Bioenergetics of
Photosynthesis

This is a volume in
CELL BIOLOGY
A series of monographs
Editors: *D. E. Buetow, I. L. Cameron, and G. M. Padilla*
A complete list of the books in this series appears at the end of the volume.

Bioenergetics of Photosynthesis

EDITED BY

Govindjee

Departments of Physiology and
Biophysics and Botany
University of Illinois
Urbana, Illinois

ACADEMIC PRESS   New York San Francisco London   1975

A Subsidiary of Harcourt Brace Jovanovich, Publishers

ACADEMIC PRESS, INC.
111 Fifth Avenue, New York, New York 10003

United Kingdom Edition published by
ACADEMIC PRESS, INC. (LONDON) LTD.
24/28 Oval Road, London NW1

Library of Congress Cataloging in Publication Data

Govindjee, Date
    Bioenergetics of photosynthesis.

    (Cell biology)
    Bibliography: p.
    1.   Photosynthesis.   2.   Bioenergetics.   I.   Title.
[DNLM: 1.   Biophysics.   2.   Photosynthesis.   QK882
G721b]
QK882.G67   581.1'3342   73-18955
ISBN 0-12-294350-3

I am grateful to my teacher and friend Eugene Rabinowitch for the guidance and training he gave me during the early part of my research career. It was he who instilled in me an interest in bioenergetics of photosynthesis. To my deep sorrow Eugene passed away on May 15, 1973 in Washington, D.C. I will always remember him as an outstanding scientist and, more importantly, as a man with deep friendship, affection, and concern for all mankind. On behalf of all the authors and myself, I dedicate this book to Eugene as a token of our appreciation for his work in photosynthesis.

Contents

1 Introduction to Photosynthesis
Govindjee and Rajni Govindjee

2 Chloroplast Structure and Function
Charles J. Arntzen and Jean-Marie Briantais

List of Contributors

Numbers in parentheses indicate the pages on which the authors' contributions begin.

Charles J. Arntzen (51), Department of Botany, University of Illinois, Urbana, Illinois.

M. Avron (373), Biochemistry Department, Weizmann Institute of Science, Rehovot, Israel.

Jean-Marie Briantais (51), Laboratoire de Photosynthèse, C.N.R.S., Gif-sur-Yvette, France.

Govindjee (1), Departments of Physiology and Biophysics and Botany, University of Illinois, Urbana, Illinois.

Rajni Govindjee (1), Department of Physiology and Biophysics, University of Illinois, Urbana, Illinois.

André T. Jagendorf (413), Division of Biological Sciences, Cornell University, Ithaca, New York.

Pierre Joliot (387), Institut de Biologie Physico-Chimique, Paris, France.

R. S. Knox (183), Department of Physics and Astronomy, The University of Rochester, Rochester, New York.

Bessel Kok (387), Martin Marietta Laboratories, Baltimore, Maryland.

J. Lavorel (223), Laboratoire de Photosynthèse, C.N.R.S., Gif-sur-Yvette, France.

Felix F. Litvin (620), Laboratory of Biophysics of Photosynthesis, Biology Department, Moscow State University, Moscow, USSR.

Satoru Murakami (555), Department of Biology, University of Tokyo, Tokyo, Japan.

Lester Packer (555), Department of Physiology—Anatomy, University of California, Berkeley, California.

George Papageorgiou (319), Department of Biology, Nuclear Research Center "Demokritos," Athens, Greece.

Kenneth Sauer (115), Department of Chemistry, Lawrence Berkeley Laboratory, Berkeley, California.

Vitaly A. Sineshchekov (620), Laboratory of Biophysics of Photosynthesis, Biology Department, Moscow State University, Moscow, USSR.

J. Torres-Pereira (555), * Department of Physiology—Anatomy, University of California, Berkeley, California.

H. T. Witt (493), Max-Volmer-Institut für Physikalische Chemie und Molekularbiologie, Technische Universität Berlin, Berlin, West Germany.

*Present address: Department of Botany, Faculty of Science, University of Luanda, Angola, Portugal.

Preface

Bioenergetics deals with the transformation of energy in biological systems. It is a branch of cell biology, biochemistry, and biophysics. In respiration, which is necessary for the survival of both plants and animals, complex organic molecules are broken down into simpler inorganic compounds. During the complete oxidation of glucose (with molecular oxygen), approximately 40% of the available chemical energy is stored in adenosine triphosphate. The energy of ATP is made available on hydrolysis and is used for various life processes (muscle contraction, biosynthesis, etc.). The complex organic molecules and oxygen, needed for respiration and therefore for life on earth, are replenished through the process of photosynthesis. In nature, the only plentiful continuously available source of energy is sunlight. In photosynthesis, this radiant energy is converted into chemical energy, with oxygen produced as a by-product, and energy stored in complex organic molecules.

The bioenergetic aspects of photosynthesis is the subject of this volume. The contributors have attempted to cover this topic in a complete and integrated fashion. Biochemical and biophysical aspects are emphasized. The work includes a review of the historical development of major concepts, analysis of experimental data, and an exposition of recent findings. The universal properties of photosynthesizing cells are emphasized; differences among green plants and photosynthetic bacteria are mentioned wherever possible. It is our hope that this volume will not only serve as a reference source for researchers but also as an introductory work for graduate students in cell biology, plant physiology, biochemistry, and biophysics.

The book begins with an introduction to photosynthesis by Govindjee and R. Govindjee (Chapter 1); followed by a discussion of the relation between chloroplast structure and function by C. Arntzen and J.-M. Briantais (Chapter 2); the primary events (light absorption, energy transfer, and primary reactions) by K. Sauer (Chapter 3); the mechanism of excitation energy migration by R. S. Knox (Chapter 4); delayed light emission by J. Lavorel (Chapter 5); chlorophyll fluorescence by G. Papageorgiou (Chapter 6); electron transport pathway by M. Avron (Chapter 7); oxygen evolution by P. Joliot and B. Kok (Chapter 8); the mechanism of photophosphorylation by A. T. Jagendorf (Chapter 9); primary acts of energy conservation in chloroplast membranes by H. T. Witt (Chapter 10); the relation of the structure of chloroplast membrane to energy coupling and ion transport by S. Murakami, J. Torres-Pereira, and L. Packer (Chapter 11); and molecular organization of chlorophyll in living cells by F. Litvin and V. A. Sineshchekov (Chapter 12). I am thankful to all the authors for providing the readers with authoritative discussions on the various aspects of the bioenergetics of photosynthesis.

Each chapter has been read and criticized by several colleagues. I have also reviewed each chapter in the book. To the following reviewers I give my most sincere thanks: P. Boyer, W. Butler, B. Chance, R. Clayton, W. Cramer, R. Dilley, D. Fleischman, N. Good, D. Krogman, B. Mayne, P. Mitchell, J. Myers, P. Nobel, R. Park, R. Pearlstein, E. Racker, K. Sauer, C. Swenberg, and L. Vernon.

My students and colleagues have also furnished advice on various chapters. I am particularly grateful to C. Arntzen, R. Emberson, Rajni Govindjee, P. Jursinic, Rita Khanna, T. Mar, P. Mohanty, R. Schooley, A. Stemler, D. VanderMeulen, and D. Wong. I am also grateful to Maarib Bazzaz, T. Wydrzynski, and Barbara Zilinskas whose research work could not get my full attention during the preparation of this volume. R. Gasanov, during his visit to our laboratory, also offered several suggestions.

The courtesy extended by the personnel in the office of the Botany Department, particularly by Shirley Langenheim and Sheila Hunt, is gratefully acknowledged. I am thankful to Sheila Hunt for typing Chapter 1, to Suzi Gomberg for typing the Subject Index, and to Alice Prickett for rendering the drawings that appear in Chapter 1. Mary Parrish helped me with the office work.

I express my deep appreciation to my wife Rajni and our children, Anita and Sanjay, who patiently endured the many weekends and evenings that were devoted to the editing of this book.

Govindjee

Bioenergetics of Photosynthesis

1

Introduction to Photosynthesis

Govindjee and Rajni Govindjee

ABBREVIATIONS

A	Intersystem intermediate
ADP	Adenosine diphosphate
AMP	Adenosine monophosphate
ATP	Adenosine triphosphate
BChl	Bacteriochlorophyll

C550	An absorption change suggested to be due to the primary electron acceptor of pigment system II.
Chl (λ)	Chlorophyll (numbers refer to the location of one of the absorption maxima, in nm)
CF	Coupling factor
Cyt (λ)	Cytochrome (numbers refer to one of the difference absorption bands)
DCMU	3-(3,4-Dichlorophenyl)-1,1-dimethylurea
DCPIP	2,6-Dichlorophenol indophenol
DPC	Diphenyl carbazide
ESR (EPR)	Electron spin resonance (electron paramagnetic resonance)
FRS	Ferredoxin reducing substance
Fd	Ferredoxin
hv	Photon of light
NAD^+	Nicotinamide adenine dinucleotide
$NADP^+$	Nicotinamide adenine dinucleotide phosphate
P430; X	Primary electron acceptor of photosystem I of chloroplasts, with a difference absorption band at 430 nm
P700	Reaction center chlorophyll of pigment system I of chloroplasts
P690	Reaction center chlorophyll of pigment system II
P870 (and P890)	Reaction center chlorophyll(s) of photosynthetic bacteria
PC	Plastocyanin
PEP	Phosphoenol pyruvate
P_i	Inorganic phosphate
PP_i	Inorganic pyrophosphate
PPNR	Photosynthetic pyridine nucleotide reductase
PQ	Plastoquinone
PS I	Photosystem I of chloroplasts (or pigment system I)
PS II	Photosystem II of chloroplasts (or pigment system II)
PSU	Photosynthetic unit
Q	Primary electron acceptor of photosystem II of chloroplasts, quencher of chlorophyll fluorescence
TMPD	N,N,N',N'-Tetramethyl-*p*-phenylenediamine
Tris	Tris(hydroxymethyl)aminomethane
UQ	Ubiquinone
X	Primary electron acceptor of pigment system I
Z	Primary electron donor of photosystem II of chloroplasts

1. GENERAL

Photosynthesis is the process by which chlorophyll-containing plants convert solar energy into photochemical energy. This energy is stored in the form of carbohydrates, providing food for man and all other heterotrophic organisms. In addition, it provides the most vitally needed supply of oxygen. The photosynthetic activity of earlier geologic eras has endowed us with massive fuel deposits. The productive potential of this process has, until recent years, seemed endless. However, the burgeoning human population, with its constantly increasing demands on both present and past products of photosynthetic activity, threatens our future survival. It is becoming more apparent that a basic understanding of the complexities of photosynthesis is necessary if we are to evaluate its true potential and wisely plan for utilization of this most fundamental metabolic activity of green plants (see Hollaender *et al.*, 1972). An understanding of this process may help us in improving its efficiency and in devising artificial photochemical systems based on it. Moreover, several basic biochemical and biophysical problems (such as mechanisms of excitation energy migration, electron transport, phosphorylation, oxidation of water molecules, and carbon fixation) can be studied with photosynthesizing organisms. The present chapter provides an introduction to the process of photosynthesis, and to the several chapters in this book dealing with the various aspects of the bioenergetics of photosynthesis.

In green plant photosynthesis, CO_2, H_2O, and light energy are the reactants, and O_2 and carbohydrates (CH_2O) are the products; this process takes place in cellular organelles called chloroplasts:

$$CO_2 + H_2O \xrightarrow[\text{chloroplasts}]{\text{light}} O_2 + (CH_2O) + 112 \text{ kcal/mole } CO_2 \qquad (1)$$

Pigment molecules, especially Chl *a*, various enzymes, and electron carriers act in a "catalytic" manner in this reaction. The overall energetics of this reaction may be looked upon in three ways. (1) In photosynthesis, a very stable arrangement of the atoms C, H, and O in CO_2 and H_2O is converted into a much less stable arrangement of the same nuclei and electrons in $(CH_2O) + O_2$; in order to drive this process the necessary energy is provided by light quanta. The total energy (ΔH) stored is 112 kcal/mole, while stored free energy (ΔF) is 120 kcal/mole, the 8 kcal/mole difference being due to the entropy term (ΔS). $[\Delta F = \Delta H - T\Delta S; + 120 = + 112 - (-8)]$. (2) A look at the total bond energy of the reactants and the products involved in the process shows a difference of ~ 105 kcal in going from H_2O and CO_2 to (CH_2O) and O_2:

$$H-O-H + O{=}C{=}O \xrightarrow{nh\nu} \begin{smallmatrix}H\\ \\H\end{smallmatrix}{>}C{=}O + O{=}O \qquad (2.1)$$

(2 × 110 kcal) (2 × 180 kcal)   (2 × 90 + 180 kcal) (115 kcal)

220 kcal + 360 kcal = 360 kcal + 115 kcal   (+ 105 kcal)

This difference is supplied by the light energy *(nh\nu)*. (3) Photosynthesis is usually considered as an oxidation–reduction reaction in which four electrons (or four H atoms) are transferred from $2H_2O$ to CO_2, oxidizing the former to O_2 and reducing the latter to (CH_2O) [see Eq. (2)]. The oxidation–reduction potential (E_0') of H_2O/O_2 couple, at pH 7.0 and under normal conditions (1 atm pressure, room temperature), is $+0.8$ V (with respect to H/H^+ being $+0.4$ V), whereas that of $CO_2/(CH_2O)$ couple is -0.4 V. Thus the difference in potential (ΔE) is 1.2 V per electron transferred. Since four electrons must be transferred from $2H_2O$ to evolve one O_2 molecule and reduce one CO_2 molecule as:

$$2H_2O \xrightarrow{nh\nu} (4H) + O_2 \qquad\qquad (2.2)$$

$$CO_2 + 4H \longrightarrow (CH_2O) + H_2O \qquad\qquad (2.3)$$

the total energy needed is ~ 110 kcal/mole (as $\Delta F = n\mathscr{F}\Delta E = 4 \times 23$ kcal/mole·Volts $\times 1.2$ V, where $\mathscr{F} =$ Faraday's constant).

Thus, the above three ways of looking at the overall energetics of photosynthesis give us approximately the same energy requirement. In terms of quanta of light needed, 120 kcal/mole means a minimum of 3 quanta of red light. However, as we shall discuss later, the mechanism of photosynthesis involves the use of 4 quanta for the accumulation of four positive equivalents (with the production of four reducing equivalents) which are needed for the evolution of one molecule of O_2 from two molecules of H_2O, and the use of another 4 quanta to transfer the four reducing equivalents to the intermediates needed for the reduction of CO_2. This is schematically shown below:

$$4(DA) \xrightarrow{4h\nu} 4(D^+A^-) \qquad\qquad (3.1)$$

$$4D^+ + 2H_2O \longrightarrow 4D + O_2 + 4H^+ \qquad\qquad (3.2)$$

$$4A^- + 4X \xrightarrow{4h\nu} 4A + 4X^- \qquad\qquad (3.3)$$

$$4H^+ + 4X^- \longrightarrow 4XH \qquad\qquad (3.4)$$

$$4XH + CO_2 \longrightarrow (CH_2O) + H_2O + 4X \qquad\qquad (3.5)$$

where $D =$ primary electron donor, $A =$ primary electron acceptor, and $X =$ a second primary electron acceptor. Thus, a *minimum* of 8 quanta are needed to evolve one O_2 and reduce one CO_2 molecule.

In addition to green plants, certain species of bacteria (e.g., green and purple) are capable of reducing CO_2 to (CH_2O). The photosynthetic bacteria differ from green plants in that they are incapable of oxidizing H_2O (perhaps due to the lack of enzymes involved in this process); instead, these organisms use substitute hydrogen (electron) donors (H_2A'). Consequently, no O_2 is evolved in bacterial photosynthesis. Van Niel (1935) proposed a general equation which describes both green plant and bacterial photosynthesis as:

$$H_2A' + CO_2 \xrightarrow[\substack{Chl\ or \\ BChl}]{h\nu} (CH_2O) + A' \qquad\qquad (4)$$

In green plants $H_2A' = H_2O$ and thus $A' = \frac{1}{2}O_2$ (not to be confused with A in Eq. 3). Equation (4) implies that the source of O_2 in photosynthesis is H_2O. This was established by Hill (1939) who discovered that isolated chloroplasts can evolve O_2 when CO_2 is replaced in a reaction mixture by ferric oxalate and ferricyanide. Furthermore, experiments in which ^{18}O-enriched water (or $C^{18}O_2$) was used as substrate (Ruben et al., 1941), or in which the ratio of $^{18}O/^{16}O$ was measured for normal H_2O, CO_2, and O_2 (Vinogradov and Teis, 1941), led to the conclusion that the isotopic composition of O_2 evolved in photosynthesis corresponds to that of H_2O, not of CO_2. These experiments clearly established that O_2 originates in H_2O.

The photosynthetic electron donor (H_2A') is different for the three major groups of photosynthetic bacteria: (1) Nonsulfur purple bacteria (Athiorhodaceae), e.g., *Rhodospirillum rubrum*; here, H_2A' is usually an organic compound. (2) Sulfur purple bacteria (Thiorhodaceae, e.g., *Chromatium* species); here, H_2A' is an inorganic sulfur compound. (3) Green sulfur bacteria (Chloroaceae, e.g., *Chlorobium thiosulfatophilum* and *Chloropseudomonas ethylicum*); these organisms can use either inorganic sulfur compounds or other organic hydrogen donors.

Unlike green plants, bacterial photosynthesis does not result in much energy storage. In fact, the oxidation of H_2S to S results in a net loss of 5 kcal/mole and the conversion of S to sulfate (SO_4^-) leads to a net storage of only 7 kcal/mole; in cases where molecular hydrogen is the donor, 25 kcal/mole are lost.

1.1. Time Sequence

Light absorption is the first act of photosynthesis occurring in 10^{-15} sec (billionth of a millionth of a second). Carbohydrate formation is the last step in photosynthesis; these enzymic reactions may take several seconds. Longer time scales are not unimportant, but they deal with regulation and growth mechanisms. Kamen (1963) has arbitrarily divided the various time scales into different eras.

1. Era of Radiation Physics: 10^{-15} to 10^{-6} sec, during which excitation processes, physical migration, and trapping of excitation energy occur (see Fig. 1):

$$Chl_g + h\nu \longrightarrow Chl_s^* \qquad \text{(Excitation)} \qquad (5.1)$$

$$Chl_s^* \longrightarrow Chl_T \qquad \text{(Intersystem crossing)} \qquad (5.2)$$

$$Chl_s^* + Chl \longrightarrow Chl + Chl_s^* \qquad \text{(Excitation energy transfer)} \qquad (5.3)$$

$$Chl_s^* \longrightarrow Chl_g + h\nu' \qquad \text{(Fluorescence)} \qquad (5.4)$$

$$Chl_s^* + T \longrightarrow T^* + Chl_g \qquad \text{(Trapping of energy)} \qquad (5.5)$$

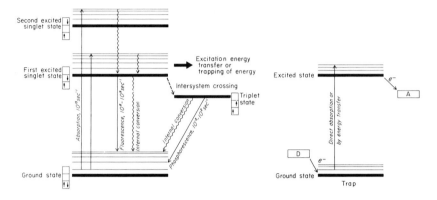

Fig. 1. Energy level diagram for "bulk" chlorophyll (left) and for the trap chlorophyll molecule (right). D: electron donor; A: electron acceptor. Small arrows in the rectangular blocks indicate spins of the electrons in the different levels.

where Chl_g = chlorophyll in ground state, Chl_s^* = chlorophyll in singlet excited state, Chl_T = chlorophyll triplet, and T = energy trap.

2. Era of Photochemistry: 10^{-10} to 10^{-3} sec in which some of the above occurs, as well as the separation of charges, or primary oxidation–reduction reactions. Three alternate ways of expressing the primary redox reaction are shown below:

$$D \cdot T \cdot {}^*A \longrightarrow \begin{cases} D^+ \cdot T \cdot A^- & (6.1) \\ D \cdot T^+ \cdot A^- \longrightarrow D^+ \cdot T \cdot A^- & (6.2) \\ D^+ \cdot T^- \cdot A \longrightarrow D^+ \cdot T \cdot A^- & (6.3) \end{cases}$$

3. Era of Biochemistry: 10^{-4} to 10^{-2} sec, in which electron carriers and enzymes are involved, leading to oxygen evolution and CO_2 fixation:

$$D^+ \cdot T \cdot A^- + \tfrac{1}{2}NADP^+ + H^+ \longrightarrow$$

$$D^+ \cdot T \cdot A + \tfrac{1}{2}NADPH + \tfrac{1}{2}H^+ \qquad \text{(Reduction of } NADP^+) \qquad (7.1)$$

$$D^+ \cdot T \cdot A \cdot + \tfrac{1}{2}H_2O \longrightarrow D \cdot T \cdot A \cdot + \tfrac{1}{4}O_2 + H^+ \qquad \text{(Oxygen evolution)} \qquad (7.2)$$

$$D^+ \cdot T \cdot A^- + ADP + P_i \longrightarrow ATP + D \cdot T \cdot A \qquad \text{(Cyclic photophosphorylation;}$$
$$\text{some ATP may also be formed}$$
$$\text{in (7.1) above)} \qquad (7.3)$$

$$CO_2 + 2NADPH + 3ATP \longrightarrow (CH_2O) +$$
$$2NADP^+ + 3ADP + 3P_i \qquad \text{(Carbon fixation)} \qquad (7.4)$$

1.2. Membrane Structure and Function

It appears that a highly ordered chloroplast membrane structure is necessary for the separation of positive and negative charges and their stabilization. The structure must somehow prevent the oxidizing and reducing equivalents from recombining to a significant extent, since recombination would result in the loss of energy as heat or light:

$$D^+\cdot T\cdot A^- \longrightarrow D\cdot T\cdot A\cdot + heat \qquad (8.1)$$

$$D^+\cdot T\cdot A^- \longrightarrow D\cdot T\cdot A + hv'' \text{ (delayed light emission)} \qquad (8.2)$$

Thus, a detailed study of the structure and composition of the chloroplast membrane is necessary for the understanding of the mechanisms of energy coupling. Two approaches have been used to elucidate the structure: (1) a morphological approach using primarily the electron microscope; and (2) a physicochemical approach. The latter can be further subdivided into two approaches: (a) nondestructive—by the use of X-ray crystallography and physical probes (e.g., fluorescent probes), and (b) destructive—by chemical analysis of the various constituents. The relation of chloroplast structure to function, as determined by combined study of electron microscopy and biochemistry, is discussed in Chapter 2 of this volume.

For photophosphorylation to occur, certain particles (CF, ATPase) attached on the outer surface of the chloroplast membranes (thylakoids) are necessary. More recently, the role of membranes has played an important part in the discussion of the mechanism of phosphorylation since it is believed that utilization of a membrane potential and/or a H^+ gradient across the thylakoid membrane provides energy for photophosphorylation (see Chapters 9 and 10 of this volume).

The complete process of photosynthesis, *in vivo*, occurs within the organelle space bounded by the outer envelope of the chloroplasts. All enzymes participating in CO_2 fixation are localized either in the matrix (stroma) of the plastid or are loosely affixed to the lamellar membranes.

1.3. Photosynthetic Units

Emerson and Arnold (1932a), using repetitive bright, short (10^{-5} sec) flashes of light with varying dark periods between the flashes, observed that the number of oxygen molecules evolved per flash increased with increasing duration of dark periods and reached a saturation value at <0.4 sec in *Chlorella* cells at $1°C$, and <0.04 sec at $25°C$. The half-time was about 0.04 sec at $1°C$, thus giving the time needed to complete the dark enzymic reactions of photosynthesis. Interestingly, they found that the reactions of the carbon fixation cycle are inhibited almost 60% by 1.14×10^{-5} M KCN and about 50% by a decrease in bicarbonate ions from 71×10^{-6} to 4.1×10^{-6} moles/liter. However, KCN and bicarbonate ions affected the flash yields (O_2/flash) quite differently: KCN did not affect this yield, whereas a decrease in bicarbonate ions reduced it by 50%. These data suggest a role for bicarbonate ions other than that in the carbon fixation cycle. We shall mention this role later. Using *Chlorella* cells with varying concentrations of Chl, Emerson and Arnold (1932b) noted that the ratio of the maximum number of oxygen molecules evolved per flash to the number of Chl molecules present was 1/2500. Thus, they suggested that for every 2500 molecules of Chl, there is present one "unit" capable of reducing one molecule of CO_2 (or evolving one molecule of O_2, as they measured O_2, not CO_2) each time the unit is activated by light (Table I). In other words, these Chl molecules must "work together to effect the reduction of one CO_2 molecule" (see Emerson, 1937). The current definition of a photosynthetic unit capable of evolving one O_2 refers to the collection of 2500 Chl molecules with its reaction center(s).

The concept of a photosynthetic unit was supported by (1) the arguments of Gaffron and Wohl (1936a, b) (also see Arnold and Kohn, 1934) who calculated that if chloroplasts did not have such units, and if individual Chl molecules had to accumulate the necessary quanta of light to evolve O_2, there would be a lag in O_2 evolution after illumination. Such a lag does not occur, although a small one, seen only with very weak light, has been observed and is due to a different reason; see Chapter 8; (2) O_2 evolution and reduction of an artificial dye (the Hill reaction) requires a minimum size of chloroplast fragments containing hundreds of Chl molecules (Thomas *et al.*, 1953); (3) the discovery of the existence of reaction centers (see Section 2.4 and Chapter 10 of this volume).

The primary photoacts of photosynthesis lead to the transfer of four hydrogen atoms (or electrons) from two H_2O to CO_2 in order to evolve one O_2 molecule. Thus, the original figure of 2400 Chl molecules/O_2 for the PSU means that 600 Chl molecules interact in transferring one H atom. It is now generally believed that each H atom is transferred from H_2O to CO_2

TABLE I

Photosynthetic Unit in Chlorella

Basis	No. of Chl molecules
O_2 evolved	2400
(H) or e^- transferred	600
hv absorbed	300

in two successive photochemical steps (see Sections 1.5 and 3.8 and Chapter 7 of this volume); therefore there must be eight primary photochemical events. The photosynthetic unit size for each of these primary photoacts can be visualized as a group of 250–300 Chl molecules (Table I). Each light quantum absorbed by any Chl molecule of the unit must be transferred through the complex until it reaches the specialized Chl *a* molecules called energy traps or reaction centers. Once the excitation energy reaches the trap, the primary photoact facilitates the transfer of an H atom (or electron) from the donor to the acceptor molecule due to the proximity of the donor and the acceptor molecules to the trap (Fig. 2). The trap molecules first undergo light-induced oxidation and subsequent reduction in the dark, leading to changes in light absorption. Alternatively, trap molecules may undergo reduction in light and oxidation in dark. Since the total number of trap molecules is extremely small (less than 1% of the total Chl), the magnitude of the light-induced absorption changes due to their redox reactions is too small to be detected by conventional spectrophotometers. However, it can be done by the use of sensitive difference spectrophotometers (see Section 2.4).

The size of a photosynthetic unit is not necessarily uniform for all plants. Schmid and Gaffron (1971) found that PSU varied between 300 and 5000 Chl/CO_2 fixed in various higher plants.

The size of the photosynthetic unit in photosynthetic bacteria is about six times smaller than in green plants. Arnold (cited by Van Niel, 1941) reported that the ratio of P_{max} (maximum number of CO_2 molecules fixed) to the number of BChl molecules present was $1/400$. If we divide 400 by the number of quanta used, this number reduces to 50. This is in fair agreement with the ratio of the number of bulk BChl molecules to that of the reaction center (also see Clayton, 1965, p. 17).

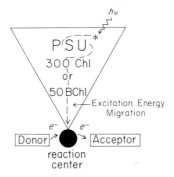

Fig. 2. Diagrammatic sketch of a photosynthetic unit (PSU). Chl: chlorophyll; BChl: bacteriochlorophyll; hv: light quantum.

1.4. Overall Quantum Yield; "Red Drop"

Quantum efficiency (the number of O_2 molecules evolved per quantum of light absorbed, or, conversely, the quantum requirement, i.e., the number of quanta required per O_2 evolved) of photosynthesis in green plants was a matter of great controversy between O. Warburg and R. Emerson for many years. Warburg reported values of the quantum requirement ranging from 2.8 to 4 quanta, whereas Emerson found it to be 8 or more quanta (see Emerson, 1958).

The quantum requirement of 4 (claimed by Warburg) does not provide sufficient energy for the hydrogen transfers and the stabilization of the various products. Our recent results (R. Govindjee *et al.*, 1968) confirm Emerson. In view of the present scheme of photosynthesis which involves two light reactions (see Section 1.5), the transfer of four hydrogens will require a minimum of 8 light quanta. A few more quanta may be needed to drive cyclic photophosphorylation. A quantum requirement of 10 would thus allow enough energy for the needed H transfer as well as the stabilization of the various products.

It has been shown that the energy absorbed by the various photosynthetic pigments is transferred to Chl *a*, ultimately reaching the reaction centers where it is converted into chemical energy. If all pigments are equally efficient in transferring energy to the reaction centers, the quantum yield of photosynthesis should remain constant as long as there is absorption by these pigments. Emerson and Lewis (1943) showed that a plot of the quantum yield of photosynthesis as a function of wavelength of light does not remain constant throughout the spectrum. There is a decline in the blue end (where absorption by carotenoids is high), and there is a drop in the red end of the spectrum, well within the red absorption band of Chl *a*. The decline of quantum yield in the red end of the spectrum has since been known as the "red drop" (Fig. 3). This can now be understood in terms of the two light reactions and the two pigment systems hypothesis—in the long-wavelength (red-drop) region, absorption by only one pigment system dominates and this is not enough for efficient photosynthesis (see below).

The quantum yield of bacterial photosynthesis (the number of CO_2 molecules fixed per quantum of light absorbed) is 0.12, the same as that in green plants. Experiments on the quantum yield of bacterial photosynthesis as a function of wavelength did not reveal any "far-red drop" equivalent to the red drop observed in green plants (unpublished experiments of the authors). This indicates the presence of one photosystem or two (or more) with overlapping absorption spectra. [For detailed discussions on photosynthetic bacteria, see Gest *et al.* (1963), Pfennig (1967) and Sybesma (1970).]

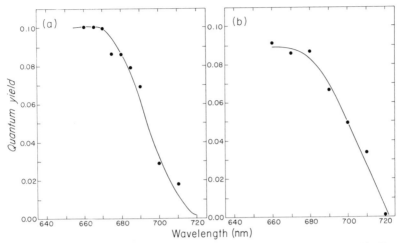

Fig. 3. Quantum yield of oxygen evolution as a function of wavelength. (a) *Navicula minima*; (b) *Chlorella pyrenoidosa*. (After Govindjee, 1960.)

1.5. Enhancement Effect; Two Pigment Systems and Two Light Reactions

Emerson *et al.* (1957) discovered that the "red drop" could be avoided, and the quantum yield increased in this region, if a supplementary light (absorbed by one of the accessory pigments, e.g., Chl *b* in green algae, phycobilins in the blue-green and red algae, and fucoxanthol in the brown algae) was added to the far-red light (absorbed by Chl *a*). This phenomenon has since been known as the "enhancement effect" (the Emerson effect, the Emerson enhancement effect, or the second Emerson effect; see Emerson and Rabinowitch, 1960).

A suggestion was made that simultaneous excitation of Chl *a* and the accessory pigments is necessary for high quantum yields, and Chl *a* alone was inefficient in photosynthesis (Emerson, 1958). In conflict with this idea was the data obtained from the Chl *a* fluorescence studies (Duysens, 1952), which showed that Chl *a* is the prime photochemical sensitizer and the accessory pigments were assigned an indirect role of energy collection and transfer to Chl *a*.

Emerson and Rabinowitch (1960) discarded the idea that Chl *b* directly sensitizes one light reaction. Instead, they suggested that two different forms of Chl *a* sensitize two different light reactions. Govindjee and Rabinowitch (1960a, b) and French *et al.* (1960) showed that the action spectra of the enhancement effect in various organisms, containing different accessory pigments, had peaks representing the effectiveness of the accessory pigment, but *in addition* there was a peak or shoulder around 670 nm which was

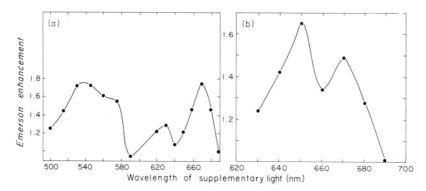

Fig. 4. Emerson enhancement (ratio of the rate of oxygen evolution in combined far-red and short-wavelength lights minus the rate in short-wavelength, and the rate in far-red light alone) as a function of short-wavelength light. (a) *Navicula minima*. (b) *Chlorella pyrenoidosa*. (After Govindjee, 1960.)

clearly in the region of Chl *a* absorption (Fig. 4). On the basis of this, the interpretation of the enhancement effect could be modified, and it was suggested that different forms of Chl *a* exist in both pigment systems. The shorter wavelength forms, e.g., Chl *a* 670 (absorbing around 670), are present in the system containing most of the accessory pigments. Duysen's fluorescence data refer mostly to these forms. The longer wavelength forms (absorbing at wavelengths > 685 nm) are present only in the other system (see review by Govindjee *et al.*, 1973). When only long-wavelength forms are excited, only one pigment system is excited and photosynthesis is low since two photosystems are required. When short-wavelength forms are excited, both pigment systems are excited, and, photosynthesis takes place normally. Excitation with short- and long-wavelength lights leads to a more balanced excitation of the two pigment systems, and photosynthesis is more efficient (enhancement effect).

It was clearly shown by Myers and French (1960a) that the synergistic or the enhancement effect in the quantum yield of photosynthesis could be brought about even when the supplementary light and the far-red light were not given simultaneously, but a few seconds apart. This implied that the product of one light reaction could exist for a few seconds and still be effective in enhancing the rate of another reaction. (See Govindjee, 1963; Fork, 1963; Myers, 1971 for further information on the enhancement effect.)

Another phenomenon related to the two "light effect" hypothesis was the discovery by Blinks (1957) of "chromatic transients" in the photosynthesis of the red alga *Porphyra perforata*. He found that when this alga was exposed to alternating red (absorbed by Chl *a*) and green (absorbed by phycoerythrin; see Section 2.1) light of equal steady-state effectiveness, the rates of O_2 "evolution" were not constant during the transition period: upon going from red to green light, an O_2 gush was observed followed by a decrease and then recovery to steady state, whereas in the transition from green to red light, no O_2 gush was observed, but a decrease in rate and then a recovery. At that time, the reason for this phenomenon was not clear. Blinks proposed two possible explanations: (1) inactivation of some of the Chl to which energy must be transferred from phycoerythrin, and its recovery; (2) altered respiration rates because the platinum electrodes used could not distinguish between light-induced O_2 evolution and uptake. He (1959, 1960) later favored the second explanation. However, Myers and French (1960b) argued against the Blinks hypothesis and suggested that the chromatic transient is closely geared to the photosynthetic sequence. The Blinks effect could, therefore, be another consequence of the two pigment systems—two light reactions hypothesis.

Hill and Bendall (1960) suggested a scheme for photosynthesis which in a highly modified form is now popularly known as the Z scheme. In the original scheme the concept of two light reactions operating in series was proposed, with cytochromes acting as intermediate electron carriers. (Independently, Kautsky et al. (1960) had also suggested the existence of two light reactions to explain the kinetics of Chl a fluorescence.) The concept of two pigment systems, based on enhancement and other effects, was added later to this scheme. Since cytochromes are thought to be a part of the intersystem chain, one would predict that the cytochromes should be oxidized by one photosystem and reduced by the other. Duysens et al. (1961; also see Duysens and Amesz, 1962) indeed showed that, in red algae, the red light (absorbed by Chl a) caused an oxidation of Cyt f, and green light (absorbed by phycobilins) caused a reduction of Cyt f. A study of the action spectra of the oxidation and reduction of cytochromes gave an idea of the composition of the two photosystems (see Duysens and Amesz, 1962), since the action spectrum of a photochemical reaction follows the percent absorption spectrum of the active pigment(s).

It has been suggested that for green plants Chl a is present in both the systems, the long-wavelength forms of Chl a (Chl a 680 and longer wavelength forms) being predominant in system I (PS I), which is weakly fluorescent at room temperature. PS I is believed to have a smaller proportion of the accessory pigments and is responsible for the oxidation of cytochromes. The other photosystem (PS II) contains relatively more of short-wavelength forms of Chl a (Chl a 670) and of the accessory pigments; PS II is believed to be responsible for the photoreduction of cytochromes. (In red and blue-green algae, a larger proportion of Chl a is present in PS I.)

After the initial characteristics of the two photosystems were defined, a great deal of evidence was presented confirming their existence, and

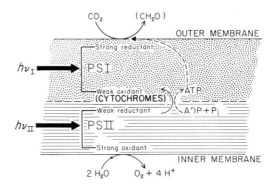

Fig. 5. Diagrammatic sketch of a portion of the thylakoid membrane of grana from higher plants. Pigment system I (PS I), upon light (hv_I) absorption produces a strong reductant and a weak oxidant. Pigment system II (PS II), upon light (hv_{II}) absorption, produces a strong oxidant and a weak reductant. Electron flow from the weak reductant to the weak oxidant is coupled to phosphorylation [conversion of adenosine diphosphate (ADP) and inorganic phosphate (P_i) to adenosine triphosphate (ATP)]. With the aid of ATP, the strong reductant (produced by PS I) reduces carbon dioxide to carbohydrate (CH_2O) in the stroma. The strong oxidant (produced by PS II) oxidizes water molecules to molecular oxygen, and H^+ ions are released on the inner membrane.

information was provided for a more detailed picture (see Section 3.1 and reviews by Duysens, 1964; Franck and Rosenberg, 1964; Robinson, 1964; Vernon and Avron, 1965; Witt, 1967; Hind and Olson, 1968; Fork and Amesz, 1969; Boardman, 1970; Goedheer, 1972).

Studies of the two photosystems have led to the hypothesis that PS I produces a strong reductant (for reducing CO_2) and a weak oxidant, and PS II is responsible for the formation of a weak reductant and a strong oxidant (responsible for the oxidation of H_2O to molecular O_2). The transfer of electrons from the weak reductant (produced by PS II) to the weak oxidant (produced by PS I) is coupled to the production of ATP (Fig. 5). As discussed by Arntzen and Briantais (Chapter 2, this volume), the PS I is, perhaps, on the outer side of the thylakoid membrane, and PS II on the inner side. [In red and blue-green algae, however, phycobilins (part of PS II) are contained in "phycobilisomes" that are attached to (or partially embedded in) the outer surface of thylakoids. Thus, the location of PS I and PS II is not yet clear in these algae.]

In photosynthetic bacteria, the Emerson enhancement effect has not been observed (Blinks and Van Niel, 1963). Thus, the existence of two pigment systems in these organisms cannot be supported from this kind of data

1.6. Overall Energetics

Duysens (1958) calculated the maximum efficiency of photosynthesis by applying the second law of thermodynamics. At low light intensities, where the efficiency is low, maximum efficiency was calculated to be about 70% in green plants. Ross and Calvin (1967) calculated the maximum amount of free energy stored, based on detailed thermodynamic arguments, to be 1.19 and 1.23 eV per photon absorbed by PS I and PS II, respectively, under an illumination of 1 klx of white light. This gives a maximum energy efficiency of about 66–68%, in agreement with Duysen's value. These values place an upper limit on the photosynthetic efficiency. For PS I, a ΔE of 1.0 eV is estimated for oxidizing P700 and reducing "X." If only 1.19 eV of free energy is available, we have only 0.2 eV of extra energy for any other use. The same may be true for PS II. Thus, from this point of view, electron transfer from H_2O to "X," which is estimated to need about 1.6 eV (assuming the Z/Z^+ couple to have an E_0' of $+1.0$ V, 0.2 V more positive than the H_2O/O_2 couple, and X/X^- to be about $-0.6V$), cannot occur with 1 quantum. This is perhaps why plants had to resort to the use of 2 quanta mechanisms.

2. PRIMARY EVENTS

2.1. Light Absorption

The various photosynthetic pigments, mentioned in Section 1, can be classified into three main groups: chlorophylls, carotenoids, and phycobilins. The function of these pigments is to provide the plants with an efficient system of absorbing light throughout the visible spectrum (see reviews by Rabinowitch and Govindjee, 1969; Govindjee and Mohanty, 1972; Govindjee and Braun, 1973). This energy is then transferred to the reaction centers, where it is utilized for the photochemical reactions (see Chapter 3 of this volume). The bulk of the pigments involved in the process of light absorption (and energy transfer) are called the light-harvesting pigments.

2.1.1. Chlorophylls

There are two kinds of Chl in higher plants and green algae: Chl a and Chl b. These are soluble in organic solvents. Chl a is the major pigment and is present in all photosynthetic organisms that evolve O_2 (bacteria do not have Chl a and are incapable of evolving O_2). Several forms of Chl a have been postulated: Chl a660, Chl a670, Chl a680, Chl a685, Chl a690, and Chl a700–720, the number indicating their respective red absorption maxima (see French, 1971; Fig. 6; and Chapter 12 of this volume). Evidence for the existence of the various forms comes from derivative spectrophotometry, low-temperature absorption measurements, and the action spectra of various photochemical reactions. The short-wavelength Chl a forms are fluorescent and are predominantly present in the PS II. The long-wavelength forms are weakly fluorescent and are predominantly present in PS I.

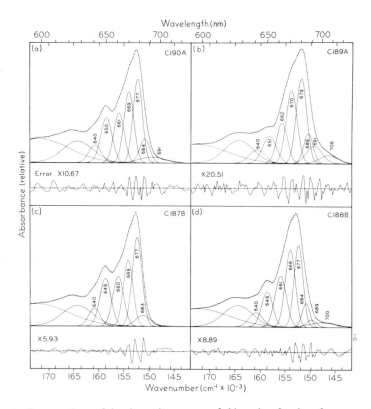

Fig. 6. Curve analyses of the absorption spectra of chloroplast fractions from stroma and grana lamellae measured at −196°C. (a) P-10K: this fraction consists of grana membranes (both pigment systems I and II); (b) P-144K: stroma lamellae containing pigment system I only; (c) F-2D: pigment system II separated from grana; (d) F-1D: pigment system I separated from grana. Note that F-2D (c) does not contain Chl *a* 689–693 and Chl *a* 700–706; the other Chl *a* and Chl *b* forms are present in all samples, although in slightly different proportions. (After R. A. Gasanov and C. S. French, personal communication, 1973.)

Chl *b* is present in all the higher plants and green algae. Its red absorption maximum occurs at 650 nm. Recently it has been shown that there are two forms of Chl *b*—Chl *b*640 and Chl *b*650. The major proportion of Chl *b* is present in PS II.

Apart from the above-listed forms of Chl, there exists a compound related to Chl, labeled Chl *c* in brown algae. The existence of yet another Chl called Chl *d* is doubtful. Its presence has been suggested in some red algae, but never proven to exist *in vivo*.

Chlorophylls *in vivo* are noncovalently bound to protein; upon treatment with organic solvents, the weak interaction is eliminated and the absorption maximum shifts to shorter wavelengths (e.g., from 675 to 660 nm for Chl *a*). (For a review of all aspects of chlorophylls, see Vernon and Seely, 1966.)

2.1.2. Carotenoids

These are the yellow and orange pigments found in almost all the photosynthetic organisms. They are soluble in organic solvents. There are two kinds of carotenoids: (1) carotenes, of which β-carotene is the most common one, are hydrocarbons; they absorb blue light; (2) carotenols are alcohols; these are commonly called xanthophylls. One of the xanthophylls (fucoxanthol), present in diatoms and other brown algae, is thought to be bound to protein *in vivo*. It is generally accepted that most of carotenes are present in PS I and xanthophylls in PS II.

2.1.3. Phycobilins

These water-soluble pigments, present in red and blue-green algae, are open-chain tetrapyrroles. There are two kinds of phycobilins: (1) phycocyanins, which predominate in the blue-green algae and account for the absorption maxima around 630 nm; (2) phycoerythrins, which predominate in the red algae, and absorb around 540 nm (see O'hEocha, 1971). Phycobilins are mainly associated with PS II, but they are present in PS I as well. ·

A working model for the composition of the two pigment systems in green plants is shown in Fig. 7.

2.1.4. Bacteriochlorophyll

The main light-harvesting pigment of photosynthetic bacteria is BChl—it is a tetrahydroporphyrin. BChl *a*, in organic solvents, has absorption maxima at 360, 590, and 760 nm. *In vivo*, the absorption is red-shifted and often has two or more maxima. For example, the purple bacterium *Chromatium* has absorption peaks at 800, 850, and 870 nm due to bulk BChl *a*. The green bacteria contain *Chlorobium* Chl, of which there are two kinds, with absorption at 725 and 750 nm; in addition they contain a small amount of BChl *a* absorbing at 810 nm (see Olson and Stanton, 1966). Some bacteria, e.g., *Rhodopseudomonas viridis*, contain BChl *b*; in these, the long-wavelength absorption maximum lies at 1025 nm.

2.2. Light Emission

2.2.1. Fluorescence

The Chl *a* molecule, upon excitation, by direct light absorption, or through the transfer of excitation energy absorbed in any accessory pigment, goes into the first excited singlet state. The potential energy of the molecule in the excited state can be dissipated in several ways (see Fig. 1). The molecule can return to its ground state through "internal conversion" with loss of the energy as heat; or it can transfer its energy to another molecule until the energy finally reaches the reaction center where it is converted into the

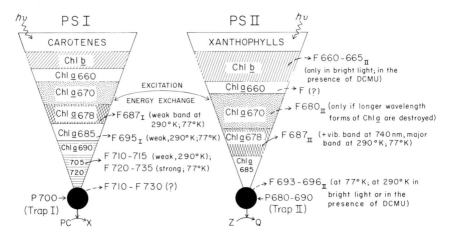

Fig. 7. A working hypothesis for the approximate distribution of the various pigments in the two pigment systems in green plants. PS I: pigment system I; PS II: pigment system II; Chl *b*: chlorophyll *b*; Chl *a*: chlorophyll *a*; the numbers indicate the approximate absorption maxima in the red end of the spectrum. The symbol F followed by numbers refers to the suggested fluorescence emission bands in nm. The conditions under which these bands are observed are listed within parentheses. Z: "primary" electron donor of PS II; Q: primary electron acceptor of PS II; PC: possible electron donor of PS I; X: primary electron acceptor of PS I; P680–P690: energy trap of PS II; P700: energy trap of PS I. DCMU: 3-(3, 4-dichlorophenyl)-1,1-dimethyl-urea. As shown in Fig. 5, the PS I is assumed to be located on the outerside of the thylakoid membrane and the PS II on the inner side such that excitation energy transfer between them is possible. (We cannot, however, discount the possibility that the two systems may be side by side on the thylakoid membrane.) In diatoms and brown algae, Chl *c* replaces Chl *b*. In red and blue-green algae, phycobilins replace Chl *b*; a large proportion of all Chl *a* forms is in PS I, and a large proportion of phycobilins is in PS II. Phycobilins are located in phycobilisomes; the physical arrangement of pigment systems in these algae is not yet clear.

chemical energy of the reaction products; alternately, the molecule can return to the ground state with the release of a quantum of light. This last process is called fluorescence (see Chapters 5 and 6 of this volume). In solution, Chl has a very high fluorescence yield (about 30%), but *in vivo* the yield is very low (about 3%), as most of the energy is used for photochemistry. The emission spectrum of Chl *a* fluorescence *in vivo* has a main band at 685 nm and a minor band at 740 nm (see Chapter 6). The excitation spectra for the various fluorescence bands indicate that most of the fluorescence at room temperature has its origin in PS II, but fluorescence in the 710–715 nm region originates also from PS I (see Govindjee *et al.*, 1973).

Most of the photosynthetic pigments (with the exception of the carotenoids) are known to fluoresce in solution. *In vivo*, however, Chl *a* fluorescence is dominant because the rest of the pigments transfer their absorbed energy to Chl *a* with a relatively high efficiency. The energy then migrates among Chl *a* molecules until it finally reaches the reaction center (for a theoretical discussion of energy migration, see Chapter 4 of this volume). Fluorescence from the trap molecules of green plants has not been measured directly, but has been inferred (Krey and Govindjee, 1964, 1966; Govindjee and Yang, 1966; Cho and Govindjee, 1970a,b) from measurements under the conditions when photochemistry is saturated (at high light intensities) or is stopped (at very low temperatures—liquid nitrogen or helium). Under such conditions, the quanta arriving at the reaction site may be given off as fluorescence or transferred back to the bulk pigments, and emitted from there. A fluorescence band at 693–696 nm has been attributed mainly to the reaction center Chl a_{II}. The correlation of various absorption bands with different fluorescence bands is discussed in Chapter 6 of this volume (also see Fig. 7).

The fluorescent property of PS II has been used extensively in studying the reactions of this photosystem. When light is turned on, Chl a fluorescence undergoes transient changes with time (see review by Govindjee and Papageorgiou, 1971). These changes have been divided into two types: the fast changes that are completed within a few seconds, and the slow changes that last several minutes. The fast changes are related to the initial photochemical events of PS II. Thus, the rise of fluorescence from a low to a high level (also known as the variable fluorescence, as opposed to the fixed or constant fluorescence observed immediately upon the onset of illumination) is suggested to indicate the reduction of a "primary" electron acceptor Q (the quencher of fluorescence, Duysens and Sweers, 1963). As Q^-* accumulates, the light quanta arriving at the reaction center can no longer drive photochemistry, and are more likely to be given off as fluorescence. When Q^- can be reoxidized by a pool of secondary acceptors or by back-reaction of PS II, fluorescence does not reach a maximum level. Ferricyanide, which can reoxidize Q^-, also lowers the fluorescence level (Malkin and Kok, 1966). When DCMU, which blocks the reoxidation of Q^- by the pool of secondary acceptors, is added, the fluorescence rises quickly to a high level and stays at that level.

Reduction of Q to Q^- occurs at 77°K as shown by the fluorescence rise at that temperature; however, this transient only partially recovers in the dark;

---

* Recently, Mauzerall (1972), using nanosecond laser flashes, has shown that Q is reduced too slowly (within microseconds) to be able to compete with singlet Chl$_a^*$, which has a lifetime of about 1 nsec; thus, Q cannot act as the primary electron acceptor of PS II. Butler (1972a) recently suggested from his data at 77°K, that this fluorescence can be governed by the "Z" (the primary electron donor) side of PS II as well. He observed that the fluorescence rise does not parallel the reduction of the primary acceptor of PS II (a component labeled C550; see discussion in Chapters 3 and 6 of this volume), but rather the oxidation of an electron donor, a high potential form of a cytochrome, labeled Cyt b559, which feeds electrons to Z when it is in the reduced state. From this it is suggested that the fluorescence rise is associated with the formation of ZQ^- rather than with Z^+Q^- (if Q can be considered as equivalent to C550). (It remains to be seen whether the same interpretation holds at room temperature where H_2O replaces Cyt b559!) Wraight and Clayton (verbal communication, 1973) have concluded that in bacterial systems, fluorescence should be taken as an indicator of the state of the reaction centers, but not as a quantitative component of a simple theory.

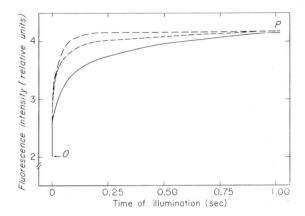

Fig. 8. Fluorescence intensity as a function of time of illumination at 77°K in isolated spinach chloroplasts. (—) first illumination; (– –) second illumination after 5 min darkness at 77°K; (–·) third illumination after 30–60 min darkness at 77°K. Actinic illumination: blue light; filters: C.S. 4-96 and C.S. 3-73; measuring wavelength: 685 nm (half-bandwidth, 6.6 nm). O: constant fluorescence; P: maximum fluorescence. (Unpublished experiments of Govindjee and Barbara Zilinskas Braun, 1972.)

a part of the change is irreversible (Fig. 8). It appears that at 77°K a portion of the primary back-reaction is blocked. Perhaps there are two pathways for the recovery of the primary reactants, one temperature-independent involving a direct reaction of Q^- with Z^+ (the oxidized form of the primary electron donor of PS II, labeled Z) within the reaction center II complex, and the other temperature-dependent via another intermediate.

Slow Chl *a* fluorescence yield changes and those changes which occur in the presence of DCMU may be related to changes in the conformation of the thylakoid membranes that often lead to changes in the "spillover" of excitation energy between PS I and II. These possibilities are discussed by Papageorgiou (Chapter 6 of this volume) and Murakami *et al.* (Chapter 11).

Photosynthetic bacteria also show transient changes in BChl fluorescence. The interpretation of the variable fluorescence in these organisms is, perhaps, the same as that given for the variable fluorescence in green plants, i.e. it represents the reduction of the primary electron acceptor(s). [For details, see Vredenberg and Duysens (1963), Clayton (1966), de Klerk *et al.* (1969), Suzuki and Takamiya (1972), and Malkin and Silberstein (1972).] In photosynthetic bacteria, fluorescence from the reaction center BChl was demonstrated by Zankel *et al.* (1968). An example of fluorescence from a reaction center preparation of *R. rubrum* is shown in Fig. 9.

2.2.2. Delayed Light Emission

Another phenomenon of light emission has been observed in photosynthetic organisms, lasting up to several minutes after the exciting light has

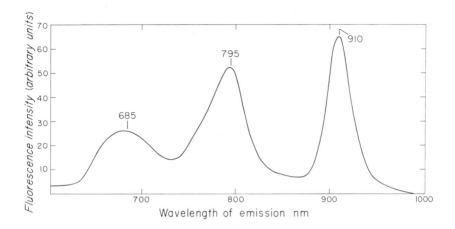

Fig. 9. Fluorescence emission spectrum of dithionite-reduced "large" reaction center preparation from *Rhodospirillum rubrum* (see W. R. Smith, 1972 for the method of preparation) at 77°K. Actinic illumination: broad-band blue light, 10^4 ergs cm^{-2} sec^{-1}; measuring conditions: C. S. 2-58 filter; halfband width, 6.6 nm; photomultiplier, RCA 7102. The peaks at 910, 795, and 685 nm are due to reaction center bacteriochlorophyll, bacteriopheophytin, and solubilized bacteriochlorophyll (?), respectively. (After Govindjee and W. Smith, unpublished observations, 1972.)

been turned off (Strehler and Arnold, 1951). This is known by various names: delayed light, afterglow, luminescence, and delayed fluorescence. The origin of part of the delayed light in green plants has been attributed to a back-reaction between the reduced primary electron acceptor (Q^-) and the oxidized primary electron donor (Z^+, or more oxidized forms of it). Any condition that favors the recombination of Q^- and Z^+ also enhances the delayed-light emission. For a detailed discussion see Chapter 5 of this volume.

2.3. Energy Transfer and Migration

As discussed earlier, the existence of photosynthetic units allows a more efficient use of the absorbed energy. This involves excitation energy migration through a maze of several hundred Chl *a* molecules, until the energy finally reaches the reaction center where it is converted into chemical energy. The process of excitation energy migration through the same kind of molecules involves *homogeneous* energy transfers. In addition, the energy absorbed by pigments other than Chl *a* in any photosystem is also transferred to Chl *a*. This kind of excitation energy transfer between different kinds of pigments is called *heterogenous* energy transfer.

Evidence for the existence of heterogeneous energy transfer comes from fluorescence measurements. Fluorescence bands ascribable to Chl *a* (acceptor) are observed, even when the excitation has been in a spectral region where Chl *a* has relatively very low absorption and where other pigments (donors:

Chl *b*, phycobilins, fucoxanthol, etc.) absorb more strongly. Also, the excitation spectrum (or the action spectrum) of Chl *a* fluorescence should theoretically follow the percent absorption spectrum of Chl *a* if there is no heterogeneous transfer; however, it remains high even in the regions of low Chl *a* absorption. This is particularly striking in the blue-green and red algae, where the action spectrum of Chl *a* fluorescence is very high in the regions of phycocyanin and phycoerythrin absorption. The efficiency of excitation energy transfer from Chl *b* to Chl *a* has been calculated to be almost 100% (Duysens, 1952; Cho and Govindjee, 1970*a*). (Under certain conditions, a small fluorescence from Chl *b* may be observed indicating a less than 100% transfer from Chl *b* to Chl *a*; see Govindjee and Briantais, 1972.)

The efficiency of excitation energy transfer from carotenoids to Chl *a* is poor (10–50% depending upon the organism), except for fucoxanthol which transfers with ~ 70% efficiency (Dutton *et al.*, 1943; Duysens, 1952; Goedheer, 1972). Phycobilins are capable of transferring the absorbed energy to Chl *a* with a very high efficiency (80–90%) (French and Young, 1952; Duysens, 1952; also see Govindjee and Mohanty, 1972) (see Fig. 10).

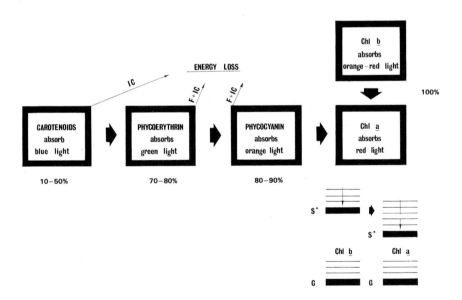

Fig. 10. Diagrammatic scheme for excitation energy transfer. F : fluorescence; IC : internal conversion; Chl *b*: chlorophyll *b*: Chl *a*: chlorophyll *a*; G: ground state; S*: excited singlet state. The numbers, in percent, refer to the efficiency of energy transfer from various pigments to Chl *a*.

Considerable overlap of the absorption band of the acceptor molecule, in this case Chl a, and the fluorescence band of the donor molecules (Chl b or phycoerythrin and phycocyanin), the appropriate orientation and the close distance of these pigments are the important conditions for the high efficiency of energy transfer. The time lag between the emission of Chl a fluorescence and the excitation of phycobilins is of the order of 0.3–0.5 nsec, whereas the lifetimes of phycoerythin and phycocyanin fluorescence are 7 and 1.8 nsec, respectively (Brody and Rabinowitch, 1957; Tomita and Rabinowitch, 1962). Thus, the energy transfer from the various pigments to Chl a can compete very effectively with the natural fluorescence of the different pigments *in vivo*. Hammond (1973) has recently measured a transfer time of 21 psec (upper limit) from Chl b to Chl a in the green alga *Chlorella*. [For literature on lifetime of excited states in green plants, see Mar *et al.* (1972), and Briantais *et al.* (1972).]

Evidence for homogeneous energy transfer (migration) comes from the fluorescence studies using polarized excitation light. Since the lifetime of the excited singlet state of Chl a is short, the fluorescence emission from Chl excited with polarized light would be highly polarized if it were from the same Chl a molecule that absorbed the light. However, experimental measurements show that the fluorescence is almost completely depolarized *in vivo* [Arnold and Meek (1956); for a recent measurement, see Mar and Govindjee (1972a)]. This could indicate that between light absorption and fluorescence emission a quantum is transferred through several molecules, each having a slightly different orientation, thus resulting in a depolarization of the fluorescence. However, most of the depolarization of fluorescence may occur in the first transfer. It may also be pointed out here that if the pigment molecules are highly aligned, information on energy migration cannot be obtained by measuring polarization of fluorescence.

The absorbed energy can migrate from one pigment molecule to another as excitation energy. The excitation energy transfer can take place by the process of exciton migration (see Chapter 4 of this volume). An exciton consists of an "excited" electron and a positive charge or a hole (left by the transfer of the electron to the excited state), the pair moving *together* from one molecule to another. The process of exciton migration involves no charge separation. Alternatively, migration of energy could be similar to that which exists in metals. In this process, electrons and holes can migrate freely (independently of each other) through conductance bands (see Arnold and Azzi, 1968). In this theory, migration of electrons and holes can occur at the same time in opposite directions in the chloroplast membrane. The electrons are finally trapped in an "electron trap," where they cause a reduction and "holes" in a "hole trap" where they cause an oxidation. As stated earlier, this in essence is the goal of the primary photochemical process, i.e., production of an oxidant and a reductant that would not recombine with each other. However, there is no definite evidence to prove the existence of extensive electron-hole migration in the photosynthetic unit.

Two hypotheses have been proposed for the interaction between the two pigment systems mentioned in Section 1.5. (1) In the *separate package* model, PS I and PS II are physically isolated from each other; there is no exchange of excitation energy between the two systems. (2) In the *spillover* model, light energy absorbed by PS II (and not used by it) spills over to PS I, but not vice versa (see Myers, 1963). More recently a dynamic picture has arisen in which spillover from PS II to PS I is controlled by light and/or concentration of salts (e.g., Mg^{2+}; see discussion by Mohanty *et al.*, 1973, and Chapters 6 and 11 of this volume). Here also there are two different schemes. In the first, Mg^{2+}, Ca^{2+} etc. cause a decrease in the "spillover" of energy from PS II to PS I (Murata, 1969). In the second scheme, these divalent ions cause an increase in the "spillover" of evergy from PS I to PS II (Sun and Sauer, 1971). Present experimental data do not allow us to choose among the two pictures.

There are also two models for excitation energy transfer among units of the same photosystem: the *lake model* and the *puddle model*. In the *lake* model, reaction centers are embedded in a statistical fashion in a lake of bulk pigments; if one reaction center is closed, excitation energy can migrate to another one. In the puddle model, each unit is isolated; there is no excitation energy exchange among the different units (see Robinson, 1967). Perhaps the truth lies somewhere in between (see discussion of a "connected model" by Lavorel and Joliot, 1972).

In photosynthetic bacteria, the evidence for excitation energy transfer is the fact that excitation of carotenoids or BChl in any of the absorption bands results in the fluorescence of the component with its absorption band toward the longest wavelength, i.e., one representing the lowest electronic level (see Olson and Stanton, 1966). However, *Rhodopseudomonas viridis* has the ability for "uphill" energy transfer from BChl *b* (at 1025 nm) to the reaction center P985 (Zankel and Clayton, 1969). (Hammond, 1973, has also recently shown the existence of uphill energy transfer from Chl *a* to phycocyanin in blue-green algae.)

2.4. Reactions at Reaction Centers

Once the energy, by whatever mechanism, reaches the reaction center it is converted into chemical energy with the production of an oxidizing and a reducing equivalent. First, the reaction center Chl (or BChl) goes into the first excited singlet state. Immediately following this event, the primary electron acceptor is reduced and the reaction center becomes oxidized, which in turn receives an electron from the primary electron donor. Thus, the reaction center is returned to its original state, but the primary electron donor and acceptor are oxidized and reduced, respectively. For a review of primary processes in bacterial photosynthesis, see Clayton (1973).

Duysens *et al.* (1956), using difference spectrophotometry, first reported a light-induced reversible change in absorption of BChl. This change has been attributed to the oxidation of reaction center BChl, since a similar decrease in absorption could be brought about by oxidizing chemicals. The quantum yield of BChl oxidation is high (close to 1.0), indicating that it plays an important role in photosynthesis.

Since this initial discovery, several investigators have looked at the light-minus-dark difference spectra in several bacteria. Parson (1968), using nanosecond laser flashes, showed that the primary photochemical reaction in bacteria was the oxidation of reaction center BChl, with a simultaneous reduction of a primary electron acceptor. It has been shown that the bleaching at 860–870 nm (representing the oxidation of the reaction center BChl) is always accompanied by a blue shift of an absorption band at 800 nm, also due to a BChl. This blue shift in the absorption band accounts for the increase in absorption at 795 nm and a decrease at 810 nm in the difference absorption spectrum (see Chapter 3 of this volume).

Recently, several investigators have isolated pigment–protein complexes from different bacterial species which are free of the bulk of light-harvesting pigments. The absorption spectra of these reaction center preparations have bands at 800 nm and at 870 nm in the near-infrared region. Upon excitation with light, the band at 870 nm bleaches totally, while that at 800 nm shifts to the blue. A detailed discussion of reaction centers is given in Chapter 3 of this volume.

Kok (1956) discovered a light-induced absorbance change at 700 nm in green plants, which he attributed to the redox reactions of Chl *a* molecules (P700), representing the energy trap or the reaction center. Subsequently, he was able to measure the corresponding absorption change in the Soret region at 433 nm, and to prove that the decrease in absorbance at 700 nm was due to oxidation of Chl *a* (Kok, 1961). Since there are two pigment systems (I and II) in green plants, we would also expect two separate energy traps. P700 has been shown to be the reaction center or the energy trap for PS I. The absorbance changes representing the energy trap of PS II remained undetected for a long time despite the efforts of several investigators. Döring *et al.* (1967, 1968, 1969) and Döring and Witt (1972) have reported an absorbance change at 435 nm and 682–690 nm which they attributed to the energy trap of PS II. The 682 nm absorption change is distinct from that of P700 because its half-life is 100 times shorter than that of P700. In addition, the change is present in chloroplast particles enriched in PS II, but not in PS I-enriched particles. (The presence of P680–690 in algae has been inferred from the appearance of a fluorescence band at 695 nm under conditions when photo-synthesis is saturated, or blocked by poisons.) The P680 change can be observed in both wet heptane-treated chloroplasts (which have very little variable fluorescence) and in Tris-washed (0.8 M, pH 8.0) chloroplasts (that also have no variable fluorescence) (Govindjee *et al.*, 1970). These data along with those presented by Döring and Witt (1972) show that the P680 change is not an artifact due to fluorescence yield changes (also see Butler, 1972b).

3. SECONDARY EVENTS

3.1. Electron Transfer in Green Plants

An acceptable model of photosynthesis must make provision for the occurrence of two distinct photoreactions, the presence of a chain of electron-transporting intermediates (redox couples), and the presence of a phosphory-lating mechanism converting ADP to ATP. In addition, it must provide for the possibility of "artificial" electron transport, in which only a part of the photosynthetic electron transport chain is used. All models proposed, save one, invoke two or more reaction centers communicating with each other by means of electron-transport chains. These are classified as models with their photoreactions coupled *in series* or *in parallel*. One exceptional model has a single reaction center that is capable of two distinct photoprocesses.

3.1.1. Coupling in Series

This is the most widely accepted model. The Hill and Bendall (1960) model (or the Z scheme), briefly discussed here, has been elaborated into a de-tailed electron path leading from water to $NADP^+$ (for details see Hind and Olson, 1968, and Chapter 7). According to this model (Fig. 11), photo-reaction II oxidizes water ($E_0' = +0.8$ V) to free oxygen, and reduces Q ($E_0' \sim 0$ V), while photoreaction I reduces a low-potential electron acceptor X($E_0' \sim -0.6$ V) and oxidizes P700 ($E_0' \sim +0.4$ V). Recently, it has been suggested that Q may be equivalent to a component producing an absorbance change at 550 nm, referred to as C550 (see Arnon *et al.*, 1971; Erixon and Butler, 1971), and similarly, X to a component referred to as P430 (see Hiyama and Ke, 1971). Oxidized P700 is reduced by the reduced Q via exer-gonic electron transport reactions that are coupled to the phosphorylation of ADP (noncyclic photophosphorylation). The carriers catalyzing the *e*-trans-port reactions are thought to be Cyt b559 (low potential), PQ, Cyt f, and PC in that order. Böhme and Cramer (1972a) have shown that there is a site of phos-phorylation between PQ and Cyt f. The low potential electron acceptor (X) for PS I can either reduce $NADP^+$ via FRS(?) and Fd, NADPH entering the carbon fixation cycle, or it can return its electron to an intermediate pool

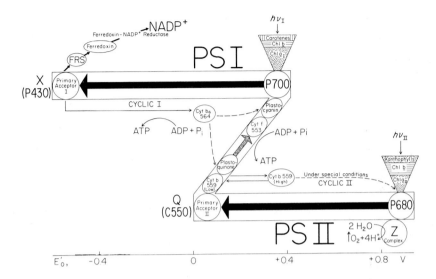

Fig. 11. The Z-scheme for electron flow in photosynthesis. The two bold horizontal arrows represent the two light reactions; all others, dark reactions. Flow of electrons from H_2O to $NADP^+$ (nicotinamide adenine dinucleotide phosphate) is designated as "noncyclic" electron flow, and from the primary acceptor of pigment system I (PS I) to the intersystem intermediates (plastoquinone, or plastocyanin) as "cyclic I". A similar cyclic flow of electrons involving only pigment system II may also exist; this could be designated as "cyclic II". E_0': oxidation–reduction potential, at pH 7, in volts; Z complex: the electron donor system for pigment system II (PS II); P680: the proposed energy trap of PS II; Chl a_{II}: bulk chlorophyll a of PS II; Chl b: chlorophyll b: Cyt b559: cytochrome b with one of its difference absorption band at 559 nm; Q (C550): electron acceptor for PS II, Q stands for the quencher of Chl a_{II} fluorescence, and C550 for a compound with a difference absorption band at 550 nm, ADP, P_i, ATP; adenosine diphosphate, inorganic phosphate, and adenosine triphosphate, respectively; Cyt f553· cytochrome f with one of its difference absorption bands at 553 nm; P700: the energy trap and the reaction center of PS I; Chl a_I: bulk Chl a of PS I: Cyt $b_6$564: cytochrome b_6 with one of its difference absorption band at 564 nm: X (P430): the PS I electron acceptor (pigment with one of its difference absorption band at 430 nm); FRS: ferredoxin reducing substance. (Data of various investigators, too numerous to mention here, are incorporated in this scheme; see text.)

carrier (probably via Cyt b_6, etc.). In the latter instance, electron transport traces a closed circuit utilizing only PS I; it is referred to as cyclic electron transport and the accompanying formation of ATP is designated cyclic photophosphorylation. (See Böhme and Cramer (1972b) for a discussion of the site of cyclic phosphorylation.)

The series model offers an explanation for the synergistic effect of the two photoreactions on the end products of chloroplast photochemistry (O_2 and NADPH). It also explains the antagonistic effect of the two photosystems on the intersystem electron carriers. The series model thus accounts for the enhancement effect observed for both oxygen evolution (discussed above) and the reduction of NADP$^+$ (R. Govindjee *et al.* 1962, 1964; Govindjee, 1963; Joliot *et al.* 1968; Avron and Ben Hayyim, 1969; Sun and Sauer, 1972) and for the reduction of Cyt f (Duysens *et al.*, 1961; Duysens and Amesz, 1962) and P700 (Kok and Gott, 1960; Kok and Hoch, 1961) by PS II and its oxidation by PS I. Since the electron carrier Q (the quencher of Chl fluorescence) is in the intersystem chain, the two photoreactions are expected to be antagonistic with respect to the Chl *a* fluorescence yield. Indeed, it is known that light absorbed by PS II causes the fluorescence yield of Chl *a* to rise, while light absorbed by PS I has the opposite effect (see Govindjee, *et al.*, 1960; Butler, 1962; Duysens and Sweers, 1963; Munday and Govindjee, 1969; Mohanty *et al.*, 1970).

A number of studies strongly support the series formulation. Some of these follow. (1) DCMU was shown to inhibit or block the electron transport from Q to other intersystem electron carriers (see Duysens, 1972). By adding DCMU and artificial electron donors, such as DCPIPH$_2$ and TMPDH$_2$, one could study PS I and its characteristics independently of PS II (Hoch and Martin, 1963). Similarly, reactions of PS II (electron flow from H$_2$O to certain Hill oxidants) have been studied by adding ferricyanide, quinones, or DCPIP, although these acceptors can also accept electrons from PS I under certain conditions (e.g., see Govindjee and Bazzaz, 1967; Selman and Bannister, 1971). Certain treatments, such as washing with 0.8 *M* Tris (at pH 8.0) (see Yamashita and Butler, 1968), irradiation with ultraviolet light (Jones and Kok, 1966) or heating chloroplasts to about 50°C for 5 min (see Yamashita and Butler, 1968; Katoh and San Pietro, 1968), result in the loss of the oxygen-evolving capacity; however, substitute electron donors (e.g., DPC and hydroxylamine) can reduce the electron acceptor and the reduction of Hill oxidants or NADP$^+$ can then be achieved. Thus, electron-flow separate from oxygen evolution can also be studied. (2) Removal of one or more components of the system results in loss of photochemical activity in partial reactions described above. Restoration of activity by the addition of the appropriate component(s) has elucidated the relative position of some of the electron carriers. Evidence from removal of carriers that block complete non-cyclic electron flow without interfering with the independent activity of either photosystem strongly implicates requisite link between the two photosystems. (3) Isolation and characterization of mutants lacking one or the other electron carrier has been very helpful in determining the location of the various electron carriers in the chain (see review by Levine, 1969). (4) Finally, detergent or mechanical fractionation of chloroplasts followed by differential centrifugation has led to the physical separation of two kinds of particles that are enriched in one or the other photosystem (see reviews by Boardman, 1970; Park and Sane, 1971; and Chapter 2 of this volume).

3.1.2. Parallel Coupling

Photosynthetic models in this classification invoke the formation of a strong reductant ($E_0 < -0.4$ V) by one or both photoreactions. This does not conflict with the energy content of a red quantum since a 700-nm photon has enough energy (1.77 eV) to span the 1.2 eV potential differences between water and $NADP^+$ (however, see Section 1.6). Although the models discussed below have not been supported by experiments as the series model has been, they will be mentioned here for completeness and for the sake of keeping an open mind toward future developments.

Gaffron (1962), in an attempt to account for the quantum requirement of 8 in bacterial photosynthesis and the photoreduction of carbon dioxide by hydrogen gas in algae, developed a scheme in which each photoreaction oxidizes cytochromes (denoted here as Y' and Z') and produces a low-potential reductant.

$$Z' + H_2O \xrightarrow{h\nu_{II}} Z'O + 2(H) \qquad (9.1)$$

$$Y' + H_2O \xrightarrow{h\nu_I} Y'O + 2(H) \qquad (9.2)$$

The oxidized cytochromes, then, react with a Mn-containing enzyme MnE (absent from photosynthetic bacteria) releasing oxygen as follows:

$$Z'O + MnEO \longrightarrow Z' + MnE + O_2 \qquad (9.3)$$

$$Y'O + MnE \longrightarrow Y' + MnEO. \qquad (9.4)$$

In photosynthetic bacteria, hydrogen donors could donate electrons to oxidized cytochromes, or there may be a cyclic reaction. Another model of parallel coupling was advanced by Govindjee et al. (1967; also see Hoch and Owens, 1963). In this model, photoreaction II produces a weak reductant, while photoreaction I supplies the ATP, or a high-energy precursor of it, by means of a cyclic electron transport. The weak reductant is assumed to reduce Fd, or another low potential intermediate, by a reverse electron flow powered by the energy-rich product of photoreaction I. In contrast to this model, McSwain and Arnon (1968) proposed that photoreaction II directly reduces Fd, while a photoreaction I-supported cyclic electron flow supplies the ATP required for CO_2 fixation. This model is based on the observation that enhancement could be demonstrated in isolated spinach chloroplasts in their laboratory only with CO_2 as the terminal electron acceptor, but not with $NADP^+$.

The existence of an enhancement effect in isolated chloroplasts with $NADP^+$ as the oxidant was, however, confirmed by Sun and Sauer (see their

1972 paper) in the presence of magnesium salt but not in its absence. Arnon *et al.* (1971) could not find this critical role of Mg salts. Sane and Park (1971) observed that when PPNR was added, enhancement was observed, but when Fd was used, it was not. We recall that in the early measurements of the enhancement effect in $NADP^+$ reduction R. Govindjee *et al.* (1962, 1964) had used PPNR and Mg^{2+} salts in their experiments. McSwain and Arnon (1972) recently confirmed the earlier work, and further showed that addition of ATP eliminated enhancement in CO_2 fixation with intact chloroplasts. Furthermore, they obtained a protein factor which, when added along with Fd, allows one to observe enhancement in $NADP^+$ reduction as with PPNR. The role of Mg^{2+} is still uncertain.

Knaff and Arnon (1969) proposed two sequential photoreactions, II_a and II_b, which are driven by the pigments of PS II. Photoreaction II_b oxidizes water and reduces an unidentified electron carrier, denoted from its difference absorption maximum as C550. Photoreaction II_a (that does not include P700) reduces Fd and oxidizes the copper protein PC. Electrons are transported from C550 to PC along a chain of carriers which includes PQ and Cyt b559. This transport is in the direction of the electrochemical gradient and it includes one phosphorylation site. In addition to the photoreactions II_a and II_b, photoreaction I carries out a cyclic electron transport through a chain of carriers that includes Chl *a* (P700), Fd, Cyt b_6, and Cyt f. There are two photophosphorylation coupling sites in this cyclic transport, one between Fd and Cyt b_6 and another between Cyt b_6 and Cyt f. Specific aspects of this model have been criticized by Esser (1972) and Bazzaz and Govindjee (1973), among others, on several grounds (also see Chapter 2 of this volume).

The parallel coupling model by Arnold and Azzi (1968) resembles the models of Govindjee and Arnon (*vide supra*) in the sense that it visualizes a Fd-reducing PS II and an auxiliary PS I-driven cyclic electron transport. The absorption of a photon by PS II results either in a reduced trap A and a Chl *a* cation, or in an oxidized trap B and a Chl *b* anion, as follows (absorption of 2 quanta in PS II are needed):

$$\text{Chl } a + A \xrightarrow{hv_{II}} \text{Chl } a^+ + A^- \tag{10.1}$$

$$\text{Chl } b + B \xrightarrow{hv_{II}} \text{Chl } b^- + B^+ \tag{10.2}$$

The ionic chlorophylls (holes and electrons) are discharged by recombination that leads to delayed light emission (hence the dependence of its intensity on the square of the excitation intensity) or to heat loss. The reduced energy trap A^- reduces Fd, while the oxidized trap B^+ decomposes water.

In general, models in which the two photoreactions of photosynthesis function with some degree of independence (e.g., the parallel coupling models) have the advantage that they do not require the synchronization of the primary photoacts for the sake of efficient photosynthesis. Such parallel models are supported by the recent work of Rurainski and Hoch (1972), who find that the addition of increasing concentration of $MgSO_4$ leads to a decreased turnover of P700 and an increased rate of $NADP^+$ reduction, i.e., there is no correlation between the number of electrons passing through P700 and those which end up in $NADP^+$; in fact, a competition is observed. On the other hand, it is difficult to adapt such models to the massive biochemical and biophysical evidence obtained in support of the series model (discussed in this and other chapters in this book).

3.1.3. The United Reaction Center

Franck and Rosenberg (1964) visualize a single reaction center performing both the photoreactions of photosynthesis. Two groups of light-collecting chlorophylls, one of which is amorphous and one semicrystalline, supply the excitation to this center. The latter group donate triplet excitation quanta, which are used for the reduction of an electron acceptor X, and the oxidation of a Cyt. This process is equivalent to photoreaction I of the series model. The amorphous Chl population supplies singlet excitation quanta, by which the reduction of the previously oxidized Cyt and the oxidation of another intermediate Z is effected. Oxidized Z extracts electrons from water to release oxygen, while reduced X provides the electrons for the reduction of carbon dioxide. These reactions are envisioned as follows:

$$X + 2Cyt \xrightarrow{Chl_T} X^{2-} + 2Cyt^+ \tag{11.1}$$

$$Z + 2Cyt^+ \xrightarrow{Chl_S^*} Z^{2+} + 2Cyt \tag{11.2}$$

$$Z^{2+} + H_2O \longrightarrow Z + 2H^+ + \tfrac{1}{2}O_2 \tag{11.3}$$

This model is now of historical importance only, since it does not provide for the chain(s) of electron transport carriers known to exist. It is mentioned here because of the possibility that PS I may operate via triplet state and PS II by singlet state; this, of course, is speculation. [See a recent review on the role of triplets by VanderMeulen and Govindjee (1973).]

3.2. Electron Transfer in Photosynthetic Bacteria

Electron flow in bacteria is thought to follow a cyclic pathway. Parson (1968) showed by means of laser flashes that in *Chromatium* chromatophore preparations, the light-induced oxidation of P890 was severalfold faster than the oxidation of a Cyt labeled C555. He also showed that the rate of rereduction of oxidized P890 after the flash corresponded to the rate of oxidation of

C555. He demonstrated that the primary photochemical reaction in bacteria is the oxidation of P890.

Following the primary event, i.e., the photooxidation of the reaction center BChl (P840, P870, or P890 depending upon the organism) and the reduction of the primary electron acceptor (X), the electrons then trace a cyclic pathway back to the oxidized reaction center via several intermediates (Fig. 12). Phosphorylation has been shown to be coupled to this electron transport. ATP (or the high-energy intermediate) thus produced supplies the energy for the reduction of NAD^+ by a reversed electron flow, coupled to substrate (i.e., external H donor) oxidation.

The identity of the intermediates involved in the electron transport is not definitely known except for the cytochromes. Through the use of absorption difference spectroscopy, light-induced oxidation of three cytochromes (C555, C552, and CC′) has been revealed by Olson and Chance (1960) in *Chromatium*. C552 photooxidation was shown to occur at low light intensities; it is auto-oxidizable under aerobic conditions. The rereduction of C552 is dependent upon the presence of oxidizable substrate (Olson and Chance 1960; Morita *et al.* 1965). The photooxidation of C555 and CC′ occurs at high light intensities, independent of C552. C555 and CC′ were linked with the cyclic electron transport (mentioned above) and C552 with a separate (?) noncyclic electron transport linked with substrate oxidation.

Morita (1968) showed that the action spectra for the oxidation of the C555 and C552 in *Chromatium* were different, suggesting the possible involvement of more than one light reaction. Cusanovitch and Kamen (1968a,b) confirmed these observations and titrated the midpoint oxidation–reduction potentials (E_0') for the various cytochromes in *Chromatium*. C555 has an E_0' (at pH 7.5) of $+ 0.32$ V, CC′ of $+ 0.18$ V, and C552 of slightly below 0.0 V. They also found that the light-minus-dark difference spectra in the near infrared region were dependent upon the ambient redox potential of the chromatophores. They suggested that the two electron transport systems were linked with different reaction centers P890 and P905 (the latter based on the presence of an increase in absorption at 905 nm in the light-minus-dark difference spectrum at low redox potentials).

Sybesma and Fowler (1968) and Fowler and Sybesma (1970) reported that in *Rhodospirillum rubrum*, C428 (C552) was oxidized at low light intensities and its rereduction was extremely slow in the absence of a substrate; however, it could be speeded up upon the addition of malate. Oxidation of C422 (C555) was shown to occur at high light intensity. They found that the action spectra for the oxidation of C422 and C428 were different. Thus they also suggested two electron transport pathways: a noncyclic pathway linked with substrate oxidation via C428, and a cyclic pathway linked with C422 and associated with phosphorylation. Whether the suggested noncyclic and cyclic electron transport pathways are mediated by different reaction centers is not certain. It has been shown that the photooxidation of both C552 and C555 is linked to the same reaction center P890 (Parson and Case 1970; Case *et al.* 1970; Dutton, 1971) in *Chromatium*, *Rhodopseudomonas viridis*, and *Rhodopseudomonas gelatinosa*. Such evidence is not available for *R. rubrum*; however, attempts at isolation of reaction center fractions from this organism has resulted in only one kind of preparation—that of P870 (Smith, 1972).

NAD$^+$ photoreduction in bacteria has been suggested (Olson and Chance, 1960; Bose and Gest, 1962; Keister and Yike, 1967; Jones and Vernon, 1969) to occur via a reversed electron transport with the help of ATP (or high-energy intermediate) produced in a light-induced cyclic electron flow. Nozaki *et al.* (1961) suggested a direct reduction of NAD$^+$ by a noncyclic electron transport. R. Govindjee and Sybesma (1970, 1972) have proposed two different pathways for NAD$^+$ reduction in *Rhodospirillum rubrum*. They suggested that succinate-supported NAD$^+$ reduction occurs by a reversed electron flow and NAD$^+$ reduction with DCPIPH$_2$ occurs directly via a noncyclic electron flow. R. Govindjee *et al.* (1974) showed that with the latter condition, low potential viologen dyes are capable of reacting with this system as inferred from light induced absorption changes of P870.

Hind and Olson (1968) also proposed a scheme of electron transport in bacteria in which there are two separate light-induced electron transport pathways. One is cyclic and is coupled to phosphorylation, thus reducing NAD$^+$ with the help of ATP, and another noncyclic pathway linked with substrate oxidation and direct NAD$^+$ reduction via Fd and flavoproteins. However, the existence of the noncyclic pathway remains to be proven (see Fig. 12 for a general scheme).

The nature of "X," the primary electron acceptor, in photosynthetic bacteria and PS I in green plants is under active investigation (see Chapter 3 of this volume for progress in this area). Dutton and Leigh (1973) have coined the term photoredoxin for the reduced form of "X" which exhibits an ESR signal centered at $g = 1.82$; it has an E_0' (at pH 7.0) of -50 mV in *R. spheroides* and -130 mV in *Chromatium*. On the basis of the EPR studies, Loach and Hales (personal communication, 1973) found that the primary electron acceptor in the photoreceptor complex preparations (from purple photosynthetic bacteria) has a midpoint potential of -0.37 V. Loach found a similar midpoint potential for quenching phototrap activity in chromatophores and whole cells when they were kept under very dark conditions and subjected to a single short pulse of light.

The presence of Fd has been shown in several photosynthetic bacteria. However, there is no conclusive evidence for its role in their electron transport pathway. Phosphorylation and cyclic electron transport have been reported in *R. rubrum* chromatophores from which Fd was absent (Horio *et al.*, 1968).

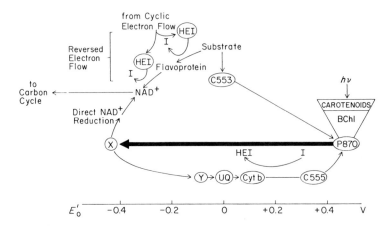

Fig. 12. A hypothetical scheme for electron flow in bacterial photosynthesis. NAD^+ (nicotinamide adenine dinucleotide) reduction may occur by reversed electron flow (see top left of scheme), or possibly by a direct reduction. E_0': oxidation–reduction, potential, at pH 7.0, in volts; P870: reaction center bacteriochlorophyll; BChl: bulk bacteriochlorophyll; X: primary electron acceptor; Y: secondary electron acceptor; UQ: ubiquinone; Cyt b: two kinds of cytochrome b; C555: cytochrome c with one of its difference absorption band at 555 nm; I: intermediate; HEI: high energy intermediate; C553: cytochrome c with one of the difference absorption band at 553nm. Whether two separate pigment systems (with their own reaction centers) sensitize two separate light reactions remains to be proven. Results of several experiments suggest the idea of a direct reduction of NAD^+. Such a direct NAD^+ reduction would be easier, if there was an electron acceptor (X) with more negative E_0' than -0.3 V. (However, an electron acceptor having a more positive E_0' than -0.3 V could also reduce NAD^+ directly under conditions when it is largely in the reduced state.)

Trebst *et al.* (1967) reported that Fd is not required for NAD^+ reduction in *R. rubrum*. Fd-linked NAD^+ reduction by molecular H_2 in *Chromatium* has, however, been reported by Weaver *et al.* (1965).

3.3. Oxygen Evolution in Green Plants

Although photosynthetic oxygen evolution was discovered 200 years ago by Priestley (see R. Hill, 1972), the oxygen evolving side of PS II is still not very well elucidated. An unknown component Z is the "primary" electron donor. Z is linked with the oxidation of H_2O through one or several steps. The nature of Z and the other components involved is not known. As noted earlier, this system is heat-labile, sensitive to ultraviolet radiation, and is (perhaps reversibly) impaired by treatment with a high molarity Tris at pH 8.0. After any of the above treatments, the oxygen-evolving capacity of the organism is lost, but substitute electron donors such as DPC or hydroxylamine can replace H_2O. We know that Mn^{2+} (see Cheniae, 1970), Cl^- (Izawa *et al.*, 1969), and bicarbonate ions (Stemler and Govindjee, 1973, 1974) are in some way involved in the oxidation of water to molecular O_2. A protein (enzyme) is implicated, but has never been isolated and purified (see Braun and Govindjee, 1972).

An important piece of information about the O_2-evolving system was provided by Allen and Franck (1955). They showed that after a period of dark adaptation, no O_2 was evolved by algae in the first flash of light; however, O_2 was evolved if this flash was preceded by weak light or another flash. Joliot *et al.* (1969) and Forbush *et al.* (1971) found that after dark adaptation the first flash of light yields no O_2, the second flash shows some, the third flash has a high yield, the fourth flash a little less, the fifth even less, then the sixth flash yields more, and the seventh flash again represents a maximum (see Fig. 1 in Chapter 8 of this volume). In this way, they found a periodic oscillation in the O_2 yield from a series of flashes, the yield reaching a peak with a period of 4. Thus, the 3rd, 7th, 11th, and so forth, flashes represented yield maxima. These oscillations eventually damped out with time, and the sequence could be repeated again by dark-adaptation of the system.

Kok *et al.* (1970) proposed a mechanism in which a light-activated accumulation of four positive charges was necessary before a molecule of oxygen could be evolved. In order to explain the occurrence of the maxima at the 3rd, 7th, and 11th flashes, they suggested that in the dark there are two stable species, such as Z and Z^+ (Z being the primary electron donor).* Thus, the Z^+ to Z^{4+} reaction needs 3 more quanta to enable the evolution of O_2; thereafter, it returns to the Z state and must build up four positive charges before the sequence is repeated (also see Mar and Govindjee, 1972b). Details of O_2 evolution are discussed by Joliot and Kok in Chapter 8 of this volume.

The basic idea is that four oxidizing equivalents must accumulate before O_2 can be evolved as follows:

$$Z \cdot P680 \cdot Q \xrightarrow{hv} Z^+ \cdot P680 \cdot Q^- \tag{12.1}$$

$$Z^+ \cdot P680 \cdot Q^- + A \longrightarrow Z^+ \cdot P680 \cdot Q + A^- \tag{12.2}$$

$$Z^+ \cdot P680 \cdot Q \xrightarrow{hv} Z^{2+} \cdot P680 \cdot Q^- \tag{12.3}$$

$$Z^{2+} \cdot P680 \cdot Q^- + A \longrightarrow Z^{2+} \cdot P680 \cdot Q + A^- \tag{12.4}$$

---

*There is some confusion regarding the nomenclature of the "primary" electron donor that donates electrons to the primary electron acceptor. The P680 itself is, in all likelihood *the* primary electron donor. For purposes of present discussion, the intermediate that donates electrons to P680$^+$ is referred to here as the primary electron donor. There are at least two intermediates between H_2O and P680—a primary donor, and a secondary donor. Unfortunately, no standard nomenclature has been accepted for these unknown intermediates. Some call it Z_1 and Z_2 (e.g., see Braun and Govindjee, 1972); Lavorel (see Chapter 5 of this volume) has used Y and Z, Papageorgiou (Chapter 6) "Z" and "S", and Avron (Chapter 7) "Z" and "E", respectively. The reader is warned that he should not transform "Z" of one author to that of another without a careful check.

$$Z^{2+} \cdot P680 \cdot Q \qquad \xrightarrow{h\nu} Z^{3+} \cdot P680 \cdot Q^- \qquad (12.5)$$

$$Z^{3+} \cdot P680 \cdot Q^- + A \longrightarrow Z^{3+} \cdot P680 \cdot Q + A^- \qquad (12.6)$$

$$Z^{3+} \cdot P680 \cdot Q \qquad \xrightarrow{h\nu} Z^{4+} \cdot P680 \cdot Q^- \qquad (12.7)$$

$$Z^{4+} \cdot P680 \cdot Q^- + A \longrightarrow Z^{4+} \cdot P680 \cdot Q + A^- \qquad (12.8)$$

$$Z^{4+} \cdot P680 \cdot Q + 2H_2O \longrightarrow O_2 + 4H^+ + Z \cdot P680 \cdot Q \qquad (12.9)$$

$$\text{Net:} \quad 2H_2O + 4A \xrightarrow{Z \cdot P680 \cdot Q \ + \ 4h\nu} O_2 + 4A^- + 4H^+ \qquad (12.10)$$

In order to explain all their kinetic data and those of Joliot, Kok and co-workers postulated the following: (1) after a dark period, the initial states are $Z^+ \cdot P680 \cdot Q$ (75%) and $Z \cdot P680 \cdot Q$ (25%); (2) there is a small possibility of "misses," i.e., the reaction will not proceed even though a PSU receives light; (3) in the light flashes used for their experiments, there was a small probability of "double hits," i.e., the reaction center moved two steps, instead of one, with one flash. With these and other assumptions (discussed by Joliot and Kok, Chapter 8, this volume), all the kinetic data were explainable. Much further work needs to be done to understand fully the mechanism of O_2 evolution.

As noted earlier, no oxygen evolution occurs in photosynthetic bacteria, as they are unable to use H_2O as an electron donor.

3.4. Photophosphorylation

Photophosphorylation was first clearly demonstrated by Arnon *et al.* (1954) and Frenkel (1954) in chloroplasts and in photosynthetic bacteria, respectively. In the Z scheme, ATP can be produced either during the exergonic flow of electrons in the noncyclic electron transport chain or via cyclic electron flow (see Section 3.1). Details of photophosphorylation are discussed by Jagendorf (Chapter 9 of this volume) and by Witt (Chapter 10).

It is generally accepted that at least one site of photophosphorylation is between PQ and Cyt f. If ADP or NH_4Cl is added to chloroplasts, PQ oxidation and Cyt f reduction is accelerated, i.e., electron flow from PQ to Cyt f is increased when ADP phosphorylation is allowed to occur, or when electron flow is uncoupled from phosphorylation with NH_4Cl. These and other data indeed suggest that one of the sites of phosphorylation is between PQ and Cyt f (see Böhme and Cramer, 1972a). From similar and other experiments, Böhme and Cramer (1972b) have further shown that electron flow from reduced Cyt b_6, that occurs via PQ, to Cyt f (cyclic) also contains an energy-coupling site; it was tentatively suggested that this coupling site is common to noncyclic electron flow. Further work is needed to establish other (suggested) site(s) of phosphorylation.

The mechanism of phosphorylation is not well understood. There are two major competing hypotheses: (1) chemical and (2) chemiosmotic (see Chapter 9 and 10 of this volume). In the former an energy-rich chemical intermediate is made and its deenergization is coupled to ATP production, whereas in the latter hypothesis, energy for ATP production is obtained by the discharge of membrane potential and/or H^+ ion gradient created across the chloroplast membrane by the electron flow. No definite conclusions can be made yet. However, the experiments of Jagendorf and Uribe (1966), in which they produced ATP in darkness from an artifically created H^+ gradient, and of Witt and coworkers (see Junge et al., 1969), in which they show that an absorption change of 515 nm (probably due to a change in membrane potential) decays faster in the presence of the uncoupler gramicidin or ADP + P_i, support Mitchell's chemiosmotic theory. [However, for different considerations, see Weber (1972, 1973).]

3.5. Carbon Fixation

After the elegant work of Benson, Bassham and Calvin (see Bassham and Calvin, 1957, and Gaffron, 1960), the path of carbon is shown as possessing the following basic features: (1) the carboxylation of a five-carbon (C_5) sugar phosphate (ribulose 1,5-diphosphate) with the production of 2 molecules of 3 phosphoglyceric acid; (2) the conversion of phosphoglyceric acid to triose phosphate (the simplest sugar) with the aid of reducing power (NADPH) and ATP made by the "light reactions" of photosynthesis; and (3) the interconversion of sugars, with the result that if one starts with 6 molecules of CO_2, 1 molecule of hexose is formed, and 6 molecules of ribulose 1,5-diphosphate are regenerated to carry on the cycle (often called the Calvin cycle; see Fig. 13). There is an additional input of an ATP molecule in the conversion of ribulose monophosphate to ribulose 1,5-diphosphate. The net reaction (for 1 glucose molecule) is:

$$6CO_2 + 12NADPH + (12H^+) + 18ATP \longrightarrow$$

$$C_6H_{12}O_6 + 12NADP^+ + 18ADP + 18P_i + 6H_2O \tag{13}$$

Recently (see review by Hatch and Slack, 1970) it has become clear that there is another pathway for CO_2 fixation, called the C_4 pathway, found in certain plants containing dimorphic (mesophyll and bundle sheath) chloroplasts. This mode is visualized by some workers as follows (see Fig. 14). In mesophyll cells, CO_2 is fixed by its addition to PEP, making oxalacetic acid. The latter is reduced to malate (in most C_4 plants called "malate formers") with the help of NADPH (produced in the light reactions). Malate is then

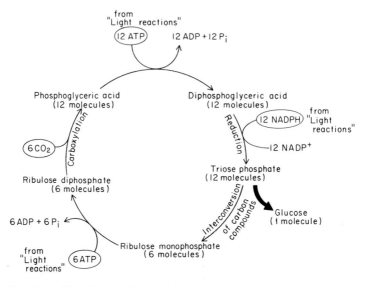

Fig. 13. A simplified diagram of the Calvin–Benson–Bassham cycle for the path of carbon fixation in photosynthesis. (See Fig. 11 for abbreviations.)

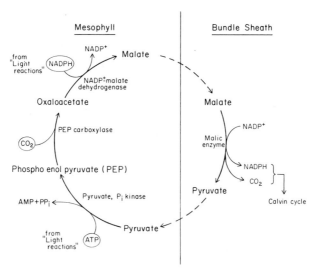

Fig. 14. A simplified diagram of the Hatch and Slack pathway for carbon fixation in plants with agranal (bundle sheath) and granal (mesophyll) chloroplasts.

translocated to the bundle sheath cells where it is converted into pyruvate and CO_2; simultaneously, $NADP^+$ is reduced to NADPH. Thus malate acts in transferring both the carbon and the reducing power from the mesophyll to bundle sheath cells. (There is also the possibility that bundle sheath cells make some of their own reducing power, and utilize CO_2 *per se* if it reaches them.) Bundle sheath cells then catalyze the usual Calvin cycle. Pyruvate is translocated back to the mesophyll cells where, with the help of an additional ATP molecule, it is converted to PEP to continue the cycle. Thus, the Hatch and Slack cycle requires an additional ATP. The above is a very simplified picture. The details of the pathway of carbon in C_4 plants (called "aspartate formers") that produce aspartate (instead of malate) is not well known yet. The path of carbon is discussed at length by Bassham (1965) and Hatch (1970) (also see Hatch *et al.*, 1970; Zelitch, 1971; Gibbs, 1971).

Almost all the photosynthetic bacteria have the ability to fix carbon by the reductive pentose (or Calvin) cycle. Evans *et al.* (1966), however, found that the green bacterium *Chloropseudomonas thiosulfatophilum* is also capable of reversing the citric acid (or Krebs) cycle. This reversal is accomplished by using reduced Fd and ATP made by light reactions. Several photosynthetic bacteria, when grown photoheterotrophically, instead of autotrophically, can obtain reducing equivalents by the oxidation of organic compounds via the citric acid cycle. Others produce it by the oxidation of externally added H_2 donors via the electron transport chain discussed above (noncyclic or by combination of cyclic with reversed electron flow). The reducing equivalents thus produced are then used for the reduction of carbon compounds (see discussion in Gregory, 1971).

Often, photosynthetic bacteria are capable of evolving hydrogen, using the hydrogenase enzyme. It is this reaction that may be of significance to the solution of our energy crisis, if one can efficiently couple the production of reducing equivalents from H_2O by green plants with bacterial hydrogen evolution (see Hollaender *et al.*, 1972).

4. SUMMARY

Photosynthesis is an oxidation–reduction process in which light energy is converted into chemical energy. In green plants, this amounts to the reduction of CO_2 into (CH_2O) and the oxidation of H_2O to molecular O_2. In photosynthetic bacteria various H donors replace H_2O, and no O_2 is evolved.

Several hundred pigment molecules (e.g., Chl *b* and Chl *a* in a green plants, or BChl in photosynthetic bacteria) somehow cooperate to perform photosynthesis. This collection of pigment molecules, along with other necessary components, is called a photosynthetic unit (PSU). In green plants, there are two pigment systems, and thus two types of PSU's. Each PSU has its own energy trap or reaction center (P700 in PS I and P680 in PS II). Certain photosynthetic bacteria possess at least one type of reaction center, labeled P870. Light energy absorbed by any of the pigment molecules in a PSU is probably transferred, as an "exciton" to the reaction centers where the primary light reaction occurs. The primary light reaction is essentially the oxidation of the reaction center and the reduction of a primary electron acceptor.

In green plants, one light reaction (I) leads to the oxidation of a cytochrome (f) and reduction of pyridine nucleotide. The other light reaction leads to the oxidation of H_2O to molecular O_2 and reduction of certain intermediate(s). The transfer of electrons from the latter to Cyt f completes the chain; this step is exergonic and is coupled to ATP production (noncyclic phosphorylation). A cyclic flow of electrons around reaction I can also lead to ATP production (cyclic photophosphorylation). ATP and reduced pyridine nucleotide produced by the two light reactions are sufficient to run the carbon fixation cycle that leads to the production of organic matter (CH_2O) from CO_2.

In photosynthetic bacteria, there is at least one light reaction that runs in a cyclic fashion producing ATP or a high-energy intermediate. By utilizing these high-energy compounds, bacteria reduce NAD^+ with external hydrogen donors (e.g., succinate), by reversed electron flow (i.e., against a potential gradient). There are, however, some indications that NAD^+ may be reduced directly by a noncyclic pathway. Whether a separate reaction center is involved here is not yet established. Photosynthetic bacteria fix carbon by various pathways.

APPENDIX I. PHOTOSYNTHESIS LITERATURE

Extensive literature exists in the area of photosynthesis research. A discussion of earlier literature appears in the most thorough and detailed treatise on photosynthesis by Rabinowitch (1945, 1951, 1956). These books are still a gold mine of information and ideas. Several more recent small books of varying emphasis and/or depth are also available. The primary photochemical reactions are discussed by Kamen (1963); the biophysical aspects of photosynthesis by Clayton (1965); carbon fixation by Bassham and Calvin (1957) and Calvin and Bassham (1962); physiological aspects by Heath (1969); and general aspects of photosynthesis by Gaffron (1960), Fogg (1968), Rabinowitch and Govindjee (1969), Zelitch (1971) and Gregory (1971). Mention should also be made of the small books by Hill and Whittingham (1953) and Rosenberg (1965).

Several valuable multiauthor books or collections of papers have appeared, many of which are proceedings of conferences held on photosynthesis.

The following is a partial list of such volumes edited by Franck and Loomis (1949), Danielli and Brown (1951), Gaffron et al. (1957), Fuller (1959), Allen (1960), Ruhland (1960), McElroy and Glass (1961), Ashida (1963), Gest et al. (1963), Kok and Jagendorf (1963), Tamiya (1963), Wurmser (1963), Krogmann and Powers (1965), Goodwin (1966), Thomas and Goedheer (1966), Olson (1967), San Pietro et al. (1967), Shibata et al. (1968), Metzner (1969), Hatch et al. (1970), Gibbs (1971), Forti et al. (1972), and Jacobi (1972). Reference is also made to three special issues of journals on photosynthesis, edited by Brown (1959), Pearlstein (1971), and Govindjee (1972).

A very valuable book on methods edited by San Pietro (1971, 1972), is available from Academic Press. Spectroscopy of chlorophyll is summarized in a book by Gurinovich et al. (1968), and all aspects of chlorophylls by Vernon and Seely (1966).

APPENDIX II. CONSTANTS, CONVERSION FACTORS, AND EQUATIONS

A. Physical Constants

Avogadro's number	N	6.023×10^{23} molecules mole^{-1}
Boltzmann's constant	k	1.3805×10^{-16} erg deg (Kelvin)$^{-1}$
Electronic charge	e	1.6021×10^{-19} C or 4.8030×10^{-10} esu
Faraday's constant	$\mathcal{F}e$	$96,494$ C g equiv^{-1}
Gas constant	R	8.314 J deg (Kelvin)$^{-1}$ mole^{-1} or 1.98 cal deg (Kelvin)$^{-1}$ mole^{-1}
Gravitational constant	g	980.665 cm sec^{-2} or 1.01325×10^{6} dyn cm^{-2}
Planck's constant	h	6.62×10^{-27} erg sec or 1.58×10^{-34} cal sec
Velocity of light *in vacuo*	c	2.997×10^{10} cm sec^{-1} or $186,000$ miles sec^{-1}

B. Conversion Factors

1 cal = 4.184×10^{7} ergs or 4.184 J (abs.) or 2.612×10^{19} eV or 2.106×10^{23} cm^{-1}.

1 cm^{-1} = 1.986×10^{-16} ergs or 1.986×10^{-23} J (abs.) or 4.747×10^{-24} cal or 1.240×10^{-4} eV.

1 einstein cm^{-2} sec^{-1} = 12×10^{14} ergs cm^{-2} sec^{-1} or $12 \times 10^{7}/\lambda$ W cm^{-2}.

1 erg = 10^{-7} J (abs.) or 2.389×10^{-8} cal or 6.242×10^{11} eV or 5.034×10^{15} cm^{-1}.

1 eV = 1.602×10^{-12} erg or 1.602×10^{-19} J (abs.) or 3.829×10^{-20} cal or 8066 cm^{-1} or photon of 1240 nm wavelength.

1 J (abs.) $= 10^7$ ergs or 0.2389 cal or 6.242×10^{18} eV or 5.034×10^{22} cm^{-1}

Temperature (T) in °Kelvin = temperature (t) in °Celsius (c) plus 273.16°C

1 W cm^{-2} $= 10^7$ ergs cm^{-2} sec^{-1} or 8.3×10^{-9} (λ) einsteins cm^{-2} sec^{-1} (λ in nm).

At any one wavelength, 1 W cm^{-2} $= 5.75 \times 10^5 \times V_{rel}$ f-c (see Fig. 3–3 for V_{rel} in R. K. Clayton, 1970).

To change some of the above units from molecular to molar units, multiply by 6.023×10^{23}. For example, 1 eV $= 3.829 \times 10^{-20}$ cal or $3.829 \times 10^{-20} \times 6.0223 \times 10^{23}$ cal mole^{-1} or 23.06 kcal mole^{-1}.

Energy of one quantum (λ in nm) $= 2 \times 10^{-9}/\lambda$ erg or $1{,}240/\lambda$ eV. The energy of 1 einstein (1 mole of quanta) is the above numbers multiplied by 6.023×10^{23}.

C. Equations

$E = h\nu = hc/\lambda$, where E = energy, h = Planck's constant, ν = frequency, c = velocity of light, and λ = wavelength of light.

$$\Phi_f = k_f/(k_f + k_d + k_p)$$

where Φ_f = quantum yield of fluorescence, k_f = rate constant (k) for fluorescence, $k_d = k$ for radiationless, and $k_p = k$ for photochemical deexcitation.

$$\Phi_f = \tau/\tau_0$$

where τ = measured lifetime of excited state and τ_0 = intrinsic lifetime when all deexcitation is by fluorescence.

$$1/\tau_0 \approx 3 \times 10^{-9}\ km^2\ \Delta k\ \varepsilon m$$

where km (cm^{-1}) = wavenumber at the peak, Δk (cm^{-1}) = half-bandwidth of the absorbance band, and εm (M^{-1} cm^{-1}) = extinction coefficient* at the absorption peak.

$OD = \log I_0/I = \log 1/T = -\log T = -\log (1-A) = \varepsilon cd$, where OD = optical density or absorbance, I_0 = incident intensity, I = transmitted intensity, T = fractional transmission, and A = fractional absorbance, ε = extinction coefficient, c = concentration, and d = optical pathlength.

$$\Delta c\ (\text{moles sec}^{-1}) = \Delta T/2.3T\ (d)\ (10^3)\ \Delta\varepsilon$$

---

*Extinction coefficient, an important property of a molecule, refers to the cross section of absorption (units of area). Therefore, the proper unit is cm^2 mole^{-1} of the material. However, one often uses the unit M^{-1} cm^{-1} for convenience in certain types of calculations. The two units are, however, equivalent as: M^{-1} cm^{-1} = moles^{-1} liter cm^{-1} = moles^{-1} (10^3) cm^3 cm^{-1} = 10^3 moles^{-1} cm^2 = 10^3 cm^2 moles^{-1}, or cm^2 mmoles^{-1}.

if $\Delta OD \ll 1$ and $\Delta T \ll T$, where Δc = change in molar concentration, ΔT = change in transmission, T = fractional transmission, d = optical path length and $\Delta \varepsilon$ = differential extinction coefficient. This equation is applied to light-induced absorption changes of various components in photosynthesis.

$$\Phi_p = \frac{\text{\# moles sec}^{-1} \text{ transformed}}{\text{\# einsteins absorbed sec}^{-1}} = \frac{\Delta c \text{ (moles sec}^{-1})}{I_0 \, A \text{ (einstein sec}^{-1})} = \frac{\Delta OD \text{ sec}^{-1}}{10^3 \, I_0 \, A \, \Delta \, \epsilon};$$

where Φ_p = quantum yield of a photochemical reaction, ΔOD is change in absorbance and other terms have the same meaning as described above.

ACKNOWLEDGMENT

We are grateful to Dr. Charles Arntzen, Dr. Warren Butler, Dr. Roderick Clayton, Dr. Richard Dilley, Dr. Ralphreed Gasanov, Dr. David Krogmann, Dr. Prasanna Mohanty, Dr. Alan Stemler, Ms. Barbara Zilinskas Braun and Mr. David VanderMeulen for their suggestions during the preparation of this manuscript. (We regret that some of the suggestions could not be incorporated in the final text.)

REFERENCES

Allen, F., and Franck, J. (1955). *Arch. Biochem. Biophys.* **58**, 510.

Allen, M. B. (ed.) (1960). "Comparative Biochemistry of Photoreactive Systems." Academic Press, New York.

Arnold, W., and Azzi, J. R. (1968). *Proc. Nat. Acad. Sci. U.S.* **61**, 29.

Arnold, W., and Kohn, H. (1934). *J. Gen. Physiol.* **15**, 391.

Arnold, W., and Meek, E. S. (1956). *Arch. Biochem. Biophys.* **60**, 82.

Arnon, D. I., Whatley, F. R., and Allen, M. B. (1954). *J. Amer. Chem. Soc.* **76**, 6324.

Arnon, D. I., Knaff, D. B., McSwain, B. D., Chain, R. K., and Tsujimoto, H. Y. (1971). *Photochem. Photobiol.* **14**, 397.

Ashida, J. (President) (1963). "Studies on Microalgae and Photosynthetic Bacteria." Jap. Soc. of Plant Physiol., Univ. of Tokyo Press, Tokyo.

Avron, M., and Ben-Hayyim, G. (1969). *Prog. Photosynthesis Res.* **3**, 1185 (For complete reference, see Metzner, 1969).

Bassham, J. A. (1965). *In* "Plant Biochemistry" (J. Bonner and J. E. Varner, eds.), Chapter 34, p. 875. Academic Press, New York.

Bassham, J. A., and M. Calvin. (1957). "The Path of Carbon in Photosynthesis." Prentice Hall, Englewood Cliffs, New Jersey.

Bazzaz, M. B., and Govindjee (1973). *Plant Sci. Lett.* **1**, 201.

Blinks, L. R. (1957). *In* "Research in Photosynthesis," p. 444 (For complete reference, see Gaffron *et al.*, 1957).

Blinks, L. R. (1959). *Plant Physiol.* **34**, 200.

Blinks, L. R. (1960). *In* "Comparative Biochemistry of Photoreactive Systems," p. 367. (For complete reference, see Allen, 1960.)

Blinks, L. R., and Van Niel, C. B. (1963). *In* "Studies on Microalgae and Photosynthetic Bacteria" (J. Ashida, ed.), pp. 297–307. Univ. Tokyo Press, Tokyo.

Boardman, N. K. (1970). *Ann. Rev. Plant Physiol.* **21**, 115.

Böhme, H., and Cramer, W. (1972a). *Biochemistry* **11**, 1155.

Böhme, H., and Cramer, W. (1972b). *Biochim. Biophys. Acta* **283**, 302.

Bose, S. K., and Gest, H. (1962). *Nature (London)* **195**, 1168.

Braun, B. Z., and Govindjee (1972). *Fed. Eur. Biochem. Soc. Lett.* **25**, 143.

Briantais, J. M., Merkelo, H., and Govindjee (1972). *Photosynthetica* **6**, 133.

Brody, S. S., and Rabinowitch, E. (1957). *Science* **125**, 555.

Brown, A. H. (ed.) (1959). Robert Emerson Memorial Issue, *Plant Physiol.* **34**, 179–363.

Butler, W. L. (1962). *Biochim. Biophys. Acta* **64**, 309.

Butler, W. L. (1972a). *Proc. Nat. Acad. Sci. U.S.* **69**, 3420.

Butler, W. L. (1972b). *Biophys. J.* **12**, 851.

Calvin, M., and Bassham, J. A. (1962). "The Photosynthesis of Carbon Compounds." Benjamin, New York.

Case, G. D., Parson, W. W., and Thornber, J. P. (1970). *Biochim. Biophys. Acta* **223**, 122.

Cheniae, G. M. (1970). *Ann. Rev. Plant Physiol.* **21**, 467.

Cho, F., and Govindjee (1970a). *Biochim. Biophys. Acta* **216**, 139.

Cho, F., and Govindjee (1970b). *Biochim. Biophys. Acta* **216**, 151.

Clayton, R. K. (1965). "Molecular Physics in Photosynthesis." Ginn (Blaisdell), Boston, Massachusetts.

Clayton, R. K. (1966). *In* "The Chlorophylls" (L. P. Vernon and G. R. Seely, eds.), p. 609. Academic Press, New York.

Clayton, R. K. (1970). "Light and Living Matter." McGraw-Hill, New York.

Clayton, R. K. (1973). *Ann. Rev. Biophys. Bioeng.* **2**, 137.

Cusanovich, M. A., and Kamen, M. D. (1968a). *Biochim. Biophys. Acta* **153**, 376.

Cusanovich, M. A., and Kamen, M. D. (1968b). *Biochim. Biophys. Acta* **153**, 418.

Danielli, J. F., and Brown, R. (Honorary Symp. Secretaries) (1951). "Carbon Dioxide Fixation and Photosynthesis." Cambridge Univ. Press, London and New York.

deKlerk, H., Govindjee, Kamen, M. D., and Lavorel, J. (1969). *Proc. Nat. Acad. Sci. U.S.* **62**, 972.

Döring, G., and Witt, H. T. (1972). *Proc. Int. Congr. Photosynthesis Res. 2nd* **1**, 39 (For complete reference, see Forti *et al.*, 1972).

Döring, G., Stiehl, H. H., and Witt, H. T. (1967). *Z. Naturforsch.* **B22**, 639.

Döring, G., Bailey, J. L., Kreutz, W., and Witt, H. T. (1968). *Naturwissenschaften* **55**, 220.

Döring, G., Renger, G., Vater, J., and Witt, H. T. (1969). *Z. Naturforsch.* **B24**, 1139.

Dutton, H. J., Manning, W. M., and Duggar, B. B. (1943). *J. Phys. Chem.* **47**, 308.

Dutton, P. L. (1971). *Biochim. Biophys. Acta* **226**, 63.

Dutton, P. L., and Leigh, J. S. (1973). *Biophys. Soc. Abstr.* p. 60a. Columbus, Ohio.

Duysens, L. N. M. (1952). Ph.D. Thesis, Univ. of Utrecht, The Netherlands.

Duysens, L. N. M. (1958). *Brookhaven Symp. Biol.* **11**, 10.

Duysens, L. N. M. (1964). *Progr. Biophys.* **14**, 1.

Duysens, L. N. M. (1972). *Biophys. J.* **12**, 858.

Duysens, L. N. M., and Amesz, J. (1962). *Biochim. Biophys. Acta* **64**, 243.

Duysens, L. N. M., and Sweers, H. E. (1963). *In* "Studies on Microalgae and Photosynthetic Bacteria" (J. Ashida ed.), p. 353. Univ. of Tokyo Press, Tokyo.

Duysens, L. N. M., Huiskamp, W. J., Vos, J. J., and Van der Hart, J. M. (1956). *Biochim. Biophys. Acta* **19**, 188.

Duysens, L. N. M., Amesz, J., and Kamp, B. M. (1961). *Nature (London)* **190**, 510.

Emerson, R. (1937). *Ann. Rev. Biochem.* **6**, 535.

Emerson, R. (1958). *Ann. Rev. Plant Physiol.* **9**, 1.

Emerson, R., and Arnold, W. (1932a). *J. Gen. Physiol.* **15**, 391.

Emerson, R., and Arnold, W. (1932b). *J. Gen. Physiol.* **16**, 191.

Emerson, R., and Lewis, C. M. (1943). *Amer. J. Bot.* **30**, 165.

Emerson, R., and Rabinowitch, E. (1960). *Plant Physiol.* **35**, 477.

Emerson, R., Chalmers, R., and Cederstrand, C. (1957). *Proc. Nat. Acad. Sci. U.S.* **43**, 133.

Erixon, K., and Butler, W. L. (1971). *Biochim. Biophys. Acta* **234**, 381.

Esser, A. F. (1972). *Biochim Biophys. Acta* **275**, 199.

Evans, M. C. W., Buchanan, B. B., and Arnon, D. I. (1966). *Proc. Nat. Acad. Sci. U.S.* **55**, 928.

Fogg, G. E. (1968). "Photosynthesis." American Elsevier, New York.

Forbush, B., Kok, B., and McGloin, M. (1971). *Photochem. Photobiol.* **14**, 307.

Fork, D. C. (1963). *Nat. Acad. Res. -Nat. Res. Council* **1145**, 352 (For complete reference, see Kok and Jagendorf, 1963).

Fork, D. C., and Amesz, J. (1969). *Ann. Rev. Plant Physiol.* **20**, 305.

Forti, G., Avron, M., and Melandri, B. A. (eds.) (1972). "Photosynthesis, Two Centuries after its Discovery by Joseph Priestley," Vol. I, Primary Reactions and Electron Transport; Vol. II, Ion Transport and Photophosphorylation; Vol. III, Photosynthesis and Evolution. *Proc. Int. Congr. Photosynthesis Res., 2nd Stresa, 1971*. Dr. W. Junk N. V. Publ., The Hague.

Fowler, C. F., and Sybesma, C. (1970). *Biochim. Biophys. Acta* **197**, 276.

Franck, J., and Loomis, W. E. (eds.) (1949). "Photosynthesis in Plants," A Monogr. of the Amer. Soc. Plant Physiol., Iowa State College Press, Ames, Iowa.

Franck, J., and Rosenberg, J. L. (1964). *J. Theoret. Biol.* **7**, 276.

French, C. S. (1971). *Proc. Nat. Acad. Sci. U.S.* **68**, 2893.

French, C. S., and Young, V. M. K. (1952). *J. Gen. Physiol.* **35**, 873.

French, C. S., Myers, J., and McLeod, G. C. (1960). *Symp. Comp. Biol.* **1**, 361.

Frenkel, A. W. (1954). *J. Amer. Chem. Soc.* **76**, 5568.

Fuller, R. C. (Chairman) (1959). *Photochem. Apparatus, Its Structure and Function, Brookhaven Symp. Biol.* No. 11. Brookhaven Nat. Lab., Upton, New York.

Gaffron, H. (1960). Energy Storage: Photosynthesis. *In* "Plant Physiology" (F. C. Steward, ed.), Vol. IB, pp. 3–277. Academic Press, New York.

Gaffron, H. (1962). *In* "Horizons of Biochemistry" (M. Kasha and B. Pullman, eds.), p. 59. Academic Press, New York.

Gaffron, H., and Wohl, K. (1936a). *Naturwissenschaften* **24**, 81.

Gaffron, H., and Wohl, K. (1936b). *Naturwissenschaften* **24**, 103.

Gaffron, H., Brown, A. H., French, C. S., Livingston, R., Rabinowitch, E. I., Strehler, B. L., and Tolbert, N. E. (eds.) (1957). "Research in Photosynthesis," *Proc. Gatlinburg Conf., 1955*. Wiley (Interscience), New York.

Gest, H., San Pietro, A., and Vernon, L. P. (eds.) (1963). "Bacterial Photosynthesis," *Symp. sponsored by the C. F. Kettering Res. Lab.* Antioch Press, Yellow Springs, Ohio.

Gibbs, M. (ed.) (1971). "Structure and Function of Chloroplasts." Springer-Verlag, Berlin and New York.

Gibbs, M. (1971). *In* "Structure and Function of Chloroplasts" (M. Gibbs, ed.), p. 169. Springer-Verlag, Berlin and New York.

Goedheer, J. H. C. (1972). *Ann. Rev. Plant Physiol.* **23**, 87.

Goodwin, T. W. (ed.) (1966). "Biochemistry of Chloroplasts," *Proc. NATO Advan. Study Inst., Aberystwyth*, 1965 Vol. I and II. Academic Press, New York.

Govindjee (1960). Ph.D. Thesis, University of Illinois at Champaign-Urbana.

Govindjee (1963). *Nat. Acad. Res. -Nat. Res. Council* **1145**, 318 (For complete reference, see Kok and Jagendorf, 1963).

Govindjee (ed.) (1972). Photosynthesis, A special issue dedicated to E. I. Rabinowitch, *Biophys. J.* **12**, 707–929.

Govindjee, and Bazzaz, M. (1967). *Photochem. Photobiol.* **6**, 885.

Govindjee, and Braun, B. Z. (1973). *In* "Physiology and Biochemistry of Algae" (W. D. P. Stewart, ed.). Blackwell Scientific Publications, Oxford (in press).

Govindjee, and Briantais, J. M. (1972). *Fed. Eur. Biochem. Soc. Lett.* **19**, 278.

Govindjee, and Mohanty, P. (1972). *In* "Biology and Taxonomy of Blue-Green Algae" (T. V. Desikachary, ed.), p. 171. Univ. of Madras, Madras, India.

Govindjee, and Papageorgiou, G. (1971). *In* "Photophysiology" (A. C. Giese, ed.), Vol. 6, p. 1. Academic Press, New York.

Govindjee, and Rabinowitch, E. (1960a). *Science* **132**, 355.

Govindjee, and Rabinowitch, E. (1960b). *Biophys. J.* **1**, 73.

Govindjee, and Yang, L. (1966). *J. Gen. Physiol.* **49**, 763.

Govindjee, Ichimura, S., Cederstrand, C., and Rabinowitch, E. (1960). *Arch. Biochem. Biophys.* **89**, 322.

Govindjee, Munday, J. C., Jr., and Papageorgiou, G. (1967). *Brookhaven Symp. Biol.* **19**, 434 (For complete reference, see Olson, 1967).

Govindjee, Döring, G., and Govindjee, R. (1970). *Biochim. Biophys. Acta* **205**, 303.

Govindjee, Papageorgiou, G., and Rabinowitch, E. (1973). *In* "Practical Fluorescence, Theory, Methods and Techniques" (G. G. Guilbault, ed.), p. 543. Dekker, New York.

Govindjee, R., and Sybesma, C. (1970). *Biochim. Biophys. Acta* **223**, 251.

Govindjee, R., and Sybesma, C. (1972). *Biophys. J.* **12**, 897.

Govindjee, R., Govindjee, and Hoch, G. (1962). *Biochem. Biophys. Res. Commun.* **9**, 222.

Govindjee, R., Govindjee, and Hoch, G. (1964). *Plant Physiol.* **39**, 10.

Govindjee, R., Rabinowitch, E. and Govindjee (1968). *Biochim. Biophys. Acta* **162**, 539.

Govindjee, R., Smith, W., and Govindjee (1974). *Photochem. Photobiol.* (in press).

Gregory, R. P. F. (1971). "Biochemistry of Photosynthesis." Wiley (Interscience), New York.

Gurinovich, G. P., Sevchenko, A. N., and Solov'ev, K. N. (1968). "Spectroscopy of Chlorophyll and Related Compounds." Izadatel' stvo Nauka i Tekhnika, Minsk. [Transl. is published by U.S. Atomic Energy Comm., Div. of Tech. Informat., as AEC to-7199 Chemistry (T1D-4500).

Hammond, J. (1973). Ph. D. thesis, Univ. of Illinois at Urbana-Champaign, Illinois.

Hatch, M. D. (1970). *In* "Photosynthesis and Photorespiration" (M. D. Hatch, C. B. Osmond, and R. O. Slayter, eds.) p. 139. Wiley, New York.

Hatch, M. D., and Slack, C. R. (1970). *Ann. Rev. Plant Physiol.* **21**, 141.

Hatch, M. D., Osmond, C. B., and Slayter, R. O. (eds.) (1970). "Photosynthesis and Photo-respiration." Wiley, New York.

Heath, O. V. S. (1969). "The Physiological Aspects of Photosynthesis." Stanford Univ. Press, Stanford, California.

Hill, R. (1939). *Proc. Roy. Soc.* **B127**, 192.

Hill, R. (1972). *Proc. Int. Congr. Photosynthesis Res. 2nd* **1**, 1 (For complete reference, see Forti *et al.*, 1972).

Hill, R., and Bendall F. (1960). *Nature (London)* **186**, 136.

Hill, R., and Whittingham, C. P. (1953). "Photosynthesis." Methuen, London.

Hind, G., and Olson, J. M. (1968). *Ann. Rev. Plant Physiol.* **19**, 249.

Hiyama, T., and Ke, B. (1971). *Arch. Biochem. Biophys.* **147**, 99.

Hoch, G., and Martin, I. (1963). *Arch. Biochem. Biophys.* **102**, 430.

Hoch, G., and Owens, O. V. H. (1963). *Nat. Acad. Sci. Nat. Res. Council* **1145**, 409.

Hollaender, A., Monty, K. J., Pearlstein, R. M., Schmidt-Bleck, F., Snyder, W. T., and Volkin, E. (1972). An inquiry into Biological Energy Conversion, Univ. of Tennessee, Knoxville, Tennessee.

Horio, T., Nishikawa, K., Horiuti, Y., and Kakeno, T. (1968). *In* "Comparative Biochemistry and Biophysics of Photosynthesis" (K. Shibata, A. Takamiya, A. T. Jagendorf, and R. C. Fuller, eds.), p. 408. Univ. Park Press, Philadelphia, Pennsylvania.

Izawa, S., Heath, R., and Hind, G. (1969). *Biochim. Biophys. Acta* **180**, 338.

Jacobi, G. (ed.) (1972) "Chloroplast Fragments Discussions of their Biophysical and Biochemical Properties." Univ. of Göttingen, Göttingen.

Jagendorf, A. T., and Uribe, E. (1966). *Proc. Nat. Acad. Sci. U.S.* **55**, 170.

Joliot, P., Joliot, A., and Kok, B. (1968). *Biochim. Biophys. Acta* **153**, 635.

Sybesma, C. (1970). *In* "Photobiology of Microorganisms" (P. Halldal, ed.), p. 57. Wiley, New York.

Sybesma, C., and Fowler, C. F. (1968). *Proc. Nat. Acad. Sci. U.S.* **61**, 1343.

Tamiya, H. (ed.) (1963). "Mechanism of Photosynthesis," *Proc. Int. Congr. Biochem., 5th.* Pergamon, Oxford.

Thomas, J. B., and Goedheer, J. C. (eds.) (1966). "Currents in Photosynthesis," *Proc. Western-Eur. Conf. Photosynthesis, 2nd, Woudschoten, The Netherlands,* 1965. Ad. Donker Publ., Rotterdam, The Netherlands.

Thomas, J. B., Blaauw, O. H., and Duysens, L. N. M. (1953). *Biochim. Biophys. Acta* **10**, 230.

Tomita, G., and Rabinowitch, E. (1962). *Biophys. J.* **2**, 483.

Trebst, A., Pistorius, E., and Baltscheffsky, H. (1967). *Biochim. Biophys. Acta* **143**, 257.

VanderMeulen, D., and Govindjee (1973). *J. Sci. Ind. Res. New Delhi* **32**, 62.

Van Niel, C. B. (1935). *Cold Spring Harbor Symp. Quant. Biol.* **3**, 138.

Van Niel, C. B. (1941). *Advan. Enzymol.* **1**, 263.

Vernon, L. P., and Avron, M. (1965). *Ann. Rev. Biochem.* **34**, 269.

Vernon, L. P., and Seely, G. R. (eds.) (1966). "The Chlorophylls." Academic Press, New York.

Vinogradov, A. P., and Teis, R. V. (1941). *C. R. Dokl. Acad. Sci. U.S.S.R.* **33**, 490.

Vredenberg, W. J., and Duysens, L. N. M. (1963). *Nature (London)* **197**, 355.

Weaver, P., Tinker, K., and Valentine, R. C. (1965). *Biochem. Biophys. Res. Commun.* **21**, 195.

Weber, G. (1972). *Proc. Nat. Acad. Sci. U.S.* **69**, 3000.

Weber, G. (1973). Paper presented at the *Conf. Mech. Energy Tranduction Biol. Sys., N. Y. Acad. Sci.* Feb. 7–9.

Witt, H. (1967). *Nobel Symp.* **5**, 261.

Wurmser, R. (President du Colloque) (1963). *La Photosynthèse, Colloq. Int. Centre Nat. Rech. Sci.* (C.N.R.S.), No. 119. Editions du C.N.R.S., 15 Quai Anatole-France, Paris, France.

Yamashita, T., and Butler, W. L. (1968). *Plant Physiol.* **43**, 1978.

Zankel, K. L., and Clayton, R. K. (1969). *Photochem. Photobiol.* **9**, 7.

Zankel, K. L., Reed, D. W., and Clayton, R. K. (1968). *Proc. Nat. Acad. Sci. U.S.* **61**, 1243.

Zelitch, I. (1971). "Photosynthesis, Photorespiration, and Plant Productivity." Academic Press, New York.

Note added in proof: During the time this volume has been in press many new papers have been published that have provided a better understanding of the process of photosynthesis. Some examples are: (1) Picosecond laser technology has permitted the measurement of lifetime of excited states in picosecond time scale [Seibert, M., and Alfano, R. R. (1974). *Biophys. J.* **14**, 269], and of the time needed (7 ± 2 psec) for excitation energy transfer plus primary oxidation reduction reaction at the reaction centers in photosynthetic bacteria [Netzel, T. L., Rentzepis, P. M., and Leigh, J. (1973). *Science* **182**, 238]. (2) Studies of structure–function relationships have lead to a tentative, but detailed, picture of how various components may be embedded in or located on the thylakoid membrane [Trebst, A. (1974). *Ann. Rev. Plant. Physiol.* **25**, 423]. (3) A light-induced electrical signal, perhaps reflecting membrane potential, which can be correlated with phosphorylation events has been measured [Witt, H. T., and Zickler, A. (1973). *FEBS Lett.* **37**, 307]. (4) The essential role of bicarbonate ions in O_2 evolution steps has been firmly established. It was shown that HCO_3^- accelerates the recovery of the reaction center II complex following light reactions so as to allow it to operate again [Stemler, A., Babcock, G. T., and Govindjee (1974). *Proc. Nat. Acad. Sci., U.S.,* in press]. (5) A phosphorylation step associated with system II has been shown to occur [Trebst, A., and Reimer, S. (1973). *Biochim. Biophys. Acta* **325**, 546; Ouitrakul, R., and Izawa, S. (1973). *Biochim. Biophys. Acta* **305**, 105]. (6) The role of C550 as a primary electron acceptor has now been challenged by several research workers—it may still be an indicator of PS II reactions. (7) Witt's group in Berlin has suggested that most of the electron flow may bypass cyt f. For an up-to-date knowledge, the reader should consult the recent issues of various journals including *Annual Review of Plant Physiology, Archives of Biochemistry and Biophysics, Biochemical Biophysical Research Communications, Biochemistry, Biochimica Biophysica Acta* (Bioenergetics), *Biophysical Journal, FEBS Letters, Journal of Biological Chemistry, Journal of Theoretical Biology, Photochemistry and Photobiology, Photosynthetica, Plant and Cell Physiology, Plant Physiology, Plant Science Letters,* and *Proceedings of National Academy of Sciences, U.S.A.*

2

Chloroplast Structure and Function

Charles J. Arntzen and Jean-Marie Briantais

ABBREVIATIONS

C_4 plants	Those plants which produce a 4-carbon dicarboxylic acid as the first product of photosynthetic CO_2 fixation (see Hatch, 1970)
C_3 plants	Those plants which produce phosphoglyceric acid (a 3-carbon compound) as the first stable product of photosynthetic CO_2 fixation (see Bassham, 1965)
CF	Coupling factor
Chl	Chlorophyll
Cyt	Cytochrome
DAB	Diaminobenzidine
DABS	Diazoniumbenzenesulfonic acid
DCMU	Dichlorophenylmethylurea
DCPIP	2,6-Dichlorophenolindophenol
DGD	Digalactosyl diglyceride
EDTA	Ethylenediaminetetraacetic acid
Fd	Ferredoxin
FRS	Ferredoxin reducing substance
MGD	Monogalactosyl diglyceride
MV	Methyl viologen
$NADP^+$	Nicotinamide adenine dinucleotide phosphate
PC	Plastocyanin
PMS	Phenazine methosulfate
PS I	Photosystem I
PS II	Photosystem II
PSU	Photosynthetic unit
P700	The reaction center Chl of PS I
RuDP carboxylase	Ribulose-1,5-diphosphate carboxylase (carboxydismutase)
TNBT	Tetranitro Blue Tetrazolium

1. INTRODUCTION

In a photosynthetic cell, the active chlorophylls are functionally organized in association with specific enzymic components within lamellar membranes. In prokaryotic organisms (photosynthetic bacteria and blue-green algae) these membranes extend throughout the cell, ranging in patterns of organization from single vesicular structures in *Rhodospirillum rubrum* to stacks of flattened vesicles in *Rhodospirillum molischianum* (Cohen-Bazire and Kunisawa, 1963, Gibbs *et al.*, 1965, Menke, 1966). In eukaryotes, the chlorophyll-bearing membranes are restricted to a specific cellular organelle: the chloroplast. This chapter will be devoted to a discussion of correlative studies of structure and function of chloroplast lamellae with emphasis on higher plants.

The overall morphology of chloroplasts may vary considerably depending on the organism studied. In the algae, morphologies differ from one long spirally wound chloroplast extending the length of the cell in some species of *Spirogyra* to a single cup-shaped chloroplast in *Chlorella*. Similarities and differences in algal chloroplast architecture have been reviewed by Kirk and Tilney-Basset (1967), and more recently by Gibbs (1970). The chloroplast of a typical higher plant is lens-shaped with a long diameter of 3–10 μm

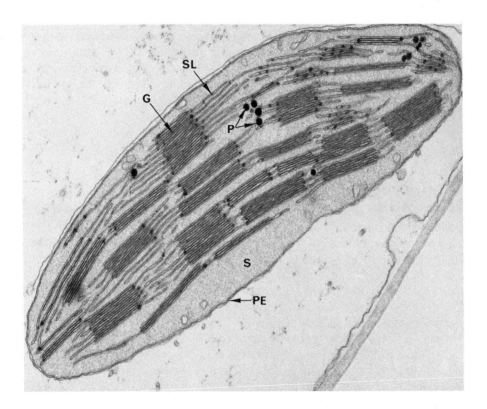

Fig. 1. A small portion of a lettuce (*Lactuca sativa*) leaf cell showing a mature chloroplast. A double membrane surrounds the chloroplast forming the plastid envelope (PE). Within the stroma (S) lie the unpaired stroma lamellae (SL), thylakoids stacked to form grana (G), and large, densely staining lipid droplets, plastoglobuli (P). The smaller, darkly staining droplets localized on the periphery of the grana stacks have been described by Sabnis *et al.* (1970); their chemical nature and physiological function, if any, remains unknown. This sample was prepared for electron microscopy by chemically fixing a portion of the intact leaf, embedding the leaf segment in plastic, and cutting ultrathin sections of the sample on a diamond knife.

(Möbius, 1920). A chloroplast from a mature leaf of *Lactuca sativa* is shown in Fig. 1. The organelle is primarily characterized by a continuous outer envelope enclosing a somewhat granular matrix (the stroma) within which the internal chloroplast membranes are embedded.

In this chapter, we will initially discuss the chloroplast envelope and stroma inclusions which have been thought to have a direct role in the biosynthesis of the internal lamellar membranes. We will then consider the structural organization of the internal lamallae with respect to two major questions. First, is there some correlation between the morphological organization of chloroplast lamellae and the function of these membranes? We will attempt to summarize those data which relate to the functional significance of stacked (grana) membranes. Second, is there any correlation between membrane substructure (as shown by electron microscopy) and the functional subunits of photosystem I and photosystem II? We will discuss the types of subunits which can be observed in chloroplast lamellae and attempt to relate this information to present knowledge on the spatial distribution of functional macromolecules within the membrane.

2. CHLOROPLAST MORPHOLOGY

2.1. The Chloroplast Envelope

The chloroplast envelope is comprised of a continuous double membrane which has a subunit structure in thin-sectioned preparations (Weier *et al.*, 1965a, 1966a). Numerous studies with isolated whole chloroplasts have indicated that the plastid envelope acts as a selective barrier to the transport of various metabolites into or out of the chloroplast. These studies are discussed in detail by Murakami *et al.* in Chapter 11 of this volume. In addition to its role as a selective boundary membrane for the chloroplast, the plastid envelope is thought to play a role in formation of new internal lamellae. Numerous reports have noted an apparent "budding" of the inner double membrane which results in the formation of vesicles within the chloroplants. These vesicles are thought to fuse to form lamellae. Much of the early literature describing vesicle formation and fusion was reviewed by Menke (1962), who concluded that there is no doubt that thylakoids can arise from the inner membrane of the plastid envelope. He did not rule out, however, a secondary site of lamellar formation not associated with the envelope. Ben-Shaul *et al.* (1964) have suggested a model for *Euglena* chloroplast membrane biosynthesis in which the inner plastid envelope membrane develops extensive invaginations to form the primary plastid lamellae. In a mutant of *Chlamydomonas reinhardii*, however, Ohad *et al.* (1967) found no evidence of plastid envelope invagination during rapid membrane biosynthesis. Moreover, a radioautographic study of membrane biosynthesis in the same algae by Goldberg and Ohad (1970) showed no movement of label from the plastid envolope to internal lamellae in the "pulse-chase" type of experiment when tritium-labeled membrane precursors were fed to the organism.

In light of the findings of Ohad and coworkers with *Chlamydomonas*, it seems pertinent that the importance of plastid envelope budding with respect to interior membrane biosynthesis be reinvestigated in other systems. Mackender and Leech (1972) have reported that the lipid composition of isolated chloroplast envelope membranes is qualitatively very different from the composition of the interior lamellae membranes. This must be taken into account in any membrane biogenesis scheme. Recent reports have described a variable pattern of differentiation along the length of the internal chloroplast lamellae (see review by Park and Sane, 1971) suggesting that a complex process of development must exist within the plastid. Furthermore, recent research with both algae and higher plants has indicated that a multistep assembly process occurs during chloroplast membrane biosynthesis (discussed in Section 3.3). Techniques are now becoming available which allow the isolation of the membranes of the plastid envelope (Mackender and Leech, 1972) and extensive subfractionation of the internal chloroplast lamellae (Arntzen *et al.*, 1972). Coupling of these procedures and use of radioactive membrane precursors may allow a specific determination of the initial site of new membrane biosynthesis and a more thorough evaluation of the role of the plastid envelope in this process.

In some plant species an extensive system of anastomizing tubules and vesicles has been shown to be contiguous with the inner limiting membrane of the chloroplast envelope (see Fig. 2 and 3). This "peripheral reticulum" has been reported to be continuous with the interior lamellar system of the chloroplast by Rosado-Alberio *et al.* (1968). There may be some question as whether their fixation procedure has induced an artifactual continuation of the membrane system, since Laetsch (1971) has reported modifications in the appearance of the peripheral reticulum by various fixatives. In spite of the possible misinterpretation of lamellar membrane–peripheral reticulum interconnections, the reports of Rosado-Alberio *et al.* and Laetsch are in agreement that the peripheral reticulum is a unique structural feature of the chloroplasts of plants possessing the C_4 dicarboxylic acid pathway for carbon fixation. There have been some reports, however, which seem to present exceptions to this rule. Bisalputra *et al.* (1969) have reported that a system of peripheral vesicles can be seen in wheat mesophyll chloroplasts (a C_3 or Calvin cycle species) and Hilliard and West (1971) and Gracen *et al.* (1972) have noted a peripheral reticulum in some varieties of *Dactylis glomerata* and *Typha latifolia* (also C_3 species). The latter authors suggest that

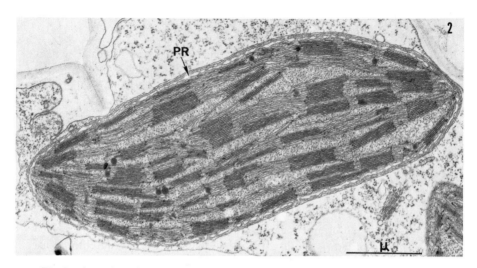

Fig. 2. A portion of mesophyll cell from the leaf of *Zea mays* showing an intact chloroplast which contains numerous grana stacks. The system of anastomizing vesicles which is continuous with the inner plastid envelope membrane is called the peripheral reticulum (PR). (Sample preparation as described for Fig. 1.) (After Chollet, 1972.)

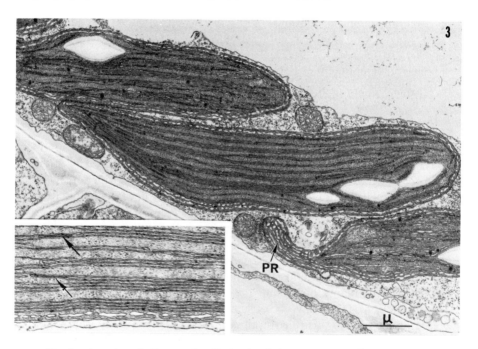

Fig. 3. A portion of a *Zea mays* bundle sheath cell showing portions of several chloroplasts. An extensive peripheral reticulum (PR) is evident. The very limited amount of membrane stacking occurring in these plastids is shown most clearly in the figure inset. Occasional regions of membrane appression are indicated by the arrows (Sample preparation as described for Fig. 1.) (After Chollet, 1972.)

there is a strong correlation between a limitation in photorespiration and the presence of a peripheral reticulum. There is also some evidence that a peripheral reticulum can be detected in the chloroplasts of C_3 plants subjected to environmental stress. Taylor and Craig (1971) have observed an extensive system of vesicles along the plastid envelope in soybean plants which were subjected to low temperatures. Chloroplasts show an extensive peripheral reticulum in some cells of drought-stressed sunflower leaves (J. S. Boyer and R. Keck, personal communication). Both soybean and sunflower plants are C_3 species.

2.2. The Chloroplast Stroma

The stroma of the chloroplast is the proteinaceous matrix which is enclosed by the chloroplast envelope (Menke, 1966). In addition to the internal chloroplast lamellar system, it contains a number of particulate structures visible by electron microscopy. These include ribosomes and strands of DNA. It is now well documented that both these components play a role in chloroplast self-regulation and replication (see Kirk, 1970; Sager, 1972). Ellipsoidal starch grains ranging in size up to 1.5 μm often lie between the internal lamellae of the chloroplast (see Fig. 2).

Many algal chloroplasts contain either round, angular, or irregularly shaped regions within the stroma which are called pyrenoids (Kirk and Tilney-Basset, 1967; Griffiths, 1970). They appear as dense, somewhat granular deposits which are occasionally transversed by the chloroplast lamellae. Recently Holdsworth (1971) has succeeded in isolating the pyrenoid of *Eremosphaera viridis*, a green alga, and has shown that it is primarily composed of two proteins. The major protein component of the isolated pyrenoid fraction was suggested to be RuDP carboxylase. This is in agreement with the findings of Goodenough and Levine (1970) and Togasaki and Levine (1970) who noted rudimentary pyrenoids in *Chlamydomonas* mutants that had low rates of RuDP carboxylase synthesis. An apparently related structure in higher plant chloroplasts is the "stroma center" discovered by Gunning (1965) in *Avena sativa* plastids. The stroma center appears as a mass of tightly packed, but regularly oriented, fibrils. Each fibril has a subunit structure which appears closely related to that of RuDP carboxylase (Gunning *et al.*, 1967). The exact function of the stroma center or the pyrenoid is still not clear. The structures may simply be a storage form for certain of the Calvin cycle enzymes.

Stroma components which are present in all plastids are the osmiophilic globules (or plastoglobuli). After osmium fixation these structures stand out as dense deposits lacking a limiting membrane (see Fig. 1, and discussion by Kirk and Tilney Basset, 1967). Separation and purification of these globules free from the chloroplast lamellae (Lichtenthaler and Sprey, 1966;

Barr *et al.*, 1967) has shown them to be composed primarily of lipophilic chloroplast quinones such as plastoquinones, vitamin K_1, α-tocopherol-quinone, and α-tocopherylquinone. Plastoglobuli increase in number in dark-grown plants, but decrease during membrane synthesis in the light (Sprey and Lichtenthaler, 1966). It has been demonstrated that chloroplasts in senescing leaves show a progressive breakdown of lamellar structure with a corresponding increase in size of the plastoglobuli (Barr and Arntzen, 1969; Lichtenthaler, 1969). It therefore appears that the plastoglobuli can function as extralamellar pools of membrane lipids which are utilized during membrane biosynthesis or which accumulate during membrane degradation.

2.3. Organization of Chloroplast Lamellae

The internal lamellar structure of higher plant chloroplasts has been studied with the electron microscope in numerous laboratories beginning with the initial investigations of Menke in 1940. By the early 1950's, it was well established that a characteristic feature of higher plant chloroplasts was the presence of small membranous disks which were stacked one upon another much like a pile of coins (Steinmann, 1952; Wolken and Palade, 1952; Frey-Wyssling and Steinmann, 1953; see review by Rabinowitch, 1956). Although the development of terminology used to describe the lamellar system has been diverse, most investigators now have adopted a system of nomenclature based on that of Menke (1962) and Weier and coworkers (1963, 1965a,b). Each grana stack is composed of two or more saclike disks termed thylakoids. The end portion of each thylakoid, which is in contact with the stroma, is called the margin. A partition is that region where two thylakoids are tightly appressed, and the loculus is the interior volume enclosed by one thylakoid. It has been recognized for many years that grana stacks are interconnected by membranous regions (the stroma lamellae) which unite thylakoids in separate stacked regions. Varying interpretations of the patterns of membranous interconnections have resulted in several models for chloroplast lamellar architecture. These have been discussed and compared by Kirk and Tilney-Basset (1967) and more recently by Park and Sane (1971). The overriding theme that has come from these studies is that each stroma lamellae is continuous with numerous thylakoids within the same grana stack and also with thylakoids in different grana stacks. (Heslop-Harrison, 1963; Wehremeyer, 1964; Weier *et al.*, 1966b; Paolillo, 1970). This extensive interconnection suggests that the loculus of the thylakoids and the space between the double stroma lamellae are all part of one single anastomosing internal chamber. While this may simply result from some unifying aspect of the original membrane biosynthesis, it does suggest, as noted by Weier *et al.* (1966b), that the internal compartment could function as a channel for diffusion of various photosynthetic substrates. Bearing in mind that most PS II is thought to be localized in the grana regions in normal mature chloroplasts of higher plants (see Section 3), this may have some practical role in the photoreactions.

One difficulty with the enthusiastic acceptance of most of the membrane models for grana organization which are discussed in the literature is that they imply a certain degree of rigidity of membrane organization. Accepting this notion makes it difficult to reconcile the results of Izawa and Good (1966a,b) who demonstrated that chloroplasts isolated in a low-salt media lose all grana stacking structure. These swollen chloroplasts appeared to be made up of continuous sheets of paired membranes. More importantly, addition of salts to the grana-free plastids resulted in restacking of the membranes to give structures that sometimes were indistinguishable from the original normally isolated chloroplasts. Murakami and Packer (1971) have extended these studies and have concluded that the area of thylakoid pairing must be a region of high hydrophobicity in the presence of cations. The experiments still do not resolve, however, how a complicated membrane system such as that described in models by Paolillo (1970), Wehremeyer (1964), Heslop-Harrison (1963) or Weier *et al.* (1966b) can reversibly change from grana stacks interconnected by stroma lamellae to parallel sheets of membranes and then back to stacked structures.

One additional aspect of this discussion should be concerned with the presence of grana stacks *in vivo*. Punnett (1966) has reported the use of a Carbowax isolation medium which allows the preparation of nongranal chloroplasts from several plant species. (These are not similar to the low-salt chloroplasts described above, but are whole, intact plastids.) Punnett contends that the grana observed in chloroplasts isolated by conventional means form as a result of damage during chloroplast isolation. The suggestion that chloroplasts are usually agranal is in agreement with certain of the earlier light microscope studies which described homogeneous chloroplasts *in vivo* (Menke, 1938; Granick and Porter, 1947; McCledon, 1954). More recently, Punnett (1971) had demonstrated that a conversion from homogeneous to granal chloroplasts in *Elodea* can be regulated by light quality and light intensity. Punnett (1966) has also stated that most electron microscopy fixatives induce grana stacking, thus indicating that most *in situ* electron microscopy preparations may not represent normal organization. In spite of these findings, however, the tendency of chloroplast membranes to organize into stacks, whether real or induced, is certainly the result of the presence of distinctive areas of membrane specialization. The fact that these grana stacking regions are different in biochemical capacities, and thus presumably in structural components, will be discussed in the following section.

The preceding discussion of grana has pertained almost entirely to higher plants. We should emphasize at this point that many algal chloroplasts do not contain stacks of disklike thylakoids. Instead, these species usually have long parallel lamellae, along which adjacent membrane pairs are usually fused for much of their length (see Kirk and Tilney-Bassett, 1967). There is no precedent for calling these pairs of appressed lamellae grana, since only two membranes are involved. There is also no evidence which would indicate whether the function (if any) of the membrane fusion in algae is related to the formation of grana stacks in higher plant chloroplasts.

3. LOCALIZATION OF PHOTOCHEMICAL ACTIVITIES ALONG THE CHLOROPLAST LAMELLAE

Spencer and Wildman in 1962 made the observation that almost all Chl fluorescence originated from the grana of isolated chloroplasts when viewed by light microscopy. They concluded that all Chl must therefore be localized in the stacked regions and not in the stroma lamellae. In later studies, Lintilhac and Park (1966) used a combined analysis of lamellar fragments by fluorescence and electron microscopy to demonstrate that all chloroplast lamellae (both grana and stoma lamellae) showed Chl fluorescence. The contradictory data from the two laboratories are probably the result of a lower fluorescence yield of Chl in the stroma lamellae (Michel and Michel-Wolwertz, 1969; Park and Sane, 1971) and an increase in fluorescence intensity in stacked regions simply due to high concentration of Chl in the tightly packed thylakoids. The low level of fluorescence by single stroma lamellae was probably not detected by light microscopy in the earlier studies (see discussion by Park and Sane, 1971).

Some of the first attempts at localizing sites of photoreductive activity within chloroplasts used a cytochemical staining procedure. Nagai, Vater, Metzner and others (see Rabinowitch, 1956) monitored the reduction of silver nitrate to metallic silver by chloroplast suspensions to demonstrate that the plastid membranes, and in particular the grana, were the primary sites of silver deposition. Somewhat later, Weier *et al.* (1966b) used TNBT as an electron acceptor for the Hill reaction and showed, by electron microscopy, that precipitated reduced diformazan was concentrated in the partition region of the grana stacks. Although the authors used this evidence to suggest that all Chl is in the grana regions, it is now possible to reinterpret this finding as showing that all PS II (which is necessary for the TNBT Hill reaction) is localized in the grana. Nir and Seligman (1970) more recently showed that photochemical oxidation of DAB results in a uniform precipitate over both grana and stroma lamellae. Since DAB acts as an electron donor to PS I in intact cells (Chua, 1972), it can be concluded that there is photochemically active Chl in both grana and stroma lamellae.

D. O. Hall *et al.* (1971, 1972) have followed Hill reaction activity in the presence of ferricyanide and copper sulfate. They have observed that the ferrocyanide produced by photoreduction precipitates as Cu ferrocyanide in large deposits visible in electron microscopy preparations. Since these deposits were localized over both grana and stroma lamellae, Hall and co-workers concluded that PS II is present in both lamellar regions. The authors also suggest, however, that there is a movement of ferrocyanide away from its initial site of reduction to another membrane crystallization point.

3.1. Mechanical Fractionation of Chloroplasts: Separation of Grana and Stroma Lamellae

The most direct means of determining the localization of photochemical activities within the grana and stroma lamellae is to isolate these two membrane regions separate from one another and characterize them biochemically. Methods to accomplish this fractionation have been developed in recent years. The techniques are based on the concept that the unpaired stroma lamellae extending between grana stacks are sheared from the more massive grana by mechanical or osmotic shock (Jacobi and Lehmann, 1969; Park and Sane, 1971). Structural evidence in support of this concept was first obtained by Gross and Packer (1967) who sonicated chloroplasts and demonstrated that the heavy fraction obtained by differential centrifugation contained mainly grana stacks, while the light fractions were composed of small vesicles. Jacobi (1969) and Jacobi and Lehmann (1969) extended these studies to demonstrate that the experimental conditions used during sonication greatly influenced the pattern of membrane disruption. They concluded that when grana structures are stabilized by high-salt concentrations and short periods of ultrasonic treatment are employed, the light vesicular membranes released are derived from stroma lamellae. This conclusion was supported by the findings of Sane *et al.* (1970) who used the French press to fragment chloroplast membranes. They used both thin sectioning and freeze-etching to demonstrate that the mechanical treatment causes breakage and vesiculation of stroma lamellae and end membranes of grana stacks. Their results were particularly convincing since use of the freeze-etch procedure allowed them to identify grana-derived membranes on the basis of membrane substructure (see Section 4.3). They could therefore conclude that none of the light vesicles released by their procedure were derived from grana partition regions.

Mechanical treatment of chloroplasts containing various amounts of stroma lamellae also provides evidence in support of the fragmentation hypothesis of Jacobi and Lehmann (1968, 1969) and Park and coworkers (Park and Sane, 1971). Sesták (1969), Goodchild and Park (1971) and Arntzen *et al.* (1971) have shown that higher amounts of Chl are released into the light fraction by French press treatment of chloroplasts isolated from young or immature leaf fragments than from mature leaves of spinach, lettuce, or radish. Since the undeveloped chloroplasts in young leaves have a greater proportion of unpaired lamellae with respect to stacked regions, these findings support the idea that mechanical treatment of chloroplasts releases vesicles which are derived from stroma lamellae.

An early suggestion that mechanically treated chloroplasts release membrane fragments which differ not only structurally, but also functionally from the remaining lamellar membranes was based on an observed difference in Chl content of the various fractions (J. A. Gross *et al.* 1966; Biehl, 1966; and Gross and Packer, 1967). A high Chl *a/b* ratio was consistently observed in the light fractions and a lower Chl *a/b* ratio was observed for the heavier membranes. In addition to the change in pigment distribution, Michel and Michel-Wolwertz (1969, 1970), using a French pressure cell to fragment chloroplasts, discovered that the light fragments separated on a sucrose gradient were greatly enriched in PS I activity, whereas the heavy membranes were somewhat enriched in PS II activity (see also Murata and Brown, 1970). Jacobi (1969) and Jacobi and Lehmann (1969) used a differential centrifugation procedure to separate light and heavy fragments from sonicated chloroplasts. Their results showed only PS I activity in the light lamellae fraction and both PS I and PS II in the heavy grana membranes. They have suggested that the unpaired stroma lamella, *in situ*, have only PS I activity, while grana stacks have both PS I and PS II. Sane *et al.* (1970), using a modification of the French press procedure of Michel and Michel-Wolwertz, have extended and refined these structure–function studies to conclude that not only the stroma lamellae but also the grana end membranes of mature, intact spinach chloroplasts contain only PS I activity, while the grana partition regions have both PS I and PS II.

Suzuki *et al.* (1970) have described the preparation of chloroplast fragments by grinding whole leaves in a low-salt buffer. The cellular homogenate was separated by differential centrifugation. The light fractions obtained had high Chl *a/b* ratios, a decreased level of accessory pigments, no Hill reaction activity, but high levels of $NADP^+$ reduction in presence of an electron donor for PS I. All of these characteristics are similar to those of the stroma lamellae particles described above. In light of the earlier demonstration of Jacobi and Perner (1962) that osmotic shock causes release of stroma lamellae fragments, it would appear that the low-salt grinding procedure of Suzuki *et al.* resulted in the release of unpaired lamellae during chloroplast preparation.

Terpstra (1970) has also reported obtaining a light fraction containing chloroplast fragments after homogenizing spinach leaves in a Braun multi-press. Samples of the homogenate were collected by differential centrifugation and then sucrose gradient separation. The light fractions (containing 3.5% of the total Chl) had PS I but not PS II activities. Terpstra suggested that these PS I particles represent chloroplast fragments which are physically separate from the major portion of the chloroplast lamellae in the chloroplast in the intact leaf. However, it seems difficult to rule out the possibility that Terpstra's PS I particles might have originated from stroma lamellae fragmentation during isolation. The highly disrupted chloroplast membranes shown in the micrographs of Terpstra (1970) would support this speculation.

Since the procedures used to isolate stroma lamellae fragments involve generation of shearing forces which would conceivably have deleterious effects on the membranes, it might be argued that the observed lack of PS II activity in the stroma lamellae fraction is an artifact of preparation. Park and coworkers (Sane *et al.*, 1970; Park and Sane, 1971; and Park *et al.*, 1971) have provided several lines of evidence regarding the composition of the fractions which indicated that this is not the case. It was found that the stroma lamellae membranes have a higher ratio of Chl a/Chl b, and grana membranes a lower ratio of Chl a/chl b than that which is observed for intact chloroplasts (Sane *et al.*, 1970). Stoma lamellae are relatively deficient in Cyt 559 and manganese, but enriched in the reaction center of PS I (P700) when compared to grana. Electrophoresis of sodium dodecyl sulfate solubilized membranes showed that stroma membranes contained a predominance of the protein complex corresponding to PS I, while grana contained complexes for both PS I and II. The light stroma lamellae fragments were found to have very little photochemically or chemically (dithionite) induced Chl fluorescence yield change (indicating a lack of PS II). These data suggest the absence rather than an inactivation of PS II in stroma lamellae.

Arntzen *et al.* (1971) have demonstrated that the energy-coupling mechanism in stroma lamellae fragments may be modified compared to that in the isolated grana stacks. While both fractions can carry out rapid rates of PMS-catalyzed cyclic photophosphorylation, only the grana demonstrated active proton accumulation (see Chapter 9 of this volume for a discussion of photophosphorylation). No light-induced proton uptake could be detected in the stroma lamellae. Proton gradient dissipating agents such as NH_4Cl or nigericin (in the presence of K^+) had only slight inhibitory effects on stroma lamellae phosphorylation, but were highly effective against grana membranes. Further evidence suggesting that stroma lamellae do not have the usual proton uptake in intact chloroplasts came from comparative studies of chloroplasts containing different relative amounts of grana and stroma membranes. These data revealed a direct relationship between the extent of grana stacking and amount of proton accumulation. All these differences suggest inherent differences in the stroma and grana membranes which are not resultant from isolation procedures.

3.2. Studies of Dimorphic Chloroplasts

It is now well recognized that certain plants which possess the C_4 dicarboxylic acid pathway of CO_2 fixation show a structural dimorphism of chloroplast structure (see Figs. 2 and 3). Hodge *et al.* (1955) and Vater (see Rabinowitch, 1956, p. 1981) were the first to show by electron microscopy that the chloroplasts of cells surrounding the vascular bundle of *Zea mays* were nearly devoid of grana stacks. The mesophyll cell chloroplasts, however, had normal grana stacks connected by unpaired stroma lamellae. Although chloroplast dimorphism is not readily obvious in all plants having the C_4 pathway for CO_2 fixation (Black and Mollenhauer, 1971; Laetsch, 1971), the structural variation is quite pronounced in certain tropical grasses such as sorghum and sugar cane and some dicots such as *Euphorbia maculata* (Laetsch, 1971). These species are of particular interest in functional studies concerning the interdependence of chloroplast structural organization and patterns of photochemical activity.

The first indication of a difference in electron transport activities in dimorphic chloroplasts came from *in situ* measurements of Hill reaction activity using TNBT chloride as the Hill electron acceptor. The light-induced blue-black reduction product was detected in the mesophyll but not the agranal bundle sheath chloroplasts of sorghum (Downton *et al.*, 1970). It should be noted that Laetsch (1971) and Laetsch and Price (1969) have shown that bundle sheath chloroplasts of young developing leaves of C_4 plants do contain grana. These grana are lost in some species during chloroplast maturation indicating that structural dimorphism of mesophyll and bundle sheath chloroplasts is due to a difference in the pattern of membrane development. Downton and Pyliotis (1971), using *in situ* histochemical staining, demonstrated that the loss of grana in sorghum bundle sheath chloroplasts during ontogeny is accompanied by a loss of Hill reaction activity.

Since the bundle sheath cells of some C_4 plants have a more rigid cell wall than the mesophyll cells, it was possible for Woo *et al.* (1970) to develop a differential grinding procedure which allowed selective breaking of the two cell types. They have used this technique to isolate separately the chloroplasts from the two cell types, thus allowing a direct study of the photochemical activities in the two types of dimorphic chloroplasts. Their results showed that agranal chloroplasts of sorghum are deficient in PS II (Hill reaction) activity, while mesophyll chloroplasts had normal PS II activities. Both chloroplast types had an active PS I. Both the histochemical results of

Downton *et al.* (1970) and the data of Woo *et al.* (1970) showed some PS II activity in *Zea mays* bundle sheath chloroplasts which have rudimentary grana stacks.

The photochemical evidence suggesting a deficiency of PS II in bundle sheath chloroplasts presented by Woo *et al.* (1970), was based on a Hill reaction using $NADP^+$ as the electron acceptor (see Chapter 7 of this volume for a discussion of electron transport in chloroplasts). Bishop *et al.* (1971a,b) and Smillie *et al.* (1971) have repeated these experiments with the same results. They have also measured Hill reaction activity using other electron acceptors, however, and have shown PS II-mediated reduction of ferricyanide or DCPIP in the agranal chloroplasts. Their conclusions from these data were that PS II is not absent from bundle sheath chloroplasts; rather there is a block between PS II and PS I that limits electron flow from water to $NADP^+$. In support of this suggestion, Smillie *et al.* (1971) have shown that $NADP^+$ reduction can be observed in bundle sheath chloroplasts of maize or sorghum if PC is added to the reaction mixture. Anderson *et al.* (1971a,b) have also demonstrated some PS II activity in sorghum bundle sheath chloroplasts (7–14 % the rate of mesophyll chloroplasts), and substantial PS II activity in maize bundle sheath chloroplasts. Bishop *et al.* (1972) demonstrated that both photooxidation and photoreduction of Cyt f (indicating a complete electron transport chain) can be detected in intact bundle sheath cells of maize. The chloroplasts of similar bundle sheath cells were shown in a parallel electron microscopy study (Smillie *et al.*, 1972) to be nearly devoid of grana stacks. The latter authors did note, however, that the ratio of PS I activity to Hill reaction activity was higher in bundle sheath chloroplasts than in plastids from mesophyll cells.

In support of the results of Smillie *et al.*, Mayne *et al.* (1971) have shown significant Hill reaction activity in bundle sheath cells from *Digitaria sanuinalis* (another C_4 plant with limited grana stacking). In this plant, the bundle sheath rates were only about one-third that observed with mesophyll cells. Arntzen *et al.* (1971), in a study of the dimorphic chloroplasts of sorghum, have measured PS II-mediated dye reduction in the presence of diphenyl carbazide (an electron donor for PS II) and PS I activity in the presence of solubilizing amounts of Triton X-100 and excess PC. Both of the procedures have been previously shown to provide accurate determinations of photochemcial activity even in partially inactivated preparations (Vernon and Shaw, 1969). Their results demonstrated that bundle sheath chloroplasts contained only about 15 % as much PS II as did mesophyll chloroplasts. On the basis of data on light-induced oxidation and reduction of a Cyt and on photochemical activities of isolated chloroplasts, Bakri (1972) has recently concluded that *Zea mays* bundle sheath chloroplasts contain PS II at about 40 % the level of mesophyll chloroplasts.

Studies of phosphorylation have tended to indicate that PS II is active in bundle sheath chloroplasts. Anderson *et al.* (1971a) found that rates of ferryicyanide-mediated noncyclic phosphorylation in the bundle sheath chloroplasts of maize were 11–19% of those observed with mesophyll chloroplasts. Arntzen *et al.* (1971) found noncyclic phosphorylation in the bundle sheath preparations to be only 10% of that of mesophyll chloroplasts from sorghum. [They did not include exogenous PC, however, which perhaps could have increased rates of electron flow to PS I in the agranal chloroplasts (Smillie *et al.*, 1971).] Polya and Osmond (1972) have measured noncyclic phosphorylation with ferricyanide as the electron acceptor for both sorghum and maize chloroplasts. When optimal conditions were attained, the PS II-mediated activity was only 7% as much as that seen in the mesophyll chloroplasts. In all of the reports of phosphorylation activity discussed above it was found that high rates of PMS mediated cyclic phosphorylation were obtained with both types of dimorphic chloroplasts.

Analysis of the composition and fluorescence characteristics of agranal chloroplasts have indicated a reduced level of PS II components. Woo *et al.* (1970) demonstrated that the quantum yield of fluorescence and fluorescence emission spectra of sorghum bundle sheath chloroplasts more closely resembled spinach detergent-derived PS I fractions than sorghum mesophyll chloroplasts. Woo *et al.* (1970) and Anderson *et al.* (1971b) showed that the sorghum agranal chloroplasts have a greatly reduced content of Cyt 559 (high potential form), low ratios of Chl/P700 and high Chl a/b ratios. These characteristics are also very similar to spinach PS I particle preparations. Mayne *et al.* (1971) have noted that bundle sheath cells of crabgrass leaves show greater P700 changes per unit Chl, less PS II-mediated delayed-light emission, and lower PS II fluorescence than the plastids isolated from mesophyll cells. (see Chapter 5 of this volume, for a discussion of delayed-light emission, and Chapter 6 for Chl fluorescence.) Bakri (1972) determined that maize bundle sheath chloroplasts are enriched in long-wavelength forms of Chl *a* (705, 693, 685) as compared with mesophyll chloroplasts. Bakri has also analyzed the various fluorescence characteristics of maize chloroplasts (fluorescence transients, degree of polarizations of fluorescence, emission and excitation spectra). These data indicate approximately three times higher ratios of PS I/PS II in bundle sheath chloroplasts as compared to the granal mesophyll chloroplasts.

The numerous studies of chloroplasts from C_4 plants described above have presented compelling evidence that agranal bundle sheath chloroplasts are at least somewhat deficient in PS II activity but have high levels of PS I. Mesophyll chloroplasts have normal levels of both photosystems (when compared to conventionally studied chloroplasts such as those from spinach). Almost all of the studies have shown that substantial levels of PS II are present in the agranal plastids, however. This is particularly true for the maize bundle sheath chloroplasts which have only occasional regions of appressed lamellae. It must therefore be concluded that there is no direct correlation between the extent of grana stacking and the level of PS II in agranal bundle sheath chloroplasts.

3.3. Correlation of Structure and Function during Membrane Biosynthesis

Studies of greening chloroplasts in several laboratories have revealed structure–function interrelationships. As will be discussed below, analysis of the pattern of onset of photochemical activities in greening plastids has revealed a stepwise pattern of development of enzymic capacities. It has become generally accepted that photosynthetic membrane biosynthesis occurs as a multistep assembly process (see reviews by Siekevitz *et al.*, 1967 and Kirk, 1970). As will be discussed below, it has been possible to ascertain the earliest time at which specific enzymic activities develop and to correlate the onset of these activities to structural changes.

In one of the earliest studies with dark-grown bean plants, Anderson and Boardman (1964) demonstrated that PS II-mediated ferricyanide photoreduction could be detected in chloroplasts isolated from leaves illuminated for 6 hr, whereas $NADP^+$ reduction was only detected in chloroplasts from leaves illuminated for more than 8 hr. Later studies in the same laboratory using etiolated peas showed the same pattern of onset of photochemical activities. In addition, it was shown that photooxidation of Cyt f, which is dependent on an active PS I, could be detected after only 30 min of greening (Boardman *et al.*, 1970). The developmental sequence occurring during membrane biosynthesis therefore appeared to include the early onset of PS I activity followed at a later time by the appearance of an active PS II. Some block in electron flow apparently limits the capacity to reduce $NADP^+$. When the greening plastids were examined by light and electron microscopy, Boardman and Anderson (1964b) and Boardman *et al.* (1970) determined that the appearance of grana stacks was closely correlated to the time at which Hill reaction activity was first detected. No stacking was observed during the early stages of greening when PS I became active. More recent results by Hiller and Boardman (1971) have indicated that photoreduction of Cyt f (indicating PS II activity) slightly follows the onset of PS I activity, but occurs well before the appearance of grana stacks. It was also observed (Boardman, 1968a) that Cyt 559 (high potential form) was not detectable in the early stages of greening but did become evident at times when grana stacking was observed. More recent data by Henningsen and Boardman (personal communication) have shown that oxygen evolution in greening barley leaves was observed ahead of the formation of Cyt 559 (high potential). The appearance of this Cyt was correlated to the time of increase in Chl content and grana stacking.

It should be noted that the Chl a/b ratio is high (>15) in dark-grown plants at early stages of greening. This ratio drops to a value near 3 at the time when grana stacks begin to form. As was discussed in an earlier section, it has been shown in several membrane fractionation studies that isolated grana have a lower Chl a/b ratio than the unpaired stromal lamellae.

Studies on the onset of photophosphorylation in greening bean (*Phaseolus vulgaris*) chloroplasts have also suggested that development of PS I precedes that of PS II. Glydenholme and Whatley (1968) showed that cyclic photophosphorylation in isolated plastids could be detected after 10 hr of plant illumination, whereas 15 hr were required to obtain noncyclic phosphorylation. Analysis of the greening tissues by electron microscopy showed only a few incipient grana at 10 hr, but more extensive stacking at 15 hr. These data suggest a correlation between development of PS II and stacked lamellae.

Rhodes and Yemm (1966) and Miller and Nobel (1972) found a correlation between the onset of grana stacking and the ability to fix CO_2 in greening barley (*Hordeum vulgare*) chloroplasts. Phung-Nhu-Hung *et al.* (1970a) have extended these studies by determining the timecourse of appearance of various photochemical activities. They have shown that very low rates of noncyclic electron flow to $NADP^+$ is present after only 2 hr of greening, while cyclic photophosphorylation catalyzed by PMS could not be detected before 4 hr. Noncyclic phosphorylation with $NADP^+$ was first observed after 10 hr. In a structural study on greening barley seedlings treated under similar conditions, Phung-Nhu-Hung *et al.* (1970b) report that grana stacking was observed after 4 hr of illumination. No mention was made of appressed lamellae occurring after a 2-hr illumination period. It therefore appears that PS I and PS II activity ($NADP^+$ Hill reaction) could be detected prior to the time of grana stacking. It is not clear at this time why there is a different pattern of appearance of cyclic phosphorylation and noncyclic electron flow in bean (Glydenholm and Whatley, 1968) and barley (Phung-Nhu-Hung, *et al.*, 1970a) seedlings greened under continuous light. It should be noted that the studies of Phung-Nhu-Hung *et al.* (1970a) on greening of barley seedlings under intermittent light (see discussion below) show that PS I activity (PMS-catalyzed cyclic phosphorylation) develops before any noncyclic electron flow could be detected. This is in agreement with the developmental pattern described above for *Phaseolus vulgaris*.

A number of greening experiments using intermittent illumination of etiolated leaves (1 msec flashes of bright light separated by 15 min dark periods) have been conducted by Sironval and coworkers (1968, 1969). They have demonstrated that repetitive flashing causes a significant increase in the Chl content of the leaf even though the total illumination period is only a fraction of a second in duration. Structural studies showed that long unpaired thylakoids are formed in the flashing regime (Sironval *et al.*, 1968, 1969 and Bradbeer *et al.*, 1970). These "primary" thylakoids sometimes show areas of of fusion after several hundred flashes. Biochemical studies of the chloroplasts from flashed leaves by Sironval *et al.* (1968) and Phung-Nhu-Hung *et al.* (1970a,b) have demonstrated that the "primary" thylakoids have a high Chl a/b ratio, an active PS I mediated cyclic phosphorylation activity, but no detectable PS II activity and a deficiency in Cyt 559. These results suggest that the development of PS I does not require stacked lamellae. It is not possible to determine whether the lack of PS II activity is the result of a limitation in structural development or is simply limited by some other aspect of membrane biosynthesis.

Argyroudi-Akoyunoglou and Akoyunoglou (1970) have shown that etiolated beans greened under intermittent light and dark cycles (2 min light, then 98 min dark, etc.) accumulate selectively Chl a. It has recently been found (Arntzen, unpublished data) that etiolated pea seedling illuminated under similar intermittent light conditions develop chloroplasts which have very few grana. These agranal chloroplasts have high Chl a/b ratios and demonstrate high specific activities in assays for PS II (ferricyanide Hill reaction), PS I (MV reduction in the presence of reduced DCPIP) and both cyclic and noncyclic phosphorylation. The results therefore show no correlation between PS II activity and the presence of grana stacks.

Oelze-Karow and Butler (1971) and DeGreef *et al.* (1971) studied the greening of etiolated bean plants exposed to far-red light. Under these conditions the biosynthetic processes of the leaf which are controlled by phytochrome are fully active, but protochlorophyll is only slowly converted to chlorophyll. They found that the greening membranes accumulate primarily Chl a. The capacity for DCMU-insensitive phosphorylation was detected after 12 hr of greening, whereas oxygen evolution and noncyclic (DCMU-sensitive) phosphorylation began after 20 hr of far-red light. Electron microscope examination of the greening leaves showed most chloroplast lamellae to exist as long unpaired primary thylakoids even after maximum rates of oxygen evolution had been attained. The authors concluded that there is no sharp correlation between structural changes and onset of photochemical activity. They suggest that grana formation is not required for oxygen evolution activity. It should be pointed out, however, that all of their micrographs do show small regions where adjacent thylakoids are fused. Since the Chl content of the plastid is very low, it may be possible that all of the active Chl is localized in these regions of limited membrane stacking. This seems unlikely, however, since the long primary thylakoids of red light-treated plants could be induced to rapidly fuse by exposure to white light indicating that they are deficient only some "stacking factor(s)."

Ohad *et al.* (1967) followed the development of photochemical activity and changes in membrane structure during chloroplast greening in the *y-1* mutant of *Chlamydomonas reinhardi*. They observed that formation of grana stacks lagged considerably behind the onset of Hill reaction activity, oxygen evolution or $NADP^+$ reduction. They concluded that fusion of thylakoids is not essential for photosynthetic activity, but they did not rule out some role of stacking in increasing the efficiency of quantum conversion.

Hoober *et al.* (1969) have also investigated the pattern of greening of *Chlamydomonas reinhardi y-1*. They have found that cells greened in the presence of 20 μg/ml chloramphenicol show rates of Chl and membrane biosynthesis which were nearly as high as those of the control even though development of photochemical activity was reduced by 65% for PS II and 50% for PS I. Under these conditions, the extent of grana stacking was reduced to about the same extent as was photochemical activity. If cyclo-heximide was added to the growth medium in place of chloramphenicol, Chl and membrane synthesis were greatly reduced, but photochemical activity and fusion of thylakoids were the same as the control on a Chl basis. These data seem to indicate that grana formation is dependent on the presence of active photosystems and not just the development of membranes containing normal levels of Chl.

3.4. Chloroplast Mutants with Altered Structure

One of the first suggestions that there is an absolute requirement for grana stacks in order to obtain PS II activity came from studies of Homann and Schmid (1967). They demonstrated that chloroplasts isolated from the yellow leaf of tobacco mutant variety *NC95* had high PS I activity but an inactive PS II. Homann and Schmid made the observation that the chloroplasts in these same leaf segments showed a lack of grana stacking and suggested a correlation between structural and functional limitations. Later studies by Schmid and Gaffron (1967) with other tobacco mutants did not confirm this correlation, however. It was found that plants with a reduced Chl content contain chloroplasts with few or no grana. These Chl-deficient plants showed rates of CO_2 fixation or O_2 evolution which were much higher than the control, thus indicating the presence of active photosystems.

Some of the most extensive studies of structure and function interrelationships in mutant plastids have been done by Goodenough and coworkers on various strains of *Chlamydomonas reinhardi* (Goodenough *et al.*, 1969; Goodenough and Staehelin, 1971). In the recent investigation, they have demonstrated that mutant *ac-31* and mutant *ac-5* grown mixotrophically contain plastids which have no appressed lamellae. Both mutants show a reduced Chl content on a per cell basis but have photochemical activities for both PS I and PS II which are higher than the control (on a Chl basis). They have unambiguously demonstrated that *in vivo* grana stacking is not a prerequisite for PS II activity.

Both structural and functional studies have been conducted on several plastome mutants of *Oenothera* (Dolzmann, 1968; Fork and Heber, 1968). Mutant *IIα*, which was shown to have an impaired PS I, was found to contain large grana stacks but few stroma lamellae. Three PS II mutants (*Iγ, Iδ*, and *I'γ*) were found to have fewer grana and/or more disorganized grana and greater amounts of unpaired membranes and vesicles. While these mutants do not reveal any direct correlation between structure and function, they do indicate that mutants lacking PS II components suffer greater disorganization of grana-stacked regions than do mutants deficient in PS I.

Highkin *et al.* (1969) have studied a Chl-deficient mutant of pea (*Pisum sativum*) which has a high Chl a/b ratio and twice the normal level of PS II activity. They have presented evidence that in the mutant both PS I and II contain fewer light-harvesting chlorophylls than the control; PS II was thought to be more severely reduced in Chl than PS I, however. Chloroplasts in this mutant had fewer grana per chloroplasts and fewer thylakoids per granum. A similar chloroplast structure was reported by Keck *et al.* (1970a) for a light-green soybean (*Glycine max*) mutant. These chloroplasts also had high Chl a/b ratio and PS I and PS II activities which were 3 to 5 times higher than controls. Keck *et al.* (1970b) report that the photosynthetic unit size of the soybean mutant did not change from that of the control. Bakri (1972) has found that an olive-green mutant of maize contains both mesophyll and bundle sheath chloroplasts which are nearly agranal; isolated chloroplasts have high PS I and PS II activities, however. In the maize, soybean and pea mutant studies, there is little correlation between a limitation in chloroplast membrane structural organization and photochemical activity. There is also little support for the concept that PS II must be localized only in grana membranes since mutants of soybean and pea had increased rates of PS II-dependent electron transport, but reduced stacking. (For a further discussion of mutants see Chapter 6 of this volume.)

3.5. Environmental Effects on Chloroplast Structure

Several studies in recent years have been concerned with modifications of cellular and chloroplast structure under conditions of plant nutrient stress. In almost all cases, moderate to severe changes in chloroplast membrane organization were observed. A few studies have attempted to correlate these changes with modifications in photochemical capacity of the lamellae.

It has been well established that manganese is an essential cofactor on the oxidizing side of PS II (Cheniae, 1970). Manganese is enriched in digitonin-derived PS II particles as compared to either PS I particles or control chloroplasts (Boardman, 1968b). Growth of plants under manganese deficiency results in development of chloroplasts which have impaired PS II activities. A relatively rapid restoration of PS II can be achieved by adding Mn^{2+} to the deficient tissue. This has been taken to indicate that most PS II components are present in the deficient plants and Mn^{2+} must be inserted into some proper site to obtain activity (see Cheniae, 1970). Possingham *et al.* (1964) have shown that the chloroplast is the only cellular organelle which shows marked structural change in Mn-deficient spinach leaves. These changes were primarily characterized by an increase in the number of thylakoids per grana stack and an almost complete loss of stroma lamellae. More severe damage was indicated by swelling of the grana thylakoids. These data suggest that the stacking of chloroplast thylakoids into grana is not dependent on a photochemically active PS II. Homann (1967) demonstrated a variable pattern of chloroplast structural changes caused by Mn deficiency in several plant species studied. He also demonstrated the persistence of grana stacks in Mn-depleted plants. Anderson and Pyliotis (1969) have also reported granal chloroplasts in Mn-deficient spinach plants.

Studies by J. O. Hall *et al.* (1972) and Baszynski *et al.* (1972) have attempted to correlate changes in chloroplast structure and function in maize seedlings deficient in either nitrogen, calcium, magnesium, phosphorus, potassium, or sulfur. No changes in PS I activity (on a Chl basis) were observed in any treatment. PS II activity was found to be constant in all treatments with the exception of an increase in specific activity in nitrogen- and sulfur-deficient plants. In both of these cases, an increase in amount of grana stacking was correlated to the increase in PS II.

Changes of light intensity have also shown to have an effect on chloroplast structure in several plant species. Reger and Krauss (1970) reported that the Chl *a/b* ratio of *Chlorella* increased with increasing light intensity during cell culture. This change was coupled to a decrease in the extent of grana stacking, but an increase in the rate of oxygen evolution (expressed on a Chl basis). Punnett (1971) has reported that light intensity and light quality affect the extent of grana stacking in *Elodea*. In this case the lack of stacking was also correlated to reduced photosynthetic enhancement (see Chapter 1 of this volume for a discussion of enhancement).

Goodchild *et al.* (1972) and Björkman *et al.* (1972) have examined the chloroplast ultrastructure of plants grown under various light intensities in the laboratory and under natural conditions. They have shown that rain-forest species and plants grown under controlled low-light conditions had greater granal development than plants grown under high light. The low-light chloroplasts contained more grana and had more thylakoids per grana stack. These changes were correlated to a lower Chl *a/b* ratio in the low-light chloroplasts. Björkman *et al.* (1972) found that the specific activities of PS I and PS II are greater in high-light as compared to the low-light chloroplasts. The change in photochemical activity was not due to a change in photosynthetic unit size (see Section 6); the authors have concluded that the increased capacity of the chloroplast reactions is governed by increased amounts of some of the electron carriers in the photosynthetic electron-transport chain.

3.6. Physiological Significance of Grana Stacking

Electron microscopic examination of green algae and higher land plants has revealed that nearly all naturally occurring healthy green cells contain chloroplasts which possess well-defined regions of lamellae appression (Kirk and Tilney-Basset, 1967). When a number of disklike membranes become stacked one upon another, the resultant structures are called grana. These grana stacks represent an unique example of membrane fusion in plant cells. It seems likely that selection for this particular pattern of membrane development during evolutionary progression must indicate a beneficial role of grana stacking in the utilization of solar energy. Numerous studies have attempted to answer the question: What is the physiological role of grana stacking?

One approach to this problem is the study of the content and of the distribution of photochemical activities within the chloroplast lamellae in hopes of finding some membrane component or photoreaction which is either wholly or partially dependent upon the presence of grana stacks. The previous sections of this chapter have presented data from numerous studies which relate to this approach. It can be immediately concluded from these studies that the PS I photoreaction and cyclic photophosphorylation do not require stacked lamellae, since high rates of these activities have been shown in the stroma lamellae isolated by mechanical fractionation of chloroplasts (Sane *et al.*, 1970; Arntzen *et al.*, 1971), in the agranal chloroplasts isolated from the bundle sheath cells of some C_4 plants (Arntzen *et al.*, 1971; Smillie *et al.*, 1971; Anderson *et al.* 1971a), and in a large number of mutants which have few or no grana stacks (Goodenough *et al.*, 1969; Goodenough and Staehelin, 1971; Highkin *et al.*, 1969; Keck *et al.*, 1970b).

The relationship between PS II and grana structure remains less precisely understood. Beginning with the data of Homann and Schmid (1967), who hypothesized that PS II activity was dependent on the presence of grana stacks, we have assembled representative data relating structure and function in Table I. The results are listed in a descending order such that the data at the top of the table strongly support the concept of Homann and Schmid, while the lower listed results demonstrate a complete lack of correlation between grana structure and PS II activity. The findings of Goodenough and

TABLE I

Selected References from the Literature that Present Evidence Relating to the Question: Are Grana Required to Obtain PS II Activity in Green Plant Chloroplasts?

Literature reference	Plant material	Sample	PS II activity	PS I activity	Whole-chain electron transport	Noncyclic	Cyclic (PMS-mediated)	Chl a/b ratio	Chemical characteristics	Correlation between PS II activity and presence of grana
			μ moles/mg Chl/hr			Photophosphorylation				
Homann and Schmid (1967)	Tobacco mutant NC 95	Yellow leaf area	18[a]	165[d]			420[e]	3.4	Increased carotenoid per Chl content in mutant	No grana detected; almost no PS II detected in chloroplasts from yellow leaf segment
		Green leaf area	210[a]	50[d]			430[e]	2.5		
Sane et al. (1970)	Mechanically fragmented spinach chloroplasts	Stroma lamellae	0[b]	169[d]				6.0	Cyt 559 content lower in stroma lamellae	PS II activity only in grana; PS I in both grana and stroma lamellae
		Grana membranes	74[b]	75[d]				2.4		

74

Reference	Chloroplast source	Plastid type						Notes
Woo et al. (1970)	Isolated chloroplasts from sorghum	Agranal bundle sheath plastids	43[d]	0[g]				Cyt 559 (high potential) absent from bundle sheath. No whole-chain electron transport but good PS I in agranal chloroplasts[h]
		Granal mesophyll plastids	38[d]	159[g]				
Arntzen et al. (1971)	Isolated chloroplasts from sorghum	Agranal bundle sheath plastids	39[b]	426[d]	23[e]	570[e]	3.9	PS II activity greatly reduced in agranal chloroplasts[h]
		Granal mesophyll plastids	271[b]	180[d]	237[e]	614[e]	2.8	
Anderson et al. (1971a)	Isolated chloroplasts from maize	Agranal bundle sheath plastids	66[b] 93[c]		14[e]	537[e]	5.4	PS II activity greatly reduced in agranal chloroplasts
		Granal mesophyll plastids	417[b] 619[c]		96[e]	520[e]	3.2	
Andersen et al. (1972)	Isolated chloroplasts from maize	Agranal bundle sheath plastids	113[b] 136[c]		0[f]			Bundle sheath chloroplasts with few or no grana have high PS II activity
		Granal mesophyll plastids	176[b] 214[c]		46[f]			

TABLE I (cont.)

Literature reference	Plant material	Sample	PS II activity	PS I activity	Whole-chain electron transport	Photophosphorylation Noncyclic	Photophosphorylation Cyclic (PMS-mediated)	Chl a/b ratio	Chemical characteristics	Correlation between PS II activity and presence of grana
			μ moles/mg Chl/hr							
Highkin et al. (1969)	Peas	Mutant variety	850[c]		190[f]			10–18		Reduced stacking but higher photochemical activity for both PS I and II in mutant chloroplasts
		Normal control	275[c]		80[f]					
Goodenough and Stachelin (1971)	Chlamy-domonas reinhardi	ac–5 mutant	335[b]	108[d]	134[f]			3		No grana stacking observed in mutant chloroplasts with full photochemical activity
		Wild type	201[b]	90[d]	101[f]					

[a] μmoles O_2 evolved/mg Chl-hr (FeCN as electron acceptor).

[b] μmoles DPIP reduced/mg Chl-hr.

[c] μmoles FeCN reduced/mg/Chl-hr.

[d] μmoles NADP reduced/mg Chl-hr. (DPIPH$_2$ as electron donor)

[e] μmoles ATP formed/mg Chl-hr.

[f] μmoles NADP reduced/mg Chl-hr (H$_2$O as electron donor).

[g] μatoms O evolved/mg Chl-hr (NADP as electron acceptor).

[h] Structural data based on figures in the literature.

coworkers with *Chamydomonas* mutants *ac-5* and *ac-31* (Goodenough *et al.*, 1969; Goodenough and Staehelin, 1971), which have high rates of PS II but no stacked membranes, provide the most conclusive evidence that occurrence of grana is not prerequisite to obtaining a photoreactive PS II. This is also supported by analysis of higher plant chloroplasts which show a greatly reduced pattern of membrane stacking but PS II activities equal to or higher than those of the controls (Highkin *et al.*, 1969; Keck *et al.*, 1970b; Bakri, 1972). It might also be noted at this point that blue-green algae, red algae, and some brown algae do not have stacked membranes, but are certainly fully competent photosynthetically (Kirk and Tilney-Basset, 1967). Furthermore, Izawa and Good (1966a, b) have demonstrated that even in higher plant chloroplasts which have stacked membranes, loss of the stacks during low-salt washing does not impair PS II-mediated oxygen evolution. The appressed membranes therefore do not seem to have a direct regulation of PS II activity.

In spite of the compelling evidence demonstrating that an active PS II does not require appressed membranes, there are also very clear data from membrane fractionation studies indicating that essentially all PS II in mature spinach chloroplasts is localized in the grana regions (Park and Sane, 1971). Unfortunately, these correlative structure–function fractionation studies have not yet been conducted on other species so it is difficult to determine whether or not the specific localization of PS II in the grana of spinach plastids is characteristic of all normally developed chloroplasts. Histochemical staining studies by Weier *et al.* (1966b) suggest that PS II is strictly localized in the grana of maize mesophyll chloroplasts and *Vicia faba* chloroplasts. The very simple question remaining is: Why is most PS II localized in partition regions of normally developed chloroplasts if appressed lamellae are not necessary for activity? (It would seem simpler to expect a uniform localization within the lamellae.) Two possible answers can be suggested at this point. First, the stroma lamellae of mature chloroplasts may represent a type of membrane which is uniquely different from the stacked lamellae in that it has different structural and functional components. This may imply that there are two different types of PS I in the chloroplast; one in the unpaired membranes and another form which exists in association with PS II in the specialized stacked membranes. Such a concept is not new; Arnon and coworkers have suggested that there are three light reactions in photosynthetic electron transport (Arnon *et al.*, 1970). In their scheme, two of the light reactions are capable of generating a strong reductant, but only one photoact is thought to involve P700.

A comparison of electron transport components localized in the grana and stroma lamellae do not support the concept of two different types of PS I in the respective fractions. Arntzen *et al.* (1972) have shown that the

Chl/P700 ratio of detergent derived grana PS I is 170, while the ratio for PS I in the stroma lamellae averages about 110 Chl/P700. Both grana and stroma membrane fractions were found to contain PC. The action spectra for $NADP^+$ reduction (in the presence of $DCPIPH_2$) for both stroma lamellae and grana-derived PS I showed a definite red shift as compared to the PS II action spectra. Isolated PS I, prepared by digitonin treatment of either grana or stroma lamellae, showed the same ultrastructural appearance and ultrafiltration characteristics. Wessels and Voorn (1972) have compared Chl *a/b* ratios, P700 content, Chl absorption maxima, action spectra, and particle size of PS I from both grana and stroma lamellae isolated by successive detergent treatments. Their results indicate a high degree of similarity in the parameters studied, irrespective of the source of the PS I. Arntzen *et al.* (1972) and Wessels and Voorn (1972) have therefore concluded that there are few, if any, functional differences between the PS I of grana and stroma lamellae. They also concluded that there is no evidence for $NADP^+$ photoreductive activity mediated by a photosystem which does not have P700 as the reaction center Chl (also see Esser, 1972).

Another possible explanation for the observation that PS II localization is limited to grana in spinach chloroplasts is that the two phenomena (stacking and PS II activity) are somehow closely linked in the process of membrane development. As was discussed above, chloroplast membrane biosynthesis is now generally accepted to occur as a multistep assembly process. As a means of summarizing this process, a generalized scheme describing the pattern of onset of photochemical activities in greening membranes in indicated in Fig. 4. Unfortunately, no single study in the literature has considered all of the various parameters of photochemical activity and changes in ultrastructure for one plant under one set of greening conditions. We have therefore left the time scale of Fig. 4 in arbitrary units to emphasize that the pattern of development of photochemical activities is relatively constant, whereas the actual time of occurrence varies according to plant species and experimental conditions.

It seems well established that greening membranes first develop PS I activity [cyclic phosphorylation (Glydenholm and Whatley, 1968), oxidation of Cyt f (Boardman *et al.*, 1970)], whereas there is a somewhat later onset of PS II [noncyclic phosphorylation (Glydenholm and Whatley, 1968), Hill reaction (Boardman *et al.*, 1970), and photoreduction of Cyt f (Hiller and Boardman, 1971)]. Cytochrome 559 (high potential) was found to be absent from dark grown plants but is found at later stages of the greening process (Hiller and Boardman, 1971; Boardman, 1968a). Chlorophyll *a/b* ratios have been observed to decline during greening from initial high values to a value near 3 for mature plastids (Kirk and Tilney-Basset, 1967). Ability of greening plastids to demonstrate light-induced proton uptake (as measured

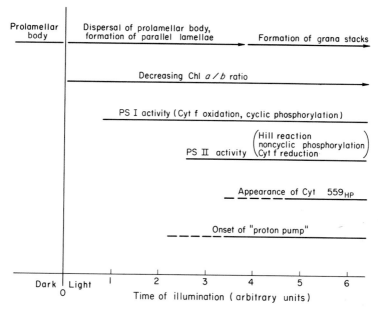

Fig. 4. Diagrammatic representation of the structural and biochemical characteristics of a "typical" dark-grown chloroplast after various periods of illumination. Arrows and lines indicate the trend of structural changes, changes in chlorophyll content, and appearance of photochemical activities. Each line is drawn to cover the representative time period over which the indicated characteristic was observed. Data used to construct this diagram were obtained from several publications (see citations in text, Sections 3.3 and 3.6). Since the time of onset of the various parameters varied in different research reports, all data were normalized on a relative time scale with arbitrary units.

by sensitivity of phosphorylation to amine uncoupling) increases during membrane development (Arntzen *et al.*, 1971). The formation of grana stacks was originally thought to be correlated to the onset of PS II (Anderson and Boardman, 1964); more recent results, however, have demonstrated a slight lag between the time of appearance of PS II activities and the onset of membrane stacking (Hiller and Boardman, 1971). We have used a dotted line in Fig. 4 to demonstrate uncertainty in the parameters of membrane synthesis which still leave some question as to the exact initial time of development (with respect to other changing parameters).

The data summarized in Fig. 4 lead to the conclusion that chloroplast membrane biosynthesis in greening tissues first results in the formation of unpaired membranes. As additional membrane components are inserted into the developing lamellae, photochemical activities appear in a stepwise fashion. One of the final steps in membrane ontogeny appears to be the complete structural development that leads to grana stack formation.

The characteristics of stroma lamellae isolated by mechanical fractionation of chloroplasts bear a remarkable resemblance to the characteristics of chloroplast membranes during the very early stages of greening. Stroma lamellae have high Chl a/b ratios, high PS I but no PS II activities, do not contain Cyt 559, and do not have an active "proton uptake" (Sane *et al.*, 1970; Arntzen *et al.*, 1971). We may speculate that stroma lamellae exist in the chloroplast as incompletely differentiated membranes. This leads to the conclusion that the early steps of membrane biosynthesis in light-grown plants may proceed in a very similar fashion to the greening process of dark-grown plants. We can also suggest that onset of PS II activity and appearance of grana stacks are closely linked, on a time scale, in light-grown plants, as was shown in Fig. 4 for the greening etioplasts. The restriction of PS II to the grana in spinach chloroplasts (Sane *et al.*, 1970) may simply be due to their parallel development during membrane biosynthesis. According to this concept, the PS I in both stroma lamellae and grana would be essentially identical with respect to functional activity since the grana thylakoids would simply be more highly developed regions of a continuous lamellar membrane system.

Recent work by Park and coworkers has been interpreted as indicating that stroma lamellae may be undergoing membrane development. Sane and Park (1971) have demonstrated that isolated stroma lamellae fragments show a decline in quantum yield for $NADP^+$ photoreduction (using reduced DCPIP as an electron donor) at wavelengths less than 700 nm. These data suggest that stroma lamellae contain Chl which is inactive in PS I reactions. Park *et al.* (1971) could not detect chemically or photochemically induced variable fluorescence in this fraction, thus indicating that the photochemically inactive Chl is not the result of damage to PS II during particle isolation. They concluded that the inactive Chl in stroma lamellae may be related to sites of membrane biosynthesis (Sane and Park, 1971). Studies by Goldberg and Ohad (1970), using radioautographic techniques, demonstrated a more rapid rate of incorporation of membrane precursors into the unpaired lamellae than into the stacked membranes of *Chlamydomonas*. This would be in support of the idea that stroma lamellae are an active site of membrane biosynthesis.

3.7. Membrane Factors Regulating Grana Stacking

We have summarized evidence in the preceding section which demonstrates that grana are not required to obtain PS II activity. Conversely, however, we may suggest that some component(s) of PS II may be required to obtain grana stacks. The PS II mutants described in Section 3.4 have been reported to have little or no grana stacking. Of the PS I mutants reported in the literature, however, there seems to be little effect on formation of grana. It should be emphasized that electron transport mediated by PS II is not required to obtain grana. Etiolated chloroplasts which were allowed to green in the presence of DCMU still showed development of stacked lamellae (Klein and Neuman, 1966, Oelze-Karow and Butler, 1971). Chloroplasts from manganese- deficient higher plants have reduced Hill reaction activity, but have numerous grana stacks (Spencer and Possingham, 1961; Possingham *et al.*, 1964; Homann, 1967). In the Mn-deficient plants, most enzymic and structural components of photoreaction II are probably present in the membrane since O_2 evolution capacity can be quickly restored to the deficient plants by adding Mn (Cheniae, 1970).

We are left with the question of which components of PS II are of importance in regulation of stacking. Chlorophyll *b* would be an attractive candidate since it is generally recognized to be a light-harvesting pigment primarily for PS II, since it is enriched in both detergent-derived PS II fractions and in mechanically derived grana preparations (Park and Sane, 1971) and since the Chl *b* concentration usually increases roughly in correlation with the onset of PS II activity and grana stacking during chloroplast greening (Anderson and Boardman, 1964). Pyliotis *et al.* (1971) have also shown that the Chl *a/b* ratio of bundle sheath and mesophyll chloroplasts of C_4 plants accurately predicts the extent of grana stacking which is present in the chloroplasts since a decreased Chl *b* content is closely correlated to reduced stacking. Suggestive evidence for a structural role of Chl *b* is contradicted by the structural studies of Goodchild *et al.* (1966) who showed that a Chl *b*-deficient mutant of barley still showed the presence of grana stacks. The number of grana/chloroplast and number of lamellae/granum were reduced in this mutant, however.

It is also tempting to speculate that the total amount of Chl in the membrane affects membrane stacking. Allen *et al.* (1972) reported that Chl content relative to protein is about 40% higher in grana as compared to stroma lamellae. Several of the algae and higher plant mutants which have photochemical activity but reduced grana stacks are characterized by their light green color. A mutant of *Chlamydomonas* (*ac-1*) does not fit this generalized scheme, however, since it has only 25% as much Chl as wild-type cells but still has the normal extent of lamellae and normal lamellar stacking (Goodenough and Levine, 1969). The Chl level, by itself, does not therefore seem to be the determinant for stacking capacity (except perhaps at very low levels of total pigment).

Other *Chlamydomonas* mutants which have high percentages of unstacked lamellae are strains *ac-115* and *ac-141* (Goodenough and Levine, 1969). Both of these mutants lack Cyt 559. They also are characterized by a high fluorescence yield (Lavorel and Levine, 1968) which is unaffected by actinic light (indicating a missing or inactive Q, the "primary" electron acceptor for PS II). Studies of greening pea chloroplast membranes by Boardman and coworkers (Boardman *et al.*, 1970) have shown that the times of illumination required to obtain detectable levels of Cyt 559 (high potential), grana stacks, and increasing amount of variable fluorescence changes are somewhat closely correlated. It should be noted that the onset of PS II activity in these pea chloroplasts is not thought to be dependent on the presence of either Cyt 559 (high potential) or grana (Hiller and Boardman, 1971). Greatly reduced levels of Cyt 559 (high potential) have also been reported in the agranal bundle sheath chloroplasts of *Sorghum* (Woo *et al.*, 1970), which are reported by some authors to have significant levels of PS II activity (Smillie *et al.*, 1972). It seems quite premature at this point to speculate that there is a causal relationship between the absence of Cyt 559 and the appearance of unstacked lamellae. We can suggest, however, that the biosynthesis of the grana "stacking factor" and the appearance of Cyt 559 (high potential) may be closely linked during membrane ontogeny (see Fig. 4).

Membrane structural organization is undoubtably controlled both by protein and lipid components. Thus it is possible that the lipid composition of stacked and unstacked membrane regions could be different. Bishop *et al.* (1971a) have shown that the agranal bundle sheath chloroplasts of maize and sorghum have a higher galactolipid content, on a Chl basis, than the mesophyll chloroplasts. Bourdu *et al.* (1972) have reported the same finding when galactolipid content was expressed on a Chl basis; on a protein basis, however, the bundle sheath chloroplasts were shown to be slightly lower in galactolipid content. Ionic lipid content was found to be higher in the agranal chloroplast fraction. Allen *et al.* (1972) have found that the relative molar amounts of glycerolipids are similar, if expressed on a membrane protein basis, in isolated grana and stroma lamellae from spinach chloroplasts.

Roughan and Boardman (1972) have recently studied the changes in lipid composition in greening pea and bean leaves. They report that MGD content increases disproportional to that of DGD during the time when grana stacking is observed. Bourdu *et al.* (1972) had previously reported that DGD/MGD ratios in mesophyll chloroplasts were lower than in the agranal bundle sheath chloroplasts of maize. Keck *et al.* (1970a) analyzed the lipid content of chloroplasts from normal soybeans and from a mutant which showed reduced grana stacking. Their results indicate a decrease in both MGD and DGD content in the chloroplasts with reduced stacking. From these studies it is not clear whether changes in galactosyl diglyceride content of chloroplast lamellae play a causal role in inducing thylakoid stacking or whether the changes are simply a secondary effect of the grana formation.

In a chloroplast which has most, if not all of the components required to obtain grana, there may be a need for an energy-requiring process to induce stacking. Erickson *et al.* (1961), in a study of the breakdown and dispersion of the prolamellae body of etiolated bean leaves, demonstrated that disk formation and grana stacking was temperature-dependent and had a high-energy (light) requirement' Butler *et al.* (1972) have shown that carbonyl cyanide *m*-chlorophenylhydrazone (an uncoupler of photophosphorylation) blocked the formation of grana stacks in greening chloroplasts. They suggest that some high-energy state or high-energy intermediate produced in the light under normal conditions is involved in lamellar fusion.

The discussion in the paragraphs above has emphasized that there must be some difference in membrane composition (presumably at the lamellar surface) in stacked regions as compared to stroma lamellae. Although we do not yet know the nature of the stacking factor(s), we must recognize that its regulation of structure is strongly affected by salts. As noted earlier, Izawa and Good (1966b) were the first to show that low-salt washing of chloroplasts caused a loss of grana structure. This effect was reversible and did not adversely affect photochemical activity. It has recently been shown that ionic concentrations probably also effect the structural organization of chloroplasts *in vivo*. Goodenough and Staehelin (1971), in a study of *Chlamydomanas* mutant *ac-31*, observed no grana stacks in examinations of normal intact cells. However, stacked lamellae were observed in chloroplasts isolated from the same cells in a high-salt medium. Moreover, if whole cells were incubated in 10% glycerol, which removed water from the cells osmotically, *in vivo* stacking was observed. The authors concluded that the *in situ* ionic environment could regulate the degree of stacking, and that ionic interactions are required for lamellar appression.

Murakami and Packer (1971) extended the studies of Izawa and Good on low-salt washed chloroplasts. They found that either divalent or monovalent cations are effective in causing reversible stacking, although the latter are less effective on a concentration basis. They suggested that the role of electrolytes is to weaken the electrostatic interaction between charged groups on the membranes. They further conclude that electrolytes decrease the water solubility of nonpolar regions such that hydrophobic interactions are the dominant forces in maintaining stacked membranes. R. A. Dilley and co-workers (personal communication) have recently shown that modification of exposed carboxyl groups on chloroplast membranes (carbodiimide treatment followed by covalent attachment of a glycine methyl ester) causes thylakoids to reassociate into grana stacks even in low salt. They interpret these results as showing that chemical modification of the charged groups on the membrane leads to a reduction in electrostatic repulsion. These data again emphasize the importance of electrostatic interaction between membranes in controlling grana stacking.

3.8. Conclusions

We have summarized evidence in the previous sections which demonstrates that appressed lamellae are not required for PS II activity. Conversely, however, it would appear that some component(s) of PS II is necessary in order to obtain grana stacks. There is no evidence, at present, which allows us to speculate as to the nature of the factor which is primarily responsible for the attraction forces between membranes. It would appear that there probably is no single "molecular glue," but rather a complex interaction between several membrane components that is regulated, at least in part, by the presence of salts.

It has been pointed out that in normal grana-containing chloroplasts, such as those of spinach, PS II is thought to be entirely localized in stacked regions of the lamellae (Sane *et al.*, 1970). A tentative explanation for this localization has been suggested on the basis of a multistep assembly of membrane constituents during lamellae biosynthesis (see Fig. 4). We suggest that initiation of new membrane biosynthesis first occurs in regions of unpaired (stroma) lamellae and that stepwise addition of membrane components leads to formation of a photochemically active and a structurally complete membrane. PS I is thought to become active very early in developing membranes. We speculate that the membrane components necessary for both PS II activity and for grana stacking are inserted into the forming membrane at a somewhat later time. It would follow that onset of PS II activity is not dependent on grana stacking, but is only fortuitously linked to the stacking process through the normal sequence of membrane biosynthetic events. This model provides a mechanism for the development of agranal chloroplasts which have an active PS II since we must only stipulate that the very last steps in membrane biosynthesis (which give rise to stacked lamellae) are somehow halted through genetic or environmental restrictions.

4. SUBSTRUCTURE OF CHLOROPLAST LAMELLAE

There is now an extensive literature describing the evidence for subunits within chloroplast lamellae. As early as 1953 Frey-Wyssling and Steinmann used shadowing techniques to demonstrate that the surface of chloroplast membranes were bumpy and therefore hypothesized that the membrane was made up of a string of globular subunits. In more recent years, globular substructure in chloroplast lamellae has also been observed in thin-sectioned material, negative stained preparations, and in freeze-fractured samples. Various aspects of these electron microscopy studies have been summarized or reviewed by several authors (Branton, 1968; Kirk, 1970; Park and Sane, 1971). There is also considerable evidence based on polarization microscopy and X-ray diffraction analysis which indicates the presence of a substructured chloroplast membrane (Branton, 1968; Kreutz, 1970). We will briefly discuss in the following sections the various types of membrane subunits visualized by electron microscopy.

4.1. Membrane Subunits: Evidence from Shadowed and Thin-Sectioned Preparations

It was observed in early studies by Steinman (1952) that isolated chloroplast membranes which were spread on an electron microscope grid and shadowed with a heavy metal appeared to have a particulate surface with a distance of about 200 Å spacing for the subunits. Park and Biggins (1964) and Park (1965) extended these studies and discovered that the subunits observed in shadowed preparations were sometimes oriented in highly ordered arrays. Each subunit had dimensions of $155 \times 180 \times 100$ Å. This particle was called a "quantasome" (possible correlations between structure and function will be discussed below). Park and coworkers, in recent papers, have emphasized that these quantasomes are on the interior surface of the thylakoid and make up the matrix of the membrane (see Park and Sane, 1971). Since the particles in the shadowed preparations are thought to be visible only on the inner surface of the vesicular thylakoid membrane, it has been proposed that they become visible as a result of distortion and differential collapse of the thylakoid membrane as it dries during sample preparation (Branton, 1968).

Electron microscopic examination of ultrathin sections of fixed chloroplast lamellae began only a short time after the initial studies of shadowed material. In some of the early reports (Hodge et al., 1955; Mühlethaler, 1960), the chloroplast membrane was interpreted as being made up of alternating layers of lipid and protein. More recent studies, however, have reported that subunits are visible in thin-sectioned lamellae (Murakami, 1964; Ueda, 1964; Hohl and Hepton, 1965; Weier et al., 1965a; Weier and Benson, 1967). Weier et al. (1965a) and Weier and Benson (1967), on the basis of studies on chloroplasts from several plants, have reported that the subunits have an electron opaque core which is 37 Å in diameter and which is surrounded on either side by a 28 Å dark rim. Ji and Benson (1968), and Ji et al. (1968), in a study of the association of structural lipids and pigments with chloroplast lamellar protein have emphasized the importance of hydrophobic lipid–protein associations. This concept is included in the globular membrane model of Weier and Benson (1967) who suggest that the globules are lipoprotein subunits made up of a protein matrix which binds Chl and lipids by hydrophobic association. It has also been suggested by Weier and Benson (1967) that four of the membrane subunits may bind together to form the quantasome observed by in shadowed preparations. It should be pointed out that all of the above-mentioned studies of subunits in sectioned lamellae have been questioned on a technical basis, since it is difficult to imagine how a series of overlapping 90 Å globules can be visualized by looking through a sample which is itself at least 400 Å thick. It must be assumed that the particles are lined up exactly one above the other to allow visualization of subunits. This problem has previously been discussed by other authors (Branton, 1968; Park and Sane, 1971).

4.2. Characterization of Surface-Bound Lamellar Subunits

Particles have been observed on the surface, or exterior face, of the thylakoids of higher plant chloroplasts by negative staining techniques (see Fig. 5). Correlative enzymic and structural studies on the surface-bound subunits have been carried out by Murakami (1968) and Moudrianakis and coworkers (see Moudrianakis, 1968). The latter authors have shown that the surface-bound particles are usually randomly distributed but occasionally appear in paracrystalline arrays with a periodicity of 150×180 Å. To identify the enzymic activity of the surface-bound particle, Howell and Moudrianakis (1967a,b) analyzed the supernatant solutions following sequential water and dilute EDTA washes of isolated lamellar membranes. They have demonstrated that 120 Å particles having a cuboid structure are released from thylakoids by water washing. The fraction containing these particles had high specific activity for RuDP carboxylase. EDTA treatment of water-washed membranes resulted in the release of a second type of particle which had Ca^{2+}-dependent ATPase activity and was shown to be a 5- or 6-sided polygon of 100 Å diameter in electron micrographs. Lamellar fragments remaining after water and EDTA washes appeared smooth by negative staining (see Figs. 6 and 7). The particle removed by EDTA washing is thought to be identical to the "coupling factor," described earlier by McCarty and Racker (1966), which

---

Fig. 5. A negatively stained, intact, immature chloroplast isolated from spinach leaves. The sample was obtained by placing a drop of isolated chloroplasts on the supportive surface of an electron microscope grid. After allowing the plastid to settle on the surface of the grid, excess liquid was removed. The chloroplast was then stained with a solution of phosphotungstic acid which dries around the chloroplast, thus revealing surface topography of the membranes. Drying of the chloroplast on the grid has resulted in the collapse of the plastid envelope (PE), giving rise to the membrane folds extending across the chloroplast. Numerous round thylakoids (T) can be seen within the plastid.

Fig. 6. Isolated, water washed spinach chloroplast thylakoids prepared for electron microscopy by negative staining. The small particulate subunits protruding from the surface of the membranes have been identified as carboxydismutase and the coupling factor (Moudrianakis, 1968). These small subunits are also visible within the chloroplast shown in Fig. 5.

Fig. 7. Smooth-surfaced spinach chloroplast thylakoids from which the suface-bound subunits were removed by EDTA washing (according to the procedure of Howell and Moudrianakis, 1967a,b).

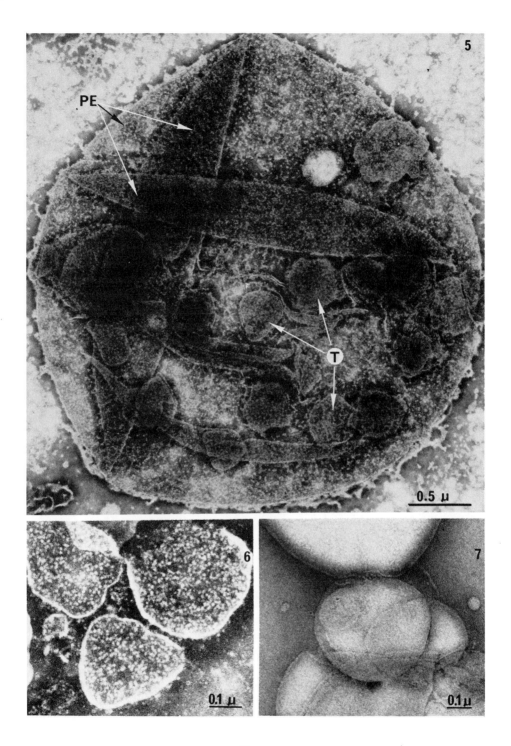

87

was suggested to catalyze the terminal steps in the energy coupling process which forms ATP. This concept is in agreement with the observation that loss of the surface particles in EDTA preparations was correlated to a loss in ability of the membranes to catalyze photophosphorylation (Howell and Moudrianakis, 1967a). It should be emphasized that the surface-bound particles on chloroplast lamellae do not participate in the electron transport activities of the plastid since "smooth," EDTA-washed membranes still carry out the Hill reaction with either dyes or NADP$^+$ as the electron acceptor (Howell and Moudrianakis, 1967b, Arntzen *et al.*, 1969).

Granules on the stroma side of the photosynthetic lamellae of *Porphyridium cruentum* were first observed by Gantt and Conti (see Gantt and Conti, 1966). Isolation of these particles free from the membrane (Gantt and Contii, 1966; Gantt and Lipschultz, 1972) has shown them to be composed of the accessory pigments phycoenythrin, R-phycoyanin, and allophycocyanin. The mechanism of energy transfer from these accessory pigments, located in a surface-bound particle, to the chlorophyll embedded within the membrane remains an interesting problem.

4.3. Interpretation of Membrane Subunits Revealed by Freeze-Fracturing

The development of the freeze-etching technique for electron microscopy (Moor, 1964) has allowed a new approach to the study of chloroplast lamellae. This procedure very clearly demonstrated the presence of subunits in the membrane (Mühlethaler *et al.*, 1965; Branton and Park, 1967). An example of this type of preparation is shown in Fig. 8. There has been considerable discussion relating to the proper interpretation of the images seen by the freeze-fracturing. In the initial studies of Mühlethaler and coworkers (1965, 1966) two different sized particles were observed. They concluded that the larger of these particles was attached to the outer surface and the smaller particle to the inner surface of a bimolecular lipid layer to give a

---

Fig. 8. A portion of an isolated maize mesophyll chloroplast prepared for electron microscopy by the freeze-etch technique (Moor, 1964). A series of longitudinal membrane fractures extending through several thylakoids of two grana stacks (G) can be seen on the right-hand side of the figure. The fracture faces exposed are: The A face: characterized by small, tightly packed, indistinct particles; the B face: characterized by large, randomly spaced particles; the C face: characterized by small, tightly packed, but very distinct particles. The stroma lamellae (SL), shown in the left central portion of the picture, have both the C fracture face and a B′ face which is characterized by the presence of very widely spaced small particles.

Samples are prepared for freeze-etching by freezing a small sample in liquid N_2. This frozen sample is fractured under a vacuum to reveal the membranes. A replica of the surface is produced by first shadowing with a heavy metal and then evaporating a layer of carbon on the fracture face. The carbon replica is then removed and examined in the electron microscope.

complete lamellar membrane. This was based on the idea that the fracture plane through the sample followed the membrane–aqueous medium interface. Branton and Park (1967), on the basis of essentially identical structural data, reinterpreted the results in a model which depicted the membrane subunits lying entirely within the membrane matrix. This conclusion was based on the idea that the fracture plane in a frozen sample would follow the hydrophobic interior of the membrane where bonding energy would be weak at the low temperature.

The concept of internal membrane fracture has been supported by the several lines of evidence. Park and Pfeifhofer (1968, 1969b) and Arntzen *et al.* (1969) have examined smooth (EDTA-washed) chloroplast lamellae from which all surface particles were removed and have demonstrated the continued presence of the freeze-etch particles in these preparations. Branton and Park (1967) noted that lipid extraction of chloroplasts completely eliminated the normal freeze-fracture surfaces, thus suggesting that the hydrophobic interior region of the membrane is the site of membrane splitting. Branton (1966) and Neushul (1970) demonstrated that large surface-bound phycobilin-containing particles, which are readily observed in cross-sectional views of *Porphyridium cruentum* chloroplast membranes, are not visible on the membrane faces after freeze fracture. Studies of spinach thylakoids by Park and Pfeifhofer (1969a) demonstrated that the exterior surfaces of the lamellae were only visible after a period of "deep-etching." Branton has summarized freeze-etch studies on artificial membranes, together with studies on red blood cell and other cellular membranes in a recent review (Branton, 1969). In all cases he concluded that the freeze fracture follows an internal region of the membrane.

Neushul (1970) and Mühlethaler and coworkers (Wehrli *et al.*, 1970; Mühlethaler, 1972) has used a double-replica technique to examine plastid membranes. In this procedure both halves of the fractured specimen can be examined to reveal the nature of both of the surfaces exposed during membrane fracture. The data obtained show that two matching membrane surfaces can be obtained after freeze-fracturing chloroplast lamellae. This is in complete agreement with the hypothesis that the fracture follows an internal region of the membrane.

The pattern of particle distribution observed on freeze-fractured chloroplast lamellae preparations is unique with respect to other membranes which have been studied by this technique. The two major size classes of particles observed in mature plant chloroplasts were reported by Branton and Park (1967) to average 175 and 110 Å in diameter. It should be noted that an earlier paper of Mühlethaler *et al.* (1965) reported chloroplast particle dimensions of 60 and 120 Å in similar preparations. The discrepancy in reported particle sizes is apparently not due to a difference in the samples examined by the two groups, but rather a difference in the way the two groups measure the particles in their micrographs (Branton, 1968). Goodenough and Staehelin (1971) have conducted a careful analysis of the size distribution of particles observed on *Chlamydomanas* chloroplast lamellae. They have stressed the fact that the subunits observed by freeze-fracturing have a broad range of sizes which show certain characteristic maxima at about 70, 100, and 160 Å when plotted as a histogram. They also point out, however, that the actual measurements are a subjective decision which may be affected by factors involved in sample preparation.

In spite of the problem of accurate determination of the size of freeze-etch particles, it is possible to give a general description of chloroplast lamellae which clearly distinguishes the different fracture faces observed by freeze-etching. Both longitudinal and cross-sectional fracture planes through grana and stroma lamellae can be seen in Fig. 8. Following the nomenclature of Branton and Park (1967) the membrane faces have been labeled as follows. Fracture face C, which is characterized by the presence of very distinct, small (110 Å), tightly packed particles, can be observed both within grana stacks and on stroma lamellae. Face A also is characterized by the presence of small, tightly packed particles; in this case the subunits are much less distinct, however. The A face is found only in stacked lamellar regions. Since the region where two thylakoids are closely appressed in a partition is thought to be a hydrophobic region (Murakami and Packer, 1971). it is likely that a freeze-fracture might follow this membrane surface. It has therefore been suggested that the A face reveals the outer surface of a thylakoid in a partition region (Branton and Park, 1967). (It should be noted that this is the only case in which a fracture face is thought to follow a membrane surface.) It has not been established whether or not the coupling factor and RuDP carboxylase are present in the partition region. If they are found in the stacked region, they must not exist as particles projecting out from the surface but rather as a more integral part of the membrane itself since no protruding subunits are evident on the A face.

Fracture face B (see Fig. 8) is perhaps the most distinctive feature of chloroplast lamellae prepared by the freeze-etch technique. It is characterized by the presence of large (175 Å diameter) subunits which are usually randomly distributed and separated from one another. These particles sometimes appear in rows and occasionally in a paracrystalline array (Branton and Park, 1967). It has been recognized recently that the large widely spaced "B" particles can only be detected within grana stacks. Phung Nhu Hung *et al.* (1970b), in a study of chloroplast greening in barley, found that the B face was only detected after the onset of grana stacking and that the large particles were localized in the stacked regions. Remy (1969) has also reported that the large particles are restricted to appressed lamellae in bean chloroplasts. Sane *et al.* (1970), in a combined membrane fractionation and structural study demonstrated that large freeze etch particles of the B face are found in grana, but not in the stroma lamellae.

The fracture plane which passes through the grana lamellae to expose the B face often follows along the same membrane into the stroma lamellae (see Fig. 8). In the unpaired lamellar region, the fracture face which is exposed under these conditions is characterized by a relatively smooth surface upon which widely dispersed, small (110–120 Å) particles are embedded. We will designate this the B′ face since it is continuous with the B face of the grana.

Goodenough and Staehelin (1971), have examined the chloroplast lamellae of *Chlamydomonas* mutants *ac-5* and *ac-31* which do not have stacked lamellae *in vivo*. They have shown that only the C face and matching B′ face (Bu face in their terminology) are present on these membranes. If the plastid membranes of mutant *ac-31* were isolated in the presence of high concentration of salt, the lamellae would form stacks. This *in vitro* stacking was accompanied by the appearance of the B face in freeze-etch preparations. Isolation of lamellae in low salts did not induce stacking and did not induce the appearance of the B face. It therefore appears that the formation of the large, widely spaced freeze-etch particle is regulated by salt concentration and membrane stacking, and that this particle therefore can only be found in partition regions. On the basis of these findings, Goodenough and Staehelin (1971). presented a chloroplast membrane model which shows the large freeze-etch particle extending through two adjacent membranes. They imply that the large particle can only form from the interaction of constituents of two membranes. This model does not agree with the data of Park and Pfeifhofer (1968, 1969b) and Arntzen *et al.* (1969) who have shown the continued presence of the B face in low-salt (EDTA-washed) chloroplast lamellae which have become unstacked and appear as small vesicles. In addition, analysis of the height of the B face subunit in higher plant chloroplasts by Branton and Park (1967) does not suggest that it is large enough to extend through two membranes. The demonstration of a reversible formation of the large freeze-etch particles in *Chlamydomonas* may simply be related to a conformational change in a single membrane which gives rise to the particle. Conformational changes of membrane constituents which are related to stacking in the presence of salts have been described by Murakami and Packer (1971).

It should be pointed out that the particulate structure of all green plant chloroplasts is probably not identical. Schwelitz *et al.* (1972), in a study of *Euglena* chloroplasts, could not detect the B fracture face (large freeze-etch particles) in regions of appressed lamellae. Robinson (1972) has reported that the chloroplasts of *Chaetomorpha* swarmers have A-, B-, C-type fracture faces in addition to a new fracture face which is completely smooth. This "naked" face was reproducibly demonstrated in regions of appressed thylakoids.

The data discussed in the beginning of this section have clearly indicated that a freeze-fracture through frozen chloroplast lamellae follows an interior region of the membrane. As has been pointed out by Kirk (1970), a segregation of the two types of membrane particles to opposing fracture faces during the membrane fracture implies that the two types of subunits are at least partially offset from one another across a center fracture plane of the membrane. This concept, together with the observation that detergents cause both a functional structural partitioning of chloroplast lamellar components, led Arntzen *et al.* (1969) to propose that the plastid membrane has a binary structure with small freeze-etch particles localized in the outer half and large freeze-etch particles in the inner half of an asymmetric membrane. This is in complete agreement with the recent double-replica freeze-etch studies of Neushul (1970) and Wehrli *et al.* (1970) which demonstrated how the two halves of the membrane fit together. Neushul has also concluded that chloroplast lamellae are asymmetrical. It should be noted that the suggestion of an asymmetric organization of membrane structural components is not unique for chloroplasts (Crane *et al.*, 1970; Wakabayashi *et al.*, 1971). Da Silva and Branton (1970) have reported that their studies of freeze-fractured cell membranes are most consistent "with a bilayer membrane model in which lipid components are locally intercalated with protein differentially associated with each half of the membrane."

4.4. Conclusions

The presence of subunits within and on the surface of chloroplast lamellae has clearly been shown. Interpretation of the freeze-etch data strongly indicates that the subunits within the membrane are arranged asymmetrically in a binary membrane. This working model is shown diagrammatically in Fig. 9. In this membrane model we have stressed the fact that the large freeze-etch particles (face B) are found only in stacked lamellar regions (Sane *et al.*, 1970). The small widely dispersed particles of the B' face are found on both stroma lamellae and grana end membranes (Sane *et al.*, 1970). Tightly packed, small freeze-etch particles (face C) are found in both grana and stroma lamellae. Our interpretation of the D surface, which shows the large freeze-etch particles slightly protruding through the interior side of a thylakoid, is based on the deep-etching studies of Park and Shumway (1968) and Park and Pfeifhofer (1969a). The particles on the D surface have been reported to sometimes occur in a paracrystalline array and are probably identical to the "quantasomes" observed in shadowed lamellar preparations.

We have, in this highly diagrammatic model, strongly emphasized an asymmetric distribution of the internal membrane subunits. In actuality there is probably a greater interdigitation of the subunits on each half of the

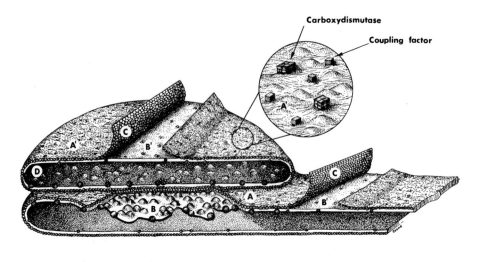

Fig. 9. A schematic representation of two appressed thylakoids and an attached portion of the stroma lamellae (patterned after earlier models of Arntzen *et al.*, 1969 and Park and Sane, 1971). The internal particulate substructure of the membrane (which is revealed by freeze-etching) is labeled to correspond to Fig. 8. It should be noted that the large membrane subunit seen on the B fracture face is only found in the partition region of a grana stack. Smaller particles are found on the B′ face of stroma lamellae and grana end membranes (Sane *et al.*, 1970). We have emphasized the binary nature of the chloroplast membrane in this figure by diagraming the small freeze etch particles (C face) on the outer half and the large freeze etch particle (B face) on the inner half of an asymmetric membrane (further discussion in text). The surface structure of the membrane, shown in the inset of the A′ face, is based on negative staining studies (see Fig. 6). The irregular surface structure of the internal (loculus) side of a thylakoid membrane is shown on the D surface (based on deep-etch studies of Park and Shumway, 1968 and Park and Pfeifhofer, 1969a).

binary membrane than we have indicated. The extent of interdigitation might possibly be related to the energy-linked change in membrane thickness reported by Murakami and Packer (1970).

5. SUBSTRUCTURAL ORGANIZATION OF MEMBRANE FUNCTIONAL COMPONENTS

In the previous section we have presented evidence for the asymmetric localization of structural components within the chloroplast lamellae. There are also several lines of evidence which suggest an asymmetrical distribution of the functional constituents of the electron transport chain within the membrane. Rabinowitch (1959, 1961) and Calvin (1959, 1961) suggested that the lamellar system spatially separates oxidizing and reducing powers. The light-induced spectral absorbance shift at 515 nm in chloroplasts suggests that a charge separation across the membrane is the result of redox reactions between asymmetrically localized electron carriers (see Chapter 10 of this volume). Mitchell (1966) has incorporated the concept of a bilateral distribution of electron carriers into his chemiosmotic energy-coupling mechanism hypothesis. Evidence for nonuniform localization of membrane constituents has been provided through the use of detergent fractionation and specific membrane probes. These latter experiments will be discussed below.

5.1. Detergent Fractionation of Chloroplast Lamellae

Detergent treatment of isolated chloroplasts is now accepted to have a twofold effect which depends on detergent concentration and time of membrane treatment. Wehremeyer (1962) showed that incubation of chloroplast lamellae with low concentrations of digitonin resulted in the release of stroma lamellae from intact grana membranes. This finding has been verified by Goodchild and Park (1971) and Wessels and Voorn (1972). Sane *et al.* (1970) and Park and Sane (1971) have stressed that mechanical fractionation of chloroplasts and the initial action of digitonin act in a similar fashion. When the only action of the detergent is to release stroma lamellae as a light fraction, a yield of no more than 7–15 % recovery of PS I in a light fraction can be expected (Sane *et al.*, 1970; Arntzen *et al.*, 1972; Wessels and Voorn, 1972). Arntzen *et al.* (1972) have shown that treatment of chloroplasts with 0.5 % digitonin for 30 min results in a secondary action of the detergent: the selective solubilization and release of PS I from both grana and stroma lamellae. Under these conditions nearly 50 % of the total chloroplast Chl can be recovered as purified PS I. Since earlier studies of Boardman and Anderson (1964a) and Ohki and Takamiya (1970) showed Chl recovery of 35 and 40 %, respectively, in fractions containing only PS I, it seems likely that these workers were removing PS I from both grana and stroma lamellae.

Treatment of chloroplast membranes with Triton X-100 at very low concentrations (0.05–0.1 % with chloroplasts at 0.3 mg Chl/ml) results in the selective release of stroma lamellae (Briantais, unpublished results). Briantais (1966) used 0.2 % Triton X-100 to prepare PS II enriched subchloroplast fragments (which did not show enhancement for O_2 evolution and hence had probably lost PS I). Higher concentrations of triton (greater than 1.0 %) result in the loss of grana by swelling and unfolding of the lamellar structure (Deamer and Crofts, 1967). For the preparation of purified PS I, Vernon *et al.* (1966) used Triton concentrations considerably higher than 1.0 %; it is likely that they were solubilizing PS I from both grana and stroma lamellae.

The mechanism of PS I release from the intact membranes has been considered with respect to lamellar structure by several workers. Anderson and Boardman (1966) suggested that PS I is loosely attached to the membrane, whereas PS II is an integral part of the structure. Briantais (1969a) showed that successive extraction of chloroplast lamellae with Triton X-100 resulted in the initial release of PS I. He concluded that in the thylakoid membranes PS I is more exposed than PS II. Henninger *et al.* (1967) postulated that digitonin causes a longitudinal splitting of lamellar membranes. This concept was supported by the work of Arntzen *et al.* (1969), who conducted a structural study of membrane subfractions derived by digitonin treatment of chloroplast lamellae. They found that the PS I fraction contained small vesicles when viewed by thin sections. The membrane thickness of these vesicles was about one-half that of the original chloroplast lamellae. Thin-sectioned samples of the PS II fraction showed long lamellae which were usually closely appressed but which were one-half the original membrane thickness when exposed as single membranous vesicles. Briantais and Giraud (unpublished results) have also found that Triton-derived PS II fractions show membranes which are approximately half as thick as the original chloroplast lamellae.

Freeze-etch examination of digitonin derived membrane fragments by Arntzen *et al.* (1969, 1972) has revealed that the PS I fraction contains only the small tightly packed freeze-etch particle (comparable to fracture face C of Figs. 8 and 9), whereas lamellar fragments enriched in PS II activity were characterized by a relative enrichment of the large freeze-etch particles on the exposed membrane faces (comparable to fracture face B of Figs. 8 and 9). They therefore concluded that digitonin causes a longitudinal splitting along the hydrophobic interior of the membrane much like the freeze-fracture of the intact chloroplasts. This splitting results in the removal of PS I particles (either as individual particles or as small aggregates of particles) from the remainder of the membrane matrix (containing PS II). Vernon *et al.* (1968, 1969) have shown that Triton-derived PS I particles appear as small strands of 70–80 Å width and variable lengths. The strands appeared to be composed of subunits. Very similar structures were observed in negative stained samples of digitonin-derived PS I (Arntzen *et al.*, 1972; Wessels and Voorn, 1972). The presence of an active PS I in these very small particles was verified by ultrafiltration techniques (Arntzen *et al.*, 1972). The latter authors demonstrated that PS I particles derived from both grana and stroma membranes can be caused to reaggregate into membranous sheets in the presence of salts. It should be noted that PS II submembrane fractions have almost always been found to occur as lamellar sheets and not as particulate subunits (Vernon *et al.*, 1968, 1969; Arntzen *et al.*, 1969, 1972). An exception to this has been the results of Vernon *et al.* (1971). These authors have subfractionated the

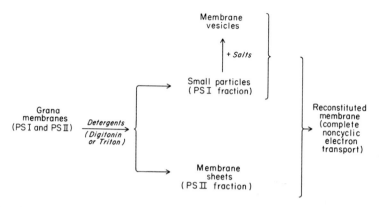

Fig. 10. A summary of the action of detergents on the structural organization of chloroplast grana membranes.

PS II membrane sheets to obtain a small amount of particulate subunits which still have PS II activity.

Evidence suggesting that removal of PS I particles from chloroplast lamellae is a reversible process has been obtained from reconstitution studies. According to Briantais (1967, 1969b), the stimulation of the O_2 burst by PS I preillumination can not be detected in isolated PS II particles, but is partially reestablished in a mixture of PS I and PS II subchloroplast fragments. This suggests that the antagonistic effect of the two photosystems on an electron pool localized between the two photosystems had been reestablished. Huzisige *et al.* (1969) were able to demonstrate the restoration of low rates of electron transfer from water to $NADP^+$ by mixing digitonin-derived PS I and II particles in the presence of plastoquinone. Arntzen *et al.* (1972) have demonstrated that digitonin-derived PS I from grana, but not PS I derived from the stroma lamellae, will reconstitute complete noncyclic electron flow with PS II fractions derived from the grana.

The action of detergents on chloroplast grana membranes is summarized in Fig. 10. As pointed out above, the data are most consistent with a membrane model in which the PS I subunits located at the membrane surface are solubilized away from the remaining portion of the membrane which has PS II activity

5.2. Selective Degradation of Membrane Constituents

Analysis of the spatial disposition of proteins within various membranes has been attempted through the use of selective digestion of specific membrane components or the use of specific labels to tag membrane surface proteins (Wallach, 1972). In the case of isolated chloroplasts, the effects of both lipolytic and proteolytic enzymes on chloroplast reactions has been studied. Okayama (1964) found that lipase treatment of chloroplasts inhibited oxygen evolution, but not PS I-mediated $NADP^+$ reduction. Gressel and Avron (1965) found that phosphorylation was inhibited by lipase treatment, although cyclic phosphorylation was less sensitive than noncyclic. Okayama *et al.* (1971) have more recently demonstrated that lipase treatment causes the apparent loss of C550, a membrane constituent closely related to the primary electron acceptor for PS II.

Protease (trypsin) treatment of isolated chloroplasts has been reported to have multiple effects. Mantai (1968) found an initial uncoupling in the treated chloroplasts. This is consistent with the idea that the coupling factor is externally located on the chloroplast lamellae, and would be very susceptible to enzymic degradation. Mantai (1969, 1970) and Selman and Bannister (1971) have reported that trypsin treatment also causes a reduction in the rate of PS II reactions, but has little or no effect on PS I.

It might initially be suggested that the experiments described above which utilize enzymic digestion of chloroplast lamellae indicate that PS II is externally located on the membrane and is therefore very susceptible to enzymic degradation. Additional results of Mantai (1968) are not in agreement with this concept, however. He has found that photochemical activities of glutaraldehyde-fixed chloroplasts (in which membrane structure is stabilized by crosslinked proteins) are not affected by either lipase or trypsin treatment, even though the enzymes still can selectively degrade the membrane constituents. He has concluded that inhibition of PS II mediated electron transport by the degradative enzymes is caused by structural disruption of the lamellae rather than inactivation of a specific component of PS II. This idea is consistent with the structural studies of lipase- and pronase-treated chloroplasts conducted by Bamberger and Park (1966). They observed distinct morphological changes in treated lamellae. The studies of enzyme-treated chloroplast lamellae therefore do not provide conclusive evidence as to the vectorial distribution of PS I and PS II components. We can only suggest from these reports that the photochemical activities of PS II are very strongly regulated by the degree of membrane organization, whereas PS I is much less affected.

Wakamatsu *et al.* (1971) have investigated the effect of chlorophyllase on isolated chloroplasts. This enzyme, which hydrolyzes chlorophyll to chlorophyllide and phytol, might be expected to cause limited structural damage to the membrane. The authors found that incubation of chloroplasts with chlorophylase severely reduced Hill reaction activity with $NADP^+$, but not with DCPIP as the electron acceptor. One suggestion from these results is that the Chl of PS I, which is involved in the reduction $NADP^+$, is more externally located in the membrane than PS II and is therefore more susceptible to enzymic digestion.

5.3. Evaluation of Membrane Structure by Immunological Techniques

Immunological techniques have proven useful in establishing the vectorial distribution of membrane constituents in chloroplast lamellae. These techniques are particularly valuable since the antibody is a large molecule which has high specificity to its antigen, and since it often inhibits the activity of the antigen. Many types of molecules have antigenic properties. The use of immunological procedures can therefore be very convenient for recognizing components of membrane surfaces.

McCarty and Racker (1966) reported the inhibition of photophosphorylation, in the presence of PMS or ferricyanide, by the antiserum made against a coupling factor removed from the chloroplasts with EDTA. By the use of the immunotechniques the CF was found to be localized on the surface of the thylakoid membrane. This is in complete agreement with the structural data obtained by Howell and Moudrianakis (1967a,b) (see Section 4.2).

Berzborn and colleagues (Berzborn et al., 1966; Berzborn, 1968, 1969a,b,c; Regitz et al., 1970) have found that vaccinating rabbits with purified chloroplast thylakoids elicits at least four different antibodies. One of these reacts with the CF as described above. The other three react with components of the electron-acceptor system of PS I (either $NADP^+$ reductase or the FRS complex). Berzborn (1969a,b) found that the antibody against $NADP^+$ reductase does not agglutinate stroma freed spinach chloroplasts in spite of the fact that a positive Coomb's test was obtained (indicating antigen–antibody interaction). Berzborn therefore suggests that both the CF and the reductase are at the lamellar surface, but the CF extends out at least 80 Å. This protrusion blocks the direct agglutination of antibodies binding to the partially hidden $NADP^+$ reductase. Further evidence supporting the idea that the reductase is in a cavity between adjacent CF's was the finding that EDTA-washed chloroplasts did agglutinate in the presence of the reductase antibody.

Regitz et al. (1970) and Regitz and Oettmeier (1972) have shown that antibodies against FRS inhibit the reduction of Fd by isolated chloroplast lamellae. This indicates that the enzyme complex is externally localized on the chloroplast membrane. Hiedemann-VanWyk and Kannangara (1971), in studies with antibodies against Fd, have concluded that Fd is also localized on the lamellar surface.

Racker et al. (1972) have found that antibodies made against a purified P700 particle inhibited the rate of MV reduction and caused the agglutination of whole chloroplasts. They have therefore concluded that P700 is externally located on the chloroplast membranes. Racker et al. (1972) and Hauska et al. (1971) have found that antibodies against either Cyt f or PC do not cause agglutination or inhibit photoreactions of chloroplast unless the plastids are sonicated in the presence of the antibodies. It was concluded that PC and

Cyt f, which act as electron carriers on the oxidizing side of PS I, are internally located in the thylakoid membranes.

The numerous reports of antibody reactions described above have all been against components of the electron transport chain which are associated with PS I. Recently Briantais and Picaud (1972) have tested the effect of preincubation of an antichloroplast serum with either PS I or PS II particles prepared by chloroplast fractionation with Triton X-100. Agglutination of chloroplast membranes in the presence of the serum was inhibited by previous incubation with PS I particles but not by the PS II lamellar fragments. Briantais and Picaud have elicited an anti-PS II antibody by injecting the Triton-derived PS II fragments into rabbits. This serum caused agglutination of the PS II particles even after a serum preincubation in the presence of chloroplast membranes. If the serum was preincubated with chloroplasts during a period of sonication, however, the agglutination capacity with PS II was reduced (Briantais, unpublished results). These studies have been interpreted to indicate that PS II is on the interior side and PS I on the exterior side of the chloroplast membrane.

Radunz *et al.* (1971) have prepared antibodies against Chl. One of their preparations indicated a surface localization of Chl since a positive Coomb's test was obtained with chloroplast lamellae. In addition, they found that one of their Chl antibodies partially inhibited PS II-mediated ferricyanide reduction, but stimulated a PS I-mediated MV reduction reaction. This may indicate that a portion of the pigments associated with PS II reactions are surface localized. Recently Braun and Govindjee (1972) have isolated an antibody active against the oxidizing side of PS II. This serum was elicited by the injection into rabbits of detergent-derived PS II subchloroplast particles. The antibody caused a low level of inhibition of PS II-mediated dye reduction in whole chloroplasts. However, this inhibition is increased twofold when the test material is PS II particles (Braun and Govindjee, unpublished data). Braun and Govindjee suggest that the limited inhibition, with whole chloroplasts, is due to the inaccessibility of the antigenic site at the membrane surface.

5.4. Site-Specific Labeling of Exposed Membrane Constituents

Dilley *et al.* (1972) have recently instigated the use of nonpermeant membrane labeling reagents for studies of chloroplast lamellae. They have reported that [^{35}S]diazonium-benzenesulfonic acid, which does not penetrate into the membrane, will react with surface-exposed lipids and proteins. When these labeled membranes were fractionated with digitonin, Dilley *et al.* (1972) found 10 to 24 times more diazonium compound bound to the PS I-enriched fraction as compared to the PS II lamellar fragments. Their results support the concept of an asymmetric distribution of functional components within the membrane with PS I being exposed at the surface of the membrane.

6. CORRELATION OF MEMBRANE SUBSTRUCTURE AND PHOTOCHEMICAL ACTIVITIES

Subunits of chloroplast lamellae have been shown in the previous sections to be of two types: those which are surface-bound and those which are an integral part of the membrane structure. There is now good agreement that the surface-bound particles are RuDP carboxylase and the CF (Moudrianakis, 1968). Correlation of the structure of the internal membrane subunits to the function of the photochemical systems residing within the membrane has not been so successful, however.

The early flashing-light experiments of Emerson and Arnold (1932) (see discussion by Rabinowitch, 1956; Gaffron, 1960) established that a large number of Chl molecules (2500) appear to cooperate in the photochemical fixation of one CO_2 molecule or evolution of one O_2 molecule (see Chapter 8 of this volume). Kohn (1936) pointed out that this large number might be divided by the number of quanta required for the evolution of one O_2 molecule to give a more proper estimate of the number of chlorophylls associated with one reductive center. This idea has been commonly accepted (e.g., see Park, 1965) and the number 250–300 chlorophylls per PSU is usually encountered in the literature. The object of numerous membrane structure–function studies has been to find, within the membrane, a morphological equivalent of the kinetically determined PSU. Of central importance in these studies has been the "quantasome" concept of Park and coworkers (see reviews by Park, 1965 and Park and Sane, 1971).

Park and Pon (1963) conducted a detailed chemical analysis of spinach chloroplast lamellae. Based on the manganese content of the membranes, they calculated a minimum molecular weight of 960,000 for a suggested lamellar subunit. When Park and Biggins (1964) discovered the existence of regularly repeating subunits within shadowed lamellar membranes, the molecular weight of a subunit was calculated to be 2.0×10^6, or twice the minimum molecular weight based on chemical determinations. From these data, Park and Biggins (1964) calculated that one lamellar subunit would contain about 230 chlorophylls and therefore hypothesized that these subunits were "quantasomes," or the morphological expression of the PSU. The quantasome concept has been of fundamental importance in stimulating and influencing numerous structure–function studies in recent years. The data obtained from these studies have sometimes argued for and sometimes against the existence of an autonomous function subunit within the membrane. The various viewpoints will be discussed below.

6.1. Isolation of Functional Membrane Subunits

Several attempts have been made to fractionate the chloroplast lamellae to isolate an individual "quantasome" particle. Park and Pon (1961), in studies of mechanically fragmented chloroplasts, maintained that aggregates of 5–6 quantasomes were fully active in ferricyanide reduction. Gross *et al.* (1964) found Hill reaction comparable to that of control chloroplast in particles containing only 2–3 quantasomes. The problem was reinvestigated by Izawa and Good (1965), however, who found that the previous workers had overlooked an uncoupling effect caused by membrane fragmentation. When this effect was recognized, lamellar disruption was found to result in a loss of photoactivity such that particles containing one or a few quantasomes were nearly inactive as compared to the controls. There is, therefore, no direct biochemical evidence for the existence of an automous functional "quantasome" unit based on fragmentation evidence.

6.2. Relationship of Surface-Bound and Internal Membrane Subunits to the "Quantasome"

Howell and Moudrianakis (1967a, b), in a combined biochemical and structural study, questioned the suggested role of the quantasome as described by Park and coworkers. They found that membrane-bound particles could be observed in a paracrystalline array on the surface of negatively stained spinach chloroplasts using the negative staining technique. The dimensions of this particulate array was approximately the same as the dimensions reported by Park and Biggins (1964) for shadowed quantasomes. In addition, Howell and Moudrianakis found that EDTA washing removed these surface particles, but did not result in loss of photoreductive capacities of the membranes. They therefore indicated that the particles organized in the paracrystalline array, which they concluded to be quantasomes, were not participating in electron transport activities of the chloroplast. This allegation was challenged by Park and Pheifhofer (1968, 1969b). They reported the continued presence of quantasomes in EDTA-washed lamellae. The differences in opinion in the two groups was not the result of variance in data but rather a difference in interpretation in the original definition of the quantasome and its suggested localization either upon or within the membrane. The distinction between these two possibilities was settled, at least partially, through the use of the freeze-etch technique for electron microscopy.

Branton and Park (1967), in their initial freeze-etch studies, suggested that the large 175 Å particles, together with the matrix material of the membrane, were identical with the quantasomes seen in shadowed preparations. This was further substantiated by the deep-etching studies of Park and Shumway

(1968) which demonstrated that the portion of the large freeze-etch particle visible on the interior (loculus) side of the membrane was identical in distribution pattern to the quantasome seen during shadowing. Analysis of EDTA washed chloroplast lamellae using freeze-etching by Park and Pheifhofer (1968, 1969b) and Arntzen *et al.* (1969) revealed the continued presence of both the small and the large freeze-etch particles in the washed membranes. It can therefore be concluded that the "quantasome," as described by Park and coworkers, is not surface bound and was not removed in the studies of Howell and Moudrianakis (1967a,b). The particles studied by Howell and Moudrianakis were undoubtedly surface-bound and primarily represent the CF, as these authors have suggested. We should emphasize that the "quantasome," which appears as a single particle with four possible subunits by heavy metal shadowing of dried lamellar membranes (Park and Biggins, 1964) is thought by Park and coworkers (see Park and Sane, 1971) to be comprised of the large (175 Å) freeze-etch particle in addition to portions of the small freeze-etch particle (C fracture face) material. Park and Pfeifhofer (1968, 1969a,b), and Park and Sane (1971) have indicated that the large freeze-etch particle is a "quantasome" with the matrix material removed and have therefore called it a "quantasome core."

Goodenough and Staehelin (1971) have reported that two mutants of *Chlamydomonas reinhardi*, which have normal photosynthetic capacities, do not have stacked lamellar membranes and do not have the large freeze-etch particle designated by Park and coworkers as the "quantasome core." Goodenough and Staehelin found that the appearance of the large freeze-etch particle could be produced by suspending the chloroplasts in higher salt concentrations. The appearance of the particles was not accompanied by any change in photochemical activities of the plastids. The authors therefore concluded that there is no direct correlation between the presence of the large freeze-etch particle (the quantasome core) and photochemical activities of the lamellae.

6.3. Fractionation of the Lamellar Membrane—Relation to the Quantasome Concept

Since the early studies of Park and Pon (1963) and Park and Biggins (1964), it has been suggested that the "quantasome" is a repeating unit within the membrane containing approximately 230 chlorophylls. It can be pointed out, however, that certain electron chain components such as Cyt f and P700 are found in concentrations of only 1 P700 or Cyt f per 400 chlorophylls and that it is now generally accepted that photosynthetic electron transport from water to $NADP^+$ is a 2-quantum process involving two photochemical systems. The data therefore suggest that the 2500 chlorophylls, determined by Emerson and Arnold (1932) to be a functional unit for CO_2 fixation, should be divided by 4 to get the size of a single "PSU" containing both PS I and II (assuming an 8-quantum requirement for fixation of one CO_2 molecule). This calculation gives a "PSU" of about 600 chlorophylls. Boardman (1968b) and Govindjee *et al.* (1967) have summarized the evidence which suggests that the "PSU" containing both PS I and PS II is comprised of two reaction center complexes with 200–300 Chl molecules in each photosystem. Detergent fractionation of chloroplasts, which allows the separate isolation of PS I and PS II (Boardman and Anderson, 1964a,b; Vernon *et al.*, 1966) has strongly supported a "separate package" model (see Myers, 1963) for the functional components of the two photosystems.

Implicit in detergent fractionation experiments is the suggestion that the two photosystems may reside in physically different regions of the membrane. Arntzen *et al.* (1969, 1972) have demonstrated that digitonin fractionation of chloroplasts results in the partitioning of both structural and functional components of the chloroplast membrane (cf. Section 4.3). They have shown that the PS I fraction obtained by detergent treatment contains only the small freeze-etch particle; the functional components of PS I are therefore not localized in the "quantasome core" described by Park and coworkers. The large freeze-etch particle (quantasome core) was found to be enriched in the detergent-derived PS II fraction. As an alternative view to the quantasome concept, Arntzen *et al.* (1969, 1972) have suggested that the small freeze-etch particle is a marker for the outermost half of a binary chloroplast membrane which contains PS I. In contrast, PS II is hypothesized to be localized within the interior half of the binary chloroplast membrane; that is, within the same region of the asymmetric membrane as the large freeze-etch particle. This hypothesis is supported by biochemical evidence for a bilateral distribution of functional constituents within the membrane, with PS II components more internally localized and PS I more externally localized (cf. Section 5). It is also interesting to note that the stroma lamellae of mature spinach chloroplasts, which have only PS I activity, have only the small freeze-etch particle (Sane *et al.*, 1970). We must quickly emphasize, however, that there is no direct evidence from membrane fractionation studies which suggests that the membrane subunits visualized by electron microscopy are actual sites of photochemical or electron transport functions. We may extend this thought by pointing out that none of the chloroplast structure–function studies reported up to the present time have provided conclusive evidence indicating a direct correlation between morphologically visible subunits lying within chloroplast lamellae and the specific photochemical activities of the membrane components.

7. SUMMARY

The stacked lamellae (grana) of higher plant chloroplasts represent a unique membranous structure in plant cells. Numerous studies have questioned the physiological significance of this pattern of membrane organization. Our summary of these works has indicated that the question is still not settled. On one hand are the reports which demonstrate that some agranal chloroplasts are capable of complete photochemical activities, thus indicating that grana are not a prerequisite for chloroplast functioning. We also can not identify a component of the chloroplast membrane which is responsible for causing membrane fusion. On the other hand, however, we must recognize that grana are found in almost all higher plants chloroplasts; it would certainly seem that they have some evolutionary advantage.

Analysis of membrane biosynthesis has indicated that PS I activity appears early in developing chloroplast lamellae. A somewhat later onset of PS II activity is often linked to the formation of grana; the structural development is not obligatory, however. It is interesting that this pattern of membrane development suggests ontogeny following phylogeny. Olson (1970) has theorized that photosynthetic bacteria, having mainly a PS I-type pigment assemblage, may be an evolutionary precursor to the present day chloroplast. Evolutionary development of PS II and more complex lamellar organization may be related to the secondary steps of membrane biosynthesis occurring in chloroplasts.

Attempts at correlating substructure of chloroplast lamellae to biochemical function have also been reviewed above. Good agreement has been reached with regard to the identification of RuDP carboxylase, the "CF," and "phycolbilisomes" (in certain algae) as being surface-bound membrane components (see Section 4.2). There is no definitive information relating to the functional properties of internal lamellar subunits, however (particles seen by thin-sectioning or freeze-etching). The "quantasome concept," which dominated the research developments in the area of chloroplast membrane substructure over the last several years, has had two major approaches. First, functional studies were based on the concept that a grouping of Chl molecules and associated proteins (enzymes) functions somewhat independently within the membrane as a "PSU" or "quantasome." This concept must now be tempered by data which indicate a high degree of interaction between reaction centers (See discussion of excitation spillover in Chapter 6 of this volume.) and cooperation between electron transport chains (Siggel et al., 1971, 1972). Second, the structural concept of a quantasome was originally formulated on the basis of repeating subunits visualized within shadowed chloroplast lamellae. Park and coworkers (see Section 4) have convincingly demonstrated that these shadowed subunits are closely related to the large freeze-etch particle (quantasome core). Correlations between the functional "quantasome" and its structural counterpart have been limited somewhat by the finding that the quantasome core is not present in certain photochemically active lamellae (see Section 6.2). It is necessary to suggest at this time that a PSU may be only a statistical unit which is not always accompanied by a structural counterpart.

Present research on the structure of chloroplast lamellae is using a wide variety of tools (immunochemistry, detergent fractionation, site-specific labelling of macromolecules, X-ray analysis (not discussed here), etc.) in addition to electron and light microscopy. There is an increasing realization of the dynamic nature of membrane constituents; structural analysis is now being related to the energy state of the membrane (see Chapter 11 of this volume). We can anticipate that future membrane models will include some aspect of this dynamic function of membranes—structural and substructural organization may soon be discussed more in terms of transitory conditions than as static models. Although the extensive studies summarized and reviewed in this chapter have not always provided answers to the fundamental questions which have been posed, we can be content with the realization that there is now a very sound body of information relating to chloroplast membrane form and function on which new and perhaps more incisive questions can be based.

Note Added in Proof

Since the completion of the literature search for this review numerous papers on chloroplast structure and function have been published, some of which relate directly to the conclusions discussed above. With respect to grana stacking, Levine and co-workers [Levine, R. P., and Duram, H. A. (1973). *Biochim. Biophys. Acta* **325**, 565, and Anderson, J. M., and Levine, R. P., (1974). *Biochim. Biophys. Acta*, in press] have discovered that two proteins, identified by disc-gel electrophoresis, are reduced in content in agranal bundle sheath chloroplasts of C_4 plants and also in algal and higher plant mutants which have limited grana stacking. These may be required to obtain stacked lamellae.

It was emphasized above that PS I and PS II are asymmetrically distributed across the chloroplast membrane such that PS II is more internally localized within the membrane. More recent data suggests that PS II components may span the membrane. Using the nonpermeant probe DABS, Giaquinta *et al.* found that oxygen evolution can be inhibited in light energized membranes. The DABS inhibition was dependent on electron flux through PS II; it was interpreted that membrane conformational changes expose some PS II components to the chemical modifier [Giaquinta, R. T., Dilley, R. A., and Anderson, B. J., (1973). *Biochem. Biophys. Res. Commun.* **52**, 1410]. Arntzen, C. J., Briantais, J.-M., Vernotte, C. and Armond, P. (1974). [*Biochim. Biophys. Acta* in press] have used the enzyme lactoperoxidase to specifically iodinate surface-exposed chloroplast membrane proteins. They have found that this treatment inhibits the function of the PS II reaction center, and have thereby suggested that this pigment–protein complex is surface localized. The data are not in disagreement with our earlier conclusion that most PS II components are buried within the membrane, however.

In studies of membrane subunits visualized by freeze-fracturing, Packer and colleagues [Wang, A. Y,-I., and Packer, L. (1973). *Biochim. Biophys. Acta* **305**, 488 ; and Torres-Pereira, J., Melhorn, R., Keith, A. D., and Packer L., (1974). *Arch. Biochem. Biophys.* **160**, 90] have demonstrated that the large freeze-etch particles of the B fracture face undergo changes in organization when chloroplasts are illuminated in the presence of a weak acid anion solution or when divalent cations are added to isolated chloroplasts. These studies have emphasized that the chloroplast membrane has a fluid matrix and that dynamic changes in the membrane can include lateral movements of macromolecular complexes visualized by electron microscopic techniques.

ACKNOWLEDGMENTS

The thoughtful suggestions of N. K. Boardman, R. A. Dilley, J. H. Hilliard, R. B. Park, and L. P. Vernon were greatly appreciated.

REFERENCES

Allen, C. F., Good, P., Trosper, T., and Park, R. B. (1972). *Biochem. Biophys. Res. Commun.* **48**, 907.

Andersen, K. S., Bain, J. M., Bishop, D. G., and Smillie, R. M. (1972) *Plant Physiol.* **49**, 461.

Anderson, J. M., and Boardman, N. K. (1964). *Aust. J. Biol. Sci.* **17**, 93.

Anderson, J. M., and Boardman, N. K. (1966). *Biochim. Biophys. Acta* **112**, 403.

Anderson, J. M., and Pyliotis, N. A. (1969). *Biochim. Biophys. Acta* **189**, 280

Anderson, J. M., Boardman, N. K., and Spencer, D. (1971a). *Biochim. Biophys. Acta* **245**, 253.

Anderson, J. M., Woo, K. C., and Boardman, N. K. (1971b). *Biochim. Biophys. Acta* **245**, 398.

Argyroudi-Akoyunglou, J. H., and Akoyunoglou, G. (1970). *Plant Physiol.* **46**, 247.

Arnon, D. I., Chain, R. K., McSwain, B. D., Tsujimoto, H. Y., and Knaff, D. B. (1970). *Proc. Nat. Acad. Sci. U.S.* **67**, 1404.

Arntzen, C. J., Dilley, R. A., and Crane, F. L. (1969). *J. Cell Biol.* **43**, 16.

Arntzen, C. J., Dilley, R. A., and Neumann, J. (1971). *Biochim. Biophys. Acta* **245**, 409.

Arntzen, C. J., Dilley, R. A., Peters, G. A., and Shaw, E. R. (1972). *Biochim. Biophys. Acta* **256**, 85.

Bakri, M. D. L. (1972). Ph.D. Thesis, Univ. of Illinois, Urbana.

Bamberger, E. S., and Park, R. B. (1966). *Plant Physiol.* **41**, 1591.

Barr, R., and Arntzen, C. J. (1969). *Plant Physiol.* **44**, 591.

Barr, R., Magree, L., and Crane, F. L. (1967). *Amer. J. Bot.* **54**, 365.

Bassham, J. A. (1965). *In* "Plant Biochemistry" (J. Bonner and J. E. Varner, eds.), p. 875. Academic Press, New York.

Baszynski, T., Brand, J., Barr, R., Krogmann, D. W., and Crane, F. L. (1972). *Plant Physiol.* **50**, 410.

Ben-Shaul, Y., Schiff, J. A., and Epstein, H. T. (1964). *Plant Physiol.* **39**, 231.

Berzborn, R. J. (1968). *Z. Naturforsch.* **236**, 1096.

Berzborn, R. J. (1969a). *Progr. Photo. Res.* **I**, 106.

Berzborn, R. J. (1969b). *Progr. Photo. Res.* **II**, 104.

Berzborn, R. J. (1969c). *Z. Naturforsch.* **246**, 436.

Berzborn, R. J., Menke, W., Trebst, A., and Pistorius, E. (1966). *Z. Naturforsch.* **21b**, 1057.

Biehl, B. (1966). *Z. Naturforsch.* **21b**, 501.

Bisalputra, T., Downton, W. J. S., and Tregunna, E. B. (1969). *Can. J. Bot.* **47**, 15.

Bishop, D. G., Andersen, K. S., and Smillie, R. M. (1971a). *Biochim. Biophys. Acta* **231**, 412.

Bishop, D. G., Andersen, K. S., and Smillie, R. M. (1971b). *Biochem. Biophys. Res. Commun.* **42**, 74.

Bishop, D. G., Andersen, K. S., and Smillie, R. M. (1972). *Plant Physiol.* **49**, 467.

Björkman, O., Boardman, N. K., Anderson, J. M., Thorne, S. W., Goodchild, D. J., and Pyliotis, N. A. (1972). *Carnegie Inst. Year Book* **71**, 115.

Black, C. C., and Mollenhauer, H. H. (1971). *Plant Physiol.* **47**, 15.

Boardman, N. K. (1968a). *In* "Comparative Biochemistry and Biophysics of Photosynthesis" (K. Shibata, A. Takamiya, A. T. Jagendorf, and R. C. Fuller, eds.), p. 206. University of Tokyo Press, Tokyo.

Boardman, N. K. (1968b). *Advan. Enzymol.* **30**, 1.

Boardman, N. K., and Anderson, J. M. (1964a). *Nature (London)* **203**, 166.

Boardman, N. K., and Anderson, J. M. (1964b). *Aust. J. Biol. Sci.* **17**, 86.

Boardman, N. K., Anderson, J. M., Kahn, A., Thorne, S. W., and Treffry, T. E. (1970). *In* "Autonomy and Biogenesis of Mitochondria and Chloroplasts" (N. K. Boardman, A. W. Linnane, and R. M. Smillie, eds.), p. 70. North Holland Publ., Amsterdam.

Bourdu, R., Brangeon, J., Costes, C., and Bazier, R. (1972). *Proc. Int. Congr. Photosynthesis Res.* **II**, 1471.

Bradbeer, J. W., Clijsters, H., Glydenholm, A. O., and Edge, H. J. W. (1970). *J. Exp.* **21**, 525.

Branton, D. (1966). *Proc. Nat. Acad. Sci. U.S.* **55**, 1048.

Branton, D. (1968). *In* "Photophysiology" (A. C. Giese, ed.), Vol. 3, p. 197. Academic Press, New York.

Branton, D. (1969). *Ann. Rev. Plant Physiol.* **20**, 209.

Branton, D., and Park, R. B. (1967). *J. Ultrastruct. Res.* **19**, 283.

Braun, B. Z., and Govindjee (1972). *FEBS Lett.* **25**, 143.

Briantais, J. M. (1966). *C. R. Acad. Sci. Paris* **263**, 1899.

Briantais, J. M. (1967). *Biochim. Biophys. Acta* **143**, 659.

Briantais, J. M. (1969a). *Physiol. Veg.* **7**, 135.

Briantais, J. M. (1969b). *Progr. Photosynthesis. Res.* **I**, 174.

Briantais, J. M., and Picaud, M. (1972). *FEBS Lett.* **20**, 100.

Butler, W. L., DeGreef, J., Roth, T. F., and Oelze-Karow, H. (1972). *Plant Physiol.* **49**, 102.

Calvin, M. (1959). *Rev. Med. Phys.* **31**, 147.

Calvin, M. (1961). *J. Theor. Biol.* **2**, 258.

Chollet, R. (1972). Ph.D. Thesis, University of Illinois, Urbana.

Cheniae, G. M. (1970). *Ann. Rev. Plant Physiol.* **21**, 467.

Chua, N. (1972). *Biochim. Biophys. Acta* **267**, 179.

Cohen-Bazire, G., and Kunisawa, R. (1963). *J. Cell Biol.* **16**, 401.

Crane, F. L., Arntzen, C. J., Hall, J. D., Ruzicka, F. J., and Dilley, R. A. (1970). *In* "Autonomy and Biogenesis of Mitochondria and Chloroplasts" (N. K. Boardman, A. W. Linnane, and R. M. Smillie, eds.), p. 53. North Holland Publ., Amsterdam.

DaSilva, P. P., and Branton, D. (1970). *J. Cell Biol.* **45**, 598.

Deamer, D. W., and Crofts, A. (1967). *J. Cell Biol.* **33**, 395.

DeGreef, J., Butler, W. L., and Roth, T. F. (1971). *Plant Physiol.* **47**, 457.

Dilley, R. A., Peters, G. A., and Shaw, E. R. (1972). *J. Memb. Biol.* **8**, 163.

Dolzmann, P. (1968). *Z. Pflanzenphysiol.* **58**, 300.

Downton, W. J. S., and Pyliotis, N. A. (1971). *Can. J. Bot.* **49**, 179.

Downton, W. J. S., Berry, J. A., and Tregunna, E. B. (1970). *Z. Pflanzenphysiol.* **63.5**, 194.

Emerson, R., and Arnold, W., (1932). *J. Gen. Physiol.* **16**, 191.

Ericksson, G., Kahn, A., Walles, B. and Von Wettstein, D. (1961). *Ber. Deut. Bot. Ges.* **74**, 221.

Esser, A. F. (1972). *Biochim. Biophys. Acta* 275, 199.

Fork, D. C., and Heber, U. W. (1968). *Plant Physiol.* **43**, 606.

Frey-Wyssling, A., and Steinmann, E. (1953). *Vierteljahressch, Naturforsch. Ges. Zuerich* **98**, 20.

Gaffron, H. (1960). *In* "Plant Physiology, A Treatise" (F. C. Steward, ed.), p. 107. Academic Press New York.

Gantt, E. and Contii, S. F., (1966). *Brookhaven Symp. Biol.* **19**, 393.

Gantt, E., and Lipschultz, C. A. (1972). *J. Cell Biol.* **54**, 313.

Gibbs, S. P. (1970). *Ann. N.Y. Acad. Sci.* **175**, 454.

Gibbs, S. P., Sistrom, W. R., and Worden, P. B. (1965). *J. Cell Biol.* **26**, 395.

Glydenholm, A. O., and Whatley, F. R. (1968). *New Phytol.* **67**, 461.

Goldberg, I., and Ohad, I. (1970). *J. Cell Biol.* **44**, 573.

Goodchild, D. J., and Park, R. B. (1971). *Biochim. Biophys. Acta* **226**, 393.

Goodchild, D. J., Highkin, H. R., and Boardman, N. K. (1966). *Exp. Cell Res.* **43**, 684.

Goodchild, D. J., Björkman, O., and Pyliotis, N. A. (1972). *Carnegie Inst. Year Book* **71**, 102.

Goodenough, U. W., and Levine, R. P. (1969). *Plant Physiol.* **44**, 990.

Goodenough, U. W., and Levine, R. P. (1970). *J. Cell Biol.* **44**, 547.

Goodenough, U. W., and Staehelin, L. A. (1971). *J. Cell Biol.* **48**, 594.

Goodenough, U. W., Armstrong, J. J., and Levine, R. P. (1969). *Plant Physiol.* **44**, 1001.

Govindjee, Papageorgiou, G., and Rabinowitch, E. (1967). *In* "Fluorescence; Theory, Instrumentation and Practice" (G. Guilbault, ed.), p. 511. Dekker, New York.

Gracen, V. E., Hilliard, J. H., Brown, R. H., and West, S. H. (1972). *Planta* **107**, 189.

Granick, S., and Porter, K. (1947). *Amer. J. Bot.* **34**, 545.

Gressel, J., and Avron, M. (1965). *Biochim. Biophys. Acta* **94**, 31.

Griffiths, D. J. (1970). *Bot. Rev.* **36**, 29.

Gross, E. L., and Packer, L. (1967). *Arch. Biochem. Biophys.* **121**, 779.

Gross, J. A., Becker, M. J., and Shefner, A. M. (1964). *Nature (London)* **203**, 1263.

Gross, J. A., Shefner, A. M., and Becker, M. J. (1966). *Nature (London)* **209**, 615.

Gunning, B. E. S. (1965). *J. Cell Biol.* **24**, 79.

Gunning, B. E. S., Steer, M. W., and Cochrane, P. (1967). *Proc. Roy. Microsc. Soc.* **2**, 378.

Hall, D. O., Edge, H., and Kalina, M. (1971). *J. Cell Sci.* **9**, 289.

Hall, D. O., Edge, H., Reeves, S. G., Stocking, C. R., and Kalina, M. (1972). *Proc. Int. Congr. Photosynthesis Res.* **II**, 701.

Hall, J. D., Barr, R., Al-Abhas, A. H., and Crane, F. L. (1972). *Plant Physiol.* **50**, 404.

Hatch, M. D. (1970). *In* "Photosynthesis and Photorespiration" (M. D. Hatch, C. B. Osmond, and R. O. slayter, eds.), p. 139. Wiley, New York.

Hauska, G. A., McCarty, R. E., Berzborn, R., and Racker, E. (1971). *J. Biol. Chem.* **246**, 3524.

Henninger, M. D., Magree, L., and Crane, F. L. (1967). *Biochim. Biophys. Acta* **131**, 119.

Heslop-Harrison, J. (1963). *Planta* **60**, 243.

Hiedemann-VanWyk, D., and Kannangara, C. G. (1971). *Z. Naturforsch.* **26b**, 46a.
Highkin, H. R., Boardman, N. K., and Goodchild, D. J. (1969). *Plant Physiol.* **44**, 1310.
Hilliard, J. H., and West, S. H. (1971). *Planta* **99**, 352.
Hiller, R. G., and Boardman, N. K. (1971). *Biochim. Biophys. Acta* **253**, 449.
Hodge, A. J., McLean, J. D., and Mercer, F. V. (1955). *J. Biophys. Biochem. Cytol.* **1**, 605.
Hohl, H. R., and Hepton, A. (1965). *J. Ultrastruct. Res.* **12**, 542.
Holdsworth, R. H. (1971). *J. Cell Biol.* **51**, 499.
Homann, P. H. (1967). *Plant Physiol.* **42**, 997.
Homann, P. H., and Schmid, G. H. (1967). *Plant Physiol.* **42**, 1619.
Hoober, K. J., Siekevitz, P., and Palade, G. E. (1969). *J. Biol. Chem.* **244**, 2621.
Howell, S. H., and Moudrianakis, E. N. (1967a). *Proc. Nat. Acad. Sci.* **58**, 1261.
Howell, S. H., and Moudrianakis, E. N. (1967b). *J. Mol. Biol.* **27**, 323.
Huzisige, H., Usiyama, H., Kikuti, T., and Azi, T. (1969). *Plant Cell Physiol.* **10**, 441.
Izawa, S., and Good, N. E. (1965). *Biochim. Biophys. Acta* **109**, 372.
Izawa, S., and Good, N. E. (1966a). *Plant Physiol.* **41**, 533.
Izawa, S., and Good, N. E. (1966b). *Plant Physiol.* **41**, 544.
Jacobi, G. (1969). *Z. Pflanzenphysiol.* **61**, 203.
Jacobi, G., and Lehmann, H. (1968). *Z. Pflanzenphysiol.* **59**, 457.
Jacobi, G., and Lehmann, H. (1969). *Progr. Photo Res.* **I**, 159.
Jacobi, G., and Perner, E. (1962). *Biochim. Biophys. Acta* **58**, 155.
Ji, T. H., and Benson, A. A. (1968). *Biochim. Biophys. Acta* **150**, 686.
Ji, T. H., Hess, J. L., and Benson, A. A. (1968). *Biochim. Biophys. Acta* **150**, 676.
Keck, R. W., Dilley, R. A., Allen, C. F., and Biggs, S. (1970a). *Plant Physiol.* **46**, 692.
Keck, R. W., Dilley, R. A., and Ke, B. (1970b). *Plant Physiol.* **46**, 699.
Kirk, J. T. O. (1970). *Ann. Rev. Plant Physiol.* **21**, 11.
Kirk, J. T. O., and Tilney-Basset, R. A. E. (1967). "The Plastids." Freeman, San Francisco, California.
Klein, S., and Neuman, J. (1966). *Plant Cell Physiol.* **7**, 115.
Kohn, H. (1936). *Nature (London)* **137**, 706.
Kreutz, W. (1970). *Advan. Bot. Res.* **3**, 54.
Laetsch, W. M. (1971). *In* "Photosynthesis and Photorespiration" (M. D. Hatch, C. B. Osmond R. D. Slayter, eds.), p. 323. Wiley (Interscience), New York.
Laetsch, W. M., and Price, I. (1969). *Amer. J. Bot.* **56**, 77.
Lavorel, J., and Levine, R. P. (1968). *Plant Physiol.* **43**, 1049.
Lichtenthaler, H. K. (1969). *Protoplasma* **68**, 315.
Lichtenthaler, H. K., and Sprey, B. (1966). *Z. Naturforsch.* **21b**, 690.
Lintilhac, P. M., and Park, R. B. (1966). *J. Cell Biol.* **28**, 582.
Mackender, R. O., and Leech, R. M. (1972). *Proc. Int. Congr. Photosynthesis Res., 2nd* **II**, 1431.
Mantai, K. E. (1968). *Carnegie Inst. Year Book* **68**, 598.
Mantai, K. E. (1969). *Biochim. Biophys. Acta* **189**, 449.
Mantai, K. E. (1970). *Plant Physiol.* **45**, 563.
Mayne, B. C., Edwards, G. E., and Black, C. C. (1971). *In* "Photosynthesis and Photorespiration" (M. D. Hatch, C. B. Osmond, and R. D. Slayter, eds,), p. 361. Wiley (Interscience), New York.
McCarty, R. E., and Racker, E. (1966). *Brookhaven Symp. Biol.* **19**, 202.
McCledon, J. (1954). *Plant Physiol.* **29**, 448.
Menke, W. (1938). *Kolloid Z.* **85**, 256.
Menke, W. (1940). *Protoplasma* **35**, 115.
Menke, W. (1962). *Ann. Rev. Plant Physiol.* **13**, 27.
Menke, W. (1966). *In* "Biochemistry of Chloroplasts" (T. W. Goodwin, ed.), pp. 3–18. Academic Press, New York.
Michel, J. M., and Michel-Wolwertz, M. R. (1969). *Progr. Photosynthesis Res.* **I**, 115.

Michel, J. M., and Michel-Wolwertz, M. R. (1970). *Photosynthetica* **4**, 146.

Miller, M., and Nobel, P. S. (1972). *Plant Physiol.* **49**, 535.

Mitchell, P. (1966). *Biol. Rev.* **41**, 445.

Möbius, M. (1920). *Ber. Deut. Bot. Ges.* **38**, 224.

Moor, H. (1964). *Z. Zellforsch.* **62**, 546.

Moudrianakis, E. N. (1968). *Fed. Proc.* **27**, 1180.

Mühlethaler, K. (1960). *Z. Wiss. Mikrosk.* **64**, 444.

Mühlethaler, K. (1966). *In* "Biochemistry of Chloroplasts" (T. W. Goodwin, ed.), Vol. I, p. 83. Academic Press, New York.

Mühlethaler, K. (1972). *Proc. Int. Congr. Photosynthesis Res., 2nd* (G. Forte, M. Avron, and A. Melandri, eds.), Vol. 2, p. 1423. Dr. W. Junk N. V. Publishers, The Hague.

Mühlethaler, K., Moor, H., and Szarkowski, J. W. (1965). *Planta* **67**, 305.

Murakami, S. (1964). *J. Electronmicrose.* **13**, 234.

Murakami, S. (1968). *In* "Comparative Biochemistry and Biophysics of Photosynthesis" (K. Shibata, A. Takamiya, A. T. Jagendorf and R. C. Fuller, eds.), p. 82. Univ. of Tokyo Press, Tokyo.

Murakami, S., and Packer, L. (1970). *J. Cell Biol.* **17**, 332.

Murakami, S., and Packer, L. (1971). *Arch. Biochem. Biophys.* **146**, 337.

Murata, N., and Brown, J. S. (1970). *Plant Physiol.* **45**, 360.

Myers, J. (1963). *Nat. Acad. Sci. Nat. Res. Council* **1145**, 301.

Neushul, M. (1970). *Amer. J. Bot.* **57**, 1231.

Nir, I., and Seligman, A. M. (1970). *J. Cell Biol.* **46**, 617.

Oelze-Karow, H., and Butler, W. L. (1971). *Plant Physiol.* **48**, 621.

Ohad, I., Siekevitz, P., and Palade, G. E. (1967). *J. Cell Biol.* **35**, 553.

Ohki, R., and Takamiya, A. (1970). *Biochim. Biophys. Acta* **197**, 240.

Okayama, S. (1964). *Plant Cell Physiol.* **5**, 145.

Okayama, S., Epel, B. L., Erixon, K., Lozier, R., and Butler, W. L. (1971). *Biochim. Biophys. Acta* **253**, 4761.

Olson, J. M. (1970). *Science* **168**, 438.

Paolillo, D. J. (1970). *J. Cell. Sci.* **6**, 243.

Park, R. B. (1965). *In* "Plant Biochemistry" (J. Bonner, and J. E. Varner, eds.), p. 124. Academic Press, New York.

Park, R. B., and Biggins, J. (1964). *Science* **144**, 1009.

Park, R. B., and Pfiefhofer, A. O. (1968). *Proc. Nat. Acad. Sci. U.S.* **60**, 337.

Park, R. B., and Pfiefhofer, A. O. (1969a). *J. Cell. Sci.* **5**, 299.

Park, R. B., and Pfiefhofer, A. O. (1969b). *J. Cell. Sci.* **5**, 313.

Park, R. B., and Pon, N. G. (1961). *J. Mol. Biol.* **3**, 1.

Park, R. B., and Pon, N. G. (1963). *J. Mol. Biol.* **6**, 105.

Park, R. B., and Sane, P. V. (1971). *Ann. Rev. Plant Physiol.* **22**, 395.

Park, R. B., and Shumway, L. K. (1968). *In* "Comparative Biochemistry and Biophysics of Photosynthesis" (K. Shibata, A. Takamiya, A. T. Jagendorf, and R. C. Fuller, eds.), p. 57. Univ. of Tokyo Press, Tokyo.

Park, R. B., Steinbach, K. E., and Sane, P. V. (1971). *Biochim. Biophys. Acta* **253**, 204.

Phung-Nhu-Hung, S., Hoarau, A., and Moyse, A. (1970a). *Z. Pflanzenphysiol.* **62**, 245.

Phung-Nhu-Hung, S., Lacourly, A., and Sarda, C. (1970b). *Z. Pflanzenphysiol.* **62**, 1.

Polya, G. M., and Osmond, C. B. (1972). *Plant Physiol.* **49**, 267.

Possingham, J. V., Vesk, M., and Mercer, F. V. (1964). *J. Ultrastruct. Res.* **11**, 68.

Punnett, T. (1966). *Brookhaven Symp. Biol.* **19**, 375.

Punnett, T. (1971). *Science* **171**, 284.

Pyliotis, N. A., Woo, K. C., and Downton, W. J. S. (1971). *In* "Photosynthesis and Photorespiration" (M. D. Hatch, C. B. Osmond, and R. O. Slayter, eds.), p. 406. Wiley (Interscience), New York.

Rabinowitch, E. I. (1956). "Photosynthesis and Related Processes" Vol. II. part 2, p. 1714–1750, 1981. Wiley (Interscience), New York.

Rabinowitch, E. I. (1959). *Discuss. Faraday Soc.* **27**, 161.

Rabinowitch, E. I. (1961). *Proc. Nat. Acad. Sci. U.S.* **47**, 1296.

Racker, E., Hauska, G., Lien, S., Berzborn, R., and Nelson, N. (1972). *Proc. Int. Congr. Photosynthesis Res., 2nd* **II**, 1970.

Radunz, A., Schmid, G. H., and Menke, W. (1971). *Z. Naturforsch.* **26b**, 435.

Reger, B. J., and Kraus, R. W. (1970). *Plant Physiol.* **46**, 568.

Regitz, G., and Oettmeier, W. (1972). *Proc. Int. Congr. Photosynthesis Res., 2nd* **I**, 499.

Regitz, G., Berzborn, R., and Trebst, A. (1970). *Planta* **91**, 8.

Remy, R. (1969). *C. R. Acad. Sci. Paris* **268**, 3057.

Rhodes, M. J. C., and Yemm, E. W. (1966). *New Phytol.* **65**, 331.

Robinson, D. G. (1972). *J. Cell. Sci.* **10**, 307.

Rosado-Alberio, J., Weier, T. E., and Stocking, C. R. (1968). *Plant Physiol.* **43**, 1325.

Roughan, P. G., and Boardman, N. K. (1972). *Plant Physiol.* **50**, 31.

Sabnis, D. D., Gordon, M., and Galston, A. W. (1970). *Plant Physiol.* **45**, 25.

Sager, R. (1972). "Cytoplasmic Genes and Organelles," p. 280. Academic Press, New York.

Sane, P. V., and Park, R. B. (1971). *Biochim. Biophys. Acta* **253**, 208.

Sane, P. V., Goodchild, D. J., and Park, R. B. (1970). *Biochim. Biophys. Acta* **216**, 162.

Schmid, G. H., and Gaffron, H. (1967). *J. Gen. Physiol.* **50**, 563.

Schwelitz, F. D., Dilley, R. A., and Crane, F. L. (1972). *Plant Physiol.* **50**, 166.

Selman, B. R., and Bannister, T. T. (1971). *Biochim. Biophys. Acta* **253**.

Sesták, Z. (1969). *Photosynthetica* **3**, 285.

Siekevitz, P., Palade, G. E., Dallner, G., Ohad, I., and Omura, T. (1967). *In* "Organizational Biosynthesis" (H. J. Vogel, J. O. Lampen, and B. Bryson, eds.), p. 331. Academic Press, New York.

Siggel, U., Renger, G., and Rumberg, B. (1971). *Proc. Int. Congr. Photosynthesis Res., 2nd* **I**, 753.

Siggel, U., Renger, G., Stiehl, H. H., and Rumberg, B. (1972). *Biochim. Biophys. Acta* **256**, 328.

Sironval, C., Bronchart, R., Michel, J. M., Brouers, M., and Kuyper, Y. (1968). *Bull. Soc. Fr. Physiol. Veg.* **14**, 195.

Sirnoval, C., Michel, J. M., Bronchart, R., and Englert-Dujardin, E. (1969). *Prog. Photosynthetic Res.* **I**, 47.

Smillie, R. M., Andersen, K. S., and Bishop, D. G. (1971). *FEBS Lett.* **13**, 318.

Smillie, R. M., Andersen, K. S., Tobin, N. F., Entsch, B., and Bishop, D. G. (1972). *Plant Physiol.* **49**, 471.

Spencer, D., and Possingham, J. V. (1961). *Biochim. Biophys. Acta* **52**, 379.

Spencer, D., and Wildman, S. G. (1962). *Aust. J. Biol. Sci.* **15**, 599.

Sprey, B., and Lichtenthaler, H. K. (1966). *Z. Naturforsch.* **21b**, 697.

Steinmann, E. (1952). *Exp. Cell Res.* **3**, 367.

Suzuki, K., Ishii, T., and Amano, H. (1970). *Sci. Rep. Saitama Univ.* **B5**, 169.

Taylor, A. O., and Craig, A. S. (1971). *Plant Physiol.* **47**, 719.

Terpstra, W. (1970). *Biochim. Biophys. Acta* **216**, 179.

Togasaki, R. K., and Levine, R. P. (1970). *J. Cell. Biol.* **44**, 531.

Ueda, K. (1964). *Cytologia* **29**, 514.

Vernon, L. P., and Shaw, E. R. (1969). *Plant Physiol.* **44**, 1645.

Vernon, L. P., Ke, B., Katoh, S., San Pietro, A., and Shaw, E. R. (1966). *Brookhaven Symp. Biol.* **19**, 102.

Vernon, L. P., Mollenhauer, H., and Shaw, E. R. (1968). *In* "Regulatory Functions of Biological Membranes" (J. Järnefelt, ed.), pp. 57–71. Elsevier, Amsterdam.

Vernon, L. P., Ke, B., Mollenhauer, H. H., and Shaw, E. R. (1969). *Progr. Photosynthesis Res.* **I**, 137.

Vernon, L. P., Shaw, E. R., Ogawa, T., and Raveed, D. (1971). *Photochem. Photobiol.* **14**, 343.

Wakabayashi, T., Korman, E. F., and Green, D. E. (1971). *Bioenergetics* **2**, 233.

Wakamatsu, K., Kobayashi, Y., and Sasa, T. (1971). *Bot. Mag. Tokyo* **84**, 101.

Wallach, D. F. H. (1972). *Biochim. Biophys. Acta* **265**, 61.

Wehremeyer, W. (1962). *Z. Naturforsch.* **17b**, 54.

Wehremeyer, W. (1964). *Planta* **62**, 272.

Wehrli, E., Mühlethaler, K., and Moor, H. (1970). *Exp. Cell Res.* **59**, 336.

Weier, T. E., and Benson, A. A. (1967). *Amer. J. Bot.* **54**, 389.

Weier, T. E., Stocking, C. R., Thomson, W. W., and Drever, H. (1963). *J. Ultrastruct. Res.* **8**, 122.

Weier, T. E., Englebrecht, A. H. P., Harrison, A., and Risley, E. B. (1965a). *J. Ultrastruct. Res.* **13**, 92.

Weier, T. E., Stocking, C. R., Bracker, C. E., and Risley, E. B. (1965b). *Amer. J. Bot.* **52**, 339.

Weier, T. E., Bisalputra, T., and Harrison, A. (1966a). *J. Ultrastruct. Res.* **15**, 38.

Weier, T. E., Stocking, C. R., and Shumway, L. K. (1966b). *Brookhaven Symp. Biol.* **19**, 353.

Wessels, J. S. C., and Voorn, G. (1972). *Proc. Int. Congr. on Photosynthesis Res., 2nd* **II**, 833.

Wolken, J. J., and Palade, G. E. (1952). *Nature (London)* **170**, 144.

Woo, K. C., Anderson, J. M., Boardman, N. K., Downton, W. J. S., Osmond, C. B., and Thorne, S. W. (1970). *Proc. Nat. Acad. Sci. U.S.* **67**, 18.

3

Primary Events and the Trapping of Energy

Kenneth Sauer

ABBREVIATIONS

BChl	Bacteriochlorophyll
BPheo	Bacteriopheophytin
CD	Circular dichroism
Chl	Chlorophyll
EPR	Electron paramagnetic resonance
Fd	Ferredoxin
FRS	Ferredoxin reducing substance
LDAO	Lauryl dimethylamine oxide
NADP$^+$	Nicotinamide adenine dinucleotide phosphate
PS I (II)	Photosystem I (II)
P700 (870)	Pigment with one of its absorption maxima at 700 nm (870 nm); reaction center of photosynthesis
SDS	Sodium dodecyl sulfate
UQ	Ubiquinone

1. INTRODUCTION

The conversion of light energy to chemical potential in photosynthesis is a form of energy transduction. It begins with the absorption of incident photons by chlorophyll or by various accessory pigments and reaches its goal with the biosynthesis of the carbohydrates, proteins, lipids, etc. that make up the complex biological organism. This chapter will treat the very earliest stages in this highly complex process—just through the point at which the first chemical intermediates appear. It will necessarily treat both the dynamics of the process and the chemical nature and arrangement of the molecules involved.

The first stage of photosynthetic energy conversion is commonly known as the "physical" part. This is by contrast with the later stages, where the influences of "chemistry" are dominant. The most penetrating experimental methods of investigating the primary processes of photosynthesis involve techniques of the physicist or physical chemist. These approaches often appear remote and mysterious to the student of biology, who may feel uncomfortable upon encountering the formalism and theoretical underpinnings that the physicist assumes as basic tools. As a result there often arises a language barrier or communication gap between the two groups. It will be a major objective of the present chapter to interpret the physical phenomena and concepts in a way that will provide insight for the biologist without being technical beyond comprehension.

The physicist approaches his task by making measurements, formulating models based on the observations, and then developing mathematical expressions to interpret the models in terms of fundamental physical laws. Here we will describe the same measurements and models, but we will use primarily diagrams and word pictures for their interpretation. The apparent complexity of the mathematical equations of the physicist should not deceive the uninitiated into overestimating their sophistication, for it is generally the quality of the model that determines how closely the physicist is able to approximate the truth.

The primary events of photosynthesis will be divided into four sequential steps and will be examined more or less in this order: (1) The absorption of light and the formation of electronic excited states of the pigment molecules; (2) the transfer of electronic excitation among the molecules in the pigment array (see also Chapter 4 of this volume); (3) trapping of the excitation at a particular location known as the reaction center; (4) the initiation of chemistry via the transfer of electrons from donors to acceptors. In addition to examining the dynamics of these steps, we will also treat the energetics of the processes and the nature of the molecular complexes in which they take place.

In order to develop a general picture in which the fundamental processes are not clouded by a host of often controversial details, it will be necessary to be selective from the vast literature on the subject. For this reason emphasis will be given to current views and a statement of outstanding problems. For comprehensive literature surveys and the historical background the reader will be referred to the many monographs and review articles on the subject. Among the most valuable of the general references which deal extensively with the primary processes are the monographs by Clayton (1965, 1970, 1971) and by Rabinowitch and Govindjee (1969). Clayton (1972) has also written an excellent summary of the current status and outstanding problems in the realm of physical processes of photosynthesis. For a comprehensive treatment of work prior to 1956, the volumes of Rabinowitch (1945, 1951, 1956) are invaluable.

2. CHLOROPHYLL SPECTROSCOPY

2.1. Chlorophylls—The Essential Pigments of Photosynthesis

The central light-absorbing pigments of photosynthesis are the chlorophylls: chlorophyll a (Chl a) in plants and algae, bacteriochlorophyll a (BChl a) or b in photosynthetic bacteria. The molecular structures of Chl a and BChl a are shown in Fig. 1. No organism is known to carry out photosynthesis without one or another of these essential pigments. In addition, there may be one or more accessory pigments that absorb light and make the resulting excitation available to the photosynthetic light reactions. Such accessory pigments include Chl b and c, *Chlorobium* chlorophyll, carotenes, xanthophylls, and phycobilins. Despite the wide occurrence of these accessory pig-

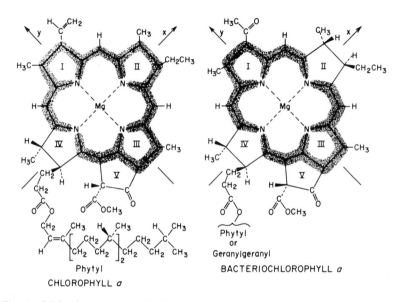

Fig. 1. Molecular structures of chlorophyll *a* and bacteriochlorophyll *a*. The porphyrin π-electrons are indicated by the shading. Absolute configurations are shown for the asymmetrically substituted porphyrin carbon atoms and for phytol. The esterifying group (phytyl or all-*trans*-geranylgeranyl) of BChl *a* varies with the bacterial source (Katz *et al.*, 1972). Chl *b* differs from Chl *a* in the replacement of —CH₃ on ring II by —C$\underset{\text{H}}{\overset{\text{O}}{<}}$. The convention for *x* and *y* directions within the plane of the porphyrin ring is indicated by the arrows.

ments and the doubtless beneficial effect of their presence in the organism for optimal performance in the natural environment, each of them is expendable as far as the essential photosynthetic energy conversion is concerned.

2.2. The Absorption of Light

Photosynthetic energy conversion is initiated by the absorption of light, either by a chlorophyll or by one of the accessory pigments. In order to understand the nature of this process we will need to examine some important aspects of photon absorption by molecules. A very readable and comprehensive account of the subject is given in the small monograph by Clayton, *Light and Living Matter*, Vol. 1: The Physical Part (1970).

The following is a simplified survey of absorption, circular dichroism, and emission spectroscopy, emphasizing those elements of particular importance in photosynthesis. The properties of the chlorophylls will be used to illustrate the general principles. In this regard we will consider spectroscopic properties that yield information about molecular interactions and associations and that can give important clues concerning the local environment. We will be especially interested in those spectroscopic states that participate in the light reactions of photosynthesis.

The absorption or capture of an incident photon by a molecule leads to a rearrangement of the distribution of electrons within its nuclear framework. The new state of the molecule is known as an excited electronic state and its energy is greater than that of the initial or ground state. The extra energy comes from the absorbed photon and is governed by the Planck relationship $\Delta\varepsilon = \varepsilon_1 - \varepsilon_0 = h\nu = hc/\lambda$, where ν is the frequency of the absorbed radiation, λ is its wavelength, h is Planck's constant of proportionality, c is the velocity of light, and ε_1 and ε_0 are the energies of the molecule in the excited (final) and ground (initial) states, respectively. The absorption process is alternatively described by an equation

$$\text{Chl} + h\nu \longrightarrow \text{Chl}^* \qquad (2.1)$$

There is actually a set of such excited electronic states, each with its own characteristic energy and electron distribution. Incident radiation is absorbed only if it has the correct energy to achieve one of these "allowed" states from the ground state of the molecule; photons with intermediate energies are transmitted by the compound and leave it unaffected. The absorption process is illustrated by an energy-level diagram (Fig. 2A). An absorption spectrum, which can be determined experimentally, is a graph of the probability or intensity of radiation absorption versus energy (usually plotted versus wavelength, λ, of the radiation). For molecules like Chl, the energy levels are rather broad because of the possibility of exciting molecular vibrations as well as the electrons themselves. This is seen in the absorption spectrum of BChl a shown in Fig. 2B. (In Fig. 2B the absorption spectrum is turned by 90° from its conventional orientation in order to relate it to the energy level diagram.)

The electronic absorption spectra of the chlorophylls consist of a set of reasonably well-defined bands extending through the ultraviolet and visible regions of the spectrum, terminating with a long-wavelength (lowest energy) band that lies between 600 and 1000 nm in the red or near-infrared. An important property of an absorption band is its strength or intensity. It can be characterized by an oscillator strength (on a scale that goes from 0 to 1 for simple, nondegenerate transitions), by an absorption cross section (i.e., the target size for an incoming photon), by a transition dipole moment, or by various other related measures. The interrelationships among these quantities are described in books on electronic absorption spectroscopy (Hanna, 1969; Sandorfy, 1964; Orchin and Jaffé, 1971). The electronic absorption bands of the chlorophylls are comparatively intense, indicating that the corresponding transitions from the ground to the excited states are strongly allowed in the presence of a flux of photons. A 1-cm cube of a solution containing only 10^{-5}

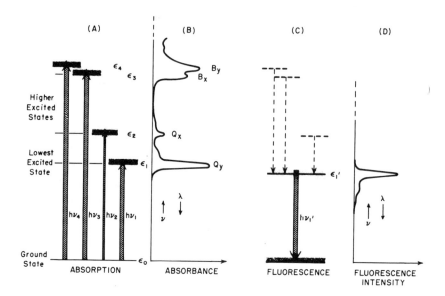

Fig. 2. Absorption and fluorescence of chlorophylls. The illustration is scaled for BChl *a*, but applies qualitatively to Chl *a* and Chl *b* as well. (A) Energy level diagram, showing spectral transitions (vertical arrows). The energy levels are broadened (shading) by vibrational sublevels that are not usually resolved in solution spectra. (B) Absorption spectrum corresponding to energy levels of part (A). This spectrum is turned 90° from the usual orientation in order to show the relationship to the energy levels. Conventions for designating the spectral transitions (Q_y, Q_x, B_x, B_y, where *x* and *y* refer to the axes shown in Fig. 1) are shown. (C) Diagram showing radiationless relaxation (dashed arrows) and fluorescence (shaded arrow). (D) Fluorescence emission spectrum corresponding to part C. Note the red shift (shorter arrow) of the fluorescence compared with the corresponding Q_y absorption illustrated in parts A and B. This Stokes shift owes to vibrational relaxation in the excited electronic state prior to fluorescence emission and in the ground electronic state after emission.

moles/liter of a chlorophyll is intensely colored. This intense absorption is, of course, of great value in providing for the efficient capture of photons for photosynthesis.

A second important property of the absorption band is its characteristic energy, $\Delta\varepsilon$ (alternatively, the wavelength, λ, or frequency, v), of the absorption maximum. These quantities are related by the Planck formula. In the case of Chl *a*, the lowest energy or longest wavelength transition occurs at about 660 nm in solution. The corresponding energy is 1.9 eV/molecule or 43 kcal/mole. Energies of this magnitude are equivalent to those required to dissociate weak covalent chemical bonds, such as that in the I_2 molecule. No bond dissociation occurs at these wavelengths for the chlorophylls, however. The excitation process results only in a change in the electron distribution within the nuclear framework. (The nuclear positions in the excited states are only slightly different from those in the ground state.) Typically, the redistribution of electron density is confined mostly to one or a few of the most labile electrons. In chlorophylls these are π-electrons from the conjugated porphyrin ring system (Fig. 1, shading). As we shall see later, these electrons are much more weakly bound in the excited molecule (C*); and in the presence of an associated molecule, A, with a strong affinity for electrons (a strong electron acceptor), an electron from the excited molecule may be transferred entirely, resulting in an ion pair

$$C \cdot A \xrightarrow{\;h\nu\;} C^* \cdot A \longrightarrow C^+ \cdot A^- \qquad (2.2)$$

This does not usually occur in simple solutions, however, where the time between collisions of C* and A is very long in comparison with the lifetime of C*.

The qualitative picture of electronic absorption described above is greatly simplified. The process can be treated (approximated, at best) using the formalism of quantum mechanics. In the quantum-mechanical picture, the properties of the electron distribution in the ground state are described by a wavefunction, ψ_0. The excited state is described by a different wavefunction, ψ_1. Each wavefunction, which varies in amplitude with the position (x, y, z) within the molecule, has the property that its square, ψ^2, gives the electron density or probability of finding the electron at each point in the molecule for the respective state.

The absorption transition itself is represented by an operation that describes how the molecule couples to the oscillating electromagnetic field associated with the incident light. The relevant quantity is the transition dipole moment, μ_{01},

$$\mu_{01} = \iiint \psi_0 \mu \psi_1 \, dx \, dy \, dz \qquad (2.3)$$

where μ is the dipole moment operator and the integration is carried out over the entire space where the wavefunctions are significantly different from zero. μ_{01}^2 is the transition dipole strength, which is a direct measure of the intensity of the transition and can be related to the area under an absorption band. This is one of the many ways in which the predictions of quantum-mechanical theory can be tested using experimentally observed properties. Even from this cursory view of a process which is much more complex when examined in detail, one can appreciate two important aspects of the absorption process:

1. The strength of the absorption depends in a very direct and sensitive way upon the wavefunctions of both the initial and final states, hence on the distribution of their constitutive electrons.

2. The transition has a vectorial character, because the dipole moment is a vectorial quantity and imparts a directionality (with respect to the molecular axis system) to the transition. A collection of molecules which are ordered, as

in a crystal or two-dimensional monolayer, will preferentially absorb incident light that is plane-polarized in the direction of the transition moments. This is the phenomenon known as *linear dichroism*. Ordinary solutions of molecules with random orientations do not exhibit linear dichroism.

The properties of two of the absorption transitions of BChl *a* are shown pictorially in Fig. 3. At each of the atoms of the porphyrin where there is appreciable π-electron density, a circle is drawn to show the magnitude of the *change* in charge density accompanying the transition. This may be either an increase or a decrease in instantaneous charge density, as indicated by the two kinds of shading. (There is no change in the zero *net* charge of this neutral molecule, only in the charge distribution.) Following the transition, certain regions of the molecule have changed their charge densities more than others. Careful analysis of the redistribution of charge permits one to establish the

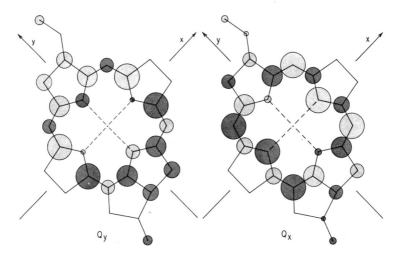

Fig. 3. Transition monopole representation of the Q_y and Q_x absorptions of BChl *a*. The dark and light shadings represent an increase or decrease, respectively, of the electron density at the various atoms. (The transition dipole is a response to the oscillating electromagnetic field of the incident light. Consequently, there is no *absolute* sign associated with the transition moment.) Note that the Q_y transition is nearly symmetric across the *y* axis, and that the principal charge displacement occurs in the *y* direction; just the opposite is true for the Q_x transition. The magnitudes of the *changes* in electronic charge are represented by the areas of the circles. Furthermore, because each contribution to a dipole moment is the product of charge times distance, charge changes farther from the center of the molecule make a correspondingly greater contribution to the transition moment than do those closer in. Taking into account the cancellation produced by adjacent increases and decreases, the reader should be able to discern that the magnitude or intensity of the Q_y transition is greater than that of the Q_x transition, in agreement with the experimental result shown in Fig. 2. The diagrams shown in Fig. 3 represent the results of theoretical calculations carried out by Dr. C. Weiss, Jr., as described by Philipson, *et al.* (1971) and Weiss (1972)

direction and *intensity* of the absorption transition. It is important to appreciate that the pictures shown in Fig. 3 are the result of theoretical calculations using molecular orbital theory. There is no good experimental way as yet to measure the details of the redistributions of charge associated with such transitions.

2.3. Circular Dichroism

Molecules which are asymmetric, and therefore not superimposable with their mirror images, possess a property known as chirality or handedness. This is the property that distinguishes a screw cut with a left-hand thread from one which is cut with a right-hand thread. The distinction is intrinsic to the screws and is not lost even if their orientations are scrambled. A left-handed screw looks left-handed from either end. In a similar way the electron density distributions or wavefunctions of chiral molecules, such as the chlorophylls, have aspects of handedness associated with them. Chlorophyll *a*, for example, has five asymmetrically substituted carbon atoms—three associated with the porphyrin ring part and two in the phytyl ester. These are shown in the molecular structure depicted in Fig. 1. If you were to examine a contour map of the electron density throughout Chl *a*, then you would easily be able to distinguish it from the map of its mirror-image molecule. (The latter molecule does not actually occur in nature, of course.) By using circularly polarized light, whose electric vector sweeps out a helix as it passes through space, to measure the absorption spectrum of chiral molecules, one can detect small differences in the absorption of left (handed) circularly polarized light from the absorption of right circularly polarized light. This difference, $\Delta A = A_L - A_R$ is known as the circular dichroism. It is a property that results from the fundamental asymmetries of the wavefunctions. CD is typically small (10^{-5} to 10^{-2}) compared with the absorption itself and is correspondingly difficult to measure. Nevertheless, sensitive CD spectrometers exist and numerous reports of CD spectra are appearing in the literature.

The CD spectrum of BChl *a* in ether solution is shown together with the normal absorption spectrum in Fig. 4. In general there will be a CD feature associated with each of the major absorption components. Notice that the CD bands can be either positive or negative. This property of the CD can be of practical advantage in enabling one to decompose or separate transitions that overlap strongly in the absorption spectrum.

By contrast with linear dichroism, even a solution of randomly oriented asymmetric molecules will exhibit CD in the absorption bands. This is because the chirality is intrinsic to the molecules and not a function of their orientation in space. For surveys of this subject the reader is referred to the monographs by Crabbé (1972) and by Velluz *et al.* (1965). The applications to photosynthesis have been reviewed recently (Sauer, 1972).

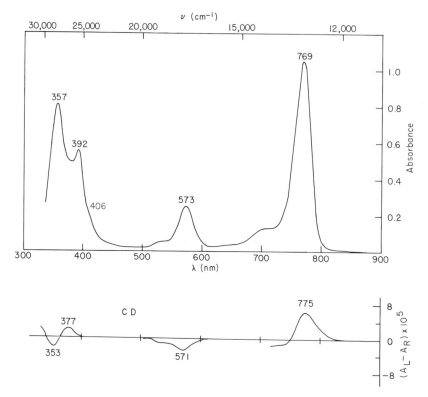

Fig. 4. Absorption and CD spectra of BChl *a* in diethyl ether at 10^{-5} moles liter^{-1} concentration. In this solvent and at this concentration the BChl *a* is monomeric. Spectra taken from Philipson and Sauer (1972).

2.4. Fluorescence

The fluorescence of the chlorophylls is discussed in several chapters in this volume (for example, see Chapter 6). It has special significance as an indicator of processes occurring very near the primary steps of energy conversion. It is important, therefore, to understand its origin and the manner in which alterations of the environment can change its characteristics.

Fluorescence is the process by which a molecule in an excited state undergoes a spontaneous decay and returns to the ground state with the simultaneous emission of radiation. It can be seen in the energy level diagram (Fig. 2C,D), and it can be represented by the equation

$$\text{Chl}^* \longrightarrow \text{Chl} + h\nu' \tag{2.4}$$

The fluorescence of Chl *a* is typical of large aromatic molecules in the following respects:

1. The fluorescence occurs only from the lowest excited state reached via absorption. Higher states, such as 2, 3, 4, etc., in Fig. 1A, undergo a very rapid relaxation to state 1 without the emission of radiation.

2. Because of thermal relaxation that occurs in the excited states prior to emission and in the ground state after emission, the maximum in the fluorescence emission spectrum typically lies a few nanometers to longer wavelength (lower energy) from the absorption maximum. This is known as the Stokes shift.

3. The measured fluorescence lifetime, τ, of Chl a in "good" solvents (see Section 2.9) is very short—only about 5 nsec (5×10^{-9} sec) (Brody and Rabinowitch, 1957; Dimitrievsky et al. 1957).

4. The proportion of excitation that results in fluorescence is about 30% for Chl a in solution (Latimer et al., 1956; Weber and Teale, 1957). Fluorescence quantum yields less than 100% indicate the presence of competing processes for deactivation of the excited state.

5. The origin of the fluorescence emission process is governed by the same kind of relationships among the wavefunctions as govern absorption. For this reason, fluorescence has essentially the same polarization as the long-wavelength absorption band.

Each of these five properties can give useful information about the excited state and its various decay processes.

Fluorescence is not the only way in which the excited state can decay. Some important competing processes are shown in the following equations:

$$Chl^* \xrightarrow{k_F} Chl + h\nu' \qquad \text{(Fluorescence)} \qquad (2.4)$$

$$Chl^* \xrightarrow{k_I} Chl + heat \qquad \text{(Internal conversion, radiationless)} \qquad (2.5)$$

$$Chl^* \xrightarrow{k_T} Chl^T + heat \qquad \text{(Intersystem crossing to triplet state)} \qquad (2.6)$$

$$Chl^* + Q \xrightarrow{k_Q} Chl + Q^* \qquad \text{(Quenching)} \qquad (2.7)$$

There may be additional processes under special circumstances, e.g., very high excitation intensities, high concentrations, photochemical reactants present, electron donors or acceptors present, etc. All of the reactions listed above are kinetically first-order in Chl* concentration. The lifetime of Chl* is determined by the rate constants for all of the reactions involved. A common way of treating this is in terms of a quantum yield. The quantum yield for fluorescence, ϕ_F, is given by the ratio of the fluorescence rate to the sum of the rates of all deactivation processes. When the processes are all first-order, the concentration of Chl* cancels out, leaving

$$\phi_F = \frac{k_F}{k_F + k_I + k_T + k_Q[Q] + \ldots} \tag{2.8}$$

The constant k_F is considered to be an intrinsic property of the molecule, governed by the quantum-mechanical properties of the ground and excited states. Its reciprocal $\tau_0 = 1/k_F$ is known as the "natural" lifetime of the excited state. This is simply the lifetime that the excited state would have in the absence of any processes competing with fluorescence. The natural lifetime can be calculated from directly measured quantities, such as the oscillator strength of the transition. Because the denominator of Eq. (2.8) above is just the reciprocal of the actual (observed) lifetime, τ, we can rewrite the equation as

$$\phi_F = \tau/\tau_0 \tag{2.9}$$

On the basis of the observed values of ϕ_F and τ given above for Chl a in solution, the reader may readily confirm that τ_0 for Chl a is about 15 nsec. This agrees closely with the value 15.2 nsec calculated from the absorption band intensity (Brody and Rabinowitch, 1957). The observed lifetime, τ, will always be less than this value in the presence of competing deactivation processes, thereby providing a very useful measure of the contribution of these competing processes. As we shall see shortly, the fluorescence yield and observed lifetimes of chlorophylls in their *in vivo* situations vary over a wide range, depending on environmental circumstances.

2.5. Triplets and Phosphorescence

The excited states that are reached directly via absorption of radiation by ground state chlorophylls are known as *singlet* states. The term singlet refers to the fact that the electron spin quantum numbers of electrons in the excited state correlate completely with those of the ground state, where all of the electrons are paired. There exists an additional important set of states, known as *triplets*, where two of the electrons have become unpaired. These triplet states are normally not produced via direct absorption of radiation by the ground state molecules; for quantum-mechanical reasons, the oscillator strengths for these transitions are usually less than 10^{-7} and the transitions are observed only under very special circumstances. The triplet states can be reached by intersystem crossing from the excited singlet states, so long as the process is energetically favorable (i.e., not uphill). This is the process represented by Eq. (2.6) above.

Triplets can be observed by their characteristic emission spectrum (phosphorescence), which has both a much longer lifetime (reflecting the very low oscillator strengths for the corresponding absorption process) and a different wavelength dependence than does fluorescence. The triplet absorption spectra (excitation to higher *triplet* states) can be measured using the techniques of flash spectroscopy suited to short-lived species. The triplets of Chl *a* and Chl *b in vitro* were first observed using flash absorption spectroscopy by Livingston *et al.* (1954). The observed lifetimes depend on the solvent, etc., but values of about 10^{-3} sec are commonly observed (Livingston, 1960). Bowers and Porter (1967) made quantitative measurements of the quantum yield of triplet formation, ϕ_T. In ether solutions, the values were 0.64 for Chl *a* and 0.88 for Chl *b*. Since the quantum yields for fluorescence in the same solvent are 0.32 and 0.12, respectively, it is apparent that these two processes together account for over 95 % of the excited singlet state decay for the chlorophylls in simple solutions.

Phosphorescence of chlorophylls is difficult to observe because of a very low quantum efficiency. Apparently the triplets return to the ground state predominantly by radiationless intersystem crossing or by quenching reactions. Emissions with decay times of the order of 10^{-3} sec have been reported for Chl *a* at 755 nm and for Chl *b* at 733 nm (Fernandez and Becker, 1959). [Longer wavelength emissions are also observed (Singh and Becker, 1960).] Because these wavelengths are greater than the corresponding fluorescence maxima, it suggests that the lowest triplet energy level lies somewhat below the first excited singlet state shown in Fig. 1. As a conconsequence, the long-lived triplet state has been a persistent dark-horse candidate for the process of trapping (extending the lifetime by decreasing the energy) of the excitation in photosynthesis (see Section 4.2).

2.6. Dichroism and Fluorescence Polarization

The orientations of the electronic transition moments of different excited states of a molecule can be determined on a relative basis by fluorescence polarization studies. In these measurements the polarized exciting light wavelength is varied throughout the region of absorption and the polarization of the fluorescence, which always comes from the first excited state, is measured. It is important that the molecules not undergo rotation during the excited state lifetimes; hence, viscous solvents are used. The measurements can be carried out on an ordinary solution containing randomly oriented molecules so long as they do not undergo rapid rotation. The information achieved is limited, however; one can only determine the angle between the absorption and the emission oscillators, and not the absolute orientation of either with respect to the molecular framework.

In order to make absolute polarization measurements one needs to use an oriented sample, such as a crystal or ordered monolayer. In this case either polarized absorption (linear dichroism) or emission can be used. Because of the very strong absorption properties of the chlorophylls, it has so far not been possible to measure spectra of single crystals. Too little light gets through to permit reliable measurements.

The most extensive studies of polarization of the chlorophyll spectra were carried out by Goedheer (1957, 1966). He studied not only fluorescence polarizations, but also the dichroism of chlorophylls incorporated into two-dimensional myelin figures. With the aid of some theoretical insight into the origins of the transitions, he was able to assign them to either the y or the x direction within the porphyrin plane (see Fig. 1). These band assignments, which are indicated in Fig. 5, have been confirmed by later studies (Gouterman, 1961; Gouterman and Stryer, 1962).

With this brief introduction to spectroscopy, we will now proceed to examine some of the properties of different chlorophylls and of the significant influences of the environment on the various spectra.

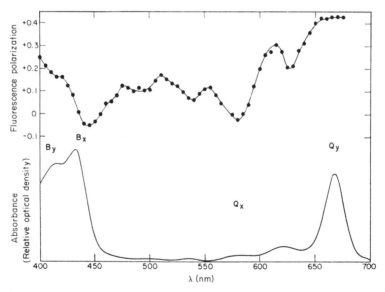

Fig. 5. Absorption spectrum of methyl chlorophyllide *a* (lower curve) and polarization spectrum of fluorescence excitation (points, and upper curve). The highest (most positive) polarization values indicate absorption oscillators parallel to the emission oscillator (e.g., B_y at 415 nm and Q_y at 660 nm) and the lowest polarizations are for perpendicular absorption and emission oscillators (e.g., B_x at 430 nm and Q_x at 580 nm). Some of the intermediate bands, such as those at 540 and 620 nm, are attributed to vibrational overtones that belong to the electronic transition whose origin is at longer wavelengths. Figure taken from Gouterman and Stryer (1962).

2.7. Comparison of the Spectra of Different Chlorophylls

Any influence which affects the electronic distribution of either the ground or the excited state of the molecule will, in general, be reflected in the absorption band intensity, wavelength position, CD strength, fluorescence efficiency, etc. In this and the following sections we will explore the consequences of some of the more important of these influences in order to uncover systematic effects. For a comprehensive survey of the spectra of the chlorophylls, the reader is referred to the following sources: Smith and Benitez (1955); Strain and Svec (1966); Goedheer (1966); and Gurinovich *et al.* (1968).

Differences in chemical structure of the chlorophylls produce spectroscopic effects that can be treated systematically. For example, the molecules Chl *a* and BChl *a* differ in two respects: (1) ring II of BChl *a* is saturated, with hydrogens on the peripheral carbon atoms, but ring II is unsaturated in Chl *a*; and (2) the acetyl substituent on ring I of BChl *a* is a vinyl group in Chl *a* (see Fig. 1). The absorption spectra of these molecules differ appreciably in the wavelengths and intensities of the bands (Fig. 6). From systematic studies of related molecules (e.g., Chl *a* with acetyl instead of vinyl on ring I), it can be demonstrated that the difference in the absorption spectra of Chl *a* and BChl *a* arises almost entirely from the saturation of ring II in the latter compound. Such spectral differences can be rationalized

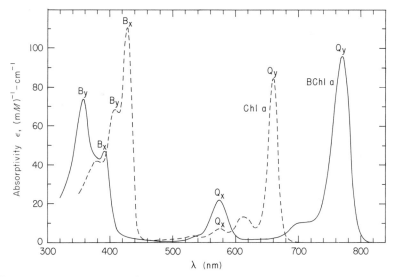

Fig. 6. Absorption spectra of BChl *a* and Chl *a* in ether solutions, compared in terms of equal molar concentrations. Millimolar absorptivities are plotted along the ordinate. Band assignments are based on polarization measurements and theoretical evaluations. The origin of the differences in intensity and wavelengths of the maxima is discussed in the text.

in terms of the electronic wavefunctions using molecular orbital theory, and we are now able to make reasonably reliable predictions of the spectral effects arising from molecular structure changes.

The most important influences are (1) the extent of saturation of the porphyrin ring system, which increases from protochlorophyll to Chl *a* or Chl *b* to BChl *a*; (2) the presence of magnesium or other atoms at the center of the porphyrin system; and (3) the occurrence of the cyclopentenone ring V in the molecule. Of lesser significance is the specific nature of the groups substituted peripherally around the ring; groups such as vinyl or acetyl which can conjugate with the porphyrin π-electron system have more influence than do alkyl substituents. Certain changes have very little, if any, detectable effect on the absorption spectrum. Examples are (1) the replacement of the phytyl esterifying group with methyl or ethyl, and (2) epimerization (reversal) of the asymmetry at the saturated carbon of ring V. As might be expected, this last change does have a readily detectable effect on the CD spectrum. In fact, systematic studies have shown that each of the asymmetric centers associated with the porphyrin ring exerts a largely independent influence on the CD. The contributions of the individual asymmetric substitution sites are, therefore, additive (Philipson *et al.* 1971).

2.8. Solvation

For a chlorophyll with a particular chemical structure, the environment of the molecule has an effect on the spectra which can be highly informative. It is usual in spectroscopy of small molecules to use as a reference point the absorption in the absence of environmental influences, i.e., the situation in the gas phase at low pressure. This kind of isolation has not yet been achieved for the chlorophylls, because of their negligible vapor pressures at moderate temperatures and their tendency to decompose at higher temperatures. When a substance is dissolved in a "good" solvent, there is a favorable interaction between the solvent and the solute which generally results in a lowering of its energy in the ground state. This may result from strong dipole–dipole interactions in the case of polar solutes in a polar solvent, or from weaker London dispersion forces in the case of nonpolar components. Other interactions, such as hydrogen bonding, can also contribute. For the same reasons the energy of the excited electronic state is lower for the molecule in solution relative to the same excited state in a vacuum. However, the electrical properties of the excited state are different from those of the ground state, as we have seen above. Strongly allowed transitions usually result in an excited state which has a dipole moment appreciably different (either greater or less) from that of the ground state. In addition, the excited state electrons are more polarizable (more loosely bound) than those of the ground state. Because both the dipole moment and the polarizability can contribute to the interaction with a solvent in solution, in general the excited state energy will be influenced to a different extent by solvation than will the ground state. As a result, the absorption bands will exhibit a red shift if the ground and excited state energies are brought closer together, and a blue shift if they separate.

Solvent effects on chlorophylls have been widely observed, the most extensive examination being that of Seely and Jensen (1965) for Chl *a*. For the red band the range of positions is from ∼ 660 to 675 nm for a set of 40 common solvents. While several influences clearly contribute, there is a readily discernible trend of increasing red shift with increasing dielectric constant of the solvent. This is the expected result if the excited state has a larger dipole moment than does the ground state.

Because of the electric dipole origin of these effects, which fall off rapidly with increasing distance, those solvent molecules which are in the immediate environment of the Chl will make the overwhelming contribution to the absorption shifts. Certain molecules, such as the strongly polar or nucleophilic compounds water, pyridine, acetone, etc., have a very strong tendency to associate with chlorophylls, probably binding at the fifth and sixth coordination positions (normal to the porphyrin ring plane) of the magnesium. In "poor" solvents (saturated hydrocarbons, benzene, carbon tetrachloride) the addition of only millimolar concentrations of a nucleophilic ligand to Chl in solution will produce a solvent shift essentially as large as if the Chl were in the nucleophilic solvent alone. This ability of spectroscopic properties to signal the immediate environment of the molecule is of great value in deducing the nature of the surroundings *in vivo*.

2.9. Aggregation

Attempts to remove the last traces of nucleophilic ligands from chlorophylls in nonpolar solvents usually results in aggregation of the chlorophylls, leading eventually to precipitation. Detailed studies of this phenomenon have been carried out by Katz and coworkers (Katz *et al.*, 1966; Ballschmiter *et al.*, 1969; Ballschmiter, 1969). It appears that, in the absence of other nucleophilic compounds, one Chl may serve as ligand for a second molecule, resulting in the formation of a dimer or, with increasing aggregation, of a precipitate. Katz *et al.* (1968) have summarized the evidence that water is commonly present as a linking molecule in these aggregates.

A new effect, known as exciton splitting, occurs in the absorption spectra of aggregated chlorophylls. This results from the coupling via coulomb interaction of the transition dipole moment of one molecule with the corresponding transition moment of one or more identical (or very similar) molecules. Exciton coupling among small and large sets of closely arrayed chromophores has been described in reviews by Kasha (1963), Tinoco (1963), and Hochstrasser and Kasha (1964). The result of the exciton coupling is that each excited electronic state is split into a set of N exciton states, where N is the number of coupled molecules in the array. What one typically observes is the formation of multiple absorption bands in regions where there was only one for the monomer. These bands are often incompletely resolved, for the energies of interaction (band splittings) are seldom much greater than the intrinsic bandwidths. Resolution may be appreciably improved by making measurements at liquid nitrogen temperature or lower (see Section 3.5). The intensities of the individual exciton components and their spectral location are not invariant, but depend sensitively on the geometry—the relative positions and orientations—of the interacting molecules. For asymmetric chromophores or for symmetric chromophores asymmetrically arranged (as for the bases in a helical nucleic acid molecule) each exciton component appears also in the CD spectrum. Both the magnitude and sign of each CD exciton component depends intimately on the geometry of the array. It is this dependence on geometry that is potentially of great value in using exciton splittings to interpret molecular arrangements and interactions in biological systems.

There is a classical analogy that may help in appreciating the origin of this exciton interaction. This is the coupled pendulum problem. If one mounts two identical pendulums near one another and sets them independently in motion, each will swing with a period or frequency determined only by the distance of its center of mass from the point of attachment. The periods of the two identical pendulums will be precisely identical, and each will swing in a constant fashion whether the second is also in motion or not. In this condition they are said to be "isolated" from one another. If one now attaches a weak elastic band between the two pendulum arms, they are observed to interact with one another to an extent determined by the strength of the coupling provided by the elastic band. (Other coupling mechanisms serve this purpose, including motion imparted to the air surrounding the pendulums or insufficient rigidity of the mount from which they are suspended.) In the coupled situation, if one pendulum is set into motion initially, then the energy will gradually be transferred to the other pendulum over a time span of many oscillations. Subsequently the oscillation amplitude will appear to move back and forth between one pendulum and the other. That there are two "resonance states" in this system can be demonstrated in the following way: If the two coupled pendulums are carefully set into motion exactly in phase with one another (parallel motion), then they will continue to oscillate at a constant characteristic frequency, ω_1. If, in a second experiment, they are started exactly 180° out-of-phase (opposed motion), then they will be observed to oscillate at a somewhat higher frequency, $\omega_2 > \omega_1$, but one that is also constant in time. The difference $\omega_2 - \omega_1$ is related to an interaction energy determined by the strength of the coupling. The coupled oscillators (pendulums) are characterized by a pair of new frequencies that are a property of *both* oscillators (including the coupling interaction). In the exciton case the molecular absorption oscillators (transition moments) are the analogs of the individual pendulums, and the coulomb interaction is the origin of the coupling. It is important to recognize that the new exciton states are properties of the entire array; we cannot assign one state to one molecule, another to the second molecule, etc.

For certain geometries (e.g., perpendicular orientation or parallel orientation of the transition moments) all of the absorption or CD intensity of a set of N molecules may appear in only one transition component and no intensity in any other. Although a number of such arrangements may exist for large numbers of coupled oscillators, these arrangements are still relatively few in comparison with the very many other possibilities (angles other than 0° and 90°) where each exciton component is present and observable. Nevertheless, counting the number of exciton components can provide us with only the *minimum* number of coupled molecules in the array. In practice, especially as it may be difficult to resolve small splittings or detect components with small amplitude, it is necessary to exercise caution and be conservative in drawing conclusions about the molecular origins of these spectroscopic effects.

It is also important to be aware that the exciton interaction is appreciable (i.e., measurable) over only relatively short distances. These are typically 20 Å or less for molecules like the chlorophylls; i.e., only five times the thickness of the porphyrin ring and less than twice its diameter. By contrast, the Förster (1965) resonance interaction (see Chapter 4) may be observed out to distances as great as 100 Å. This distinction becomes significant in connection with the analysis of mechanisms of delocalization of electronic excitation (see Section 3.7 of this chapter, and Chapter 4).

Chlorophyll molecules provide excellent examples of the usefulness of the analysis of exciton interactions, particularly with the ability to observe them in CD as well as in absorption spectra. In suitable solvents, such as carbon tetrachloride, it is possible to prepare solutions that contain a high proportion of dimers of chlorophylls. The assignment of these species as dimers is confirmed by measurements of equilibrium (Sauer *et al.*, 1966), apparent molecular weights, and infrared and nuclear magnetic resonance spectra (Katz, *et al.*, 1963; Closs *et al.*, 1963). Absorption spectra of these solutions of (bacterio)chlorophyll confirm the presence of dimers in the appearance of two exciton components (one, a long-wavelength shoulder). Furthermore, CD spectra clearly show the presence of two bands of opposite

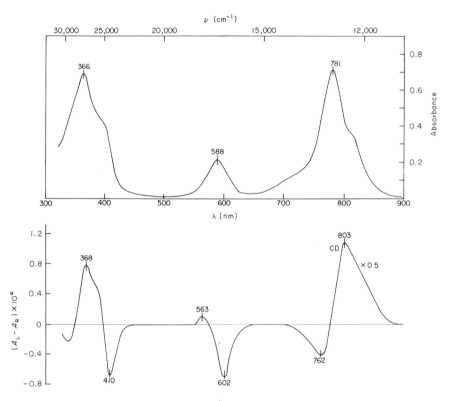

Fig. 7. Absorption and CD spectra of BChl *a* in CCl₄ at 10^{-3} moles liter^{-1} concentration. Under these conditions the BChl *a* is largely dimeric. Spectra taken from Dratz *et al.* (1966). CD amplitude has been multiplied by 0.5 at wavelengths longer than 700 nm. Relative to the absorption, the CD amplitudes of the dimers are severalfold greater than those for the monomers, shown in Fig. 4.

sign where only one is present in monomeric chlorophyll solutions (Dratz *et al.*, 1967). Figure 7 shows the absorption and CD spectra of a solution of BChl *a* in carbon tetrachloride, under conditions where the BChl *a* molecules occur largely as dimers. The consequences of the exciton interaction can be observed as a splitting of the absorption bands (note the new "shoulder" on the long-wavelength or Q_y band) and, much more dramatically, in the multiplicity of the CD bands. Compare the spectra of Fig. 7 with those in Fig. 4 for BChl *a* in ether, a solvent where the monomer predominates. Inspection of the CD spectra will demonstrate that the *amplitudes* of CD exciton components of the dimeric BChl *a* are several times larger than the CD bands of the monomeric form of this molecule. The effect of the exciton interaction on the energy levels of some of the electronic states of BChl *a* is illustrated schematically in Fig. 8.

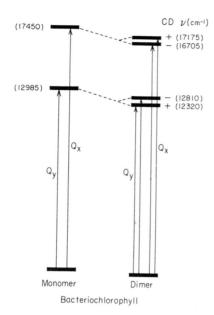

Fig. 8. Energy level diagram for BChl a, showing the splitting of excited state levels resulting from exciton interaction in the dimers. The various energy levels are characterized by their frequencies, in cm^{-1}. This diagram illustrates the origin of the differences between the spectra shown in Figs. 4 and 7.

Chlorophyll crystals provide an interesting example of the usefulness of the CD spectra as a supplement to absorption spectra. Microcrystalline suspensions of Chl a in saturated hydrocarbon solvents exhibit a very sharp absorption band, strongly red-shifted to about 740 nm (Jacobs et al., 1957).* This band remains unresolved even at liquid nitrogen temperatures. The CD spectrum in this region exhibits two large components of opposite sign (Dratz et al., 1966). This indicates the presence of at least two exciton states, albeit nearly degenerate, for Chl a crystals. Even so, why should there be only two components in a spectrum of crystals where there is potentially a very large number of interacting molecules? Analysis of this problem shows that for a periodic array of molecules, as in a molecular crystal, the maximum number of allowed transition components is equal to the number of mole-

---

* Jacobs et al. (1957) also show the spectrum of a microcrystalline suspension of methyl bacteriochlorophyllide a with an absorption maximum at 840 nm. We have obtained identical results with suspensions of BChl a crystals in isooctane. In an alternative procedure, a solution of 0.9 mg of BChl a in 150 μl acetone was diluted with 2 ml isooctane and allowed to stand overnight at 4°C. The resulting microcrystalline suspensions exhibited a pronounced absorption maximum at 940 nm that was somewhat larger than peaks at 840 and 774 nm. The 940 nm band had a pronounced tail extending to beyond 1000 nm. No irreversible changes were noted upon dissolution of the crystals by dilution with acetone (J. Ku and K. Sauer, unpublished results).

cules in the unit cell (Hochstrasser and Kasha, 1964). The detailed X-ray crystallography of Chl *a* has not been determined; however, it is known to have a similar crystal habit and unit cell dimensions to that of methyl pheophorbide a (Chl *a* minus Mg and with phytyl replaced by methyl). Recent analysis of the single-crystal X-ray diffraction pattern of methyl pheophorbide *a* shows it to contain two molecules per unit cell (Fischer *et al.*, 1972). It is apparently the predominant interaction between two molecules in the unit cell that we observe in the CD spectrum of the Chl *a* microcrystals.

The origin of the very pronounced red shift of the long-wavelength absorption band in the crystals is obscure. Although the effect of an environment of other molecules of Chl *a* cannot be duplicated by a colorless solvent, nothing we know from solution spectra would lead us to expect such a large absorption band shift. It would appear that there are effects operating in the crystal that have no counterpart in solutions of either the monomeric or aggregated pigment.

The fluorescence properties of the chlorophylls are affected even more dramatically by aggregation. Watson and Livingston (1950) observed a strong quenching of Chl *a* fluorescence with increasing concentration in ether, and Weber (1960) has shown that the quenching behavior is consistent with a model in which the excitation is transferred to nonfluorescent dimers. A similar quenching occurs as a function of surface concentration of chlorophylls in a two-dimensional monolayer at an air–water interface (Tweet *et al.*, 1964). Pronounced fluorescence depolarization as a consequence of electronic excitation transfer is also seen at high surface concentrations in Chl-containing monolayers (Trosper *et al.*, 1968).

2.10. The Usefulness of Spectroscopic Measurements

Before going on to examine the spectra of the chlorophylls *in vivo*, it will be worthwhile to present a brief summary of the different spectroscopic parameters and the kind of information that can be deduced from them. The following abbreviated list is drawn from the more detailed descriptions just completed.

1. Wavelength shifts of absorption maxima provide information about the electrical characteristics of the surrounding matrix. The inclusion of Chl *a* crystals, for example, in photosynthetic membranes would be readily apparent from the extreme red shift of the long-wavelength absorption maximum.

2. Multiple bands that disappear upon extraction into a "good" solvent are indicative of plural molecular sites or of coupling between closely associated chlorophylls. CD spectra provide even more dramatic evidence than do absorption spectra.

3. Polarization of absorption (dichroism) or fluorescence may be indicative of net molecular orientations within the surrounding matrix. Fluorescence depolarization may occur as a consequence of excitation transfer within a pigment array.

4. Decreased fluorescence lifetimes or efficiencies can occur as a result of chlorophyll aggregation, the presence of specific excited state quenchers or competition from photochemistry.

5. Transient species, such as the triplet states, give rise to their own characteristic absorption or emission spectra.

6. Bleaching or disappearance of the absorption of pigments provides direct information about the photochemical fates of the responsible molecules. As we shall see in the next section, this has been a most valuable tool in elucidating the primary photochemical processes of photosynthesis.

These are but a sampling of the many ways in which electronic spectroscopy can furnish useful information about photochemical kinetics and molecular structure. Many new techniques are emerging and give promise of expanding the horizons of our search. Some of the most exciting of these are X-ray or electron-induced electron emission spectroscopy, resonance Raman spectroscopy (Lutz, 1972) double resonance methods and high-intensity, ultrashort pulse laser spectroscopy (Seibert et al., 1973). The imaginative application of these approaches to photosynthesis research can be expected in the near future.

3. CHLOROPHYLLS IN PHOTOSYNTHETIC MEMBRANES

3.1. Spectra of Chlorophylls in the Biological Environment

Measurements of in vivo absorption and fluorescence spectra of entire photosynthetic organisms (leaves, algae, bacteria, etc.), of organelles, membrane preparations, and of subunits have been made with increasing sophistication during the past 40 years. The following general observations are noteworthy.

1. Chlorophyll absorption bands are much more complex in vivo than they are in solution. This is most readily apparent for BChl in the photosynthetic bacteria, where a number of well-resolved components are observed in the near-infrared (Wassink et al., 1939; Clayton, 1963a). In algae and higher plants the presence of as many as 8 to 10 spectral components has been reported in in vivo absorption spectra, especially at low temperatures (Brown and French, 1959; Frei, 1962; Butler, 1966; Butler and Hopkins, 1970a,b).

2. Fluorescence spectra also exhibit composite character, although generally of lesser complexity. In some cases (especially for higher plants

and algae), new components appear only at very low temperatures (Brody, 1958; Goedheer, 1961; Butler, 1961; Govindjee *et al.*, 1967).

3. Extraction of the photosynthetic membranes with organic solvents invariably leads to a dramatic simplification of the structure of both the absorption and fluorescence spectra. There is no good evidence that the pigments *in vivo* are *chemically* any different from the Chl *a*, Chl *b*, BChl, etc. molecules found in the extracts (Clayton, 1966a).

4. The wavelength maxima are generally appreciably red-shifted from those in solution, but not nearly so much as for the crystals. There is no evidence to support the earlier notion that Chl in the membranes is in a crystalline environment.

5. Neither absorption nor fluorescence is strongly polarized, supporting the idea that the pigments are not ordered in a simple periodic lattice (Goedheer, 1957; R. A. Olson *et al.*, 1961, 1962). The polarization evidence is not entirely negative, however. There is some indication from these studies of an underlying order of a more complex character (Morita and Miyazaki, 1971; Breton and Roux, 1971; Becker *et al.*, 1973).

6. Both the absorption and fluorescence are known to be kinetically heterogeneous. Small portions of the absorption may disappear or shift in wavelength upon photoexcitation (Duysens, 1952; Kok, 1956). A large portion of the fluorescence exhibits a change in its efficiency upon photoexcitation ("live" fluorescence), but another portion does not ("dead" fluorescence) (Clayton, 1969). See Section 4.3 and a review by Govindjee and Papageorgiou (1971) for a fuller description of these forms of fluorescence. The shapes of absorption envelopes can be altered appreciably by varying the growth conditions or in response to genetic mutation. Such changes occur dramatically in the course of the greening of seedlings germinated in the dark and then transferred to the light (Shibata, 1957).

These observations support a picture in which the chlorophylls occur *in vivo* in a variety of environments. Much of the work during the past 15 years has been aimed at determining the number and nature of the different environments, what role is played by the pigments in each environment, and how the individual components are coupled together in the natural photosynthetic membrane.

3.2. Photosynthetic Membranes—Structure and Organization

Photosynthetic bacteria and chloroplasts from higher plants and algae are filled with internal membranes or vesicles. These can be readily visualized by electron microscopy (Park, 1966; Branton and Park, 1967; see also Chapter 2 of this volume), and they are known to contain the Chl pigments and the sites of photochemical activity (Schachman *et al.*, 1952; Park and

Pon, 1961; Lintilhac and Park, 1966). Preparations of these isolated membranes and their fragments from broken cells have been subjected to chemical and spectrophotometric analysis. In addition to the Chl and carotenoid pigments, the membranes contain a variety of cytochromes, quinones, nonheme iron (including bound Fd, in the case of chloroplast membranes), proteins, and colorless lipids (Park and Biggins, 1964; Oelze and Drews, 1972). The membranes are typically of the order of 5–6% by weight of (bacterio)chlorophyll, which corresponds to an average concentration greater than $0.05 \ M$. For such a complex and heterogeneous system it must be recognized that the physical and chemical properties of the chlorophylls *in vivo* may bear little relationship to those of chlorophylls in dilute solutions, in homogeneous thin films or in pure crystals.

Numerous early, and even recent, models for photosynthetic membranes incorporate pigment monolayers, linear arrays, quasicrystalline aggregates, homogeneous solid solutions or other readily visualized arrangements. Although such arrangements may be convenient for purposes of carrying out theoretical calculations, it is the purpose of this section to present evidence that such models are quite unrealistic. The picture that emerges from recent investigations is one in which the membrane is a composite of individual macromolecular subunits arranged in a regular, ordered fashion (Branton and Park, 1967). The major subunits include reaction center complexes, small chlorophyll proteins, electron transport proteins such as the cytochromes and Fd, and the components responsible for coupling of phosphorylation. The overall structure also includes the carotenoids, quinones, colorless lipids and proteins and, in the case of higher plants, the apparatus involved in oxygen evolution. The manner in which these individual components are associated in functional photosynthetic membranes is one of the most active and challenging areas of current investigation.

3.3. Reaction Centers

Ever since the pioneering and highly imaginative studies of Emerson and Arnold (1932a,b), it has been known that the concentration of the sites of photochemical activity in photosynthetic organisms corresponds to only a small fraction (typically less than 1% in chloroplasts; of the order of 1% in bacteria) of the chlorophyll present. The initial studies described the maximum photochemical yield of O_2 evolved (or CO_2 fixed) resulting from a brief (μsec) saturating flash of actinic light. Confirmation of this picture was obtained subsequently when active membrane components were identified on the basis of photoinduced absorption changes (Duysens, 1952; Kok, 1956), and the content of cytochromes, quinones, Fd, Mn, etc. was determined quantitatively. The studies of photoinduced reversible absorption changes, in particular, have revealed the participation of key components in the photosynthetic light reactions; e.g., a small portion of the BChl known as P870 (Vredenberg and Duysens, 1963) and a similar small portion of Chl *a* known as P700 in the case of higher plants (Kok and Hoch, 1961). (The numbers refer to the wavelengths of maximum absorption change in the red end of the spectrum and are somewhat variable, depending on the organism, method of sample preparation, etc.) The photoinduced bleaching of these components is due to the reversible oxidation of a small portion of the (bacterio)chlorophyll.

A most important advance in our ability to examine the details of the bacterial membrane organization occurred when it was discovered that chemical treatments (e.g., strong oxidants, detergents) or illumination at high intensity of a carotenoidless mutant results in the selective destruction of the major, chemically inactive BChl and in the consequent exposure of the photochemically active sites. These are now generally known as reaction centers or reaction center complexes, and they are free of the overburden of chemically inactive, light-absorbing pigments, called the antenna or light-harvesting pigments (Clayton, 1963b; 1966a; Loach *et al.*, 1963; Kuntz *et al.*, 1964). The reaction center preparations exhibit photoinduced absorption changes and unpaired electron (EPR signal) production characteristic of the intact chromatophores or bacteria. The absorption spectra and the photoinduced absorption changes characteristic of one of the earliest and most widely studied of these reaction center preparations are shown in Fig. 9.

The isolation and purification of a reaction center complex was first reported by Reed and Clayton (1968). They treated chromatophores of *Rhodopseudomonas speroides, R-26* (carotenoidless) mutant with Triton X-100, a nonionic detergent, followed by sedimentation in a sucrose density gradient and polyacrylamide gel filtration. The reaction center complex was shown by Reed (1969) to have a molecular weight of about 650,000 daltons, relatively large amounts of ubiquinone, iron, and copper, together with P870, P800, and cytochromes (see table). Recently procedural variations have produced an active reaction center particle with lower molecular weight (Clayton and Wang, 1971). In addition to the photo-chemically active BChl absorption changes, the reaction center particles exhibit an EPR signal at $g = 2.0025$ that is stoichiometric with and kinetically similar to $P870^+$ (Bolton *et al.*, 1969).

By using the zwitterionic detergent LDAO, Feher (1971) was able to isolate an active reaction center particle with a molecular weight under 100,000 daltons. Each reaction center P870 was associated with about 1 molecule of UQ and 1 nonheme iron. Clayton and Yau (1972) reported the absence of cytochromes from a similar LDAO preparation and, by contrast with the previous report, the presence of only 0.1 molecule of UQ per P870; however, it is possible that a portion of the quinone had been lost owing to

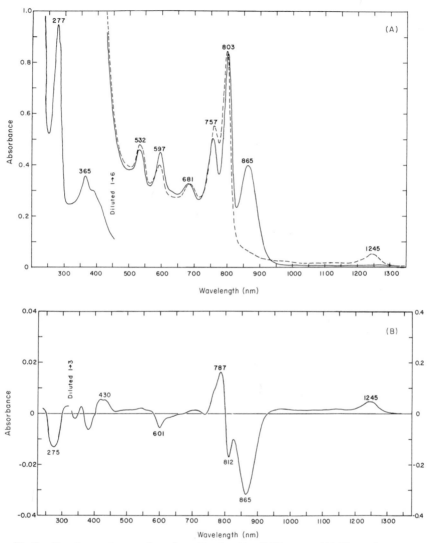

Fig. 9. Reaction center complexes from *R. spheroides, R-26* mutant. (A) Absorption spectra taken in the "dark" (solid curve) and under cross-illumination by a beam of 800 nm actinic "light" (dashed curve). (B) Difference spectrum, "light" minus "dark," between the illuminated and unilluminated samples. Figure is taken from Reed (1969).

degradation during the extraction (R. K. Clayton, personal communication). Further treatment of the LDAO particles with the anionic detergent SDS split them into two subunits; one with a molecular weight of 44,000 daltons retained the P870 activity (Feher *et al.*, 1971; Okamura *et al.*, 1974). This is the smallest active subunit so far reported. We shall return to discuss its

Comparison of Properties of Isolated Photosynthetic Bacterial Reaction Centers Free of Antenna Bacteriochlorophyll

Organism	Detergent	MW (kdaltons)	Cyt 562 / P865	Cyt 552 / P865	UQ / P865	Fe / P865	Reference
R. spheroides (R-26)	Triton (1%)	650	2	1	1.3	16	Reed (1969); Reed et al. (1970)
	Triton (2.3%)	440	0.8	0	7.4	5.4	Feher (1971); Feher et al. (1972)
	LDAO	70				3.2	
	LDAO + EDTA				1	0.8	
	SDS + LDAO	37			1	0.2–0.3	
R. spheroides (wild-type)	LDAO	~100	0	0	<0.1		Clayton and Yau (1972)
	SDS	200	<0.02	0.90	3.3	20	Slooten (1972a)
	SDS/AUT	120	<0.02	0.25	1.2		
R. spheroides (strain Y)	CTAB	150		1.5	8	0.5 (nonheme)	Reiss-Husson and Jolchine (1972)
R. rubrum (G-9)	Triton		Cyt 422 / P870: 0.5				Gingras and Jolchine (1969)
	LDAO		<0.1				Wang and Clayton (1973)
R. rubrum (wild-type)	SDS	80–100	≥1.0		UQ / P870: 1.7		Smith et al. (1972)
	SDS(0.05%)	35	~0				
R. viridis	SDS	110	Cyt 558 / P960: 2	Cyt 553 / P960: 5			Thornber et al. (1969); Thornber and Olson (1971)

properties below. At this stage in the elucidation of reaction center properties and stoichiometry it is useful to establish and denote a distinction between the small *reaction center particles* obtained by Clayton and Wang (1971), by Feher *et al.* (1971) and at low concentrations of SDS by Smith *et al.* (1972), and the larger *reaction center complexes* that include cytochromes and larger amounts of UQ and nonheme iron (Reed, 1969; Thornber and Olson, 1971; Slooten, 1972a; Reiss-Husson and Jolchine, 1972; Smith *et al.*, 1972).

Following the initial report by Reed and Clayton, a number of other organisms were investigated. A summary of some of the properties of reaction centers from several of these preparations is given in the table. Reaction centers have now been isolated from several species of photosynthetic bacteria, including some wild-type species with the normal complement of carotenoids. Various detergents have been used and a wide range of molecular weights is seen. Similarly, a considerable variation in the content of demonstrated or potential electron transport cofactors (cytochromes, UQ, nonheme iron) is found. In each case a few molecules of pheophytinized BChl are associated with the reaction center; however, their role in the primary photochemistry is not yet clear. In each of the cases described in Table I, the bulk or antenna BChl is completely absent in the reaction center preparation. This results in a decrease of about 30-fold in the overall BChl content compared with the intact chromatophores.

Organisms other than the purple photosynthetic bacteria cited in the table have thus far resisted such a dramatic enrichment in reaction center pigments relative to the antenna. Detergent solubilized preparations from *Chromatium* D exhibit molecular weights of about 500,000 daltons, but still contain about 40 antenna BChl molecules per active center (Garcia *et al.*, 1966; Thornber, 1970). Still larger photoactive particles, with molecular weights in excess of 1.5×10^6 daltons, have been prepared without the use of detergents from the green bacteria *Chloropseudomonas ethylica* (Fowler *et al.*, 1971) and *Chlorobium thiosulfatophilium* (Fowler *et al.*, 1973; J. M. Olson *et al.*, 1973). Particles of much smaller molecular weight were obtained using simultaneous treatment at high (alkaline) pH with urea and Triton X-100 (AUT) from *R. spheroides* (MW 160,000 daltons) (Loach *et al.*, 1970a) and from *Rhodospirillum rubrum* (MW 100,000 daltons) (Loach *et al.*, 1970b). Nearly all of the antenna pigments, BChl, and carotenoids, are retained in these preparations together with the photochemical reaction centers. In an interesting extension of these studies, Loach and Hall (1972) report that column electrophoresis removes most of the iron from these preparations, leaving a ratio Fe/P865 = 0.2 to 0.3. This is of special relevance with regard to the possibility that the primary electron acceptor contains nonheme iron (see Section 4.5). Feher *et al.* (1972) and Okamura *et al.* (1974) report a similarly low value for the iron content of the 44,000-dalton particle from *R. spheroides R-26* mutant, obtained by combined LDAO/SDS treatment.

Attempts to prepare exposed reaction centers of P700 from higher plants have been frustrated by the inability to remove all of the antenna pigment, although enrichments of 10- to 20-fold have been reported. Yamamoto and Vernon (1969; Vernon *et al.*, 1969) obtained a preparation (HP700) of particles with a Chl/P700 ratio of 30:1 using extraction with organic solvents (acetone:hexane) followed by suspension with Triton. Sane and Park (1970) have used low temperature acetone extraction to achieve a preparation with a 15:1 ratio of Chl:P700. Other particles have been prepared from chloroplasts and from a blue-green alga by Thornber and co-workers using SDS treatment followed by hydroxylapatite chromatography (Thornber and Olson, 1971; Dietrich and Thornber, 1971); however, these have larger contents of antenna Chl *a*. Vernon *et al.* (1971) used Triton to achieve an enriched particle with PS II activity from chloroplasts.

We shall return to consider the photochemical properties of these reaction centers in Section 4.

3.4. Chlorophylls in the Reaction Centers

A major feature of photoinduced absorption changes in purple photosynthetic bacteria, chromatophores, and reaction centers is the bleaching of absorption around 870, 600, and 375 nm, associated with the major BChl absorption bands, and accompanying increases in absorption at 1250, 1140, 980, and 420–450 nm (Clayton, 1962a). These are now thought to result from the one-electron oxidation of reaction center BChl (see Section 4.4). At the same time there occurs a blue shift of 3–4 nm in the position of a second absorption peak at 800 nm. Clayton (1966a) and Clayton and Sistrom (1966) made extensive studies of the relationship of these components and reached the conclusion that the absorption at 800 and at 870 nm results from BChl components of the same reaction center. Because the absorption bands occur in the ratio 2:1 in a variety of bacterial reaction centers and because oxidative bleaching resulted in a loss of at most one-third of the total absorption residing in the two components, the view was put forward that P870 represents a reactive oxidizable BChl molecule, and the P800 represents two other BChl molecules that participate in the reaction center but do not become oxidized. (However, see below for revised stoichiometry.) Not only is the light absorbed at 800 nm transferred efficiently to the reaction center (Clayton and Sistrom, 1966), but fluorescence from the reaction center is distinctly less polarized when excited at 800 than at 870 nm (Ebrey and Clayton, 1969). This evidence supports the idea that the excitation at 800 nm produces electronic states that are distinct from, but closely coupled to, those reached directly via absorption at 870 nm.

CD spectra provide support for strong coupling among several associated BChl molecules in the reaction centers. Sauer *et al.* (1968) observed evidence of exciton interaction in the CD spectra of reaction center particles from the *R-26* mutant of *R. spheroides*. The CD spectrum indicated the presence of at least two energy levels (states) in the 800 nm absorption band and one more at 865 nm for the reduced reaction center preparation. Upon chemical oxidation or illumination with actinic light, the CD spectrum changed in a way consistent with a smaller number of interacting molecules. The absorption and CD changes at room temperature are illustrated in Fig. 10. Recently Reed and Ke (1973) have extended these measurements to low temperature (77°K) where they observe a resolution of the long-wavelength CD feature in reduced reaction centers into two (+) components at 877 and 888 nm. The associated absorption band shifts to 893 nm at low temperature, but remains unresolved. In the chemically oxidized reaction centers at 77°K the long-wavelength absorption is greatly attenuated, and both of the long-wavelength CD components are absent. Furthermore, careful investigation

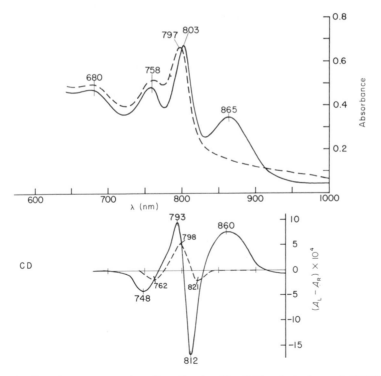

Fig. 10. Reaction center complexes from *R. spheroides*, *R-26* mutant, at room temperature. Absorption spectra (top) and CD spectra (bottom) for samples either in the dark (solid curves) or under cross-illumination by an actinic beam. (Sauer *et al.*, 1968.)

of the ratio of BChl to BPheo in purified reaction centers shows it to be very close to 2.0 (Reed and Peters, 1972; Straley *et al.*, 1973) and leads to the conclusion that reaction centers contain an even number of BChl molecules. This evidence, taken together, supports the idea that at least four BChl molecules are strongly coupled in the (reduced) reaction center particles, and the resulting exciton interaction gives rise to four states in the CD spectra (see Section 2.9). Because each state giving rise to an absorption or CD component results from the coupling of all four molecules, it is not appropriate in this model to assign one molecule to each absorption or CD band. Upon oxidation, the four coupled molecules lose one electron to the acceptor, and an altered exciton interaction is left among the remaining three BChl molecules.

Reaction center preparations from *R. rubrum* and *Rhodopseudomonas viridis* exhibit room temperature absorption and CD properties similar to those of *R. spheroides* (Philipson and Sauer, 1973), apart from the long-wavelength shifts owing to the presence of BChl *b* instead of BChl *a* in *R. viridis*. At 77°K the absorption spectrum of reduced reaction centers from *R. viridis* exhibits two well-resolved long-wavelength components at 929 and 987 nm. These may be the counterparts of the CD components in *R. spheroides* observed by Reed and Ke (1973). A low temperature absorption spectrum of reduced reaction centers from *R. rubrum* exhibits only a single long-wavelength absorption band at 890 nm (Philipson and Sauer, 1973), as was observed by Reed and Ke (1973) for the low-temperature absorption of *R. spheroides* reaction centers.

The particles from *Chromatium* (Thornber's fraction A) containing substantial antenna BChl, exhibit similar changes in the room temperature CD spectrum upon illumination (Philipson and Sauer, 1973). Oxidation of the reaction center/antenna pigment complexes results in bleaching of only one of the exciton components; the others merely shift in wavelength and exhibit altered exciton coupling.

Several recent studies support the exciton model involving strong coupling between several BChl molecules in the reaction centers. EPR spectra of both protonated and deuterated algae and bacteria characteristically show EPR line narrowing for the signal at 2.0025 in comparison with the EPR of chemically oxidized Chl or BChl in solution (Norris *et al.*, 1971). The explanation put forward for this line narrowing is that the positive charge of the oxidized pigment *in vivo* is delocalized, probably over two molecules. Similar conclusions were reached by McElroy *et al.* (1972). This interpretation has been confirmed by subsequent studies involving electron-nuclear double resonance (ENDOR) spectroscopy (Feher *et al.*, 1973; Norris *et al.*, 1973). These results are clearly at variance with a picture in which a single P870 molecule becomes oxidized in the absence of strong coupling with at least one other molecule.

In the case of PS I from higher plants, recent studies of the HP700 preparation of Yamamoto and Vernon (1969) using CD spectroscopy (Philipson *et al.*, 1972) confirm the earlier observation of Amesz *et al.* (1967), Döring *et al.* (1968), Murata and Takamiya (1969) and Vernon *et al.* (1969) that the P700 absorption change is coupled to a weaker photobleaching at about 680 nm. Photoinduced CD changes of about equal magnitudes but of opposite sign occur at the two wavelengths (Philipson *et al.*, 1972). Although the P700 oxidation has been shown to correspond to a one-electron transition, the decrease in absorption intensity is apparently distributed, albeit unequally, between two components of an exciton pair. In this case it seems clear that there is no "P700 molecule" per se.

In principle, it should be possible to count the number of coupled BChl molecules in purified reaction center preparations from the number of components in the absorption or CD spectra. In practice it is possible to determine only a minimum number, as mentioned in Section 2.3, because some of the transitions may be degenerate and unresolved, or the strengths of one or another of the transitions may be fortuitously very small owing to the particular molecular configuration. For reaction centers from *R. spheroides* and from *R. viridis* we estimate a minimum of four interacting BChl molecules in the reduced state. Other estimates range from 2 to 5. Clayton (1966a) originally concluded that the number was 3, on the basis of observed extinction coefficients in reaction center complex preparations and because nearly one-third of the absorption at 800 and 870 nm was lost upon light or chemical oxidation. Furthermore, alcohol extracts of reduced and of oxidized reaction center complexes recovered BChl in the ratio 3:2. A similar stoichiometry was proposed for reaction center complexes prepared from *R. spheroides*, strain Y (Jolchine *et al.*, 1969), *R. rubrum* (*G-9* mutant) (Gingras and Jolchine, 1969) and *R. viridis* (Thornber and Olson, 1971; Thornber, 1971).

Several workers, in addition to Reed and Ke (1973) and Straley *et al.* (1973) cited above, have recently presented evidence that supports a larger number of BChl molecules in the reaction centers. Feher (1971) analyzed the metal content of highly purified LDAO reaction center particles from *R. spheroides* and found a ratio of 5:1 for Mg/Fe. The possibility that this was also the ratio of BChl to the primary acceptor led Feher to suggest that the previously accepted values be reexamined. Clayton *et al.* (1972a) analyze the case for 3, 4, or 5 BChl per reaction center on the basis that BPheo is an integral part of the reaction center (Yau, 1971). Values greater than 3 BChl per reaction center inevitably lead to quantum yields greater than unity for certain substrate oxidations and to a loss in correspondence between oxidized P870 and the number of unpaired electrons as measured by EPR. However, these discrepancies may result, in part, from incorrect values assumed for the extinction coefficient of P870. On the other hand, numbers less than 4 BChl per reaction center are difficult to reconcile with the absorption and CD spectra (Reed and Ke, 1973; Philipson and Sauer, 1973). The evidence of Straley *et al.* (1973) supporting the presence of 4 BChl and 2 BPheo in bacterial reactions centers is perhaps the strongest available at the time of this writing.

3.5. Bacteriochlorophyll and Chlorophyll Proteins

In addition to the various particles containing active reaction centers with or without accompanying antenna pigments, a variety of (bacterio)-chlorophyll proteins without any measurable photochemical activity have been isolated. These fall into two distinct classes: (1) water-soluble pigment proteins that are obtained by sonication or other methods of cell rupture followed by fractionation of the cellular debris, and (2) pigment complexes obtainable only through the use of detergents, which are required both to excise the complexes from the internal membranes and to provide them with sufficient hydrophilic character to become water-soluble.

A prime example of the first type is the water-soluble BChl protein isolable from sonicates of the green bacteria *C. ethylica* or *Chlorobium thiosulfatophilum* (J. M. Olson and Romano, 1962; J. M. Olson, 1971). This substance has been highly purified and is readily crystallizable. It has a molecular weight of 152,000 daltons and consists of four equivalent sub-units, each containing 5 BChl molecules (Thornber and Olson, 1968). The BChl are apparently well covered by a protein coat and are in close proximity to one another within the subunits, giving rise to pronounced exciton splittings in low-temperature absorption and CD spectra (Fig. 11). The spectroscopic properties of the BChl protein do not, however, closely resemble those of the membrane-bound antenna BChl associated with green bacterial reaction center particles (Fowler *et al.*, 1971; J. M. Olson *et al.* 1973). Despite the conclusion that it is not an integral part of the photo-chemically active vesicles of the green bacteria, the water-soluble BChl protein is able readily to transfer electronic excitation to the photosynthetic reaction centers in order to drive the light reactions.

Several interesting water-soluble Chl proteins have been isolated from higher plants. One obtained from *Chenopodium album* or from cauliflower has a molecular weight of 78,000 daltons and contains 6 Chl *a* and 1 or 2 Chl *b* (Yakushiji *et al.*, 1963; Takamiya, 1971; T. Murata *et al.*, 1971). Upon illumination the pigments of the Chl protein from *Chenopodium* undergo an irreversible photooxidation. The Chl protein from cauliflower is light-insensitive. Other small Chl proteins have been isolated from *Lepidium*; however, their function in these organisms is not known (Takamiya, 1971; T. Murata and Murata, 1971). Frozen, aged, and thawed spinach chloro-

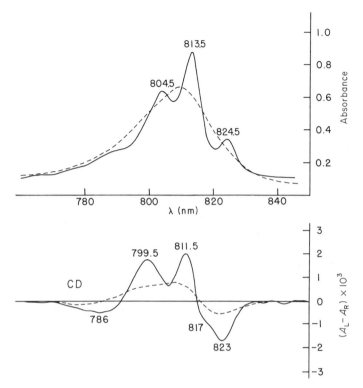

Fig. 11. BChl–protein complex from *C. ethylica*; absorption and CD spectra taken for the same sample at room temperature (dashed curves) and at 77°K (solid curves). Both an increase in resolution of the components and a considerable enhancement of the CD features are evident at the lower temperature. Curves taken from Philipson and Sauer (1972).

plasts release two small water-soluble Chl proteins that contain plasto-quinones and carotenoids as well (Tachiki *et al.*, 1969).

The second class of inactive pigment proteins requires solubilizing detergents for their isolation. Two BChl proteins obtained from *Chromatium* D, fractions B and C, using SDS have molecular weights of about 100,000 and contain about 10 BChl molecules together with carotenoids (Thornber, 1970). A much smaller pigment protein with molecular weight about 14,000 and containing one molecule of BPheo (presumably converted from BChl during an acid-extraction step) was obtained from *R. spheroides* using SDS (Fraker and Kaplan, 1972). A BChl protein of about 9000 molecular weight and associated with the antenna BChl has been isolated by Clayton and Clayton (1972). This is the smallest of the pigment proteins isolated to date.

Chlorophyll proteins isolated from higher plant chloroplasts incubated with SDS are of two types (Kung and Thornber, 1971). One has a molecular weight of 100,000 and a Chl a/b ratio $5:1$; the other has a molecular weight of 35,000 and a Chl a/b ratio $1:1$. These are thought to derive from PS I and II respectively.

It is tempting to postulate that these detergent-solubilized (bacterio)-chlorophyll proteins without reaction centers represent subunits of the antenna pigment systems. In several cases they are obtained in yields corresponding to more than half the (bacterio)chlorophyll of the membranes prior to detergent treatment. Nevertheless, in most cases it remains to be determined that the pigment proteins exist in a similar configuration within the active membranes and that the detergents have not caused a major rearrangement of the constituents. This note of caution applies equally to the reaction center preparations described in Section 3.4.

3.6. Models of Photosynthetic Lamellae

The preparation and elucidation of an almost bewildering variety of reaction center particles, reaction centers with associated bulk pigments and inactive (bacterio)chlorophyll proteins has led to the formulation of various models for the photosynthetic lamellae. Thornber (1970) has envisioned a set of basic components or building blocks which are suitably arranged to give both the light-harvesting properties and the photochemical activity of the photosynthetic units. The building blocks consist of a set of reaction centers containing the primary electron donor and acceptor molecules, cytochromes serving as secondary donors, and (bacterio)chlorophyll proteins serving as the antenna. The basic units of the latter may contain between 2 and 20 pigment molecules encased in protein. These basic units are presumed to be held together by lipid components that are extracted by the solubilizing detergents.

The two-dimensional photosynthetic lamellae of higher plants are thought to contain repetitions of the photosynthetic units in a closely packed array (Branton and Park, 1967). Similar models for bacterial membranes or vesicles are envisioned by Reed *et al.* (1970) and by Slooten (1972a). The picture that emerges from these ideas is akin to the tesselated or pebble mosaic floors characteristic of ancient pavements uncovered in the Near East. Imagine a characteristic and well-defined unit pattern compounded from a set of differently colored or shaped pebbles, and then imagine this compound unit repeated extensively in each direction on the surface. Mosaic models with varying degrees of rigidity have been presented for a wide variety of subcellular membranes (Green *et al.*, 1967; Singer and Nicolson, 1972). This pebble mosaic model, illustrated in a conjectural fashion in Fig. 12, has features which should command the attention of those

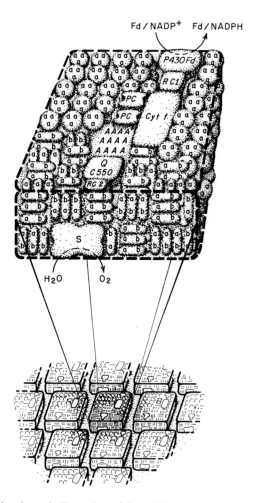

Fig. 12. A highly schematic illustration of the pebble mosaic model. At the top is an expanded view of a single unit, consisting of an integrated array of different "pebbles": the electron transport cofactors [ferredoxin (Fd), Cytochrome f (Cyt f), plastocyanin (PC), intersystem intermediates, plastoquinone (A), electron acceptor of PS II (Q), etc.], reaction centers (RC 1 and RC 2) and specific pigment-protein subunits (Chl a proteins and Chl $a + b$ proteins). Carotenes and uncharged colorless lipids may be interspersed between components as a kind of glue; charged sulfo- and phospholipids may occur at the peripheral surface of the unit in contact with the aqueous surroundings. The model omits such additional features as phosphorylation coupling factors, for which the stoichiometry and placement are even less well known. At the bottom is pictured a portion of a single thylakoid membrane, which consists of a two-dimensional mosaic of the individual units. Electron micrographs show that the arrays are not always so regular as indicated here. This variability may control the extent of electronic excitation energy transfer from one unit to its neighbors.

who seek to understand the essential properties of active photosynthetic membranes. We will examine some of its consequences for the phenomenon of electronic excitation transfer in Section 3.7.

Other models with lesser degrees of internal organization have been postulated by Kreutz (1970) and by Weier and Benson (1967). These differ in placing all or part of the pigments in a kind of two-dimensional layer or sheet with interspersed carotenoids or colorless lipids, and do not emphasize the packaging of the chlorophylls in individual small protein subunits. Numerous earlier models exist, but they are even more speculative because they did not profit from the recent studies of membrane substructure.

Support for rather specific interactions of particular proteins with reaction centers or light-harvesting pigments comes from studies of protein composition following extensive detergent treatment (Remy, 1971; Fraker and Kaplan, 1972; Clayton and Haselkorn, 1972). Mutants which are missing the light reaction or reaction center activities are found to be deficient also in specific proteins (Gregory *et al.*, 1971; Segen and Gibson, 1971; Clayton and Haselkorn, 1972).

The membranes or thylakoids of chloroplasts are enclosed vesicles, for which the inside and outside surfaces are distinguishable using the electron microscope (Branton and Park, 1967). (See Chapter 2 by Arntzen and Briantais, this volume.) In addition, labeling studies using diazonium compounds (Dilley *et al.*, 1972) or antibodies (Briantais and Picaud, 1972) indicate that these faces differ in their activity, with access to PS I on the outside and PS II on the inside of the chloroplast thylakoids.

3.7. Excitation Transfer and Trapping—The Pebble Mosaic Model

Within a few picoseconds (1 psec $= 10^{-12}$ sec) following the absorption of a photon by a Chl anywhere in an antenna of up to 500 molecules, the resulting electronic excitation arrives at the photochemical trap (see Section 4.3). This rapid transfer of excitation places severe demands on the properties of such an array. The mechanisms invoked for the excitation transfer range from the diffusion of an exciton wave throughout a more or less homogeneous array of closely spaced chlorophyll, to a migration via a Förster mechanism (resonance transfer) involving the hopping of the excitation to neighboring molecules more or less at random (random walk). The purpose of this section is to point out aspects of this problem which have not been given sufficient consideration in any theoretical model to date. A detailed description of the models that have been studied and of their predictions is given by Knox in Chapter 4 of this volume.

Most of the models of excitation transfer have used a simple picture of a regular repeating lattice (in 1, 2, or 3 dimensions) of identical pigment molecules (Duysens, 1964; Bay and Pearlstein, 1963; Robinson, 1966; Hoch and Knox, 1968; Knox, 1968). The photochemical traps are assumed to be dispersed more or less uniformly in this array (the lake model), although some models consider only isolated photosynthetic units containing a single trap (the isolated puddles model). Evidence gathered from studies of fluorescence yields as a function of the fraction of reaction centers in the oxidized condition (and presumably not capable of acting as traps) suggest that excitation can pass relatively easily from one such unit to another in photosynthetic bacteria and PS II of algae (Clayton, 1966c, 1967; Duysens, 1966a; Borisov, 1967; Tumerman and Sorokin, 1967).

Several models have considered heterogeneous arrays in which molecules which are closer to the traps have somewhat lower energy excited states than do those molecules farther away (Duysens, 1966b; Seely, 1973). In certain arrangements calculations suggest that the speed of trapping can be increased in this way as much as 10-fold over that of equivalent homogeneous arrays (Borisov and Fetisova, 1971).

A different kind of picture is suggested by the pebble mosaic model described in Section 3.6 and illustrated in Fig. 12. The isolation of pigment protein complexes containing a small number (2 to 7) of strongly coupled (bacterio)chlorophyll molecules suggests that the antenna may consist of a repeating network of such pigment proteins. This model is *spatially* heterogeneous, in the sense that the antenna array consists of repeating units of *groups* of closely spaced chlorophylls in Chl protein subunits. The consequences of this model cannot be appreciated by simply reordering the lattice points in the previous studies, however, for there is ample evidence that the underlying mechanism of molecular interaction is quite different for the closely grouped molecules within each protein subunit from what it is for interaction between adjacent subunit groups.

The absorption and CD spectra of Olson's BChl protein from green bacteria (Philipson and Sauer, 1972) and of both a Chl a/b protein and a Chl a protein from higher plants (Murata and Murata, 1971; Murata, Philipson and Sauer, unpublished results) suggest that the (bacterio)-chlorophyll molecules within these proteins are relatively strongly coupled. The resulting exciton states are properties of the molecular group. In these exciton states it is appropriate to consider the electronic excitation to be spread over all of the chlorophylls in the group, probably in a time shorter than the 10^{-13} sec characteristic of thermal relaxation in such molecules. Furthermore, there is probably a similarly rapid transition (intersystem crossing) to the lowest energy state(s) in the manifold of exciton levels. Evidence to support this view comes from the low-temperature fluorescence spectrum of the BChl protein (Olson, 1971), where it is seen that the fluorescence spectrum is the mirror image not of the entire absorption manifold, but only of the long-wavelength exciton component. Although the fluorescence lifetime undoubtedly is much longer (ca. 10^{-9} sec) than the relaxation lifetimes that we are concerned with above, the fluorescence result emphasizes the conclusion that we should not consider these closely coupled pigment molecules to act as individual entities for purposes of excitation transfer calculations within or between photosynthetic units.

It is reasonable to suppose, and consistent with the meager experimental data available, that excitation transfer *between* adjacent BChl protein or Chl protein molecules in the antennas occurs via the very weak coupling typically invoked in the Förster mechanism. The rate of such transfer varies inversely as the sixth power of the distance between centers and is proportional to the overlap between the emission spectrum of the donor group and the absorption spectrum of the acceptor group.

While an arrangement such as that described above has not been explored by quantitative calculations, we can predict some of its features relative to that of a homogeneous array of regularly spaced, individual chlorophylls. Because intergroup transfer will be much slower than the intragroup delocalization, the effect of grouping the pigments is to reduce the number of Förster steps required to reach the trap by a factor N_g, where N_g is the number of pigment molecules in each closely coupled group. The time required to reach the trap is therefore shortened by a similar factor. On the other hand, the overlap between the fluorescence and absorption envelopes may be reduced for the groups as compared with the separated molecules (monomers). The fluorescence comes only from the lowest state(s) of the manifold, whereas the absorption is distributed (albeit unequally) among all of the states. The relevant properties of the groups need to be known in greater detail, however. Emissive transition probabilities for the groups and their orientation with respect to adjacent groups play a strong role in the transfer probability (Knox and Chang, 1971). The focus of interaction between the protein sheaths of adjacent Chl proteins provide a plausible rationale for supposing that the antenna consists of a "regular" array of these subunits with some well-defined, as opposed to a random, mutual orientation.

It will be a challenge for the future both to theoreticians and to experimentalists to explore the properties and consequences of the pebble mosaic model of photosynthetic membranes.

3.8. Synopsis of Photosynthetic Membrane Organization

Before going on to discuss the photochemistry of the reaction centers, let us first summarize some of the contributions that electronic spectroscopy has made to our knowledge of the function and arrangement of the chlorophyll pigments in photosynthesis.

1. The chlorophylls *in vivo* are bound into lipoprotein membranes and occur there in several distinct spectroscopic and functional forms.

2. Antenna chlorophyll molecules, typically 99% of the total, serve to absorb incident photons and to transfer the resulting electronic excitation to the reaction centers.

3. Reaction center chlorophylls, roughly 1% of the total, participate directly in the primary photochemistry. The reaction centers utilize the excitation energy to pump electrons from donors to acceptors in the electron transport chain.

4. Small BChl-containing proteins with the general characteristics of either the antenna or the reaction center BChl have been isolated from photosynthetic bacteria. These proteins have a small number (2 to 7) of BChl per molecule or subunit.

5. Photosynthetic membranes probably consist of assemblies of antenna chlorophyll proteins, reaction center complexes, electron transport proteins and other smaller molecules packaged into integrated units that are repeated in a more or less regular array distributed over the membrane surface.

6. In the pebble mosaic model, photon absorption results in electronic excitation that is very rapidly delocalized via strong exciton interaction among several pigment molecules within each antenna chlorophyll protein. The excitation is then transferred by a slower hopping process to other antenna chlorophyll protein molecules until it eventually reaches the reaction center.

4. PRIMARY PHOTOCHEMISTRY OF PHOTOSYNTHESIS

4.1. Definition of Photosynthetic Energy Trapping

For the purposes of this section we shall define photosynthetic energy trapping as the process that results in the disappearance of (singlet) electronic excitation in a potentially productive manner. Dissipative processes, such as fluorescence and thermal degradation, are not included in the definition. The onset of the "chemical" stage occurs upon the appearance of the first of a series of oxidized and reduced intermediates. It is important to recognize that there is an unexplored region lying between the trapping of electronic excitation and the initial electron transfer reactions. Speculations have populated this region with triplet states and other metastable species, but there is little evidence at present to support their direct participation in the energy conversion process. The process of trapping is, therefore, defined operationally in terms of certain initial and final states. The precise boundary between the physics and chemistry of the process may prove difficult to locate, however, unless the model is carefully specified.

An alternative way of looking at the trapping process is to consider the motion of electrons and holes (missing electrons) in the activated system.

The absorption of radiation by the photosynthetic pigments results in the promotion or excitation of a bound electron to a higher energy level and leaves a hole behind in the original level. During excitation migration this electron and its associated hole move together among the molecular arrays (e.g., between adjacent Chl proteins) in a correlated fashion. One feature that is characteristic of this correlation is that the excited electron always has the opposite spin to that of the remaining electron in the orbital containing the hole. When trapping occurs this spin correlation is lost. The electron and the hole eventually move independently of one another, either in conduction bands or among a set of donor and acceptor molecules. The free energy of the electronic excited states has then become converted into stored chemical potential, and we have entered the region of chemistry.

Recent reviews emphasizing various aspects of the photochemical trapping have been written by Fork and Amesz (1970), by Witt (1971) and by Clayton (1972). Tributsch (1971) has presented a formulation of the trapping process in electrochemical terminology.

4.2. Primary Photochemical Processes

The experimental evidence suggests that the first events consequent upon trapping of the electronic excitation are oxidation–reduction (electron transfer) reactions. The evidence varies in detail, depending on whether photosynthetic bacteria or higher plants are studied. It is worthwhile starting with some generalizations to provide a framework for interpreting the detailed observations.

The earliest detectable event appears to be the transfer of an electron from a Chl or BChl (C) in the reaction center complex to a primary acceptor (A_1), according to the equation

$$C^* \cdot A_1 \longrightarrow C^+ \cdot A_1^- \tag{4.1}$$

This process is very rapid (< 20 nsec) and proceeds with a quantum efficiency of the order of unity. It is temperature-independent and occurs even at $1°K$. The reaction can be observed by characteristic changes in the absorption spectrum (P870 or P700) and by the distinctive EPR signal of the oxidized pigments.

Reaction (4.1) competes very favorably with the other mechanisms of singlet excited state decay listed in Section 2.4. As a consequence, photosynthetically active reaction centers exhibit a low yield of fluorescence, short fluorescence decay times, low yields of triplet formation, etc. relative to chlorophylls in solution. We will see in Section 4.3 how measurements of these quantities can be used to deduce information about the kinetics of the primary photochemistry.

Following the primary event, several secondary processes may occur, either separately or together.

1. The primary acceptor may pass the electron to one or more secondary acceptors (A_2):

$$C^+ \cdot A_1^- \cdot A_2 \longrightarrow C^+ \cdot A_1 \cdot A_2^- \qquad (4.2)$$

2. An electron may be transferred from a donor (D_1) associated with the reaction center complex:

$$D_1 \cdot C^+ \cdot A_1^- \longrightarrow D_1^+ \cdot C \cdot A_1^- \qquad (4.3)$$

3. The process may reverse to regenerate the pigment excited state, leading to delayed fluorescence (Strehler and Arnold, 1951)

$$^+(D_2 \cdot D_1 \cdot C) \cdot (A_1 \cdot A_2)^- \longrightarrow D_2 \cdot D_1 \cdot C^* \cdot A_1 \cdot A_2 \qquad (4.4)$$
$$\downarrow$$
$$D_2 \cdot D_1 \cdot C \cdot A_1 \cdot A_2 + hv'$$

4. The reduced and oxidized species may be coupled via an external (cyclic) electron flow pathway

$$^+(D_2 \cdot D_1 \cdot C) \cdot (A_1 \cdot A_2)^- \longrightarrow D \cdot C \cdot A + \text{heat} \qquad (4.5)$$

The process represented by steps 1 and 2 are extended through other electron carriers in noncyclic electron transport. The ultimate electron donor in higher plant photosynthesis is H_2O and the electron acceptor is CO_2 via Fd and $NADP^+$. Process 4 may also result in the production of stored chemical potential via a cyclic phosphorylation step.

While the events described above have been extensively documented, several additional intermediate stages have been proposed but have not yet been convincingly demonstrated. For example, the triplet state of Chl has been invoked as an intermediate in the trapping process prior to the initial electron transfer step (Robinson, 1963, 1966; Franck and Rosenberg, 1964). Although evidence is available supporting the occurrence of the Chl triplet in photosynthetic organelles or membranes, the triplet state is observed only when the primary electron transfer step is blocked (Moraw and Witt, 1961; Rikhireva et al., 1968; Dutton et al., 1972, 1973; Borisov et al., 1970). We cannot, however, rule out the possibility that triplets do participate in the normal course of the photosynthetic energy conversion, but that their lifetimes are too short or concentrations too low to have been observed.

An alternative mechanism of photochemical trapping invokes the separation of electrons and holes into conduction bands characteristic of large molecular arrays (Calvin, 1958; Arnold, 1965). Such a delocalization of charges over a large matrix would provide a favorable entropy barrier against their wasteful recombination. The hole and electron would productively be captured by an electron donor and an acceptor, respectively, associated with the molecular array. Subsequent events would occur much as in the simpler mechanism presented previously. Photoconductivity and light-induced dielectric changes have been observed for dried films of chloroplasts (Arnold and Sherwood, 1957) or chromatophores (Arnold and Clayton, 1960). Recently, high-frequency (10^{10} Hz) photoconduction has been observed using an electrodeless (microwave cavity) technique (Blumenfeld *et al.*, 1970; Bogomolni, 1972). Bogomolni made measurements of the Hall effect of this photoconductivity. He obtained evidence for both positive and negative carriers in both chromatophores and chloroplasts, but the quantum yield for their formation was less than 10^{-2}. The nonphotosynthetic *PM-8* mutant of *R. spheroides* did not exhibit evidence of positive carriers, however.

Reaction center preparations do not contain the antenna BChl molecules, and there is only a very small number of molecular sites over which charge delocalization can occur. From the ability of these reaction center preparations to carry out normal photochemical trapping it would appear that electron conduction bands typical of organic semiconductors are not necessary for photosynthetic energy conversion.

4.3. Trapping Times Deduced from Fluorescence Measurements

Once a reaction center becomes activated, it will not be able to utilize additional excitation until C^+ and A_1^- [Eq. (4.1)] are returned to their dark states. During this dead time no further photochemistry is possible, and the additional absorbed energy is lost via heat or emitted radiation. The refractory period lasts until the secondary donor, D_1, transfers an electron to C^+ *and* the reduced primary acceptor, A_1^-, is able to transfer an electron to a secondary acceptor, A_2.

$$D_1 \cdot C^+ \cdot A_1^- \cdot A_2 \longrightarrow\longrightarrow D_1^+ \cdot C \cdot A_1 \cdot A_2^- \qquad (4.6)$$

The available evidence (Sections 4.4 and 4.5) suggests that there is only one oxidizable (bacterio)chlorophyll and one primary acceptor molecule in each reaction center complex. However, the secondary donors and acceptors can, in many cases, transfer or accept more than one electron upon successive flashes or continuing illumination (Parson, 1969; Case and Parson, 1971). Once these secondary pools are emptied or filled, additional refractory periods or kinetic rate-limiting steps will be observed.

As the primary trapping sites become progressively filled there is a concomitantly greater probability that absorbed photons will be released as fluorescence. As a consequence, an increase in fluorescence efficiency (i.e., the *ratio* of emitted to absorbed light intensities) is observed with increasing intensities of actinic light (Duysens and Sweers, 1963; Vredenberg and Duysens, 1963). Fluorescence, thus, becomes a relatively more important pathway for excitation loss as the chemical steps saturate. We expect, on the basis of Eq. (2.9), that an increased fluorescence efficiency will be accompanied by an increased fluorescence lifetime. Studies by Tumerman and Sorokin (1967), and by Müller *et al.* (1969) of the fluorescence lifetime of Chl *a* in *Chlorella* did indeed show a strong intensity dependence. The lifetime increased from a value of 0.35 nsec at low intensity to 1.9 nsec at saturation. [This observation appears to resolve earlier discrepancies in lifetime values (Brody and Rabinowitch, 1957; Tomita and Rabinowitch, 1962; Butler and Norris, 1963; W. J. Nicholson and Fortoul, 1967), assuming that the different experiments were carried out at different exciting light intensities.] Mutants with nonfunctional reaction centers did not show any significant variation in fluorescence lifetime over the same intensity range (Müller *et al.*, 1969).

Two significant properties of this *in vivo* fluorescence are worth noting. The minimum fluorescence efficiency, extrapolated to zero actinic intensity, is distinctly greater than zero, and the saturation value is distinctly less than that for the Chl *a* or BChl in solution. From the observed lifetime of 0.35 nsec at low intensity (Müller, *et al.*, 1969) and the intrinsic lifetime of 15 nsec for Chl *a*, we can calculate [Eq. (2)] an expected quantum efficiency of 0.023. This is in excellent agreement with the observed value of 0.024 for *Chlorella* at low intensities (Latimer *et al.*, 1956). Clayton (1969) has interpreted this as predominantly "dead" fluorescence which is insensitive to the condition of the photochemical traps. The variable fluorescence is designated "live" fluorescence, by contrast. Clayton attributed these two forms to different Chl pigments or Chl in different situations in the lamellae.

The live fluorescence has been extensively studied, both in higher plants and in bacteria, as an indication of the condition of the photochemical traps. Clayton (1967) has provided a detailed analysis of the quantitative aspects of this correspondence. The live fluorescence of higher plants appears to originate in PS II, and largely from the Chl *a* antenna pigments. Fluorescence from PS I can be observed readily at low temperatures (Goedheer, 1964, 1965); however, its presence at room temperature is greatly reduced. Because of strong emission band overlap and the difficulties of isolating pure, unaltered PS I particles, it is difficult to document the presence of PS I fluorescence at room temperature (Goedheer, 1972). No live fluorescence of PS I has been reported. The virtual absence of both live and dead fluorescence from PS I at room temperature is not understood at present, but it suggests a rather different reaction center configuration from those of PS II or photosynthetic bacteria.

Zankel *et al.* (1968) have studied reaction centers isolated from *R. spheroides* and have observed a "live" fluorescence that appears to originate from the reaction center pigments themselves. On the basis of the very low quantum yield, approximately 4×10^{-4}, one can calculate a lifetime of 7 psec (7×10^{-12} sec) for the electronic excited state of the BChl of the reaction centers. The lifetime of the live fluorescence was investigated for intact bacteria using a sensitive phase fluorometer by Borisov and Godik (1970). They concluded that excitation reaches the open traps within 30–50 psec, which was at the limit imposed by their instrument. At the other extreme, the dead fluorescence from the bacteria when the traps were closed had a lifetime of 1.35 nsec. Since the quantum yield for photochemistry at low actinic intensities is of the order of unity in these preparations, the photochemical trapping must occur within a few picoseconds following absorption of a photon.

4.4. P870 and P700—Primary Electron Donors

The first experimental evidence for the direct participation of BChl or Chl in the primary photochemistry was the observation by Duysens (1952) of a photoinduced, reversible bleaching of a small portion (ca. 1 %) of the absorption spectrum of BChl in intact bacterial cells or chromatophores. Although the difference spectrum varies somewhat from one organism to another, we shall refer to this species as P870, after the long-wavelength component of the bleaching. Shortly thereafter, Kok (1956; Kok and Hoch, 1961) reported an analogous species, P700, characteristic of higher plants and algae. P700 is now known to be associated with PS I. In a search for a corresponding species associated with PS II in oxygen-evolving organisms, Döring *et al.* (1967, 1968, 1969) uncovered a rapidly reversible bleaching at 680 nm; however, they concluded that its role may be an indirect one and not reflect the oxidation of a PS II reaction center Chl (Witt, 1971). Floyd *et al.* (1971) showed that a portion of the P680 decays with kinetics corresponding to the oxidation of Cyt b559 at low temperature. They favored placing P680 as an active component of the PS II reaction center. Butler (1972a) has written a detailed analysis of this question, concluding that there is no good evidence at present to rule out P680 as the primary electron donor of PS II.

At about the same time as the photoinduced absorption changes P870 and P700 were first observed, studies using EPR techniques uncovered the photoproduction of unpaired electrons in chloroplasts and in photosynthetic bacteria. Sharp, structureless EPR signals were observed to appear upon illumination and to decay in the dark (Commoner *et al.*, 1956; Calvin and Sogo, 1957). At first it proved difficult to establish the relationship between the absorption changes and the EPR signals; however, extensive studies carried out in several laboratories during the past few years have established a common origin for them. We are now reasonably confident that the bleaching of P870 or P700 and the appearance of the EPR signal (signal 1, in the case of higher plants) both witness the one-electron oxidation of (bacterio)chlorophyll in the reaction center. Kohl (1972) has written a comprehensive review of the properties of the EPR signals associated with the photosynthetic reactions.

Both P870$^+$ and P700$^+$ and the corresponding EPR signals in chromatophores and chloroplasts can be produced chemically in the dark through the use of an oxidant such as ferricyanide. Each active pigment undergoes a one-electron oxidation, and the reversible midpoint potentials are $+0.44$ V for P870/P870$^+$ (Kuntz et al., 1964) and $+0.43$.V for P700/P700$^+$ (Kok, 1961). The light-induced changes disappear progressively as the pigments are oxidized chemically, as shown in Fig. 13A. In *Chromatium* a somewhat higher potential of 0.49 V for P870 has been reported by Cusanovich et al. (1968). In spinach chloroplasts in the presence of ferricyanide a new photo-induced EPR signal has been observed at 77°K by Malkin and Bearden (1973). Preliminary measurements indicate that it is associated with the reaction center Chl of PS II. Its reduction potential appears to be greater than $+0.54$ V.

Loach and Sekura (1967) showed that the decay kinetics of the photo-induced absorption changes owing to P870 are identical to those of the EPR signal. Further kinetic and stoichiometric evidence supporting the identification of P870$^+$ as oxidized BChl and responsible for the EPR signal was reported by Bolton et al. (1969) and by McElroy et al. (1969). Figure 14 illustrates this similar kinetic behavior for reaction centers from *R. spheroides*. Bolton et al. (1969) reported that the quantum yield for the formation of P870$^+$ and of the EPR signal are both unity; however, a more recent molar extinction determination for P870$^+$ by Straley et al. (1973) leads to a 20 % lower value for the quantum yield for P870$^+$. This difference is of the order of the experimental uncertainty in the EPR measurements. In the case of higher plants, studies of Weaver (1968) and of Warden and Bolton (1972) demonstrated the identity of the kinetics of P700 bleaching and EPR signal 1, at the same time resolving some apparent earlier discrepancies for this species.

Under high intensity, laser-pulse illumination the onset times of the P870 and P700 absorption changes are too rapid to measure using currently available techniques of sensitive flash absorption spectrometry. The experiments show that they are faster than 50 nsec for P870 (Seibert et al., 1971) and faster than 20 nsec for P700 (Witt and Wolff, 1970). These limits are still very much longer than the trapping time of 7 psec (7×10^{-12} sec)

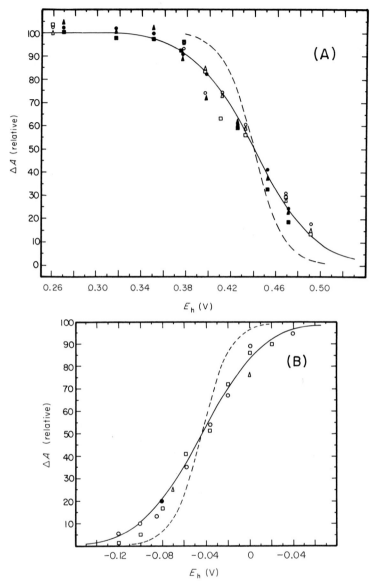

Fig. 13. Oxidation reduction titrations of the light-induced P870 absorption change in chromatophores of *R. rubrum*. The two figures are taken from Kuntz *et al.* (1964). (A) Oxidation of P870 by added ferricyanide (open symbols: ○, 865 nm; △, 810 nm; □, 792 nm), followed by a reductive titration with sodium dithionite (solid symbols). Measurements were made at three different wavelengths characteristic of the primary photochemical oxidation. The solid line is the theoretical curve for a one-electron oxidation with a midpoint potential of +0.439 V. The dashed line is the theoretical curve for a two-electron oxidation. (B) Titration of P870 at low potentials, presumably the reduction of the electron acceptor. The experimental procedure was varied for the different symbols (see original reference for details). The solid line is the theoretical curve for a one-electron reduction with a midpoint potential of −0.044 V; dashed line is for a two-electron reduction.

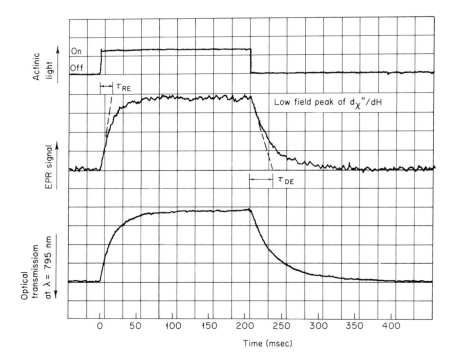

Fig. 14. Kinetics of the EPR and absorption changes at 795 nm for reaction centers of *R. spheroides, R-26* mutant, upon actinic illumination. The same sample was used for both experiments. Rise times at 77°K, which vary with incident intensity from 14 ± 3 msec (at 10^5 erg cm^{-2} sec^{-1}) to 30 ± 5 msec (at 10^4 erg cm^{-2} sec^{-1}), agree within $\pm 10\%$ for the two measurements. Decay times are 30 ± 3 msec for both measurements at 77°K, and exhibit no significant change down to 1.7°K. The figure and the numerical results are taken from McElroy *et al.* (1969).

estimated by Zankel *et al.* (1968) on the basis of low fluorescence yields in reaction center preparations.

Bleaching at 864 nm in reaction centers of *R. spheroides R-26* was recently reported to occur 7 ± 2 psec following activation at 530 nm using a single pulse from a mode-locked laser (Netzel *et al.*, 1973). Since BPheo is the predominant species absorbing at 530 nm, this result demonstrates that excitation is transferred from BPheo to P870 in 7 psec and the reaction center BChl (ground state) becomes bleached. At the time of this writing it is not known whether the process goes beyond the excited electronic state to BChl$^+$ in times of a few psec.

Further evidence supporting the primacy of both the absorption and EPR signals comes from their characteristically high quantum yields and virtual independence of temperature. For bacterial systems both the EPR signal (Sogo *et al.*, 1959) and the P870 absorption change (Arnold and Clayton, 1960) can be induced rapidly by light at low temperatures, even down to

1°K. At room temperature the quantum yields are close to unity for both the P870 change (Loach and Sekura, 1968; Parson, 1968; Beugeling, 1968; Bolton *et al.*, 1969; Clayton *et al.*, 1972a,b; Slooten, 1972a) and for the EPR signal (Loach and Walsh, 1969; Weaver and Weaver, 1969; Loach and Hall, 1972) for chromatophores and reaction center preparations. The quantum yield of the P870 bleaching is changed hardly at all at 1°K (Clayton, 1962b).

The wide range of evidence cited above provides strong support for the proposal that the primary electron donor associated with the photosynthetic light reactions is (bacterio)chlorophyll in a special environment. In addition to the other chlorophylls present in the reaction centers (Section 3.4), one important feature of this special environment is the close proximity of a good electron acceptor.

4.5. Primary Electron Acceptors

The primary electron acceptor is defined as the first component to receive an electron from the primary donor, now presumed to be P870 or P700. It must be recognized at the outset that the primary acceptor may prove to be more than one molecule (e.g., a pool of acceptors), or there may be several different primary acceptors for a given donor (parallel pathways). Furthermore, the specific primary acceptor(s) may change depending on reaction conditions (e.g., electrochemical potential) or other chemical treatments which the photosynthetic materials have undergone.

A variety of compounds have been proposed for the primary electron acceptor in reaction centers. These include quinones, nonheme iron proteins, pteridines, cytochromes, a second chlorophyll molecule, and other components identified only in terms of absorbance changes or EPR signals. This is one of the most active areas of current research.

The difference between the electrochemical potentials of oxidized (bacterio)chlorophyll and the reduced primary acceptor defines the chemical potential produced from the absorbed photon. Ross and Calvin (1967) present arguments based on the second law of thermodynamics that limit this difference to about 1.2 eV for organisms containing Chl *a* and about 0.8 eV for photosynthetic bacteria. Since the oxidation of P700 and of P870 each exhibits a midpoint potential of about $+0.44$ V, this places lower limits for the potentials of the primary acceptors of about -0.76 V and -0.36 V in plants (PS I) and bacteria, respectively.

If an external chemical reducing agent added to the photosynthetic materials is able to reduce the primary acceptor

$$C \cdot A + e^- \longrightarrow C \cdot A^- \tag{4.7}$$

then P870 or P700 are no longer able to transfer electrons upon illumination and the associated absorption and EPR changes disappear. The state

C·A⁻ is inactive photochemically and exhibits a high fluorescence efficiency (Clayton, 1966b). Kuntz *et al.* (1964) found a midpoint potential of -0.04 V for a reversible, one-electron reductive titration of P800 (coupled to P870) in the bacterium *R. rubrum* (Fig. 13B). A similar value, -0.03 V, was found by Nicolson and Clayton (1969) for isolated reaction centers from *R. spheroides*. Both of these investigations used steady illumination. Using flash spectroscopy following brief (10 nsec) laser pulses, Dutton (1971) and Seibert and De Vault (1971) found values near -0.14 V for *Chromatium* D and for *R. gelatinosa*, and obtained evidence that a small portion of the light reactions was operative down to -0.35 V. From similar studies on *R. spheroides*, Dutton and Jackson (1972) observed a value of -0.025 V, in excellent agreement with that of Nicolson and Clayton (1969) using steady illumination. Although these potentials are measured for different organisms and, therefore, need not be the same, it is important to recognize that measurements made using steady illumination may differ from those using brief flashes even for the same organism. The origin of this difference is kinetic rather than thermodynamic, and we shall return to consider the underlying reasons shortly.

In the case of chloroplasts from higher plants, a series of studies using added weak electron acceptors (Kok *et al.*, 1965; Zweig and Avron, 1965) point to a potential of about -0.50 V for the primary acceptor. In some cases added electron acceptors with midpoint potentials below -0.60 V could be reduced to some extent by chloroplasts (Kok *et al.*, 1965; Black, 1966). These studies were all carried out using steady illumination; however, they did not involve reversible titrations of an endogenous chloroplast electron acceptor. It is possible for an electron donor to reduce an acceptor with a more negative *midpoint* potential if the concentration ratios of oxidized and reduced forms of the respective molecules favor the transfer. For one-electron carriers, in accordance with the Nernst equation, each order of magnitude change in the concentration ratio adds (or subtracts) 0.06 V to its *effective* potential. As a consequence, one must be cautious in interpreting the results of such studies.

In none of these cases does the outside limit for the potential of the primary acceptors appear to challenge the thermodynamic constraints. As more definite information becomes available, it will be interesting to learn what is the actual thermodynamic efficiency of photosynthetic energy conversion.

Attempts have been made to isolate and characterize the primary acceptor of chloroplasts. Ferredoxin has been proposed to serve this role, particularly by Arnon (1969). The midpoint potential of soluble spinach Fd, which is -0.45 V, would seem to make it too weak a reductant to account for the photoreduction of the viologen or dipyridyl dyes (Kok *et al.*, 1965; Black, 1966).

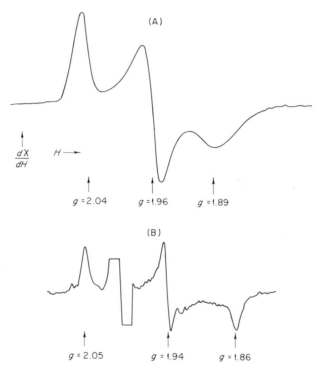

Fig. 15. EPR spectra of species obtained from spinach chloroplasts. (A) Soluble spinach ferredoxin, chemically reduced with sodium dithionite and measured at 20°K (Malkin and Bearden, 1971). (B) Whole-spinach chloroplasts, illuminated using 715 nm light and measured at 25°K. The spectrum of an unilluminated (dark) sample has been subtracted (Bearden and Malkin, 1972b). The large signal at $g = 2.0025$ owes to P700$^+$. The features at $g = 1.86$, 1.94, and 2.05 are to be compared with the EPR spectrum of reduced ferredoxin in part (A).

The primary acceptor is membrane-bound both in plants and in bacteria; the P870 and P700 absorption changes survive cell breakage and extensive washing. Recent studies by Hiyama and Ke (1971a,b) and Ke (1972) have identified an absorption change with a maximum at 430 nm in chloroplast particles enriched in PS I activity and having the characteristics of the primary acceptor. It couples to P700 and exhibits a one-electron reduction with a midpoint potential at about -0.47 V. The rise time for this absorption change is 100 nsec or less following a brief laser flash. Observations have been reported only for detergent-treated or sonicated membranes so far, and there remains the possibility that this acceptor is not the physiological primary one.

Membrane-bound Fd (iron–sulfur protein) has been found in chloroplasts and has been implicated as the primary acceptor in higher plants on the basis of low-temperature EPR spectroscopy by Malkin and Bearden (1971; Bearden and Malkin, 1972a). In this case whole chloroplasts were studied, but it was necessary to cool the samples to 25°K in order to assay the reduced acceptor. The evidence consists of several EPR features, typical of those observed for the nonheme iron, labile sulfur proteins of which Fd is an example. A comparison of these EPR signals is shown in Fig. 15. The EPR spectrum appears upon illumination of chloroplasts at 77°K or room temperature (Leigh and Dutton, 1972; Evans *et al.*, 1972), or upon chemical reduction using dithionite.* Yang and Blumberg (1972) report that the amount of Fd signal measured at 1.5°K is only 10–20% of the P700 EPR signal amplitude; however, Bearden and Malkin (1972b) find a ratio close to unity in a set of carefully documented quantitative measurements at 20°K.

Neither kinetic nor redox titration studies have been possible using the low-temperature EPR assay in order to establish the role of this component under physiological conditions. Because of the apparent requirement for a primary acceptor with a midpoint potential of -0.50 V, the membrane-bound "Fd" may have a somewhat lower potential than does the soluble Fd, or it may be a chemically distinct species with the observed EPR spectrum. Such a modification or chemical distinction would be required to reconcile the EPR results with the P430 of Hiyama and Ke (1971b), for soluble Fd does not exhibit the absorption difference spectrum reported for P430 (Fig. 16). Ke (1973) has recently reviewed this subject in some detail.

Several components extractable from chloroplast lamellae or from blue-green algae have been proposed as primary acceptors for PS 1. A FRS has been reported by Yocum and San Pietro (1969, 1970, 1972). Its presence is required for photoreductions involving Fd as an electron carrier (including $NADP^+$ reduction) by lamellar fragments separated from sonicated chloroplasts; however, its absence does not affect the primary photooxidation of P700 in FRS-depleted chloroplast fragments (Yocum and San Pietro, 1972). A component (S_{L-eth}) solubilized from chloroplast membranes by low-temperature diethyl ether extraction exhibits similar properties, and serves to reverse the inhibition caused by an antibody that blocks Fd-mediated $NADP^+$ reduction (Regitz *et al.*, 1970; Trebst, 1972; Regitz and Oettmeier, 1972). Inhibition by the same antibody can be overcome by FRS (Regitz *et al.*, 1970; Yocum and San Pietro, 1970). The two components, FRS and S_{L-eth}, have similar, but not identical, UV absorption spectra. The molecular weight of FRS is about 4000, whereas that of S_{L-eth} is about 300,000.

---

* Dithionite is commonly used as a "strong" reductant; however, its effective potential is difficult to deduce because of a complex dependence on pH and because of the instability of the species in solution.

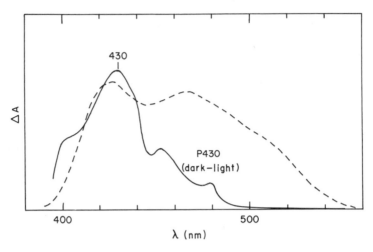

Fig. 16. Comparison of the dark-minus-light difference spectrum of the P430 of Hiyama and Ke (1971b) with the oxidized-minus-reduced difference spectrum of spinach ferredoxin. The two curves have been arbitrarily normalized near 430 nm for purposes of comparison; however, Hiyama and Ke estimate that $\Delta\varepsilon_{430}$ (light–dark) for P430 is 12 ± 3 mM^{-1} cm^{-1}, whereas the $\Delta\varepsilon_{430}$ (oxidized–reduced) for ferredoxin is only about 3 mM^{-1} cm^{-1}.

A component with similar properties has been separated from blue-green algae by Honeycutt and Krogmann (1972). Despite the apparent role of these compounds in the electron transport pathway between P700 and soluble Fd, there is no convincing evidence at present that any of them serves as the primary acceptor of PS I.

The search for the primary acceptor in photosynthetic bacteria has resulted in similar complexities. Feher (1971) reported an EPR spectrum of three broad lines that accompanies the sharp signal of BChl$^+$ of *R. spheroides* reaction centers at 1.4°K (Fig. 17). The broad lines, which are generally reminiscent of nonheme iron, are produced in the light and they decay in the dark with the 30 msec time constant of the BChl$^+$ EPR signal. The preparation of reaction centers used in this particular study by Feher contained one iron per P870. From experiments in which the potential was established prior to freezing, Dutton and Leigh (1973) concluded that the midpoint potential of this light-induced EPR signal at 10°K is -0.05 V and that it results from a one-electron photoreduction. A similar low-temperature EPR signal was observed for subchromatophore particles from *Chromatium* (Leigh and Dutton, 1972). In this organism it exhibited a midpoint potential of -0.13 V. These values are in good agreement with those deduced for the primary electron acceptors on the basis of titrations of the P870 or P800 absorption changes (see above).

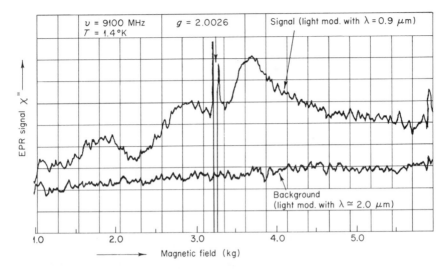

Fig. 17. The light-induced EPR signal at 1.4°K for *R. spheroides* (*R-26* mutant) reaction centers using either actinic light (upper trace) or infrared radiation that is not actinic (lower trace), taken from Feher (1971). Note, in comparison with the traces shown in Figs. 15 and 18, that this scan is over a much wider range of magnetic field, and that the spectrum is shown in the absorption mode rather than in the more conventional first derivative mode.

Subsequently Loach and Hall (1972), working with a preparation of photoreceptor subunits from *R. rubrum* containing active P870 but only 0.30 equiv of Fe per reaction center, observed a rather narrow EPR signal at $g = 2.0050$ which was kinetically similar to the $BChl^+$ signal at $g = 2.0025$. This observation was corroborated by Feher *et al.* (1972) using a new low-iron preparation of active reaction centers from *R. spheroides* obtained by more extensive detergent treatment of their previous preparation (Fig. 18). By measuring this EPR spectrum also at higher microwave frequencies, these workers were able to demonstrate a striking similarity to the EPR signal of the radical anion of UQ. They postulate that UQ is not the physiological primary acceptor, but has adopted this role by virtue of the removal of the iron-containing species. Ke (1969) had previously examined absorbance changes at 280 nm attributable to UQ in untreated *Chromatium* chromatophores. On the basis of the disappearance of these changes in the presence of reducing agents that did not abolish the P890 changes, he concluded that it behaved like a secondary rather than a primary electron acceptor from BChl. Reed *et al.* (1969) came to a similar conclusion for reaction centers from *R. spheroides*. Clayton *et al.* (1972a) presented evidence for a one-electron acceptor prior to UQ in reaction centers from *R. spher-*

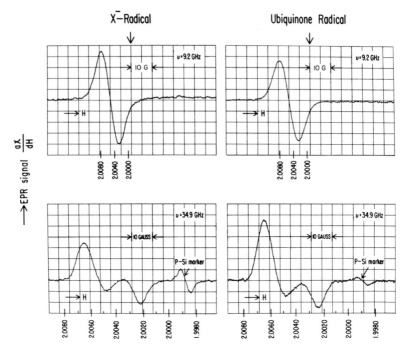

Fig. 18. The light-induced EPR signal for a subunit of *R. spheroides* (*R-26* mutant) reaction centers with low iron content (spectra at the left) and the EPR of the radical of ubiquinone Q-50 (spectra at the right) ($T = 1.3°K$). Curves shown at the top and bottom in each case were measured at different microwave frequencies. The higher microwave frequency provides a much better resolution of the components of the *g* tensor. Figures are reproduced from Feher *et al.* (1972).

oides, confirming the conclusion of Ke and of Reed *et al.*. A new spectroscopic component that exhibits absorption changes at 450, 320, and 270 nm has been studied and proposed as a candidate for the primary acceptor (Clayton and Straley, 1970, 1972; Slooten, 1972b). Its difference spectrum resembles that of ubisemiquinone radical.

With such a plethora of possible primary acceptors, it is refreshing that some previous candidates can now apparently be ruled out. Reed and Mayne (1971) have looked for pteridines in chromatophores and reaction centers prepared from the *R-26* mutant of *R. spheroides* and have found a convincingly insufficient amount present for it to remain a serious candidate.

Rather than leave wide open the question of why there should be such apparently contradictory evidence concerning the nature of the primary acceptor, it is worthwhile first to examine some possible limitations inherent in the experimental findings. The distinction between primary and secondary electron acceptors is important to appreciate. Primacy is determined by which of several acceptors receives the electron first from the chlorophyll. If the electron is transferred quickly to the secondary acceptor [see Eq. (4.2)] or if there is rapid equilibration between the primary and secondary acceptors, then both of them will appear to change with the kinetics of the $C \longrightarrow C^+$ reaction. There are still several ways in which the distinction could be made, however. By using very short, high-intensity light pulses the primary process [Eq. (4.1)] may be speeded up sufficiently to separate it from the secondary $A_1^- + A_2 \longrightarrow A_1 + A_2^-$ reaction, as in the studies of the 280 nm absorption change by Ke (1969). Alternatively, an inhibitor of this last reaction that still permits electrons to be transferred to A_1, can serve to distinguish between them. Clayton et al. (1972b) report that o-phenanthroline serves further to characterize UQ as a secondary acceptor in R. spheroides reaction centers. However, in this approach, as with the extensive detergent treatment, there remains the possibility that the normal pathways are altered and non-physiological ones are exposed by the chemical treatment.

A second complication arises in the event of rapid equilibration of primary and secondary electron acceptors. Chemical titrations or photochemical processes occurring on a long time scale will sense only some average property, e.g., midpoint potential, of the heterogeneous pool. Even when the addition of substances accelerates or retards the rates of the coupled reactions to the same extent, one cannot be certain that the primary and secondary acceptors do not equilibrate at a still more rapid rate. Reductive titrations, such as the disappearance of the P870 change, will then appear to involve two (or more) electrons, as was observed for whole cells and chromatophores of R. rubrum by Loach (1966). For the bacterial chromatophores of R. gelatinosum, but not for those of Chromatium, Dutton (1971) found titration evidence of two primary acceptors. One has a midpoint potential of -0.14 V, but the other is still active at -0.35 V. He considered the possibility that these participate in two different reaction centers in R. gelatinosum.

The primary reactants of PS II of higher plants are even less well understood. The case for the fast absorption change of P680 reflecting the primary electron donor was described in Section 4.4. A component that appears to control the variable fluorescence of PS II, designated Q by Duysens and Sweers (1963), has been implicated as the primary acceptor. Under some circumstances it behaves similarly to a component, C550, that is seen in difference absorption spectra (Knaff and Arnon, 1969; Erixon and Butler, 1971).

Recent experiments of Mauzerall (1972) showed, however, that the fluorescence rise controlled by Q is delayed following a brief laser flash. The delay is 25 nsec following the first flash after prolonged darkness, but it increases to 3 μsec for subsequent flashes in a closely spaced series. These delays are sufficiently long to provide evidence against Q being the primary acceptor of PS II. Furthermore, Butler (1972b) has found that the onset of variable fluorescence from chloroplasts at $-196°C$ is distinctly slower than the photoreduction of C550. The kinetics of the change in variable fluorescence are identical to those of Cyt b559, which serves as an electron donor to PS II at $-196°C$. Studies of Vermeglio and Mathis (1973a,b) show that the oxidation of Cyt b559 exhibits poor correlation with C550 reduction at low temperatures. The authors conclude that a second component can compete with Cyt b559 in donating electrons to PS II under these conditions. The lack of correspondence between Q and the primary acceptor of PS II has been suggested by others (Govindjee *et al.*, 1970; Okayama and Butler, 1972), and an excellent review of this complex subject is given by Butler (1973). At present the weight of evidence favors C550 as either the primary electron acceptor of PS II or else as an indicator of phenomenological changes associated with the primary electron transfer step. Part of the C550 change observed at room temperature is, in addition, an indicator of changes in membrane potential associated with the thylakoids. The variable fluorescence appears to be influenced by the state of the primary and secondary donors as well as by the primary acceptor.

Because of the many and sometimes conflicting reports of the nature of the primary process in photosynthetic energy conversion, it is useful to outline a current picture of some of the important features discussed in the preceding paragraphs. This picture must be considered provisional, owing to the uncertainties of some of the conclusions.

1. The immediate consequence of excitation trapping is a one-electron transfer reaction that occurs with a quantum yield near unity.

2. The primary electron transfer occurs in a time less than 20–50 nsec and is temperature-independent down to 1°K.

3. The primary electron donor is a Chl or BChl molecule that shares its (photo)oxidized state with one to three additional (bacterio)chlorophylls in the reaction center complex. The midpoint potential for this oxidation is about $+0.44$ V.

4. The identity of the primary electron acceptor molecule(s) is uncertain, but Fd-like nonheme iron proteins or quinones are popular candidates. The reduction of the primary acceptor occurs with a midpoint potential of -0.03 V or less in photosynthetic bacteria and -0.5 V or less in PS I of higher plants.

4.6. ——And Beyond

With this discussion of the nature of primary electron donors and acceptors we have reached the limit of the territory that we set out to explore. We can see pathways leading off in various directions; in many cases their topography is described in other chapters in this volume. For example, we have said relatively little here about the processes associated with oxygen evolution in higher plants and algae, which is central to the discussions of other chapters.

New insights will help to open fresh avenues of approach to photosynthetic energy conversion. To cite just one example, recent studies of Parson and coworkers (Case and Parson, 1971; Callis *et al.*, 1972) show that the storage of free energy in the primary process in photosynthetic bacteria is primarily an entropy effect, and has little if any contribution from enthalpy or chemical bond energy sources. One way to interpret this finding is to view the primary process in terms of a photoinduced rearrangement of the electron distribution of the molecules in the reaction center complex, in a way that does not result in an increase in the enthalpy of the system. The chemical potential associated with this charge redistribution may arise in a fashion analogous to that for solid-state photovoltaic devices (solar cells). The EPR evidence that indicates the delocalization of the unpaired electron over more than one chlorophyll molecule in photoxidized reaction centers (Section 3.4) may be a manifestation of this entropy contribution. On the basis of absorption and CD spectroscopy, all reaction centers studied so far appear to contain more than one (bacterio)chlorophyll molecule.

In most of the work that has been described in this chapter, more questions have been raised than have been answered. Although we have come a long way in the past 40 years, the next generation need have no fear that the major outstanding problems are all but wrapped up. We still are very far from knowing nature's secret for converting solar energy into useful chemical potential on such an enormous scale. At the same time there is reason to hope that we can improve on the present utilization of sunlight. Photosynthesis makes use of only a fraction of 1% of the intensity at suitable wavelengths falling on the surface of the earth. This is an important direction that we must explore in order to conserve our dwindling resources of conventional (including fissionable) fuels.

5. SUMMARY

In the preceding pages a broad map has been sketched, covering the processes of absorption of radiation, excitation transfer, and photochemical trapping. Although much detailed information is still missing, it is worthwhile presenting a brief summary of those areas that we feel confident about and those which are largely unknown.

1. The excitation of molecules of chlorophyll (Chl *a*, BChl *a*, etc.) is prerequisite to photosynthetic energy conversion. While other molecules may serve for the initial absorption of radiation, excitation which is not transferred to one of the principal chlorophyll pigments is not available for photochemistry.

2. The sites at which the photochemical trapping occurs are located in lipoprotein membranes. In these membranes they occur at a relatively low concentration (1% or less) in comparison with the total chlorophyll of the antennas.

3. Quantum yields for photon utilization can approach 100%, which demonstrates that excitation can move from the many antenna molecules to the few traps with very little loss.

4. The isolation of chlorophyll proteins, reaction center particles and electron transport proteins in good yields has led to the formulation of a membrane model—the pebble mosaic model—in which these and other components constitute subunits of a photosynthetic unit. The membranes are visualized as larger arrays formed by repetition of these basic units in an interacting network.

5. Excitation transfer occurs via two distinct mechanisms; exciton delocalization within small closely coupled groups of chlorophylls in the chlorophyll proteins, and resonance or Förster transfer between the chlorophyll proteins. (The latter could also occur, at least in part, via exciton interaction if the pigment molecules of adjacent chlorophyll proteins are sufficiently closely coupled.)

6. Trapping occurs when the electronic excitation has been converted to separate oxidized and reduced species in the reaction centers. The molecule oxidized is a special reaction center chlorophyll; the specific nature of the molecule reduced is uncertain, but nonheme iron proteins or quinones are current popular candidates.

7. There is a span of time, between 10^{-13} and 10^{-8} sec, following photon absorption about which we know almost nothing. There may be essential steps in this interval involving conduction electrons, triplet states, or other entities that will only be confirmed by a new generation of experiments.

8. Finally, nothing that we are reasonably certain of in the current body of photosynthetic knowledge leads us to require processes outside our current experience in physics or chemistry. Quantum yields are comfortingly at or less than unity; neither the first nor the second law of thermodynamics is under attack; tested mechanisms of excitation transfer are adequate to account for the efficiency of trapping; and the chemical properties of the

primary donors and acceptors are reasonable ones for the molecules proposed.

In other words, we need not look for totally new phenomena to be at the heart of this fundamental biological process. It appears that the best set of tools for elucidating photosynthetic energy conversion will include solid foundations in chemistry and physics—and an adventurous mind.

ACKNOWLEDGMENTS

My colleagues and students at Berkeley, especially Drs. Larry Vickery, Vicki Sato, Jerry Babcock, and Ken Philipson, have provided very helpful suggestions toward improving this chapter. Drs. Roderick K. Clayton and John S. Leigh, Jr., suggested valuable clarifications in the text and provided me with new information on several topics. Those difficulties that remain for the reader probably occur because I failed to heed all of their advice. I should like to thank the several authors who very kindly authorized inclusion of their original figures or sent me copies of manuscripts in advance of publication.

Support for the research of the author's group, some of which is described in this chapter, came partly from the U.S. Atomic Energy Commission and partly through a grant from the National Science Foundation (GB-24317).

REFERENCES

Amesz, J., Fork, D. C., and Nooteboom, W. (1967). *Stud. Biophys. Berlin* **5**, 175.
Arnold, W. (1965). *J. Phys. Chem.* **69**, 788.
Arnold, W., and Clayton, R. K. (1960). *Proc. Nat. Acad. Sci. U.S.* **46**, 769.
Arnold, W., and Sherwood, H. K. (1957). *Proc. Nat. Acad. Sci. U.S.* **43**, 105.
Arnon, D. I. (1969). *Naturwissenschaften* **56**, 295.
Ballschmiter, K. (1969). *Angew. Chem. Int. Ed.* **8**, 617.
Ballschmiter, K., Truesdell, K., and Katz, J. J. (1969). *Biochim. Biophys. Acta* **184**, 604.
Bay, Z., and Pearlstein, R. M. (1963). *Proc. Nat. Acad. Sci. U.S.* **50**, 962.
Bearden, A. J., and Malkin, R. (1972a). *Biochem. Biophys. Res. Commun.* **46**, 1299.
Bearden, A. J., and Malkin, R. (1972b). *Biochim. Biophys. Acta* **283**, 456.
Becker, J. F., Geacintov, N. E., Van Nostrand, F., and Van Metter, R. (1973). *Biochem. Biophys. Res. Commun.* **51**, 597.
Beugeling, T. (1968). *Biochim. Biophys. Acta* **153**, 143.
Black, C. C. (1966). *Biochim. Biophys. Acta* **120**, 332.
Blumenfeld, L. A., Kafalieva, D. N., Livshits, V. A., Solov'ev, I. S., and Chetverikov, A. G. (1970). *Dokl. Akad. Nauk SSSR* **193**, 700.
Bogomolni, R. A. (1972). Ph.D. Thesis, Univ. of California, Berkeley, Lawrence Berkeley Lab. Rep. LBL-1036.
Bolton, J. R., Clayton, R. K., and Reed, D. W. (1969). *Photochem. Photobiol.* **9**, 209.
Borisov, A. Yu. (1967). *Dokl. Akad. Nauk SSSR* **173**, 208.
Borisov, A. Yu., and Fetisova, Z. G. (1971). *Mol. Biol.* **5**, 509.
Borisov, A. Yu., and Godik, V. I. (1970). *Biochim. Biophys. Acta* **223**, 441.
Borisov, A. Yu., Godik, V. I., and Chibisov, A. K. (1970). *Mol. Biol.* **4**, 500.
Bowers, P. G., and Porter, G. (1967). *Proc. Roy. Soc.* **A296**, 435.
Branton, D., and Park, R. B. (1967). *J. Ultrastruct. Res.* **19**, 283.

Breton, J., and Roux, E. (1971). *Biochem. Biophys. Res. Commun.* **45**, 557.

Briantais, J. -M., and Picaud, M. (1972). *FEBS Lett.* **20**, 100.

Brody, S. S. (1958). *Science* **128**, 838.

Brody, S. S., and Rabinowitch, E. (1957). *Science* **125**, 555.

Brown, J. S., and French, C. S. (1959). *Plant Physiol.* **34**, 305.

Butler, W. L. (1961). *Arch. Biochem. Biophys.* **93**, 413.

Butler, W. L. (1966). *In* "The Chlorophylls" (L. P. Vernon and G. R. Seely, eds.), p. 343. Academic Press, New York.

Butler, W. L. (1972a). *Biophys. J.* **12**, 851.

Butler, W. L. (1972b). *Proc. Nat. Acad. Sci. U.S.* **69**, 3420.

Butler, W. L. (1973). *Accts. Chem. Res.* **6**, 177.

Butler, W. L., and Hopkins, D. W. (1970a). *Photochem. Photobiol.* **12**, 439.

Butler, W. L., and Hopkins, D. W. (1970b). *Photochem. Photobiol.* **12**, 451.

Butler, W. L., and Norris, K. H. (1963). *Biochim. Biophys. Acta* **66**, 72.

Callis, J. B., Parson, W. W., and Gouterman, M. (1972). *Biochim. Biophys. Acta* **267**, 348.

Calvin, M. (1958). *Brookhaven Symp. Biol.* **11**, 160.

Calvin, M., and Sogo, P. B. (1957). *Science* **125**, 499.

Case, G. D., and Parson, W. W. (1971). *Biochim. Biophys. Acta* **253**, 187.

Clayton, R. K. (1962a). *Photochem. Photobiol.* **1**, 201.

Clayton, R. K. (1962b). *Photochem. Photobiol.* **1**, 305.

Clayton, R. K. (1963a). *In* "Bacterial Photosynthesis" (H. Gest, A. San Pietro, and L. P. Vernon, eds.), p. 495. Antioch Press, Yellow Springs, Ohio.

Clayton, R. K. (1963b). *Biochim. Biophys. Acta* **75**, 312.

Clayton, R. K. (1965). "Molecular Physics in Photosynthesis." Ginn (Blaisdell), Boston, Massachusetts.

Clayton, R. K. (1966a). *Photochem. Photobiol.* **5**, 669.

Clayton, R. K. (1966b). *Photochem. Photobiol.* **5**, 679.

Clayton, R. K. (1966c). *Photochem. Photobiol.* **5**, 807.

Clayton, R. K. (1967). *J. Theoret. Biol.* **14**, 173.

Clayton, R. K. (1969). *Biophys. J.* **9**, 60.

Clayton, R. K. (1970). "Light and Living Matter," Vol. 1, The Physical Part. McGraw-Hill, New York.

Clayton, R. K. (1971). "Light and Living Matter," Vol. 2, The Biological Part. McGraw-Hill, New York.

Clayton, R. K. (1972). *Proc. Nat. Acad. Sci. U.S.* **69**, 44.

Clayton, R. K., and Clayton, B. J. (1972). *Biochim. Biophys. Acta* **283**, 492.

Clayton, R. K., and Haselkorn, R. (1972). *J. Mol. Biol.* **68**, 97.

Clayton, R. K., and Sistrom, W. R. (1966). *Photochem. Photobiol.* **5**, 661.

Clayton, R. K., and Straley, S. C. (1970). *Biochem. Biophys. Res. Commun.* **39**, 1114.

Clayton, R. K., and Straley, S. C. (1972). *Biophys. J.* **12**, 1221.

Clayton, R. K., and Wang, R. T. (1971). *Methods Enzymol.* **23**, 696.

Clayton, R. K., and Yau, H. F. (1972). *Biophys. J.* **12**, 867.

Clayton, R. K., Fleming, H., and Szuts, E. Z. (1972a). *Biophys. J.* **12**, 46.

Clayton, R. K., Szuts, E. Z., and Fleming, H. (1972b). *Biophys. J.* **12**, 64.

Closs, G. L., Katz, J. J., Pennington, F. C., Thomas, M. R., and Strain, H. H. (1963). *J. Amer. Chem. Soc.* **85**, 3809.

Commoner, B., Heise, J. J., and Townsend, J. (1956). *Proc. Nat. Acad. Sci. U.S.* **42**, 710.

Crabbé, P. (1972). "ORD and CD in Chemistry and Biochemistry." Academic Press, New York.

Cusanovich, M. A., Bartsch, R. G., and Kamen, M. D. (1968). *Biochim. Biophys. Acta* **153**, 397.

Dietrich, W. E., Jr., and Thornber, J. P. (1971). *Biochim. Biophys. Acta* **245**, 482.

Dilley, R. A., Peters, G. A., and Shaw, E. R. (1972). *J. Membrane Biol.* **8**, 163.

Dimitrievsky, O. D., Ermolaev, V. L., and Terenin, A. N. (1957). *Dokl. Akad. Nauk SSSR* **114**, 751.

Döring, G., Stiehl, H. H., and Witt, H. T. (1967). *Z. Naturforsch.* **22b**, 639.

Döring, G., Bailey, J. L., Kreutz, W., Weikard, J., and Witt, H. T. (1968). *Naturwissenschaften* **55**, 219.

Döring, G., Renger, G., Vater, J., and Witt, H. T. (1969). *Z. Naturforsch.* **24b**, 1139.

Dratz, E. A., Schultz, A. J., and Sauer, K. (1967). *Brookhaven Symp. Biol.* **19**, 303.

Dutton, P. L. (1971). *Biochim. Biophys. Acta* **226**, 63.

Dutton, P. L., and Jackson, J. B. (1972). *Eur. J. Biochem.* **30**, 495.

Dutton, P. L., and Leigh, J. S. (1973). *Biochim. Biophys. Acta* **314**, 178.

Dutton, P. L., Leigh, J. S., and Seibert, M. (1972). *Biochem. Biophys. Res. Commun.* **46**, 406.

Dutton, P. L., Leigh, J. S., Jr., and Reed, D. W. (1973). *Biochim. Biophys. Acta* **292**, 654.

Duysens, L. N. M. (1952). Thesis, Univ. of Utrecht.

Duysens, L. N. M. (1964). *Progr. Biophys. Mol. Biol.* **14**, 1.

Duysens, L. N. M. (1966a). *Brookhaven Symp. Biol.* **19**, 71.

Duysens, L. N. M. (1966b). *In* "Currents in Photosynthesis" (J. B. Thomas and J. C. Goedheer, eds.), p. 263. Ad Donker, Rotterdam.

Duysens, L. N. M., and Sweers, H. E. (1963). *In* "Studies on Microalgae and Photosynthetic Bacteria," p. 353. Univ. of Tokyo Press, Tokyo.

Ebrey, T. G., and Clayton, R. K. (1969). *Photochem. Photobiol.* **10**, 109.

Emerson, R., and Arnold, W. (1932a). *J. Gen. Physiol.* **15**, 391.

Emerson, R., and Arnold, W. (1932b). *J. Gen. Physiol.* **16**, 191.

Erixon, K., and Butler, W. L. (1971). *Biochim. Biophys. Acta* **234**, 381.

Evans, M. C. W., Telfer, A., and Lord, A. V. (1972). *Biochim. Biophys. Acta* **267**, 530.

Feher, G. (1971). *Photochem. Photobiol.* **14**, 373.

Feher, G., Okamura, M. Y., Raymond, J. A. and Steiner, L. A. (1971). *Biophys. Soc. Abstr. New Orleans* **11**, 38a.

Feher, G., Okamura, M. Y. and McElroy, J. D. (1972). *Biochim. Biophys. Acta* **267**, 222.

Feher, G., Hoff, A. J., Isaacson, R. A. and McElroy, J. D. (1973). *Biophys. Soc. Abstr.* **13**, 61a.

Fernandez, J., and Becker, R. S. (1959). *J. Chem. Phys.* **31**, 467.

Fischer, M. S., Templeton, D. H., Zalkin, A., and Calvin, M. (1972). *J. Amer. Chem. Soc.* **94**, 3613.

Floyd, R. A., Chance, B., and De Vault, D. (1971). *Biochim. Biophys. Acta* **226**, 103.

Förster, Th. (1965). *In* "Modern Quantum Chemistry" (O. Sinanoglu, ed.), Vol. 3, p. 93. Academic Press, New York.

Fork, D. C., and Amesz, J. (1970). *In* "Photophysiology" (A. C. Giese, ed.), Vol. V, pp. 97–126. Academic Press, New York.

Fowler, C. F., Nugent, N. A., and Fuller, R. C. (1971). *Proc. Nat. Acad. Sci. U.S.* **68**, 2278.

Fowler, C. F., Gray, B. H., Nugent, N. A., and Fuller, R. C. (1973), *Biochim. Biophys. Acta* **292**, 692.

Fraker, P. J., and Kaplan, S. (1972). *J. Biol. Chem.* **247**, 2732.

Franck, J., and Rosenberg, J. L. (1964). *J. Theoret. Biol.* **7**, 276.

Frei, Y. F. (1962). *Biochim. Biophys. Acta* **57**, 82.

Garcia, A., Vernon, L. P., and Mollenhauer, H. (1966). *Biochemistry* **5**, 2399.

Gingras, G., and Jolchine, G. (1969). *Prog. Photosynthesis Res.* **1**, 209.

Goedheer, J. C. (1957). Thesis, Univ. of Utrecht.

Goedheer, J. C. (1961). *Biochim. Biophys. Acta* **53**, 420.

Goedheer, J. C. (1964). *Biochim. Biophys. Acta* **88**, 304.

Goedheer, J. C. (1965). *Biochim. Biophys. Acta* **102**, 73.

Goedheer, J. C. (1966). *In* "The Chlorophylls" (L. P. Vernon and G. R. Seely, eds.), p. 147. Academic Press, New York.

Goedheer, J. C. (1972). *Ann. Rev. Plant Physiol.* **23**, 87.

Gouterman, M. (1961). *J. Mol. Spectrosc.* **6**, 138.

Gouterman, M., and Stryer, L. (1962). *J. Chem. Phys.* **37**, 2260.

Govindjee, and Papageorgiou, G. (1971). *In* "Photophysiology" (A. C. Giese, ed.), p. 1. Academic Press, New York.

Govindjee, Papageorgiou, G., and Rabinowitch, E. (1967). *In* "Fluorescence Theory, Instrumentation and Practice" (G. Guilbault, ed.), pp. 511–564. Dekker, New York.

Govindjee, Döring, G., and Govindjee, R. (1970). *Biochim. Biophys. Acta* **205**, 303.

Green, D. *et al.* (1967). *Arch. Biochem. Biophys.* **119**, 312.

Gregory, R. P. F., Raps, S., and Bertsch, W. (1971). *Biochim. Biophys. Acta* **234**, 330.

Gurinovich, G. P., Sevchenko, A. N., and Solov'ev, K. N. (1968). "Spectroscopy of Chlorophyll and Related Compounds." Izdatel'stvo Nauka i Tekhnika, Minsk (*English Transl.; U.S. At. Energy Comm.*, AEC-tr-7199, 1971).

Hanna, M. W. (1969). "Quantum Mechanics in Chemistry," 2nd ed., Chapter 4. Benjamin, New York.

Hiyama, T., and Ke, B. (1971a). *Proc. Nat. Acad. Sci. U.S.* **68**, 1010.

Hiyama, T., and Ke, B. (1971b). *Arch. Biochem. Biophys.* **147**, 99.

Hoch, G., and Knox, R. S. (1968). *In* "Photophysiology" (A. Giese, ed.), Vol. 3, p. 225. Academic Press, New York.

Hochstrasser, R. M., and Kasha, M. (1964). *Photochem. Photobiol.* **3**, 317.

Honeycutt, R. C., and Krogmann, D. W. (1972). *Biochim. Biophys. Acta* **256**, 467.

Jacobs, E. E., Holt, A. S., Kromhout, R., and Rabinowitch, E. (1957). *Arch. Biochem. Biophys.* **72**, 495.

Jolchine, G., Reiss-Husson, F., and Kamen, M. D. (1969). *Proc. Nat. Acad. Sci. U.S.* **64**, 650.

Kasha, M. (1963). *Radiat. Res.* **20**, 55.

Katz, J. J., Closs, G. L., Pennington, F. C., Thomas, M. R. and Strain, H. H. (1963). *J. Amer. Chem. Soc.* **85**, 3801.

Katz, J. J., Dougherty, R. C., and Boucher, L. J. (1966). *In* "The Chlorophylls" (L. P. Vernon and G. R. Seely, eds.), p. 185. Academic Press, New York.

Katz, J. J., Ballschmiter, K., Garcia-Morin, M., Strain, H. H., and Uphaus, R. A. (1968). *Proc. Nat. Acad. Sci. U.S.* **60**, 100.

Katz, J. J., Strain, H. H., Harkness, A. L., Studier, M. H., Svec, W. A., Janson, T. R., and Cope B. T. (1972). *J. Amer. Chem. Soc.* **94**, 7938.

Ke, B. (1969). *Biochim. Biophys. Acta* **172**, 583.

Ke, B. (1972). *Arch. Biochem. Biophys.* **152**, 70.

Ke, B. (1973). *Biochim. Biophys. Acta* **301**, 1.

Knaff, D. B., and Arnon, D. I. (1969). *Proc. Nat. Acad. Sci. U.S.* **63**, 963.

Knox, R. S. (1968). *J. Theoret. Biol.* **21**, 244.

Knox, R. S., and Chang, J. C. (1971). Abstracts. *Conf. Primary Photochem. Photosynthesis, Argonne, Ill.* p. 13.

Kohl, D. H. (1972). *In* "Biological Application of Electron Spin Resonance" (H. M. Swartz, J. R. Bolton, and D. C. Borg, eds.), p. 213. Academic Press, New York.

Kok, B. (1956). *Biochim. Biophys. Acta* **22**, 399.

Kok, B. (1961). *Biochim. Biophys. Acta* **48**, 527.

Kok, B., and Hoch, G. (1961), *In* "Light and Life," (W. D. McElroy and B. Glass, eds.), p. 397. Johns Hopkins Press, Baltimore, Maryland.

Kok, B., Rurainski, H. J., and Owens, O. von H. (1965). *Biochim. Biophys. Acta* **109**, 347.

Kreutz, W. (1970). *Advan. Bot. Res.* **3**, 53.

Kung, S. D., and Thornber, J. P. (1971). *Biochim. Biophys. Acta* **253**, 285.

Kuntz, I. D. Jr., Loach, P. A., and Calvin, M. (1964). *Biophys. J.* **4**, 227.

Latimer, P., Bannister, T. T., and Rabinowitch, E. I. (1956). *Science* **124**, 585.

Leigh, J. S. Jr., and Dutton, P. L. (1972). *Biochem. Biophys. Res. Commun.* **46**, 414.

Lintilhac, P. M., and Park, R. B. (1966). *J. Cell Biol.* **28**, 582.

Livingston, R. (1960). *Radiat. Res. Suppl.* **2**, 196.

Livingston, R., Porter, G., and Windsor, M. (1954). *Nature (London)* **173**, 485.

Loach, P. A. (1966). *Biochemistry* **5**, 592.

Loach, P. A., and Hall, R. L. (1972). *Proc. Nat. Acad. Sci. U.S.* **69**, 786.

Loach, P. A., and Sekura, D. L. (1967). *Photochem. Photobiol.* **6**, 381.

Loach, P. A. and Sekura, D. L. (1968). *Biochemistry* **7**, 2642.

Loach, P. A., and Walsh, K. (1969). *Biochemistry* **8**, 1908.

Loach, P. A., Androes, G. M., Maksim, A. F., and Calvin, M. (1963). *Photochem. Photobiol.* **2**, 443.

Loach, P. A., Sekura, D. L., Hadsell, R. M., and Stemer, A. (1970a). *Biochemistry* **9**, 724.

Loach, P. A., Hadsell, R. M., Sekura, D. L., and Stemer, A. (1970b). *Biochemistry* **9**, 3127.

Lutz, M. (1972). *C. R. Acad. Sci. Paris* [B] **275**, 497.

Malkin, R., and Bearden, A. J. (1971). *Proc. Nat. Acad. Sci. U.S.* **68**, 16.

Malkin, R., and Bearden, A. J. (1973). *Proc. Nat. Acad. Sci. U.S.* **70**, 294.

Mauzerall, D. (1972). *Proc. Nat. Acad. Sci. U.S.* **69**, 1358.

McElroy, J. D., Feher, G., and Mauzerall, D. C. (1969). *Biochim. Biophys. Acta* **172**, 180.

McElroy, J. D., Feher, G., and Mauzerall, D. C. (1972). *Biochim. Biophys. Acta* **267**, 363.

Moraw, R., and Witt, H. T. (1961). *Z. Phys. Chem. (Frankfurt)* **29**, 25.

Morita, S., and Miyazaki, T. (1971). *Biochim. Biophys. Acta* **245**, 151.

Müller, A., Lumry, R., and Walker, M. S. (1969). *Photochem. Photobiol.* **9**, 113.

Murata, N., and Takamiya, A. (1969). *Plant Cell Physiol.* **10**, 193.

Murata, T., and Murata, N. (1971). *Carnegie Inst. Wash. Yearbook* **70**, 504.

Murata, T., Toda, F., Uchino, K., and Yakushiji, E. (1971). *Biochim. Biophys. Acta* **245**, 208.

Netzel, T. L., Rentzepis, P. M., and Leigh, J. (1973). *Science* **182**, 238.

Nicolson, G. L., and Clayton, R. K. (1969). *Photochem. Photobiol.* **9**, 395.

Nicholson, W. J., and Fortoul, J. I. (1967). *Biochim. Biophys. Acta* **143**, 577.

Norris, J. R., Uphaus, R. A., Crespi, H. L., and Katz, J. J. (1971). *Proc. Nat. Acad. Sci. U.S.* **68**, 625.

Norris, J. R., Druyan, M. E., and Katz, J. J. (1973). *J. Amer. Chem. Soc.* **95**, 1680.

Oelze, J., and Drews, G. (1972). *Biochim. Biophys. Acta* **265**, 209.

Okamura, M. Y., Steiner, L. A., and Feher, G. (1974). *Biochemistry* **13**, 1394.

Okayama, S., and Butler, W. L. (1972). *Biochim. Biophys. Acta* **267**, 523.

Olson, J. M. (1971). *Methods Enzymol.* **23**, 636.

Olson, J. M., and Romano, C. A. (1962). *Biochim. Biophys. Acta* **59**, 726.

Olson, J. M., Philipson, K. D., and Sauer, K. (1973). *Biochim. Biophys. Acta* **292**, 206.

Olson, R. A., Butler, W. L., and Jennings, W. H. (1961). *Biochim. Biophys. Acta* **54**, 615.

Olson, R. A., Butler, W. L., and Jennings, W. H. (1962). *Biochim. Biophys. Acta* **58**, 144.

Orchin, M., and Jaffé, H. H. (1971). "Symmetry, Orbitals and Spectra," Chapter 8. Wiley (Interscience), New York.

Park, R. B. (1966). *In* "The Chlorophylls" (L. P. Vernon and G. R. Seely, eds.), p. 283. Academic Press, New York.

Park, R. B., and Biggins, J. (1964). *Science* **144**, 1009.

Park, R. B., and Pon, N. G. (1961). *J. Mol. Biol.* **3**, 1.

Parson, W. W. (1968). *Biochim. Biophys. Acta* **153**, 248.

Parson, W. W. (1969). *Biochim. Biophys. Acta* **189**, 397.

Philipson, K. D., and Sauer, K. (1972). *Biochemistry* **11**, 1880.

Philipson, K. D., and Sauer, K. (1973). *Biochemistry* **12**, 535.

Philipson, K. D., Tsai, S. C., and Sauer, K. (1971). *J. Phys. Chem.* **75**, 1440.

Philipson, K. D., Sato, V. L., and Sauer, K. (1972). *Biochemistry* **11**, 4591.

Rabinowitch, E. I. (1945, 1951, 1956). "Photosynthesis," Vols. I and II (Parts 1 and 2). Wiley (Interscience), New York.

Rabinowitch, E., and Govindjee. (1969). "Photosynthesis." Wiley, New York.

Reed, D. W. (1969). *J. Biol. Chem.* **244**, 4936.

Reed, D. W., and Clayton, R. K. (1968). *Biochem. Biophys. Res. Commun.* **30**, 471.

Reed, D. W., and Ke, B. (1973). *J. Biol. Chem.* **248**, 3041.

Reed, D. W., and Mayne, B. C. (1971). *Biochim. Biophys. Acta* **226**, 477.

Reed, D. W., and Peters, G. A. (1972). *J. Biol. Chem.* **247**, 7148.

Reed, D. W., Zankel, K. L., and Clayton, R. K. (1969). *Proc. Nat. Acad. Sci. U.S.* **63**, 42.

Reed, D. W., Raveed, D., and Israel, H. W. (1970). *Biochim. Biophys. Acta* **223**, 281.

Regitz, G., and Oettmeier, W. (1972). *Proc. Int. Congr. Photosynthesis Res., 2nd, Stresa* (G. Forti, M. Avron, and A. Melandri, eds.), p. 499. Dr. W. Junk, Publ., The Hague.

Regitz, G., Berzborn, R., and Trebst, A. (1970). *Planta* **91**, 8.

Reiss-Husson, F., and Jolchine, G. (1972). *Biochim. Biophys. Acta* **256**, 440.

Remy, R. (1971). *FEBS Lett.* **13**, 313.

Rikhireva, G. I., Sibel'dina, L. A., Gribova, Z. P., Marinov, B. S., Kayushin, L. P., and Krasnovsky, A. A. (1968). *Dokl. Akad. Nauk SSSR* **181**, 1485.

Robinson, G. W. (1963). *Proc. Nat. Acad. Sci. U.S.* **49**, 521.

Robinson, G. W. (1966). *Brookhaven Symp. Biol.* **19**, 16.

Ross, R. T., and Calvin, M. (1967). *Biophys. J.* **7**, 595.

Sandorfy, C. (1964). "Electronic Spectra and Quantum Chemistry," Chapter 5. Prentice-Hall, Englewood Cliffs, New Jersey.

Sane, P. V., and Park, R. B. (1970). *Biochem. Biophys. Res. Commun.* **41**, 206.

Sauer, K. (1972). *Methods Enzymol.* **24**, 206.

Sauer, K., Lindsay Smith, J. R., and Schultz, A. J. (1966). *J. Amer. Chem. Soc.* **88**, 2681.

Sauer, K., Dratz, E. A., and Coyne, L. (1968). *Proc. Nat. Acad. Sci. U.S.* **61**, 17.

Schachman, H. K., Pardee, A. B., and Stanier, R. Y. (1952). *Arch. Biochem. Biophys.* **38**, 245.

Seely, G. R. (1973). *J. Theoret. Biol.* **40**, 173, 189.

Seely, G. R., and Jensen, R. G. (1965). *Spectrochim. Acta* **21**, 1835.

Segen, B. J., and Gibson, K. D. (1971). *J. Bacteriol.* **105**, 701.

Seibert, M., and De Vault, D. (1971). *Biochim. Biophys. Acta* **253**, 396.

Seibert, M., Dutton, P. L., and De Vault, D. (1971). *Biochim. Biophys. Acta* **226**, 189.

Seibert, M., Alfano, R. R., and Shapiro, S. L. (1973). *Biochim. Biophys. Acta* **292**, 493.

Shibata, K. (1957). *J. Biochem.* **44**, 147.

Singer, S. J., and Nicolson, G. L. (1972). *Science* **175**, 720.

Singh, I. S., and Becker, R. S. (1960). *J. Amer. Chem. Soc.* **82**, 2083.

Slooten, L. (1972a). *Biochim. Biophys. Acta* **256**, 452.

Slooten, L. (1972b). *Biochim. Biophys. Acta* **275**, 208.

Smith, J. H. C., and Benitez, A. (1955). *In* "Modern Methods of Plant Analysis" (K. Paech and M. V. Tracey, eds.), Vol. IV, p. 142. Springer Verlag, Berlin and New York.

Smith, W. R., Jr., Sybesma, C., and Dus, K. (1972). *Biochim. Biophys. Acta* **267**, 609.

Sogo, P. B., Jost, M., and Calvin, M. (1959). *Radiat. Res. Suppl.* **1**, 511.

Straley, S. C., Parson, W. W., Mauzerall, D. C., and Clayton, R. K. (1973). *Biochim. Biophys. Acta* **305**, 597.

Strain, H. H., and Svec, W. A. (1966). *In* "The Chlorophylls" (L. P. Vernon and G. R. Seely, eds.), p. 21. Academic Press, New York.

Strehler, B. L., and Arnold, W. (1951). *J. Gen. Physiol.* **34**, 809.

Tachiki, K. H., Parlette, P. R., and Pon, N. G. (1969). *Biochem. Biophys. Res. Commun.* **34**, 162.

Takamiya, A. (1971). *Methods Enzymol.* **23**, 603.

Thornber, J. P. (1970). *Biochemistry* **9**, 2688.

Thornber, J. P. (1971). *Methods Enzymol.* **23**, 688.

Thornber, J. P., and Olson, J. M. (1968). *Biochemistry* **7**, 2242.

Thornber, J. P., and Olson, J. M. (1971). *Photochem. Photobiol.* **14**, 329.

Thornber, J. P., Olson, J. M., Williams, D. M., and Clayton, M. L. (1969). *Biochim. Biophys. Acta* **172**, 351.

Tinoco, I. Jr. (1963). *Radiat. Res.* **20**, 133.

Tomita, G., and Rabinowitch, E. (1962). *Biophys. J.* **2**, 483.

Trebst, A. (1972). *In Proc. Int. Congr. Photosynthesis Res.*, *2nd, Stresa* (G. Forti, M. Avron, and A. Melandri, eds.), p. 399. Dr. W. Junk, Publ., The Hague.

Tributsch, H. (1971). *Bioenergetics* **2**, 249.

Trosper, T., Park, R. B., and Sauer, K. (1968). *Photochem. Photobiol.* **7**, 451.

Tumerman, L. A., and Sorokin, E. M. (1967). *Mol. Biol.* **1**, 628.

Tweet, A. G., Bellamy, W. D., and Gaines, G. L. Jr., (1964). *J. Chem. Phys.* **41**, 2068.

Velluz, L., Legrand, M., and Grosjean, M. (1965). "Optical Circular Dichroism." Academic Press, New York.

Vermeglio, A., and Mathis, P. (1973a). *Biochim. Biophys. Acta* **292**, 763.

Vermeglio, A., and Mathis, P. (1973b). *Biochim. Biophys. Acta* **314**, 57.

Vernon, L. P., Yamamoto, H. Y., and Ogawa, T. (1969). *Proc. Nat. Acad. Sci. U.S.* **63**, 911.

Vernon, L. P., Shaw, E. R., Ogawa, T., and Raveed, D. (1971). *Photochem. Photobiol.* **14**, 343.

Vredenberg, W. J., and Duysens, L. N. M. (1963). *Nature (London)* **197**, 355.

Wang, R. T., and Clayton, R. K. (1973). *Photochem. Photobiol.* **17**, 57.

Warden, J. T., and Bolton, J. R. (1972). *J. Amer. Chem. Soc.* **94**, 4351.

Wassink, E. C., Katz, E., and Dorrestein, R. (1939). *Enzymologia* **7**, 113.

Watson, A., and Livingston, R. L. (1950). *J. Chem. Phys.* **18**, 802.

Weaver, E. C. (1968). *Photochem. Photobiol.* **7**, 93.

Weaver, E. C., and Weaver, H. E. (1969). *Science* **165**, 906.

Weber, G. (1960). *In* "Comparative Biochemistry of Photoreactive Systems" (M. B. Allen, ed.), p. 393. Academic Press, New York.

Weber, G., and Teale, F. W. J. (1957). *Trans. Faraday Soc.* **53**, 646.

Weier, T. E., and Benson, A. A. (1967). *Amer. J. Bot.* **54**, 389.

Weiss, C. Jr. (1972). *J. Mol. Spectrosc.* **44**, 37.

Witt, H. T. (1971). *Quart. Rev. Biophys.* **4**, 365.

Witt, K., and Wolff, Ch. (1970). *Z. Naturforsch.* **25b**, 387.

Yakushiji, E., Uchino, K., Sugimura, Y., Shiratori, I., and Takamiya, F. (1963). *Biochim. Biophys. Acta* **75**, 293.

Yamamoto, H. Y., and Vernon, L. P. (1969). *Biochemistry* **8**, 4131.

Yang, C. S., and Blumberg, W. E. (1972). *Biochem. Biophys. Res. Commun.* **46**, 422.

Yau, H. F. (1971). *Photochem. Photobiol.* **14**, 475.

Yocum, C. F., and San Pietro, A. (1969). *Biochem. Biophys. Res. Commun.* **36**, 614.

Yocum, C. F., and San Pietro, A. (1970). *Arch. Biochem. Biophys.* **140**, 152.

Yocum, C. F., and San Pietro, A. (1972). *In Proc. Int. Congr. Photosynthesis Res.*, *2nd, Stresa* (G. Forti, M. Avron, and A. Melandri, eds.), p. 477. Dr. W. Junk, Publ., The Hague.

Zankel, K. L., Reed, D. W., and Clayton, R. K. (1968). *Proc. Nat. Acad. Sci. U.S.* **61**, 1243.

Zweig, G., and Avron, M. (1965). *Biochem. Biophys. Res. Commun.* **19**, 397.

4

Excitation Energy Transfer and Migration: Theoretical Considerations*

R. S. Knox

1. EXCITED ELECTRONIC STATES IN CONDENSED SYSTEMS

1.1. Excitons in Solids

"Excitons transport electronic energy through solids with no corresponding mass and charge transport." The principles on which this statement is based were laid out in the 1930's in a sequence of papers (Frenkel, 1931,

---

*Preparation of this manuscript was supported in part by the National Science Foundation, under Grant GU-4040.

1936; Peierls, 1932; Wannier, 1937; Mott, 1938; Franck and Teller, 1938), and the subject lay dormant for about ten years. During the 1950's a determined search for unambiguous exciton transport of energy resulted largely in a further detailed understanding of absorption and emission spectra of all sorts of solids. Especially worthy of note are Davydov's (1948) theory and the pioneering work of Gross and Nikitine on semiconductor spectra (see, e.g., Knox, 1963). The main difficulty in demonstrating exciton transport in the pure sense stated above lay in separating the exciton and photon excitation of impurities used as detectors, and in the natural tendency for excitons to ionize thermally. Some progress was made in organic systems, however (Simpson, 1956), where excitons did appear able to diffuse into the crystal to distances inaccessible to the strongly absorbed photons.

In the 1960's both high-resolution spectroscopy and high-intensity sources were brought into play and exciton motion was demonstrated unambiguously. But before describing the experiments a few terms should be defined for readers not familiar with solid-state terminology. The Frenkel, or *tight-binding*, or *molecular* exciton is an excited state of a condensed system which very closely resembles a molecular excited state, either as a coherent delocalized combination of states or as a localized excitation which hops from site to site. The main characteristic of any Frenkel exciton is its correspon-

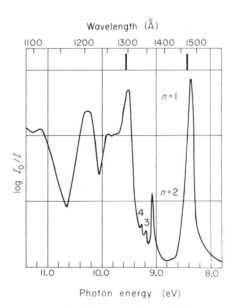

Fig. 1.   Absorption spectrum of solid xenon (after Baldini, 1962). The two vertical lines indicate the positions of two atomic resonance lines which correspond to the two highest peaks. Lines $n = 2$, 3, etc. are Wannier exciton peaks (see text).

dence with some atomic or molecular state, as seen directly in a comparison of spectra. The *Wannier*, or *Mott*, or *weak-binding*, or *effective-mass* exciton has no immediate counterpart in a molecular constituent of the solid. It is an extreme charge-transfer state and occurs in materials of high dielectric constant where the lowest bound excited state is an exciton with a very large radius. In at least one remarkable case, both types of exciton may be seen in a single absorption spectrum (solid xenon, Fig. 1). The peaks at 8.4 and 9.5 eV are Frenkel states corresponding to atomic lines at 8.45 and 9.6 eV, while the lines labeled $n = 2, 3$, etc. are Wannier states unknown in the atomic spectrum.

The classification according to spectra is partially valid with respect to mobility. All Wannier excitons are expected to be highly mobile, reflecting the easy motion of the electron and hole from which they are basically constructed. Most Frenkel excitons are expected to have low mobility, because their size is comparable to lattice constants, and, since they prevail in molecular solids, they interact with molecular vibrations as well as lattice vibrations. However, there are numerous cases in which Frenkel excitons could be considered "highly mobile," as will be seen below.

1.2. Coherent Excitons

To discuss coherence of excitons it will be necessary to introduce the following notation. In a condensed system of N molecules, a state in which the jth molecule is in its excited state α at energy E_α may be written as Φ_j^α. This is a many-particle state, the details of which will not concern us here. If there is any coupling between two states with excitation on different molecules, Φ_j^α is not likely to be a stationary state, and we may find a linear combination Ψ_K^α to be a better description:

$$\Psi_K^\alpha = \sum_{j=1}^{N} c_{K_j} \Phi_j^\alpha \tag{1.1}$$

where the coefficients c_{K_j} are determined by the details of the system. The simplest example of Eq. (1.1) is the pair of electronic states in a dimer ($N = 2$),

$$\Psi_{\pm}^\alpha = \sqrt{\tfrac{1}{2}}\Phi_1^\alpha \pm \sqrt{\tfrac{1}{2}}\Phi_2^\alpha \tag{1.2}$$

The two states have different energies, and exciton coherence (the correlation of the phase of the excitation on different molecules) was first recognized by these spectral effects. In crystals it was seen in the limited sense that the splitting, there known as Davydov splitting, implied coherence over at least two molecules (Davydov, 1948; Knox, 1963, Chapter 3). In dye aggregates in solution the Scheibe (1939) polymers are the best known example of spectral coherence effects, and the relevant theory was developed by McRae and Kasha (1958).

From the point of view of exciton dynamics, coherence is much more subtle. The simple observation that excitation moves from one locale to another at any particular diffusive rate does not establish coherence (Avakian *et al.*, 1968). The first experiment sufficient to show coherence was the magnetostark effect observation in CdS by Thomas and Hopfield (1961). The method was to show that a true exciton velocity **v** existed, which was done by observing that the electric field causing the Stark effect was being augmented by an amount $\mathbf{v} \times \mathbf{B}/c$, when the exciton was in the presence of a magnetic field B. (Here c is the speed of light in vacuum, which appears as a result of using Gaussian units.) The magnitude of the velocity agreed with the theoretical value computed assuming coherent Wannier states, and its direction agreed with that predicted from conservation of momentum in the exciton–photon interaction.

Much theoretical subtlety is required if observable effects of coherence other than simple Davydov splittings are to be predicted. A full density-matrix formalism is required because the elementary coherent states [Eq. (1.1)] are no more stationary than the elementary localized states from which they are built. (See Section 4 for comments on recent theory.)

1.3. Incoherent Excitons

The wavelength of the light which is absorbed and converted into excitons is much larger than both the absorption length and the intermolecular spacings in most cases. Typically, these are $\lambda = 5000$ Å, $\alpha^{-1} = 100$ Å for a strong absorber, and $a_0 = 10$ Å, respectively. One therefore expects some coherence in all excitons at the time of their creation, and it probably does exist for a brief instant. The success of Thomas and Hopfield's method is due to the long lifetime of the particular exciton produced, and the method is accordingly restricted. In most situations we can safely assume that unless (1) observations are made in picoseconds or less, or (2) the exciton line-width is considerably smaller than 0.5 cm^{-1}, the initial coherence associated with the photon's spatial extent will be washed out by vibrational broadening. In most instances it is well to assume that the light is depositing localized excitations like raisins in a pudding, which then diffuse about. The process which localizes these excitations is strong scattering by vibrations and ir-regularities, which simultaneously give the observed width to the absorption line.

Simpson's (1956) observations on anthracene were interpreted on this localized model, and he concluded that singlet excitons were moving about with a diffusion length of about 450 Å. A diffusion equation governs such a process, i.e.,

$$\frac{\partial \rho(\mathbf{r}, t)}{\partial t} = D \nabla^2 \rho(\mathbf{r}, t) - \frac{1}{\tau}\rho(\mathbf{r}, t) + E(\mathbf{r}, t) - \sum_i k_i(\mathbf{r} - \mathbf{r}_i)\rho(\mathbf{r}, t) \quad (1.3)$$

where D is the diffusion constant, $\rho(\mathbf{r}, t)$ the probability density of excitons at location $\mathbf{r}$ at time t, τ is the exciton lifetime in the absence of traps, $E(\mathbf{r}, t)$ is the volume rate of production of excitons at time t, and the last term on the right is a trapping term representing the total rate at which all traps i capture excitation energy, and is written in a form which allows variable rates of trapping $k_i(\mathbf{r} - \mathbf{r}_i)$ depending on the distance from the exciton to the trap locations $(\mathbf{r}_i)$. Quite often this term may be replaced by a trapping boundary condition (see, e.g., Pearlstein, 1966, 1972b). The dynamics of the exciton are determined by D and the trapping term. Förster (1948) first showed how D relates to the optical properties of molecules, in his famous paper on resonance transfer of energy between molecules. (It is interesting that at the time, he was merely trying to find a high-density continuation of his low-density theory of polarization quenching.)

A decisive demonstration of exciton diffusion came when Kepler *et al.* (1963) produced massive numbers of triplet excitons in anthracene crystals by laser excitation, and essentially followed their motion by looking at the delayed fluorescence produced by collisions among them. To appreciate this process one must first call that the triplets have a relatively long lifetime ($\sim 10^{-2}$ sec in anthracene; see Avakian and Merrifield, 1968); then it must be noted that the energy of two triplet excitons, each at 1.8 eV, is sufficient to produce one singlet exciton which fluoresces at 3.1 eV (in the blue). Exciton "fusion" of this type has now been studied wifely, and the exciton diffusion constant (actually a tensor in organic crystals, which are anisotropic) has been measured with precision. Exciton "fission" is also observed in which one singlet breaks into two triplets. Recent reviews of this subject are by Avakian and Merrifield (1968), Robinson (1970), and, with special reference to magnetic field effects, by Avakian and Suna (1971) and Swenberg and Geacintov (1973). A more general review of excitons in organic crystals is that of Birks (1970a).

1.4. Excitons *in Vivo*

Nearly every bit of the light energy absorbed by living systems spends part of its time as an exciton. There seem to be only two clear-cut cases where this fact may have very much significance. The first is in photosynthetic cells, where efficiency in energy gathering is required and the exciton acts as a directable energy package. The second is in polynucleotides, where the migration of high photon energy to particular base pairs may have a damaging effect. In neither of these cases are the usual advantages of the solid state available. Spectral resolution is low, temperatures may not be varied appreciably without damage of or irreversible changes in the *in vivo* state, and electric and magnetic fields produce numerous interfering side effects. Transfer along polynucleotide chains can, of course, be studied *in vitro*. Recent reviews are by Guéron and Shulman (1968) and Eisinger and Lamola (1971). Since the present volume is concerned with bioenergetics of photosynthesis we will limit our considerations to photosynthetic systems.

At the same time that the theory of the exciton was being developed, Emerson and Arnold (1932) were performing their classic experiments which indicated, on kinetic grounds, that many chlorophyll molecules were involved in collecting the light for each primary reaction in the photosynthetic process. This kinetic concept led, after many years of development, to the physical or physiological concept of the photosynthetic unit (PSU), in which Emerson and Arnold's superabundance of chlorophylls were thought of as divided into groups, each associated with a center at which reactions take place, these groups being built of 50 to 300 chlorophyll molecules whose function was to channel the energy to the reaction center. The PSU concept is reviewed by Rabinowitch (1956, Chapters 32 and 34), Duysens (1964), and Clayton (1965). During the last several years chlorophyll–protein complexes identified with the reaction center itself have been isolated in several bacteria and prepared in relatively high concentrations in some algae (see, e.g., Reed and Clayton, 1968; Sane and Park, 1970; Thornber and Olson, 1971).

The first serious theoretical discussion relating to Emerson and Arnold's work was that of Franck and Teller (1938), who asked whether by random walk the excitons could, on the average, reach the reaction centers during their normal lifetimes. Their answer was negative. Theory therefore did little to help develop the physical PSU concept, until Duysens (1952) provided quantitative data on transfer probabilities and a careful analysis of fluorescence data. Franck and Teller had assumed results from one-dimensional random walk theory, which effectively stretched out the path needed to reach the reaction centers. Detailed discussions may be found in the reviews by Duysens (1964) and Robinson (1967). During the past decade much theoretical effort has been put into theoretical models of the PSU and related random walk problems, starting with the work of Bay and Pearlstein (1963) and Pearlstein (1966, 1967). Some of it will be discussed in this article.

The reader has by now noticed the bias of this chapter toward theoretical work. Our purpose will be to discuss recent developments in excitation transfer and migration, mainly theoretical, with the purpose of making a frank assessment of the impact theory is having, and with the goal of identifying some areas into which it could move. There is presently a tremendous disparity between the power of the mathematical techniques available and the range of problems on which they may be realistically focused.

By excitation energy *transfer* we will generally mean two-molecule (donor–acceptor) processes. By excitation energy *migration* we will mean any motion, coherent or incoherent, involving more than two molecules. In most cases of practical interest migration is a sequence of transfers.

2. ENERGY TRANSFER

Before approaching energy migration in general it is necessary to deal with its simplest form: transfer between two molecules. There is no need for a general theoretical summary because it is hard to surpass Förster's (1965) treatment. Our plan of attack is as follows: the "very weak coupling" theory, which corresponds to the incoherent exciton case, will be introduced in a semiclassical (but rigorous) way. Its applicability to all phenomena of interest to photosynthesis is adopted as a working hypothesis. This is done not only for simplicity but also because we believe it is sufficient for most purposes. Any observable manifestations of Förster's weak coupling (intermediate coupling in the terminology of some authors) or strong coupling cases will be highlighted by this approach.

2.1. Theory of Förster Resonance Transfer Process

Consider a molecule, the donor (D), whose natural excited state lifetime in a given environment with no acceptor present is τ_0. Let the fluorescence emission spectrum $F_D(\omega)$ be normalized on an angular frequency scale ($\omega = 2\pi v$, where v is the ordinary frequency). Then the rate of radiation can be written

$$k_r = \frac{1}{\tau_0} = \frac{1}{\tau_0}\int F_D(\omega)\,d\omega \qquad (2.1)$$

Consider next an acceptor molecule A, which may be of the same type, located at a distance R, and initially unexcited. If the acceptor is at a great distance from D (such as position A_3 in Fig. 2), it can be considered solely as an absorber of electromagnetic radiation. This occurs when $R > \lambdabar$, where $\lambdabar = (\lambda/2\pi n)$, λ is the wavelength of the radiation in the medium, and n is the index of refraction (see, e.g., Jackson 1962, Section 9.2). Under these conditions A presents only the usual absorption cross section, $\sigma_A(\omega)$, and the amount of energy transferred to A can be calculated by introducing the

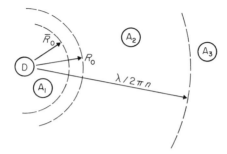

Fig. 2. Various donor (D) – acceptor (A) configurations in relation to the characteristic lengths discussed in the text. A_3 is in the far zone of D, while A_1 and A_2 are in its near zone. A_1 is in what might be called the Förster zone. The donor and acceptor may be molecules of the same type.

factor $\sigma_A/4\pi R^2$ into Eq. (2.1):

$$k_t(R \gg \tilde{\lambda}) = \frac{1}{\tau_0} \int F_D(\omega) \left[\frac{\sigma_A(\omega)}{4\pi R^2} \right] d\omega \qquad (2.2)$$

The cross section, taken here to be in centimeter-gram-second units, is related to the molar absorption coefficient by σ_A (cm^2) = $0.3825 \times 10^{-20} \epsilon_A$ (liter mole^{-1} cm^{-1}). Since typical molecular optical cross sections at their peaks are of the order of a few square angstroms, and $\tilde{\lambda}$ is several hundred angstroms, the transfer rate given by Eq. (2.2) is minuscule. It is presented for the purpose of displaying the R^{-2} dependence of the rate, which may be understood as follows: in the far zone, the electric field of the radiating donor falls off as R^{-1}. This field induces a transition dipole on the acceptor which is proportional to the field, and hence to R^{-1}. But the work done on the acceptor, which is the energy transferred to it, is proportional to the product of the field and dipole moment, and hence goes as R^{-2}.

When the acceptor is moved closer to the donor, such as to position A_2 or A_1 in Fig. 2, the only part of the immediately preceding argument which changes is the size of the field. It is increased sharply by the emergence of terms of the form $\tilde{\lambda}^2 R^{-3}$, which were negligible when R was greater than $\tilde{\lambda}$. Therefore the two factors of R^{-1} each acquire a multiplier $\tilde{\lambda}^2/R^2$ and the transfer rate becomes:

$$k_t(R \ll \tilde{\lambda}) \approx \frac{1}{\tau_0} \int F_D(\omega) \left(\frac{\tilde{\lambda}}{R} \right)^4 \left[\frac{\sigma_A(\omega)}{4\pi R^2} \right] d\omega \qquad (2.3)$$

The only reason this expression is not exact is that the appropriate angular factors have not been included. When they are, and an average over all orientations of the dipole moment of A is taken, and an average over all acceptor positions at distance R (or an average over all donor orientations) is made, (2.3) becomes

$$k_t = \frac{1}{\tau_0} \cdot \frac{1}{R^6} \cdot \left[\frac{3}{4\pi} \int \lambda^4 F_D(\omega) \sigma_A(\omega) \, d\omega \right], \tag{2.4}$$

a particularly compact form of the Förster result. The numerical coefficient has been chosen to correspond to an acceptor absorption cross section measured in a random solution.

The near-zone transfer rate, Eq. (2.4), applies to virtually the entire region* inside the sphere of radius λ (Fig. 2), but in much of this region, such as at A_2, the rate is still smaller than the total rate of radiation from the donor. At a distance such that R^6 becomes equal to the square brackets in Eq. (2.4), the transfer rate becomes equal to the radiative rate, and this characteristic distance is known as R_0:

$$R_0 \equiv \left[\frac{3}{4\pi} \int \lambda^4 F_D(\omega) \sigma_A(\omega) \, d\omega \right]^{1/6} \tag{2.5}$$

All the quantities necessary to calculate k_t may be obtained from the spectra of isolated donors and acceptors, but frequently it is found convenient to work with the observed donor lifetime τ instead of the natural lifetime τ_0, which can be obtained spectrally using the Ladenburg relation (see, e.g., Strickler and Berg, 1962). Since $\tau = \phi_F \tau_0$, where ϕ_F is the fluorescence yield in the absence of the acceptor, Eq. (2.4) may be rewritten as

$$k_t = \frac{1}{\tau} \frac{1}{R^6} \left[\phi_F \frac{3}{4\pi} \int \lambda^4 F_D(\omega) \sigma_A(\omega) \, d\omega \right] \tag{2.6}$$

and a new characteristic distance $\overline{R}_0$ is introduced, defined as the sixth root of the square bracket of Eq. (2.6), and so $\overline{R}_0 = \phi_F^{1/6} R_0$. This $\overline{R}_0$ is the acceptor distance at which transfer competes equally with the *total* donor deexcitation rate. One therefore has the choice between two representations of k_t:

$$k_t = \frac{1}{\tau_0} \left(\frac{R_0}{R} \right)^6 = \frac{1}{\tau} \left(\frac{\overline{R}_0}{R} \right)^6 \tag{2.7}$$

The distinction between R_0 and $\overline{R}_0$ is important, but there is no apparent agreement in the literature about which one is meant by the symbol R_0.

---

*There is a region near λ in which an R^{-4} transfer rate is applicable, but it is difficult to explore.

Using the wrong one can cause an error in the implied k_t by a factor of ϕ_F or $1/\phi_F$. Förster's original R_0 is what we are calling $\overline{R}_0$; Duysens (1964) redefined R_0 as we have done it, arguing that it is more intrinsic to the donor–acceptor pair. The Duysens definition has generally been accepted in the theoretical photosynthesis literature.

In a treatment paralleling the present one, we have presented a graphical representation of the parameters R_0, $\overline{R}_0$, and λ, and their relation to the various rates (Knox, 1973).

Some simplifications and approximations have been made in deriving Eqs. (2.4)–(2.7). One simplification is the angular average, which will be discussed in Section 2.3. One approximation is the assumption that the donor and acceptor transition moments always act like dipoles; this will also be discussed in Section 2.3. Another approximation is that the wavefunctions of the two molecules do not overlap. If they do, and if there is already dipolar interaction, this interaction is very strong, and the pair should be treated more as a dimer. If they overlap and there is no appreciable dipolar interaction, higher multipoles and exchange effects will become important (Dexter, 1953). We will not consider either case here. Finally, the validity of the incoherent excitation model is an implicit assumption, introduced at the outset when a definite molecule, the donor, was taken to be excited. To proceed with this point of view, we must assume that it makes sense to speak of the probability that one or the other molecule is excited, and see how it changes in time. If $\rho_D(t)$ is the probability that the donor is excited, and $\rho_A(t)$ is the probability that the acceptor is excited, then the incoherent model is taken to imply that they satisfy the equations

$$\frac{\partial \rho_D(t)}{\partial t} = -\frac{1}{\tau_D}\rho_D(t) - F_{AD}\rho_D(t) + F_{DA}\rho_A(t)$$

$$\frac{\partial \rho_A(t)}{\partial t} = -\frac{1}{\tau_A}\rho_A(t) - F_{DA}\rho_A(t) + F_{AD}\rho_D(t)$$

$$(2.8)$$

where τ_D and τ_A are the observed unimolecular lifetimes of donor and acceptor, respectively, and F_{IJ} is the Förster rate constant k_t for transfer from J to I.

2.2. Measurement of the Rates

2.2.1. Fluorescence and Polarization Quenching

If the donor and acceptor are distinct molecules and have appropriate spectra (Fig. 3), it is clear that the transfer rate F_{DA} (acceptor to donor) will be negligible compared to F_{AD} (donor to acceptor). Therefore, D will lose energy through simple transfers to all its neighbors, and only the first of Eqs.

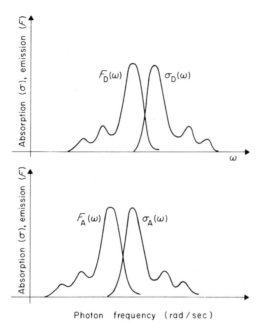

Fig. 3. Donor and acceptor spectra particularly suitable for efficient irreversible transfer. F_D overlaps σ_A well, but F_A and σ_D are widely separated.

(2.8) is needed to determine ρ_D (as $F_{DA} \to 0$):

$$\frac{\partial \rho_D}{\partial t} = - \left(\frac{1}{\tau_D} + \sum_i F_{A_iD} \right) \rho_D(t) = - \frac{1}{\tau_{\text{eff}}} \rho_D(t)$$

The effective lifetime τ_{eff} will be smaller as more acceptors are brought in closer to the donor. Since $\phi_F = \tau_{\text{eff}}/\tau_0$, the donor fluorescence yield drops with increasing concentration of acceptors in solution; at a characteristic acceptor concentration $\bar{c}_0 = (\frac{4}{3}\pi\bar{R}_0^3)^{-1}$ it has dropped to about half. Measurement of $\bar{c}_0$, or fitting the data to a more precise theoretical curve, yields $\bar{R}_0$. Direct measurements of the transfer rate, using known donor and acceptor distances in specially prepared "molecular stick" configurations, verifying the R-dependence of F_{AD}, were performed by Stryer and Haugland (1967).

The simple quenching method fails as a means of measuring $\bar{R}_0$ for transfer between like molecules, which is the case of most interest to us. The reason is the complete symmetry of the situation: if D "quenches" D then the "quencher" has the same fluorescence characteristics as the donor and will merely reemit the energy as if no transfer had taken place. However, the

set of excitations on the molecules originally excited constitute an anisotropic source of radiation because the absorbed light had no polarization
component in the direction of propagation. To established notation and a
basis for further discussion the most elementary polarization theory will
be developed here (Perrin, 1926).

Light is incident in the y direction on a collection of N molecules whose
emission is by dipole radiation and whose transition moments are
$\mu_j(j = 1, \ldots, N)$.* The incident light may be polarized in the z direction
(polarization unit vector z) or it may be unpolarized. In the standard configuration for measurement of the degree of polarization p, emission intensity
outward along the x axis is measured in two polarizations specified by the
unit vectors z, parallel to the incident polarization plane, and y, perpendicular to it. These intensities will be called I_z and I_y, respectively, and p is
defined by

$$p = \frac{I_z - I_y}{I_z + I_y} \tag{2.9}$$

As is well known, the contribution of molecule j to the intensities I_z and I_y
will be proportional to $(\mu_j \cdot z)^2$ and $(\mu_j \cdot y)^2$, respectively, provided j is excited
in the first place. Suppose the probability that j is excited is given by ρ_j, then,
the intensities are

$$I_z = C\sum_j \rho_j(\mu_j \cdot z)^2$$
$$I_y = C\sum_j \rho_j(\mu_j \cdot y)^2 \tag{2.10}$$

where the sum runs over all molecules and C is a proportionality constant
common to all molecules, which we are assuming to be identical; and the
degree of polarization is

$$p = \frac{\sum_j \rho_j[(\mu_j \cdot z)^2 - (\mu_j \cdot y)^2]}{\sum_j \rho_j[(\mu_j \cdot z)^2 + (\mu_j \cdot y)^2]} \tag{2.11}$$

---

*The transition moment is the matrix element of the vector er between the ground and
excited states of the molecule, where e is the electronic charge and r is the sum of the position
coordinates of the electrons in the molecule. If the light has a polarization vector in some
direction s, and if the excited state is nondegenerate, which is the case in most complex molecules of interest here, the probability of absorption or emission by molecule j is proportional
to $(\mu_j \cdot s)^2$ (see, e.g., Hochstrasser, 1971, p. 8).

In practice, the polarization is usually measured in a steady state so a time average must be made over the intensities in Eq. (2.10). Either ρ_j or μ_j or both may have a time dependence. The central problem of polarization quenching theory is to know ρ_j and μ_j and perform the time averages, and the sums, which are really space averages, in the proper way. A few useful special cases will indicate the complexity of the problem.

2.2.1.1. Emission from a Randomly Excited Array. Let us express μ_j in standard spherical coordinates,

$$\mu_j = \mu_j(\cos \phi_j \sin \theta_j \mathbf{x} + \sin \phi_j \sin \theta_j \mathbf{y} + \cos \theta_j \mathbf{z}) \qquad (2.12)$$

where $\mathbf{x}$, $\mathbf{y}$, and $\mathbf{z}$ are the usual Cartesian unit vectors fixed in the system and μ_j is the magnitude of μ_j. Then $(\mu_j \cdot \mathbf{z})^2 = \mu_j^2 \cos^2 \theta_j$ and $(\mu_j \cdot \mathbf{y})^2 = \mu_j^2 \sin^2 \phi_j \sin^2 \theta_j$. If the array is randomly excited, ρ_j is a constant and the sum over j corresponds to a spherical average of $\cos^2 \theta$ in the case of I_z and $\sin^2 \phi \sin^2 \theta$ in the case of I_y. These two averages are both equal to $\frac{1}{3}$ and therefore $p = 0$.

2.2.1.2. Emission from a Fixed Array Excited by Polarized Light. Here the exciting light makes a nonrandom excitation and ρ_j is not constant. Molecule j will be excited with a probability proportional to $(\xi_j \cdot \mathbf{z})^2$, where ξ_j is the absorption transition dipole moment. Assuming at first that $\xi_j = \mu_j$, which is the case when absorption and emission involve the same electronic excited state, we have (omitting common constants)

$$I_z = \sum_j (\mu_j \cdot \mathbf{z})^2 (\mu_j \cdot \mathbf{z})^2$$

$$I_y = \sum_j (\mu_j \cdot \mathbf{y})^2 (\mu_j \cdot \mathbf{z})^2 \qquad (2.13)$$

which means that now I_z is an average of $\cos^4 \theta$ and I_y is an average of $\sin^2 \phi \sin^2 \theta \cos^2 \theta$. These are, respectively, $\frac{1}{5}$ and $\frac{1}{15}$ so $p = \frac{1}{2}$. If ξ and μ are not parallel, but are known to have an angle α between them, the averaging process becomes more tedious. Levshin (1925) showed that the polarization is changed to $(3 \cos^2 \alpha - 1)/(\cos^2 \alpha + 3)$.

2.2.1.3. Emission from Molecules Undergoing Rotation. If the emitting molecule can rotate physically during the fluorescence lifetime τ, then at the time of emission its contribution to p will be changed. An average of $(\mu \cdot \mathbf{z})^2$ and $(\mu \cdot \mathbf{y})^2$ over this lifetime is required. Perrin (1929) showed that the polarization in this case is given by

$$\left(\frac{1}{p} - \frac{1}{3} \right) = \left(\frac{1}{p_0} - \frac{1}{3} \right) \frac{2}{3u - 1} \qquad (2.14)$$

where p_0 is the polarization in the absence of rotation and u is the average value of $\cos^2 \theta$ for unit vector starting from time zero at $\theta = 0$ and performing a random walk, the average being taken over the lifetime. The time dependence of $\cos^2 \theta$ is

$$\cos^2 \theta(t) = \frac{1}{3}(1 + 2 \exp [-6\mathscr{R}t])   \tag{2.15}$$

where $\mathscr{R} = k_B T/8\pi a^3 \eta$ is the Einstein rotational Brownian motion rate, k_B is Boltzmann's constant, T is the absolute temperature, a is the molecular radius, and η is the viscosity of the solution. For a solution with $\eta = 10$ cP, with $a = 10$ Å, and $T = 300°$ K, the rate $6\mathscr{R}$ is about 10^8 sec^{-1} which is comparable with fluorescence lifetimes. Therefore, by varying solvents and temperatures, the polarization may be "frozen in" or the effects of the rotation displayed. The average of Eq. (2.15) over the lifetime is

$$u = \tau^{-1} \int_0^\infty e^{-t/\tau} \cos^2 \theta(t) \, dt = (1 + 2\mathscr{R}\tau)/(1 + 6\mathscr{R}\tau)   \tag{2.16}$$

It can be seen by comparing Eqs. (2.16) and (2.14) that when $k_B T$ is very large, so that a lot of rotation occurs during τ, the average u goes to $\frac{1}{3}$ and p becomes zero. On the other hand, when $k_B T$ is small, u remains equal to 1 and p remains equal to p_0.

The theory of the polarization of fluorescence from complex molecules for which there is no single rate of Brownian rotational relaxation is understandably much more complicated. The simple theory whose main results are described above holds only for spherically symmetric molecules. Weber et al. (1971) have developed a detailed phenomenological approach for the general case of asymmetric rotators with absorption and emission oscillators not necessarily parallel. A qualitative summary has been published (Weber, 1972).

2.2.1.4.   Case of Unpolarized Excitation.   In this case the degree of polarization may still be nonzero because the exciting light still has no polarization component parallel to its propagation direction. Among the results which change are that p becomes $\frac{1}{3}$ in the simple case, above where formerly it was $\frac{1}{2}$; and Eq. (2.14) becomes

$$\left(\frac{1}{p} + \frac{1}{3}\right) = \left(\frac{1}{p_0} + \frac{1}{3}\right)\frac{2}{3u - 1}   \tag{2.17}$$

(Weber, 1952).

2.2.1.5.   Case of Excitation Transfer.   Again let us consider the simplest case, where absorbing and emitting dipoles are parallel, and assume no rotational depolarization. We ask what p will be measured if the excitation can be transferred to like molecules. In the rotational depolarization case the time dependence of ρ_j was just a simple decay and that of $\mu_j(t)$ was approximated by Brownian rotation. Now, ρ_j is complicated and μ_j is fixed. The time dependence of ρ_j in the simplest possible case is determined by a pair of equations like (2.8), where D is the molecule whose contribution to Eq. (2.11) we are trying to compute, and A is a neighbor. A statistical distribution of possible nearest neighbor locations can be employed, but then we must know (i) does the neighbor contribute to the polarization when it emits? and (ii) what happens at moderate and high densities when there is on the average more than one neighbor within a distance $\overline{R}_0$? The answer to (i) is *yes*, but when we average over the set of nearest neighbors of all donors whose dipoles are in some given direction, say μ_D, not only is the transfer rate independent of μ_D but so also is the resulting polarization of emission from the neighbors; or at least it was shown by Galanin (1950) that these neighbors contribute only about 0.02 to the value of p. The contributions of the acceptors to the degree of polarization must be considered anew in any theory applying to nonrandom solutions. Attempts to answer (ii) have led to a very large literature. A fairly complete review is contained in the author's treatment (Knox, 1968b) but further references may be found in, e.g., Ore and Eriksen (1971). A detailed comparison and consolidation of several theories was given by Craver (1971a), who found that a single curve fits all

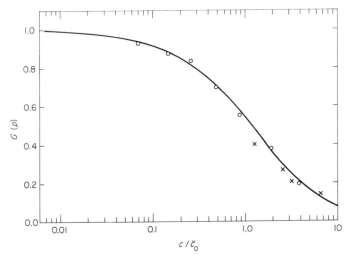

Fig. 4. A "universal curve" proposed by Craver for the theory of concentration quenching of fluorescence polarization. The ordinate is $G(p)/G(p_0)$, where $G(p) = (1/p - 1/3)^{-1}$ and p_0 is the value of p at zero concentration. The ordinate is equal to p/p_0 for all practical purposes. The abscissa is the concentration in units of $\overline{c}_0 = (4\pi \overline{R}_0^3/3)^{-1}$. Experimental points shown correspond to Weber's data on phenol in propylene glycol (Weber 1960a). (o) thick films, 5 mm; (x) thin films, 30 μm; (—) theory.

theories quite well, to within 1% at low concentrations, 5% at the critical concentration $\bar{c}_0 = (4\pi \bar{R}_0^3/3)^{-1}$, and 25% at high concentrations. This average curve, shown in Fig. 4, is taken from Table I of Craver and Knox (1971).

2.2.2. Discussion of Data on Chlorophyll a

Any theoretical interpretation of energy migration processes in chlorophyll (Chl) *in vivo* must rely on some kind of information on pair transfer rates. In this section we present a collection of results from polarization quenching in Chl *a* solutions which should provide almost any theory with a good parameter! The situation is summarized in Table I. Each entry is the value of Förster's $\bar{R}_0$ as deduced by the author quoted, or as modified (see notes and below). The values *should* be comparable with each other since in most cases there was randomness in two or three dimensions or there was no reason to suspect otherwise, and virtually the same theoretical analysis has been applied to each.

The measurement by Seely (1970) is certainly the most complex represented in the table, but his analysis is extensive and should be producing a reliable $\bar{R}_0$. We can only speculate that a very subtle steric correlation existed which kept the relative dipole orientations slightly biased toward the notches in the transfer patterns (see Section 2.3), if Seely's value of $\bar{R}_0$ is to be considered compatible with the others. Skipping to the high values: in the monogalactolipid solutions, Trosper *et al.* (1968) may have likewise been the victims of subtle correlations which emphasized the lobes of the transfer patterns. Goedheer's 92 Å, revised by us from his 87 Å (Knox, 1968b), may be somewhat affected by difficulties in determining the limiting

<table>
<tr><td colspan="3" align="right">**TABLE I**</td></tr>
<tr><td colspan="3">The Förster Parameter $\bar{R}_0$ as Deduced from Fluorescence Depolarization Data (Chlorophyll *a*).</td></tr>
<tr><td>Data by</td><td>Solvent</td><td>$\bar{R}_0(\text{Å})^a$ Notes</td></tr>
<tr><td>Seely (1970)</td><td>(Polymer attachment)</td><td>42 → 46?   —[b]</td></tr>
<tr><td>Trosper *et al.* (1968)</td><td>Oleyl alcohol (monolayer)</td><td>62 ± 11   —[c]</td></tr>
<tr><td>Tweet *et al.* (1964)</td><td>Monolayer on water</td><td>65 ± 8   —</td></tr>
<tr><td>Weber (1960b)</td><td>Paraffin</td><td>60, 74   —[d]</td></tr>
<tr><td>Kelly and Porter (1970)</td><td>Lecithin (solid)</td><td>73   —[e]</td></tr>
<tr><td>Goedheer (1957)</td><td>Castor oil</td><td>92 ± 7   —</td></tr>
<tr><td>Trosper *et al.* (1968)</td><td>Monogalactolipid</td><td>96 ± 22   —[c]</td></tr>
</table>

[a]The numbers are not necessarily those originally published by the quoted authors. (see notes and text).

[b]Higher figure proposed by Craver (1971a).

[c]As modified by Craver (1971b).

[d]The two figures represent results obtained with different excitation (see text).

[e]Not deduced from polarization (see text).

value of p at low concentrations, as pointed out privately by R. Pearlstein who suggests that Goedheer's data may predict an even higher $\overline{R}_0$. This leaves the central group of values, ranging from 62 to 73 Å, all of which require clarifying remarks. The value quoted for Trosper et al. has been revised upward by Craver (1971b) on the basis of a theory specifically designed for two-dimensional solutions. The next entry (65 Å) is corrected downward from the quoted value to account for missing π in the Förster formula originally used (Tweet et al. 1964) (the author thanks A. G. Tweet for this observation). Weber's two values have been deduced by us from his data and they correspond to two different conditions of excitation: 74 Å using 366 nm only, and 60 Å using 436 nm in addition to 366 nm. This unusual result deserves further experimental study. The reasons we do not list Weber's quoted value of 36 Å have been given before (Knox, 1968b). Finally, Kelly and Porter's 73 Å was deduced by them from ordinary quenching, and its interpretation as a true intermolecular transfer parameter is in some doubt. Kelly and Patterson (1971) deduce 80 Å from polarization quenching in Chl b.

Some small variations in $\overline{R}_0$ might be expected from variations in fluorescence yields and from differing indexes of refraction. $\overline{R}_0$ would probably be most sensitive to changes in the overlap between emission and absorption, although one expects these two curves to shift roughly equally as the solvent changes, leaving the overlap integral unchanged. To illustrate some "sensitivities" we have constructed Table II, which starts out with the simple theoretical Förster parameter as computed from Eqs. (2.5) and (2.7) (Duysens, 1964). The second line shows that a shift of either the absorption or emission curve to optimally increase the overlap, as might happen at a trap, leads to an increase of 20% in R_0. This means tripling the transfer rate. Another enormous jump occurs if we are dealing with optimal alignment of the dipoles, as discussed in the next section.

If the numbers in Table I are in fact comparable in all respects, then we have an uncertainty about the transfer rate between Chl a molecules which amounts to a factor of more than 100.

TABLE II

The Förster Parameters R_0 and $\overline{R}_0$ Calculated from Fluorescence and Emission Spectra of Chlorophyll a in Ether[a]

Spectral overlap	Dipolar orientation	R_0(Å)	$\overline{R}_0$(Å)
As in ether	Random	70	58
Complete	Random	84	70
Complete	Coaxial alignment	113	94

[a]Throughout, the fluorescence yield is taken to be $\phi_F = 0.33$. "Complete" overlap refers to an artificial situation in which the emission spectrum is shifted to produce a maximal transfer rate.

2.2.3. "Uphill" Transfer

The capability for transfer of excitation energy from donor to acceptor depends only on the degree of overlap and the total strengths of the absorption and emission characteristics, and not on the relative positions of the peaks. This is due to the fact that transfer is always occurring in a manner which conserves energy, and the heat bath of the donor may supply energy as well as accept it. To be sure, large overlaps are more likely to occur when the acceptor is red-shifted relative to the donor, as sketched in Fig. 3. But Zankel and Clayton (1969) have shown that measurable transfer from a donor (B850) to a blue-shifted acceptor (B800) exists in *Rhodopseudomonas spheroides*. The uphill transfer rate is about half the downhill rate, according to their analysis. The height of the hill climbed is about 3.6 times the room temperature value of $k_B T$.

2.3. Anisotropy and Size Effects

2.3.1. Dipole–Dipole Interaction Patterns

Few individual transfer rates are properly represented by Eqs. (2.4) and (2.7) even when all conditions for validity are met. These are averages over ensembles of donor–acceptor pairs. If the details of the dipolar interactions are maintained without taking such averages, the transfer rate is actually found to contain a factor

$$f = \tfrac{3}{2} \left[\mu_D \cdot \mu_A - 3(\mu_D \cdot \mathbf{r})(\mu_A \cdot \mathbf{r}) \right]^2 = \tfrac{3}{2} \kappa^2 \qquad (2.18)$$

(Förster, 1948), where μ_D, μ_A, and $\mathbf{r}$ are unit vectors parallel to the donor transition dipole moment, acceptor transition dipole moment, and the vector separating the two molecules, respectively. A spherical average over *any two* of the three unit vectors makes $\kappa^2 = \tfrac{2}{3}$ and therefore $f = 1$. In viscous solvents, the average has little meaning for any single donor–acceptor pair. It must be applied only when it is certain that the superposition of a large number of donor environments, chosen at random, is in fact random. Even then, it must be applied at the right point in the calculation. Fairly extensive discussions of this matter as it affects polarization quenching theory are given by Knox (1968b), Craver (1971a), and Craver and Knox (1971). We conclude that for random solutions the average may be safely taken first.

The information submerged by the averaging process is illustrated in Fig. 5, which contains polar plots of the Förster transfer rate in units of the average rate. In both parts of the figure the donor's dipole is taken to be fixed and vertical, and the acceptor is at a fixed distance and at a polar

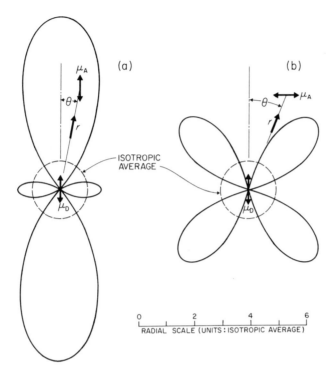

Fig. 5. Examples of the angular dependence of the Förster rate. (a) Transition moments parallel, in plane. (b) Transition moments perpendicular, in plane. The solid curve is the function f [see Eq. (2.18)] and the dashed circle is the usual Förster angular average.

angle θ. In Fig. 5a the acceptor dipole is kept parallel to the donor dipole. The angular factor f is then equal to $\frac{3}{2}(1 - 3\cos^2\theta)^2$, which is plotted as a solid curve. Now it may easily be seen that the lobes of the cloverleaf (Fig. 5b) in some way fill in the notches in the other pattern (Fig. 5a) so that an average over just the one vector μ_A already might appear smooth. What is not so easily appreciated is the depth of the notches in three dimensions. The small dashed circle is the curve $f = 1$, which is the relative size of the transfer rate one is accepting if he uses the "$\kappa^2 = \frac{2}{3}$" average. Figure 5 dramatically illustrates how in an ordered system individual rates can be larger, or as small as one wants.

2.3.2. "Monopole" Effects

It has long been recognized that molecules having transition dipole moments which are spread out over large π electron systems may have interactions at small distances which are not well approximated by the dipole–dipole interaction (London, 1942). If the intermolecular distance is comparable with the size of the π-electron system, each dipole "sees" the internal structure of the other. Golebiowski and Witkowski (1959) first pointed out that this correction could be made in the Förster theory using London's method, but no numerical work seems to have been done prior to Chang's thesis (1972). In Fig. 6 we show some of her results on Chl *a* which were obtained using a full quantum mechanical description of the excited states to compute London's "monopoles," which are just the pieces of the transition dipoles distributed over the π-electron sites.

There is a correction to the dipole–dipole rate which amounts to $+50\%$ or -30%, depending on orientation, at distances of 15 Å. The correction persists even at large distances when the dipole moments are coaxial (case I). An especially interesting comparison is between cases II and III, which theoretically have the same rate in the dipole approximation; in fact, they do not have precisely the same rate because when the porphyrin rings are coplanar (case II), some of the monopoles are closer together

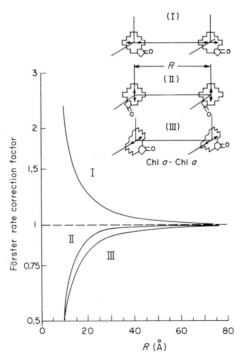

Fig. 6. Effect of molecular size on the rate of transfer between chlorophyll *a* molecules in the relative positions shown (after Chang, 1972). The ordinate is the enhancement (or reduction) factor on a logarithmic scale and the abscissa is the intermolecular separation in angstroms.

and the interaction is enhanced. Perhaps the most important point is that the rate is enhanced in exactly that situation where it is already largest (compare case I with Fig. 5a, $\theta = 0°$). The anisotropy at distances smaller than 15 Å is therefore very sharply increased. It is expected that a similar situation prevails in all the porphyrin derivatives. Needless to say, the correction factors of Fig. 6 will alter the pure R^{-6} dependence of the rate, at small distances, but it still varies rapidly with R.

3. RANDOM WALKS OF EXCITONS ON LATTICES

3.1. Physical Description of the Problem

The goal of a theory of antenna pigments which deliver energy to a trap is to relate measured observables with a proposed or known structure of the antenna itself. Although in many model organic systems the trap itself fluoresces and provides information, in the case of the photosynthetic unit the traps do not usually give such a direct display of the energy they receive, although the fact that they receive energy may be inferred by bleaching of their characteristic absorption (see Chapter 3). Therefore, *in vivo*, primary sources of information on the travel of the exciton remain its own fluorescence and fluorescence polarization. [For a discussion of Chl *a* fluorescence and its relation to photosynthetic processes, see reviews by Govindjee *et al.* (1967), Govindjee and Papageorgiou (1971), and Goedheer (1972)].

As we remarked earlier, polarization memory is lost in a random solution as soon as the excitation has left its original site. In a system with some degree of order, the appearance of nonzero polarization may mean either a limited amount of transfer *or* considerable transfer among an ordered group of molecules which by their orientation tend to restore or maintain the polarization. For example, in photosynthetic systems it seems clear that migration and trapping take place, so the observed polarization of about 0.01 to 0.06 (Arnold and Meek, 1956; recent work by Mar and Govindjee, 1972) might be due to either a small group of nontransferring molecules or to partial order within the diffusion length of the excitions, or both. On the basis of model calculations, Hemenger and Pearlstein (1971) find that p should remain close to zero ($\sim 10^{-3}$) *in vivo*. Variation of external parameters is beginning to produce valuable information about this problem. Geacintov *et al.* (1971) have found a marked effect of magnetic fields on the polarization in *Chlorella* and several other organisms and cells, and Mar and Govindjee (1972) find an effect due to the addition of poisons in *Chlorella* and *Porphyridium*. It is clear that the need for a theory of fluorescence polarization in partially ordered systems will soon be felt. In this section we review the theory of fluorescence quenching and try to indicate which elements of the theory can be applied to the polarization problem.

To discuss excition migration in detail we will continue to use the working hypothesis that the incoherent limit is appropriate, in which case we follow the time dependence of the excitation of the molecules by expanding the set of equations (2.8) to cover as much of the system as is necessary. This is the method of Bay and Pearlstein (1963) and of Pearlstein (1966, 1967). It is virtually identical to using a random walk picture (Pearlstein, 1966; Robinson, 1967; Knox, 1968a), and all the power of random walk methods can therefore be employed where necessary.

3.2. Kinetics of Exciton Migration

3.2.1. The Stern–Volmer Equation for Excitons

It is tempting to use straightforward kinetics in dealing with excitons but the results are not highly accurate. To illustrate this, consider a dense solid solution of molecules at a concentration $[B]$ in which a low density $[B^*]$ of excitons exists, $[B^*] \ll [B]$. The time development of $[B^*]$ is determined by

$$\frac{d[B^*]}{dt} = CI - \frac{[B^*]}{\tau} - k[B^*][X] \qquad (3.1)$$

where I is the intensity of the light and C has appropriate factors and dimensions to make CI a volume rate of exciton production; $1/\tau$ is the monomolecular decay rate, $[X]$ is a density of traps, and k is a bimolecular rate constant for exciton–trap collisions resulting in loss of an exciton. The assumption of a distribution of traps all of which are relatively accessible to all lattice sites is a "lake" model, as contrasted to a model in which each trap has a specific group of bulk chlorophylls which can provide energy to it alone (the "puddle" model). We consider $[X]$ constant or slowly varying. As is well known, the fluorescence yield in the steady state is found from this equation to be of the Stern–Volmer form,

$$\phi_F = \frac{\phi_F^0}{1 + \tau k[X]} \qquad (3.2)$$

where $\phi_F^0 = \tau/\tau_0$ and τ_0 is the radiative lifetime of the exciton. A derivation of (3.2) may be found, e.g., in the paper of Vredenberg and Duysens (1963), who verified the linear dependence of ϕ_F^0/ϕ_F on $[X]$ in *Rhodospirillum rubrum* (see also Clayton, 1966).

The size of the rate constant k is determined by the rates of at least two separate processes, one the average rate of flow of the excitons to the vicinity of the traps, and the other the rate of capture by the trap once the excitation has arrived. In ordinary simple solution kinetics, the two rates appear as factors, one a velocity of relative approach of $[B^*]$ and $[X]$ and the other a capture across section, so $k = v(B^*)\sigma_x$, where $v(B^*)$ and σ_x are the two quantities just described. In our case, it is reasonable to take $v(B^*) = cFa$ and $\sigma_x = a^2$, where c is the coordination number of the lattice (number of nearest neighbors), a is the lattice constant, and F is the Förster transfer rate between two bulk molecules (see e.g., Knox, 1973). Therefore $k = cFa^3 \approx cF[B]^{-1}$ and the fluorescence yield becomes

$$\phi_F = \frac{\phi_F^0}{1 + cF\tau \dfrac{[X]}{[B]}} \qquad (3.3)$$

This equation can be given the following partially correct interpretation: the denominator includes two terms giving the relative probabilities of fluorescence and trapping during each step of a random walk, since $cF\tau$ is the number of steps taken during the exciton's lifetime and $[X]/[B]$ is the probability that a trap will be encountered. This is only "partially correct" because at low trap densities the exciton's random walk takes it back over many molecules already visited. The quantity $cF\tau$ should be replaced by the number of *different* sites visited during the lifetime of the exciton. For several regular lattices this quantity has been computed (see Montroll, 1964). A convenient way to modify Eq. (3.3) is to replace c by a (smaller) effective coordination number c'. A table of values of c' is found in Knox (1973). Exact calculations using the theory of the next section show that in many practical cases, the use of the low-density effective coordination number c' is an overcorrection; traps at moderate densities such as 1/30 to 1/3000, where trapping competes effectively with fluorescence, also compete rather successfully with "revisiting." In view of the fact that Eq. (3.3) is in any event only a rough approximation for any biological application, it may be used with confidence for calculating orders of magnitude provided that the actual coordination number is replaced by the average of c and c' (see Section 2).

As the traps are bleached, $[X]/[B]$ is reduced to zero and ϕ_F is observed to approach ϕ_F^0 in such a way that ϕ_F^0/ϕ_F is linear in the bleaching fraction (Vredenberg and Duysens, 1963):

$$\phi_F = \frac{\phi_F^0}{1 + cF\tau(1 - \eta) \dfrac{[X]}{[B]}} \qquad (3.4)$$

Here η is the fraction of traps bleached.

The simplest "puddle" model fluorescence yield can be derived directly from Eq. (3.3) by first applying it, as an approximation, to the case of a system of N molecules with one trap, in which case $[X]/[B] = 1/N$:

$$\phi_F \text{ (one unit open)} = \frac{\phi_F^0}{1 + \dfrac{cF\tau}{N}} \tag{3.5}$$

In this model each trap has associated with it a group of molecules (the puddle) from which the excitation cannot escape, so when occupied each unit has a fluorescence yield ϕ_F (closed) $= \phi_F^0$. When a fraction η of the traps are bleached,

$$\phi_F = (1 - \eta) \frac{\phi_F^0}{1 + \dfrac{cF\tau}{N}} + \eta \phi_F^0 \tag{3.6}$$

The dependence of ϕ_F^0/ϕ_F is far from linear in η and is essentially hyperbolic. Further evidence for the "lake model" are the lifetime measurements of Tumerman and Sorokin (1967) and Briantais et al. (1972), who find only a single lifetime as opposed to the two which would be associated with the two components of the yield on a puddle model.

Kinetic models allowing partial interunit transfer have been introduced by Joliot and Joliot (1964), Clayton (1967), Paillotin (1971, 1972), and others. Our purpose here is only to see how well the simple kinetic model works at $\eta = 0$, for which the models coincide. The two models also coincide at $\eta = 1$, where the serious assumption must be made that ϕ_F^0 is the *in vivo* saturation value of ϕ_F; no "dead chlorophyll" or similar model assumptions can be added without introducing more parameters. Generally the saturation value is about 0.10 and ϕ_F ($\eta = 0$) is about 0.02. So $1 + cF\tau[X]/[B]$ is 5, or

$$F \approx \frac{4}{c\tau[X]/[B]}$$

An average coordination number, chosen according to the discussion above, is not far from 4, so the implied Förster transfer rate is simply $[B]/\tau[X]$ or N/τ, where N is the number of bulk molecules per trap. In order of magnitude, then, since τ (chosen arbitrarily as the lifetime of Chl a in solution) is 5 nsec, $F \approx 6 \times 10^{10} \text{sec}^{-1}$ in a unit with 300 antenna molecules per trap; so the average spacing implied by the Förster rate is $R \approx 27$ Å when $\overline{R}_0 = 70$ Å. This value of F is much too small, and the implied spacing much too large, in view of the estimates based on Chl concentrations in the lamellae (see, e.g., Bay and Pearlstein, 1963).

The preceding is just another presentation, essentially, of the problem Pearlstein (1966) discovered after his extensive calculations. Transfer to the trap is very efficient, almost too efficient, so the absolute values of the yield and the ratio of saturated to unsaturated yields probably has to be understood by extending the theoretical model, to include the effect of trap-limited kinetics on the fluorescence yield.

As an example of one of the more interesting recent contributions to this problem, we may mention the paper of Borisov and Godik (1972), who assert on the basis of phase fluorometry measurements that in certain bacteria the lifetime of the fluorescence which they consider "photosynthetic" (as opposed to "background") actually decreases as saturation is approached, and has a very small value at all intensities (less than 1 nsec). They conclude that the Förster mechanism is to be questioned; but it should be clear from our discussion above that a small lifetime might even improve the theoretical situation. It is also possible that transfer between the two fluorescing systems has an effect on the phase shift measurement (cf. Section 4.2). The results of Borisov and Godik are, moreover, in experimental contradiction with those of Govindjee et al. (1972), who find that the lifetime increases along with the fluorescence yield in roughly the same range of intensity. Although the bacterial species were not identical, both groups worked with the genera *Chloropseudomonas* and *Rhodospirillum*.

Lavorel and Joliot (1972) have observed that as trap bleaching progresses, the bleached and unbleached units may condense together in such a way that "islets" of fluorescing and nonfluorescing units are formed. The arrangement of the islets of nonfluorescing (trapping) units may be such that the "lake" of the lake model is divided into puddles. This clearly can happen in one dimension, where the authors are able to give numerical illustrations of the fact that the fluorescence yield depends not only on η itself but on the spatial distribution of bleached units. Whether this idea will be important in higher dimensions, in which the topology may allow the excitation to flow around trapping islets, may be seen in their two-dimensional calculations reportedly in progress.

We do not wish to minimize the importance nor even the inevitability of making model refinements to improve on the theoretical situation, especially those such as Joliot's and Clayton's which have direct access to experimental parameters. However, we do not believe that the internal structure of the exciton flow system has yet been studied quantitatively enough to be considered a closed issue. As we saw earlier, Chl pair configurations can have very low associated transfer rates as well as very large ones, compared with the average which is always assumed. Even the value of $\overline{R}_0$ is not certain within a fairly wide range. We will limit ourselves here to a discussion of the primary excitation flow system in the low bleaching ($\eta = 0$) limit.

3.2.2. Master Equation and Random Walk Approaches

Although the kinetic method just described gives a good qualitative and partially quantitative picture, it is possible to be much more specific about fluorescence yields in systems with linear coupling. The method is an application (Knox, 1968a) to the Förster master equations (Förster, 1948) of the inverse matrix method of Montroll and Shuler (1958). It consists of making each molecule a kinetic unit and collecting all of the decay and transfer processes into a single matrix G which appears in the master equations:

$$\frac{\partial \rho_i}{\partial t} = -\sum_{j=1}^{N} G_{ij}\rho_j + E_i \tag{3.7}$$

Here ρ_i is the probability that molecule i is in the excited state, N is the total number of molecules, E_i is the rate of external excitation of molecule i, and G_{ij} is a general set of constant coupling coefficients; for example, in practice, when $i \neq j$ the matrix element G_{ij} is $-F_{ij}$, where F_{ij} is the rate of transfer from molecule j to molecule i computed with (say) the Förster theory. Also, when $i = j$ the diagonal element G_{ij} includes not only all the outgoing transfer rates $\Sigma_j F_{ji}$ but also an abitrary monomolecular decay rate $1/\tau_i$ which can vary from molecule to molecule. Examples of the transfer matrix are given by Knox (1968a) and by Pearlstein (1972a).

Recently a very extensive development of the master equation approach has been presented by Hemenger et al. (1972). In addition to providing an elegant formation in terms of Green's functions for a perfect lattice, they provide precise quenching formulas including specific back-transfer parameters and a demonstration that except at very high concentrations, the degree of randomness of trap locations does not affect the rates.

The steady-state solutions of Eq. (3.7) are most conveniently written in terms of the matrix inverse to G.

$$\rho_j(\text{steady state}) = \sum_{k=1}^{N} (G^{-1})_{jk} E_k \tag{3.8}$$

from which the fluorescence yield follows at once (Knox, 1968a, 1973; Appendix of this chapter):

$$\phi_F = \sum_{j=1}^{N} \sum_{k=1}^{N} \frac{1}{\tau_{0j}} (G^{-1})_{jk} f_k \tag{3.9}$$

where f_k is the fraction of light absorbed by molecule k and $1/\tau_{0j}$ is the natural radiative rate of molecule j. This equation has been applied to lattices in which comparable simple kinetics of the type described in the last section could be used, and the exact results lie between curves computed with the true coordination number and an effective coordination number (Knox, 1973).

Equation (3.8) has also been applied to model systems in which the dipolar form of the Förster interaction is not averaged (Knox, 1973). In a periodic square array with one trap for every 25 sites, the trapping time can be smaller by as much as a factor of $\frac{1}{3}$ or larger by a factor of 8 than that calculated with the Förster average ($\kappa^2 = \frac{2}{3}$). If the molecules are fairly tightly packed, so that the monopole effects become important, these factors change to $\frac{1}{6}$ and 16. Hemenger et al. (1972) have discussed general conditions under which anisotropy may be expected to affect the kinetic dimensionality of trapping rates, and they conclude that only extreme anisotropy does so.

Equation (3.8) may also be used to set up a rigorous theory of depolarization by transfer. Since E_k is the rate of external excitation of molecule k, it is proportional to $(\mu_k \cdot \mathbf{z})^2$ in the notation of Section 2, and the intensity equations (2.10) become

$$
\begin{aligned}
I_z &= C\sum_j \sum_k (\mu_j \cdot \mathbf{z})^2 (G^{-1})_{jk} (\mu_k \cdot \mathbf{z})^2 \\
I_y &= C\sum_j \sum_k (\mu_j \cdot \mathbf{y})^2 (G^{-1})_{jk} (\mu_k \cdot \mathbf{z})^2
\end{aligned}
\tag{3.10}
$$

These may in turn be inserted into Eq. (2.9) and an exact calculation made in principle. The role of G^{-1} as a coupler between absorbers and emitters of radiation is even more striking here than in the fluorescence yield equation. Where there is no transfer, G^{-1} is diagonal and there is complete correlation between the absorbers (k) and emitters (j). When there is widespread transfer, G^{-1} is more homogeneous, off-diagonal, and independent of j and k, or so it would seem, because in the presence of appreciable transfer in random dipolar arrays the two sums in I_z and I_y may be done independently to yield $p = 0$.

Several calculations of trapping time in the photosynthetic unit have been done explicitly in terms of random walk theory (Robinson, 1967; Montroll, 1969; Sanders et al., 1971). The master equation is a rather specialized form of this approach which is more easily related to practical kinetic considerations, such as finite lifetimes of the bulk chlorophylls. For the most part, the connection between the master equation results and random walk results is a matter of interconnecting parameters such as jump times and rate constants (Knox, 1968a; Sanders et al., 1971), but Lakatos-Lindenberg et al. (1972) have observed that under certain conditions an oscillation can occur in the one-dimensional random walk case which does not occur in the master equation solution. They believe that this oscillation will generally be unimportant and the two approaches identical in their predictions as long as one has nonlocalized initial conditions, nonzero probability per step that the walker stays at the same (bulk) site, or other randomizing processes such as variable stepping times and hopes to distant neighbors. Lakatos-Lindenberg *et al.* give a very thorough discussion of the connection between the two points of view.

Seely (1972, 1973) has used the probability-matrix method (Robinson, 1967) to calculate trapping times in several theoretical model systems designed to show the effect of variable transfer rates and transition dipole orientations. In particular, he shows that if different spectral forms of Chl are arranged in bands or layers about the reaction center, the irreversibility introduced because of favorable spectral overlapping can increase the trapping rate by a factor of 4 or 5. His calculations essentially add a very specific molecular model to the earlier kinetic considerations of Borisov (1967).

3.3. Dynamics of Exciton Migration

It is now well established that excitons in molecular crystals, moving in essentially the same way as we have described such motion in the photosynthetic unit, may collide and destroy each other. The product is radiation or a new exciton, usually accompanied by heat to conserve energy. The existence of the inverse process (fission of an exciton, with the help of thermal energy, into two excitons) is also well established. Much of the precise work on triplet exciton collisions has been done by the du Pont group (for a review of theirs and other work see Avakian and Merrifield, 1968). A recent comprehensive review of exciton dynamics is by Swenberg and Geacintov (1973).

Exciton collisions add a new decay term $-\gamma[B^*]^2$ to Eq. (3.1) and a corresponding term in $\rho_i\rho_j$ to the master equation (3.7). In crystalline anthracene the value of γ for singlet and triplet excitons is in the range of 10^{-8} to 10^{-11} cm^3 sec^{-1}, and so we see that to compete with the natural decay rate of 10^9 sec^{-1}, the product $\gamma[B^*]$ must be 10^9 sec^{-1} or $[B^*] \sim 10^{17}$ to 10^{20} cm^{-3}. This is a very high exciton density, one which is seldom encountered in photosynthetic research where perhaps 10^{10} to 10^{11} cm^{-3} is common. It seems fairly certain, therefore, that collisions are not an important decay mode in the bulk. However, for three reasons they are worthy of mention here: their role in a model of delayed light production, their probable role in limiting coherent energy storage modes, and the quenching of singlets by triplets.

The existence of delayed light in anthracene, explained fully by exciton collisions (see Avakian and Merrifield, 1968) suggested that the phenomenon of delayed light in photosynthesis (Strehler and Arnold, 1951) might have a similar origin. Noting, however, that the low-exciton densities mentioned above preclude a simple interpretation, Stacy *et al.* (1971) suggested a somewhat more complex model depending on feedback of triplet energy from traps to the bulk. They were unable to observe in *Chlorella* the large and characteristic magnetic field effects on exciton–exciton collision probabilities which are observed in the polyacene crystals. Nonetheless, their paper is important as a reminder that triplet states may still have a role to play in photosynthesis, despite the general failure of models based on triplets in earlier days (see Clayton, 1965, pp. 178–182), and despite some recent measurements on intact chloroplasts which indicate very short lifetimes (~ 30 nsec) for Chl triplets (Breton and Mathis, 1970; Mathis, 1971).

Collisions would probably prevent the establishment of the interesting storage modes proposed by Fröhlich (1968, 1969) as we have argued in detail elsewhere (Knox, 1969). Although the Fröhlich theory includes bimolecular exciton terms, they are exciton-number conserving, and dissipative terms such as $-\gamma[\text{B}^*]^2$ are ignored.

The third bulk exciton collision phenomenon is a singlet–triplet fusion process. We have recently shown (Rahman and Knox, 1973) that resonance transfer of energy from an excited singlet to a triplet, raising the latter to a higher triplet, is a highly efficient process which is due to the Förster interaction. The triplet–triplet transition is strongly dipole allowed, and the corresponding Förster radius in chl a $\overline{R}_0^{st}$ (analogous to $\overline{R}_0$ of Section 2) is 39 Å. This means that triplets, wherever they occur in the bulk, can act as very efficient quenchers of singlets. They therefore carry with them a region of destruction which covers an equivalent of more than 75 sites on a simple cubic lattice of lattice constant 15 Å.

4. SPECIAL TOPICS

4.1. The Approach to Strong Coupling

The oldest theory of transfer of electronic excitation energy was that of the Perrins (J. Perrin, 1925, 1927; F. Perrin, 1932). It was based on the same dipole–dipole interaction $J = \mu^2/R^3$ that causes the Förster transfer, but it assumed coherence between the excitations on the donor and acceptor molecules. Förster recognized that the coherence assumption, not the form of the interaction, was at the root of the difficulties of the Perrins' theory. The quantitative difficulty was the prediction of much too large a characteristic R_0. The transition from a theory of coherent transfer was made and carefully explained in a series of publications by Förster (1946, 1948, 1951). The physical arguments are as follows: if the excited acceptor state has a sufficiently long lifetime, compared with $\hbar/J$, then immediate transfer back to the donor must be considered and a self-consistent quantum-mechanical state constructed. But the excited acceptor state, because of a great multiplicity of vibrational states, has a very short lifetime against transitions into neighboring states of nearly the same energy but whose phase need have no relation to the initial one. The lifetime against this loss of phase is of the order of magnitude of $\hbar/\Delta E$, where ΔE is a vibrational bandwith. Consequently, when $\Delta E > J$ the incoherent theory is appropriate. Only when there are no effects which disturb the phase relation between the donor and acceptor states will the R^{-3} interaction be observable, so $\Delta E > J$ is a weak condition. Even if J and ΔE are comparable, or J is somewhat larger than ΔE, there is no guarantee that energy is being transferred at a rate related to R^{-3}. It is sometimes said that R^{-3} transfer has been observed as a Davydov splitting; however, this only shows coherence of the wavefunction over a unit cell or over the donor–acceptor pair, and transfer in the usual sense of an energy transporting process is unnecessary in the first place!

Over the past 10 or 12 years there have been some attempts to reintroduce coherent excitation transfer into photosynthesis theory, either in terms of an exciton model of the unit (Robinson, 1964, 1967; Pearlstein, 1968) or as a correction for special effects in donor–acceptor transfer (e.g., Bauer *et al.*, 1972). In the former case, we know of no firmly predicted observables which do not depend on arbitrary parameters and rate constants and by which the suitability of the coherent model could be tested; on the other hand, this same statement applies to the incoherent model. Indeed, at a fixed temperature it is unlikely that the two models can be distinguished in principle, as was argued some time ago by Avakian *et al.* (1968). The situation is much better in the case of molecular crystals, where diffusion of excitons is well characterized as a tensor quantity (Mulder, 1968) and the theory of transfer to traps from the bulk, while not without its difficulties, seems firmly based on the incoherent model at ordinary temperatures (Powell and Kepler, 1969; Birks, 1970b; Pearlstein *et al.*, 1971).

The specific references to temperature in the preceding paragraph are prompted by the recent appearance of two very formal treatments (Haken and Reineker, 1968, 1971; Grover and Silbey, 1971) which combine the coherent and incoherent limits. In Haken and Reineker's treatment, for example, the diffusion coefficient in a linear chain model contains explicit contributions from both types of motion:

$$D = R^2 \left[F + \frac{\frac{1}{2}(J/\hbar)^2}{F + 2\Gamma} \right] \tag{4.1}$$

where R is the lattice constant, J is the transition dipole–dipole interaction, Γ is the halfwidth of the absorption band of the molecule, and F is the transfer rate in the purely incoherent limit, called $2\gamma_1$ by Haken and

Reineker. Whether the first or second term is dominant depends on the proportionality constant between F and J^2, which in turn depends heavily on the extent of overlap between the absorption and emission spectra of the molecules and other temperature-dependent factors. However, it is clear that D itself is basically proportional to J^2; and that Eq. (4.1) reduces to one of the usual two limiting cases when one term or the other is dropped. When $F = 0$ the diffusion constant is proportional to J^2/Γ, which is essentially the product of the square of the group velocity ($v \propto J$) and a scattering lifetime ($\tau \propto 1/\Gamma$) appearing in transport theories of the exciton diffusion (Agranovich and Konobeev, 1968). When J is formally set equal to zero in Eq. (4.1), the one-dimensional form of the incoherent transfer diffusion constant (Pearlstein, 1966) is obtained, $D = R^2 F$. Haken and Reineker have applied the theory to excitons in solid anthracene and conclude that the first term dominates at high temperatures, while the second ("coherent") term dominates at low temperatures. Agranovich and Konobeev had earlier placed an upper limit of $T = 100°K$ on the validity of their calculations in the coherent limit.

Two additional theoretical treatments have been published recently, one by Hemenger, Lakatos-Lindenberg, and Pearlstein (1974), which applies the Grover-Silbey theory to a lattice with traps, and one by Kenkre and Knox (1974), which is an extension of Förster's theory to include coherence effects through the use of a non-Markoffian rate equation.

Recent low-temperature studies on *Chlorella* bearing on exciton transport (Cho and Govindjee, 1970, Mar *et al.*, 1972) might possibly be analyzed in terms of these new theoretical approaches, but the temperature effect is hardly as straightforward as in crystalline materials. The broad lines, relatively large spacings, and disorder in transition dipole orientations which probably exist in the photosynthetic unit chlorophyll array, along with the fact that the ambient temperature is well above the region for coherent excitons in ordered crystals, make it seem quite unlikely that coherent excitons have any real importance in the photosynthetic mechanism, as Duysens (1952) surmised. Of course we refer to coherence over relatively large distances, not to the coherence which is associated with the clusters of chlorophyll in the chlorophyll-protein complexes of the pebble-mosaic model (see Chapter 3). From our *transport* point of view these clusters are the single molecules of exciton theory.

4.2. Excitation Transfer and Phase Fluorometry

The finite lifetime of molecular excitations produces a lag between absorption and emission, and this lag can be used to measure the lifetime (Bailey and Rollefson, 1953). The well-known relationship between the lifetime, τ, the angular frequency ω at which the incident light is modulated, and the phase of the emitted light's modulation with respect to that of the incident light is

$$\tan \delta = -\omega\tau \tag{4.2}$$

The method can be extended to the case of two uncoupled absorbers of lifetime τ_1 and τ_2 each of which absorb a certain fraction of the incident light given by f_1 and f_2, respectively; the result is (see, e.g., Pearlstein, 1966)

$$\tan \delta = \frac{(1 + \tan^2 \delta_2)f_1\phi_1 + (1 + \tan^2 \delta_1)f_2\phi_2 \tan \delta_2}{(1 + \tan^2 \delta_2)f_1\phi_1 + (1 + \tan^2 \delta_1)f_2\phi_2} \tag{4.3}$$

or an equivalent form given by Tumerman and Sorokin (1967). In Eq. (4.3), δ is the observed phase shift and δ_1 and δ_2 are the phase shifts which would be measured if the two components were measured separately, i.e., $\tan \delta_i = -\omega\tau_i$.

Usually if one were interested in measuring two lifetimes he would not go out of his way to mix the two molecules or subsystems in the first place. Therefore, Eq. (4.3) is usually applied to situations in which systems 1 and 2 are inseparably mixed and to obtain the correct answer one must hope excitation transfer is absent or small. The derivation of Eq. (4.3) even without transfer is tedious, and *with* transfer the result is prohibitively complicated for ordinary algebraic manipulation. We present here a method which puts the theory of $\tan \delta$ in a computer-ready form and allows its calculation for any arbitrarily large system of arbitrarily coupled emitters (provided the kinetics are linear). The formalism is also applied to an important configuration from which a transfer rate can be deduced directly from phase shift measurements.

4.2.1. Theory: The Complex Transfer Matrix

Consider a sytem containing N subsystems, each with N_k molecules of a single absorbing type. Illumination of intensity I_k falls on subsystem k and the subsystem has an absorption coefficient B_k. (While a single source of illumination is generally presumed, the various subsystems might absorb at different wavelengths in practice, or some might be in the shadow of others, so that all the I_k need not be the same.) Now if $n_k(t)$ is the number of excited molecules at time t in subsystem k, in the presence of illumination, transfer, and decay the most general linear Markoffian equation for the n_k is

$$\frac{dn_k}{dt} = I_k B_k N_k - \sum_{l=1}^{N} G_{kl} n_l(t) \tag{4.4}$$

The coefficients G_{kl} are defined in the same way as those appearing in Eq. (3.7), this time referring more generally to subsystems which may be larger than single molecules. Let us suppose that any one of the three quantities I_k, B_k, or N_k is modulated (simultaneously for all subsystems) in such a way that Eq. (4.5) holds for all k:

$$\frac{dn_k}{dt} = I_k^0 B_k^0 N_k^0 (1 + \lambda \cos \omega t) - \sum_{l=1}^{N} G_{kl} n_l(t) \qquad (4.5)$$

It is shown in the Appendix that the phase of the emission relative to that of the excitation modulation is given by

$$\delta = \arg \sum_{j=1}^{N} \sum_{k=1}^{N} A_j (\mathscr{G}^{-1})_{jk} f_k \qquad (4.6)$$

where A_j is the natural radiative rate of subsystem j, f_k is the fraction of light absorbed by subsystem k on the average:

$$f_k = \frac{I_k^0 B_k^0 N_k^0}{\sum\limits_{l=1}^{N} I_l^0 B_l^0 N_l^0} \qquad (4.7)$$

and $\mathscr{G}$ is a complex transfer matrix

$$\mathscr{G} = G + i\omega \, \mathbf{1} \qquad (4.8)$$

in which $i = \sqrt{-1}$, $\mathbf{1}$ is the unit matrix, and G consists of the array of rate constants G_{kl}.

4.2.2. Application to a Two-Component System

Consider the two-component system illustrated in Fig. 7. Both components absorb, in relative amounts f_1 and $f_2 (f_1 + f_2 = 1)$, and transfer occurs at a rate F from subsystem 1 to subsystem 2. Only the latter's radiation is observed in the output ($A_1 = 0$, $A_2 \neq 0$). The complex transfer matrix, Eq. (4.8), has the form

$$\mathscr{G} = \begin{pmatrix} k_1 + F + i\omega & 0 \\ -F & k_2 + i\omega \end{pmatrix} \qquad (4.9)$$

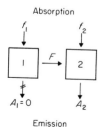

Absorption

Emission

Fig. 7. Schematic of a coupled system in which subsystems 1 and 2 absorb fractions f_1 and f_2, respectively, of the total absorbed light, but only subsystem 2 emits. Excitation transfer proceeds with a rate constant F from 1 to 2.

where k_1 and k_2 are the total decay constants for uncoupled subsystems 1 and 2, respectively (k_1 does not include F but k_2 includes A_2). Upon inverting the matrix and using Eq. (4.6), we obtain

$$\delta_{f_1 f_2} = \arg \left[A_2 \frac{F}{(k_2 + i\omega)(k_1 + F + i\omega)} f_1 + A_2 \frac{1}{k_2' + i\omega} f_2 \right] \quad (4.10)$$

where $\delta_{f_1 f_2}$ is the phase shift predicted when the absorbed fractions are f_1 and f_2. If these absorbed fractions can be varied at will, by using selective filters or other means, then all three rate constants can be measured directly in terms of three phase shifts. From Eq. (4.10) we have, for three special cases,

$$\delta_{01} = -\tan^{-1}(\omega/k_2)$$

$$\delta_{10} = -\tan^{-1}[\omega/(k_1 + F)] + \delta_{01} \quad (4.11)$$

$$\delta_{\frac{1}{2}\frac{1}{2}} = \tan^{-1}[\omega/(k_1 + 2F)] + \delta_{10}$$

and so

$$F = \omega[\cot(\delta_{\frac{1}{2}\frac{1}{2}} - \delta_{10}) + \cot(\delta_{10} - \delta_{01})] \quad (4.12)$$

obviously, k_2 can be obtained directly from δ_{01}, and k_1 can be extracted easily. This result shows that phase fluorometry can be used to measure a transfer rate directly, provided the absorption fractions can be varied; the theory is not necessarily a hopeless algebraic jungle, and if it shows any danger of becoming one, calculations can be performed directly by computer, using the complex transfer matrix and Eq. (4.6).

The derivation of Eq. (4.12) was stimulated by conversations with Profs. Govindjee and C. Swenberg during a discussion of the possibility of learning the rate of excitation transfer from carotene (subsystem 1) to bacteriochlorophyll (subsystem 2). It has to be admitted that the extraction of F is not as simple when both subsystems contribute to the fluorescence and their emissions cannot be separated efficiently. The appropriate term to be added to Eq. (4.10) inside the brackets is $A_1(k_1 + F + i\omega)^{-1} f_1$, and the phase shifts become dependent on A_1/A_2. By then setting $F = 0$ the Pearlstein and Tumerman–Sorokin results, Eq. (4.3), can be obtained.

5. SUMMARY

In this chapter several particular aspects of the theory of excitation transfer and migration have been reviewed. In the introduction a comparison between the situation in solids, for which exciton theory is developed, and the more complex case of biological systems is given. Section 2 is concerned with the quantitative description and measurement of the rate of pairwise excitation transfer between like molecules, with particular attention to chlorophyll *a*. For this molecule a comparison of a series of published measurements is made, and it is argued that considerable analysis will be required before the rate is known accurately. At present there is a possible experimental spread of over a factor of 50. The effects on the rate of relative orientations of the donor and acceptor transition dipole moments, usually submerged by angular averages, are illustrated explicitly. Corrections to the rate due to the finite extent of the porphyrin ring, recently calculated by J. Chang, are described. These corrections show that the anisotropy of the rate at small intermolecular separations is even greater than that predicted for point dipoles. Section 3 reviews some of the most elementary means by which pairwise rates can be used to construct a theory of excitation migration, and it is emphasized that the variability of the pairwise rate is a two-edged sword: the fluorescence yields computed with simple models may be unreliable, but on the other hand models with built-in efficiency (such as those of Seely) can be constructed without too many arbitrary parameters. Section 4 embraces two special topics. The first is the question of whether excitation energy transport proceeds in a coherent or incoherent manner, and it is argued that the question is moot in photosynthetic systems as long as one is careful to distinguish between the local coherence present in chlorophyll–protein complexes and the character of the actual transport of energy over large distances. Finally, we present a new theory of phase fluorometry which may be useful in determining rates of energy transfer between fluorescing species.

ACKNOWLEDGMENTS

The author would like to record his sincere gratitude to Drs. Govindjee, R. P. Hemenger, K. Lakatos-Lindenberg, R. Pearlstein, and C. Swenberg for their assistance in correcting and recommending improvements in the first draft of the manuscript.

REFERENCES

Agranovich, V. M., and Konobeev, Yu. V. (1968). *Phys. Status Solidi* **27**, 435.
Arnold, W., and Meek, E. S. (1956). *Arch. Biochem. Biophys.* **60**, 82.
Avakian, P., and Merrifield, R. E. (1968). *Mol. Cryst.* **5**, 37.
Avakian, P., and Suna, A. (1971). *Mat. Res. Bull.* **6**, 891.
Avakian, P., Ern, V., Merrifield, R. E., and Suna, A. (1968). *Phys. Rev.* **165**, 974.
Bailey, E. A., and Rollefson, G. K. (1953). *J. Chem. Phys.* **21**, 1315.
Baldini, G. (1962). *Phys. Rev.* **128**, 1562.
Bauer, R. K., Szalay, L., and Tombacz, E. (1972). *Biophys. J.* **12**, 731.
Bay, Z., and Pearlstein, R. M. (1963). *Proc. Nat. Acad. Sci. (US)* **50** 1071.

Birks, J. B. (1970a). "Photophysics of Aromatic Molecules," Chap. 11. Wiley-Interscience, New York.

Birks, J. B. (1970b). *J. Phys. (B)* **3**, 1704.

Borisov, A. Yu (1967). *Biophysics* **12**, 727.

Borisov, A. Yu., and Godik, V. I. (1972). *J. Bioenergetics* **3**, 211.

Breton, J., and Mathis, P. (1970). *Compt. Rend. Acad. Sci. (Paris)* **271**, 1094.

Briantais, J. M., Merkelo, H., and Govindjee (1972). *Photosynthetica* **6**, 133.

Chang, J. C. (1972). Ph.D. Thesis, University of Rochester, Rochester, N. Y.

Cho, F., and Govindjee (1970). *Biochim. Biophys. Acta* **205**, 371; *216*, 139.

Clayton, R. K. (1965). "Molecular Physics in Photosynthesis," Blaisdell, New York.

Clayton, R. K. (1966). *Photochem. Photobiol.* **5**, 807.

Clayton, R. K. (1967). *J. Theoret. Biol.* **14**, 173.

Craver, F. W. (1971a). Ph.D. Thesis, University of Rochester, Rochester, New York.

Craver, F. W. (1971b). *Molec. Phys.* **72**, 403.

Craver, F. W., and Knox, R. S. (1971). *Molec. Phys.* **72**, 385.

Davydov, A. S. (1948). *Zh. Eksp. Teor. Fiz.* **18**, 210.

Dexter, D. L. (1953). *J. Chem. Phys.* **21**, 836.

Duysens, L. N. M. (1952). Thesis, the University of Utrecht, Utrecht, The Netherlands.

Duysens, L. N. M. (1964). *Prog. Biophys.* **14**, 1.

Eisinger, J., and Lamola, A. A. (1971). *In* "Excited States of Proteins and Nucleic Acids" (R. F. Steiner and I. Weinryb, eds.), p. 107. Plenum, New York.

Emerson, R., and Arnold, W. (1932). *J. Gen. Physiol.* **15**, 391; **16**, 191.

Förster, Th. W. (1946). *Naturwissenschaften* **33**, 166.

Förster, Th. W. (1948). *Ann. Physik* [6] **2**, 55.

Förster, Th. W. (1951). "Fluoreszenz Organischer Verbindungen," Van den Hoek and Rupprecht, Göttingen.

Förster, Th. W. (1965). *In* "Modern Quantum Chemistry, Part III: Action of Light and Organic Molecules" (O. Sinanoglu, ed.), p. 93. Academic, New York.

Franck, J., and Teller, E. (1938). *J. Chem. Phys.* **6**, 861.

Frenkel, J. I. (1931). *Phys. Rev.* **37**, 17, 1276.

Frenkel, J. I. (1936). *Physik. Z. Sowjet.* **9**, 158.

Fröhlich, H. (1968). *Nature* **219**, 743.

Fröhlich, H. (1969). *Nature* **221**, 976.

Galanin, M. D. (1950). *Tr. Inst. Fiz. Akad. Nauk USSR* **5**, 341.

Geacintov, N., Van Nostrand, F., Pope, M., and Tinkel, J. B. (1971). *Biochim. Biophys. Acta* **226**, 486.

Goedheer, J. C. (1957). Ph.D. Thesis, The University of Utrecht, Utrecht, The Netherlands.

Goedheer, J. C. (1972). *Ann. Rev. Plant Physiol.* **23**, 87.

Golebiowski, A., and Witkowski, A. (1959). *Rocznicki Chem.* **33**, 1443.

Govindjee and Papageorgiou, G. (1971). In *Photophysiology: Current Topics in Photobiology and Photochemistry*, Vol. 6 (A. C. Giese, ed.), Academic, New York, pp. 2–46.

Govindjee, Papageorgiou, G., and Rabinowitch, E. (1967). *In* "Fluorescence Theory, Instrumentation, and Practice" (G. G. Guilbault, ed.), Marcel Dekker, New York, pp. 511–564.

Govindjee, Hammond, J. H., and Merkelo, H. (1972). *Biophys. J.* **12**, 809.

Grover, M., and Silbey, R. (1971). *J. Chem. Phys.* **54**, 4843.

Guéron, M., and Shulman, R. G. (1968). *Ann. Rev. Biochem.* **37**, 571.

Haken, H., and Reineker, P. (1968). *In* "Excitons, Magnons and Phonons in Molecular Crystals" (A. B. Zahlan, ed.), pp. 185–194. Cambridge Univ. Press, London.

Haken, H., and Reineker, P. (1971). *Z. Physik.* **249**, 253.

Hemenger, R. P., and Pearlstein, R. M. (1971). Abstracts of the Conference on Primary Photochemistry of Photosynthesis, Argonne, Ill., Nov. 17–19, 1971, p. 14 (unpublished).

Hemenger, R. P., Pearlstein, R. M., and Lakatos-Lindenberg, K. (1972). *J. Math. Phys.* **13**, 1056.

Hemenger, R. P., Lakatos-Lindenberg, K., and Pearlstein, R. M. (1974). *J. Chem. Phys.* (in press).

Hochstrasser, R. M. (1971). *In* "Excited States of Proteins and Nucleic Acids" (R. F. Steiner and I. Weinryb, eds.), pp. 1–29 Plenum, New York.

Jackson, J. D. (1962). "Classical Electrodynamics," Wiley, New York.

Joliot, A., and Joliot, P. (1964). *Compt. Rend. Acad. Sci.* (*Paris*) **258**, 4622.

Kelly, A., and Patterson, L. K. (1971). *Proc. Roy. Soc.* (*London*) **A324**, 117.

Kelly, A., and Porter, G. (1970). *Proc. Roy. Soc.* (*London*) **A315**, 149.

Kenkre, V. M., and Knox, R. S. (1974). *Phys. Rev. B* (in press).

Kepler, R. G., Caris, J. C. Avakian, P., and Abramson, E. (1963). *Phys. Rev. Lett.* **10**, 400.

Knox, R. S. (1963). *Solid State Phys. Suppl.* **5**.

Knox, R. S. (1968a). *J. Theoret. Biol.* **21**, 244.

Knox, R. S. (1968b). *Physica* 39, 361.

Knox, R. S. (1969). *Nature* **221**, 263.

Knox, R. S. (1973). "Primary Molecular Events in Photobiology," pp. 45–79. *Proc. NATO Advanced Study Inst. Photobiol. Badia Fiesolana, 1972.* Elsevier, Amsterdam.

Lakatos-Lindenberg, K., Hemenger, R. P., and Pearlstein, R. M. (1972). *J. Chem. Phys.* **56**, 4852.

Lavorel, J., and Joliot, P. (1972). *Biophys. J.* **12**, 815.

Levshin, V. L. (1925). *Z. Physik* **32**, 307.

London, F. (1942). *J. Phys. Chem.* **46**, 305.

Mar, T., and Govindjee (1972). *Proc. Int. Congr. Photosynthesis Res. 2nd, The Hague* **1**, 271.

Mar, T., Govindjee, Singhal, G. S., and Merkelo, H. (1972). *Biophys. J.* **12**, 797.

Mathis, P. (1971). *Abstracts of the Conference on Primary Photochemistry of Photosynthesis, Argonne, Ill.*, Nov. 17–19, 1971, p. 15 (unpublished).

McRae, E. G., and Kasha, M. (1958). *J. Chem. Phys.* **28**, 721.

Montroll, E. W. (1964). *Proc. Symp. Appl. Math. (Am. Math. Soc.)* **16**, 193.

Montroll, E. W. (1969). *J. Math. Phys.* **10**, 753.

Montroll, E. W., and Shuler, K. (1958). *Adv. Chem. Phys.* **1**, 361.

Mott, N. F. (1938). *Trans. Faraday Soc.* **34**, 500.

Mulder, B. J. (1968). *Philips Res. Rpt. Suppl.* **4**, 1.

Ore, A., and Eriksen, E. L; (1971). *Phys. Norveg.* **5**, 57 (1971), and references therein.

Paillotin, G. (1971), *Proc. Int. Cong. Photosynthesis Res., 2nd, The Hauge*, **1**, 331.

Paillotin, G. (1972). *J. Theoret. Biol.* **36**, 223.

Pearlstein, R. M. (1966). Ph.D. Thesis, University of Maryland, College Park, Maryland.

Pearlstein, R. M. (1967). *Brookhaven Symp. Biol. (BNL 608)* **19**, 8.

Pearlstein, R. M. (1968). *Photochem. Photobiol.* **8**, 341.

Pearlstein, R. M. (1972a). *J. Chem. Phys.* **56**, 2431.

Pearlstein, R. M. (1972b). *Bull. Am. Phys. Soc.* [2] **17**, 31.

Pearlstein, R. M., Hemenger, R. P., and Lakatos-Lindenberg, K. (1971). *Phys. Rev. Lett.* **27**, 1509.

Peierls, R. E. (1932). *Ann. Physik* [5] **13**, 905.

Perrin, F. (1926). *J. Phys Radium* 7, 390.

Perrin, F. (1929). *Ann. Phy.* (*Paris*) **12**, 169.

Perrin, F. (1932). *Ann. Chim. Phys.* **17**, 283.

Perrin, J. (1925). *Conseil Chim. Solvay, 2nd, Bruxelles, 1924* Gauthier-Villars, Paris, p. 322.

Perrin, J. (1927). *Compt. Rend. Acad. Sci.* (*Paris*) **184**, 1097.

Powell, R. C., and Kepler R. G. (1969). *Phys. Rev. Lett.* **22**, 636.

Rabinowitch, E. (1956). "Photosynthesis and Related Processes," Vol. 2, Part 2, Interscience, New York.

Rahman, T. S., and Knox, R. S. (1973), *Phys. Status Solidi (b)* **58**, 715.

Reed, D. W., and Clayton, R. K. (1968). *Biochem. Biophys. Res. Comm.* **30**, 471.

Robinson, G. W. (1964). *Ann. Rev. Phys. Chem.* **15**, 311.

Robinson, G. W. (1967). *Brookhaven Symp. Biol. (BNL 608)* **19**, 16.

Robinson, G. W. (1970). *Ann. Rev. Phys. Chem.* **21**, 429.

Sanders, J. W., Ruijgrok, Th. W., and ten Bosch, J. J. (1971). *J. Math. Phys.* **12**, 534.

Sane, P. V., and Park, R. B. (1970). *Biochem. Biophys. Res. Commun.* **41**, 206.

Scheibe, G. (1939). *Z. Angew. Chem.* **52**, 631.

Seely, G. R. (1970). *J. Phys. Chem.* **74**, 219.

Seely, G. R. (1972). *Proc. Int. Congr. Photosynthesis Res. 2nd, The Hague* **1**, 341.

Seely, G. R. (1973). *J. Theoret. Biol.* **40**, 173, 189.

Simpson, O. (1956). *Proc. Roy. Soc. (London)* **A238**, 402.

Stacy, W. T., Mar, T., Swenberg, C., and Govindjee (1971). *Photochem. Photobiol.* **14**, 197.

Strehler, B. L., and Arnold, W. (1951). *J. Gen. Physiol.* **34**, 807.

Strickler, S. J., and Berg, R. A. (1962). *J. Chem. Phys.* **37**, 814.

Stryer, L., and Haugland, R. P. (1967). *Proc. Nat. Acad. Sci. (US)* **58**, 719.

Swenberg, C. E., and Geacintov, N. E. (1973). *In* "Organic Photophysics" (J. B. Birks, ed.), Chapter 10. Wiley-Interscience, New York.

Thomas, D. G., and Hopfield, J. J. (1961). *Phys. Rev.* **124**, 657.

Thornber, J. P., and Olson, J. M. (1971). *Photochem. Photobiol.* **14**, 329.

Trosper, T., Park, R. B., and Sauer, K. (1968). *Photochem. Photobiol.* **7**, 451.

Tumerman, L. A., and Sorokin, E. M. (1967). *Mol. Biol.* **1**, 628.

Tweet, A. G., Bellamy, W. D., and Gaines, G. L., Jr. (1964). *J. Chem. Phys.* **41**, 2068.

Vredenberg, W. J., and Duysens, L. N. M. (1963). *Nature* **197**, 355.

Wannier, G. H. (1937). *Phys. Rev.* **52**, 191.

Weber, G. (1952). *Biochem. J.* **51**, 145.

Weber, G. (1960a). *Biochem. J.* **75**, 335.

Weber, G. (1960b). *In* "Comparative Biochemistry of Photoreactive Systems" (M. B. Allen, ed.), Academic, New York, 1960.

Weber, G. (1972). *Ann. Rev. Biophys. Bioeng.* **1**, 553.

Weber, G., Borris, D. P., DeRobertis, E., Barrantes, F. J., La Torre, J. L., and Llorente DeCarlin, M. C. (1971). *Mol. Pharmacol.* **7**, 530.

Zankel, K. L., and Clayton, R. K. (1969). *Photochem. Photobiol.* **9**, 7.

APPENDIX: Theory of Phase Fluorometry

To derive Eq. (4.6), we begin by rewriting (4.5) in complex form, with $I_k^0 B_k^0 N_k^0$ taken to be real:

$$\frac{dn_k}{dt} = I_k^0 B_k^0 N_k^0 (1 + \lambda e^{i\omega t}) - \sum_{l=1}^{N} G_{kl} n_l(t) \tag{A.1}$$

The solutions will be complex, and we will seek those of the form

$$n_k(t) = n_k^0(t) + \lambda n_k^1(t) e^{i\omega t} \tag{A.2}$$

where $n_k^0(t)$ is the solution of the "homogeneous" problem produced by setting $\lambda = 0$. The quantities n_k^1 are complex but can be taken to be time-independent,* and they satisfy the following equation, which is found by substituting (A.2) into (A.1)

$$\sum_{l=1}^{N}(G_{kl} + i\omega\delta_{kl})n_{l}^{1} = I_{k}^{0}B_{k}^{0}N_{k}^{0} \tag{A.3}$$

where δ_{kl} is the Dirac δ function ($\delta_{kl} = 1$ if $k = l$, and zero otherwise). Equation (A.3) may be solved formally by introducing the complex matrix defined in Eq. (4.8), with the result

$$n_{l}^{1} = \sum_{k=1}^{N} (\mathcal{G}^{-1})_{kl}I_{k}^{0}B_{k}^{0}N_{k}^{0} \tag{A.4}$$

Therefore, if we have a quasi-steady state in which n_{k}^{0} is also a constant, given by the solution of Eq. (A.1) with $\lambda = 0$,

$$n_{l}^{0} = \sum_{k=1}^{N} (G^{-1})_{lk}I_{k}^{0}B_{k}^{0}N_{k}^{0} \tag{A.5}$$

the total emission using Eqs. (A.2), (A.4), and (A.5), is

$$J(t) = \mathrm{Re} \sum_{j=1}^{N} A_{j}n_{j}(t) = \mathrm{Re} \sum_{j=1}^{N}\sum_{k=1}^{N} A_{j}[(G^{-1})_{jk} + \lambda(\mathcal{G}^{-1})_{jk}e^{i\omega t}]I_{k}^{0}B_{k}^{0}N_{k}^{0} \tag{A.6}$$

The observable in phase fluorometry is the phase shift δ, defined by comparison with the phase of the modulation:

$$J(t) = J_{0} + \lambda J_{1} \cos (\omega t + \delta) = \mathrm{Re}(J_{0} + \lambda J_{1}e^{i\delta}e^{i\omega t}) \tag{A.7}$$

where J_{0} and J_{1} are real and J_{0} is the average emission. A direct comparison of the second term in the extreme right of Eq. (A.7) with that in Eq. (A.6) yields

$$J_{1}e^{i\delta} = \sum_{j=1}^{N}\sum_{k=1}^{N} A_{j}(\mathcal{G}^{-1})_{jk}I_{k}^{0}B_{k}^{0}N_{k}^{0} \tag{A.8}$$

This equation may be divided by $\sum_{l=1}^{N}I_{l}^{0}B_{l}^{0}N_{l}^{0}$, a real quantity, and then when the arguments of the left- and right-hand sides are equated, the basic result given in Eq. (4.6) is obtained. In the same notation, a few more steps show that the average fluorescence yield $J_{0}/\sum_{l=1}^{N}I_{l}^{0}B_{l}^{0}N_{l}^{0}$ is given by

$$\phi_{F} = \sum_{l=1}^{N}\sum_{k=1}^{N} A_{l}(G^{-1})_{lk}f_{k}$$

[see Eq. (3.9)].

---

*This is not an approximation; it means that all other time dependence has decayed out or is found in other Fourier components of the solution, for which similar equations would hold.

5

Luminescence

J. Lavorel

SYMBOLS AND ABBREVIATIONS*

1. Luminescence

L	Luminescence intensity
$\bar{L}$	Average luminescence intensity over a phosphoroscope period
$\Sigma L(\ )$	Light sum or time integral of L over limits specified within parentheses; in particular, $\Sigma L(\infty)$ is the total light sum for the whole decay of L
LX	Stimulated luminescence under perturbation X, with $X =$

	A	acid transition
	AB	acid–base transition
	E	variable electric field
	O	addition of organic substance
	S	salt jump
	T	temperature jump

I, t_I	intensity of actinic light, time interval of action of actinic light
t_L	any of the succeeding time intervals after actinic illumination when luminescence can be observed

The experimental protocol is noted as a luminescence symbol followed by:

$$[t_I | t_{L_1} | t_{L_2} | t_{L_3}]_u (A = a, B = b, \ldots)$$

where

t_{L_1}	time interval before measurement of luminescence
t_{L_2}	time interval when luminescence is measured
t_{L_3}	time interval after measurement of luminescence (t_{L_1} may be the minimum t_L imposed for technical reasons; in the phosphoroscope methods, t_{L_1} and t_{L_3} are the occultation intervals during a cycle)
u	units in which time intervals are expressed
A, a, etc.	factor A has been given the value a, etc.

Variants:

$[n^* | \ldots]$	observation occurs after n short actinic flashes
$[|t_{L_{1a}}, A, t_{L_{1b}}, B, \ldots|]$	= actinic illumination followed by perturbation A, after $t_{L_{1a}}$, plus perturbation B, after $t_{L_{1b}}$, etc.

2. Fluorescence

F	Fluorescence intensity
ϕ	Total fluorescence yield
$\Delta\phi$	Fluorescence yield of the variable part
ϕ_0	Fluorescence yield of the constant part (ϕ and $\Delta\phi$ are conveniently measured in units of ϕ_0)
(ϕ)	Fluorescence yield of the luminescence emission
$\varphi(i)$	Fluorescence yield per photosynthetic unit belonging to a cluster of size i.

*It will be easier to read this chapter if the reader first masters the definition of symbols listed here.

3. Mechanistic symbols

ϵ_F, ϵ_L	Singlet state exciton (chlorophyll) arising from light absorption or luminescence recombination process
J	Rate of injection of ϵ_L in the photosynthetic units
τ	Reciprocal of (quasi) first-order decay of a kinetic component (within a limited time range)
C_+^-, C^-, C_+, C	Forms of the photosystem II center ZChlQ after photochemical transformation and resulting from the protonation equilibria

4. Abbreviations

Chl	Chlorophyll
CNP	2-Chloro-6-nitrophenol
CCCP	Carbonylcyanide-m-chlorophenylhydrazone
DCMU	3-(3,4-Dichlorophenyl)-1,1-dimethylurea
DCPIP	2,6-Dichlorophenol indophenol
DNB	2,4-Dinitrobenzene
DNP	2,4-Dinitrophenol
DPSC	1,4-Diphenylsemicarbazide
Fd	Ferredoxin
Gmcd	Gramicidin
IOX	Ioxynil or 3,5-diiodo-4-hydroxybenzonitrile
MV	Methyl viologen
NADP$^+$	Nicotinamide adenine dinucleotide phosphate
Nig	Nigericin
OP	o-Phenanthroline
PQ	Plastoquinone
PS	Photosystem
PSU	Photosynthetic unit
PU	Phenylurethane
Vmc	Valinomycin

1. INTRODUCTION

Quantum conversion in the photosynthetic apparatus results from the photochemical functioning of a "vectorial" device which utilizes the energy of light collected by the Chl "antenna" to separate electron-hole pairs. The charge separation is stabilized first, by the reduction and oxidation of a primary electron acceptor and electron donor permanently bound to the "reactions center" Chl, and second, by protonation equilibria of the primary donor and acceptor; these stabilization steps condition the noncyclic electron flow through the photosynthetic chain. They are accompanied by the buildup and decay of a "gradient" (or difference) of photoelectric potential and the formation of a pH gradient across the membrane. The direction of these gradients with respect to the inner and outer phases of the thylakoid indicates that the activated center is a photochemical "dipole" with the positive end pointing toward the inside of the thylakoid. Luminescence (or delayed light, delayed fluorescence, etc.) results from the decay of singlet state (Chl) excitons which have been "injected" in the "collector" Chl molecules by the spontaneous "recombination" of electron-hole pairs of the PS II dipoles; the activation energy of the recombination process reflects the extent of stabilization; further differentiation of the stabilization energy is a consequence of a four-step accumulation of positive charges on the donor side of the centers; "luminescence excitons" are not easily distinguishable from "fluorescence excitons" created by light absorption and, as the latter, they are subject to radiative or nonradiative decays, migration, and trapping by "open" centers.

This synthetic definition points to the complexity of luminescence and to its intimate connection with several important areas in photosynthesis. It is also intended to emphasize the accelerated, extensive, and intensive development of luminescence studies in the photosynthetic community, dating from the year 1951, when this unexpected phenomenon was discovered by Strehler and Arnold (1951).

In this chapter, my approach has developed midway between impassive compilation on my right and egocentric contemplation on my left (although admittedly a little more on the left side!). It is understood, of course, that I have had to be selective in the presentation of facts and ideas. In consequence of the above definition of luminescence, two remarks must be made. First, kinetics is basic in luminescence studies, hence I have frequently used qualitative kinetic arguments throughout and devoted a major section to the quantitative discussion of kinetic results (Section 6). Second, luminescence has been more and more a source of pertinent arguments in several areas: fluorescence (see Chapter 6), oxygen evolution (see Chapter 8), and properties of the thylakoid membrane (see Chapter 11). I have frequently used results and arguments from these areas and discussed the corresponding mechanisms rather freely and independently.

2. METHODS

2.1. Definition and Symbols

A number of symbols are necessary. They have been listed above under "Symbols and Abbreviations"; however, they will always be defined the first time they appear in the text. I will introduce here the most frequently used symbols concerning the quantities being measured and the main parameters. From the operational standpoint, the observed luminescence intensity L is a function of the duration of the "actinic" period t_I, during which the sample is absorbing the light of intensity I, of a series of successive time intervals t_L, which define an occultation or waiting period, an observation period, etc., and of various experimental factors. I will use two other symbols for qualifying luminescence: $\Sigma L(t_L)$, the time integral of L over the interval t_L and $\overline{L}$ the mean value of L over a phosphoroscope period. The main aspects of luminescence, i.e., decay, induction (and transients), and light sum are thus characterized by three quantities: L, $\overline{L}$, and ΣL. I have found it useful to introduce a condensed notation for specifying luminescence signals as a function of the main parameters, for instance:

$$L[t_I|t_{L1}|t_{L2}]_u(A = a, B = b, \ldots)$$

where t_{L1} is the waiting or dark interval preceding the observation of L during the interval t_{L2}, the time being expressed in unit u, and a, b are numerical values assumed by factors A, B, etc. The symbol ΣL should replace L whenever t_{L2} is not small as compared to the mean lifetime of the decay during t_{L2}. This precise but cumbersome notation will only be used when necessary, with variations in connotation and freedom in writing (for instance, $L[|2|]_{msec}$ is meant for L observed at 2 msec from the origin of the decay, the missing information being irrelevant for the purpose, or not found in the published source).

2.2. Methods

Much of the difficulty and uncertainty encountered in compiling and interpreting luminescence results stems from the wide choice of time parameters and other factors permitted by an extreme variety of methods, and from the variation in kinetic nature of the signal as seen through different experimental devices. Thus, while it will be important to clarify as much as possible the strict interrelation between methodology and interpretation, I have not attempted an exhaustive compilation of all the methods and, instead, have pointed to the principal operational characteristics, possibilities, and recent technical advances in the most common families of methods.

2.2.1. Performance of the Various Methods

2.2.1.1. Methodological. 2.2.1.1.1. Time resolution. The minimum value of t_{L_1} in $[|t_{L_1}|]$; the L curve being highly polyphasic, nothing can be inferred on the decay during the "blind" period from the actually observed decay ("head cutoff" effect).

2.2.1.1.2. Nature of the actinic period $[t_I||](I, \ldots)$. The light flash or pulse may induce a single photochemical "turnover," or several turnovers in such a manner that the significant factor is the light energy ["reciprocity" law, $L = f(It_I)$] or, besides, that t_I is independently a factor controlling side- or back-reactions, filling of "pools," etc. $(L = f(It_I, t_I)$.

2.2.1.1.3. Repetitive and sequential mode, $\overline{L}$ *or* $L\ [n^*|\ |]$. Although the distinction is somewhat arbitrary, the more recent sequential mode [sequence of isolated short saturating flashes (n^*)] is less prone to various kinetic effects which may weaken the significance of the results than the more popular repetitive mode (phosphoroscope). Experimental factors affecting $\overline{L}$ are often difficult to foresee and interpret owing to the "averaging" effect over the successive phases of the decay. The time period, $t_I + t_L(\max)$, may interfere with several turnover times of the photosynthetic system giving rise to various "convolution" or "memory" effects in the sense that $\overline{L}$ at period n may not be independent of what occurred at periods n-1, n-2, etc. However, the induction curve of $L = f(t) = f(nt_I)$ gives invaluable insight into the functioning of large pools in the system.

2.2.1.1.4. Perturbation factors. A more or less abrupt change of a factor X during the L decay may introduce, aside from the result which one looks for, an interesting "differentiation" effect resulting from the simultaneous variation of X and t_L. [The "thermoluminescence" or "glow curve" method is an extreme instance of perturbation where the rate of change of X (= temperature) is relatively low].

2.2.1.2. Technical. 2.2.1.2.1. Signal-to-noise ratio (S/N). Depending on the method, S/N improvement may be obtained by simply setting a large time constant in the measuring instrument or it may have to be done by repetition and statistical averaging (see Witt, 1971), with (Lavorel, 1971) or without changing the sample.

2.2.1.2.2. Light source. With continuous light, the occulting device* sets the time resolution and also the "contrast" ratio, i.e., I (during the actinic period)/I (during the dark period). Pulsed light sources have their built-in occulting device, but some suffer from a poor contrast ratio in the time immediately following the flash.

2.2.1.2.3. Measure of ΣL. If the light sum is to be measured up to its maximum value $(t_L = \infty)$, the instrument must allow variation of t_L over a very large range and must be able to monitor a very large span of signal sizes.

2.2.1.2.4. Pretreatments and perturbations. Some methods are specifically designed for performing such operation before or during the normal light–dark cycle.

2.2.2. Flow Method

It was introduced in the field by Strehler and Arnold (1951). Minimum $t_L\ [t_L(\min)]$ is $\cong$ 10 msec. All the time parameters are transformed into

---

*For instance, rotating sector, mechanical shutter, etc.

distance and section parameters. $L[|t_L|]$ may be obtained by monitoring the signal at a variable distance from the actinic chamber, the simultaneous measurement of the fluorescence intensity F may be realized with very little artifact. S/N improvement is easily achieved by the time constant principle (and also through the use of large volumes). By nature, it is also a "single shot" method (see Section 2.2.4) permitting a variety of pretreatments. The simplicity and convenience of operation is somewhat offset by the need to use large volumes of suspension, a poor temperature control, a medium light collection capacity, and uncertainty in time definition due to imperfect turbulent flow (Lavorel, 1969a).

2.2.3. Phosphoroscope Method

This group of methods works on the repetitive mode. It was introduced by Arthur and Strehler (1957). One may subdivide it according to the technique of occultation.

2.2.3.1. Mechanical. This is the most popular "Becquerel" type. The cycle $[t_I/t_{L_1}/t_{L_2}/t_{L_3}]$ is produced by a pair of rotating sectors with openings at fixed angular distance, the observation interval t_{L_2} being delimited by two occultation intervals, t_{L_1} and t_{L_3}. $t_L(min) \cong 0.1$ msec. One must notice that the ratio of the time parameters is usually kept constant and that the maximum value of the ratio is $\cong 10$. The measurement of $\overline{L} = f(t)$, whether as induction phase or steady state, is the most frequent use of this method; however, the decay curve of L may also be displayed on an oscilloscope during t_{L_2} (Sweetser et al., 1961). Decay curves of a similar type (but actually of different nature) were formerly obtained by Arthur and Strehler, with fixed sector geometry, but variable rotation speed. The average I is evidently constant but It_I per cycle is changing and the consequences mentioned in Section 2.2.1.1.2 are likely to occur. This is plainly visible when both procedures for scanning the L decay are used simultaneously (see Bertsch et al., 1969). The "Becquerel" type has many advantages which explains its widespread use: simplicity of operation, economy of biological material and biochemicals, good light collection capacity, and high yield of signal per unit of time (if one is interested in that), but in my opinion, it is too limited in the choice of the time parameters.

2.2.3.2. Other Techniques. Improvements in the time resolution have been obtained with the help of special techniques. Occultation with a thin rotating sector of a focused laser beam achieves $t_L(min) \cong 3 \mu sec$; addition of a second sector actuated by a step motor allows one to work on the sequential mode (laser phosphoroscope; Lavorel, 1971). By synchronizing short saturating flashes with a high-speed sector in order to cut off the tail of the flash, Zankel (1971) has obtained single turnover operation—which is hardly possible with the laser phosphoroscope—at the cost, however, of a lesser time resolution (t_L (min) $\cong$ 50 μsec). This method belongs actually to the sequential type. Very promising techniques for rapidly switching a photomultiplier on and off in conjunction with electronic flashes (Ruby, 1968; Stacy et al., 1971) have been used to get better time resolution and light saturation and this trend is likely to be intensified in the near future. Recently, the highest time resolution (t_L(min) $\cong$ 250 nsec) has been achieved by Haug et al. (1972) by occultation of a continuous laser beam with a fast rotating sector.

2.2.4. "Single-Shot" Method

This method, a special case of the sequential type, is characterized by a unique cycle $[t_I | t_L]$ with no limitation in the upper range of the two parameters, t_I and t_L. It was introduced by Mayne (1968). This method is intended for studying the slow component of the L decay. It is simple, has and needs high light collection efficiency, does not require a high time resolution (however, t_L(min) $\cong$ 20 msec is easily obtained with a mechanical shutter).

2.2.5. "Jump" Method

Although, from the technical standpoint, this method appears as a variation of the preceding types, its purpose is to "trigger" changes in luminescence by an appropriate perturbation occurring during t_L. The perturbation is chemical (acid–base transition, salt addition, etc.) or physical (temperature, external electric field). In a single-shot apparatus, the perturbation is applied by injection of a suitable solution into the sample cuvette either before the observation interval (Mayne, 1968) or during this interval (Mar and Govindjee (1971)). The speed of mixing is not very high ($\cong$ 0.1 sec). Faster mixing is achieved with a flow method (of the "stopped-flow" type): the two liquid volumes to be mixed are expelled from syringes and forced into turbulent mixing by a Y-shaped tubing which transfers the liquid into the measuring cuvette (Barber and Kraan, 1970). The mixing time is $\cong$ 70 msec and the transfer time after mixing $\cong$ 50 msec. The appropriate notation for the sequence of events, for instance in an acid–base transition experiment will be $[|t_{L_1} A t_{L_2} B t_{L_3}|]$, i.e., the waiting period is divided into three intervals t_{L_1}, t_{L_2}, and t_{L_3}, the acid phase (A) starts at the end of t_{L_1} and the base phase (B) at the end of t_{L_2}. With the phosphoroscope method, likewise, a perturbation on $\overline{L}$ may be observed while the apparatus is running with the actinic light turned on (Barber, 1972).

2.2.6. Sequential Method

As already noted, there is a similarity between this method and the phosphoroscope method. The difference is that, in the sequential mode, as introduced by Barbieri *et al.* (1970), the outcome of each cycle is monitored separately with special attention directed to the transient changes of L decay from one cycle to the next. Short saturating electronic flashes are used in order to insure single photochemical turnover per cycle. The photomultiplier does not need to be protected from the actinic flash if $t_L(\min)$ is limited to $\cong 20$ msec (see Zankel (1971) for better time resolution).

3. FLUORESCENCE AND LUMINESCENCE

Following the discovery of luminescence, two fundamental points concerning its significance were soon recognized:

1. Luminescence, as well as "prompt" fluorescence, results from the radiative deactivation of the first singlet excited state of Chl a. This was demonstrated by the identity between luminescence and fluorescence spectra (Strehler and Arnold, 1951; Arnold and Davidson, 1954).

2. Luminescence is intimately related to the primary events of the photochemical energy conversion in the photosynthetic apparatus. A number of convincing arguments were produced in favor of this proposition. The action spectra for excitation of luminescence and fluorescence were seen by Arnold and Thompson (1956) to be identical in many photosynthetic organisms (green, blue-green, and red algae, photosynthetic bacteria, chloroplasts of higher plants). Luminescence requires the functional integrity of the photochemical apparatus as evidenced by the disappearance of both photosynthetic activity and luminescence after thermal denaturation (Strehler and Arnold, 1951) and Mn deficiency (Kessler, *et al.*, 1957). A number of physical or chemical factors were found to act with a striking parallelism on photosynthesis and luminescence (Arthur and Strehler, 1957).

These two fundamental properties led Arthur and Strehler (1957) to propose that the origin of luminescence was in the back-reaction (or recombination) of the primary photooxidized and photoreduced products of the light reaction. This section and Section 4 are devoted to the experimental developments and elaboration of concepts which, in large part, stemmed from these early discoveries and hypotheses.

Let us first consider the common origin of fluorescence and luminescence. In current language, we may say that both emissions testify to the radiative decay of singlet state excitons of identical nature but different origin. Accordingly, I will distinguish them as luminescence (ϵ_L) or fluorescence (ϵ_F) excitons. The term "exciton" is used here in a loose sense for a quantum of electronic excitation (singlet) migrating throughout the collector(s) of the photosynthetic unit(s) (PSU). In spite of their different origin, ϵ_F and ϵ_L are expected to obey the same law of decay, or very similar ones. Taking into account the conclusion that luminescence (see Section 4) and the variable part of fluorescence ($\Delta\phi$) are associated with the functioning of PS II, one may conjecture that luminescence and fluorescence have a common yield factor, that luminescence, being an intensity ($(h\nu)$ sec^{-1}), has also a rate factor and that factors influencing the fluorescence yield (e.g., reduced primary electron acceptor of PS II Q$^-$) should also affect luminescence. The ideas were expressed independently by three authors: Duysens (1963), Butler (1966), and Lavorel (1966).

3.1. The Yield Factor and the Rate Factor: "The *L* Relation"

As a consequence, I have proposed (Lavorel, 1968) that *L*, the luminescence intensity, be expressed in close analogy with *F*, the fluorescence intensity. For the latter quantity, the (quantum) yield ϕ is defined as a proportionality factor between *F* and *I*, the absorbed light intensity, or:

$$F = \phi I \tag{1}$$

Accordingly, *L* should be defined by the relation:

$$L = (\phi)J \tag{2}$$

The "*L* relation" [Eq. (2)] simply says that a flux *J* of ϵ_L is injected into the PSU collector and is subsequently expressed as a light flux *L*, with quantum yield (ϕ). Several points are raised by the *L* relation, which I will now consider. They concern the respective significance of ϕ and (ϕ) in Eqs. (1) and (2), the possibility of genuine effects of delaying the decay of excited states and redistributing them in the photosynthetic apparatus, and the kinetic factors controlling ϕ, (ϕ), and *J*.

3.1.1. Separate vs Connected Models of the Photosynthetic Unit

ϵ_F and ϵ_L are, in a sense, "probes" which "see" the state of the chlorophyll territory. The question, then, is to know whether, in spite of the fact that they are close relatives, they actually both see the same thing. The answer is in part dependent upon which model one chooses for the PSU, more precisely, upon the "connectedness" of the PSU's depending on the extent and freedom of migration of the excitons (see Lavorel and Joliot, 1972).

1. The case is trivial if the PSU's are separated (no exciton exchange between units). According to the recombination hypothesis, which I have extended in a more precise sense in keeping with modern evidence, *J* should be a function of the recombination rate of the "activated" PS II centres (Lavorel, 1968). Following the modern terminology, the two opposite processes are expressed as:

$$\text{PS II photoreaction } ZChlQ + \epsilon_F \longrightarrow {}^+ZChlQ^- \tag{3}$$
$$\text{Luminescence} \qquad\qquad {}^+ZChlQ^- \longrightarrow ZChlQ + \epsilon_L \tag{4}$$

(Equations (3) and (4) are simplified forms expressing, as we shall see, a more complex situation; I will use the simplified forms when dealing with general aspects of the processes and where no confusion or ambiguity can result.) Note also that ϵ_F stands in Eq. (3) of its own right: its definition does not imply that it only decays by fluorescence. Then, Eqs. (1) and (2) are to be replaced by:

$$F = \varphi(1)I[(C_+^-)] + \varphi(0)I[(C)] \tag{5}$$
$$L = \varphi(0)J[(C_+^-)] \tag{6}$$

where $\varphi(1)$ is the fluorescence yield of the unit in its closed state (P state, nonphotoactive center: $^+ZChlQ^-$) and likewise $\varphi(0)$ for the open state (0 state, photoactive center; $ZChlQ$); $I[(C_+^-)]$, $I[(C)]$ are proportional to the concentrations (C_+^-) of "closed" centers ($^+ZChlQ^-$) or (C) of "open" centers, respectively, and $J[(C_+^-)]$ is a function (simple presumably) of (C_+^-). It is already apparent that ϕ and (ϕ) in the general Eqs. (1) and (2) *cannot* be the same (although they may have identical emission wavelength dependence): when performing the recombination the closed unit becomes open and the only chance of escape of ϵ_L as luminescence is given by the yields $\varphi(0)$ hence, Eq. (6). In terms of rate constants, the "reduced" yields φ are expressed as:

$$\varphi(0) = \frac{f}{p + f + d} \tag{7}$$

$$\varphi(1) = \frac{f}{f + d} \tag{8}$$

where p, f, and d are the first-order rate constants of exciton decay by photochemical trapping [Eq. (3)], radiative and nonradiative processes, respectively. Less trivial is the delay effect* (see Appendix I). Assuming, as a special simple condition, that the PS II centers are isolated from the PS photosynthetic chain (no electron flow) and on account of the recombination reaction [Eq. (4)] being first-order (see Section 4.1) with rate constant τ_L^{-1}, i.e., $J[(C_+^-)] = (1/\tau_L)(C_+^-)$, the total light sum is:

$$\Sigma L(\infty) = r\frac{f}{f + d + (1 - r)p}(C_+^-)_0 = r\frac{\varphi(1)\varphi(0)}{(1 - r)\varphi(1) + r\varphi(0)}(C_+^-)_0 \tag{9}$$

where $(C_+^-)_0$ is the value of (C_+^-) at the time origin and r the exciton yields of the recombination reaction (see Joliot *et al.*, 1971); and the lifetime of

---

*A "delay" in the expression of ϵ_L as luminescence will occur because ϵ_L has some probability of being immediately retrapped by the (now open) center which has just produced it (or by neighboring open centers as well, if the PSU's are connected); this effect is similar to the well known increase of τ in fluorescence lifetime measurements due to reabsorption of fluorescence radiation within the volume of the sample.

closed center, which incidentally is also in the present case that of both ϕ and L, is:

$$\tau_{LD} = \tau_L \left(1 - \frac{\mathrm{rp}}{p + f + d} \right)^{-1} = \tau_L \frac{\varphi(1)}{(1 - r)\varphi(1) + r\varphi(0)} \qquad (10)$$

As $\varphi(0) \ll \varphi(1)$, we see that the "actual" lifetime of the L decay (τ_{LD}) may be much longer than expected from its mechanistic, "natural" value (τ_L). The ratio τ_{LD}/τ_L, which is a measure of the delay effect, may be ≥ 100 and has probably some significance with respect to the wide spectrum of lifetime of luminescence.

2. Experimental evidence (for instance, kinetics of the photochemical rise of $\Delta\phi$) rules out the separate model, at least as a two-states, first-order system. The alternative—in fact, not the only one (see Lavorel, 1972)— is the connected model where exciton exchange between units is allowed, with the consequence that the yield per unit φ is not only dependent upon the unit being open or closed, but also on the state of the neighboring (connected) units. This is one aspect of what Lavorel and Joliot (1972) have called the *îlot* effect. Accordingly, ϕ for a closed unit is a function of the size i of the cluster (*îlot*) of which it is a member, and of a distribution parameter d_i (shape of the cluster and position of the unit in the cluster). The sad point is that, except for a one-dimensional array of units, d_i is almost impossible to evaluate (qualitatively, one can only say that φ, on the average, is larger for a compact cluster than for an elongated or branched cluster of the same size and that it is also larger, in a given cluster, for units in a central position than for those in a peripheral position). Equations (5) and (6) are, therefore, replaced by:

$$F = \sum_i \alpha(i) \sum_{d_i} \varphi(i, d_i) I\left[(C_+^-) \right] \qquad (11)$$

$$L = \sum_i \alpha(i) \sum_{d_i} \varphi(i - 1, d_{i-1}) J\left[(C_+^-, i, d_i) \right] \qquad (12)$$

where $\alpha(i)$ is the proportion of units belonging to clusters of size i. In this form, the L relation is obviously intractable, and one feels the urge to look for situations where matters get simpler.

1. In the P state (all units closed) obviously $\varphi(\infty)[\text{connected}] = \varphi(1)$ [separate], hence (ϕ) can be factorized in the L relation and J calculated. I will assume this to be true when looking at the L decay for very small t_L, under light saturation (see Section 6.1.1).

2. Close to the 0 state (all units open), with a few randomly distributed, and therefore isolated, units, the situation is not much different from that of the separate model. The delay effect operates, with the same con-

sequences as above [Eqs. (9) and (10)]. Note, however, that $\varphi(1)$ [connected] $< \varphi(1)$ [separate], because an ϵ_F created by light absorption in a closed unit has an additional type of decay, besides "inside" trapping, and nonradiative deactivation, namely, "edge" trapping by a neighboring open unit. Again the simple L relation applies and this situation would be worthy of thorough exploration: in fact the slow component(s) of L belongs to it, which presumably explains the relatively simple kinetics observed there (see Section 6.1.3); the emission spectrum should be that of $\varphi(0)$ and an interesting comparison could be made with the spectra of the constant part (ϕ_0) and of the variable part $(\Delta\phi)$ of fluorescence.

The edge trapping as noted above by neighboring open units is characteristic of the connected model and gives rise to a "redistribution" effect* which, like the delay effect, is a simple consequence of the recombination hypothesis and of the L relation. Qualitatively, the redistribution effect is a function of the number of open units: it should increase in importance as the L decay proceeds and affect—again, in a manner almost impossible to evaluate quantitatively—the delay effect. Both effects should in principle contribute to the kinetic peculiarities of the L decay (Lavorel, 1969a,b).

3.1.2. Type of Exciton Movement

The speed of the exciton migration (for a theoretical discussion, see Chapter 4) should also condition its ability to probe the Chl territories. Fast traveling means blurred landscapes! On the one hand, if we are dealing with fast excitons (with low probability of trapping per visit to the photochemical centers), the same average picture should be seen by ϵ_L and ϵ_F. On the other hand, with slow excitons, enjoying a leisurely random walk throughout the green PSU's, the exact place of birth of the exciton might not be without consequence. In particular, one may wonder whether ϵ_L sees as large a territory as ϵ_F does and whether the spatial range of ϵ_L is not rather restricted to the neighborhood of centers. The question is not futile, if the holochrome composition of the PSU is not homogeneous (Borisov, 1967; Seely, 1972; Paillotin, 1972).

3.1.3. The Variable and Constant Parts of Fluorescence

Neglecting all of the above difficulties and assuming that the L relation can be factorized, it remains to be decided which "macroscopic" fluorescence

*In a system of connected PSU's, an ϵ_L produced by recombination in a center may be retrapped in another, neighboring center. Repetition of this process in effect produces a "redistribution" of the activated state C_+^- among the centers. If recombination *is* deactivation [see Eq. (45)], redistribution modifies the S state of the centers (Z side of PS II) and should be of consequence in the kinetics of the deactivation process, as noted by Joliot *et al.* (1971) and Lavorel (1971).

yield should be used to calculate J. Will it be the variable yield $\Delta\phi$ or the total yield $\phi(=\phi_0 + \Delta\phi)$? In other words, within the present context, the question is to know exactly which part of the fluorescence belongs to PS II. Clayton (1969) has pointed out the importance of this choice when using the L relation and concluded that most of ϕ_0 was actually "dead" fluorescence, i.e., not indicative of PS II activity. On the scale of fluorescence levels, Clayton puts the true zero of "live" or PS II fluorescence a few percent below the experimental 0 level. There are also some theoretical reasons to believe that Clayton's conclusion is right (Lavorel and Joliot, 1972). As several authors (including myself) have set $(\phi) = \phi$ instead of the presumably more correct

$$(\phi) = \Delta\phi \tag{13}$$

I shall call the resulting values of $J:J_T$ (with the choice ϕ) and J_V (with the choice $\Delta\phi$). We note that:

$$J_V = \left(1 + \frac{1}{\Delta\phi}\right)J_T \tag{14}$$

($\Delta\phi$ is, for practical purposes, measured in units of ϕ_0) and therefore we expect that J_T will not behave as J_V at the end of the L decay when $\Delta\phi$ is small.

3.1.4. The Kinetic Factors of the L Relation

This question is not solely of theoretical interest. Some authors have taken the L relation as predicting that luminescence varies always in the same direction as fluorescence. Obviously, the L relation does *not* say that. The simple reason is that (ϕ) and J are functions of several variables (kinetic factors) which may determine them quite differently. The problem is best understood by differentiation of the L relation: changing a factor X by a small amount dX induces a variation dL according to

$$dL = \left[(\phi)'_X J + (\phi)J'_X\right] dX \tag{15}$$

(primed quantities with subscript X are derivatives with respect to X). A strict parallelism between L and ϕ is only expected when $J'_X \cong 0$. The case is frequent but far from general; this can be a priori understood by knowing that J is a more complicated function than (ϕ), hence, when X is varied, screening or compensating effects may operate in J that will "buffer" it, while (ϕ) will experience a larger shift (see Clayton's experiment, Section 3.2.2.).

3.2. Experimental Evidence and the L Relation

I shall select here a few conspicuous cases where Eq. (15) assumes a simple form, where the effect of the most notorious factor Q^- common to

$\Delta\phi$ [henceforth, according to Eq. (13)] and J is visible and where apparent discrepancies are easily and significantly explained. Some examples are taken here for illustrative purposes and will be dealt with thoroughly in Section 6.

3.2.1. Dependence on Wavelength of Emission $(X = \lambda)$

J'_X in Eq. (15) is zero and we are back to the identity of F and L emission spectra. It is worth mentioning that the identity of these spectra has been confirmed by many investigators: Azzi (1966), Lavorel (1969a) with exact simultaneous measurements of F and L, Ruby (1971), Zankel (1971) in the t_L range of $\cong 100$ μsec. An exception was found only by Bonaventura and Kindergan (1971) and it should await verification before further comments are made. The simple property $J'_\lambda = 0$ has been used in my laboratory to compare plots of L and F at 715 nm vs L and F at 685 nm, the slope of which eliminates the J factor; the two wavelengths were chosen to emphasize (see Lavorel, 1962) the respective role of ϕ or $\Delta\phi$ in the L relation. Figure 1*

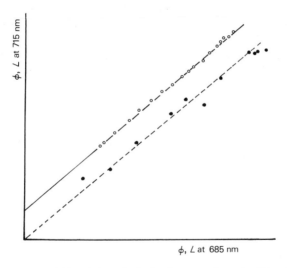

Fig. 1. Fluorescence yield, ϕ (○), or L, the luminescence intensity (●) at 715 nm vs same respective quantities at 685 nm in *Chlorella*. The independent variable is t_I, the time spent in light, for ϕ (photochemical O–P rise) or I, the intensity of actinic light, for L measured at $[5/0.02/3]_{msec}$ (i.e., light period = 5 msec, waiting period = 20 μsec, measurement period = 3 msec). The units are arbitrary; however, they stand in the same ratio for both quantities. (Etienne *et al.*, 1970).

---

*It is seen in Fig. 1 that the ratio of *total* F intensities at the two chosen wavelengths is variable, whereas the ratio of *variable* F intensities is constant. The fact that the ratio of L intensities is constant and identical within experimental errors to the ratio of *variable* F intensities is a good indication that, in the L relation, $(\phi) = \Delta\phi$. Another consequence is that the zero level of PS II fluorescence must lie very close to the 0 level (zero level of $\Delta\phi$).

indicates that the two slopes are identical. This result is in good agreement
with Clayton's conjecture concerning the zero level of PS II fluorescence
(Etienne *et al.*, 1970).

3.2.2. *Dependence on Concentration of Reduced Acceptor* (Q^-) $(X = Q^-)$

Clayton (1969) has observed two cases of striking parallelism in variations
of $\Delta\phi$ and $\overline{L}$ which seems to be clearly correlated with changes in the concentration of Q^-: the first is observed at low light intensity upon the addition
of ferricyanide (see Fig. 26). It is known that DCMU, by interrupting the
electron flow between Q^- and A (i.e., PQ), will increase (Q^-) when its
steady-state value is not maximum and that, on the contrary, a PS II acceptor,
such as ferricyanide, will decrease (Q^-) when its steady-state value is maximum. Hence Eq. (15) will be approximately satisfied provided that in both
cases $\Delta\phi \cdot J_X' < \Delta\phi_X' J$. The DCMU effect would be the simplest to explain
since we already know that $\Delta\phi$ is small. For the ferricyanide effect, we would
have to assume that J is well "buffered"; remembering that we are dealing
here with a transient effect, J would respond with a larger time lag to an
abrupt change in (Q^-) than $\Delta\phi$ would. As we shall see in Section 6, this
conclusion is only partly correct.

I have tried numerous types of plots for data obtained with the flow
method on simultaneous variations of L and ϕ. One of them (Fig. 2) points
to the following relation:

$$L^{\frac{1}{2}} \sim \phi \tag{16}$$

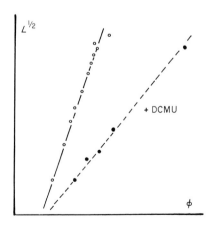

Fig. 2. The square root of the luminescence intensity $(L^{\frac{1}{2}})$ vs ϕ, the fluorescence yield,
in the absence ($\bigcirc$) and in the presence ($\bullet$) of 10^{-5} M CMU in *Chlorella*. Simultaneous
measurements of L and ϕ by a flow method in the time range of 10^{-2} sec after actinic illumination. The independent variable is I, the intensity of actinic light. (Lavorel, 1966.)

("~" meaning "a linear function of"). Since (Q⁻) is changing simultaneously with L and ϕ, Eq. (16) could be interpreted to mean that both J and
$\Delta\phi$ are proportional to (Q⁻), which is theoretically fairly correct for J and
quite approximate for $\Delta\phi$ (see Joliot and Joliot, 1964). Therefore, I believe
that this relation can only be used as a qualitative argument. Incidentally,
it is interesting to note that DCMU changes the slope (Fig. 2) in a manner
suggesting that this poison decreases J, the recombination rate.

Extreme instances, where L is zero and ϕ maximum, are easily explained
from basic principles and without contradicting the L relation: J will be
zero whenever the primary electron acceptor Q is missing or nonfunctioning
[*Chlamydomonas* mutant *ac 115* (Lavorel and Levine, 1968), destruction of
C 550 (Q?) by lipase treatment (Butler, 1971)], or the centers are forced
into the reduced nonphotoactive state ZChlQ⁻ [*Scenedesmus* "adapted"
to hydrogen (Lavorel, 1969)]. (Clayton and Bertsch (1965) reported the
absence of delayed light in a mutant of *Rhodopseudomonas spheroides* that
lacks functioning reaction centers.)

Two simple rules are consequences of Q being a factor in both $\Delta\phi$ and
J: (1) no L without photoinduced $\Delta\phi$; and (2) every factor modifying $\Delta\phi$
must modify L. To my knowledge, the only exception to rule 1 is found in
bacterial chromatophores where Fleischman (1971) observed a chemiluminescence induced by a redox transition. So far, no equivalent treatment
has been found to produce ⁺ZChlQ⁻ in the dark, in chloroplasts. With
respect to rule 1, it may be recalled that the acid–base induced L was formerly thought not to require preillumination (Mayne and Clayton, 1966).

3.2.3. Dependence on Time ($X = t$)

Again in qualitative agreement with the L relation, one always finds that,
during the decay, J, whether calculated as J_V or J_T, behaves as expected,
i.e., as a decreasing function of t_L. For very short t_L, when first-order
kinetics for the components of L and the factors in the L relation are legitimate approximations, Eq. (15) predicts for the time constants the following
relation:

$$\tau_L^{-1} = \tau_{\Delta\phi}^{-1} + \tau_J^{-1} \qquad (17)$$

where the τ's are time constants and the subscripts refer to the whole process
(L) or its kinetic components ($\Delta\phi$ and J). (See Appendix I). We shall see
later (Section 6.2.1) that Eq. (17) is fairly well verified by Zankel's (1971)
results.

In this section, I have shown that the L relation occupies incongruously
a key position: if taken mechanistically, it is too complicated and thus use-
less; if taken phenomenologically, it is inexact or at least approximate. It
must, however, be recognized that it has had some heuristic virtues, and is
still useful in this sense.

4. LUMINESCENCE AND STATES OF PHOTOSYSTEM II CENTERS

Luminescence is a PS II property. This proposition is not only consistent with all earlier findings when considered in the light of the two systems theory (see Section 3), but is also strongly supported by a number of recent results which I shall now discuss. The first clear indication in this sense was found independently by Goedheer (1962, 1963) and Bertsch (1962). They demonstrated an antagonistic effect of light I and light II on luminescence in correspondence with the two main PS II responses: variable fluorescence and oxygen emission. Other arguments were found later, among the more striking are: absence of *L* in mutants devoid of PS II activity (Bertsch *et al.*, 1967; Lavorel and Levine, 1968; Haug *et al.*, 1972), identity of the action spectra of *L* and PS II activity (Ruby, 1971), absence of *L* in PS I subchloroplast particles (Lurie *et al.*, 1972; Vernon *et al.*, 1972), and finally all the arguments presented in Section 3 in support of the *L* relation. However, the significant contribution toward the understanding of how the PS II centers are involved in the recombination has come from results obtained in Joliot's group (Barbieri *et al.*, 1970). They will therefore stand in a prominent place during this discussion.

4.1. Interdependence of the Two Primary Substrates of Luminescence

A basic problem concerning the recombination hypothesis is to decide whether ^+Z and Q^- recombine as a bimolecular or as a monomolecular process. There is much evidence pointing to the kinetic independence of the two "sides" of the PS II centers (fluorescence, oxygen emission). That they more or less ignore each other is not only a fortunate circumstance, but, also, an absolute prerequisite for a good efficiency of this charge separation device. And it must be recognized that in it may lie one of the most important virtues of the vectorial model first proposed by Mitchell (1966; also see 1967) and later elaborated by Witt's (1971) group, among others. The kinetic independence of Z and Q has probably played some role in enticing some authors to write $L \sim (^+Z)(Q^-)$, another reason being that luminescence, from the kinetic point of view, "seems" to be a bimolecular process (see Section 6.3.2.1). I think that one cannot overstress the point that at the photochemical stage, the concept of a "complex" (center, trap, etc.), i.e., an entity where a primary donor (Z) and a primary acceptor (Q) are permanent residents, in close contact with the photochemical effector (Chl), is even more central than the vectorial concept. In other words, even if the centers act cooperatively (connected PSU) at the photophysical stage, they must, for reasons of efficiency, work noncooperatively as independent entities at the photochemical stage. [For the functional independence of centers seen from the Z side, see Kok *et al.* (1970).] The point that the photosynthetic chains may (or must) work cooperatively during the later stages through several agencies (electric field, pH gradients, common pool of PQ, etc., see Witt, 1971) is another but different matter. If, then, we adhere to the recombination hypothesis we must recognize that: (1) among the four possible forms (in a simplified version) of the PS II centers, i.e., ZChlQ, $^+$ZChlQ$^-$, ZChlQ$^-$, and $^+$ZChlQ, the *unique primary* substrate of luminescence is $^+$ZChlQ$^-$; and (2) the recombination process is *monomolecular*, i.e.,

$$L \sim (^+ZChlQ^-) = (C_+^-) \qquad (18)$$

At this point, I must add that if L is of second-order (or, for that matter, of any order) we must look for and find pertinent reasons.

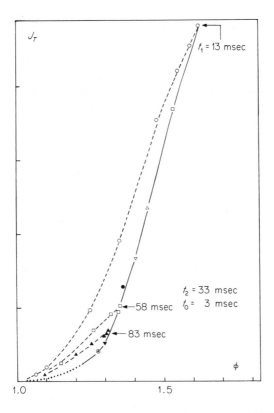

Fig. 3. The ratio of luminescence yield to the total fluorescence yield (J_T) as a function of ϕ, the total fluorescence yield (normalized at 0 level, $\phi_0 = 1.0$) in *Chlorella*. Protocol: flow method, measurements at $[33|t_L|3]_{msec}$ (i.e., light period = 33 msec, waiting period = t_L, measurement period = 3 msec). Continuous line: intensity of actinic light kept constant and maximum [I(max)], t_L increasing from $\cong 10$ to $\cong 150$ msec, toward the origin. Dashed line: t_L kept constant as indicated, I decreasing from I(max), toward the origin (Lavorel, 1969a.)

4.2. The "Z" Side (Donor)

The experiment depicted in Fig. 3 was perhaps the first kinetic indication that Q^-, which was the central point in the L relation, was not alone in determining the L decay, and that, according to the recombination hypothesis, one should look also towards the Z side. More pertinent was the reflection by Joliot (see Barbieri *et al.*, 1970) that luminescence as well as "deactivation" of the O_2 emission mechanism in the dark, on account of their similar decays in the 1–100 sec time range, bespoke the presence of species or states of considerable lifetime, common to both processes.

4.2.1. Oscillations of the Medium Decay of L

The above prompted a fruitful comparison of O_2 yield and L (or rather ΣL) induced by a sequence of brief, saturating flashes (Barbieri *et al.*, 1970). Figure 4 depicts such a comparison; one must note that the luminescence signal here is $\Sigma L[n^*|20|220]_{msec}$. The similarity of the two oscillatory patterns up to the 6th flash (period of four, and damping) and the phase advance by one step of the L "sequence" as compared to the O_2 "sequence" are noteworthy. The finding of a very small luminescence at the first flash confirmed an earlier report by Jones (1967). In Kok's theory of the O_2 evolution

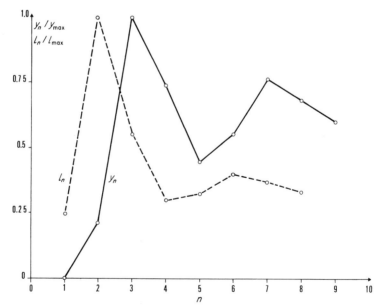

Fig. 4. Comparison of sequences of O_2 emission (y_n, continuous line) and luminescence (l_n, dashed line) in *Chlorella* dark-adapted for 3 mn. Flashes spaced 0.24 sec apart. Luminescence is measured at $t_L = 0.24$ sec after the flash. (Barbieri *et al.*, 1970.)

mechanism, the O_2 yield at the nth flash reveals the population of S_3 states which were formed after the (n-1)th flash; therefore, the O_2 sequence shows that the S_3 concentration becomes maximum *after* the 2nd, 6th, etc. flash (see Chapter 8). Since the maximas of ΣL also occur at these 2nd, 6th, etc. flash, one must agree with Joliot that the S_3 state is actively involved in luminescence. This conclusion was strengthened in a later report (Joliot *et al.*, 1971) showing that, with a low concentration of hydroxylamine, the L sequence was shifted by about two steps, in precisely the same fashion as was demonstrated by Bouges (1971a,b) for the O_2 sequence. Two other significant results were that, from $t_L \cong 1$ sec and up, L seemed to run parallel to the *rate* of S_3 and, after a while, to the *concentration* of S_2 (see Fig. 18) and that in general, the long tail of the $\Delta\phi$ decay, i.e., $\Delta\phi[n^*|2|]_{sec}$, followed a sequence almost identical to that of the sum $(S_2) + (S_3)$. The first of these two results were taken to mean that S_2 also was a luminescence substrate, although less active than S_3.

4.2.2. Oscillations of the Fast Decay of L

Using the same sequential mode with much better time resolution, Zankel (1971) was able to show that L had several fast components, and that two of them, with $\tau = 35$ μsec and $\cong 200$ μsec, respectively, were oscillating in *exact* synchrony with the O_2 sequence, i.e., with maximum amplitude at the 3rd, 7th, etc. flash (Fig. 5). The different behavior of the L decay in the medium and fast t_L range was traced back by Zankel to a "phase inversion" occurring at $t_L \cong 6$ msec when comparing $L[2^*| \ |]$ and $L[3^*| \ |]$ (Fig. 6). For the moment, let us only examine the possible reasons why the amplitude of the various components oscillates. (I shall come back to the significance of their number in Section 6.) To explain the phase inversion, Zankel reasons that, from one state to the next, one has to consider a

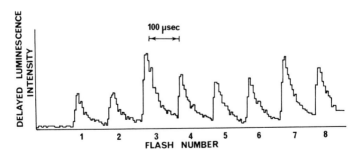

Fig. 5. Sequence of luminescence decays in dark-adapted chloroplasts. 3 μsec saturating flashes (1–8) spaced 1 sec apart. Each decay runs from $\cong 65$ to $\cong 160$ μsec after the flash. Note maxima at 3rd and 7th flashes (Zankel, 1971.)

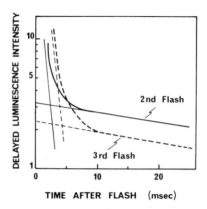

Fig. 6. (Delayed) luminescence as a function of time after flash. Note the phase inversion at $t_L = 6$ msec between the second (continuous line) and third (dashed line) flash. Chloroplasts. Conditions as in Fig. 5. Compare the situation at long t_L with the L sequence in Fig. 4. (Zankel, 1971.)

stabilization lifetime τ_{S_n} according to the following scheme:

$$S_n \xrightarrow{(h\nu)} S_n^* \xrightarrow{\tau_{S_n}^{-1}} S_{n+1} \tag{19}$$

In other words, the flash results in the actual accumulation of S_{n+1} *after* a delay of the order of τ_{S_n}. This time may be taken to be the estimated delay between excitation and O_2 evolution ($\cong 0.1$–1 msec). If the unstabilized forms S_n^* has the same luminescence properties as the stabilized form S_n, then the two series of results are in agreement and Joliot's conclusion that S_3 is the most "luminous" state is preserved. An alternative explanation, which I favor (see also Kraan, 1971), may be examined. Actually, it all depends on the nature of the stabilization process [Eq. (19)] and on the estimation of τ_{S_n}. One must remember that the "experimental" 4 S cycle results from a "mechanistic" 5 S cycle where two states, S_4 and S_0, have been telescoped into a single one (usually written S_0), due to a fast discharge, too fast to be seen directly in ordinary O_2 measurements, of the four positive equivalents (or holes) to produce one O_2 molecule. Obviously, this discharge, cannot be faster than the stabilization of S_3; it may even be of a different nature, for instance, we can describe this step as:

$$S_3 \xrightarrow{h\nu} S_3^* \xrightarrow{\tau_{S_3}^{-1}} S_4 \xrightarrow{\tau_{O_2}^{-1}} S_0 + O_2 \tag{20}$$

Furthermore, if τ_{O_2} is not negligible as compared to τ_{S_3}, we might as well consider S_4 as a luminescence substrate and attribute to it the luminescence components seen in the $t_L \cong 100\ \mu sec$ range. We shall see in Section 5 that there are good reasons to believe that S_4 is even more "luminous" than S_3.

4.3. The "Q" Side (Acceptor)

After discussing the "*L* relation" approach (Section 3), I noted its inherent ambiguity. The "substrate" approach which is fully expanded in this section, was not absent altogether, but it was impeded by mathematical difficulties concerning the definition of $\Delta\phi$ and J as function of Q^-. In a series of elegant experiments, Bennoun (1970) circumvented these difficulties. He was reinvestigating the well-known inhibitory effect of DCMU on the $\Delta\phi$ decay (Duysens and Sweers, 1963). A puzzling point was that, at concentrations higher than necessary to bring about complete inhibition of O_2 evolution, a slow decay of $\Delta\phi$ persisted (Lavorel, 1966). Bennoun (1970) observed that, when DCMU acted in the presence of hydroxylamine (low concentrations), the slow decay was halted somewhere in between the *P* and *O* levels and without any change in speed; at $2 \times 10^{-5} M [NH_2OH]$, the decay was completely halted, staying at the *P* level. Since, in the same concentration range, NH_2OH was known to act as competitive electron donor to the Z side, it became evident that the incomplete decay of $\Delta\phi$ was caused by the accumulation of ZChlQ⁻, due to a fast reduction of ⁺Z by NH_2OH, and that, consequently, the slow decay itself in presence of DCMU alone could only occur through the recombination pathway [Eq. (4) in Section 3.1.1]. If so, luminescence which is only partly inhibited in presence of DCMU (see Section 6.2.2.6 and 6.3.2.2), should be completely suppressed in presence of both DCMU and $2 \times 10^{-5} M$ NH_2OH and, indeed, Bennoun observed that it was (at least, for $t_L > 0.3$ sec). Similar results and conclusion were independently presented by Mohanty *et al.* (1971) and Stacy *et al.* (1971). One may notice that these experiments are also in agreement with the conclusions of the hydrogen "adaptation" experiment mentioned in Section 3.2.2, but these are more elegant and, as we shall see when analysing the kinetic aspects of these experiments (see Section 6.2.2.3), they have far-reaching consequences.

The conclusion of this section is simple. We are at least in possession of very strong arguments—qualitative, but unambiguous—in favor of the recombination hypothesis. We were looking for *the* substrate. We have found several. Just how many are there? At this point, I will venture to say that there are four in number, with luminescence ability arranged in the following order:

$$S_4 > S_3 > S_2 > S_1 \tag{21}$$

5. LUMINESCENCE AND ELECTROOSMOTIC PROPERTIES OF THE PHOTOSYNTHETIC MEMBRANE

So far, we have paid little attention to the energetic aspects of luminescence. We know from classical kinetics that the speed of almost all processes is limited by a Boltzmann factor involving an activation energy term. Accordingly, we must write Eq. (18) in the following explicit form (see Arnold and Azzi, 1968):

$$L = (\phi)v \exp{(-\Delta E/KT)}(C_+^-) \qquad (22)$$

where, v is a "frequency" factor, ΔE is the activation energy, K the Boltzmann constant and T the absolute temperature (energy will be expressed in (molar) electron-volts; at room temperature, $KT \simeq 0.025$ eV). C_+^-, hence forth, will stand for any of the luminescence substrates which we have recognized in Section 4: Eq. (22) is thus a simplified version of a more complex equation which need not be written in full. The activation energy problem has given rise to two kinds of developments in luminescence studies: (1) For ordinary chemical processes, there is but one way to change the Boltzmann factor, this is by varying temperature T, which, consequently allows us to measure ΔE; we shall discuss it in Section 6. (2) One may naturally ask what is the order of magnitude of ΔE: according to the recombination hypothesis, we can obtain an estimate of ΔE as the difference between the energy of the first singlet–singlet transition of Chl (1.8 eV) and that of the two terminal redox couples, Q^-/Q and H_2O/O_2 ($\cong 0.8$ eV). This 1 eV "gap" is quite impressive (1 eV $= 23.06$ kcal mole^{-1}) and has been a potent motivation for speculations concerning the mechanism of luminescence (see Section 7).

There is, however, another trend of research which unexpectedly but finally had to be connected with the activation energy problem. It may be traced back to two main sources. First, in the light of Mitchell's (1966) chemiosmotic theory, several important experimental results (Neumann and Jagendorf, 1964; Jagendorf and Uribe, 1966) came out, suggesting that the "high energy state" which results in ATP production might reveal another type of stored energy, besides redox potential energy, in the form of a proton concentration gradient across the thylakoid membrane. Second, Mayne (1967) made the surprising discovery that various substances known to "uncouple" phosphorylation from the photosynthetic electron flow (thereby increasing the latter rate) were inhibitory to luminescence. These two approaches merged when Mayne and Clayton (1966) observed that an artificial pH gradient (ΔpH) set across the membrane induced, in addition to the ATP burst of the Jagendorf and Uribe experiment, a surge of luminescence. This merger has subsequently resulted in an important experimental and theoretical development which we shall now discuss.

We must ask how all this is related to the ΔE problem. The idea came to Barber (Barber and Kraan, 1970) and to Crofts (quoted by Fleischman, 1971), that the activation energy "barrier" might be lowered by an amount V (Fig. 7), on account of a membrane potential V set across the membrane

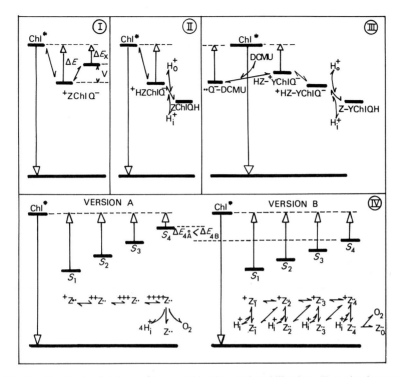

Fig. 7. Energy level scheme for recombination and stabilization. (I) Activation energy hypothesis: recombination of the photoelectric dipole, $^+ZChlQ^-$, (upward arrow) brings the reaction center chlorophyll into the excited state, Chl* before emission of a quantum of luminescence (downward arrow); the activation energy, ΔE, is changed into ΔE_X by modification V of the electroosmotic potential of the membrane. (II) Protonation hypothesis: the photoelectric dipole, $^+HZChlQ^-$, is stabilized in photochemical dipole, ZChlQH, through protonation equilibria with the external (H_0^+) and internal (H_i^+) proton pools. (III) Extended protonation hypothesis with complete photoelectric dipole $HZ^{\pm}YChlQ^-$ and additional stabilization by complex formation with DCMU. Note that the DCMU complex is also able to recombine. (IV) Modulation of activation energy by S states: each horizontal bar marked S_1, S_2, etc., is meant for the photoelectric dipole in state S_1, S_2, etc.; version A is a 4-step deprotonation of Z side ($^{n+}Z\cdot\cdot$ is a symbol for the Z side of the dipole carrying n^+ charges, equivalent to n protons); version B is a 1-step deprotonation of Z side (since deprotonation may now occur at each step, the symbols $^+Z_{ii}$ are used for the Z side of the dipole in state S_n); note that the activation energy is expected to be smaller in version A than in version B and the difference in activation energy (version A – version B) to increase from S_1 to S_4.

and of the vectorial character of the center, i.e., its proper "orientation" within the membrane. It must be noticed that this hypothesis amounts to saying that in this kind of system, the Boltzmann factor may be independently modified by changing temperature, as in ordinary homogeneous chemical systems, or by changing the activation energy itself and that this unique property of bioenergetic systems is strictly related to and explained through their heterogeneous (i.e., structured) nature. The Barber Crofts hypothesis will be generalized, remembering that the reaction center, as a photochemical dipole, is able to store and transduct by "coupling" electrical as well as chemical energy and that, consequently, it may be acted upon by an electrical potential gradient as well as by chemical potential gradients.

This field will be surveyed in order of increasing complexity, which means, not in the chronological order.

5.1. Electric Field

Recently, Arnold and Azzi (1971) discovered an effect which should have aroused considerable interest (but apparently has not). When applying an external variable electric field across a preilluminated chloroplast suspension, an extra luminescence signal LE is produced. (Henceforth, LX will designate an artificially induced luminescence, the symbol X being characteristic of the external agent.) The effect can be observed at a field amplitude of $\cong 100$ V cm^{-1} and frequency from $\cong 10^2$ up to $\cong 10^4$ Hz, where it saturates. Interestingly, LE displays a frequency doubling property and its amplitude is fairly proportional to the third power of the field amplitude (Fig. 8).

I believe that the explanation is simple (see Appendix II) and may be taken as an introductory and fundamental point in this section. As the C_+^- dipoles are statistically oriented with respect to the external electric field, at any

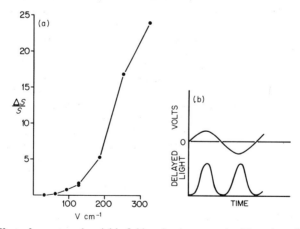

Fig. 8. Effect of an external variable field on luminescence in chloroplasts. Luminescence is observed $\cong 30$ sec after a saturating flash. Frequency of field = 60 Hz. (a) Dependence of the relative stimulated luminescence ($\Delta S/S$) upon the field strength. (b) Frequency doubling of stimulated luminescence; field strength = 314 V cm^{-1}. (Arnold and Azzi, 1971.)

instant, half of them, roughly speaking, will have their ΔE *increased* by an amount $F|X|d$ ($F = 1$ Faraday, $|X|$ is the absolute instant value of the field strength, d is the dipole length), whereas the other half will have their ΔE *decreased* by just the same amount. Therefore, again roughly speaking, only the second half will be induced to luminesce more than in the absence of the field, the total effect giving a time and field strength dependence as follows:

$$LE(t) = L \frac{\sinh[(Fd/RT)X_0 \sin(\omega t)]}{(Fd/RT)X_0 \sin(\omega t)} \tag{23}$$

which leads to frequency doubling; Eq. (23) also shows that LE is a strongly increasing function of $|X|$ (although not exactly of the third power). The frequency saturation probably reflects the increased electrical impedance of the electrolyte in the suspension medium as ω is raised. I believe that the LE effect is in urgent need of confirmation in view of its importance and significance, and, if confirmed, should find considerable application and extension. For instance, it would permit one to monitor luminescence, as a modulated signal, while the actinic light is *on*. It would be also interesting to look for an LH signal, i.e., induced by a variable magnetic field, owing to Faraday's law of induction applied to the dipoles considered as part of microscopic electronic and ionic "circuits" embedded in the membrane.

5.2. Diffusion Potential

Soon after Mayne and Clayton's demonstration of a ΔpH-induced luminescence, Miles and Jagendorf (1969) and independently Barber and Kraan (1970) observed that addition of salts to a preilluminated chloroplast suspension likewise induced an increase in luminescence (LS). LS decays in a few seconds (see Fig. 9). The effect is somewhat nonspecific and is effected by so-called "uncouplers", ionophorous substances and PS II inhibitors. A significant difference was pointed out by Miles and Jagendorf: unlike the pH jump, the salt jump did not induce any ATP formation, indicating that luminescence and photophosphorylation were not related in the same way to the elusive "high energy state."

5.2.1. Goldman Equation

The idea behind the Barber and Kraan experiment was the following. If the ΔpH-induced luminescence had to be explained by the lowering of ΔE brought about by a change in "protonmotive force" (Mitchell, 1966), then a similar effect was expected when a "diffusion" potential difference across the membrane was established by a sudden addition of salt. Diffusion potentials are well known to electrochemists and electrophysiologists: whenever two ions of opposite signs and unequal mobility diffuse, an electric

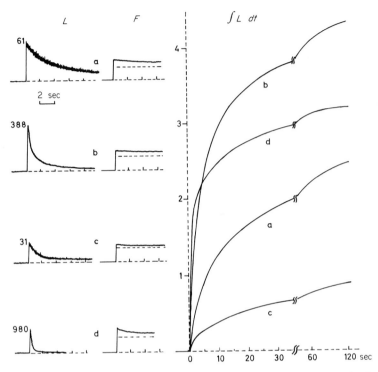

Fig. 9. Perturbation experiments in chloroplasts. Left: luminescence intensity L or LX (note that scales of ordinate are different; actual numerical values at initial peak are indicated). Middle: fluorescence intensity F (dashed line is F level after 2 mn darkness, approximating level of constant F). Right: Sum of light or time integral of L, ΣL or ΣLX. ΣL is control, ΣLX is experiment with perturbation X. Initial suspension medium: sucrose, tricine pH 7.8. Protocol: $[5\ (\text{pH }7.8)|X_1,\ 3,\ X_2|]_s$ (i.e., light period = 5 sec in initial medium, followed by perturbation X_1, waiting period = 3 sec and perturbation X_2; then measurement period) with: a = $\Sigma L{:}X_1 = X_2 = $ buffer pH 7.8; b = $\Sigma LS{:}X_1 = $ buffer pH 7.8, $X_2 = $ NaCl[1] 0,3 M; c = $\Sigma LA{:}X_1 = $ buffer pH 7.8, $X_2 = $ succinic ac. pH 4.2; d = $\Sigma LAB{:}X_1 = $ succinic ac. pH 4,2, $X_2 = $ Tris pH 8.7. (Kraan *et al.*, 1970; Kraan, 1971.)

potential gradient develops of such signs and magnitude as to slow down the faster ion and speed up the slower one. This state of matter or "electro-diffusion" is described exactly by a system of partial differential equations: the Nernst–Planck equation and the Poisson equation, which unfortunately is quite difficult to solve in general. A particular solution was derived by Goldman (1943) for use in (physiological) membranes. Under restrictive conditions (constant field across the membrane, no net electric current) which seem to be fairly well met for LS, the Goldman equation predicts that the electric potential difference resulting from unequal distribution of monovalent ions on both sides of the membrane is (see Appendix III):

$$V_{i-o} = \frac{RT}{F} \ln \left[\frac{\sum_j pC_j(C_j)_o + \sum_j pA_j(A_j)_i}{\sum_j pC_j(C_j)_i + \sum_j pA_j(A_j)_o} \right] \quad (24)$$

where subscripts i and o refer to the inner and outer phase of the thylakoid in our case, the summation is over all j ionic species (cations: C_j, anions: A_j) and pC_j and pA_j are permeability coefficients (analogous to the mobility coefficients of "free" electrodiffusion). Note that the potential difference V_{i-o}, as defined, is positive if the potential of the inner phase is larger than that of the outer one.

In a salt jump experiment, for instance with KCl, Eq. (24) assumes a simpler form:

$$V_{i-o} = \frac{RT}{F} \ln \left[\frac{pK(K^+)_o + a}{pCl(Cl^-)_o + b} \right]$$

where a and b are constants representing the contribution of all other ions present in the system before the jump. If, further, a and b are relatively small,

$$V_{i-o} = \frac{RT}{F} \ln \frac{pK(K^+)_o}{pCl(Cl^-)_o} \quad (26)$$

The thylakoid membrane is "notoriously" less permeable to anions than to cations, hence V_{i-o} is very likely to be positive ("inside"). According to the ΔE hypothesis (Barber–Crofts), the new activation energy (ΔE_s) will be

$$\Delta E_s = \Delta E - FV_{i-o} \quad (27)$$

which is very likely *less* than ΔE (see Fig. 7). Combining Eqs. (2), (6), and (7), the L equation will read:

$$LS = (\phi)v \exp \left(-\frac{\Delta E}{RT} \right) \frac{pK(K^+)_o + a}{pCl(Cl^-)_o + b} (C_+^-) \quad (28)$$

I note for future reference [compare Eqs. (21)–(27)] that LS/L is temperature-independent, except for the permeability coefficients (however, since they enter as a ratio, their temperature effect should be small). This point would be interesting to check experimentally. We will group, following Barber and Kraan, all the nonrelevant terms for LS into a constant α and write in general, in place of Eq. (28):

$$LS = \alpha \frac{pC(C^+)_o + a}{pA(A^-)_o + b} \quad (29)$$

5.2.2. Experimental Evidence

The above relation for *LS* has been shown to be in fairly good agreement with the experiments of Barber and Kraan (1970), Barber and Varley (1972), and Barber (1972).

5.2.2.1. Permeability. Most experiments in *LS* and *LX*, in general, were made with stopped-flow devices, $t_L \cong 10$ sec, i.e., when *L* itself is very low. This was a nice condition for the salt-jump effect to show up. Equation (29) predicts that *LS* should increase when β, defined by Barber as the ratio of the anion permeability to the cation permeability ($=$ pA/pC), decreases. For instance, β is much lower for potassium benzoate (with a bulky anion) than for KCl and, indeed, *LS* is $\cong 19$ times larger with potassium benzoate than with KCl (Krann, 1971). Valinomycin (inducing a high permeability to K^+) specifically enhances *LS* for potassium benzoate (Fig. 10).

5.2.2.2. Competition. When several salts are added simultaneously, Eq. (29) has to be modified accordingly and this affords another type of verification: under suitable conditions (constant cation concentration), a hyperbolic relation is easily derived between *LS* and the two anion concentrations in the mixture. This relation is verified in general, except for benzoate (Fig. 11), but the discrepancy is reasonably well accounted for assuming a nonnegligible permeation of the undissociated benzoic acid. In a similar vein, successive additions of the same salt produce decreasing *LS* signals. Incidentally, such competition experiments were used by Barber and Varley (1972) to estimate membrane permeabilities for various ions and the effect of ionophorous agents on them.

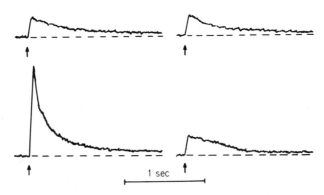

Fig. 10. Luminescence stimulated by addition of benzoate salts (0.3 *M*) in control (top) or after pretreatment with Vmc ($3.3.10^{-6}$ *M*) (bottom) in chloroplasts. Addition of K benzoate (left) or Na benzoate (right). Initial suspension medium sucrose, Tris, pH 7.2. Protocol: $[20|10,S|]_{sec}$ (i.e., light period $=$ 20 sec, waiting period $=$ 10 sec, followed by addition of salt and measurement period). (Kraan, 1971.)

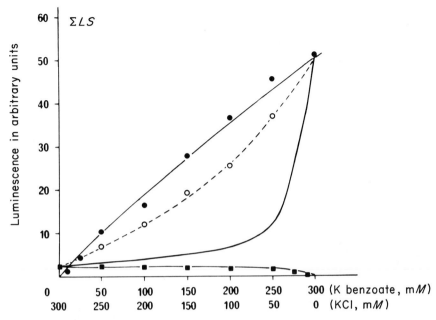

Fig. 11. Luminescence stimulated by addition of KCl (■), potassium benzoate (●) or both salts in competition (at constant $[K^+] = 300$ mM) (○) in chloroplasts. Solid line without dots is theoretical curve for the competition experiment. Protocol similar to that of Fig. 10 ([5|10, S|]$_{sec}$). (Barber and Varley, 1972.)

5.2.2.3. Luminescence in mV Scale. An excellent verification of the Goldman equation was made by Barber (1972) who expressed V_{i-0} as a function of LS [substitution of Eq. (29) into Eq. (24)]. $\overline{LS}$ was measured with a phosphoroscope at $t_L \cong 1$ msec (Fig. 12). This is equivalent to the calibration of luminescence in electric potential units. The zero of the voltage scale was set by the addition of Vmc (K^+ being already present in the medium) which should "short circuit" the photoelectric field (see Appendix III). One might argue that the photoelectric field has not disappeared at $t_L \ll 1$ msec, if only for reason of large differences in rates. (The photochemical charge separation is presumably much faster than the diffusion of K^+ across the membrane, even assisted by Vmc; besides, K^+ has to go back and forth at each phosphoroscope cycle.) Nevertheless, the results are impressive: the diffusion potential is quite independent of $\overline{L}$ (which may vary from one chloroplast preparation to the next) and the same observation applies to its modification by Vmc. We see that the photoelectric potential at $t_L \cong 1$ msec, as measured on this scale, is in the range of 50–80 mV [to be compared with its maximum value $\cong 200$ mV, as calculated from the 515 nm absorbance change (see Witt, 1971)]. [The inhibition of $\overline{L}$, under similar conditions, by Vmc was observed earlier by Wraight and Crofts (1971).]

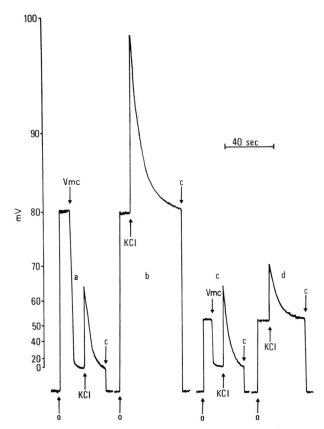

Fig. 12.   Luminescence stimulated with KCl, with or without Vmc in chloroplasts. Additions where indicated by arrows. a, b and c, d are different batches of chloroplasts. See text for mV scale. Protocol: steady-state signal in phosphoroscope method, $[1.5 \| 0.3 \|]_{msec}$ (i.e., light period = 1.5 msec, measurement period = 0.3 msec). (Barber, 1972.)

5.3. pH Transition

The discovery of a luminescence induced by a pH gradient (*LAB*)* by Mayne and Clayton (1966) was particularly exciting, because it first looked like a true "chemiluminescence" (see also Fleischman, 1971, and Section 3.2.3), although the inhibitory effect of DCMU had some curious "smell". It was subsequently recognized by Mayne (1968) that preillumination was necessary in order for *LAB* to be manifested. This field has been since extensively explored, notably by Kraan *et al.* (1970) and Kraan (1971), and has given rise to a most important hypothesis, which I shall discuss in detail.

*A, B stands for "acid", "base", respectively; although the order of letters (AB) implies a pH shift in the positive direction (which is the case in many experiments), the generic symbol *L*AB will be used irrespective of the sign of the pH shift.

5.3.1. Properties of Luminescence Induced by Acid–Base Transition (LAB)

A priori, there should be nothing unusual about H^+; after all, it is another ion and its effect on luminescence should have been included above. Whatever has been said there also applies to H^+ and this should be kept in mind in considering its specific effects. However, LAB has to be discussed separately, because the photosynthetic chain is "transducing" energy both to electrons and protons.

LAB was studied by Kraan and collaborators with the same experimental procedure as for LS. The following factors were found to affect quantitatively the size and shape of the signal: t_L, t_{LA} (i.e., t_L spent in the acid phase of the process), pH of the acid phase, size of ΔpH, and chemical composition of buffer. The main results were:

1. LAB is the strongest of all signals under identical conditions (Fig. 9). In addition, another signal LA was observed when shifting the pH of the chloroplast suspension to very low values (pH = 4); its properties seem to reflect, in part, a destructive effect on the membrane.

2. LAB is, more than any other similar signal, attended by sizeable effects on $\Delta\phi$ (Figs. 9 and 13); as a rule $\Delta\phi$ decreases upon lowering the pH starting from near neutrality and increases with pH change in the opposite direction [see also de Kouchkovsky (1972) and Papageorgiou and Govindjee (1971)].

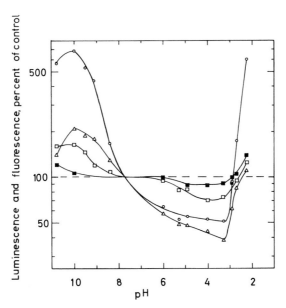

Fig. 13. Effect of size of pH jump on luminescence (○), light-sum (△), variable fluorescence (□) and constant fluorescence (■) in chloroplasts. Protocol similar to that of Fig. 9. Initially, pH was 7.8. (Kraan, 1971.)

3. *L*AB may be larger or lower than the control depending on the size and sign of ΔpH (Fig. 13, which also displays the narrow pH range of the special *LA* signal). Starting from pH 7.8, it is seen that luminescence is enhanced by shifting to the basic side, and, on the contrary, depressed by shifting to the (medium) acid side. Operationaly this depression may not actually be an inhibition but a manifestation of the "head cut-off effect" (t_{LA} is long, see Section 2.2.1.1.2). Over restricted pH ranges, *L*AB is fairly close to an exponential function of ΔpH.

4. DCMU has a complex effect: *L*AB is decreased and the ΔpH-induced change of $\Delta\phi$ is suppressed.

5.3.2. *Kraan's Hypothesis: Protonation Equilibria in the PS II Center*

Rather than insisting on any fashionable; pH-induced "structural" change, which, in particular for *LA*, is not altogether ruled out. Kraan points directly to the important concept: there must be a translocation of protons from the outer to the inner phase, more or less concomitant to the opposite translocation of electrons driven by the PS II centers; experimental evidence is clear on that point (light-induced ΔpH, henceforth: Δ*pH) and the two processes might as well occur in close connection to each other. Figure 14 depicts schematically the hypothesis. Accordingly, following Kraan's nomenclature, we have to consider two *separate* equilibria:

$$\left.\begin{array}{c} Q^- + H^+ \rightleftharpoons QH \\ ZH^+ \rightleftharpoons Z + H^+ \end{array}\right\} \tag{30}$$

occurring after the charge separation and only Q^- and ZH^+ must be considered as substrates of luminescence. Essentially, this is the "protonated" version of the recombination hypothesis. In addition, Kraan assumes that ZH^+ is identical or in equilibrium with that entity in S_3 which carries three positive charges and he includes a pH-sensitive equilibrium with the PQ pool on the Q side:

$$Q^- + H^+ + \tfrac{1}{2}PQ \rightleftharpoons Q + \tfrac{1}{2}PQH_2 \tag{31}$$

this process being blocked by DCMU.

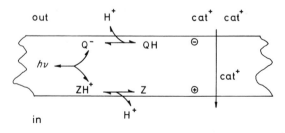

Fig. 14. Scheme of the PS II protonation hypothesis. See text. (Kraan *et al.*, 1970.)

The main evidence in favor of this scheme—which obviously works in the right direction—is, according to Kraan, the exponential dependence of LAB on ΔpH. Indeed, one may write (see Appendix IV):

$$L \sim \left\{ 1 + \frac{(H^+)_o}{K_Q} + \frac{K_Z}{(H^+)_i} + \frac{K_Z}{K_Q} \cdot \frac{(H^+)_o}{(H^+)_i} \right\}^{-1} \tag{32}$$

where subscripts o and i have the usual meaning and K_Q, K_Z are the dissociation equilibrium constants in Eq. (30). If the last term is dominant in Eq. (32), it follows that

$$L \sim \exp\left[-\frac{1}{2.3}(pH_i - pH_o) \right] = \exp\left(-\frac{1}{2.3}\Delta pH \right) \tag{33}$$

The DCMU effect is mainly the result of blocking of the Q/PQ equilibrium [Eq. (31)]: the lowering of Q^-, hence of $\Delta\phi$ by acidification is no longer possible and, during the base phase of LAB, Q^- is no longer formed from PQH_2.

5.3.3. Extension of the PS II Protonation Hypothesis

Kraan's hypothesis raises several interesting points and needs, in my opinion, some extension.

5.3.3.1. *L is Monomolecular.* Kraan's reasoning is entirely built on the assumption that L is bimolecular which, as suggested in Section 4.1. is wrong.* However, this turns out to be without consequence to Eq. (30), because (see Appendix IV) the more complicated system of equilibria involving the four possible forms of the center: C, C_+^-, C^-, C_+ (where C is HZChlQ, C_+^- is $^+$HZClQ, etc.), with the *four* corresponding equilibrium constants K_Q^+, K_Q, K_Z^-, K_Z, happens to give under reasonable assumptions, just the same result. These "reasonable assumptions" might nevertheless be open to question.

5.3.3.2. *Titration Curve.* Dropping, as we did, three terms in Eq. (32) is presumably not correct, when dealing with pH's departing from neutrality. Then, leaving Eq. (32) as it is, we may look at Fig. 13 with a slightly different interpretation in mind. I believe that the complex pattern is reminiscent of the titration curve of a zwitterionic substance, suggesting a pK_Q in the range 8–9 and pK_Z somewhere below 4. [QH has also been considered recently as a weak acid by Wraight *et al.* (1972).]

5.3.3.3. *Stabilization.* It is instructive to bring together Crofts' hypothesis and Kraan's hypothesis (Fig. 7) and consider them from the point

---

*The point that the L decay may sometimes display a *second-order* behavior does not mean necessarily that the recombination is *bimolecular*. As indicated in Section 4.1., the high photochemical efficiency of the reaction center cannot be understood if the primary electron acceptor and donor are freely diffusing species.

of view of the activation energy of recombination; we see two successive steps of stabilization: (1) from the excited Chl effector to the *photoelectric* dipole, and (2) from the latter to the *photochemical* dipole. One may notice that the two kinds of dipole will not respond in the same manner to the cooperative factors of the photosynthetic apparatus considered in Section 4.1, and, in particular, that the transition (photoelectric $\longrightarrow$ photochemical) is attended with a loss of sensitivity to the electric field because, only, the species C_+^- has the character of an electric dipole and is acted upon by the electric field. The kinship of the two processes is also reflected in an attenuated temperature dependence, expected from the structure of Eq. (32), which reminds us of what was previously said for Eq. (28). Here also, experimental verification would be required.

5.3.3.4. The S States. The energy level scheme helps one to go one step further: I would like to speculate that the four dipoles corresponding to S_4, S_3, S_2, S_1 have their energy levels arranged in such order as to satisfy (see Fig. 7) the inequalities postulated in Eq. (21). This would be a natural assumption if the Z "head" of the dipole is physically able to accumulate up to four positive charges. There is unfortunately too little information available at the present to go any further in the discussion. However, at least two alternative schemes suggest themselves concerning the protonation state on the Z side (Fig. 7). If, for instance, a Δ*pH sequence would show a periodicity of four in synchrony with the O_2 sequence, version A of the hypothesis (Fig. 7) would be demonstrated. The energy level arrangement of the S dipoles is not the same in version B, and here no periodicity in the Δ*pH sequence is expected. In parallel with the Δ*pH sequence, it would also be interesting to look for a $\Delta\psi$ or $\Delta515$ sequence (photoelectric membrane potential): an oscillatory pattern is again expected for version A as compared to version B.

An obvious extension for further investigation of the protonation stabilization and of its exact significance for the Z side would be to use the isotopic effect and reexamine the whole field of luminescence (as well as fluorescence and O_2 emission) when protons are replaced by deuterons.

5.4. Oscillations

Recently, Hardt and Malkin (1973) have shown that the various *LX*, when produced in the sequential mode (at $t_L \cong 1$–10 sec), also oscillate just like the O_2 sequence, with maxima at the 3rd, 7th, etc. flashes. This brought to a clear-cut conclusion an old controversy in the field of luminescence, which I did not mention so far. Were there several *distinct* forms of luminescence arising from *distinct* substrates and processes? Hardt and

Malkin's results have the unescapable conclusion that there is but *one* luminescence and that it may be "triggered" (Mayne, 1968) by various external factors. (Other less stringent arguments will also be discussed in Section 6.4.2.)

Hardt and Malkin have also contributed to studies of *LS*, *L*AB, *L*A, but their results are concerned with kinetics and will be mentioned in Section 6. Lastly, Hardt and Malkin (1972b) have discovered a new type of triggered luminescence (in symbol, *L*0) by the addition of various organic substances. They have tried to correlate *L*0 with the dielectric constant, the solubility in the membrane and rate of penetration into it of the agents; the correlation seems, on the average, significant. They ascribe the effect to a facilitation of detrapping of charged entities resulting from a decrease in the dielectric constant. *L*0 is probably an important phenomenon which deserves further study. At this point, one may only say that, besides the dielectric constant, the dipole moment of these agents could also be a significant parameter.

In concluding this section, I wish to stress again the paramount importance of the activation energy concept in luminescence, as first noted by Arnold (1966). Activation energy means stabilization and we have seen here that two consecutive steps of stabilization were operating in succession during the energy transduction from photons to electrons plus protons. This section also brings to an end the qualitative survey of luminescence studies. "Qualitative" is not derogatory, it often means "yes" or "no," but, ultimately, every theory has to pass the test of quantitative kinetic analysis.

6. KINETICS OF LUMINESCENCE

Section 6 has been referred to so much in the above sections, that, I think, it does not need more introductory comments. First, I will discuss the main consequences of luminescence being a rate and, second, review and discuss decay and induction time courses, together with their various factors.

6.1. Kinetic Signification of L

As evidenced in the structure of the L relation [Eq. (2) in section 3], the luminescence intensity (or flux) L is, to a constant factor r (see Section 3.1.1.1), the rate of decay by recombination of C_+^- (with collective meaning of this symbol):

$$L = \left[-\frac{d(C_+^-)}{dt} \right]_L \tag{34}$$

L, in general, is a part of the total decay rate of C_+^-:

$$-\frac{d(C_+^-)}{dt} = \left[-\frac{d(C_+^-)}{dt}\right]_L + \sum_k \left[-\frac{d(C_+^-)}{dt}\right]_k \tag{35}$$

where the summation is over all the k processes in competition with L—in fact, mostly electron and proton flows.

Depending on the relative contribution of L in the total decay, two simple situations may occur.

1.
$$\left[-\frac{d(C_+^-)}{dt}\right]_L \ll -\frac{d(C_+^-)}{dt}$$

I shall call this the "leakage type." Obviously, in this case, luminescence has a negligible effect on the kinetics of C_+^- and on the law of its decay; however, it allows one to "monitor" the time course of C_+^- (Lavorel, 1971). Therefore, in the leakage type, the decay of C_+^- has to be calculated separately and L is some function of the kind discussed previously; for instance, we would have to write Eq. (22) as (including explicit functions of t_L):

$$L(t_L) = (\phi)_{t_L} v \exp\left[-\Delta E_{t_L}/RT\right](C_+^-)_{t_L} \tag{36}$$

where it is understood that $(\phi)_{t_L}$, ΔE_{t_L} and $(C_+^-)_{t_L}$ are functions of t_L.

2.
$$\left[-\frac{d(C_+^-)}{dt}\right]_L \cong -\frac{d(C_+^-)}{dt}$$

For reasons that will become clear, I shall call this the "deactivation type." In this case, recombination is not negligible; it is, in fact, *the* only process of decay for C_+^- and we have to write:

$$L(t_L) = -\frac{d[C_+^-(t_L)]}{dt} \tag{37}$$

Consequently, if both decays $[L$ and $(C_+^-)]$ are known experimentally, they must verify Eq. (36), and, the kinetic law for (C_+^-) may be obtained by time integration of the kinetic law of L.

To be complete, I shall speak of "intermediate type" when none of the above relations are valid.

The t_L scale is a good means of appreciating the respective importance of the above defined types of L (see Fig. 15). At short t_L, the leakage type is likely to be predominant, for the obvious reason of efficiency of the photochemical act. Conversely, at long t_L, there are no forbidding reasons not to consider the deactivation type as the main one. The intermediate type would then occur around some critical time "t_L (crit)" which according to this reasoning should be identified with one of the turnover τ's of the PS

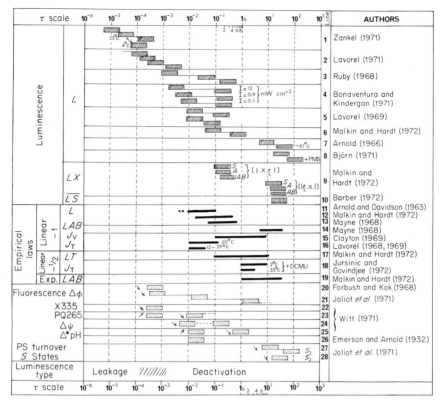

Fig. 15. The τ scale. Log scale of time. L = luminescence intensity; LX = luminescence intensity stimulated by X (= S = addition of salt, = A = addition of acid, = AB = acid–base transition, = T = temperature jump); $\overline{LX}$ = average of same type of signal over a phosphoroscope period; J_V, J_T: see L relation, using variable, total fluorescence, respectively; $\Delta\phi$ = variable fluorescence yield; X335 = absorbance change at 335 nm due to a hypothetical carrier between photoreaction II and plastoquinone; PQ 265 = absorbance change at 265 nm due to plastoquinone; $\Delta\Psi$ = change in membrane potential; Δ*pH = light induced pH change. Monophasic (exponential) kinetic components are represented by rectangles extending from $0.5\,\tau$ to $2.\tau$, the area below the decay curve between these limits being $\simeq 50\%$ of the total area. $\uparrow$:rise, $\downarrow$:decay. Line 9: $[\,|,X,t_L|\,]$ = decay as a function of time following perturbation, $[\,|t_L,X,|\,]$ = decay as function of time before perturbation.

II centers: the lifetime for the fast oxidation of Q^- seems a fair approximation (Fig. 15). We may also expect that, in blocked or otherwise nonfunctional systems, the deactivation type will be predominant.

A confusing point should be mentioned. The fact that L has to be a "small" process has nothing to do with the speed of its decay. Table I shows several criteria of rapidity for various kinetic processes; it is seen that, for

TABLE I

Properties of Some Kinetic Laws[a]

LAW $X = f(t)$	Initial rate constant (k) $\left(-\frac{1}{X}\frac{dX}{dt}\right)_{t=0}$ f	$\left(-\frac{dt}{dX}\frac{dX}{dt^2}\right)_{t=0}$ f'	Time constant (τ) f (Multiexponential analysis)	f'	Half-life $(t_{1/2})$ $X(t_{1/2}) = \frac{1}{2}X(0)$ f	$X'(t_{1/2}) = \frac{1}{2}X'(0)$ f'	$\dfrac{(k)f'}{(k)f}$	$\dfrac{(\tau)f'}{(\tau)f}$	$\dfrac{(t_{1/2})f'}{(t_{1/2})f}$
1st order $X = \exp(-kt)$	k	k	k^{-1}	k^{-1}	$\ln(2)k^{-1}$	$\ln(2)k^{-1}$	1	1	1
2nd order (biequimolecular, linear (-1)) $X = (1 + kt)^{-1}$ $\{=f[X(0)]\}$	k $\{=f[X(0)]\}$	$2k$	(nd)	(nd)	k^{-1}	$(\sqrt{2} - 1)k^{-1}$	2	(nd)	2.41
(binonequimolecular) $X = \dfrac{\alpha \exp(-kt)}{1 - \beta \exp(-kt)}$ $\alpha + \beta = 1$	k/α	$(1 + \beta)k/\alpha$	$(nk)^{-1}$ $n = 1,\infty^b$	$(nk)^{-1}$ $n = 1,\infty^b$	—	—	$1 + \beta$	>1	—
3rd Order (linear $(-\frac{1}{2})$) $X = (1 + kt)^{-1/2}$ $\{=f[X(0)]\}$	$k/2$ $\{=f[X(0)]\}$	$3k/2$	(nd)	(nd)	$3k^{-1}$	$(\sqrt[3]{4} - 1)k^{-1}$	3	—	5.11
General polyphasic (multiexponential) $X = \Sigma\alpha_i \exp(-k_i t)$ $\Sigma\alpha_i = 1$ $\{=f[X(0)]\}$	$\Sigma\alpha_i k_i$	$\dfrac{\Sigma\alpha_i k_i^2}{\Sigma\alpha_i k_i}$	k_i^{-1}	k_i^{-1}	—	—	$\dfrac{\Sigma\alpha_i k_i^2}{(\Sigma\alpha_i k_i)^2}$ ≥ 1	—	—

[a] f = timecourse of the quantity X defining the process; f' = timecourse of its rate. nd: the quantity is not defined. (—) means no general expression.

[b] τ of successive components.

the functions we are interested in (positive monotonously decreasing), the rate of decay of the time derivative is always equal to or larger than that of the function itself and, precisely, that the equality holds only for the first-order process.

In the deactivation type, Eq. (37) shows clearly the importance of $\Sigma L(\infty)$, the total time integral of L, since it provides a measure of C_+^- present at $t_L = 0$. I should stress again the importance of the "head cutoff effect", hence the need for exploring very short t_L, in a correct measurement of ΣL; examples for this point will be given below.

It is interesting to point out that Eq. (37) for the deactivation type, is the kinetic counterpart of the "substrate" approach, equivalent to the L relation for the "exciton" approach. Therefore, the delay and redistribution effects discussed in Section 2.1.1 should or might also be taken into consideration as factors in the decay law of C_+^-.

A last remark is of practical importance. The numerical analysis of kinetic data in the literature is too often unsatisfactory. We still use graphical approximative procedures of another epoch, while modern means do exist for a rational estimation of kinetic parameters and, above all, for appreciation of goodness of fit between the experimental set and the model function (see for instance, Isenberg and Dyson, 1969). I believe, like Cope (1964), that, for unorthodox kinetics, the analysis in terms of exponential components (polyphasic type) is a valid procedure, corresponding to the Fourier decomposition of periodic phenomena. It does not "mean" a priori anything mechanistically; it may, however, suggest a mechanism and, above all, allow a precise comparison of kinetic behavior under different experimental conditions. I suggest that a "spectral" display (amplitude vs lifetime) similar to Fig. 15 be systematically used to qualify complex kinetic phenomena.

6.2. Decays

The above definition of t_L (crit) should lead to classification of decays into two types. However, for practical and technical reasons, I will distinguish three types: fast, medium and slow, with arbitrariness in t_L; this actually is of no consequence if only orders of magnitude are considered. Throughout this part, Fig. 15 should be frequently consulted for easy comparison of lifetimes in luminescence and connected areas, of validity of empirical laws, etc. I will simply refer to it as the "τ scale."

6.2.1. Fast Decays

The time domain will be defined as $t_L < t_L$ (crit) ($\cong$ a few 100 μsec to 1 msec). As mentioned above and seen on the τ scale, t_L (crit) is an important parameter for PS II and may be called its turnover time: events occurring at this time (fast decay of $\Delta\phi$ and reduction of plastoquinone) are characteristic of electron transfer from PS II to PS I. The electric field is still there, we deal here with a leakage luminescence.

Few data are available so far in this area. We discussed in Section 4.2.2 the qualitative aspects of the oscillatory pattern which Zankel (1971) reported. Beyond t_L (min) $\cong 50\,\mu s$ and up to $t_L \cong 100\,\mu sec$, he observed three first-order components, which I will refer to as $z1, z2, z3$, with lifetimes of 10, 35, and 200 μsec. Actually, only $z2$ (35 μsec) could be clearly qualified as first-order for reason of limitation of t_L range. No oscillation was found for the $\Delta\phi$ decay sequence. The following properties were seen:

1. $z1$ (10 μsec) was only seen at the first flash and disappeared in the presence of DCMU. I note that, due to the "head cutoff effect", these properties cannot be taken as proof that, either at the second, etc. flashes or in presence of DCMU, this component is non-existent: all one can say is that it might have been missed.

2. $z2$ (35 μsec), as well as $z3$, oscillates as already mentioned; it is partly inhibited in amplitude (α) by DCMU, but the lifetime is not changed. The activation energies for α and τ are 0.25 and 0.20 eV, respectively.

3. $z3$ (200 μs) is altogether eliminated in the presence of DCMU. We may thus agree, with Zankel, that it reflects the oxidation decay of Q (toward PQ) associated with the decay of $\Delta\phi$. At this point, it is interesting to note that we are in a situation where Eq. (17) (Section 3) is legitimate: since in this case $\tau_J = \tau_{\Delta\phi}$, we would find $\tau_{\Delta\phi} \cong 2.\tau_L$, which is in fair agreement with actual data ($\tau_{\Delta\phi} \cong 600\,\mu sec$ as compared to $2 \times 200\,\mu sec$, see τ scale).

I have recently extended Zankel's data down to t_L (min) $\cong 6\mu sec$ (unpublished) and found results in good agreement with his. An interesting point (Fig. 16) is that the second component ($\cong z2$) is quite prominent *after* treatment with a high concentration of NH_2OH (destructive to the Z side) and that it is inhibited by DCMU, with the appearance of a shorter component ($1D$).

In keeping with the ideas developed in the preceding sections, I would like to propose the following scheme (see Fig. 7,IV):

$$z1: {}^{(n)+}Z{-}^{+}YChlQ^{-} \rightleftharpoons {}^{(n+1)+}Z{-}YChlQ^{-} \tag{38}$$

The distinction between Y the true primary donor, and Z, the "hole" accumulating device was postulated by Bennoun (1971a,b) and Renger (1972) (also see Kraan, 1971). The kinetic component $z1$ is to be considered as a step in the stabilization of the photoelectric dipole. After destructive treatment with NH_2OH, this component disappears because, Z being cut off from the center (Bennoun), we are left with YChlQ.

$$z2: {}^{(n+1)+}Z{-}YChlQ^{-} + H^{+} \rightleftharpoons {}^{(n+1)+}Z{-}YChlQH \tag{39}$$

(Q protonation in Kraan's hypothesis). This step is still seen with the

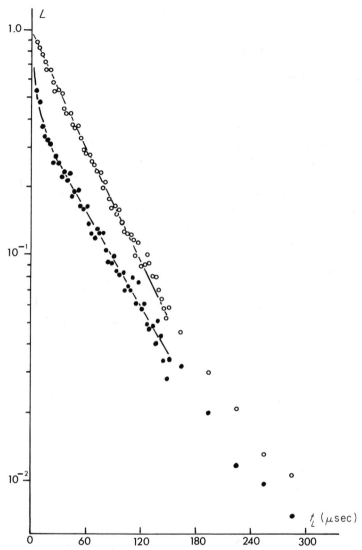

Fig. 16. Fast decay of luminescence intensity (L) in the presence of NH_2OH $10^{-3}\,M$ (○) or NH_2OH $10^{-3}\,M$ + DCMU $10^{-5}\,M$ (●) in *Chlorella*. 1 μsec xenon laser flash. First point recorded at $t_L = 6\,\mu$sec. Each point is the average of signal over 100 samples, each having received 100 flashes. (Lavorel, 1972.)

"cutoff center" YChlQ (action of NH_2OH), but it is depressed in the presence of DCMU, while component $1D$ appears. Therefore, one may assume another process in competition with the $z2$ step [Eq. (39)] as follows:

$$1D: {}^{(n+1)+}Z\text{–}YChlQ^- + DCMU \rightleftharpoons {}^{(n+1)+}Z\text{–}YChlQ^-\text{–}DCMU \quad (40)$$

The stability of the DCMU complex would be sufficient to account for the DCMU inhibition of the following (classical) step:

$$z3: {}^{(n+1)+}Z-YChlQH + \tfrac{1}{2}PQ \rightleftharpoons {}^{(n+1)+}Z-YChlQ + \tfrac{1}{2}PQH_2 \quad (41)$$

We have to remember that all these components are subject to amplitude "modulations" of two kinds: the first, on account of the validity of the L relation near the "P" state [hence, Eq. (17)], the other resulting from variation in the distribution of the S_n states with their different ΔE_S [see Fig. 7,IV and Eq. (35)]. The second modulation obviously explains the inhibition of the oscillatory pattern by DCMU observed by Zankel.

Very recently, Haug *et al.* (1972) reported analysis of L at a still shorter t_L range, where the junction with the tail of the fluorescence decay immediately following the actinic flash can be observed. Figure 17 shows the almost vertical drop of the prompt fluorescence signal during the "closure" of the fast shutter, from which $\tau_F \cong 2$ nsec may be calculated; at $t_L \cong 0.3\,\mu$sec, the fastest L component with $\tau_L \cong 34\,\mu$sec appears almost flat and horizontal as seen at this time scale. The simplicity of this pattern is noteworthy: it clearly indicates that fluorescence is actually "prompt" as compared to luminescence and well separated chronologically—and mechanistically—from the latter. Under these experimental conditions, extrapolation at

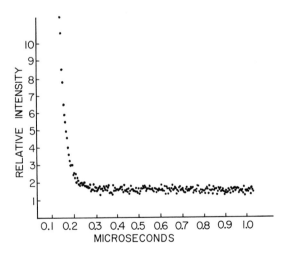

Fig. 17. Junction of fast component of luminescence decay (near horizontal section) with fluorescence decay at the closure of shutter (near vertical Section) in *Scenedesmus*. Protocol: steady-state signal in phosphoroscope method, $[10^3|0.25|10^3|0.25]_{\mu sec}$ (i.e., light period = 1 msec; occultation period = 0.25 μsec, measurement period = 1 msec, occultation period = 0.25 μsec). (Haug *et al.*, 1972.)

$t_L = 0$ seems warranted and Haug *et al.* are able to calculate the ratio L/F as 1/165. These beautiful results firmly establish the long-sought-for distinction between luminescence and fluorescence (see Arnold and Davidson, 1963).

Although luminescence is qualified as a "leakage" within this t_L range, it may still be significant in accounting for the "misses" in the functioning of the O_2 emission mechanism. Finally, a puzzling point is raised when comparing the simple kinetic behavior of L decays in the case of Fig. 16 with the report by Mauzerall (1972) that the "photochemical" rise of $\Delta\phi$ takes up to 10 μsec to get to its maximum.

6.2.2. Medium Decays

The t_L domain will be from $\cong 100$ μsec up to $\cong 10$ sec. The choice of such a large span on the τ scale is justified technically, because many results were obtained in this area under comparable conditions and methodology, and theoretically, because several important events correlated with luminescence are occurring in this time range: slow decays of $\Delta\phi$, decay of ψ, rise and fall of Δ *pH, emptying of the PQ pool and, mainly, onset of O_2 deactivation (see τ scale). From now on, we cannot decide a priori whether we are dealing with any of the three luminescence types which have been defined above.

This area furnishes the most abundant crop of data and has to be surveyed with many subdivisions.

6.2.2.1. Distribution of Lifetimes. This distribution is displayed on the τ scale. It represents the components resulting from analysis of a limited set of decay curves as polyphasic decays (sums of exponentials), either as given in the literature or estimated by me. It includes heterogeneous material, such as very low temperature or perturbation data. It is evident that no simple pattern emerges from this compilation. Besides the complexity of luminescence kinetics, this situation probably results from many circumstantial factors, such as diversity of temperature (lines 5–7) or actinic intensity (line 4). There is possibly a "blind" area in the range of $t_L \cong 1$–10 sec of technical origin (maximum t_L of fast methods, low level signals).

6.2.2.2. Empirical Laws. Many authors have endeavored to fit their data with simple mathematical equations (resulting, after rearrangement, in linear plots). The t_L domains of validity of these empirical laws, according to what is published, has been indicated on the τ scale. They concern both L and J (or their various equivalents in perturbation experiments). For the moment, one may only note that these laws are rather few in number, that they generally are in agreement among themselves (within homogeneous categories) and that two main types seem to occur:

$$\text{linear } (-1): X^{-1} \sim t_L \tag{42a}$$

$$\text{linear } (-\tfrac{1}{2}): X^{-1/2} \sim t_L \tag{42b}$$

i.e., either the reciprocal of the quantity (X) [Eq. (42a)] or the reciprocal of its square root [Eq. (42b)] is a linear function of t_L. These facts are probably significant and I shall frequently refer to them below and in Section 7.

6.2.2.3. Correlated Empirical Laws. By this I mean the (few) published cases where the kinetics of luminescence *and* of its suspected substrate can be or have been quantitatively correlated. Obviously, according to the law [Eq. (37)], the deactivation type would be the easiest situation that one may hope to deal with. Two cases are known.

1. Recall (see Section 4.3) that Bennoun (1970) concluded, from his qualitative comparison of $\Delta\phi$ and L in the presence of DCMU with or without NH_2OH, that Q^- in presence of DCMU could only decay through the recombination process. By applying an integral relation between $\Delta\phi$ and Q^- (see Malkin (1966)), which is independent of any PSU model and permits a photochemical "titration" of Q^- when the curves $\Delta\phi = f(t_I, t_L)$ are known, Bennoun could also show that the decay of Q^- was linear (-1) [Eq. (42a)], i.e.:

$$(Q^-)^{-1} = (Q^-)_0^{-1}(1 + kt_L) \tag{43}$$

Now, it can be simply shown that Bennoun's conclusion was also quantitatively correct. If it is, then luminescence is of the deactivation type and, according to Eq. (37), L obeys a linear $(-\tfrac{1}{2})$ law (see Appendix V), i.e.:

$$(L)^{-1/2} = (L)_0^{-1/2}(1 + kt_L) \tag{44}$$

I have plotted Bennoun's luminescence data according to Eq. (44) and compared them with his Q^- data plotted according to Eq. (43) (Fig. 18). The agreement is excellent. Since the L data were not taken at the same temperature as the Q^- data, the proof is not complete, for both k's in Eqs. (43) and (44) should be identical. However, the Arrhenius plot (Fig. 18) shows that the point obtained from L fits nicely with the two other points obtained from Q^-. Hence, we have at last a clear-cut proof that the deactivation type does exist. In the DCMU system, we can write:

$$\text{Luminescence} = \text{deactivation} \tag{45}$$

and read it in both directions.

An unexpected confirmation is found from data collected by Jursinic and Govindjee (1972) also on DCMU-treated algae. They found (see τ scale)

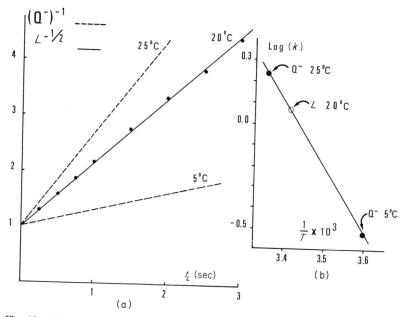

Fig. 18. Linear (-1) law for (Q^-), the concentration of reduced primary acceptor of photoreaction II and linear $(-\frac{1}{2})$ law for L, the luminescence intensity for *Chlorella* in the presence of $2.10^{-5}\,M$ DCMU. (Data from Bennoun, 1970.) (a) $1/(Q^-)$ (from Bennoun's Fig. 1B) (dashed lines) and $1/\sqrt{L}$ (calculated from Bennoun's Fig. 4) (continuous line). (b) Arrhenius plot of k from $1/(Q^-)$ and $1/\sqrt{L}$ plots [see text, Eqs. (43) and (44)].

that J_T was linear $(-\frac{1}{2})$. Assuming that $\Delta\phi$ is nearly proportional to (Q^-) and that, as in Bennoun's experiment, (Q^-) is linear (-1), then it is found that, provided $\Delta\phi < 1$, J_T had to be linear $(-\frac{1}{2})$ (see Appendix V), Incidentally, this result would indicate that the L relation can be factorized and that it is not such bad as approximation after all.

2. In section 4.2.1 we mentioned various correlations found by Joliot *et al.* (1971) between the kinetics of luminescence and that of the S states in normal algae (Fig. 19). A quantitative check has not been made but still the comparison is suggestive and supports Joliot's conclusion that deactivation is closely associated with luminescence. I dare say that again, here, Eq. (45) closely applies and that we have for $t_L \cong 1 - 100$ sec predominantly a deactivation-type luminescence (whence the name).

6.2.2.4. Light Energy or Intensity. It was indicated in Section 2.2 that in general $L[t_I||t_L] = f(I, t_I, t_L)$, where both light parameters may occur as It_I (reciprocity law). We may expect this law to be verified for small t_I when large pools (PQ, Δ*pH) do not have the time to be filled or modified.

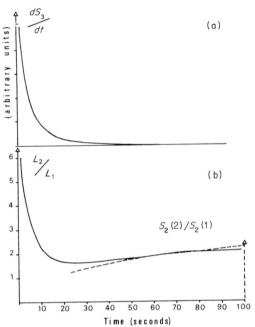

Fig. 19. Comparison of deactivation and luminescence in *Chlorella*. (a) Rate of decay of S_3. (b) Ratio of L after 2 (L_2) or 1 (L_1) flash in continuous line; ratio of S_2 measured 30 sec after 2 [$S_2(2)$] or 1 [$S_2(1)$] flash in dashed line. (Joliot *et al.*, 1971.)

Looking at the τ scale, we see that the upper limit of validity of the reciprocity law is in the range of $t_I \cong 0.1–1$ msec [$t_I(\text{crit})$]. Experiments confirm this point (Lavorel, 1971). The light-curve, whether vs. energy or intensity, depends also on t_L in a complex manner. L signals in general saturate when the light factor increases. It seems appropriate to discuss the light curves with reference to a critical intensity, analogous to $t_L(\text{crit})$, and a convenient as well as reasonable choice is the light condition where the photochemical part of $\Delta\phi$ saturates [Ic (crit)]. Table II contains experimental information concerning the light dependence of luminescence: types of empirical laws, quantitative or qualitative appreciation of the light condition with reference to t_I (crit), etc. Several remarks are in order.

1. For light conditions below Ic (crit) and t_I smaller than t_I (crit), there is a general agreement: one sees a shift from a I^2 to a I law. Jones (1967) who first observed the I^2 law took it as an argument in favor of the two-quantum electron-hole model (see Section 7); I observed, however, that the L relation was a simpler explanation (Lavorel (1969a)). For larger t_I, the situation is less simple: Kraan (1971) found that L was $\sim I^2$ at $t_L \cong$ 40 msec, but $\sim I$ at $t_L \cong 3$ sec; this probably reflects the fact that the two

factors of the L relation do not decay at the same rate: the decay of J is faster than that of $\Delta\phi$ (see Fig. 20) and, at $t_L = 3$ sec, $\Delta\phi$ is the predominant factor.

2. The role of large pools is plainly evident when light conditions are above Ic crit. Comparison of data in Table II is instructive, whereas light saturation obtains both for very short ($\cong 20\,\mu\text{sec}$) and very long ($\cong 3$ sec)t_L's, one finds a component ($t_L \cong 30 - 40$ msec) which keeps increasing with both I or t_I. The τ scale suggests that the Δ *pH is a good candidate.

3. The general shape of the light curve is sigmoidal (I^2, then I, then saturation). In other words, it is bound to be of any fractional order in I in a transition region situated in between the I region and saturation. Clayton (1969) found evidence for a $I^{1/2}$ law in a limited region which, however, when compared to the light curve given by Stacy et al. (1971), is precisely the transition region. This is unfortunate, because Clayton's belief in a bimolecular mechanism (see Section 7) relied in large part on the $I^{1/2}$ law.

6.2.2.5. *Temperature.* Three ranges of temperature have to be distinguished for their operational or theoretical significance: low range, physiological range, and range of thermal inactivation of PS II. The low range will be considered separately below (see Section 6.4.1).

6.2.2.5.1. *Physiological temperatures.* The picture is quite involved in this temperature range. Figure 20 shows the decay curves of $L[33\|3]_{\text{ms}}$, $\Delta\phi$, and J_T at 12 and 29°C; $\Delta\phi$ has an anomalous phase inversion which reflects in part into L, the result being that J_T is temperature-independent for $t_L < 50$ msec. In the presence of DCMU and at longer times ($L[5-10\|]$ and $t_L(\text{min}) = 1$ sec), Jursinic and Govindjee (1972) also found an anomaly for the Arrhenius plot of k (from $J_T^{-1/2}$) in the range 10–15°C. We recall that the possibility of anomalous temperature dependence was forseen for the electric potential effect [Eq. (27)] and the protonation effect [Eq. (31)]. They also found in T jump experiments ($t_L = 1$ sec, $T = 24 \longrightarrow 45$°C) that the size of the extra luminescence LT was independent of the light intensity in a range where L itself was not yet saturated. In Section 2.2.1.1.4, I mentioned the differentiation effect inherent in perturbation procedures on kinetic phenomena. The reason is that, while temperature has been changed by a certain ΔT, time has also increased by a certain Δt_L; therefore:

$$LT = \left(\frac{\partial L}{\partial T}\right)_{t_L} \Delta T + \left(\frac{\partial L}{\partial t_L}\right)_T \Delta t_L \tag{46}$$

It can be shown that, contrary to their result, LT/L should be independent of I (see Appendix V) unless the activation energy is itself *dependent* on I. We have thus an additional indirect evidence for the effect of the Δ *pH on

TABLE II

Light Factors of Luminescence

Authors	Signal[a] (method)	Type of Law[b] I^2	Type of Law[b] I	Type of Law[b] $I^{1/n}$	Saturation vs I_{crit}^c	It_i law[b]	Observations[d]
Arnold (1955)	ΣL (mF)	+				+ < 0.003 mW cm^{-2}	
Jones (1967)	L (mF)		+ +				No preillumination after 1* $\left.\right\rbrace t_L \cong 0.1$ sec
Mayne (1968)	LAB (mJ)		0.01–0.15 mW cm^{-2}				
Clayton (1969)	J_v (mP)			+ $n = 2$ 0.01–0.25 mW cm^{-2}			
Bonaventura and Kindergan (1971)	L (mP)						$\alpha_{1/\alpha_2}{}^e$ 0.1–12 mW cm^{-2}

Reference	Signal					Decay $[\alpha 1|\alpha 2|\ldots]$
Lavorel (1971)	L	+			$+\,(\cong I_{crit})$	$[4.4|0.02|0.03]_{msec}$
	ΣL	+			$-$	$[4.4|0.02|30]_{msec}$
	(mP)					
Ruby (1971)	L		+			<0.004 mI cm^{-2}
	(mP)		$n = 2$			From* background $I\,\}\,t_L \cong 1$ msec
Stacy	$\dfrac{L}{\overline{L}}$	$+$ <0.001 mW cm^{-2}	$+$ $n = 1.26$ >0.01 mW cm^{-2}			$[|0.2|30]_{msec}$
et al. (1971)	(mP)					$[3.5|1.5|5]_{msec}$
Kraan (1971)	L	$+$ $\cong 0.1$ mW cm^{-2}	$+$ $\cong 0.2$ mW cm^{-2}	$-$		$[5|0|04]_{sec}$
	(m$\overline{S}$)		$+$ $\cong 0.1$ mW cm^{-2}	$+$ 0.2mW cm^{-2} $(\cong I_{crit})$		$[5|3]_{sec}$

[a]Signals: L = luminescence; $\overline{L}$ = L averaged over a phosphoroscope period; ΣL = light sum; LAB = L stimulated by acid–base transition; J_V = see L relation with variable fluorescence. Methods: mF = flow; mJ = jump; mP = phosphoroscope; m$\overline{S}$ = single shot; * = flash.

[b]I = light intensity; t_I = time interval of action of light.

[c]I_{crit} = light saturation of $\Delta\phi$ (see text).

[d]See Symbols and Abbreviations.

[e]$\alpha 1$, $\alpha 2$ = 1st, 2nd exponential component.

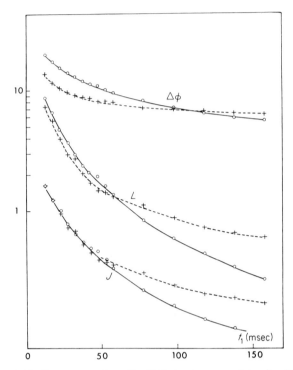

Fig. 20. Variable fluorescence intensity ($\Delta\phi$), luminescence intensity (L) and ratio of L to ϕ (total fluorescence intensity) (J) in *Chlorella* with flow method at $[33|t_L|3]_{msec}$ (i.e., light period = 33 msec, variable waiting period, measurement period = 3 msec) for two temperatures: 12°C (continuous line) and 29°C (dashed line). Each group of curves has been shifted on log scale of ordinate for best reading. (Lavorel, 1969.)

the PS II protonation equilibria. The same differentiation effect predicts, as in Malkin and Hardt's experiments (see τ scale), that LT must be linear $(-\frac{1}{2})$ if L is linear (-1), without the need of postulating different mechanisms for L and LT (see Appendix V).

   6.2.2.5.2 Thermal denaturation. The irreversible thermal inhibition in the 45–50°C range has been confirmed by many investigators since the earliest studies. This inhibition goes along with the thermal destruction of Z (in Z–YChlQ). Actually, as shown by Ruby (1971), a fast component $(t_L \cong 1$ msec) resists destruction. It is interesting to note the similarity between thermal denaturation and chemical agents destructive to Z from this point of view: they also are unable to destroy the fast component of L (see Fig. 16 for the effect of NH$_2$OH).

   6.2.2.6. Other Factors. Various chemical treatments modify luminescence. We have already discussed the effects of two PS II inhibitors, DCMU

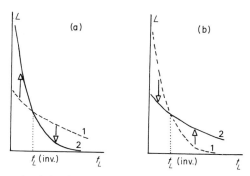

Fig. 21. Phase inversion in luminescence (L) decay (schematic): (a) positive or (b) negative when going from condition 1 to condition 2.

and NH_2OH. It is quite difficult to survey and summarize the effects of such treatments and to present them synthetically and independently of the kinetic aspect of luminescence being studied (decay, induction). In fact, the same substance may have altogether opposite effects depending on how luminescence is monitored. This situation essentially stems from the polyphasicity of L and the more or less independent behavior of its successive phases. The importance of the choice of the time parameters $[|t_{L_1}|t_{L_2}]$ in appreciating the effect of a chemical treatment is most clearly seen in the phase-inversion effect which very often attends such treatments (Fig. 21): depending on whether t_{L_1} is smaller or larger than t_L (inv), the inversion point, the treatment may result in increased or decreased luminescence signals. (Henceforth, phase inversion will be qualified "positive" when the amplitude α is *increased* before t_L and decreased there after, and "negative" for the opposite situation). Obviously, monitoring the entire L decay is the best way of appreciating the effect and judging its significance. Another difficulty is that very often several chemical treatments are purposely superposed (see Section 6.2.2.3), which makes it impossible to survey systematically and separately each case.

6.2.2.6.1. Inhibitors. Table III is a summary of data concerning the the effect of PS II inhibitors and other chemical agents. Looking at luminescence only, the distinction between inhibitors per se and other agents which I shall examine below is not always clear.

There is fairly general agreement concerning the effect of DCMU: in strong light, α of the fast component is depressed, whereas that of the medium components is not modified or even increased; the inversion point depends on experimental conditions; the concentration range ($\cong 10^{-5}$ M) where the effect appears is consistent with what is known of the effect on PS II activity. We have already discussed some of the effects of DCMU; as they are quite complex, it is worth considering here what may be expected a priori.

TABLE III

Effect of Inhibitors on Luminescence

Inhibitors[a]	Authors	(met.) Signal[b]	Material[c]	C(M) × 10^{5d}	Effect at t_L^e(s) 10^{-3}	10^{-2}	10^{-1}	1	10^1	10^2	Line	Observations
DCMU[f]	Sweetser et al. (1961)	L (mP)	a	9							1	I sat. ↘; CMU; ≅ OP
	Bertsch et al. (1963)	L (mP) LAB	a	0.1							2	CMU
	Mayne and Clayton (1966)	L (mJ)	cp								3	
	Lavorel (1968)	L (mF)	a	1							4	on J_T; 1*.
	Lavorel (1968)	L (mF)	a	≅ 0.1							5	1*; ≅ PU
	Miles and Jagen- Dorf (1969)	LX (mJ)	cp	> 0.1 0.5							6	X = S, A, AB; 50% inhib.
	Björn (1971)	L (m$\bar{S}$)	a								7	I at 730 nm
	Bonaventura and Kindergan (1971)	L (mP)	a								8	
	Itoh et al. (1971a)	$\bar{L}$ (mP)	cp								9	≅ CMU
	Mohanty et al. (1971)	L (mP)	a	1							10	
	Ruby (1971)	L (mP)	a	1							11	
	Wells et al. (1972)	L (mP)	cp								12	On DAD effect (cf. Fecy^{3+})

Compound	Reference	Scale[a]	Type[c]	C(M)[d]	Effect[e]	Line	Notes
NH_2OH[g]	Wraight and Crofts (1971)	L / (mP)	cp			13	
	Arthur and Strehler (1957)	L / (mP)	a			14	
	Bertsch et al. (1963)	L / (mP)	cp	> 0.1 / > 1.0		15	
	Lavorel (1968)	L / (mF)	a	40		16	50% inhib.
	Mohanty et al. (1971)	L / (mP)	a	100		17	+ DCMU = further
	Stacy et al. (1971)	L / (mP)	a			18	+ MV = further
OP	Bertsch et al. (1963)	L / (mP)	a	> 0.0 / > 1		19	(cf. also line 1)
NaN_2	Bertsch et al. (1963)	L / (mP)	a	> 1		20	
CNP	Sweetser et al. (1961)	L / (mP)	a	1		21	
	Bertsch et al. (1963)	L / (mP)	a	< 10 / > 20		22	50% inhib.
DNB	Lavorel (1968)	L / (mF)	a	< 1 / 30		23	also ϕ_o, ϕ
IOX	Lavorel (1968)	L / (mF)	a	< 1 / > 1		24	ϕ_o, ϕ

[a] See Symbols and Abbreviations
[b] For abbreviations, see Table II.
[c] a = algae; cp = chloroplasts.
[d] C(M) = molar concentration.
[e] ↗ = stimulation; → = inhibition; ‖ = approximate limit of effect on log t_L scale t_L = time interval after light.
[f] See also Figs. 2, 16, and 26, and Sections 3.2.2., 4.2.2., 4.3., 6.2.1., 6.2.2.3.-6., 6.2.3., and 6.3.
[g] See also Fig. 16 and Sections 4.2.1., 4.3., 6.2.1., 6.2.2.6., and 6.4.2..

1. Stimulation: From the stand point of the L relation, an increase of luminescence is expected, since, for the same t_L, $\Delta\phi$ is larger in the presence of DCMU; this is especially true when I is low (Clayton, 1969).

2. Inhibition: However, there is also an effect on J which is apparent in Figs. 2–16 and which we ascribed (see Section 6.2.1) to formation of a complex with Q^-. Besides interfering with the electron flow to PQ, blocking of PS II activity has two consequences, namely, lack of formation of the higher S states (S_3, S_4) and decrease of the $\Delta*pH$ (Wraight and Crofts, 1971). These two indirect effects are evidently predominant in the repetitive mode.

The phase inversion seen under DCMU presumably reflects this complex situation: at short t_L, the inhibitory effects must be more important; at longer t_L, both the relative increase of S_1 and S_2 and a larger $\Delta\phi$ should compensate the former effects. We have discussed in Section 4 triggered luminescence in terms of destabilization. Again, various results may be obtained. Inasmuch as luminescent states are still formed but decay slowly in the presence of DCMU, LX may be increased: this has been observed by Hardt and Malkin (1972a), provided t_L is kept short and I is low. On the other hand, with higher value of I and longer t_L, the absence of S_3 and S_4 plus shifting of the Q–PQ equilibrium (see Section 6.3.2) should result in overall inhibition (Mayne, 1967, Kraan, 1971). Other inhibitors (o-phenanthroline, phenylurethane in Table III) act like DCMU.

As Table III shows, the stimulation of L at short t_L and complete inhibition at $t_L > 20$ msec by NH_2OH is very well documented. We recall the conclusions obtained above from various experimental results that a high concentration of NH_2OH, upon disconnecting Z from the PS II center abolishes its four-step functioning; furthermore, since $^+YChlQ^-$ is "shorter" than the complete dipole, $^+HZ–YChlQ^-$, one may understand that it is less stable and that it should decay faster. When NH_2OH is acting at lower concentration as a substrate (but not destructive to Z), the effects, as studied by Bennoun (1971a,b), are a little more involved, although consistent with the above scheme. The experiments and results are summarized by the following inequalities:

$$L[1*||]\ (DCMU, NH_2OH) > L[1*||]\ (DCMU) \qquad (47a)$$

$$L[1*||]\ (DCMU) > L[10*||]\ (DCMU, NH_2OH) \qquad (47b)$$

with $t_L < 20$ msec, DCMU $= 2.10^{-5}M$ and $NH_2OH = 4.10^{-5}M$.* Equation (47a) is in support of Bennoun's hypothesis where Y is the primary electron donor, distinct from Z. The affinity of Y is larger to NH_2OH than to

---

*Equations (47a) and (47b) say that L induced by one flash is larger with both DCMU and NH_2OH than with DCMU alone, but that L induced by one flash with DCMU is larger than L after the 10th flash with both DCMU and NH_2OH.

Z and the discharge of the hole from ^+Y is slower to NH_2OH than to Z. In addition (see Bouges, 1971b), the formation of the NH_2OH-Y complex is slow and limiting under repeated excitation, resulting in the accumulation of inactive forms $YChlQ^-$ (DCMU is present) and low luminescence [Eq. (47b)]. As the NH_2OH concentration is increased, the stimulating effect in $L[1*||]$ [Eq. (47a)] diminishes and a phase inversion occurs which Bennoun ascribes to a special site of action of the poison. However, I believe that the two types of action outlined above (namely, (a) competition between Z and NH_2OH for ^+Y, and (b) destructive splitting of the bound Z–Y) are themselves sufficient to account for the results.

Some other effects can be briefly mentioned. DNB at high concentration acts as an artificial trap competing with the centers for capture, and a general inhibition of ϕ_0, $\Delta\phi$ and L results (Lavorel, 1969a). At a lower concentration, it shows a stimulating effect. The same thing is observed with chloronitrophenol (Bertsch et al., 1963) and IOX. It is not clear as yet if this stimulation has to be explained by a modification of the physicochemical environment of the centers or by a kinetic effect (phase inversion). It may be significant that, in the case of DNB, the stimulation seems to occur concomitantly with a specific suppression of the "I" to "P" (thermal) phase of the $\Delta\phi$ rise (Etienne and Lavergne, 1972).

6.2.2.6.2. Electron donors or acceptors. The "ferricyanide effect" has played a most important role in the development of the "electron-hole" model of luminescence and, from this point of view, deserves special attention. This effect has been thoroughly studied in chloroplasts by Bertsch and his group (1969). It can be described as a strongly positive phase inversion at t_L (inv) $\cong$ 4 msec, the short component of L being enhanced by a factor of 3–4. Other electron acceptors (MV, DCPIP) act qualitatively in the same way, while the combination NADP + Fd brings about a uniform increase of α and p-benzoquinone a uniform decrease.

The conclusion seems straightforward: luminescence is enhanced whenever a noncyclic electron flow is "activated" by the presence of electron acceptors. In effect, on comparing L from *in situ* chloroplasts to L in chloroplast suspensions, it is clear that addition of ferricyanide to the latter does restore to some extent the luminescence pattern seen in the intact material. This seems to stand as a definitive argument in favor of the electron-hole model—and particularly its "type *I*," two-quantum version (Bertsch and Lurie, 1971). This extremely sophisticated device is depicted in Fig. 22. According to this model, the mobile charges, which are responsible for the fast component of luminescence, are only active in the process if the photochemical traps (t_{H_2O} and t_A) are constantly emptied during light-induced noncyclic electron flow.

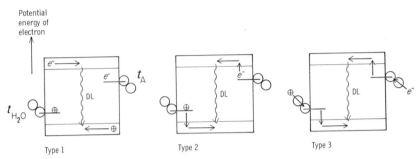

Fig. 22. Two quantum electron-hole model. Type 1:2 hv form 2 pairs e^-, $\oplus$; after trapping at the primary donor (t_{H_2O}) and acceptor (t_A) sites, one pair is left for annihilation and luminescence (L) production; no energy of activation. Type 2: same, except that L-producing pair results from thermal activation of charges in the primary sites. Type 3: same, except that e^- and $\oplus$ need further activation energy for going "upstream" in the electron transport chain; e^- is moving in the conduction band (top rectangle), $\oplus$ in the valence band (bottom rectangle) of the chlorophyll aggregate. (Bertsch and Lurie, 1971).

At first sight, we seem to be in trouble with the recombination model which I endeavor to illustrate here. However, there are several experimental results which do not quite fit in with the electron-hole picture. First, recall the instantaneous inhibitory effect of ferricyanide observed by Clayton (1969) (Section 2.2.2). Second, Bertsch and Lurie (1971), when studying this sort of effects on normal (N) or high Tris-treated (HT) chloroplasts, observe the following.

1. HT as compared to N results in a positive phase inversion at $t_L(\text{inv}) \cong$ 1.3 msec (Fig. 23), which they ascribe to a Tris-induced cyclic electron flow "around" the PS II centers—not to be confused with the recombination reaction. This additional assumption must be considered in the light of the restoration of a variable fluorescence by DCMU (see Yamashita and Butler, 1968) and the inhibition of $\overline{L}$ by the same poison in HT chloroplasts (Itoh *et al.*, 1971b). One then has to assume that DCMU also blocks this Tris-induced side pathway. (Also see Mohanty *et al.*, 1972).

2. Addition of an electron donor (DPSC) results in a negative phase inversion at t_L (inv) < 1 msec as compared to HT alone (Fig. 24); in other words, we are again qualitatively in the situation of N chloroplasts. Bertsch and Lurie explain this effect by the interruption of the cyclic path (competition with DPSC) and the lack of emptying of the t_A traps (no electron acceptor present).

3. Addition of an electron acceptor (NADP$^+$) to combination β (above), results again in a positive phase inversion at t_L (inv) $\cong$ 3.5 msec (with respect to 2) (Fig. 24). Since the noncyclic electron flow is now activated, luminescence of type I should increase.

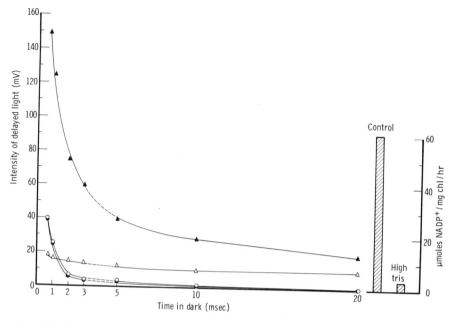

Fig. 23. Luminescence from normal or Tris-treated (HT) chloroplasts, in the absence or presence of $NADP^+$. Decay measured during phosphoroscope cycle. ($\triangle$) Control; ($\blacktriangle$) control + $NADP^+$; ($\bigcirc$) HT; ($\bullet$) HT + $NADP^+$. (Bertsch and Lurie, 1971.)

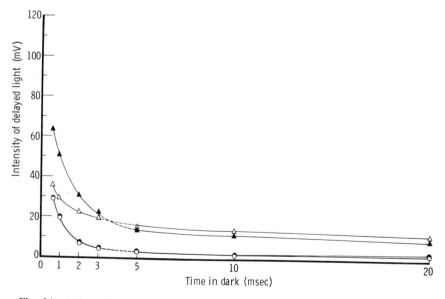

Fig. 24. Effect of electron donor and acceptor on luminescence of Tris-treated (HT) chloroplasts. ($\bigcirc$) Control; ($\bullet$) $NADP^+$; ($\triangle$) DPSC; ($\blacktriangle$) DPSC + $NADP^+$. See Fig. 23 and list of abbreviations. (Bertsch and Lurie, 1971.)

281

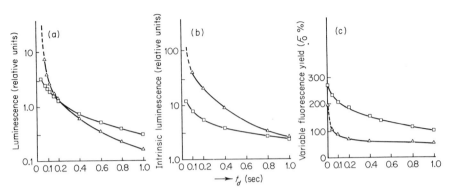

Fig. 25. Ferricyanide effect in chloroplasts. Control (□) or in the presence of 5.0 ×
10^{-5} M potassium ferricyanide (△). (a) Luminescence intensity, L; (b) $L/\Delta\phi$; (c) variable
fluorescence intensity, $\Delta\phi$. Conditions as in Fig. 9, except for protocol: $[5|0.04|]$ sec (i.e.,
light period = 5 sec, waiting period = 40 msec, variable measurement period). (Kraan,
1971.)

All of the above-described effects are explained equally well in the recombination hypothesis. A priori, it seems difficult to account for an increase
of luminescence of N chloroplasts in the presence of an electron acceptor
(ferricyanide effect), since the result must be a decrease of the concentration
of Q^-. A look at Fig. 25, where Kraan (1971) has monitored both L and ϕ
(and calculated J_V), suggest that, in a sense, ferricyanide has, roughly speak-
ing, an effect just opposite to that of DCMU (see also Clayton, 1969). Positive phase inversion means both accelerated decay and low luminescence at
long t_L. The first aspect is clearly explained by equipartition of the S states
and buildup of Δ^*pH resulting from increased electron flow, the second
one goes simply along with a faster decay of $\Delta\phi$ resulting from a faster
oxidation of Q^-. Both aspects are indeed the reverse of the DCMU pattern.
[A similar analysis for the ferricyanide effect is given by Stacy *et al.*
(1971).]

We also have to remember that a special type of phase inversion occurs
when Z is disconnected. This probably accounts for point 1 above and should
be likened to similar effects already mentioned (NH_2OH effect at high concentration and thermal denaturation of PS II).

6.2.2.6.3. Ionophores. The data on this class of compounds are summarized in Table IV; they are in general consistent with what we have already
discussed in Section 5. By facilitating various ionic movements through the
thylakoid membrane, they tend to stabilize the dipoles (either photochemical
or photoelectric) and thus inhibit luminescence. The details of their effect
will be best demonstrated below when dealing with induction and transients.

The dynamic aspect of ionophores is clearly displayed in the observation of Kraan (1971): Gmcd (10^{-6} M) depresses selectively the fast and medium components of the decay, L being inhibited by $\cong 50\%$ at $t_L \cong 30$ msec and not at all at $t_L \cong 1$ sec. According to Kraan, this indicates the limited lifetime of the light-induced electric field.

To summarize the survey of the action of chemicals, one might say that either a simple situation arises from competition between Z and an artificial donor for ^+Y, from complete elimination of $Z(NH_2OH)$, from artificial stabilization brought about by collapse of electric or ionic gradients (ionophores), or a more complex situation occurs where luminescence is affected in opposite directions by changes in turnover rate resulting in perturbation of the S states population, variation in contents of large pools and variation in the concentration of Q^- (ferricyanide and DCMU effects).

6.2.3. Slow Decay

The t_L domain is defined as larger than $\cong 10$ sec. In this area, the main relaxations of the photosynthetic apparatus are well over, except for the deactivation of S_2 and S_3 (and perhaps even more slowly, S_1). In this respect, the correlated empirical laws discussed in Section 6.2.2.3 should also belong to this area except for the fact that, assuming luminescence to be of the deactivation type, time differentiation (see Table I) has the effect of shifting the L decay to the left side of the τ scale as compared to the corresponding S decay.

Data in this area are scarce for obvious technical reasons (very low signals requiring time integration or time-consuming photon counting). The study by Björn (1971) of a far-red, long-lived luminescence must be mentioned. The decay was followed up to 6 mn after 730 nm illumination. It has a most unusual behavior: a minimum is reached at $t_L \cong 30$ sec, then a maximum at $t_L \cong 1$–1.5 mn, followed by a first-order decay ($\tau = 105$ sec). PMS simplifies this pattern by suppressing the maximum and inducing a faster first-order decay ($\tau \cong 33$ sec). For low enough concentration of PMS, the light sum is however conserved. From this effect and that of various other chemicals, Björn concludes that this slow component of L reflects the buildup of a large pool, presumably of Δ^*pH, which he can estimate as corresponding to $\cong 10^5$ Chl molecules, i.e., a size comparable to a single thylakoid.

Although of a different origin, the relaxation decays following the maximum of LX in perturbation experiments might be mentioned here. They are simple exponential decays (see τ scale) ranging from $\tau \cong 0.2$–0.3 sec for LS, LAB, LA (Malkin and Hardt, 1972) to $\tau \cong 28$ sec for $\overline{LS}$ (Barber, 1972). In the latter case, $S = KCl$ and the relation is kinetically controlled by the slow entry of Cl^- into the inner phase of the thylakoid, since τ is not modified after the addition of Vmc (increasing K^+ permeability).

TABLE IV

Effect of Ionophores, etc. on Luminescence

Agents or conditions[a]	Authors	Signal (met.)[b]	Material[b]	Effect	Observations	Line
Gmcd	Mayne (1967)	$\frac{L}{(mP)}$	cp	↗	10^{-5} mg liter^{-1} = 50% inhib; compl.[c] Hill rm.	1
	Kraan (1971)	$\frac{LX}{(mJ)}$	cp	↗	X = AB; higher inhib. at low pH (A), long t_{LA}	2
CCCP	Clayton (1969)	$\frac{L}{(mP)}$	cp	↗	at $\bar{L}$st; 10^{-7} M	3
	Itoh et al. (1971a)	$\frac{L}{L}$ (mP)	cp	↗	10^{-7}–10^{-6} M; inhib. slow phase > inhib. fast phase	4
FCCP	Wells et al. (1972)	$\frac{L}{(mP)}$	cp	↗	10^{-6} M; on both L and Fecy^{3+} effect	5
	Wraight and Crofts (1972)	$\frac{L}{(mP)}$	cp	↗	≅ 1 mg liter^{-1} = 50% inhib. on slow phase	6
DNP	Arthur and Strehler (1957)	$\frac{L}{(mP)}$	a	↗	t_L ≅ 10–360 msec	7
	Sweetser et al. (1961)	$\frac{L}{(mP)}$	cp	↗↘	t_L ≅ 5–22 msec and > 1 sec $\bigg\}$ t_L ≅ 0.2–1 sec $\bigg\}$ 4 × 10^{-5} M	8
	Lavorel (1969b)	$\frac{L}{L}$ (mF)	a	↗	[d] 2 × 10^{-4} M = 50% inhib; 1*; t_L = 7–37 msec	9
Vmc[e]	Wraight and Crofts (1972)	$\frac{L}{L}$ (mP)	cp	↗	On fast phase; max ↘ at pH 7–8	10
Nig	Wraight and Crofts (1972)	$\frac{L}{(mP)}$	cp	↗	½↘ at pH 5.9 and 8.7 ≅ 1 mg liter^{-1} = 50% inhib. on slow phase	11

Compound	Reference	Symbol	cp	Hill reaction	Notes	
CH$_3$NH$_2$	Mayne (1967)	$\bar{L}$ (mP)	cp	↗	$10^{-3}\ M = 50\%$ inhib.; compl. Hill rn.	12
	Itoh et al. (1971a)	$\bar{L}$ (mP)	cp	↗	Specifically on slow phase	13
	Itoh et al. (1971b)	$\bar{L}$ (mP)	cp	↗	On stimulation of L induced by e donors in HT chloroplasts.	14
NH$_3$	Mayne (1967)	$\bar{L}$ (mP)	cp	↗ $2 \times 10^{-3}\ M$	Compl. Hill rn.	15
	Wraight and Crofts (1972)	$\bar{L}$ (mP)	cp	↗	$0.1\ M = 50\%$ inhib. on slow phase	16
	Zankel (1971)	L (mP)	cp	↗	on $z2$ and $z3$ (see Section 6.2.1)	17
~P cond.	Mayne (1967)	L (m$\bar{S}$)	cp	↗	when P_i added to cp(ADP, Mg^{2+},...)	18
Inhibitors ~P	Mayne (1967)	L (mP)	cp	✕ (0.02 M)	= Phloridzin	19
		L (mP)	cp	↗ (0.08 M)	= DiO9; when added to syst. line 18	20
	Björn (1971)	L (m$\bar{S}$)	a	✕	= Desaspidin; $t_L > 10$ sec	21

[a] See Symbols and Abbreviations. ~P cond. = phosphorylating conditions; inhibitors ~P = inhibitors of phosphorylation.

[b] For abbreviations, see Tables II and III.

[c] complementary action: Hill reaction ↗, L ↘.

[d] weakly ↘.

[e] See also Figs. 10–12.

6.3. Induction and Transients

We shall now consider relatively slow kinetic phenomena which appear most clearly and simply when monitoring $\overline{L}$ (phosphoroscope methods). We have seen that this signal is produced in the repetitive mode through compression of the time scale and limitation of the time response of the detecting instrument (averaging). This change of time scale is ideally suited for monitoring and studying slow changes in the photosynthetic apparatus which indirectly affect luminescence, when the material is subjected to a light–dark transition (induction) or to a varying external factor (transients). However, as I already pointed out in Section 2.2 some caution is required in interpreting $\overline{L}$ data; it is important to stress this point again.

$\overline{L}$ is in fact a truncated ΣL, with the inherent effect of blurring out the relative contributions of the various components of the L decay. Let us take as an illustrative example the case of a two-exponential decays:

$$L = \alpha_1 \exp(-t_L/\tau_1) + \alpha_2 \exp(-t_L/\tau_2) \qquad (48)$$

If τ_1, τ_2 are both $\ll t_L$ (max),

$$\overline{L} = \alpha_1\tau_1 + \alpha_2\tau_2 \qquad (49)$$

$\overline{L}$ is thus a function of four a priori independent variables and any change in the signal will by necessity be interpreted with some arbitrariness. Admittedly, the above example was intended to dramatize the situation and, for instance, increasing the frequency of the phosphoroscope will tend to transform the meaning of $\overline{L}$ to

$$\overline{L} \cong \alpha_1 + \alpha_2$$

which now is simpler. Furthermore, the kinetic characteristics of the induction phases (rapidity, sensitivity to various treatments, etc.) may often somehow attenuate the ambiguity. Another important difference between the repetitive and sequential modes concerns the average state of the centers and results from either truncation (t_L (max) smaller than the photochemical turnover time) or, less often, non saturating light. The consequence is that the maximum amount of luminescence substrate $^+HZChlQ^-$ is never reached, because some inactive form is accumulating, either as $ZChlQ^-$ in the case of truncation or as $HZChlQ$ in the case of nonsaturation

6.3.1. Induction

As was reported earlier by Brugger (1957), the induction of $\overline{L}$ in algae bears some similarity to that of F. In chloroplasts, it has been generally observed that the two induction curves are not as a rule parallel and that the steady-state value is reached sooner for F than for $\overline{L}$. The induction phase has been studied more recently and its variations in the presence of various chemical additives have given further insight into the interplay of internal factors controlling luminescence.

Two successive phases during the development of induction were seen by Wraight and Crofts (1971) and Itoh *et al.* (1971a,b): a first photochemical fast phase ($\cong 0.1$ sec) more or less concomitant to the rise in $\Delta\phi$ and a slower phase which reaches full amplitude after $\cong 10$ sec. The shape of the curves are not found exactly identical, but the properties of the two phases as observed with various treatments are in general agreement.

In the presence of electron acceptors, the slow phase is much amplified, whereas the fast phase is essentially unmodified (Fig. 26). The slow phase is inhibited by many ionophores or uncoupling agents, by DCMU and by other treatments (sonication, Triton X-100) (see Tables IV and V). The effect of Vmc (plus KCl) is noteworthy: it depresses specifically the fast phase (but fails to abolish it altogether). From the latter effect and its pH dependence, Wraight and Crofts conclude that the fast phase is associated with the electric component of the "protonmotive force," while, for reasons of its slow kinetics and general sensitivity to ionophores, they attribute the slow phase to the pH component of the same force. With Kraan (1971) (see Section 5), they insist on the role of the pH-sensitive redox equilibria on the two sides of PS II in controlling the activation energy term of L (as discussed above, one might speak of pH-sensitive stabilization step in addition to a direct, saltlike effect of Δ^*pH on the activation energy). Nearly the same conclusions are proposed by Itoh *et al.*; their experiments with HT chloroplasts and the effects of electron donors suggest to them more precisely that the first phase is intimately associated with the primary charge separation in PS II centers.

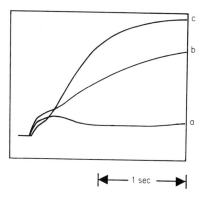

Fig. 26. Induction curve of luminescence in chloroplasts; (a) control; (b) $2 \times 10^{-4} M$ ferricyanide; (c) $2.5 \times 10^{-4} M$ DAD. Protocol: average signal in phosphoroscope method at $[3|0.3|1|0.3]_{msec}$ (i.e., light period = 3 msec; occultation period = 0.3 msec; measurement period = 1 msec occultation period = 0.3 msec). (Wraight and Crofts, 1971.)

Both groups have simultaneously monitored the induction phases of F and $\overline{L}$ and their results deserve more scrutiny. Some cases of parallelism (in the presence of electron acceptors) seem to be reasonably well accounted for by the L relation. But the effects on the F induction and steady-state level are far from simple. For instance, Vmc (Wraight and Crofts) or methylamine (Itoh *et al.*) both inhibit the slow phase of $\overline{L}$ and slightly stimulate F.

An unconventional type of induction has been studied by Ruby (1971). Schematically, the experiment is: $n\{[\text{background light}] \ L[1*\|t_L]\}$. The induction curve at $t_L \cong 1$ msec after a long dark period is reminiscent of that of F; for short dark incubation ($\cong 2$ mn), the induction shows only a fast first-order rise; after cessation of background light, a slow second-order decay of L is observed. This last feature probably has some relevance to the linear (-1) decay of L discussed above.

6.3.2. *Transients*

Transients encompass a large class of kinetic effects. We have already encountered several of them and discussed them separately because of their specific characteristics and significance, chromatic transients in Section 4 and perturbation experiments in Section 5. We will be concerned here with the time course of transients induced by chemical treatments (DCMU, ferricyanide, etc) of which we already know the final, steady-state outcome. The analysis will be limited to the few sets of systematic data available, mainly those of Clayton (1969), which were qualitatively discussed in Section 3.2.2 in support of the L relation.

Clayton measured transients induced on $\overline{L}$ by such chemical treatments after the light steady-state was obtained (Fig. 27). The salient point in these experiments is that both ferricyanide and DCMU show a transient effect just opposite to the steady-state effect which many people have observed.

6.3.2.1. Ferricyanide. The conditions of Fig. 27 are $\overline{L}[|500\|]_{\text{msec}}$ ($I = 3.1$ mW cm^{-2}). Addition of $5.10\,M$ ferricyanide causes a drop of $\overline{L}$, followed by a slow rise to the initial level; the curve of $\Delta\phi$ is almost exactly parallel. When however, luminescence is monitored at shorter t_L, $[|5|]_{\text{msec}}$, other conditions being identical, the drop in $\overline{L}$ is immediately followed by a rise above the initial level; in other words, the normal ferricyanide effect (stimulation) obtains after a short negative lag. Clayton explains the different behavior of $\overline{L}$ at short and large t_L by a loss of "memory" inherent to second-order processes. Indeed, for such processes, $(X)^{-1} = (X)_0^{-1} + kt$ [(X) = concentration of substrate, k = second-order rate constant] and, with large enough values of t, (X) is practically independent of $(X)_0$. However, the reason is probably different, since we have seen above how and why the phase inversion attending the ferricyanide effect (Fig. 25) had to produce

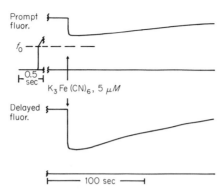

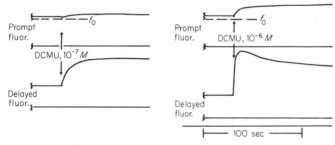

Fig. 27. Transients in fluorescence (prompt fluor.) and luminescence (delayed fluor.) (average signal in phosphoroscope method) after the addition of ferricyanide (top) or DCMU (bottom). (For details, see text. (Clayton, 1969.))

opposite behavior at short and long t_L. I must add also that Clayton's "bimolecular" hypothesis rests on too weak an experimental basis (see Section 6.2.2.4) to be considered seriously.

6.3.2.2. DCMU. The conditions of the experiment are similar to the above ferricyanide experiment except for a much weaker actinic light ($I = 0.082$ mW cm^{-2}) and intermediate t_L, $[|250|]$ $_\text{msec}$. The positive fast transient is best seen at a high concentration of DCMU (Fig. 27): it is immediately followed by the normal inhibitory effect, whereas at lower concentration the latter effect does not show (a similar positive transient has probably been seen by Wraight and Crofts (1971), as indicated by a positive ill-recorded perturbation in their curves). A very interesting phenomenon was observed by Clayton when, a few seconds after interrupting the phosphoroscope in the dark phase, DCMU was added: both L and $\Delta\phi$, which are then in the "tail" of their decay, show a conspicuous positive transient. [I have independently observed this effect on $\Delta\phi$ (Lavorel, 1969b)]. One must stress the fact that this effect is occurring in the dark; it is most simply explained according to the hypothesis of complex formation between Q^- and DCMU (Section 6.2.1, above). Complex formation would shift the Q/PQ equilibrium towards the Q^- side and this should result in a transient increase in fluorescence, if Q^-–DCMU is as "fluorescent" as Q^- alone.

6.3.2.3. Fast and Slow Effects. The slow or steady-state effects of ferricyanide and DCMU have already been discussed above and do not need further comment. Except that it is natural to see them develop rather slowly, since they correspond mainly to the readjustment of the large pools of protons and PQ in the photosynthetic apparatus.

The fast transients in Clayton's experiments seem to be simple consequences of the truncation and nonsaturation effects which I mentioned above. In the case of truncation, and strong light, as with ferricyanide, the reduction of the turnover time (oxidation of Q^-) should decrease the proportion of the nonphotoactive forms of the centers, among which is the luminescence substrate; in the case of nonsaturation, purposely chosen by Clayton for the DCMU experiment, augmentation of the turnover time has the opposite effect of a relative increase of the nonphotoactive forms. On this point, Clayton's analysis has exactly the same conclusion.

6.4. Special Topics

This review would not be complete without mentioning other types of luminescence measurements, such as "glow curves", and the important quantity ΣL taken as a function of t_L or as a total sum.

6.4.1. Low Temperatures

At low enough temperatures, all thermal reactions are "frozen" while the photochemical steps can still make the charge separation. In such case, the luminescent substrates are preserved as metastable states and will not show up until, by raising the temperature, the Boltzmann factor, $\exp(-\Delta E/RT)$ has reached a high enough value. It is seen qualitatively that the critical temperature of destabilization is lower the smaller the activation energy. This is the classical method of thermoluminescence or "glow curves" (see Appendix V). It has been applied to preilluminated samples of photosynthetic materials and has given rise to interesting results. Not unlike some crystals, such samples evolved successive peaks of luminescence as the temperature was progressively raised, revealing traps of different "depths." The data are summarized in Table V. The activation energies of these various traps are in agreement with the range of values one might expect (see Section 5 and Section 1). Their existence was generally taken as proof of metastable states resulting from "abortive" functioning of the photosynthetic apparatus. One

TABLE V

Characteristics of Low-Temperature Luminescence

Authors	Type[a]	0.1	0.2	0.3	0.4	0.5	0.6	0.7	0.8	0.9	Peak T (°C)	DCMU	O²	T.D.[b]	Observations
Arnold and Azzi (1968)	Z										−155				Not relevant to photosynthesis
	A					0.46–0.53[c]					−6	+		+	
	B						0.52–0.6				+30	o		+	heating rate = 3° sec⁻¹
	C						0.56–0.64				+52	+		+	
Shuvalov and Litvin (1969)	I	0.16									−160	++	+		Triplet
	II														
	III			0.35							−15	++	+		
	IV									0.9	+20	o	o		

[a] Authors' denominations.

[b] T.D. = thermal denaturation.

[c] Depending on the choice of frequency factor in glow-curve equation [see Appendix V, Eq. (51)].

may propose to relate them to the S states of PS II (see Arnold and Azzi, 1971). Their number is right, taking the A, B, C components of Arnold and Azzi (1968), for correspondence with S_3, S_2, S_1, which are the states which could be formed in the described experimental conditions (long dark pretreatment, low temperature, illumination, glow curve). In any case, since some of them are able to glow below 0°C, one may wonder whether they give evidence in favor of the photoelectric dipoles rather than the photochemical dipoles (at what temperature does protonation stabilization stop operating?)

6.4.2. Light Sum

ΣL has been used very little in the literature. Its significance is however quite interesting. This was showed above for the quantity $\Sigma L(\infty)$ in the case of dispersed separated closed units, near the "O" state (Section 2.1.1), and, more generally, in the case of luminescence of the deactivation type, where the time function $\Sigma L(t_L)$ is seen to mirror the decay of C_+^- [see the time integral of Eq. (37)]. It should certainly be interesting to plot this function in sequence experiments and compare it with the decays of the various S states (an equivalent operation is done in Fig. 19).

Kraan *et al.* (1970) are among the few who have used ΣL for a demonstrative purpose (see Fig. 9). They noticed that L and the various LX show much wider variations among themselves than the corresponding ΣL. This fact they took in favor of the "triggering" interpretation of luminescence in perturbation experiments, as opposed to the hypothesis of distinct substrates and processes (see Section 5.4). Their demonstration could have had even stronger value, had their recordings started at a shorter t_L. Obviously the "head cut-off effect" is a quite serious source of error when estimating $\Sigma L(\infty)$. An interesting example follows. I reported earlier (1971) that $\Sigma L(\infty)$ in presence of NH_2OH was approximately half the control value [the summation being started at $t_L(\min) \cong 27\ \mu sec$] and speculated on its significance. This was erroneous, because, more recently with, $t_L(\min) \cong 6\ \mu sec$, $\Sigma L(\infty)$ (NH_2OH) turned out to be of the same order of magnitude as the control value (and perhaps larger).

With the end of this important section, we are ready to attempt to draw some conclusions on the mechanism of luminescence. All the data presented and the arguments developed showed in a more precise and detailed form, if not always more quantitative, the role of the numerous factors which control or modulate the recombination process of the primary back reaction. Several quantitative interpretations are still pending, such as the linear (-1) and linear $(-\frac{1}{2})$ empirical laws. Perhaps the main point in this section is the recognition of two main types of luminescence and the demonstration, in two limited cases, of the existence of the deactivation type. In my opinion, any generalization would be premature, but the question is unescapable: is Eq.

(45)—which identifies luminescence and deactivation—the exception rather than the rule?

7. A MECHANISM AND SEVERAL PROBLEMS

In concluding this chapter, I believe that a fairly coherent picture of luminescence phenomena in plants can be projected out of the survey and discussion of present evidence. It is not a "one-man" picture and I might say that it is most agreeable and, in a sense, more convincing, to see that it results from the convergence of contributions from many workers and groups in the field. In preparing this chapter, I have had the privilege of getting acquainted with many ideas, of understanding them more deeply than I had before, and I have, in the hope of reaching a maximum of coherence, combined, adjusted, systematized and sometimes extended them with a few ideas of my own. It is said that theories are just good for being disproved by experiment. In this sense, I have not tried to escape the challenge and, instead, have proposed several experimental tests which might help in correcting or refining the picture.

It will probably be useful to summarize the key hypotheses which stand as the foundation of this theory.

H1. Luminescence results from recombination of charges in the photoelectric PS II dipole, $^+YChlQ^-$; as such, recombination is a monomolecular process.

H2. ε_L is almost indistinguishable from ε_F.

H3. The kinetics of luminescence is dominated by the activation energy of recombination (Arnold and Azzi) and the Boltzmann factor may be modified by artificially induced changes of the electroosmotic state of the membrane (Barber and Crofts) as a consequence of the vectorial character of the PS II centers (Mitchell). Stabilization, i.e., increase of activation energy, normally proceeds along the following steps: (i) photoelectric dipoles, $^+YChlQ^-$ and $^+HZ–YChlQ^-$ (Bennoun); (ii) photochemical dipole Z–YChlQH through protonation equilibria (Kraan); and (iii) translocation of protons across the membrane and electron flow through the photosynthetic chain.

H3'. NH_2OH splits the $Z(O_2$ evolving head) off the dipole; DCMU affords a parasitic stabilized state by complexing with Q^-, this state is still able to recombine and luminesce.

H4. Luminescence is modulated by the S state of the O_2 evolution mechanism (Joliot); this modulation is best seen as an increased destabilization from S_1 to S_4.

H5. The Q $\longleftrightarrow$ PQ equilibrium is pH-sensitive and operates along the whole time scale of luminescence and fluorescence decays.

H6. There is a critical time in the τ scale of photosynthetic events which separate two main types of luminescence: leakage and deactivation.

It is fair and appropriate, at this point, to compare this proposal with several other postulated mechanisms. The "electron-hole" model (Bertsch) has some similarity to the present mechanism (in H3); its main weakness, in my opinion, is its two-quantum version (type I) which postulates that the activation of electron flow must always result in increased luminescence. The "triplet exciton" model, further extended to a "bimolecular mechanism" by Stacy and coworkers was first proposed in a naïve attempt to overcome the activation energy problem (Lavorel); the least one can say is that it is waiting for decisive evidence on the crucial point of the effect of magnetic fields on the triplet–triplet annihilation process.

Ominous problems are still obstructing the way. Some of them are in the realm of experimental projects, others may be tentatively attacked on the basis of present evidence.

7.1. What Do the Empirical Linear (-1) or Linear ($-\frac{1}{2}$) Laws Mean?

I believe that, on the one hand, the linear (-1) decay of Q^- and the resulting linear ($-\frac{1}{2}$) decay of L in the presence of DCMU, and the linear (-1) decay of L in the normal case, on the other hand, are not unrelated phenomena. In both cases, we are in a range of the τ scale where relaxation of the electroosmotic state of the membrane is occurring, either in the form of decay of $\Delta\psi$ or of Δ*pH. One may thus look for a general explanation whereby the recombination process is an important factor limiting the rate of relaxation of $\Delta\psi$ or is itself strongly conditioned, through the protonation equilibria, by the decay of Δ*pH.

In Appendix 6, two models are proposed to this aim. It will be noticed that they rely heavily on an unorthodox type of kinetic law (the Elovich–Cope law), the main virtue of which is its ability to cope with rate processes in heterogeneous media and more precisely transfer of charges across an energy barrier. The general linear (-1) decay of L is depicted as resulting from the discharge of the thylakoîd membrane considered as a capacitor, the original feature being that the recombination of C_+^- change both the charge of the capacitor and its leakage resistance. In the case of the DCMU system, an explanation is sought in a rate limitation of the recombination process through the functioning of a "proton translocator" which is subjected to a diode-like law of conduction of charges. Recourse to the Elovich–Cope law is a reasonable explanation why in this case the decay of Q^- is second-order, although definitely not bimolecular [$t_{1/2}$ of

decay rather independent of initial concentration of Q^- (Bennoun, 1971a)].
The role played by H_i^+ in this model is consistent with Bennoun's finding
that the decay of Q^- in the DCMU system is faster under acid conditions.

A similar type of model, where the decay of ψ, assumed to be exponential,
and the electronic conduction through an interface are determinant, has
been independently considered by Tributsch (1972). It has been successfully
applied to fit data of Stacy et al. (1971) on L decay in the presence of NH_2OH;
it also finds some support in an empirical multiexponential decay law
verified in the t_L range of $10^{-4}-10^{-3}$ sec (see τ scale, line 2).

7.2 What Are the Deprotonation Steps of the Z Side in the Four S States?

This question cannot presently be answered for lack of pertinent experimental evidence*, but at least extreme situations may occur which could be
tested by $\Delta *pH$ and $\Delta 515$ sequence experiments. The outcome should be of
consequence for the relative extent of stabilization of the S states (see
Fig. 7). One must note, however, that the explanation offered above for the
linear (-1) laws would not depend much on the result; for instance, if the
first version proved to be true (i.e., Z from S_1 to S_3 is carrying stable positive
charges), nOH^- could be fixed on $^{n(+)}Z$ liberating an equivalent amount of
nH^+ into the inner phase of the thylakoîd.

7.3. Is the Deactivation Type of Luminescence Exceptional?

This problem is crucial not only for luminescence, but also for O_2 emission and fluorescence. It is obvious that, if the recombination, beyond
the critical time in H6, is of the "deactivation" type, then the luminescence,
O_2 deactivation and $\Delta\phi$ decays must be strictly related to each other. There
are two lines of evidence which seem to contradict a generalization of the
deactivation type; they both express a state of crisis which has developed
lately concerning Duysens'Q hypothesis.

(a) The decay of $\Delta\phi$, if it is interpreted as reoxidation of Q^- and if Q is the
primary PS II acceptor, is too slow as compared to the Emerson–Arnold
period ($\cong 20$ msec), i.e., the limiting turnover time of the whole photosynthetic chain (Lavorel, 1966). Forbush and Kok (1968) showed that only
$\frac{1}{3}$ of Q react rapidly with PQ ($=A$), with $\tau \cong 0.6$ msec. A further step was
taken by Joliot et al. (1971) who stated that the maximum τ for Q^- reoxidation
was $\cong 1$ sec and showed that, beyond this time, $\Delta\phi$ was well correlated with

---

*When completing this manuscript, it came to my knowledge that a $\Delta *pH$ sequence
experiment had been performed by Fowler (1972) (see Chapter 8).

(S_2) + (S_3) although for an unknown reason. Consequently, they had to look for a reductant other than Q^- to account for luminescence at longer time within the recombination hypothesis.

(b) Flashing light conditions, either of the repetitive (Emerson and Arnold) or sequential type (Joliot), seem to promote a maximum efficiency of light utilization by the photosynthetic apparatus when the flash period is in the range of 0.1–1.0 sec. One must notice that the maximum O_2 yield for flash obtained under this optimal situation has direct consequence on the determination of the PSU size.

The only weak or uncertain point is ultimately the assumption that, under optimal conditions in flashing-light experiments, each cycle involves the maximum amount of photoactive centers, i.e., ZChlQ, in simplified form. I believe that this assumption is open to question and that, unfortunately, the unequivocal evidence might be difficult to find. Indeed, Lemasson and Barbieri (1972) remarked that their deactivation curves were by necessity inaccurate, on account of neglect of the form $ZChlQ^-$. Moreover, when extending the Emerson–Arnold (1932) experiment to longer flashing periods, as did Renger (1972), one finds a progressive drop of the yield, due to the onset of deactivation, as one might have expected. Thus, nobody can ascertain that the maximum of the yield vs flash period curve signifies that all centers are at work, in the form ZChlQ.*

Therefore, it is permissible to contemplate the possibility that, even in the above-defined optimal conditions, a certain steady-state proportion of centers are in a nonphotoactive state, say $ZChlQ^-$ in simplified form. Several consequences (α) and arguments (β) concerning this "idle centers" hypothesis follow:

$\alpha 1$. The maximum PS II yield and the size of the PSU are respectively under- and overestimated. The answer might be obtained in flash experiments with Z-split systems (YChlQ) in the presence of very efficient artificial donors.

$\alpha 2$. Idle centers, on account of their slow independent decays, must "spoil" or damp the four step periodicity of the O_2 sequence (this is an instance of the truncation effect). Note that in weak continuous light there are no idle centers (maximum quantum yield).

$\alpha 3$. Luminescence is not distinguishable from O_2 deactivation, all factors affecting the former must affect in a similar way the latter, and reciprocally.

---

*The maximum yield in flashing light should be obtained when all centers have relaxed to the photoactive state, ZChlQ, during the dark period between flashes, provided that S_3 has not been deactivated. Since deactivation proceeds at a nonnegligible rate, the maximum in Renger's experiment might indicate a competition between relaxation and deactivation; hence, it does not prove that all centers are at work.

Since S_0 has been observed to accumulate as the predominant state under special experimental conditions, nothing forbids us from postulating that S_1 is also a (weak) luminescence substrate.

$\alpha 4$. Although the function $\Delta\phi = f[(Q^-)]$ need not be simple, Duysens' hypothesis is valid. The whole decay of $\Delta\phi$ is controlled by Q^-, and by PQH_2, through the pH-sensitive $Q \longleftrightarrow PQ$ equilibrium.

$\beta 1$. The hypothesis is consistent with the proposed mechanism of luminescence, in particular with H1, H4, and H5.

$\beta 2$. It is easily seen that the $\Delta *pH$ "buffers" counteract the forward PS II to PS I electron flow, on account of the pH-sensitive redox equilibria at both ends of the "internal" electron transport chain ["back pressure," in the terminology of Witt (1971)]. Therefore, the overall PS II–PS I equilibrium constant:

$$\frac{(Q)(P)}{(Q^-)(P^+)} = K'_{II}K'_{I} \frac{(H^+)_0}{(H^+)_i} = K_{II.I} \tag{53}$$

should be rather low under light-induced electron flow, as was observed by Joliot *et al.* (1968) and Delrieu (1969). As the $\Delta *pH$ relaxes, the back pressure is also released, and the electrons in the PQ pool are sparingly and slowly fed back to the recombination sites for luminescence and O_2 deactivation. This is a likely explanation for the sluggishness of deactivation and the slow component of luminescence.

$\beta 3$. Even when the decay of $\Delta\phi$ is almost completed, DCMU acts according to H3' to produce a transient rise of fluorescence (see Section 6.3.2), which shows that $\Delta\phi$ is still under control of Q^-.

$\beta 4$. The decay of S_3 (Bouges, 1971a), like that of Q^-, in the presence of DCMU (see above) is speeded up by lowering the external pH; this is consistent with a destabilization according to H3, if this pH change also results in acidification of the inner phase of the thylakoid.

$\beta 5$. Isolated chloroplasts as compared to whole cells show as a rule a slower deactivation and a lower luminescence. This is qualitatively expected for the deactivation type luminescence [see Eq. (36)]. One may guess that the membrane has become more leaky to H^+ compared to the intact state, due to preparative procedure, and that, consequently, a smaller $\Delta *pH$, as outlined above to explain the linear (-1) decay, results in a slower recombination.

Finally, one may wonder whether deactivation and luminescence, which, in this hypothesis, seem to bespeak a nonnegligible energy storage, might not have a role as a safety device in strong light or otherwise indicate some sort of energy transduction mechanism. But this is altogether pure speculation.

8. SUMMARY

In this chapter, results and ideas concerning luminescence phenomena in photosynthesizing plants are reviewed and discussed. The central concept is that of recombination of charges. Recombination is the exact reversal of primary charge separation (which is the result of quantum conversion by the reaction center) and represents a normally small loss in the quantum efficiency of photosynthesis. Once created by quantum conversion, the electron-hole pair is assumed to constitute a photoelectric dipole $^+YChlQ^-$ (where Chl is the reaction center chlorophyll or photochemical effector, ^+Y the oxidized primary donor and Q^- the reduced primary acceptor). Recombination is impeded by a large activation energy (of the order of 1.0 eV) and is in competition with several efficient stabilization steps, which further increase the activation energy along the recombination path. These steps are the early events which separate in time the quantum conversion from the electron and proton flow throughout the photosynthetic chain. When recombination occurs, it liberates a quantum of energy resulting in the creation of a luminescence exciton, ϵ_L. Not unlike the fluorescence exciton, ϵ_F, created by light absorption among the collector Chl molecules of the photosynthetic units (PSU's), ϵ_L during its brief life span (10^{-9} to 10^{-8} sec) will migrate randomly throughout the Chl territories and will disappear by being trapped by an open center or by radiative or nonradiative decay. Luminescence is the radiative decay of ϵ_L.

In Section 2, the methods for monitoring L, the luminescence intensity, are surveyed. Attention is paid to the numerous experimental factors which characterize each method and to the interplay between methodology and interpretation. In all methods, luminescence is monitored in a time interval, t_L, following the absorption of light by the photosynthetic cells or organelles. Most methods are limited in the minimum value of t_L accessible to the analysis; as a consequence, important and significant details in the decay of L is ignored. The light period is characterized mainly by its duration, t_I, and the intensity of the actinic beam, I; depending on both factors, a single light period may result in one or several photochemical turnovers and, consequently, different L patterns. When using repetitive (phosphoroscope) or sequential methods, several light–dark cycles are concatenated; one likely drawback is a memory effect in the sense that the light–dark period may be too short as compared to the time needed for the complete relaxation of the luminescence system.

Section 3 is a reflection on the postulate of a quasi-identity between the two types of excitons, ϵ_F and ϵ_L. This postulate is founded on many experimental results indicating that luminescence, as variable fluorescence, is a

PS II quantity. It finds a quantitative expression in the so-called L relation:

$$L = (\phi)J$$

where two factors are recognized: J, the rate factor, is the net influx of ϵ_L, resulting from the recombination process, injected in the PSU's; (ϕ), the yield factor, is the probability of expression of ϵ_L as luminescence. A consequence of this point of view is the delay effect resulting from repeated recombination and trapping acts; it should in effect delay the decay of the ϵ_L population. If, further, exciton exchange between units is allowed (connected PSU's), (ϕ) is expected to bear some relationship to the macroscopic fluorescence yield, ϕ, or rather to its variable part, $\Delta\phi$. Although no simple expression can be given for this relationship, it may be grasped qualitatively by recognizing that at least one factor Q, the "primary" PS II electron acceptor, is a factor common to (ϕ), J, and $\Delta\phi$. Cases of experimental parallelism between the kinetic behavior of fluorescence and luminescence are given as illustration and discussed.

Whereas the exciton approach to the recombination hypothesis was the object of Section 3, the substrate approach was considered in Section 4. As a consequence, ^+Z, as well as Q^-, is to be considered as a kinetic factor of L. This point of view is borne out by the striking parallelism between the oscillating patterns seen in sequence experiments (with short electronic flashes fired in succession, several hundred milliseconds apart) for both oxygen evolution and luminescence. Analysis of the oxygen sequence has given rise to the so-called four-S state theory of the Z side of PS II. Following this theory, four states, S_1 to S_4, corresponding presumably to four successive oxidation states of Z, result from four photochemical acts on the same reaction center and must be successively occupied before one O_2 molecule is evolved. The luminescence sequence suggests that all four states are involved as substrates for the recombination process and that their efficiency for luminescence production decreases in the order:

$$S_4 > S_3 > S_2 > S_1$$

Furthermore, another aspect of recombination is evidenced in fluorescence experiments in the presence of DCMU and NH_2OH at low concentration. The latter, by scavenging the positive charge on the Z side, acts in competition to recombination; consequently, it stops the $\Delta\phi$ decay in the dark (reoxidation of Q^-) and inhibits luminescence.

Stabilization or activation energy is the central theme of Section 5. Like other processes, luminescence is governed by a Boltzmann factor, exp $(-\Delta E/RT)$ (where ΔE is the activation energy, R a constant, and T the absolute temperature). But, unlike in ordinary homogeneous systems, the Boltzmann factor of the recombination process may be changed not only

by varying temperature, but also by varying ΔE itself. These unique properties stem from the vectorial character of the photoelectric dipole, i.e.,
its precise orientation within the photosynthetic membrane, with its positive-
head pointing inside the thylakoid. The vectorial concept is suggested by
many experiments: the electric potential of the membrane, resulting from
the oriented photoelectric dipoles, is thought to be responsible for the small
absorption change observed in the carotenoid bands (see Chapter 10); it can
be modified by external agents in such a way as to decrease the activation
energy of recombination, thus stimulating luminescence. These external
agents include variable external electric fields and diffusion potential set
across the membrane by unequal permeability (further modified by ionophorous agents) of ions. According to this picture, one may understand the
decreasing luminescence ability of the S states mentioned in Section 4, since
ΔE is expected to decrease as positive charges are accumulated on Z by
consecutive photochemical acts. A large activation energy barrier prevents
most dipoles to recombine, hence it stabilizes the charge separation. Further
stabilization is obtained by protonation of the Q side and deprotonation
of the Z side. The stabilization through protonation equilibria is clearly
suggested by the finding that an external pH transition may trigger an extra
luminescence. This step is also the likely explanation for the light-induced
pH gradient observed in chloroplast suspensions. In facts, it must be
considered in addition to quantum conversion, as the PS II contribution
to the electron *plus* proton transduction properties of the photosynthetic
membrane.

In Section 6, the kinetic aspects of luminescence are reviewed and discussed. On general grounds, as a consequence of the recombination
hypothesis and as regards the kinetic significance, two types of luminescence
must be distinguished. If recombination is negligible as compared to
stabilization and electron–proton transduction, luminescence is of the
leakage type. L is then proportional to the concentration of C_+^- (the photoelectric dipole) and the latter is completely controlled by the main processes
of stabilization and transduction. This situation is likely to occur for small
values of t_L, not larger than 1–10 msec. If, on the contrary, recombination
is the only process whereby C_+^- disappears, luminescence is of the deactivation type and L obeys the relation:

$$L = - \frac{d(C_+^-)}{dt_L}$$

This probably occurs at t_L of the order of seconds, concomitantly with the
deactivation of the O_2 evolving system, i.e., the spontaneous dark-trans-
formation of the higher S states into the lower ones. The deactivation type

is also demonstrated to occur in nonfunctional systems, such as in the presence of DCMU. Most results published in the literature are concerned with L decay in a t_L domain when either an intermediate type of luminescence or the deactivation type are dominant. The effects of many physical and chemical factors (light intensity, temperature, inhibitors, external electron acceptors and donors) have been studied. They are never simple. If light energy is delivered in a short enough time, saturation of luminescence occurs, indicating that all reaction centers are transformed by light into C_+^-; however, for longer time of action of light, L may keep increasing, in particular on account of the slow buildup of the light-induced pH gradient. The temperature factor may be studied in perturbation experiments. Glow curves (slow perturbation) indicate several activation energies, corresponding presumably to several of the S states; in jump experiments, the activation energy is seen to have an anomalous behavior in some temperature range, which probably reflects the complex interplay of the electroosmotic potential of the membrane with the activation energy of recombination. Among the effects of chemicals, those of DCMU and ferricyanide are noteworthy, DCMU has both an inhibitory effect on luminescence, by preventing the population of the higher S states and by inhibiting the light-induced pH gradient, and a stimulating effect, by increasing the luminescence yield factor (ϕ). The effect of ferricyanide is more or less opposite to that of DCMU. When monitoring the time-average of the L signal with a phosphoroscope method, one may study the induction phase (transition from the dark-adapted to the light-adapted state) or transients (following a sudden change in one external factor). Such studies clearly show, notably in the presence of ionophorous agents, the dependence of luminescence on the electroosmotic state of the membrane.

In Section 7 various arguments in favor of the recombination theory are summarized, other postulated mechanisms are compared to the latter theory, and some problems among those which are as yet unsolved are tentatively attacked. Several models are proposed to account for some of the empirical laws which the L decays are often found to obey; these models make use of unorthodox kinetic principles which seem particularly well suited to describe the behavior of the photoelectric dipole in its heterogeneous, structured environment. In an attempt to extend the deactivation type to the slow components of the L decay and to correlate it with the deactivation of the O_2 emission system and the decay of variable fluorescence, the "idle centers" hypothesis is considered; following this hypothesis, it is argued that, in many experimental situations with periodic light excitation, not all centers are at work in the photoactive state. Several observed facts are discussed in the light of this hypothesis.

APPENDIX I

The Delay Effect (Separate PSU's)

The kinetic equations are:

$$C_+^- \xrightarrow{\frac{k_L}{p}} (C + \epsilon_L), \; k_L = \tau_L^{-1}$$

$$\epsilon_L \xrightarrow{f} h\upsilon \tag{1a}$$

$$\epsilon_L \xrightarrow{d} \text{heat}$$

C_+^- and C are the luminescence substrate and the photoactive form of the reaction center, respectively; k_L; p, f, and d are first-order rate constants for recombination, trapping—this process is first-order in the case of separate PSU's—radiative deactivation and nonradiative deactivation, respectively. C_+^- is a form of reaction center of PS II, and ϵ_L is singlet exciton from luminescence. With $(C_+^-) = x$, $(\epsilon_L) = y$, and exciton yield r for recombination, this is equivalent to the differential equations:

$$\left. \begin{array}{l} \dot{x} = dx/dt = -k_L x + py \\ \dot{y} = -(p + f + d)y + rk_L x \end{array} \right\} \tag{1b}$$

It is legitimate (see below) to set $\dot{y} \cong 0$. Hence,

$$y = \frac{rk_L}{p + f + d} x \tag{1c}$$

and the decay of C_+^- is first-order with time constant:

$$\tau_{LD} = \tau_L \left(1 - \frac{rp}{p + f + d} \right)^{-1} \tag{1d}$$

Equations 1c and 1d, with the appropriate order of magnitude of rate constants i.e., $k_L \ll p, f$, or d, and reasonable value of $r (\gg 0)$, a posteriori justify the assumption $\dot{y} \cong 0$.

The total light sum (ΣL) is $\int_0^\infty f y \, dt$, or:

$$\Sigma L(\infty) = \int_0^\infty f \frac{rk_L}{p + f + d} x_0 \exp(-t/\tau_{LD}) \, dt \tag{1e}$$

$$= r\frac{f}{f + d + (1 - r)p}(C_+^-)_0$$

It is also interesting to give τ_{LD}, $\Sigma L(\infty)$ and $L(=fy)$ in terms of the reduced quantum yields $\varphi(0) = f/(p + f + d)$ and $\varphi(1) = f/(f + d)$:

$$\tau_{LD} = \tau_L\frac{\varphi(1)}{(1 - r)\varphi(1) + r\varphi(0)} \tag{1f}$$

$$\Sigma L(\infty)\,L = r\frac{\varphi(1)\varphi(0)}{(1 - r)\varphi(1) + r\varphi(0)}(C_+^-)_0 \tag{1g}$$

$$L = \varphi(0)rk_L(C_+^-) \tag{1h}$$

Equation (1h) should be compared with Eq. (6) in the text, whence $J(C_+^-) = rk_L(C_+^-)$.

L Relation near P State (i.e., When Fluorescence Yield is Maximum)

$\tau_{\Delta\phi}$ and τ_J being the shortest time constant of $\Delta\phi$ and J, L is given by:

$$\begin{aligned} L &= L_0 \exp(-\tau_{\Delta\phi}^{-1}t) \exp(-\tau_J^{-1}t) \\ &= L_0 \exp(-\tau_L^{-1}t) \end{aligned} \tag{1i}$$

which results in Eq. (17):

$$\tau_L^{-1} = \tau_{\Delta\phi}^{-1} + \tau_J^{-1}$$

APPENDIX II

Effect of a Variable Electric Field

If θ is the angle between the electric field (vector) X and the dipole $^+$ZChlQ$^-$ (vector), the positive direction being defined from $^+$Z to Q$^-$, the activation energy is modified by:

$$-\frac{Fd}{RT}X \cos \theta \tag{2a}$$

(with $X = X_0 \sin(\omega t)$ and ω the angular frequency of the field). The probability of this orientation being $(\frac{1}{2}) \sin \theta\, d\theta$, the average effect over all orientations is (because of the Boltzmann factor):

$$LE = L \int_{\theta=0}^{\pi} \exp\left(\frac{Fd}{RT}X \cos \theta\right) \frac{1}{2}\sin \theta\, d\theta \tag{2b}$$

or

$$LE = L\frac{\sinh\left(\dfrac{Fd}{RT}X\right)}{\dfrac{Fd}{RT}X} \tag{2c}$$

which is an even function of X, hence, frequency doubling (see Fig. 8). The relative increase in luminescence is:

$$\frac{LE - L}{L} = \frac{1}{3}\left[\frac{Fd}{RT}X\right]^2 + \frac{1}{5}\left[\frac{Fd}{RT}X\right]^4 + \ldots \tag{2d}$$

This theory neglects the simultaneous variations of L and X (see the differentiation effect in Appendix 5). The dependence of the local field X on the external field has not been stated explicitly.

APPENDIX III

The Goldman Equation (see Goldman, 1943)

Three partial differential equations are in general required to solve electrodiffusion problems.

The Nernst–Planck Equation

The *flux* j_k, number of particles of species k crossing the unit surface per unit time, is related to the (mean) *velocity* V_k and the *concentration* C_k of particles by:

$$j_k = C_k V_k \tag{3a}$$

(j and V are vectors; here we only consider the one-dimensional case).
The *velocity* is proportional to the *force* F_k acting on the particle:

$$V_k = U_k F_k \tag{3b}$$

where U_k is the *mobility* of the particle (a positive scalar).
One may set F_k equal to minus the *gradient* of the thermodynamic *potential* μ_k. In the general case (charged particle):

$$\mu_k = \mu_k^0 + RT \ln (C_k) + Z_k F\psi \tag{3c}$$

where Z_k is the charge of the particle (in unit valence), F a constant ($=1$ Faraday), and ψ the electric potential. From Eqs. (3a), 3(b), (3c), we get the

Nernst–Planck equation:

$$j_k = C_k U_k \left[-\frac{\partial u_k}{\partial x} \right] = C_k U_k \left[-\frac{RT}{C_k} \frac{\partial C_k}{\partial x} + Z_k FE \right]$$

$$= -U_k RT \frac{\partial C_k}{\partial x} + C_k U_k Z_k FE \tag{3d}$$

where E = electrical field strength. (It is understood that j, C, and E are, in general, functions of x and t.) Equation (3d) in terms of the corresponding *current density* $i_k = Z_k F j_k$, is (after some rearrangements):

$$i_k = -U_k Z_k FRT \frac{\partial C_k}{\partial x} + C_k U_k Z_k^2 F^2 E \tag{3e}$$

(Note that J has always the opposite sign (direction) to the concentration gradient, while i has always the same sign (direction) as the electric field.)

The Poisson Equation

$$\frac{\partial E}{\partial x} = \frac{F}{\epsilon} \sum_k Z_k C_k \tag{3f}$$

where ϵ = electric permittivity of the medium and summation is overall k ionic species. As a consequence of Poisson's equation, the field is constant in an *electroneutral* region (i.e., where $\sum_k Z_k C_k = 0$).

The Continuity Equation

$$\frac{\partial j_k}{\partial x} + \frac{\partial C_k}{\partial t} = 0 \tag{3g}$$

This equation is important for non-steady-state problems. It is a *conservation law*: the time variation of the local concentration results only from the balance of incoming and outcoming fluxes at this place.

The Goldman solution assumes: (1) steady-state conditions; (2) constant field within the membrane; and (3) no net electric current through the membrane. The consequence of (1) and (2) is that only the Nernst–Planck equation [Eq. (3e)] needs to be considered, with E = constant and the partial derivative replaced by the total derivative. The general solution is:

$$C_k(x) = \frac{i_k}{F^2 E} \cdot \frac{1}{U_k Z_k^2} + \alpha \exp \left[\frac{Z_k F}{RT} Ex \right] \tag{3h}$$

where α is a constant to be determined from the boundary condition. We will take the x axis directed toward the inside with origin at the outside boundary, i.e., subscript "0" at $x = 0$, "i" at $x = d$ (d is the width of the membrane). Taking the difference of Eq. (3h) at $x = d$ and $x = 0$, we get:

$$\alpha = \frac{(C_k)_i - (C_k)_o}{\exp\left[\dfrac{Z_k F}{RT} Ed\right] - 1} \tag{3i}$$

substituting α into Eq. (3h) and solving for i_k we have for $x = o$

$$\frac{i_k}{F^2 E} = \left\{ U_k Z_k^2 (C_k)_o \exp\left[\frac{Z_k F}{RT} Ed\right] - U_k Z_k^2 (C_k)_i \right\} \left\{ \exp\left[\frac{Z_k F}{RT} Ed\right] - 1 \right\} \tag{3j}$$

Assumption (3) implies summation of Eq. (3j) for all ions and setting it equal to zero. The result is simple only if all species have the same valency (we take $Z_k = \pm 1$). Furthermore, assuming with Goldman that near the boundaries the concentrations in the membrane C_k and in the liquid phase $\overline{C}_k$ (cation or $\overline{A}_k$ for anion) are related by a partition coefficient ($C_k = \beta_k \cdot (\overline{C}_k$ or $\overline{A}_k)$), this operation results in:

$$\exp\left[\frac{F}{RT} Ed\right] \left\{ \sum_k pC_k (\overline{C}_k)_o + \sum_k pA_k (\overline{A}_k)_i \right\} -$$

$$\left\{ \sum_k pC_k (\overline{C}_k)_i + \sum_k pA_k (\overline{A}_k) \right\} \tag{3k}$$

Rearranging, taking the logarithm and remembering that $E = d\psi/dx = -V_{i-o}/d$, we obtain the Goldman equation [Eq. (23)]. The "permeability" coefficients pC_k, etc are defined by

$$P_{C_k} = U_{C_k} \beta_{C_k}, \text{ etc.} \tag{3l}$$

A simple case of practical importance is that of an ion the permeability of which is much larger than that of any other ion: for instance, K^+ in the presence of Vmc. In such a case, V_{i-o} takes on the value of the "resting" potential of that ion

$$V_{i-o} = \frac{RT}{F} \ln\left[\frac{(K^+)_o}{(K^+)_i}\right] \tag{3m}$$

This potential difference is of magnitude and sign such as to neutralize, under *steady-state* conditions, any preexisting membrane potential.

APPENDIX IV

Charge Separation and Protonation Equilibria of Photosystem II Centers

A hypothetical complete kinetic scheme is depicted in Fig. 28.

Step 1: charge separation and recombination.

Steps 2, 2', 3, and 3' are, except for the subscripts "i" and "o" of H^+, the classical steps of acid–base equilibria of a zwitterionic substance; K_Z^-, K_Z, K_Q^+, K_Q are not independent:

$$K_Q^+ K_Z^- = K_Q K_Z \qquad (4a)$$

It is assumed that protonation stabilization through the above steps is fast compared to the Q–PQ equilibrium (steps 4 and 4') and not much perturbed by the latter.

Step 5: transition to higher S state. Note that the whole scheme depends upon which version (A or B in Fig. 7) is chosen for the deprotonation mechanism on the Z side. As written, the scheme directly applies to version B. For version A, it applies to S_4 with suitable modification of the stoichiometry for H^+; for the other states, steps 2, 2' etc may be alternatively written as stabilization equilibria involving $(OH^-)_i$. Note also that depending upon the version and the state, step 5 is either an identity or a reaction whereby O_2 is liberated and H_2O is reloaded on Z.

The main path is believed to be 1, 2, 3, 4, and 5.

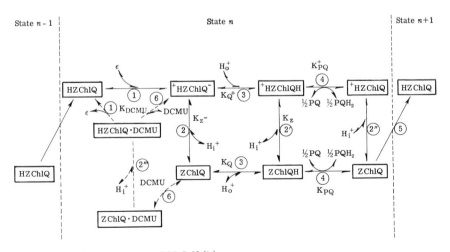

BOP. 5-28 (fe)

Fig. 28. Kinetic scheme of PS II charge separation protonation stabilization, electron flow to PQ and complexation with DCMU. For details, see text and Appendix IV.

In the DCMU system, steps 3 and beyond are assumed to be quasi in-existent due to fast complexation. Note that in this case recombination is mainly through step 1'.

After destruction by light of a certain amount (C_T) of HZChlQ:

$$(C_T) = (C_+^-) + (C_+) + (C^-) + (C) \tag{4b}$$

From the conservation equation (4b), relation (4a), and any two of the four equilibrium reactions, 2, 2', 3, and 3', the concentration of a given form may be calculated as:

$$(C_+^-) = (C_T) \cdot \left\{ 1 + \frac{(H^+)_o}{K_Q^+} + \frac{K_Z^-}{(H^+)_i} + \frac{K_Z}{K_Q} \frac{(H^+)_o}{(H^+)_i} \right\}^{-1} \tag{4c}$$

If relation (4a) is solved with the assumption that the Z and Q sides "ignore" themselves, i.e., if we state that $K_Q^+ = K_Q$ and $K_Z^- = K_Z$, Eq. (4d) simplifies into:

$$(C_+^-) = (C_T) \cdot \left\{ \left[1 + \frac{(H^+)_o}{K_Q} \right] \left[1 + \frac{K_Z}{(H^+)_i} \right] \right\}^{-1} \tag{4d}$$

which is formally what one may deduce from Kraan's simplified hypothesis (bimolecular reaction, two separate equilibria). In this case, of course, (C_+^-), etc. are to be replaced by $(^+HZ)(Q^-)$, etc.

APPENDIX V

Luminescence of the Deactivation Type in the DCMU System

Experiment of Bennoun (1970)

If luminescence is of the deactivation type, L is equal to the time deriv-ative of the substrate Q^- [Eq. (37)]. Since (Q^-) is linear (-1) [Eq. (43)],

$$L = \frac{d(Q^-)}{dt} = -k(Q^-)_o \frac{1}{(1 + kt_L)^2} \tag{5a}$$

hence Eq. (44).

Experiment of Jursinic and Govindjee (1972)

Since L is linear $(-\frac{1}{2})$ [Eq. (44)] and assuming that $\Delta\phi$ is nearly linear (-1), it follows from the L relation that J_V is also linear (-1). From the definition of J_V and J_T (Eq. 14):

$$J_T = \frac{\Delta\phi}{\Delta\phi + 1} J_V \tag{5b}$$

and J_T will be linear $(-\frac{1}{2})$, provided that $\Delta\phi \ll 1$.

The Differentiation Effect

The basic equation, of which Eq. (46) is a special case, should be used for any (small) perturbation effect whenever the rate of jump $\beta = dX/dt$ is not very large:

$$\Delta L = \left\{ \left(\frac{\partial L}{\partial X} \right)_{t_L} \beta + \left(\frac{\partial L}{\partial t_L} \right)_X \right\} \Delta t_L \tag{5c}$$

in what follows, $X = T$ and $(C_+^-) = x(t_L)$, a function of t_L.

Experiment of Jursinic and Govindjee (1972)

L is of the deactivation type, thus, from Eq. (37):

$$L = -\frac{d(C_+^-)}{dt_L} = k(T)x(t_L) \tag{5d}$$

where temperature dependence arises from the factor $k(T)$; therefore

$$LT = \Delta L = \left\{ \frac{dk(T)}{dT} x(t_L)\beta + k(T)\frac{dx(t_L)}{dt_L} \right\} \Delta t_L \tag{5e}$$

or

$$\frac{\Delta L}{L} = \left\{ \frac{dk(T)}{dT} \frac{1}{k(T)}\beta - k(T) \right\} \Delta t_L \tag{5f}$$

The right member in Eq. (5f) is only a function of ΔE and should not depend on I. To account qualitatively for the result: ΔL not dependent on I, i.e., $\Delta L/L$ dependent on I (L is not light-saturated), one may assume that ΔE is I-dependent, which might arise if $\Delta *pH$ is included in the activation energy factor.

Experiment of Malkin and Hardt (1972)

L is (empirically) linear (-1), i.e., precisely:

$$L = L_o(1 + k(T)t_L)^{-1} \tag{5g}$$

hence

$$LT = L_o \left\{ -(1 + k(T)t_L)^{-2}\frac{dk(T)}{dT}\beta t_L - (1 + k(T)t_L)^{-2}k(T) \right\} \Delta t_L \tag{5h}$$

or

$$(LT)^{\frac{1}{2}} = \left\{ L_o \left[-\frac{dk(T)}{dT}\beta t_L - k(T) \right] \Delta t_L \right\}^{\frac{1}{2}} (1 + k(T)t_L)^{-1} \tag{5i}$$

Glow Curves

L must be written explicitly as a function of T and t_L [cf. Eq. (36)] ; a simplifying assumption—which might be open to question—is that the only time-dependent factor is $(C_+^-) = x(t_L)$

$$L = \left[-\frac{d(C_+^-)}{dt_L}\right]_L = (\phi)v \exp(-\Delta E/RT)x(t_L) \qquad (5j)$$

The peak of the glow curve obtains when $\Delta L/\Delta t_L = 0$, i.e.:

$$\frac{\Delta L}{\Delta t_L} = (\phi)v \exp(-\Delta E/RT)\left\{\frac{\Delta E}{RT^2}x(t_L)\beta + \frac{dx(t_L)}{dt_L}\right\} = 0 \qquad (5k)$$

If one may assume that $[-dx(t_L)/dt_L]_L = -dx(t_L)/dt_L$ (deactivation type), the above condition is satisfied at a temperature T such that:

$$(\Delta E/RT) \exp(\Delta E/RT) = (\phi)vT/\beta \qquad (5l)$$

which, except for the (ϕ) factor, is the classical formula of Randall and Wilkins (see Arnold and Azzi, 1968).

APPENDIX VI

The Elovich Law and Cope's Approximation (see Cope, 1964)

A variety of processes in the field of heterogeneous kinetics are found to obey the empirical Elovich law:

$$-\dot{x} = -\frac{dx}{dt} = m \exp(nx) \qquad (6a)$$

where the variable x is dimensionless and m and n are constants. By integration:

$$\frac{1}{n}\exp(-nx) = mt + \frac{1}{n}\exp(-nx_0) \qquad (6b)$$

By substitution of Eq. (6b) into (6a), it is seen that the rate of decay is linear (-1):

$$(-\dot{x})^{-1} = nt + \frac{1}{m}\exp(-nx_0) \qquad (6c)$$

From a hypothetical scheme of redox catalysis (which we need not consider in detail here) Cope derived the law:

$$-\dot{x} = k\frac{X}{C - X} \qquad (6d)$$

where X is the concentration of substrate, k and C are constants, or in reduced form (with appropriate redefinition of k):

$$-\dot{x} = k\,\frac{x}{1-x} \qquad (6e)$$

Cope noted that, if $0.1 \ll x \ll 0.9$, $\ln[x/(1-x)]$ is fairly well approximated by $2x - 1$ or taking the inverse functions:

$$\frac{x}{1-x} \cong \exp(2x - 1) \qquad (6f)$$

Substitution of Eq. (6f) into Eq. (6e) gives a law of the Elovich type:

$$-\dot{x} = [k\exp(-1)]\,\exp(2x) \qquad (6g)$$

Equations (6c)–(6e) show that both x and $-\dot{x}$ can be approximated by linear (-1) laws

$$(x)^{-1} = (x_0)^{-1} + 2kt$$

$$(-\dot{x})^{-1} = k^{-1}[(x_0)^{-1} - 1] + 2t \qquad (6h)$$

It is interesting to note that the Cope law is highly polyphasic, as shown by expansion of the right member of Eq. (6e) in powers of x.

Justification of the Elovich–Cope Decay Law

One may propose several models where a kinetic variable has to obey the fundamental Eq. (6e). I shall give two of them. These models have in common the assumption that rate limitation is somehow determined by the existence of one or several interphases, that the rate of change of the concentration in one compartment is proportional to the flux of the substance through the corresponding interphase, the classical mass action law is completely dropped, except for the expression of chemical equilibria, but this may as well be considered as a consequence of the definition of thermodynamic potentials.

The "Leaky" Capacitor

This is the simplest model (Fig. 29). The discharge of a capacitor C through a resistor R is given in terms of the charge Q by the differential equation:

$$\frac{Q}{C} + R\frac{dQ}{dt} = 0 \qquad (6i)$$

Let us consider the case where R is the "leakage" resistance of the dielectric which constitues the capacitor and further let us assume that R is a decreasing function of the charge Q:

$$R = R_0 - \alpha Q \qquad (6j)$$

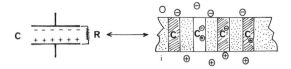

Fig. 29. The leaky capacitor. Electrical equivalent scheme (left). Application to the thylakoid (right); only the recombination path C_+^- (open block) is assumed to offer low resistance for the flow of charge from **i** to **O**; $\oplus$ and $\ominus$ represent unbalance of electric charges. For details, see text and Appendix VI.

where R_0 and α are constants (it is understood that R_0, α, and Q are such that R is never zero or negative). Taking x as the dimensionless variable $\alpha Q/R_0$ and $R_0 C = \tau_0$, Eqs. (6i)–(6j) lead to:

$$-\dot{x} = \tau_0^{-1} \frac{x}{1 - x} \tag{6k}$$

which is identical with Cope's equation.

The assumption, implicit in Eq. (6j), is quite reasonable when this model is applied to the thylakoîd membrane considered as a capacitor. One may assume that C_+^- are sites of low resistance through the "dielectric" of the capacitor, i.e., through the membrane; the surface density of these sites is obviously a function of the momentary charge of the capacitor and, through the protonation equilibria relations such as Eq. (4c) (Appendix 4), it may be taken as proportional to this charge. The way Q is affecting R in Eq. (6j) is a consequence of this reasoning. The conclusion of this first model is that the rate of discharge of the leaky capacitor is linear (-1).

The Proton Translocator

This second model has some definite similarity with the chemical cells considered in Mitchell's theory. Let us consider that two redox surface bound, H^+-sensitive equilibria exist at corresponding sites on the two sides of the membrane (Fig. 30). Each one is specified by the redox potential

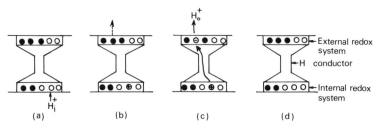

Fig. 30. The proton translocator. At rest [(a) and (d)] each redox system is poised (fixed ratio of reduced ($\bullet$) to oxidized ($\bigcirc$) forms). Four imaginary steps in translocation of H^+ are depicted from (a) to (d). Note that, except for the fact that one H^+ has been translocated, configuration (a) and (d) are identical. For details, see text and Appendix VI.

formula:

$$E = E^0 + \frac{RT}{F} \ln \left[\frac{(\text{ox}) \cdot (\text{H}^+)}{(\text{red})} \right] \tag{6l}$$

Let us assume that these two redox systems are separately poised [the ratio (ox)/(red) is maintained constant], that they are linked by a "hydrogen conductor" that will "couple" the external and internal values of E and (H^+) and, further, that the flux through the translocator obeys the diode (or Tafel) equation:

$$I = I_S[\exp(FV/RT) - 1] \cong I_S \exp(FV/RT) \tag{6m}$$

where I is the proton flux, and V is taken as the difference between the two terminal redox potentials [Eq. (6l)]. The simplifying assumption in Eq. (6m) signifies that one is not attempting to describe the terminal part of the decay where the exponential term is not longer large as compared to 1. The introduction of the Tafel formula may be justified on the ground that it gives an adequate empirical law for charge transfer across liquid–solid interphases (see Cope, 1964). It is also apparent that, given an adequate poise of the two redox systems, the translocator has in effect a diode-like behavior for proton conduction.

As a consequence, the flux j_H, taken positive in the inside-outside sense, is given by:

$$j_H = j_H^0 \left[\frac{(\text{ox})}{K(\text{red})} \right]_i \left[\frac{K(\text{red})}{(\text{ox})} \right]_0 \left[\frac{(\text{H}^+)_i}{(\text{H}^+)_0} \right] \tag{6n}$$

Now, it is seen, as a special application of the conservation Eq. (3g) (Appendix III), that the rate of variation of the excess internal proton concentration $-d[\delta(\text{H}^+)_i]/dt$ must be proportional to j_H and that the ratio $(\text{H}^+)_i/(\text{H}^+)_0$ must vary as $\delta(\text{H}^+)_i/[C - \delta(\text{H}^+)_i]$, C being a constant. For this second model, the conclusion again is that $\delta(\text{H}^+)_i$ decays according to the Elovich–Cope law.

The Linear (-1) Decay of L

As indicated on the τ scale, this law is mainly verified for the t_L range of 10^{-2} to 1 sec. It is also seen that many events associate with L are occurring during this time range. Thus one would hardly expect a simple kinetic behavior. For the same reason, the following explanation of the linear (-1) law must be regarded as highly tentative. It might so happen that, in this t_L range, the recombination pathway constitutes the principal exit route for the positive charges accumulated in the inner phase of the thylakoid and that the rate of recombination, i.e., L, is precisely equal to the rate of discharge of the membrane. If this is true, then the law simply follows from model 1 (the leaky capacitor).

The Linear (-1) Decay of Q^- in Bennoun's System (1970)

We recall that, in this case, L is of the deactivation type. From this fact and the empirical law, it follows that

$$(C_T) = (C_T)_0 (1 + kt)^{-1} \tag{6o}$$

$$L = h(C_+^-) = -\frac{d(C_T)}{dt} \tag{6p}$$

h and k being constants. Carrying the time differentiation in Eq. (6p) explicitly according to Eq. (6o), we find:

$$\frac{(C_+^-)}{(C_T)} = \frac{h^{-1}k}{1 + kt} \tag{7q}$$

This same ratio may also be expressed as a function of the proton gradient [Eq. (4e) of Appendix 4] with a suitable modification to take into account that C_+^- is practically present only in the form of the DCMU complex (H3'). Accordingly, we see that:

$$\frac{(H^+)_i}{(H^+)_i + K_Z} = \frac{k'}{1 + kt} \tag{7r}$$

This amounts to assuming that $(H^+)_i$ itself is linear (-1). In this case, model 2 (the proton translocator) is a possible explanation [together with the set of Eq. (6h)]. Obviously in the DCMU system, we expect that the pH differential across the membrane is mainly of PS I origin. The decay of Q^- is in this case rate-limited, through the protonation equilibrium of the Z side, by the rate of proton escape. One may tentatively assign to PQ the role of a proton translocator in the sense of this model.

ACKNOWLEDGMENTS

I sincerely appreciate the help that Dr. Govindjee and Dr. B. Mayne gave me in bringing my text into its present form. My thanks are due to Dr. G. Renger, who corrected several mathematical errors in the Appendixes, and to Mrs. Y. Tsacas and Mrs. F. Rosengard, who carried the burden of the material preparation of my manuscript.

REFERENCES

Arnold, W. (1955). *In* "The Luminescence of Biological Systems," (Franck H. Johnson, ed.), pp. 47–50. The American Association for the Advancement of Science, Washington, D.C.
Arnold, W. (1966). *Science* **154**, 1046.
Arnold, W., and Azzi, J. R. (1968). *Proc. Nat. Acad. Sci. U.S.* **61**, 29.
Arnold, W., and Azzi, J. (1971). *Photochem. Photobiol.* **14**, 233.
Arnold, W. and Davidson; J. B. (1954). *J. Gen. Physiol.* **37**, 677.

Arnold, W., and Davidson, J. B. (1963). *Nat. Acad. Sci. Nat. Res. Council Pub.* **1145**, 698.
Arnold, W., and Thompson, J. (1956). *J. Gen. Physiol.* **39**, 311.
Arthur, W. E., and Strehler, B. L. (1957). *Arch. Biochem. Biophys.* **70**, 507.
Azzi, J. R. (1966). Oak Ridge National Laboratory Tech. Memo no. 1534.
Barber, J. (1972). *Biochim. Biophys. Acta* **275**, 105.
Barber, J., and Kraan, G. P. B. (1970). *Biochim. Biophys. Acta* **197**, 49.
Barber, J., and Varley, W. J. (1972). *J. Exp. Bot.* **23**(74), 216.
Barbieri, G., Delosme, R., and Joliot, P. (1970). *Photochem. Photobiol* **12**, 197.
Bennoun, P. (1970). *Biochim. Biophys. Acta* **216**, 357.
Bennoun, P. (1971a). Thèse Doctorat d'Etat, Faculté des Sciences, Paris.
Bennoun, P. (1971b). *Compt. Rend. Acad. Sci.* **273**, 2654.
Bertsch, W. (1962). *Proc. Nat. Acad. Sci. U.S.* **48**, 2000.
Bertsch, W., and Lurie, S. (1971). *Photochem. Photobiol.* **14**, 251.
Bertsch, W. F., Davidson, J. B., and Azzi, J. R. (1963). *In* "Photosynthetic Mechanisms of Green Plants," pp. 701–710. The Committee on Photobiology of the National Academy of Sciences, National Research Council, Washington, D.C.
Bertsch, W., Azzi, J. R., and Davidson, J. B. (1967). *Biochim. Biophys. Acta* **143**, 129.
Bertsch, W., West, J., and Hill, R. (1969). *Biochim. Biophys. Acta* **172**, 525.
Björn, L. O. (1971). *Photochem. Photobiol.* **13**, 5.
Bonaventura, C., and Kindergan, M. (1971). *Biochim. Biophys. Acta* **234**, 249.
Borisov, A. Yu. (1967). *Biofizika* **12**, 630. (English transl. *Biophysics* **12**, 727).
Bouges, B. (1971a). Thèse Doctorat 3ᵉ cycle, Faculté des Sciences, Paris.
Bouges, B. (1971b). *Biochim. Biophys. Acta* **234**, 103.
Brugger, J. E. (1957). *Res. Photosyn. Pap. Discuss. Gatlinburg Confer. 1955*, pp. 134–141.
Butler, W. L. (1966). *Curr Top Bioenerg* **1**, 49.
Butler, W. L. (1971). *FEBS Lett.* **19**, 125.
Clayton, R. K. (1969). *Biophys. J.* **9**, 60.
Clayton, R. K., and Bertsch, W. F. (1965). *Biochem. Biophys. Res. Commun.* **18**, 415.
Cope, F. W. (1964). *Proc. Nat. Acad. Sci. U.S.* **51**, 809.
Delrieu, M. J. (1969). *Progr. Photosynthesis Res.* **2**, 1110.
Duysens, L. N. M. (1963). *Colloq. Int. Cent. Nat. Rech. Sci. 1962* No. 119, p. 158.
Duysens, L. N. M., and Sweers, H. E. (1963). *In* "Studies on Microalgae and Photosynthetic Bacteria" (Japanese Society of Plant Physiologists, ed.), pp. 353–372. The University of Tokyo Press, Tokyo.
Emerson, R. and Arnold, W. (1932). *J. Gen. Physiol.* **15**, 391; **16**, 191.
Etienne, A. L., and Lavergne, J. (1972). *Biochim. Biophys. Acta* **283**, 268.
Etienne, A. L., Lavergne, J., and Lavorel, J. (1970). Unpublished observations.
Fleischman, D. E. (1971). *Photochem. Photobiol.* **14**, 277.
Forbush, B. and Kok, B. (1968). *Biochim. Biophys. Acta* **162**, 243.
Fowler, C. F. (1972). Paper presented at the Annual Meeting of the Biophysical Society of America, Columbus, Ohio, *Abstract*, p. 64a.
Goedheer, J. C. (1962). *Biochim. Biophys. Acta* **64**, 29.
Goedheer, J. C. (1963). *Biochim. Biophys. Acta* **66**, 6.
Goldman, D. E., (1943). *J. Gen. Physiol.* **27**, 37.
Hardt, H., and Malkin, S., (1972a). *Biochem. Biophys. Res. Commun.* **46**, 668.
Hardt, H., and Malkin, S. (1972b). *Biochim. Biophys. Acta* **267**, 588.
Hardt, H., and Malkin, S. (1973). *Photochem. Photobiol.* **17**, 433.
Haug, A., Jaquet, D. D., and Beall, H. C. (1972). *Biochim. Biophys. Acta* **283**, 92.
Isenberg, I., and Dyson, R. D. (1969). *Biophys. J.* **9**, 1337.
Itoh, S., Katoh, S., and Takamiya, A. (1971a). *Biochim. Biophys. Acta* **245**, 121.
Itoh, S., Murata, N., and Takamiya, A. (1971b). *Biochim. Biophys. Acta* **245**, 109.

Jagendorf, A. T., and Uribe, E. (1966). *Proc. Nat. Acad. Sci.* **55**, 170.
Joliot, A., and Joliot, P. (1964). *Compt. Rend. Acad. Sci.* **258**, 4622.
Joliot, P., Joliot, A., and Kok, B. (1968). *Biochim. Biophys. Acta* **153**, 635.
Joliot, P., Joliot, A., Bouges, B., and Barbieri, G. (1971). *Photochem. Photobiol.* **14**, 287.
Jones, L. W. (1967). *Proc. Nat. Acad. Sci.* **58**(1), 75.
Jursinic, P., and Govindjee (1972). *Photochem. Photobiol.* **15**, 331.
Kessler, E., Arthur, W., and Brugger, J. E. (1957). *Arch. Biochem. Biophys.* **71**(2), 326.
Kok, B., Forbush, B., and McGloin, M. (1970). *Photochem. Photobiol.* **11**, 457.
de Kouchkovsky, Y. (1972). *Proc. Int. Congr. Photosyn. Res. 2nd, Stresa* **1**, 233.
Kraan, G. P. B. (1971). Ph.D. Thesis, University of Leiden, Reider, The Netherlands.
Kraan, G. P. B., Amesz, J., Velthuys, B. R., and Steemers, R. G. (1970). *Biochim. Biophys. Acta* **223**, 129.
Lavorel, J. (1962). *Biochim. Biophys. Acta* **60**, 510.
Lavorel, J. (1966). *Proc. West Eur. Confer. Photosyn. 2nd Woudschoten, Zeist, The Netherlands, 1965*, p. 39.
Lavorel, J. (1968). *Biochim. Biophys. Acta* **153**, 727.
Lavorel, J. (1969a). *Progr. Photosynthesis Res.* **2**, 883.
Lavorel, J. (1969b). In *Biophysical Aspects of Photosynthesis*, meeting of the British Biophysical Society, p. 3.
Lavorel, J. (1971). *Photochem. Photobiol.* **14**, 261.
Lavorel, J. (1972). *Compt. Rend. Acad. Sci.* **274**, 2909.
Lavorel, J., and Joliot, P. (1972). *Biophys. J.* **12**, 815.
Lavorel, J., and Levine, R. P. (1968). *Plant Physiol.* **43**, 1049.
Lemasson, C., and Barbieri, G. (1972). *Proc. Int. Congr. Photosyn. Res., 2nd, Stresa* **1**, 107.
Lurie, S., Cohen, W., and Bertsch, W. (1972). *Proc. Int. Congr. Photosyn. Res., 2nd, Stresa, 1971* **197**.
Malkin, S. (1966). *Biochim. Biophys. Acta* **126**, 433.
Malkin, S., and Hardt, H. (1972). *Proc. Int. Congr. Photosyn. Res., 2nd, Stresa* **1** 253.
Mar, T., and Govindjee, (1971). *Biochim. Biophys. Acta* **226**, 200.
Mauzerall, D. (1972). *Proc. Nat. Acad. Sci.* **69**, 1358.
Mayne, B. C. (1967). *Photochem. Photobiol.* **6**, 189.
Mayne, B. C. (1968). *Photochem. Photobiol.* **8**, 107.
Mayne, B. C., and Clayton, R. K. (1966). *Proc. Nat. Acad. Sci.* **55**, 494.
Miles, C. D., and Jagendorf, A. T. (1969). *Arch. Biochem. Biophys.* **129**, 711.
Mitchell, P. (1966). *Biol. Rev. Cambridge Phil. Soc.* **41**, 445.
Mitchell, P. (1967). *Federation Proc.* **26**, 1370.
Mohanty, P., Mar, T., and Govindjee. (1971). *Biochim. Biophys. Acta* **253**, 213.
Mohanty, P., Braun, B. Z., and Govindjee. (1972). *FEBS Lett.* **20**, 273.
Neumann, J., and Jagendorf, A. T. (1964). *Archiv. Biochem. Biophys.* **107**, 109.
Paillotin, G. (1972). *J. Theoret. Biol.* **36**, 223.
Papageorgiou, G., and Govindjee (1971). *Biochim. Biophys. Acta* **234**, 428.
Renger, G. (1972). *Physiol. Veg.* **10**, 329.
Ruby, R. H. (1968). *Photochem. Photobiol.* **8**, 299.
Ruby, R. H. (1971). *Photochem. Photobiol.* **13**, 97.
Seely, G. R. (1972). *Proc. Int. Congr. Photosyn. Res., 2nd, Stresa* **1**, 341.
Shuvalov, V. A., and Litvin, F. F. (1969). *Mol. Biol.* (Russian), **3**, 59.
Stacy, W. T., Mar, T., Swenberg, C. E., and Govindjee. (1971). *Photochem. Photobiol.* **14**, 197.
Strehler, B. L., and Arnold, W. (1951). *J. Gen. Physiol.* **34**, 809.
Sweetser, P. B., Todd, C. W., and Hersh, R. T. (1961). *Biochim. Biophys. Acta* **51**, 509.

Tributsch, H. (1972). *Photochem. Photobiol.* **16**, 261.

Vernon, L. P., Klein, S., White, F. G., Shaw, E. R., and Mayne, B. C. (1972). *Proc. Int. Congr. Photosyn. Res., 2nd, Stresa* **1**, 801.

Wells, R., Bertsch, W., and Cohen, W. (1972). *Proc. Int. Congr. Photosyn. 2nd Stresa* **1**, 207.

Witt, H. T. (1971). *Quart. Rev. Biophys.* **4**, 365.

Wraight, C. A., and Crofts, A. R. (1971). *Eur. J. Biochem.* **19**, 386.

Wraight, C. A., Kraan, G. P. B., and Gerrits, N. M. (1972). *Proc. Int. Congr. Photosynthesis Res., 2nd, Stresa* **1**, 951.

Yamashita, T., and Butler, W. L. (1968). *Plant. Physiol.* **43**, 1978.

Zankel, K. L. (1971). *Biochim. Biophys. Acta* **245**, 373.

6

Chlorophyll Fluorescence: An Intrinsic Probe of Photosynthesis

George Papageorgiou

ABBREVIATIONS

A	Intersystem intermediates
ADP	Adenosine diphosphate
ANS	1-Anilino-8-naphthalene sulfonate
ATP	Adenosine triphosphate

C550	Absorbance decrease at 550 nm (547 nm at 77°K) accompanying electron flow through PS II
CCCP	Carbonyl cyanide *m*-chlorophenylhydrazone
Chl	Chlorophyll
Cyt	Cytochrome
DAD	Diaminodurene
DCMU	3-(3,4-Dichlorophenyl)-1,1-dimethylurea
DCPIP	2,6-Dichlorophenol indophenol
EDTA	Ethylenediaminetetraacetate
FCCP	Carbonyl cyanide *p*-trifluoromethoxyphenylhydrazone
Fd	Ferredoxin
$NADP^+$	Nicotinamide adenine dinucleotide phosphate
PC	Plastocyanin
PMS	Phenazine methosulfate
PQ	Plastoquinone
PS I	Photosystem I
PS II	Photosystem II
PSU	Photosynthetic unit
Q	Quencher of fluorescence; also "primary" electron acceptor for PS II
P680	Chl *a* serving as energy trap for PS II
P700	Chl *a* serving as energy trap for PS I
X	Primary electron acceptor for PS I
X_E	Phosphorylating electrochemical potential
Z	Primary electron donor to PS II
S	Water-splitting catalyst

1. INTRODUCTION

1.1. General

A portion of the light intercepted by a plant is absorbed by the photosynthetic pigments, creating a supply of singlet electronic excitation energy. The greater part of this energy (about 85 % under optimal conditions) is used in photosynthesis. The remainder is lost as heat, or it is radiated as fluorescence. At physiological temperatures, fluorescence emanates mostly from the Chl *a* of PS II, with a main band maximum at 685 nm. Exceptional are the biliprotein-containing algae (Rhodophyta and Cyanophyta), in which the Chl *a* fluorescence is accompanied by the fluorescence of the biliproteins. At low temperature, the fluorescence of the plants originates from the Chl *a* of both photosystems, with band maxima mainly at 684 nm and 697 nm for PS II, and at 720–740 nm for PS I.

Measurable fluorescence parameters are the fluorescence spectrum, the excitation spectrum, the degree of polarization, the mean lifetime of fluorescence, and the quantum yield of fluorescence (for recent reviews on this subject, see Govindjee *et al.*, 1967, and Goedheer, 1972). The first two are static properties, more suitable for the description of a system held invariant during the measurement. In theory, the degree of fluorescence polarization and the mean lifetime of fluorescence can be correlated with the dynamic properties of the fluorescing system (e.g., rotation of the excited fluorescer, attack by excitation quenchers; see Weber, 1972). However, they are difficult to measure, particularly in biological preparations, and hence they have not been much used in kinetic studies.

The relative fluorescence yield is the most convenient parameter for kinetic studies (for an account of the techniques usually employed see Govindjee and Papageorgiou, 1971). In a photosynthetic organism, or in isolated photosynthetic tissue, the yield of Chl a fluorescence (which is proportional to the measured intensity for a given optical geometry) can be considered as an index of the momentary supply of the singlet Chl a excitation of PS II. At a constant rate of light absorption, this supply is controlled by the following deexcitation processes:

1. The light-induced electron transport across the reaction centers of PS II. This is the only significant process from the point of view of photosynthesis, and it consumes the greatest part of the PS II excitation.

2. The dissipation of the excitation as heat. This may occur at points of imperfection, such as dimeric or aggregated Chl with low-lying excited singlet states. Alternatively, it may occur by means of quenching of the singlet- or the triplet-excited population by endogenous quenchers, such as molecular oxygen.

3. The emission of Chl a fluorescence. At ordinary temperature this is emitted mostly by Chl a678 of PS II. It is a linear function of the rate of photosynthesis only under very special circumstances (cf. Sections 1.2 and 3.1).

4. The transfer of excess excitation from the fluorescing Chl a of PS II to the weakly fluorescing Chl a of PS I. This process is known as the "spillover" of the excitation energy; it is subject to control by metal cations and possibly by other factors (Section 4.2).

Before proceeding to the next section, we would like to emphasize two often overlooked points. First, with the exception of the light-induced electron transport, the processes that limit the momentary supply of Chl a excitation do not necessarily operate through the reaction center of PS II. Excitation may be quenched either at the reaction center, or at the antenna pigments that convey electronic excitation to this center (Amesz and Fork, 1967). Second, in a dynamic system, such as an active thylakoid, where light generates both electron and ion transport, the rates of all the deexcitation processes are variable with time.

1.2. Chlorophyll a Fluorescence of Constant and of Variable Yield

Spectroscopic evidence suggests that the fluorescent Chl of the green plants, Chl a, is integrated in the membrane of the thylakoid in the form of a small number of distinct molecular associations. These have been resolved by (1) low temperature absorption and fluorescence spectrophotometry (see, e.g., Cho, 1969; Cho and Govindjee 1970a–c); (2) matching the absorption spectrum of the *in vivo* pigments with a superposition of model curves (French, 1972); (3) spectroscopic analysis of the Chl *a* forms as they appear successively in a greening plant (Litvin, 1972); (4) second derivative low-temperature spectrofluorometry (Sineshchekov, 1972); and (5) fourth derivative low-temperature absorption spectrophotometry (Butler and Hopkins, 1970). The superior resolution of low-temperature spectrofluorometry is illustrated in Fig. 1, where the room temperature and the liquid nitrogen temperature fluorescence spectra of the green alga *Chlorella pyrenoidosa* are displayed. At room temperature, only the main Chl *a* fluorescence band at 688 nm, and its satellite band at about 740 nm are evident. At low temperature, four distinct fluorescence components, located at 686, 698, 718, and 725 nm, are resolved. The major forms of Chl *a in vivo* that are identified by these techniques, designated by their approximate red absorption band maxima (e.g., Chl *a*670), are listed in Table I.

With the exception of P680 and P700, which function at the reaction centers of PS II and PS I, most of the *in vivo* chlorophylls are participants of both photosystems. This is borne out by the analysis of the absorption spectra of PS I and PS II fractions into model Gaussian and Lorentzian curves (French, 1972). Of the PS II Chl population, Chl *a*678 emits the largest fraction of the room temperature fluorescence. Chl *b*, and Chl *a*670

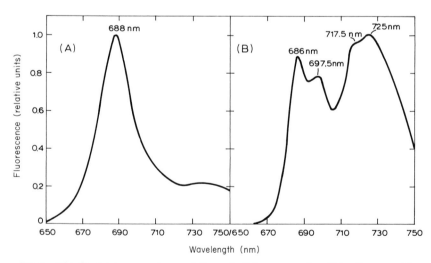

Fig. 1. The fluorescence spectrum of the unicellular green alga *Chlorella pyrenoidosa*; (A) of a cell suspension at room temperature; (B) of cells frozen with liquid nitrogen (77°K). (Redrawn from Cho, 1969.)

TABLE I

Chlorophyll *a* Forms *in vivo* Denoted by Their Approximate Red Absorption Band Maxima; Half-Bandwidths, *Tentative* Photosystem Assignments and Fluorescence Band Maxima[a]

Chl *a* form	Half-bandwidth (nm)	Photosystem assignment		Fluorescence band[b]
Chl *a*662	11	PS II	PS I	—
Chl *a*670	10	PS II	PS I	F681
Chl *a*678[c]	10	PS II	PS I	F687
Chl *a*683	11	PS II		F693
Chl *a*692	13	PS II	PS I	—
Chl *a*695	—		PS I	F715
Chl *a*705	18		PS I	F735

[a]See Cho (1969); Butler and Hopkins (1970); and French (1972).

[b]With the exception of F687, all other fluorescence bands are resolvable only under special conditions, e.g., low temperature or partial extraction of the pigments.

[c]This band is resolved further to constituent bands absorbing at 676 nm and ot 679 nm in the fourth derivative of the absorption spectrum of Chl *a in vivo* at $-196°C$ (French, 1972).

transfer their excitation entirely to Chl *a*678, and only after the latter is extracted, Chl *a*670 becomes fluorescent (Cho and Govindjee, 1970c). The chlorophylls of PS I fluoresce weakly, but their fluorescence output which is resolvable by special methods (photosystem fractionation, low temperature) appears to be quite insensitive to the photochemical activity at the reaction center of PS I. Experimental proof for this was obtained by Vredenberg and Slooten (1967), who showed that the fluorescence of isolated PS I particles remains constant under conditions in which electrons are transferred from P700 to $NADP^+$.

The emergence of new fluorescence bands on freezing the photosynthetic tissue (Fig. 1) indicates that ordinarily the *in vivo* Chl *a* can be distinguished into fluorescing (Chl *a*678), and into weakly fluorescing or nonfluorescing subpopulations. Further evidence in support of this is as follows.

1. Direct excitation of Chl *a* generates less fluorescence than sensitization by excitation of the accessory pigments (Duysens, 1952; French and Young, 1952). Presumably, the accessory pigments transfer their excitation preferentially to the fluorescent Chl *a*.

2. The quantum yield of Chl *a* fluorescence calculated from the ratio of the actual fluorescence lifetime ($\tau = 1.6$ nsec, Merkelo *et al.*, 1969) to the natural fluorescence lifetime ($\tau_0 = 15.2$ nsec, Brody, 1957) is about 0.1. On the other hand, the directly measured quantum yield is less than 0.03 (Latimer *et al.*, 1957). The discrepancy arises from the ability of the fluorescence lifetime method to select for the fluorescent Chl *a* only, while the direct method is indiscriminate, taking into account both the fluorescent and the nonfluorescent components

3. Fractionation of chloroplasts either with detergents (Boardman *et al.*, 1966; Briantais, 1967; Vredenberg and Slooten, 1967), or with the French press (Michel and Michel-Wolwertz, 1969, 1970; Park *et al.*, 1971), provides a heavy subchloroplast fraction containing the pigments and activities of both photosystems, and a light fraction composed almost exclusively of PS I particles (see Chapter 2 of this volume). The heavy fraction is as much as six times more fluorescent than the light fraction.

The portion of the Chl *a* excitation which is reemitted as fluorescence varies in response to the rate of photosynthesis. In general, but not always, weak Chl *a* fluorescence typifies a vigorous photosynthesis and strong Chl *a* fluorescence a weak or inhibited photosynthesis. Not all of the Chl *a* emission, however, is sensitive to the photosynthetic electron transport. The kinetic pattern of the Chl *a* fluorescence of dark-adapted photosynthetic organisms, during a period of continuous illumination, led Lavorel (1959) and Clayton (1969) to emphasize the distinction between fluorescence of variable and fluorescence of constant yield.

The variable Chl *a* fluorescence, which originates from the pigment population of PS II, is sensitive both to the rate of electron transport through the reaction centers of this photosystem, and to the changes in the thylakoid membrane ultrastructure that accompany the phosphorylating electron transport. At the very early stage of an illumination period (i.e., for about $\frac{1}{2}$ sec), the quantum yield of the variable fluorescence (Φ_F) and of the rate of oxygen evolution (Φ_{PS}) add up to a constant. If the fraction of the Chl *a* excitation lost by thermal degradation can be set proportional $(1-a/a)$ to the quantum yield of fluorescence, then Φ_F and Φ_{PS} can be linked by a "complementarity" equation (Lavorel and Joliot, 1972).

$$\Phi_{PS} + \frac{1}{a}\Phi_F = 1 \tag{1}$$

As will become apparent later, this equation breaks down at longer periods of illumination. Additive, rather than multiplicative terms are needed to account for the nonphotochemical and nonfluorescent Chl *a* excitation losses.

Variable Chl *a* fluorescence is always associated with the heavy subchloroplast fraction (PS I + PS II), rather than with the light fraction (PS I), suggesting the granum as its likely cytological locus (Park *et al.*, 1971). Characteristically also, mutants blocked at their PS II function show a diminished amplitude of the light-induced variable Chl *a* fluorescence (cf. Section 5). On the other hand, the constant yield fluorescence may originate from the pigments of PS I, or it may contain a substantial proportion of their contribution (Lavorel and Joliot, 1972). There exists no evidence to refute this statement, which gains some support from the observation that at 720 nm, where the emissions of both photosystems overlap, the share of the constant yield component is higher than at 685 nm, where PS II is emitting (Lavorel, 1962; Govindjee and Briantais, 1972). PS II pigments, however, may also contribute to the constant yield fluorescence, if the trapping of the exciton at the reaction center is not 100% efficient (Mar et al., 1972; Briantais et al., 1972).

2. THE REACTION CENTER COMPLEX OF PHOTOSYSTEM II

The reaction center complex of PS II utilizes singlet Chl a electronic excitation to perform an oxidoreduction against the redox potential gradient. For this task, it is endowed with a photoreceptor chromophore (Chl a) and with a redox donor-acceptor couple (Z, Q), whose chemical nature remains unknown. Two mechanisms have been advanced to account for the function of the reaction center complex of PS II. The first visualizes the excited photoreceptor as a sensitizer facilitating, but not mediating, a direct electron exchange between Z and Q (Döring et al., 1969). In the second, the photoreceptor Chl a molecule is itself oxidized as a result of the primary photoconversion (Kautsky et al., 1960; Duysens and Sweers, 1963; Butler, 1972b). The two alternatives can be rendered as follows:

$$ZChlaQ \xrightarrow{hv} ZChla^*Q \longrightarrow Z^+ChlaQ^- \tag{2}$$

$$ZChlaQ \xrightarrow{hv} ZChla^*Q \longrightarrow ZChla^+Q^- \longrightarrow Z^+ChlaQ^- \tag{3}$$

The intimate association of the components of this reaction center is evidenced by: (1) the occurrence of a photoinduced electron transport from Z to Q at $-196°C$ (Erixon and Butler, 1971a; Okayama and Butler, 1972a); (2) the fact that all known inhibitors, or inhibitory treatments, that block the photosynthetic electron transport, act either before Z or after Q (see Katoh, 1972, and references cited there).

2.1. The Pigment 680

Employing short periodic actinic flashes (390–500 nm; width 20 μsec; repetition frequency 10 Hz), Döring et al. (1967) were able to detect a fast absorbance change located within the main red absorption band of Chl a in vivo (see Chapter 10 of this volume). This change (a decrease) decays with a time constant of 200 μsec, a rate of comparable magnitude to the rate of the slow component of the dark reversal of the photobleaching of P700 as reported by Haehnel and Witt (1972). The same reaction was observed at $-196°$ by Floyd et al. (1971). The spectrum of this absorbance change is typical of Chl a in vivo, having major bands at 682 and 435 nm, and a minor band at 640 nm (Döring et al., 1969; see Fig. 6 in Chapter 10 of this volume).

The location of these bands on the long-wavelength side of the absorption bands of the PS II bulk Chl *a* suggests that they may originate from the photoactive Chl *a* that serves as the excitation energy trap of the PS II reaction center. In analogy to the designation P700 for the excitation trap of PS I, this Chl *a* form was labeled as P680 (Floyd *et al.*, 1971). Further evidence in support of the identification of P680 as the excitation trap of PS II is (Döring *et al.*, 1968, 1969):

(1) The 682 nm absorbance change is present only in the PS II-enriched subchloroplast fractions; (2) DCMU at 0.1 μM reduces both the fast absorbance change at 682 nm and the O_2 signal to one-half their normal amplitude; (3) the P680 turnover is faster than the photoreduction of PQ, an electron carrier operating close to the reaction center of PS II.

The possibility that the 682 nm absorbance decrease may arise from a Chl *a* fluorescence artifact was recently discussed by Butler (1972a). Because of the extensive overlap of the absorption and fluorescence spectrum of Chl *a*, a flash-induced increase in the fluorescence yield will be seen by the absorbance monitoring apparatus as a "bleaching." The optical arrangement of Döring, however, minimized the interference from the fluorescence signal in two ways: first by the large distance between sample and photodetector (small solid angle of observation), and second by the intervention of appropriate interference filters before the photodetector. The reality of the 682-nm absorbance change gains a firmer footing from the occurrence of a companion bleaching in the Soret band (435 nm; Fig. 6 in Chapter 10 of this volume), which obeys the same kinetics.

Whether P680 is an oxidoreducible compound like P700 or a sensitizer which, when excited, enables Z to donate an electron to Q, is now a matter of controversy. Döring *et al.* (1967, 1969) favor the second alternative because the bleaching of P680 is faster than the photoreduction of PQ, a portion of which may function as the primary electron acceptor of PS II (component X-335; see Witt 1971). Furthermore, the P680 signal occurs in photosynthetic tissue deprived of its electron donor to PS II by Tris-washing, or by extraction with wet heptane (Govindjee *et al.*, 1970). In the latter instance, the much reduced variable fluorescence of the extracted material makes more remote the possibility of a substantial fluorescence artifact in the P680 signal (R. Govindjee *et al.*, 1970).

The phototransformation of P680 to a sensitizer of weaker absorptivity can be expressed as follows:

$$P680 \underset{t \simeq 200\,\mu\text{sec}}{\overset{t \simeq 10\,\mu\text{sec}}{\rightleftharpoons}} P680' \qquad (4)$$

On the assumption of a sensitizer of zero absorptivity, Butler (1972a) calculated from Döring's data 1 P680 per 10^4 chlorophylls, a figure considerably lower than the accepted relative abundance (1 : 300) of the reaction center pigments. As Butler (1972b) suggested, the dark recovery of P680' may have a fast component with a time constant of the order of 10 μsec, which was not detected by Döring. It is possible, however, that the absorptivity of P680' is not zero as assumed, and hence the real ratio of P680 : Chl is higher than 1 : 10^4.

Butler (1972a) has strongly questioned the "sensitizer hypothesis" pointing out that Tris-washed chloroplasts retain sufficient amounts of electron donor (Cyt b559) to accomplish the photoreduction of Q. Supporting evidence for this is the increase of fluorescence to its maximal yield, when the reoxidation of Q by the intersystem pool oxidants has been blocked with DCMU (Yamashita and Butler, 1968a). As shown by Mohanty et al. (1972), DCMU has this effect only when it is added during the dark period that precedes the excitation of fluorescence, since only then it causes the reaction centers to accumulate in the state Z^+ChlQ^-. Added in the light, DCMU generates no increase in fluorescence, because the reaction centers are already oxidized by the intersystem oxidants to the state Z^+ChlQ. Denoting the intersystem oxidants, collectively, as A, and the DCMU-introduced block between Q and A by two vertical bars, we may write for these alternatives:

$$ZChlQA \xrightarrow{DCMU} ZChlQ \parallel A \xrightarrow{h\nu} Z^+ChlQ^- \parallel A \qquad (5)$$

$$ZChlQA \xrightarrow{h\nu} Z^+ChlQ^-A \longrightarrow Z^+ChlQA^- \xrightarrow{DCMU} Z^+ChlQ \parallel A \qquad (6)$$

It appears, then, that in this system at least, the Chl a fluorescence rises in response to the oxidation of the electron donor, rather, than in response to the reduction of the electron acceptor of PS II (Butler, 1972b). As shown below, this oxidation is not the primary photoinduced electron transfer reaction.

$$ZChlQ \xrightarrow{h\nu} ZChl^*Q \longrightarrow ZChl^+Q^- \longrightarrow Z^+ChlQ^- \qquad (7)$$

This notion is consistent with Mauzerall's (1972) conclusion that the Chl a fluorescence rise is too slow to be accounted for by a primary photochemical oxidoreduction. This conclusion is based on the observation that it takes 20 μsec for the fluorescence of Chlorella to rise to its maximal yield, following a 10-nsec wide saturating light flash, while the time constant of the primary electron transfer step (reduction of Q) must approximate the lifetime of Chl a fluorescence. It must be pointed out, however, that, while the reduction of the reaction center chromophore may control the rate of fluorescence rise, its level is determined in response to the relative abundance of the high yield fluorescent species $ZChlQ^-$, which exists in a redox equilibrium with the pool of intersystem oxidants (see Section 3.1).

TABLE II
Properties of the Reaction Center Chromophores of Photosystems II and I

Property	P680 (PS II)	P700 (PS I)
Chemical characterization	Chl $a^{a,b}$	Chl a^c
Difference absorption bands, nm	435, 640, 682[b]	438, 660, 682–703[c,d]
Fluorescence band maximum	693–698[e,f]	720 (?)
Type of reaction	Sensitizer,[b] or electron carrier[g]	Electron carrier[c]
Rise time of the absorbance change	$\leq$ 20 nsec[d]	20 nsec[d]
Half-life of the photobleached state	200 μsec[b]	20 μsec,[h] 200 μsec[h]
Midpoint redox potential	≤ -50 mV[i]	450 mV[d]
Temperature of inactivation	55°C[b]	65°C[b]
Sensitivity to aging	$\tau \simeq 95$ h (0°C)[d]	No[d]
Sensitivity to DCMU	$C = 0.1\ \mu M^d$	No[d]

[a]Döring *et al.* (1967). [b]Döring *et al.* (1969).
[c]Kok and Hoch (1961). [d]Witt (1971).
[e]Krey and Govindjee (1964). [f]Cho and Govindjee (1970a).
[g]Butler (1972a). [h]Haehnel and Witt (1972).
[i]Erixon and Butler (1970a).

It has been suggested that a small fluorescence band at 693–698 nm may originate from the reaction center chromophore of PS II (Krey and Govindjee, 1964, 1966). This component has been detected in the difference spectra obtained by subtracting the fluorescence of algae excited with nonsaturating light from the fluorescence excited with saturating light. In effect, its presence in the difference spectrum means that the 693-nm fluorescence band contains a higher share of the variable fluorescence, a conclusion supported also by the occurrence of this emission in the DCMU-minus-normal difference fluorescence spectra. Excitation spectra of algae frozen to 4°K (Cho, 1969; Cho and Govindjee, 1970b) indicate that the 693-nm fluorescence band originates from a Chl *a* form absorbing at 686 nm. This fluorescence band, and the C550 response (cf. Section 2.3) are lost after exhaustive extraction of lyophilized chloroplasts with hexane, but reconstitution of the extracted material with added β-carotene and PQ-A regenerates them (Okayama and Butler, 1972b). These results, and the absence of the 693 nm emission from the PS I-enriched chloroplasts suggest P680 as the origin of the 693-nm fluorescence band. Some properties of the photosynthetic reaction center chromophores are listed in Table II.

2.2. The Primary Electron Donor to Photosystem II

One of the least known links of the photosynthetic electron transport chain is the electron donor to the reaction center of PS II, the entity that is usually denoted as Z. The only spectroscopic property that has been assigned to this compound is a light-induced absorbance change at 320 nm (Vater *et al.*, 1968); however, this change may also be due to PQ (Stiehl and Witt, 1969). Several other properties of Z have been deduced from its interactions with the water-splitting enzyme S, and the primary electron acceptor of PS II, Q. Following excitation, P680 donates an electron to Q (in less than 60 nsec; Mauzerall and Malley, 1971), and abstracts an electron from Z. The second process is much slower than the first, requiring about 20 μsec to reach completion (Mauzerall, 1972). Oxidized Z receives an electron from the oxidoreducible enzyme S,* which can accumulate sequentially up to four oxidizing equivalents. S, in consequence, can exist in the forms S_0, S_1, S_2, S_3, and S_4, where the subscripts signify the number of positive charges per molecule (Joliot *et al.*, 1969; Kok *et al.*, 1970; Forbush *et al.*, 1971). *Here we must recognize, however, that Z and S may be identical, since they have not been discriminated experimentally.* In this case, S denotes a complex of Z molecules capable of accumulating up to four positive charges. Reduced Q discharges its electron to the intermediate pool of oxidants (pool A), which serves as an electron sink, while the four times oxidized S (S_4) reacts with two molecules of water to release one molecule of oxygen. The reactions at, and near, the reaction center of PS II, are as follows:

$$ZChlQ \xrightarrow{h\nu} ZChl^*Q \tag{8}$$

$$ZChl^*Q \longrightarrow ZChl^+Q^- \tag{9}$$

$$ZChl^+Q^- \longrightarrow Z^+ChlQ^- \tag{10}$$

$$Z^+ChlQ^- + A \longrightarrow Z^+ChlQ + A^- \tag{11}$$

$$S_{n-1} + Z^+ChlQ \longrightarrow ZChlQ + S_n \tag{12}$$

$$S_4 + 2H_2O \longrightarrow S_0 + 4H^+ + 0_2 \tag{13}$$

After long adaptation to darkness, only the forms S_0 (25%) and S_1 (75%) are present (see Chapter 8 of this volume). The gradual redistribution of the water-splitting enzyme among its four oxidation forms, during continuous illumination, gives rise to the activation phenomena in the rate of oxygen evolution and fluorescence (Delosme *et al.*, 1959; Joliot, 1965b, 1968; Bannister and Rice, 1968; cf. Section 3.1, and Chapter 8 of this volume). The activation phase may be eliminated if, prior to the continuous illumination, the dark-adapted tissue is given two bright light flashes, by which the state S_3 of the water-splitting enzyme becomes populated (Joliot, 1968; Forbush *et al.*, 1971). S_4 reacts with water very fast, its concentration being practically zero at all times.

---

* Lavorel (Chapter 5 of this volume), however, has used "Y" for the primary electron donor, and Z for the component that accumulates four positive charges.

The question whether S and Z are metalloproteins, and more specifically ferro- or manganoproteins, remains still unanswered. Extreme Mn deficiency impairs the transfer of electrons from water to Q, as deduced by the disappearance of the variable fluorescence of Chl *a* (Homann, 1968b; Cheniae and Martin, 1970). The loss of the variable fluorescence, however, can also arise from an increased spillover to the non- or weakly fluorescent PS I, or even from the direct quenching by some quencher such as O_2 whose penetration to the pigment bed has been facilitated by the removal of the bound Mn (cf. Section 4.2). Thus, the reported suppression of the low-temperature emission of P680 (F693), when the bound Mn is extracted with hydroxylamine (Mohanty *et al.*, 1971a), can be attributed either directly to a structural change due to the removal of Mn, or to a facilitated diffusion of oxygen to the Chl *a* site.

The correlation of the water-splitting catalyst with the high potential form of Cyt b559, which serves as an electron donor to PS II only when the capacity of the chloroplasts to photooxidize water is inhibited (Erixon and Butler, 1971c), encounters several difficulties. The midpoint potential of this Cyt ($E_0' = 450$ mV; Erixon *et al.*, 1972) is less than the 815 mV needed for the homolytic cleavage of the O—H bond. Bendall and Sofrova (1971) estimated the oxidizing potential of the process $S_3 \longrightarrow S_4$ to be much higher than that of the $S_0 \longrightarrow S_1$, but it is questionable whether four Cyt b559 molecules can be viewed as a single entity that can be oxidized in four consecutive steps. Recent evidence suggests that the photooxidation of water persists under conditions at which the high potential Cyt b559 is not detectable. Addition of FCCP, or hydroxylamine, to isolated chloroplasts converts the high potential form to the low potential one faster than it affects the electron transport from water to methyl viologen (Cramer and Böhme, 1972). About 30% of the normal rate of oxygen evolution with added Hill oxidants still remains when, following trypsin digestion, or the addition of 2,4,6-triiodophenol, Cyt b559 is no longer detectable (Cox and Bendall, 1972).

While this evidence weighs against the involvement of the high potential Cyt b559 in the water-splitting process, at low temperature this Cyt is apparently the primary electron donor to PS II, since its photooxidation is kinetically related to the rise of Chl *a* fluorescence (Knaff and Arnon, 1969a; Erixon and Butler, 1971a; Erixon *et al.*, 1972; Butler, 1972b). At room temperature, Cyt b559 is photooxidized only in Tris-washed chloroplasts buffered at high pH (Knaff and Arnon, 1969c; 1971a; Arnon *et al.*, 1971; Bendall and Sofrova, 1971). At physiological conditions, Cyt b559 may function as an intermediate electron carrier, operating between the PS II photoreaction and Cyt f. Evidence for this is derived both from work with *Chlamydomonas* mutants (Levine and Gorman, 1966; Levine, 1969), and from the antagonistic effect of red and far-red light on the oxidation state of

Cyt b559 (Cramer and Butler, 1967; Hind, 1968; Ben-Hayyim and Avron, 1970). In isolated grana stack particles, Cyt b559 may function on a side pathway to PS II, since its oxidation–reduction rates are too slow to account for the rate of NADP$^+$ reduction (Fork, 1972).

2.3. The "Primary" Electron Acceptor

Until recently, only one spectroscopic property was assigned to the primary electron acceptor Q of the reaction center complex of PS II, namely, the capacity of its oxidized form to quench the variable fluorescence of Chl a (cf. Section 3). Research carried out in the laboratories of Arnon and of Butler, however, indicates that the photoreduction of Q is accompanied with an absorbance decrease at about 550 nm, at room temperature, and at 546 nm at $-189°C$, provided that Cyt f is maintained in its oxidized state. This discovery affords a second, Q-dependent, spectroscopic variable by which the properties of Q can be studied. The evidence that led to the identification of the 550-nm absorbance change with the photoreduction of Q is the following.

In 1969, Knaff and Arnon (1969b) discovered a light-induced absorbance decrease at about 550 nm in isolated chloroplasts, which is observable on the condition that the interference from the α-band of Cyt f has been eliminated by the addition of ferricyanide. The 550-nm absorbance change was attributed to a hitherto unkown component of the intermediate electron transport chain, which was given the designation C550. Several lines of evidence support this assignment. C550 is detectable in the PS II-rich subchloroplast fractions obtained either with the digitonin (Knaff and Arnon, 1969b; Arnon et al., 1971), or with the French press procedure (Erixon and Butler, 1971c). It is detectable in the PS I-lacking mutant No. 8 of *Scenedesmus obliquus*, but not in the PS II-minus mutant No. 11 of the same species. Strong reductants, such as sodium dithionite and borohydride, but not ascorbate, simulate the light-induced absorbance decrease at 550 nm (Erixon and Butler, 1971c). The early red drop at 675–680 nm of the action spectrum of the photoreduction of C550 at low temperature indicates a PS II-driven process. On the other hand, C550 is photooxidized by PS I, since far-red light is more effective in that (Knaff and Arnon, 1971b; Bendal and Sofrova, 1971).

C550 must function very close to the primary photochemical conversion of PS II, since its absorbance change is detectable even at low temperature (84°K; Knaff and Arnon, 1969b). DCMU and o-phenanthroline, that block the electron transport just after Q, do not inhibit the light-induced spectral response of C550, but they delay its dark reoxidation by the intermediate pool oxidants. The 550-nm absorbance change is absent from Tris-washed chloroplasts, which are known to be impaired on the electron donor side of PS II. On restoring, however, electron donation to PS II, by means of the artificial electron donor benzidine, the 550-nm absorbance change is also restored (Arnon *et al.*, 1971).

While undoubtedly the light-induced absorbance decrease at 550 nm is associated with the photochemical reduction of Q, it is far from certain that C550 and Q are identical, or even that C550 is a particular chemical compound, as its symbol intends to imply. Contrary to what happens at low temperature, at room temperature and in the absence of ferricyanide the quantum yield of the C550 photoreduction does not fall off in the far-red region, as does that of the Q-dependent variable fluorescence increment (Ben-Hayyim and Malkin, 1972). Thus while C550 is an excellent indicator of the primary electron acceptor of PS II at low temperature, it is not so at room temperature where both photosystems appear to contribute to it. Butler (1972c) has shown that the absorbance changes at 518 nm, and at 550 nm are partly suppressed by the combined action of valinomycin and nigericin, while the further addition of DCMU abolishes the 518 nm change, but not the residual 550 nm change. Since the 518 nm absorbance change is an indicator of a light-induced transmembrane potential (Witt, 1971), Butler infers that the latter accounts for part of the room temperature C550 signal.

Of crucial importance for the C550 response is the structural integrity of the thylakoid membrane, since treatments that perturb the membrane ultrastructure, such as irradiation with ultraviolet light, digestion with lipase, and heat treatment abolish it (Butler and Okayama, 1971; Erixon and Butler, 1971b). Characteristically, the C550 response which has been eliminated by an exhaustive extraction of the photosynthetic tissue with hexane, reappears when β-carotene (not an intermediate electron carrier) is added to the extracted sample (Okayama and Butler, 1972b). Butler (1972a) has suggested that the C550 signal is a spectral response isomorphic in many respects with the light-induced fluorescence rise, which manifests the existence of Q. It is possible, that both C550 and Q are spectroscopic expressions of changes in the structural detail in the neighborhood of the primary electron acceptor of PS II. As discussed above, the photobleaching of C550 appears to consist of two contributions, one responsive to the electron transport through PS II, and the other to the electrical component of the photoinduced transmembrane potential that generates the 518-nm absorbance change (Butler, 1972c), Significantly, the variable yield fluorescence is responsive, also, both to the photosynthetic electron transport, and to the chemical component (pH gradient) of the transmembrane potential (see Section 4.1).

During a dark interval that follows an illumination period, the yield of the variable Chl *a* fluorescence decays, in a process which may, or may not, be indicative of a dark reoxidation of the primary electron acceptor of PS II.

Zankel (1971), using an ultrasonically modulated excitation beam at 14 MHz to exclude a large portion of the delayed light, measured a fast first-order decay with a time constant of 100 μsec, followed by a slower decay process. This decay occurs at the same rate as the dark reversal of the P680 signal (Döring *et al.*, 1969), suggesting that at that time interval the control of the variable fluorescence yield may be exerted by the primary electron donor rather than by the primary electron acceptor of PS II. Slower dark decay components of the variable Chl *a* fluorescence yield have been reported by several investigators. Decay processes with time constants of 0.6–4 msec, and of 0.1–0.8 sec have been attributed to the reoxidation of Q by the pool of intersystem oxidants (Joliot, 1965a; Forbush and Kok, 1968; Heath, 1970; Joliot *et al.*, 1971; cf. Section 3.1). Slower decay modes, with time constants of 5 sec and of 30–40 sec, may correspond to a back flow of electrons from Q to the oxidized intermediates that exist on the water side of PS II (Heath, 1970; Joliot *et al.*, 1971).

Bennoun (1970), and independently Mohanty *et al.* (1971a), discovered that the light-induced increment of the variable Chl *a* fluorescence stays undiminishingly high in the darkness when DCMU and hydroxylamine are present. With the electron transport to the intersystem oxidants blocked by DCMU, the decay of the variable Chl *a* fluorescence occurs by means of recombinations such as:

$$Z^+ChlQ^- \longrightarrow ZChlQ \tag{14}$$

$$Z^+ChlQ^- \longrightarrow ZChl^*Q \longrightarrow ZChlQ + \text{delayed light} \tag{15}$$

Hydroxylamine competes with Q for electron donation to Z^+, promoting the accumulation of the species $ZChlQ^-$, which cannot recombine to $ZChl^*Q$ and emit delayed light (see Chapter 5 of this volume). Hydroxylamine, however, was found to quench only the long-term delayed light (Bennoun, 1970), and to have no effect on the delayed light of the millisecond range (Stacy *et al.*, 1971).

The midpoint redox potential of Q can be determined either in respect to the variable Chl·*a* fluorescence, or in respect to the 550-nm absorbance change with suspensions of chloroplasts poised with suitable redox buffers. In either case, care must be exercised to limit the measurement only to Q, i.e., to exclude interference from oxygen and from the intermediate pool carriers. Redox titrations of the variable Chl *a* fluorescence and of the 546 nm absorbance change at low temperature gave $E_0' = -50$ mV (Erixon and Butler, 1971a).

In spite of the accumulated information on its spectral properties and about its interactions with other electron carriers of the photosynthetic electron transport chain, we are still unable to assign Q, unequivocally, to a

TABLE III
Properties of the Chl *a* Fluorescence Quencher Q (C550)

1. Oxidoreducible substance; $E_0 = -50$ mV; one electron carrier[a]
2. Oxidized in darkness by the pool of intersystem oxidants[b] and the pre-PS II oxidants[c]; the dark reoxidation is inhibited in the presence of both DCMU and NH_2OH[d, c]
3. Quencher of the variable fluorescence of Chl *a in vivo*[b]
4. Difference absorption band at 77°K, associated with the oxidized form, located at 546 nm[f]
5. The associated spectroscopic changes (absorption, fluorescence) are labile to lipase, exhaustive extraction by hexane, ultraviolet light, and heat treatment[g, h]

[a]Erixon and Butler (1970a).   [b]Duysens and Sweers (1963).
[c]Murata *et al* (1966b).   [d]Bennoun (1970).
[e] Mohanty *et al.* (1971a).   [f]Knaff and Arnon (1969b).
[g]Butler and Okayama (1971).   [h]Erixon and Butler (1971b).

given class of chemical compounds. If the absorbance change at 550 nm is a property of Q per se, then porphyrins such as chlorophylls and cytochromes must be ruled out for the lack of a companion absorption change in the Soret region (400–500 nm; Bendall and Sofrova, 1971). Plastoquinone (including plastochromanol, as proposed by Kohl *et al.*, 1969), is also unlikely because of its relatively slow photoreduction ($t_{1/2} \simeq 20$ msec; Stiehl and Witt, 1969). A spectral component (X-335), whose light-induced absorbance rise at 320 nm decays with a time constant of 0.6 msec has been suggested by Stiehl and Witt (1968), and Witt (1971), as a possible link between PS II and PQ. Its chemical nature is unknown. Table III lists some of the established properties of this crucial link of the photosynthetic electron transport chain.

3. THE KINETICS OF CHLOROPHYLL *a* FLUORESCENCE

When light is turned on a plant, photosynthesis does not set in at once in full force. On the way to an overall steady state, the photosynthetic apparatus passes through several transitory stages. These events, known as the induction phenomena, have been studied mostly in respect to the rate of oxygen evolution and Chl *a* fluorescence. The former variable reflects the rate of electron transport through PS II, the latter the momentary density of electronic excitation of PS II.

Traditionally, the induction of Chl *a* fluorescence, or the "Kautsky effect" after Kautsky (1931) who first observed it visually, has been distinguished into a fast change, and into a slow change (Wassink, 1951; Rabinowitch, 1956; for a recent review, see Govindjee and Papageorgiou, 1971). Lasting only a few seconds, the fast fluorescence change in the case of isolated higher plant chloroplasts (class II) consists of a biphasic rise from an initial level O, (F_0) to a plateau $P(F_\infty$; Fig. 2A, lower curve). O represents the fluorescence

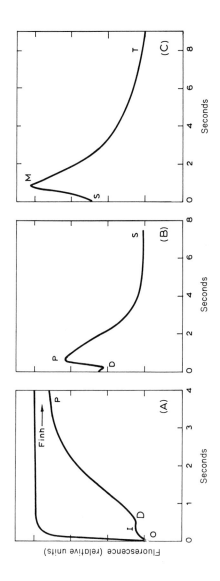

Fig. 2. The various stages of the fluorescence timecourse of Chl *a in vivo.* (A) The fast rise in isolated broken (class II) spinach chloroplasts in the absence (lower curve) and in the presence (upper curve) of the electron transport inhibitor DCMU. (B) The fast rise, and the ensuing fast decay of Chl *a* fluorescence in the red alga *Porphyridium cruentum* (redrawn from Mohanty *et al.,* 1971b). (C) The slow fluorescence change in the green alga *Chlorella pyrenoidosa* (Papageorgiou and Govindjee, unpublished experiment).

level at a physiological state in which all the intersystem intermediates are oxidized. At *P*, a steady state has been attained, in which electrons are fed to the intermediate pool by PS II, and abstracted by PS I at equal rates. When the intersystem electron transport has been inhibited by a poison such as DCMU, the fluorescence of Chl *a* rises very fast to a higher plateau (Fig. 2A, upper curve). In algae, in intact higher plant leaves, and in isolated higher plant chloroplasts that retain their outer membrane (class I chloroplasts; Krause, 1972), the rise *OP* is followed by a somewhat slower decay to a lower level *S*. This is illustrated in Fig. 2B in the case of the unicellular rhodophyte *Porphyridium cruentum*. The slow change *SMT*, that ensues the *PS* decay, lasts for several minutes. It has been documented mostly in algae, and in some cases in higher plant leaves, especially under the influence of an environmental stress, such as a CO_2-rich atmosphere. A typical time course of the slow fluorescence change in *Chlorella* is shown in Fig. 2C. The blue-green algae differ in that their fluorescence decay along *MT* is very slow, with *M* representing effectively a plateau (Papageorgiou and Govindjee, 1967; Mohanty, 1972). Figure 3 illustrates the slow fluorescence change in the blue-green alga *Anacystis nidulans* induced with various intensities of excitation.

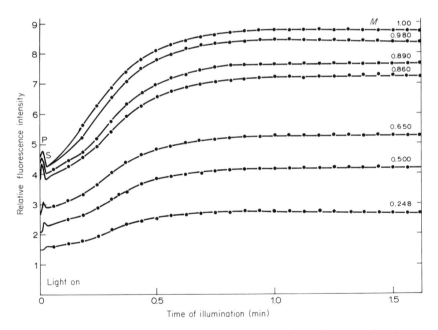

Fig. 3.   The slow change of Chl *a* fluorescence of the unicellular blue-green alga *Anacystis nidulans* at various intensities of 633 nm excitation. Maximal excitation intensity, $1.0 = 29$ kergs cm^{-2} sec^{-1}. A small fast fluorescence transient is discernible near the left-hand ordinate. (Reproduced from Mohanty and Govindjee, 1973b.)

The fast fluorescence rise along OP (Fig. 2A) coincides with a slowdown of the electron flow through PS II, as a result of the gradual reduction of the intersystem intermediates. The decay PS and the slow fluorescence change SMT reflect processes in which the ultrastructure of the thylakoid membrane plays a fundamental role. These processes are related to the transmembrane ionic fluxes on the way to the establishment of the prephosphorylation high energy state.

3.1. The Fast Fluorescence Rise

On shining light on dark-adapted photosynthetic tissue, the fluorescence of Chl a reaches very fast (probably within an interval comparable to its lifetime) the initial level O. This level is the constant yield component, which is insensitive to the physicochemical processes that take place in the thylakoid membrane. The fluorescence rise OI is almost synchronous with a rise in the rate of oxygen evolution, and it is known as the "activation phase." Its underlying physical cause is the accumulation of the water-splitting catalyst in the S_3 state (cf. Section 2.1; also Chapter 8 of this volume). The ensuing decline ID is more pronounced in anaerobic systems. Since it is sensitized by far-red light, it has been attributed to the oxidation of the intermediate pool and Q by PS I (Munday, 1968; Munday and Govindjee, 1969a–c). Following D, the Chl a fluorescence rises to P in a process that is synchronous with a decay in the rate of oxygen evolution, which in algae reaches a minimum at the time of P (Delosme et al., 1959; Joliot, 1968; Bannister and Rice, 1968). The rise DP accounts for the major part of the variable fluorescence; the latter (i.e., the amplitude $P–O$) can be up to five times greater than the constant yield fluorescence (e.g., see Munday and Govindjee, 1969a).

To rationalize the fast fluorescence rise, Duysens and Sweers (1963) postulated an electron carrier Q as the primary oxidant (electron acceptor) of the reaction center of PS II (cf. Section 2.3). Oxidized Q is supposed to be a very effective quencher of the exciton that is trapped at the reaction center, since it makes use of this energy to abstract an electron from the primary donor against the electrochemical gradient. Reduced Q (QH), on the other hand, is not a fluorescence quencher. An exciton arriving at such a reaction center has four possible routes open to it: (1) fluorescence from the reaction center chromophore; (2) radiationless deexcitation to the ground state; (3) transfer to another trap (vide infra); and (4) return to the bulk chlorophylls, wherefrom it may be emitted as fluorescence or degraded by a nonradiative process. Accordingly, the variable fluorescence consists of the combined emissions of Chl $a678$ and P680 (but perhaps more of the former). Similar hypotheses, relating the Chl a fluorescence output to the rate of the photosynthetic electron transport, have been advanced earlier by Ornstein et al. (1938), and by Kautsky et al. (1960), but the Duysens and Sweers hypothesis gained the widest acclaim primarily because it is built within the conceptual framework of the two photosystems and two photoreaction mechanisms of photosynthesis.

According to the present understanding of the Duysens and Sweers hypothesis, the amplitude $P–O$ measures the redox potential difference spanned by the Q/QH couple. This communicates primarily with the pool of intersystem intermediates, but it is also influenced by the redox state of intermediates existing after PS I and before PS II (Munday and Govindjee, 1969a,b; Murata *et al.*, 1966a). Although, as suggested by Mauzerall (1972) and by Butler (1972b; cf. Section 2.1), the rise rate of the variable fluorescence may not be under the control of the primary photoreduction in PS II (i.e., of the reaction $Chl*Q \longrightarrow Chl^+Q^-$), nevertheless its magnitude reflects the redox potential of the Q/QH couple.

The reactions that donate or abstract electrons from Q can be formulated as follows:

$$Z \underset{k_2}{\overset{k_1 I}{\rightleftharpoons}} Q \underset{k_4}{\overset{k_3}{\rightleftharpoons}} A \tag{16}$$

According to this scheme, the photoreduction of Q can be accelerated either by means of a very high intensity of excitation ($k_1 I \gg k_3$; Morin, 1964), or by blocking the flow of electrons between Q and A with DCMU. Such conditions generate maximal amplitudes of the variable fluorescence. However, even higher amplitudes are possible by reducing Q chemically with sodium dithionite (Homann, 1969), as the back-flow of electrons to the oxidizing side of PS II is eliminated. This back-flow is toward Z (Murata *et al.*, 1966b), rather than to molecular oxygen (Malkin, 1968), since it ceases when Z is kept reduced with hydroxylamine (Bennoun, 1970; Mohanty *et al.*, 1971a). Denoting, after Homann (1969), the total room temperature fluorescence at the O level as F_0, and as F_∞, F_{inh}, F_{ox}, and F_{red}, the P levels in the absence of any additions, with DCMU, with a Hill oxidant, and with a strong reductant, respectively, we may rank the variable fluorescence amplitudes of isolated chloroplasts as follows:

$$F_{ox} - F_0 < F_\infty - F_0 < F_{inh} - F_0 < F_{red} - F_0 \tag{17}$$

Large amounts of sodium dithionite, however, under PS II illumination may cause an irreversible decrease of Chl a fluorescence as a result of a reductive destruction of the chlorophylls (Karapetyan and Klimov, 1972). Also, at saturating light intensities $F_\infty \simeq F_{inh}$.

The fast fluorescence rise to P does not obey a first order exponential law, but it is rather sigmoidal (Morin, 1964; Delosme, 1967). An exponential rise is predicted by a system of independent PSU's each of which reduces Q in a 1-quantum step. To account for the sigmoidal fluorescence rise on the basis of an independent unit system, Morin (1964) postulated a series of two photoreactions, with different photochemical rate constants, by which the reduction of Q is achieved. More recently Doschek and Kok (1972) restated this concept and suggested two possible reaction alternatives, one of which requires Q to be a two-electron acceptor, and another which considers two very fast interacting acceptors (Q and R):

$$Q \xrightarrow{hv} Q^- \xrightarrow{hv} Q^{2-} \tag{18}$$

$$QR \xrightarrow{hv} Q^-R \longrightarrow QR^- \xrightarrow{hv} Q^-R^- \tag{19}$$

The sigmoidal fluorescence rise to P can also be accounted for by model systems in which closed PSU's are allowed to impart their excitation to other, closed or open, units. Extending the analysis of an experiment in which the rate of oxygen evolution in weak light was determined as a function of the fraction of open PS II reaction centers (flash yield of oxygen), Joliot and Joliot (1964) derived the following relationships for the variable fluorescence F, and for the rate of oxygen evolution V

$$F = \frac{kI[QH]}{1 - p[QH]} \qquad V = \frac{kI[Q]}{1 - p[QH]} \tag{20}$$

Here, k is a rate constant, I the excitation intensity, and p the probability of excitation exchange between a closed reaction center and another either open or closed. In isolated spinach chloroplasts and in *Chlorella pyrenoidosa* p was estimated to be about 0.5–0.6 (see also Forbush and Kok, 1968). The nonlinear correlation between the variable fluorescence and [QH] derives from the increased probability of photochemical trapping of the absorbed excitation because of the interunit excitation exchange. This exchange may, also, account for the slow photoreduction of the intersystem intermediates in the presence of unsaturating concentrations of CMU, which manifests itself as a slow rise of Chl a fluorescence (Murata *et al.*, 1966a).

The constancy of the probability p for the interunit excitation exchange has been questioned on theoretical grounds by Lavorel (1967) and Lavorel and Joliot (1972), and on experimental grounds by Elgersma (1972). Using very low excitation intensities (200 ergs cm^{-2} sec^{-1}), Elgersma obtained a first-order rise of the variable Chl a fluorescence, as it would be predicted by a system of independent PS II units ($p = 0$). Recently, Lavorel and Joliot (1972), proposed a model of connected PS II units to explain the sigmoidal OP rise without any recourse to a probability p of fixed value. This model recognizes the fact that a cluster of occupied PSU's will tend to grow at its edges, because the light-harvesting Chl territory of the neighboring units has increased. A larger islet (*îlot*) of occupied PSU's would lead to an increased fluorescence since it predicts an increased lifetime for the exciton (i.e., smaller probability of hitting a quenching center). According to this model, the fluorescence will rise in the beginning slowly, but then, as the islets of occupied units begin to expand and merge, it will rise more rapidly. Introduction of quenching centers in the Chl bed, e.g., by adding the nonreducible quencher *m*-dinitrobenzene, limits the islet growth and converts the sigmoidal rise to exponential. This result is comparable with that of Elgersma (*vide supra*), since in either case the abundance of excitation is less than required for rapid islet growth.

The gradual fluorescence rise along OP evidences a reversible interaction between Q and A, since a unidirectional electron donation predicts a sharp rise once all A is reduced. Because of this reversibility, the variable Chl *a* fluorescence is influenced by the combined reduced populations of Q and A. A relative measure of the size of the combined populations is the area bound by the fluorescence rise curve and its asymptote (i.e., the integral $\int \triangle F(t)dt$; Malkin and Kok, 1966; Murata *et al.*, 1966a; de Kouchkovsky and Joliot, 1967; Forbush and Kok, 1968). When the communication between Q and A is broken by a poison such as DCMU, or when the pool A has been depleted by extraction (dry *n*-heptane; R. Govindjee *et al.*, 1970), this area becomes minimal. Exhaustive extraction of lyophilized chloroplasts with hexane destroys the C550 absorbance response and the capacity to photoevolve oxygen with ferricyanide as a Hill oxidant. Both these functions are restored by adding simultaneously β-carotene and PQ-A to the suspension (Okayama and Butler, 1972b). In the presence of small quantities of added oxidants, such as ferricyanide and $NADP^+$, the integral $\int \triangle F(t)dt$ correlates with the amount of oxidant added (Malkin and Kok, 1966; Heath and Packer, 1968). As it might be expected from the foregoing, large amounts of added Hill oxidants obliterate the Chl *a* fluorescence rise in isolated chloroplasts.

The biphasic form of the fast fluorescence rise in isolated chloroplasts suggests the existence of two kinetically distinct subpools of intersystem oxidants. In total, the ratio of A carriers to Q is about 17:1 to 18:1 (Joliot, 1965a,b; Malkin and Kok, 1966; Forbush and Kok, 1968). One-third of these belong to the fast-reacting subpool A_1 ($t_{1/2} = 4$ msec), the remaining two-thirds to the slow reacting subpool A_2 ($t_{1/2} = 100$ msec). Whether the mode of interaction of Q with the two subpools is sequential (Murata *et al.*, 1966a) or branched (Forbush and Kok, 1968) is not decided yet.

3.2. The Fast Fluorescence Decay

The fluorescence rise OP in algae, in intact higher plant green tissue, and in isolated class I chloroplasts is succeeded by a somewhat slower decay to the quasi-steady level S (Fig. 2B), that coincides with a rise in the rate of oxygen evolution (Bannister and Rice, 1968). Phenomenologically, therefore, the decay *PS* signals an increased photochemical utilization of the available electronic excitation at the expense of the fraction emitted as fluorescence. Closer scrutiny of these phenomena, however, suggests that this may not be a sufficient cause. In 1943, for example, Kautsky and Franck reported the disappearance of the *PS* decay in oxygen-depleted algae, a result confirmed recently by Munday (1968), Munday and Govindjee (1969a), and by U. Franck and his co-workers (U. Franck and Hoffmann, 1969; U. Franck *et al.*, 1969; Schreiber *et al.*, 1971). This is illustrated in Fig. 4, which depicts the effect of progressing anaerobiosis on the *PS* decay of the green alga *Chlorella pyrenoidosa*. Deep anaerobiosis ($\simeq 10$ ppm oxygen) destroys the *OP* rise as well. In such systems, a high initial *O* level ($F_0 \simeq F_\infty$) is obtained, followed by a protracted decay (Schreiber *et al.*, 1972) which has been attributed to an anaerobiosis-induced inactivation of PS II units. Munday and Govindjee 1969b,c), and Mohanty *et al.* (1970), noted a far-red light stimulation of the Chl *a* fluorescence during the fast (*OPS*) change. This result cannot be explained on the hypothesis that Q is the only determinant of the variable fluorescence yield, since the oxidation of Q by the far-red light ought to suppress the fluorescence. The normal response, i.e., the far-red light suppression of Chl *a* fluorescence, becomes predominant later during the stage of the slow (*SMT*) fluorescence change.

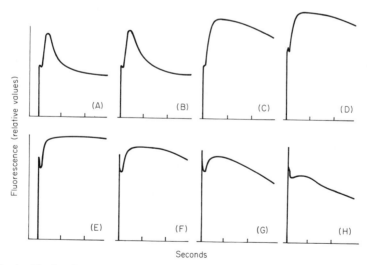

Fig. 4. The fast fluorescence transient of *Chlorella pyrenoidosa* at various stages of progressing anaerobiosis. (A) normal *Chlorella*; (B) *Chlorella* bubbled for 60 min with high purity nitrogen; (C) after 120 min of nitrogen bubbling; (D) after 150 min of nitrogen bubbling; (E) after 180 min of nitrogen bubbling; (F) (G) (H) at conditions of deeper anaerobiosis. (Redrawn from Franck *et al.*, 1969.)

Unlike the fluorescence rise OP, the fluorescence decay PS (Fig. 2B) is not determined exclusively by the ratio Q/QH. Although the inhibition of the PS decay by DCMU underlines the importance of the redox potential of this couple, other evidence suggests that the decay does not reflect a mere reversal of the process that led to the OP rise. The decay PS is more susceptible to physiological strain than the rise OP. In addition to its higher sensitivity to anaerobiosis, it is also more severely inhibited by low temperature, suggesting a larger apparent activation energy for the P to S decay than for O to P rise (U. Franck et al., 1969). Recognizing the distinct characteristics of the decay PS, Lavorel (1959), and Duysens and Sweers (1963) attributed it, partly, to the conversion of the photochemical and photoactive Chl a to a nonphotochemical and nonphotoactive form. The active Chl a is responsible for the fluorescence rise OP and the concomitant decay of the rate of oxygen evolution. The inactive form reverts to the active after dark incubation. As it is to be detailed in Section 4.1, we now have evidence relating the decay PS to the light-induced ion fluxes, and specifically to the acidification of the thylakoid interior, that leads to the prephosphorylation high-energy state.

3.3. The Slow Fluorescence Change

The fast fluorescence transient OPS in algae is succeeded by a second slower change, during which the fluorescence yield of Chl a rises to a maximum M, and then declines slowly to a low terminal level T (second wave of fluorescence induction, cf. Fig. 2C). At the end of this change, the rate of oxygen evolution, and the quantum yield of Chl a fluorescence have attained steady-state values. Sometimes higher plant leaves, especially when maintained in a CO_2-enriched atmosphere, are also capable of a slow fluorescence transient (McAlister and Myers, 1940; Frank et al., 1941), but usually they show only the fast change. Only the fluorescence of Chl a suffers from the induction phenomenon. The phycobilin fluorescence of the cyanophytes and the rhodophytes does not respond to the fluctuations of the Chl a fluorescence yield (French and Young, 1952; Govindjee et al., 1966; Mohanty et al., 1971b). The induction processes, therefore, exert no influence, neither directly on the emission of the phycobilins, nor on their capacity to transfer their excitation to Chl a.

The slow fluorescence change in algae is a complicated phenomenon, consisting of at least two distinguishable, and under certain conditions separable processes, namely, the process that leads to the S to M rise, and the process that leads to the M to T decay. This decay is particularly slow in the blue-green algae (Papageorgiou and Govindjee, 1967, 1968a; Mohanty,

1972). In the green algae, it is slower at high pH (Papageorgiou and Govindjee, 1971), while it is eliminated altogether by the carboxydismutase inhibitor cyanide anion (Wassink and Katz, 1939), and by the photophosphorylation uncoupler FCCP (Papageorgiou and Govindjee, 1968b, 1969). The physical mechanism that underlies the MT decay of fluorescence will be discussed in section 4.3.

Apart from their temporal separation, the fast (OPS) and the slow (SMT) fluorescence changes are distinguishable in several other respects. The typical for the fast change mirror image symmetry (complementarity) between the time courses of the rate of oxygen evolution and Chl a fluorescence breaks down during the slow change. Along the fluorescence rise from S to M, the rate of oxygen evolution also rises, and then stays approximately constant, while the fluorescence decays along MT (Papageorgiou, 1968). Other properties that distinguish the fast and the slow fluorescence transients are: (a) the saturation of the slow rise SM at a different excitation intensity than the rise OP (Papageorgiou and Govindjee, 1968a,b; 1969; Mohanty et al., 1971b); (b) the insensitivity of the slow change to the oxygen content of the gas phase over the sample (Wassink and Katz, 1939; Kautsky and Eberlein, 1939), as contrasted to the sensitivity of the fast change (Kautsky and U. Franck, 1943); (c) the insensitivity of the fast change to cyanide and the effective elimination by the same poison of the MT decay (Wassink and Katz, 1939).

Although the slow fluorescence change requires electron transport supported by the photosynthetic pigments, the above related phenomenology suggests that the reduction level of the Q/QH couple is not the only determinant of the Chl a fluorescence yield. Cyclic electron transport around PS I can also generate slow changes in the Chl a fluorescence yield as testified by by the following observations. (1) At very high light intensities, DCMU does not block the slow change in green algae, presumably because of the appreciable contribution of the DCMU-insensitive cyclic electron transport (Bannister and Rice, 1968). (2) Cyclic electron transport accounts, most probably, for the slow fluorescence change of DCMU-poisoned blue-green algae (Govindjee et al., 1966; Papageorgiou and Govindjee, 1968a; Duysens and Talens, 1969; Mohanty and Govindjee, 1973), and the PS II-minus mutants ac-115 and ac-141 of Chlamydomonas reinhardii. On the other hand, the Cyt f-minus mutant ac-206, and the P700-minus mutant ac-80a of the same alga, in which both the cyclic and the noncyclic transports are blocked, are devoid of the slow fluorescence change (Bannister and Rice, 1968). An interesting Chlamydomonas mutant (F-54; Sato et al., 1971), blocked at the terminal state of ATP formation, has not been studied as yet for its Chl a fluorescence properties.

Various observations suggest that the physical mechanism which under-lies the slow fluorescence change is due, in part at least, to the light-requiring adaptive process that controls the distribution of excitation between the pigment populations of PS II and PS I. This evidence will be discussed in section 4.3.

4. STRUCTURAL CONTROL OF THE YIELD OF CHLOROPHYLL *a* FLUORESCENCE

Active thylakoids are dynamic entities whose structural and functional characteristics respond to environmental stimuli. The massive ionic move-ment across the thylakoid membranes in illuminated chloroplasts affects both the shape (configuration) of the thylakoids and the ultrastructure (con-formation) of their membranous envelopes (see Chapter 11 of this volume). As Witt and his associates have shown (Chapter 10 of this volume), illumina-tion displaces electrons to the exterior of the lamella and sets up an electric field of 10^5 V/cm across it (Junge and Witt, 1968; Schliephake *et al.*, 1968, for a recent review see Witt, 1971).In response to the field force, protons are translocated from the surrounding stroma to the thylakoid interior, while metal cations (primarily K^+ and Mg^+) and water move out to preserve the electrical and osmotic equilibrium (Jagendorf and Hind, 1963; Dilley and Vernon, 1965; Nobel, 1967).

The ultrastructural changes that the light-induced electroosmotic im-balance entails find expression in the spectroscopic properties of the chloro-phylls, inasmuch as these pigments are major building blocks of the thylakoid membrane. The fluorescence parameters of Chl *a* (e.g., the quantum yield and the lifetime) are particularly sensitive to the structural perturbations. The first to hint on a connection between phosphorylation and Chl *a* fluores-cence was Strehler (1953), who demonstrated a change in the ATP content of *Chlorella* cells concurrent with the slow fluorescence induction. More recently, Papageorgiou (1968) and Papageorgiou and Govindjee (1968a,b; 1969) employed uncouplers of the photophosphorylation to establish a clear correlation between the slow change of Chl *a* fluorescence and the physico-chemical events that precede the ADP phosphorylation step. These results have been confirmed and extended by the work of Bannister and Rice (1968), and of Mohanty (1972); for a recent detailed review the reader is referred to Govindjee and Papageorgiou (1971). Apart from the photosynthetic electron transport, therefore, the variable portion of the Chl *a* fluorescence responds to the ion-dependent ultrastructural changes of the thylakoid membrane. In this section, the evidence for these effects, and their implications on the mechanism of photosynthesis are discussed.

4.1. The X_E-Linked Quenching of Chlorophyll a Fluorescence

In 1969, Murata and Sugahara observed an uncoupler sensitive "fluorescence lowering" on the addition of PMS and ascorbate to DCMU-poisoned spinach chloroplasts. (A PMS-induced fluorescence lowering in DCMU-poisoned chloroplasts was reported earlier by L. Yang and Govindjee; see Govindjee *et al.*, 1967, p. 553.) Since PMS is a catalyst of the cyclic electron transport, the fluorescence lowering was interpreted as signaling the development of the high energy state that leads to the cyclic photophosphorylation. Arnon *et al.* (1965) and Heath and Packer (1968) also reported that PMS suppresses the fluorescence of unpoisoned chloroplasts, but here the lowering can be due, to a large extent, to an acceleration of the noncyclic electron transport by the added oxidant. The X_E(high-energy state)-linked fluorescence lowering is unequivocally demonstrable only when the interference of Q has been eliminated, either by poisoning with DCMU, or by employing saturating illumination, or by both. Nevertheless, it has been shown that the noncyclic electron transport, also, gives rise to an energy-dependent fluorescence quenching. In unpoisoned systems, the X_E-linked effect can be resolved from the Q-linked one on the basis of the slower kinetics of the former, and of its different pH optimum. This is at pH 8.5, while optimal quenching by Q occurs at pH 6.5 (Wraight and Crofts, 1970).

Figure 5 illustrates a fluorescence-lowering experiment performed with with a DCMU-poisoned suspension of the unicellular cyanophyte prokaryote *Anacystis nidulans*. On the addition of a small amount of PMS, which is kept

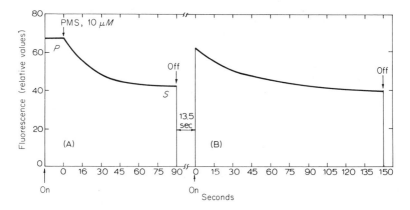

Fig. 5. The PMS-induced Chl a fluorescence lowering in a DCMU-poisoned (10 μM) suspension of the blue-green alga *Anacystis nidulans*, in the presence of 1 mM ascorbic acid. (A) the fluorescence lowering generated by the addition of 10 μM PMS; (B) the fluorescence lowering of the same sample after a dark rest of 13.5 sec. (Experiment performed in the author's laboratory.)

reduced with 1 mM ascorbate, the high fluorescence yield decays to a level lower by 35%, reflecting the buildup of the high energy state (curve *A*). Characteristically, the fluorescence lowering is reversed after a short dark rest, since a second illumination reproduces its typical time course (curve *B*).

Fluorescence lowering induced with PMS in the absence of ascorbate (Mohanty, 1972; Mohanty and Govindjee, 1973), as well as in the presence of ascorbate (Papageorgiou and Tsimilli-Michael, unpublished experiments) has been observed also with the chlorophyte eukaryote *Chlorella pyrenoidosa*. Figure 6 illustrates the effect of oxidized PMS on the fluorescence yield of DCMU poisoned *Chlorella*. The maximal fluorescence lowering (ca. 75%) is obtained at about 50 μM PMS. Characteristically, the PMS-induced effect is reversed in darkness, or by the addition of FCCP, but only partly. This author believes that the incomplete reversibility may indicate a direct, Stern–Volmer type, quenching of the Chl a fluorescence by PMS, which is added to the X_E-linked quenching.

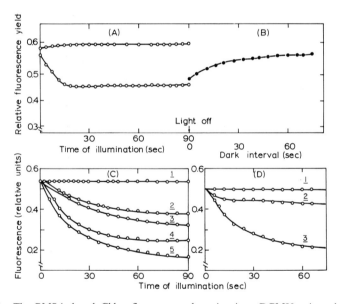

Fig. 6. The PMS-induced Chl a fluorescence lowering in a DCMU-poisoned (10 μM) suspension of the green alga *Chlorella pyrenoidosa*, in the absence of added ascorbate, (A) upper curve, 10 μM DCMU; lower curve, 10 μM DCMU + 25 μM PMS. (B) dark recovery of the fluorescence lowering as measured by weak 1/10 sec flashes. (C) fluorescence lowering induced by various concentrations of PMS, in the presence of (1) 10 μM DCMU; (2) + 10 μM PMS; (3) + 20 μM PMS; (4) + 30 μM PMS; (5) + 50 μM PMS. (D) fluorescence timecourses with (1) 10 μM DCMU alone; (2) 10 μM DCMU, 4 μM FCCP, and 30 μM PMS; and (3) 10 μM DCMU and 30 μM PMS. (Reproduced from Mohanty, 1972.)

The thylakoid membrane is known to become energized under the combined influence of two kinds of photoinduced potentials that set up across lamella: an electric potential due to a charge gradient, and a chemical potential due to concentration gradients. Of these two, it is the chemical component, and in particular the protonation of the thylakoid interior, that contributes mostly to the fluorescence lowering (Wraight and Crofts, 1970). This has been deduced from the more effective inhibition of the fluorescence lowering by uncouplers which obliterate the pH gradient by means of an uncontrolled proton transport (dianemycin, nigericin, NH_4Cl), than by uncouplers that dissipate the charge gradient (valinomycin in the presence of K^+). Preventing the light-induced uptake by having the chloroplasts suspended in a polygalacturonate solution, or by washing the coupling factor away with EDTA, results in the destruction of the fluorescence lowering response caused by the addition of DAD. Both effects are regenerated either by adding salts to the polygalacturonate inhibited system, or by adding dicyclohexylcarbodiimide to the system devoid of the coupling factor (Cohen and Sherman, 1971). Using PMS, however, instead of DAD, Mohanty et al. (1972) were unable to induce a fluorescence lowering in EDTA-washed chloroplasts.

Although related to the protonation of the thylakoid interior, we can hardly interpret the fluorescence lowering as a quenching of the Chl a fluorescence by protons, since these cannot accept electronic energy. It is likely, however, that protonation modifies the ultrastructure of the membrane in such a manner as to facilitate the diffusion of quenchers of electronic excitation (such as molecular oxygen) to the pigment sites. Figure 7 illustrates an experiment designed to test this possibility (Papageorgiou et al., 1972a). Curves A and B depict the PMS-ascorbate induced fluorescence lowering of a normal and of a gently deoxygenated suspension of spinach chloroplasts, respectively. In the oxygen-poor sample (5% of the original oxygen content) the fluorescence lowering is less by more than 50%. On the other hand, a similarly deoxygenated suspension, that was subsequently allowed to reequilibrate with the atmosphere, is capable of the full extent of the PMS-induced fluorescence lowering (curve C). The importance of the X_E-promoted changes in the membrane ultrastructure is emphasized further by the disappearance of the fluorescence lowering in glutaraldehyde-fixed chloroplasts (Mohanty et al., 1972). Although such chloroplasts retain 40–50% of their proton pumping ability (West and Packer, 1970), this may not suffice for a conformation change of the glutaraldehyde crosslinked membrane elements.

The kinetics and the oxygen requirement of the X_E-linked fluorescence lowering resemble similar features of the P to S decay of the fast fluorescence

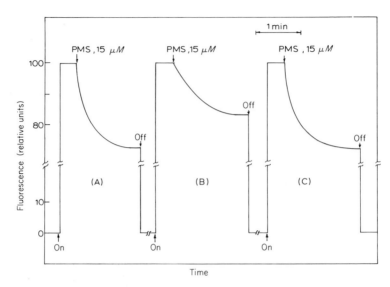

Fig. 7. The PMS-induced fluorescence lowering in isolated aerobic and anaerobic spinach chloroplasts poisoned with 10 μM DCMU. (A) aerobic suspension; (B) chloroplasts made anaerobic by repetitive freeze-thaw-pump cycles; (C) chloroplasts made anaerobic as in (B), and then left at $0°-4°C$ to reequilibrate with the atmosphere (Papageorgiou *et al.*, 1972a).

transient in algae (cf. Section 3.2). Indeed, this parallelism points to the light-driven protonation of the thylakoids as one of the generic causes of the decay along *PS*. (The other cause is probably the phototransformation of the photosynthetic apparatus from the light I state to the light II state; cf. Section 4.3.) We must note, however, that while the *PS* decay is sensitive to DCMU, the X_E-linked fluorescence lowering is most unequivocally expressed in the presence of this metabolic poison, because it eliminates the influence of the noncyclic electron transport on the fluorescence yield. This may not be a contradiction, however, because of the added cofactors (PMS, DAD) which make the contribution of the cyclic electron transport to X_E the most significant.

The kinetic records of Chl *a* fluorescence and oxygen evolution (Bannister and Rice, 1968) indicate that the decay *PS* sets in only after the rate of oxygen evolution by algae begins to rise, following a transition from a state of dark adaptation to a state of continuous illumination. This may emphasize the importance of the local (i.e., in the lamella) concentration of oxygen for this phase of the fast fluorescence transient. In the absence of molecular oxygen, and possibly of other quenchers, the high energy state has no effect on the excitation of Chl *a*; hence, the *PS* decay is absent from a deoxygenated suspension of algae, as observed originally by Kautsky and Franck (1943). Pertinent to this interpretation is the observation by Mohanty *et al.* (1972)

that PMS lowers both the yield and the lifetime of chloroplast fluorescence. The simultaneous suppression of these two magnitudes is a criterion for the dynamic mode of fluorescence quenching, in which the quencher attacks selectively only the excited population of the fluorescer (Förster, 1951; Weller, 1961; Vaughan and Weber, 1970).

Oxidants, and more generally compounds having an electron-deficient moiety in their molecular structure (e.g., the aromatic ring of the nitroaromatic compounds), are known to be effective quenchers of excited porphyrins in solution (Livingston and Ke, 1950; Whitten et al., 1968). Quinones have been shown to quench both the variable and the constant fluorescence of Chl a in vivo, acting not only through the Q/QH couple, but also directly on the pigment bed, in a manner influenced by their solubility in a lipid phase (Amesz and Fork, 1967). Molecular oxygen has a low-lying "σ singlet" (762 nm), which can accept electronic excitation from porphyrins. In the presence of singlet oxygen acceptor, the excitation donor can be quenched in a nondestructive manner (Wilson and Hastings, 1970; Gurinovich et al., 1971). Carotenoids are likely acceptors of singlet oxygen, owing to their involvement in the "xanthophyll cycle." It has been shown that the zeaxanthin of higher plant leaves incorporates molecular oxygen, in a light-dependent process, to form the mono- and the diepoxide. This, by means of a reductive deoxygenation in the presence of H^+, reverts to the zeaxanthin again (Yamamoto and Chichester, 1965; Yamamoto and Takeguchi, 1972). The fact that epoxy carotenoids occur only in oxygen-evolving organisms lends support to their postulated origin from carotenoids that serve as oxygen acceptors. A possible mechanism which utilizes the xanthophyll cycle to relieve Chl from excess excitation is as follows:

$$Chl^* + O_2 \longrightarrow Chl + O_2^* \tag{21}$$

$$Carotenoid + O_2^* \longrightarrow epoxy\ carotenoid \tag{22}$$

$$Epoxy\ carotenoid + 2H^+ + 2e \longrightarrow carotenoid + H_2O \tag{23}$$

Finally, due to its ground-state paramagnetic character, oxygen is expected to promote the crossover to the Chl triplet; because of its long lifetime, this state is particularly vulnerable to excitation quenchers. Mathis (1969) reported a rapid transfer of triplet excitation from Chl a to carotenoids (β-carotene, lutein) in digitonin micelles. Evidence obtained in Witt's laboratory (Witt, 1971; Wolff and Witt, 1972) suggests that the triplet Chl a transfers its excitation to a special carotenoid, whose triplet population becomes manifest by its absorbance at 520 nm. This effect occurs at light intensities well above light saturation, and it is considered as an additional valve for the dissipation of the potentially destructive excess excitation of Chl a. Duysens et al. (1972) reported that the fluorescence which is excited by high intensity flashes is of lower yield than the fluorescence excited with less intense flashes. Since the quenching effect of the intense light is speeded up in the presence of oxygen, it was attributed to the conversion of the singlet-excited Chl *a* to the triplet state, a process favored by the proximity of the paramagnetic ground-state oxygen.

4.2. Effects of Metal Cations on Chlorophyll *a* Fluorescence

Electrolytes are known to influence the morphology and the function of chloroplasts (see Chapter 11 of this volume). Water molecules move in or out of the thylakoid, in response to the light-induced electroosmotic imbalance, depending on the anion content of the medium. Thus, when weak acid anions (e.g., acetate) are present, illuminated chloroplasts appear shrunk relative to their state in darkness, while with strong acid anions they appear swollen (Packer, 1962, 1972; Crofts *et al.* 1967). Illumination also generates ion-dependent processes that influence the spatial arrangement of the thylakoids and their association into grana (Izawa and Good, 1966; Murakami and Packer, 1970a,b). In addition to the shape (configuration) changes of the thylakoids, the ultrastructure (conformation) of their membrane is also susceptible to light-induced or salt-induced changes (Murakami and Packer, 1969, 1970a,b, 1971). On illumination, or on acidification in darkness, the membrane becomes thinner by roughly 20%. This thinning has its origin in the protonation of negatively charged groups (carboxyls, phosphates), by which the membrane-bound charge is lowered. Macroscopically, the thinning of the membrane becomes manifest at an increase in the 90° light scattering, while the volume changes of the thylakoid show up mostly as turbidity (0° light-scattering) changes.

The divalent, and less so the monovalent, metal cations enhance the hydrophobic character of the thylakoid membrane and promote the hydrophobic association of thylakoids into grana. Increased hydrophobicity may also contribute, among other things, to the larger fluorescence output from the membrane-bound extrinsic fluorescence probe ANS (Murakami and Packer, 1971). Significantly, this fluorescence enhancement is less (but not absent) in glutaraldehyde-fixed preparations.

As an amphipathic molecule, however, ANS probes the lipid–water interfaces of the lamella, rather than the hydrophobic regions where the chlorophylls are embedded. Divalent metal cations produce ultrastructural changes in those regions as well, since they have been shown to enhance the accessibility of the pigment bed itself to added excitation quenchers (Papageorgiou, 1972; Papageorgiou *et al.*, 1972b). Figure 8 illustrates an experiment bearing this out. The presence of 1 *M* $MgCl_2$, is shown to promote greatly the quenching by nitrobenzene of the fluorescence of *Chlorella* fragments.

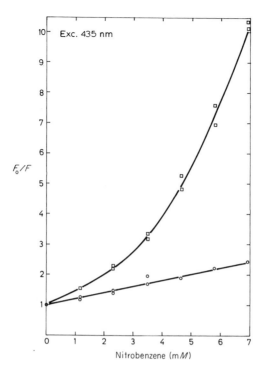

Fig. 8. Stern–Volmer plots of the quenching of the fluorescence of *Chlorella pyrenoidosa* fragments with nitrobenzene. Lower curve, without added $MgCl_2$; upper curve, in the presence of 1 M $MgCl_2$ (Papageorgiou, 1972).

According to the classical Stern–Volmer (1919) equation, the ratio of the fluorescence intensities in the absence (F_0) and in the presence of an added quencher (F) is proportional to the collision rate constant k for quencher–fluorescer encounters:

$$(F_0/F) = 1 + \gamma k \tau C \qquad (24)$$

where γ is the probability of effective collisions, τ the lifetime of the unquenched fluorescence, and C the concentration of the quencher. As Fig. 8 indicates, the presence of $MgCl_2$ dramatically enhances the quenching ability of nitrobenzene. Two causes, dependent on the membrane ultrastructure, can most likely account for this enhancement. First, the preferential migration of nitrobenzene to the lipid phases of the lamella, a process favored by the Mg^{2+}-induced increase in the hydrophobicity of the lamella; and second, an increased rate of excitation exchange among the chlorophylls, which will increase the probability of an exciton visiting a quenching center. In either case, the slope of the plot of F_0/F against the quencher concentration [see eq. (24)] can be interpreted as an indicator of the membrane ultrastructure in the neighborhood of the chlorophylls. As in many other electrolyte-dependent phenomena, the anion of the added salt is unimportant, and salts of monovalent metal cations are less effective.

The ultrastructure-mediated electrolyte control of the photosynthetic process operates both at the level of enzymic reactions, and at the level of the supply of electronic excitation energy to the reaction centers. At the bio-chemical level, salts, and especially those of the alkaline earths, are required for the coupling of the electron transport to the photophosphorylation (Jagendorf and Smith, 1962; Shavit and Avron, 1967; Gross *et al.*, 1969; Walz *et al.*, 1971). At the level of the distribution of the excitation energy salts regulate the transfer of excitation from one photosystem to the other. We shall discuss the second type of electrolyte control in some detail, since our knowledge about it derives mostly from fluorimetric evidence.

Homann (1969) was the first to show that on adding Mg^{2+} to a DCMU-poisoned suspension of *Phytolacca americana* chloroplasts the amplitude of the variable Chl *a* fluorescence increases,* while the constant yield fluorescence remains unchanged. A similar experiment with isolated spinach chloroplasts is shown in Fig. 9. In it, the inhibition by small amounts of EDTA is also illustrated. Fluorimetric titration of the intermediate pool of unpoisoned *Phytolacca americana* chloroplasts, with and without added Mg^{2+}, by measuring the area between the fluorescence curve and its asymptote, showed that the added salt exerts no effect on the size of the pool (Homann, 1969). At the end of the Mg^{2+}-induced rise, longer lifetimes of the Chl *a* fluorescence have been measured, than in the absence of added Mg^{2+} (Mohanty *et al.*, 1973). This important result implies a dynamic quenching process, which operates exclusively on the excited Chl *a* population, when the thylakoids are depleted of their Mg^{2+}.

Working with fragmented *Euglena* chloroplasts, Brody *et al.* (1966) noted that when preparations in salt solutions of high ionic strength (up to 1.5 *M*) are frozen to the temperature of liquid nitrogen, the far-red fluorescence band at 715–735 nm appears to be suppressed, and the main short red band at 685 nm enhanced relative to the fluorescence spectrum of preparations in solutions of low ionic strength. This phenomenon was ascribed to salt-induced changes in the conformation of the proteins to which the pigments are attached, which lead to a decrease in the probability of energy transfer from the Chl *a* form that emits at 685 nm, to that which emits at 715–735 nm.

---

* Gross (1973) has observed that 10 m*M* monovalent cations are required in addition to divalent cations in order to observe this phenomenon. The monovalent ions cause a decrease in fluorescence that is reversed by divalent ions and high concentrations of monovalent cations.

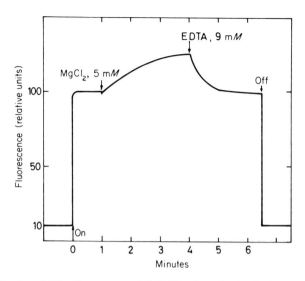

Fig. 9. The rise of Chl *a* fluorescence of DCMU-poisoned spinach chloroplasts generated by the addition of 5 m*M* MgCl$_2$, and its reversal by 9 m*M* EDTA. (Experiment performed in the author's laboratory.)

Both these phenomena, namely, the greater fluorescence output and the spectral redistribution of fluorescence, which are generated by the addition of metal cations to suspensions of isolated chloroplasts, were studied in detail by Murata and his co-workers (Murata 1969b, 1971a, and b; Murata *et al.*, 1970) and by Mohanty *et al.* (1973). Using much lower concentrations (3–5 m*M*) of added Mg^{2+} salts than Brody, Murata observed an intensification of the PS II fluorescence at low temperature (F685 and F696), accompanied by a suppression of the PS I fluorescence (F735). The magnitude of this effect is illustrated by the ratios of the fluorescence peaks, calculated from Murata's data (1969b); in the absence of Mg^{2+}, F735/F685 = 2.42 and F735/F696 = 2.09, while in its presence F735/F685 = 1.54 and F735/F696 = 1.15. The Mg^{2+}-affected redistribution of the excitation among the antenna pigments of PS II and PS I is also reflected in the rates of electron transport across the respective reaction centers. Thus, the rate of the photoreduction of NADP$^+$ by DCPIPH$_2$ (photoreaction I) in the presence of Mg^{2+} is only 0.7 of the rate in its absence (Murata, 1969b; Ben-Hayyim and Avron, 1971). On the other hand, Mg^{2+} stimulates the electron transport from water to DCPIP (photoreaction II) by a factor of 1.2 (Murata, 1969b), and from water to ferricyanide by a factor of 1.7 (Ben-Hayyim and Avron, 1971). Salts of other alkaline earths, and Mn^{2+}, induce the same fluorescence responses as Mg^{2+}, but Al^{3+} is less effective, and Zn^{2+} and Cd^{2+} are not effective at all (Murata *et al.*, 1970). Much higher concentrations of monovalent metal and alkylammonium cations are required to produce the same effects as Mg^{2+} (Homann, 1969; Murata, 1971a).

Murata (1969b) rationalized these results on the hypothesis that Mg^{2+}, once inside the thylakoid, inhibits the spillover of the excess excitation from PS II to PS I, enhancing thus the rates of fluorescence and photochemistry of the former photosystem at the expense of the latter. Other relevant observations, however, are not as readily accountable by this hypothesis. The assumption, for example, that the rate of $NADP^+$ reduction reports directly on the rate of electron flux through P700 has been challenged recently by Rurainski and Hoch (1972). Direct measurements of the rates of $NADP^+$ photoreduction and electron flux through P700 show the first to increase and the second to decrease, when Mg^{2+} is added to the chloroplast suspension. Mg^{2+} enhances the electron transport rate from water to $NADP^+$, a process that involves both photoreactions (Avron and Ben-Hayyim, 1971; also see Rurainski *et al.*, 1971). At a concentration of 7.5 mM, added Mg^{2+} optimizes both the rate of this process, and the Emerson enhancement effect (Sun and Sauer, 1972). This optimization is not readily interpretable on the basis of Murata's hypothesis, because this cation would be expected to inhibit, rather than facilitate the efficient utilization of the excess excitation of one photosystem by the other.

As an alternate hypothesis, Sun and Sauer (1972) proposed that Mg^{2+}, and other divalents, promote rather than suppress the exchange of electronic excitation between the two photosystems. According to this view, the enhanced PS II fluorescence and electron transport across PS II is a consequence of a facilitated excitation transfer from PS I to PS II when Mg^{2+} is present. This hypothesis presupposes a model in which PS II and PS I have equal absorbances in the region from 620 to 680 nm, a requirement that is not supported by the absorption spectra of physically separated photosystems (Anderson and Boardman, 1966).

The physical basis of these phenomena is the cation-regulated ultrastructure of the thylakoid membrane, as several lines of evidence appear to suggest. The Mg^{2+}-induced and the Na^+-induced fluorescence rise in chloroplasts follows similar kinetics as the associated changes in the 90° light scattering (Murata, 1971b); the latter has been shown to manifest changes in the membrane thickness (Murakami and Packer, 1969, 1971). These phenomena are restricted to the grana portion of the chloroplast, where both photosystems are to be found, and they are absent from the stroma thylakoids which contain only PS I. Furthermore, neither isolated PS I and PS II particles, nor glutaraldehyde-fixed chloroplasts show the electrolyte-induced fluorescence changes (Murata, 1971b; Mohanty *et al.* 1972).

It has been suggested that the conformation change brought about by the metal cations succeeds in diminishing the PS II to PS I spillover by increasing the spatial separation of their respective pigment populations (Duysens,

1969, 1972; Murata, 1969b; Govindjee and Papageorgiou, 1971; Myers, 1971). This mechanism derives support from recent electron microscopic evidence, which positions PS I on the outer, and PS II on the inner layer of the thylakoid membrane (Arntzen *et al.*, 1969; Park and Sane, 1971; Chapter 2, this volume). According to the R^{-6} law of the resonance excitation exchange (Förster, 1951; cf. Chapter 4 of this volume), a greater separation will lead to a diminished spillover. Assuming a complete cessation of the intersystem excitation exchange when Mg^{2+} is present, Mohanty *et al.* (1973) calculated a quantum efficiency of 0.12 for the spillover process in isolated spinach chloroplasts. It is plausible, however, to accept that metal cations affect not only the spacing of the membrane layers, but also the detailed fine structure of each layer, as might be suggested by the cation-induced enhancement of the nitrobenzene quenching of Chl *a* fluorescence (Papageorgiou, 1972; Papageorgiou *et al.*, 1972b; Papageorgiou and Argoudelis, 1973).

The changes in the membrane ultrastructure that are brought about by low concentrations of divalent metal cations differ from those due to the protonation of the thylakoid interior in several respects. In the first place, protonation quenches the fluorescence (cf. Section 4.1), while added metal cations enhance it. In suspensions of whole algal cells, high concentrations of salts (e.g., 270 m*M* KCl; de Kouchkovsky, 1972; Mohanty and Govindjee, 1971) have been found to suppress the fluorescence, but here the true effect of the cation on the thylakoid membrane is interfered with from the osmotic contraction of the whole cell. In isolated chloroplasts, with or without Mg^{2+}, the extent of the X_E-linked fluorescence lowering caused by cofactors of the cyclic electron transport is the same (Cohen and Sherman, 1971; Mohanty *et al.*, 1972). Contrary to the change induced by metal cations, the fluorescence lowering affects all three Chl *a* fluorescence bands at low temperature (F685, F696, and F735). This signifies that the spillover of electronic excitation from one photosystem to the other is not inhibited by the PMS-induced high-energy state (Mohanty *et al.*, 1972). The light-scattering changes initiated by the addition of PMS to isolated higher plant chloroplasts are slower than the PMS-induced fluorescence lowering (Murata, 1971b). Finally, the fluorescence change promoted by metal cations is either a dark, or a phototriggered, process in contrast to the PMS-induced fluorescence lowering which is a photochemical process (Papageorgiou and Isaakidou, unpublished observations).

4.3. The Light I State and Light II State Interconversions

In whole organisms, phenomena related to the metal cation control of the distribution of the electronic excitation of Chl *a* can be effected by illumination with light absorbed preferentially by the one or the other photosystem. Several minutes of illumination with PS II light suffices to transform the photosynthetic organism into a state (light II state) in which fewer quanta are delivered to the PS II, than at the beginning of the illumination period. Prolonged darkness, or prolonged illumination with PS I light, on the other hand, lead to a state (light I state) in which a larger share of the available excitation ends up in PS II. Away from light saturation, the fraction of quanta delivered to PS II in the light II state is 0.9 of what is delivered to the same photosystem in the light I state. At any given value of the rate of oxygen evolution, i.e., at any given fraction of open PS II traps, light I state gives off more fluorescence since a higher share of the excitation is delivered to the fluorescent Chl *a* of PS II (Bonaventura, 1969; Bonaventura and Myers, 1969; Murata, 1970).

The phototransformation of a dark-adapted photosynthetic organism from the light I state to the light II state is represented in the kinetics of the slow fluorescence change. At the point *M* (cf. Fig. 2C) the light I state predominates. More quanta are delivered to PS II and this results in a more intense fluorescence. Continuous illumination with PS II light slowly transforms the organism to the light II state, in a process manifested macroscopically by the fluorescence decay along *MT*. At the light II state, corresponding to the terminal level T, fewer quanta end up in PS II, and hence the emitted fluorescence is weaker.

Murata (1970) was the first to correlate explicity the light I to light II state interconversions with the metal cation control of the distribution of the Chl *a* excitation between the two photosystems. At the light II state (level *T*), the thylakoids are depleted from their metal cations, and the spillover to the nonfluorescing, or weakly fluorescing, PS I proceeds uninhibited. At the light I state, the thylakoids are rich in metal cations; consequently the spillover is blocked and the output of fluorescence from PS II is enhanced. Supporting this interpretation is the intensification of the far-red fluorescence, and the suppression of the short red one, in *Porphyridium cruentum* cells that have been preilluminated with light absorbed mainly by PS II. Murata (1969a) interpreted this result as suggesting an increased excitation spillover from PS II to the far-red fluorescing PS I. The distribution of metal cations on either side of the thylakoid membrane may also account for the reported suppression of fluorescence when salts are added to suspensions of *Chlorella* (de Kouchkovsky, 1972; Mohanty and Govindjee, 1971, 1973), if one assumes that salts penetrate to the stroma, but not to the interior of the thylakoid. This state resembles, somehow, the metal cation rich state of the stroma of illuminated chloroplasts.

Mechanically damaged cells (scratched epidermis of spinach or tomato leaf, punctured *Nitella obtusa* cells) are devoid of the slow fluorescence change, and in the case of the alga, also of the concurrent light-induced decrease of the negative electric potential that exists between the vacuole sap and suspension medium. Because of this Vredenberg (1970) disputes the importance of the thylakoid envelope for the light I to the light II state transformations and lays emphasis on the chloroplast and on the cell membrane. This interpretation, however, requires a clearer description of the site of the mechanical damage.

The transformations from one light state to the other can be viewed as a short range physiological adaptation of the plants to the quality of the light field in which they chance to exist. Without a requirement for companion changes in the chemical composition of the light-harvesting apparatus, this mechanism enables the plant to channel the excess excitation away from that photosystem which happens to be in a condition of overload. This short-range adaptation is to be contrasted with the long-range adaptive processes, with which the plants adjust their pigment composition to meet the physiological strains imposed by the quality and the quantity of the illuminating light. As an example of the latter mechanism, we may refer to the higher pigmentation of algae grown at low-light intensities, relative to algae grown in abundant light (see, for example, Ghosh and Govindjee, 1966).

TABLE IV

Tentative Assignments of the Induction Phases of Chl a Fluorescence to the Photochemical and Physiological Events That Intervene between a State of Dark Adaptation and a State of Light Adaptation[a]

Fluorescence induction phase	Photochemical or physiological process[b]
Activation rise, OI	Photoconversion of the water-splitting catalyst to the S_3 state; reduction of Q^{c-e}
Fluorescence decay, ID	Reoxidation of Q by PS I[f]
Fluorescence rise, DP	Photoreduction of the pool of intersystem electron carriers[f-h]
Fluorescence decay, PS	pH-promoted quenching of Chl a fluorescence by excitation quenchers, such as molecular oxygen[i]; photoconversion of light I to light II state[j-l]
Slow fluorescence rise, SM	Photoconversion of light II to light I state[l]
Slow fluorescence decay, MT	Photoconversion of light I to light II state[j,k]

[a]Cf. Fig. 3.

[b]At any phase of the fluorescence induction time course, the Chl a fluorescence yield is controlled by more than one photophysiological processes. Accordingly, the assignments of Table IV refer to the processes which exert the predominant influence in each phase.

[c]Joliot (1965b).   [d]Joliot (1968).

[e]Kok *et al.* (1970).   [f]Munday and Govindjee (1969a).

[g]Kautsky *et al.* (1960).   [h]Duysens and Sweers (1963).

[i]Papageorgiou *et al.* (1972a).   [j]Bonaventura and Myers (1969).

[k]Murata (1970).   [l]Mohanty *et al.* (1972).

Concluding this section, we would like to summarize the tentative assignments of the consecutive phases of the fluorescence time course of Chl
a in vivo to the various physicochemical and physiological processes that take
a photosynthesizing plant from the state of adaptation to darkness to that of
adaptation to light. This is done in Table IV.

5. FLUORESCENCE PROPERTIES OF PHOTOSYNTHETIC MUTANTS

Photosynthetic mutants are plants with genetically impaired photosynthetic apparatus (Kirk and Tilney-Basset, 1967; Levine, 1968, 1969). The
mutation may affect the structure and the arrangement of the thylakoids, the
pigment content and composition, the electron transport sequence, the
mechanism of photophosphorylation, and the CO_2 fixation pathway. Point
defects in the genome result in simple phenotypic responses (one or a few
missing components), while extranuclear mutations, such as in the plastome,
cause grosser abnormalities. Photosynthetic mutants have been induced
and studied in the genera *Chlamydomonas, Scenedesmus, Euglena, Chlorella,
Cyanidium, Oenothera, Hordeum,* and *Zea* (Table V).

To have an effect on the fluorescence properties of Chl *a in vivo,* the
mutation must concern either the electron transport across PS II, or the
pigment composition. In the first instance, its effect will be reflected primarily
in the kinetics of the fast fluorescence change; in the second, it will also
influence the shape of the excitation and emission spectra, and possibly the
slow change of the fluorescence yield. Lesions on either side of the PS II
photoreaction lead to a reduced output of variable fluorescence, operationally defined as the area bound by the fluorescence intensity ordinate, the trace
from *O* to *P,* and the asymptote of the latter (cf. Fig. 2A). Restoration of the
variable fluorescence by direct electron donation to the PS II reaction center
complex (Yamashita and Butler, 1968a and b) is taken to signify a lesion on
the electron donor side of PS II. A third alternative is the direct elimination
of Q, which destroys all expression of a Q-linked control of the Chl *a*
fluorescence yield. It is not certain, however, whether a missing Q means
"high" or "low" fluorescence. In general, the presence of an active Q can
always be demonstrated by the fluorescence increment produced on the
addition of sodium dithionite, which chemically reduces Q to QH.

Photosynthetic mutants with missing Q, or having their electron transport
blocked right after Q, have been described by several investigators. These
mutants are more fluorescent than their wild-type counterparts, a property
that has been put to use in screening for mutant colonies from agar plates
(Bennoun and Levine, 1967; Garnier, 1967). The absence of Q-linked control
of the fluorescence level leads to a high F_0 and to a small amplitude of variable
fluorescence.

TABLE V Photosynthetic Mutants

Species	Strain	Type of mutation	Affected components or functions
Chlamydomonas reinhardii	*ac-115*; *ac-141*	Nuclear	Cyt b559, PQ
	ac-21	Nuclear	M[e]
	ac-206	Nuclear	Cyt c553
	ac-208	Nuclear	PC
	ac-80 α	Nuclear	P700
	ac-20	Nuclear	RuDP carboxylase
	F-54	Nuclear	ATP synthesis
	F-60	Nuclear	Phosphoribulosekinase
	lfd-2; *lfd-13*; *lfd-15*; *lfd-17*; *lfd-27*; *lfd-29*	Nuclear	Cyt b559 (high potential)
Scenedesmus obliquus[f]	*ScD₃-α*; *ScD₃-4*; *ScD₃-5*; *ScD₃-10*; *ScD₃-11*; *ScD₃-15*; *ScD₃-40*; *ScD₃-42*; *ScD₃-67*	Unknown	Less total Chl; altered carotenoid profile; reduced PQ-A/Chl ratio; blocked PS II electron transport
Scenedesmus obliquus[f]	*ScD₃-8*	Unknown	P700
	ScD₃-26; *ScD₃-50*	Unknown	Blocked PS I electron transport, and diminished PS II transport
Chlorella pyrenoidosa[g]	*G41, G43, G44, G45*	Unknown	Diminished or no Chl b
Cyanidium caldarium		Unknown	Phycocyanin
Euglena gracilis[i]		Unknown	PS I-induced oxidation of Q
Oenothera hookeri[j]	*Iα, Iγ, Iσ, IIγ*	Plastome	Q, intersystem intermediates
Oenothera suaveolens[j]	*IIα*	Plastome	Cyt f, P700
Hordeum vulgare[k,l]		Nuclear	Less Chl a Chl b missing
Zea mays	(*olive necrotic 8147*)[m]	Nuclear	Stomataless leaves[n]; less total Chl and carotenoids; preponderance of PS I pigments and function
Nicotiana tabacuum[o]	*Su/su*; *Su/su yellow-green*; *Su/su yellow*	Extra nuclear	Less Chl; less dense lamellation
Nicotiana tabacuum[o]	*NC 95 yellow-green*	Extranuclear	Less Chl; less dense lamellation
Nicotiana tabacuum[o]	*NC 95 yellow*	Extranuclear	No grana

[a] Levine (1968).
[b] Levine (1969).
[c] Epel and Levine (1971).
[d] Epel et al. (1972).
[e] Unidentified intersystem intermediate
[f] Bishop and Wong (1971).
[g] Allen (1959).
[h] Volk and Bishop (1968).
[i] Russel et al. (1969).
[j] Fork and Heber (1968).
[k] Boardman and Highkin (1966).
[l] Boardman and Thorne (1968).
[m] Bazzaz et al. (1974).
[n] Observation by C. J. Arntzen (personal communication to Govindjee).
[o] Homann and Schmid (1967).

Chlamydomonas reinhardii strains blocked near the primary photoreaction of PS II have been described by Levine and co-workers (Levine, 1963; Smillie and Levine, 1963a,b; Gorman and Levine, 1965, 1966; Levine *et al.*, 1966; Lavorel and Levine, 1969). The mutant strains *ac-115* and *ac-141* lack Cyt b559 and have a reduced PQ content. Their fluorescence timecourse resembles that of DCMU-poisoned wild-type cells, and on protracted illumination they show a slow rise of the Chl *a* fluorescence yield (slow induction; Bannister and Rice, 1968). Clearly, this illustrates the independence of the slow fluorescence change from a functional Q. The mutant strain *ac-141*, as well as a new mutant strain, *Chlamydomonas reinhardii hfd-91*, were recently reported to lack the C550 absorbance response and to be devoid of the high potential Cyt b559. Chloroplast fragments isolated from the mutant *hfd-91* are highly fluorescent and they do not respond to the addition of ferricyanide or sodium dithionite, treatments that suppress or enhance, respectively, the fluorescence of chloroplast fragments derived from wild-type *Chlamydomonas*. Possibly, then, either Q or the reaction center of PS II is missing from these mutants (Epel and Butler, 1972).

Mutants, which on a Chl basis are more fluorescent than their wild-type counterparts, have been also described in the genera *Euglena* (Russell *et al.*, 1969), *Oenothera* (Fork and Heber, 1968), and *Scenedesmus* (Bishop, 1962; Butler and Bishop, 1963; Senger and Bishop, 1966; Pratt and Bishop, 1968; Bishop and Wong, 1971; see Table V). The high fluorescence output of these mutants may suggest that the processes which reoxidize the photoreduced Q have been impaired by the mutation. The *Scenedesmus obliquus* mutants that have been isolated and studied by Bishop and his co-workers have an altered carotenoid distribution, less total Chl, and less PQ-A, and in one case C550 is missing (mutant ScD_3-11; Erixon and Butler, 1971c). The loss of several components indicates that these mutations cause the deletion of discreet segments of the PS II apparatus, rather than the deletion of a single component. However, the major forms of the *in vivo* chlorophylls are present in their normal ratios, and no significant change in the chloroplast structure has been detected. The mutations appear to have introduced blocks on the electron acceptor side of PS II, since these mutants are unable to photoreduce DCPIP with the artificial electron donor diphenylcarbazide. Their variable fluorescence is about one-tenth of that in the wild-type *Scenedesmus*.

Six-low-fluorescence *Chlamydomonas reinhardii* mutant strains (*lfd 2, lfd 13, lfd 15, lfd 17, lfd 27*, and *lfd 29*), blocked on the electron donor side of PS II have been isolated and studied by Epel and Levine (1971), and Epel *et al.*, (1972). These mutants, although similar to normal *Chlamydomonas* in total Chl content, Chl *a* : Chl *b* ratio, and Chl *a* fluorescence excitation spectra do not photoevolve oxygen. Electron donors to PS II (*p*-phenylenediamine, *p*-aminophenol, and *p*-hydroquinone) restore the photoinduced,

DCMU-sensitive, electron transport to $NADP^+$ in chloroplast fragments derived from these mutants. The existence of a functional PS II reaction center is evidenced further by the presence of the C550 and the Cyt b559 absorbance responses in these fragments. Compared to the wild-type *Chlamydomonas*, these mutants were found to contain only one-half of the high potential (ascorbate reducible) Cyt b559. According to Epel *et al.* (1972), this may suggest the existence of two distinct pools of the high potential Cyt b559, which operate on the water side of PS II.

Various photosynthetic mutants with modified light-harvesting systems are known, but not all of them have been studied in respect to their fluorescence properties. One of the most investigated cases is that of a barley mutant (*Hordeum vulgare* No. 2), which lacks Chl *b* entirely and has a diminished Chl *a* content. Although its chloroplasts appear disorganized, with fewer grana per chloroplast and fewer thylakoids per granum, they are nevertheless capable of a 40% faster electron transport and of a higher light intensity saturation than normal barley chloroplasts (Boardman and Highkin, 1966). The Chl *a* fluorescence yield of the mutant *Hordeum* chloroplasts, measured against fluorescein, is 3 to 4 times lower than in the wild type when the Chl *a* is excited directly, and 5 to 6 times lower when it is excited via Chl *b*. Low temperature (77°K) fluorescence spectroscopy shows a deficiency in the Chl *a* form that emits at 696 nm (P680), while the bands corresponding to Chl *b* are missing from the excitation spectrum. The mutant is characterized by a low O-level fluorescence and an insignificant light-induced fluorescence increment. Reduction of Q by dithionite raises the steady-state fluorescence 2 to 3 times, indicating the presence of an active Q. Due to the preponderance of the PS I pigments, however, Q is kept mostly oxidized (Boardman and Thorne, 1968).

A lethal mutant strain of *Zea mays* (*olive necrotic 8147*) was recently studied in respect to the fluorescence, and the structural and biochemical characteristics of its chloroplasts (Bakri, 1972; Bazzaz *et al.*, 1974). Compared to leaves from normal *Zea mays*, the leaves of this mutant have less total Chl (30%) and less carotenoids (70%), and the chloroplasts of the mutant have smaller PSU's. In addition, they lack stomata, because of which the plants develop necrotic symptoms within 10–15 days (Arntzen, personal communication to Govindjee). Structurally, the bundle sheath chloroplasts of the mutant resemble those of normal maize, but the mesophyll chloroplasts have few and incompletely formed grana. Functionally, the mutant is characterized by higher saturation rates of both PS I and PS II in the mutant, but there is a preponderance of the PS I activity over that of PS II, which correlates with the higher ratios of PS I/PS II pigments and of constant over variable fluorescence, relative to chloroplasts from normal maize. Although the greatly distorted chloroplast phenotype can be due to a direct effect of the mutation, it is also possible that the mutation influences the phenotype indirectly by preventing the complete greening of the mutant leaves. It has been known for some time that, during the greening stage of etiolated higher plant leaves, PS I predominates over PS II in pigmentation and in function (Akoyunoglou *et al.*, 1966; Argyroudi-Akoyunoglou *et al.*, 1972; Argyroudi-Akoyunoglou and Akoyunoglou, 1973).

The photosynthetic mutants of *Nicotiana tabacuum* (*aurea Su/su, Su/su variegated yellow-green and yellow,* and *NC 95 variegated yellow-green and yellow*; Homann and Schmid, 1967) have less total Chl and altered Chl *a*/Chl *b* and Chl/carotenoid ratios. With the exception of the *NC 95 yellow*, they are characterized by higher CO_2 fixation rates and higher levels of light saturation, resembling in this respect the Chl-poor barley mutant of Boardman and Highkin (1966). On the other hand, the Chl *b*-minus mutants of *Chlorella pyrenoidosa* (*G41, G44,* and *G45*) carry out a less efficient photosynthesis than the wild-type *Chlorella* (Allen, 1959). Also, the phycocyaninless mutant of *Cyanidium caldarium*, studied by Volk and Bishop (1968), does not differ from normal *Cyanidium* in the rate of photosynthesis and photoreduction of *p*-benzoquinone at low light intensity.

6. FLUORESCENCE PROPERTIES OF MANGANESE-DEFICIENT PHOTOSYNTHETIC TISSUE

Mn is an indispensible plant micronutrient (Pirson, 1937; Pirson *et al.*, 1952; Eyster *et al.*, 1958). It is also an important link in the photosynthetic electron transport chain, operating between the site of water decomposition and the reaction center of PS II (Kessler *et al.*, 1957; Kessler, 1968a,b, 1970; Homann, 1967; Schmid and Gaffron, 1968), while it has no effect on the electron transport through PS I (Cheniae and Martin, 1966; Böhme and Trebst, 1969). There are four, tightly bound on protein, Mn atoms associated with each PS II center, which are distributed in two sequentially functioning subpools (Cheniae and Martin, 1970, 1971).

Mn deficiency can be induced either by extraction from the fully developed green tissue, or by culturing the plants in a Mn-poor medium. Cation chelating compounds, such as EDTA, hydroxyquinoline, and *o*-phenanthroline, fail to detach the functional Mn. This Mn, however, can be released by pronase treatment (Homann, 1967; Donnat and Briantais, 1967), by chaotropic agents and Triton X-100 (Cheniae and Martin, 1966), and by high concentrations of Tris (Homann, 1967; Cheniae and Martin, 1970). Most advantageous appears to be the release of bound Mn by heating the chloroplasts, or the intact leaves, since this treatment avoids the simultaneous extraction of several chloroplast components, and causes the least structural damage (Cheniae and Martin, 1966; Homann, 1968b).

Plants grown on Mn-poor media differ in more than one respect from the normal plants. Growth deficiency leads to chlorosis, altered Chl a/b ratio, slower metabolism, and structural disorganization of the chloroplasts (Richter, 1961; Pirson *et al.*, 1952; Eyster *et al.*, 1958; Anderson and Thorne, 1968; Constantopoulos, 1970; Possingham *et al.*, 1964; Teichler-Zallen, 1969; Gavalas and Clark 1971). As pointed out by Homann (1968a), these secondary effects may interfere with and obscure the functional differences which arise from the absence of Mn.

6.1. Fluorescence Spectra and Fluorescence Yield

Compared to the low-temperature fluorescence spectra of Mn-sufficient samples, those of the deficient ones are characterized by a prominent far-red band (720–740 nm), and by a suppressed 693 nm emission (Cheniae and Martin, 1966; Itoh *et al.*, 1969; Heath and Hind, 1970). Apart from the structural and constitutional effects due to the Mn deficiency, the enhanced far-red emission may result from an increased excitation spillover from PS II to PS I, since Mn has been shown to be most effective in suppressing this spillover (Murata *et al.*, 1970). Although this cation is not an integral part of the PS II reaction center, the weaker 693 nm fluorescence of the deficient samples suggests that Mn exerts an influence on it (Cheniae, 1970).

The fluorescence yield of spinach chloroplasts made deficient by Tris extraction at pH 7.8 (Itoh *et al.*, 1969), and of *Euglena* grown on a Mn-free heterotrophic medium (Heath and Hind, 1970) is smaller than the corresponding yield of the control samples. On the other hand, Anderson and Thorne (1968) found the Mn-deficient chloroplasts isolated from autotrophically grown spinach to be more fluorescent than those from Mn-sufficient plants. These discrepant results underscore the overriding importance of the secondary structural effects on the fluorescence of Chl a *in vivo*. The severe structural changes of the autotrophically grown spinach, such as the diminished leaf area and the altered Chl a/b ratio, hardly allow the unequivocal correlation of the Mn deficiency to the increased fluorescence emission.

6.2. Fluorescence Kinetics

Mn deficiency is accompanied by a suppression of the variable Chl a fluorescence. This may result from an inhibition either of the electron supply to Q, or from a faster withdrawal of electrons from it. Either effect can cause a narrowing of the interval in which the redox potential of Q/QH oscillates. The alternative of an electron transport block on the reducing side of PS II in Mn-deficient plants has been espoused by Anderson and Thorne (1968) on the basis of a higher F_0 and a smaller $F_\infty - F_0$ observed with chloroplasts from growth-deficient spinach. This interpretation has been challenged on the grounds that F_0 is critically dependent on structure and, therefore, not a valid measure to compare the reduction levels of Q in the deficient and the control samples. In fact, the accumulated evidence favors the assignment of the functional Mn to the oxidizing side of PS II. This evidence, fluorimetric or other, is as follows.

1. In contrast to the difference $F_\infty - F_0$, $F_{inh} - F_0$ and F_{inh} vary little with progressing Mn deficiency, so long as the inner subpool (the one closer the PS II reaction center) of the functional Mn is unaffected. In the absence of electron transport inhibitors, $F_\infty - F_0$ and the rate of the *OP* rise fall off as the Mn deficiency increases (Homann, 1968a,b; Cheniae and Martin, 1970). Were Mn an intersystem carrier, the kinetics of Chl *a* fluorescence of Mn-deficient samples, with and without DCMU, would be indistinguishable. When the inner subpool of functional Mn is depleted, F_{inh} and $F_{inh} - F_0$ become smaller signifying an incomplete reduction of Q, possibly because of a recombination of the photoproducts of PS II (Homann, 1968b; Cheniae and Martin, 1970).

2. As long as the inner Mn subpool is functional, the variable Chl *a* fluorescence can be partially restored by added electron donors that supply electrons directly to the inner pool (Yamashita and Butler, 1969; Cheniae and Martin, 1970; Lozier *et al.*, 1971).

3. Several hydrogen-adapted algae (*Ankistrodesmus braunii, Chlorella vulgaris,* and *Scenedesmus obliquus*) have been shown to suffer no adverse effect in their growth, or in Chl content, when they are cultured in Mn-free media. Under similar conditions, nonadapted strains have their growth retarded and tend to chlorosis (Kessler *et al.*, 1957; Kessler, 1968a, 1970).

Kessler (1968b, 1970) found the Mn deficiency to abolish the fast and slow fluorescence induction features in *Chlorella vulgaris* and *Ankistrodesmus braunii,* both in air-adapted and hydrogen-adapted cultures. Addition of Mn^{2+} restores the typical "two-wave" induction pattern in aerobic *Chlorella,* while in the hydrogen-adapted and Mn-sufficient *Ankistrodesmus* and *Chlorella* only the fast fluorescence transient is present.

7. SUMMARY

We have attempted, in this chapter, to correlate the complex but typical fluctuations of the variable fluorescence of Chl *a in vivo* to the physicochemical and physiological processes that occur at, or near, the lamella, as a green plant is taken from a state of dark adaptation to a state of light adaptation. On the basis of the experimental evidence, which has been accumulated mostly in the last decade, we can now assign the various features of the Chl *a* fluorescence timecourse to few, mutually interacting processes. These processes are initiated simultaneously at the onset of illumination, but they advance to their respective steady states with very different rates. Arranged in a decreasing rate order, these processes are:

1. The adjustment of the electron carriers in the neighborhood of the PS II reaction center (electron donor and acceptor groups) to the light environment. This gives rise to the activation phenomena, expressed by the *OI* phase of the fluorescence timecourse (Fig. 3).

2. The adjustment of the intersystem electron carriers to the light environment, which becomes manifest in the fluorescence rise along *IDP*. At this stage, and since the other processes did not have a chance to advance yet, the output of the variable fluorescence is directly responsive to the rate of the photosynthetic electron transport.

3. The dissipation of excess Chl *a* excitation by endogenous quenchers, such as molecular oxygen, whose penetration to the pigment bed is structurally regulated. This process (along with that of paragraph 4) contributes to the fluorescence decay along *PS*. Its biological importance may rest in the dissipation of the potentially destructive excess of Chl *a* excitation. Since this effect requires, in addition to the electron transport across PS II, the capacity to develop and to maintain a pH difference across the thylakoid membrane, a higher degree of structural integrity is necessary than required for the fluorescence rise along *OP*. Hence, the *PS* fluorescence decay is typical only of intact plants and of class I chloroplasts (i.e., isolated chloroplasts retaining, to a large proportion, their outer envelopes).

4. The gradual shift in the distribution of the Chl *a* excitation from a state that favors the PS II pigment population, at the beginning of the illumination period, to a state that favors the overflow of PS II excitation to PS I, at the end of a protracted illumination. The distribution of divalent metal cations, particularly of Mg^{2+}, on either side of the thylakoid membrane as regulated by the light-driven electron transport appears to be the generic cause of the reapportionment of the Chl *a* excitation. These effects are manifest in the slow change of Chl *a* fluorescence. Its appearance in intact cells and class I chloroplasts only, rather than in isolated broken chloroplasts, underscores the importance of chloroplast integrity.

Finally, we must call attention to the fact that, when exposed to light stresses of longer duration, plants undergo more drastic changes, in which the pigment composition, the chloroplast structure, and the disposition of chloroplasts in the leaf are modified.

ACKNOWLEDGMENT

The author is grateful to Drs. Govindjee, W. L. Butler, and P. K. Mohanty for reading the manuscript and suggesting many improvements.

REFERENCES

Akoyunoglou, G., Argyroudi-Akoyunoglou, J. H., Michel-Wolwertz, M. R., and Sironval, C. (1966). *Physiol. Plant.* **19**, 1101.

Allen, M. B. (1959). *Brookhaven Symp. Biol.* **11**, 339.

Amesz, J., and Fork, D. C. (1967). *Biochim. Biophys. Acta* **143**, 97.

Anderson, J. M., and Boardman, N. K. (1966). *Biochim. Biophys. Acta* **112**, 403.

Anderson, J. M., and Thorne, S. W. (1968). *Biochim. Biophys. Acta* **162**, 122.

Argyroudi-Akoyunoglou, J. H., and Akoyunoglou, G. (1973). *Photochem. Photobiol.* **18**, 219.

Argyroudi-Akoyunoglou, J. H., Feleki, Z., and Akoyunoglou, G. (1972). *Proc. Int. Congr. Photosynthesis Res., 2nd* (G. Forti, M. Avron, and A. Melandri, eds.), Vol. 3, pp. 2417–2426, Dr. W. Junk, N. V. The Hague.

Arnon, D. I., Tsujimoto, H. Y., and McSwain, B. D. (1965). *Proc. Natl. Acad. Sci. U.S.* **54**, 927.

Arnon, D. I., Knaff, D. B., McSwain, B. D., Chain, R. K., and Tsujimoto, H. Y. (1971). *Photochem. Photobiol.* **14**, 397.

Arntzen, C. J., Dilley, R. A., and Crane, F. L. (1969). *J. Cell. Biol.* **43**, 16.

Bannister, T. T., and Rice, G. (1968). *Biochim. Biophys. Acta* **162**, 555.

Bakri, M. D. L. (1972). Ph.D. Thesis, Univ. of Illinois, Urbana.

Bazzaz, M. B., Govindjee, and Paolillo, D. J. (1974). *Z. Pflanzen-physiol.* **72**, 181

Bendall, D. S., and Sofrova, D. S. (1971). *Biochim. Biophys. Acta* **234**, 371.

Ben-Hayyim, G., and Avron, M. (1970). *Eur. J. Biochem.* **14**, 205.

Ben-Hayyim, G., and Avron, M. (1971). *Photochem. Photobiol.* **14**, 389.

Ben-Hayyim, G., and Malkin, S. (1972). See Argyroudi-Akoyunoglou *et al.* (1972), pp. 61–72.

Bennoun, P. (1970). *Biochim. Biophys. Acta* **216**, 357.

Bennoun, P., and Levine, R. P. (1967). *Plant Physiol.* **42**, 1284.

Bishop, N. I. (1962). *Nature (London)* **195**, 55.

Bishop, N. I., and Wong, J. (1971). *Biochim. Biophys. Acta* **234**, 433.

Boardman, N. K., and Highkin, H. R. (1966). *Biochim. Biophys. Acta* **126**, 189.

Boardman, N. K., and Thorne, S. W. (1968). *Biochim. Biophys. Acta* **153**, 448.

Boardman, N. K., Thorne, S. W., and Anderson, J. M. (1966). *Proc. Nat. Acad. Sci. U.S.* **56**, 586.

Böhme, H., and Trebst, A. (1969). *Biochim. Biophys. Acta* **180**, 137.

Bonaventura, C. (1969). Ph.D. Thesis, Univ. of Texas, Austin.

Bonaventura, C., and Myers, J. (1969). *Biochim. Biophys. Acta* **189**, 366.

Briantais, J. M. (1967). *Biochim. Biophys. Acta* **143**, 650.

Briantais, J. M., Merkelo, H., and Govindjee (1972). *Photosynthetica* **6**, 133.

Brody, S. S. (1957). *Rev. Sci. Instrum.* **28**, 1021.

Brody, S. S., Ziegelmair, C. A., Samuels, A., and Brody, M. (1966). *Plant Physiol.* **41**, 1709.

Butler, W. L. (1972a). *Biophys. J.* **12**, 851.

Butler, W. L. (1972b). *Proc. Nat. Acad. Sci. U.S.* **69**, 3420.

Butler, W. L. (1972c). *FEBS Lett.* **20**, 334.

Butler, W. L., and Bishop, N. I. (1963). *Nat. Acad. Sci. Nat. Res. Council Publ.* **1145**, 91.

Butler, W. L., and Hopkins, D. W. (1970). *Photochem. Photobiol.* **12**, 439.

Butler, W. L., and Okayama, S. (1971). *Biochim. Biophys. Acta* **245**, 237.

Cheniae, G. M. (1970). *Ann. Rev. Plant Physiol.* **21**, 467.

Cheniae, G. M., and Martin, I. F. (1966). *Brookhaven Symp. Biol.* **19**, 406.

Cheniae, G. M., and Martin, I. F. (1970). *Biochim. Biophys. Acta* **197**, 219.

Cheniae, G. M., and Martin, I. F. (1971). *Plant Physiol.* **47**, 568.

Cho, F. (1969). Ph.D. Thesis, Univ. of Illinois, Urbana.

Cho, F., and Govindjee (1970a). *Biochim. Biophys. Acta* **205**, 371.

Cho, F., and Govindjee (1970b). *Biochim. Biophys. Acta* **216**, 139.

Cho, F., and Govindjee (1970c). *Biochim. Biophys. Acta* **216**, 151.

Clayton, R. K. (1969). *Biophys. J.* **9**, 60.

Cohen, W. S., and Sherman, L. A. (1971). *FEBS Lett.* **16**, 319.

Constantopoulos, G. (1970). *Plant. Physiol.* **45**, 76.

Cox, R. D., and Bendall, D. S. (1972). *Biochim. Biophys. Acta* **283**, 124.

Cramer, W. A., and Böhme, H. (1972). *Biochim. Biophys. Acta* **256**, 358.

Cramer, W. A., and Butler, W. L. (1967). *Biochim. Biophys. Acta* **143**, 332.

Crofts, A. R., Deamer, D. W., and Packer, L. (1967). *Biochim. Biophys. Acta* **131**, 97.

Delosme, R. (1967). *Biochim. Biophys. Acta* **143**, 108.

Delosme, R., Joliot, P., and Lavorel, J. (1959). *C. R. Acad. Sci. Paris* **249**, 1409.

Dilley, R. A., and Vernon, L. P. (1965). *Arch. Biochem. Biophys.* **111**, 365.

Donnat, P., and Briantais, J. M. (1967). *C. R. Acad. Sci. Paris* **265**, 21.

Döring, G., Stiehl, H. H., and Witt, H. T. (1967). *Z. Naturforsch.* **22b**, 639.

Döring, G., Bailey, J. L., Kreutz, W., and Witt, H. T. (1968). *Naturwissenschaften* **55**, 220.

Döring, G., Renger, G., Vater, J., and Witt, H. T. (1969). *Z. Naturforsch.* **24b**, 1140.

Doschek, W. W., and Kok, B. (1972). *Biophys. J.* **12**, 832.

Duysens, L. N. M. (1952). Ph.D. Thesis, Univ. of Utrecht.

Duysens, L. N. M. (1969). Structure, Function, and Control Mechanisms in Photosynthetic Organelles. *Gordon Res. Conf., Plymouth, New Hampshire.*

Duysens, L. N. M. (1972). *Biophys. J.* **12**, 858.

Duysens, L. N. M., and Sweers, H. E. (1963). *In* "Studies on Microalgae and Photosynthetic Bacteria" (Jap. Soc. Plant Physiol., eds.), pp. 353–372. Univ. of Tokyo Press, Tokyo.

Duysens, L. N. M., and Talens, A. (1969). *Prog. Photosynthesis Res.* **2**, 1073.

Duysens, L. N. M., Van der Schatte, T. E., and Den Haan, G. A. (1972). *Int. Congr. Photobiol., 4th, Bochum, Germany, Abstr.* p. 277.

Elgersma, O. (1972). *Int. Biophys. Congr., 4th, Abstr. Contributed Papers* Vol. I, p. 325, Acad. of Sci. of the U.S.S.R.

Epel, B. L., and Butler, W. L. (1972). *Biophys. J.* **12**, 922.

Epel, B. L., and Levine, R. P. (1971). *Biochim. Biophys. Acta* **226**, 154.

Epel, B. L., Butler, W. L., and Levine, R. P. (1972). *Biochim. Biophys. Acta* **275**, 325.

Erixon, K., and Butler, W. L. (1971a). *Biochim. Biophys. Acta* **234**, 381.

Erixon, K., and Butler, W. L. (1971b). *Biochim. Biophys. Acta* **253**, 488.

Erixon, K., and Butler, W. L. (1971c). *Photochem. Photobiol.* **14**, 427.

Erixon, K., Lozier, R., and Butler, W. L. (1972). *Biochim. Biophys. Acta* **267**, 375.

Eyster, H. C., Brown, T. E., Tanner, H. A., and Hodd, S. L. (1958). *Plant Physiol.* **33**, 235.

Floyd, R. A., Chance, B., and Devault, D. (1971). *Biochim. Biophys. Acta* **226**, 103.

Forbush, B., and Kok, B. (1968). *Biochim. Biophys. Acta* **162**, 243.

Forbush, B., Kok, B., and McGloin, M. (1971). *Photochem. Photobiol.* **14**, 307.

Fork, D. C. (1972). *Biophys. J.* **12**, 909.

Fork, D. C., and Heber, U. (1968). *Plant Physiol.* **43**, 606.

Förster, Th. (1951). "Fluoreszenz Organischer Verbindungen." Vandenhoeck E. Ruprecht, Göttingen.

Franck, J., French, C. S., and Puck, T. T. (1941). *J. Phys. Chem.* **45**, 978.

Franck, U. F., and Hoffmann, N. (1969). *Progr. Photosynthesis Res.* **2**, 899.

Franck, U. F., Hoffmann, N., Arenz, H., and Schreiber, U. (1969). *Ber. Bunsenges. Phys. Chem.* **73**, 871.

French, C. S. (1972). *Proc. Nat. Acad. Sci. U.S.* **68**, 2893.

French, C. S., and Young, V. K. (1952). *J. Gen. Physiol.* **35**, 873.

Gavalas, N. A., and Clark, H. E. (1971). *Plant Physiol.* **47**, 139.

Garnier, J. (1967). *C. R. Acad. Sci. Paris* **265**, 874.

Ghosh, A. K., and Govindjee (1966). *Biophys. J.* **6**, 611.

Goedheer, J. C. (1972). *Ann. Rev. Plant. Physiol.* **23**, 87.

Gorman, D. S., and Levine, R. P. (1965). *Proc. Nat. Acad. Sci. U.S.* **54**, 1665.

Gorman, D. S., and Levine, R. P. (1966). *Plant Physiol.* **41**, 1968.

Govindjee, and Briantais, J. M. (1972). *FEBS Lett.* **19**, 278.

Govindjee, and Papageorgiou, G. (1971). *In* "Photophysiology" (A. C. Giese, ed.), Vol. VI, pp. 1–46. Academic Press, New York.

Govindjee, Döring, G., and Govindjee, R. (1970). *Biochim. Biophys. Acta* **205**, 303.

Govindjee, Munday, J. C., Jr., and Papageorgiou, G. (1966). *Brookhaven Symp. Biol.* **19**, 434.

Govindjee, Papageorgiou, G., and Rabinowitch, E. I. (1967). *In* "Fluorescence: Theory, Instrumentation, and Practice" (G. G. Guilbault, ed.), pp. 511–564. Dekker, New York.

Govindjee, R., Govindjee, Lavorel, J., and Briantais, J. M. (1970). *Biochim. Biophys. Acta* **205**, 361.

Gross, E. (1973). *Biophys. Soc. Abstr.* 64a.

Gross, E., Dilley, R. A., and San Pietro, A. (1969). *Arch. Biochem. Biophys.* **134**, 450.

Gurinovich, G. P., Sevchenko, A. N., and Solov'ev, K. N. (1971). *In* "Spectroscopy of Chlorophylls and Related Compounds," pp. 455–466. Publ. AEC-tr-7199.

Haehnel, W., and Witt, H. T. (1972). See Argyroudi-Akoyunoglou *et al.* (1972), Vol. I, pp. 469–476.

Heath, R. L. (1970). *Biophys. J.* **10**, 1173.

Heath, R. L., and Hind, G. (1970). *Biochim. Biophys. Acta* **189**, 222.

Heath, R. L., and Packer, L. (1968). *Arch. Biochem. Biophys.* **125**, 1019.

Hind, G. (1968). *Photochem. Photobiol.* **7**, 369.

Homann, P. H. (1967). *Plant Physiol.* **42**, 997.

Homann, P. H. (1968a). *Biochim. Biophys. Acta* **162**, 545.

Homann, P. H. (1968b). *Biochem. Biophys. Res. Commun.* **33**, 229.

Homann, P. H. (1969). *Plant Physiol.* **44**, 932.

Homann, P. H., and Schmid, G. (1967). *Plant Physiol.* **42**, 1619.

Itoh, M., Yamashita, K., Nishi, T., Konishi, K., and Shibata, K. (1969). *Biochim. Biophys. Acta* **180**, 509.

Izawa, S., and Good, N. E. (1966). *Plant Physiol.* **41**, 544.

Jagendorf, A. T., and Hind, G. (1963). *Nat. Acad. Sci. -Nat. Res. Council Publ.* **1145**, 599.

Jagendorf, A. T., and Smith, M. (1962). *Plant. Physiol.* **37**, 135.

Joliot, A., and Joliot, P. (1964). *C.R. Acad. Sci. Paris* **258**, 4622.

Joliot, P. (1965a). *Biochim. Biophys. Acta* **102**, 116.

Joliot, P. (1965b). *Biochim. Biophys. Acta* **102**, 135.

Joliot, P. (1968). *Photochem. Photobiol.* **8**, 451.

Joliot, P., Barbieri, G., and Chabaud, R. (1969). *Photochem. Photobiol.* **10**, 309.

Joliot, P., Joliot, A., Bouges, B., and Barbieri, G. (1971). *Photochem. Photobiol.* **14**, 287.

Junge, W., and Witt, H. T. (1968). *Z. Naturforsch.* **23b**, 244.

Karapetyan, N. V., and Klimov, V. V. (1972). See Elgersma (1972), Vol. I, p. 324.

Katoh, S. (1972). *Plant Cell Physiol.* **13**, 273.

Kautsky, H. (1931). *Naturwissenschaften* **19**, 964.

Kautsky, H., and Eberlein, R. (1939). *Biochem. Z.* **305**, 137.

Kautsky, H., and Franck, U. (1943). *Biochem. Z.* **315**, 139.

Kautsky, H., Appel, W., and Amann, H. (1960). *Biochem. Z.* **332**, 277.

Kessler, E. (1968a). *Planta* **81**, 264.

Kessler, E. (1968b). *Arch. Mikrobiol.* **63**, 7.

Kessler, E. (1970). *Planta* **92**, 222.

Kessler, E., Arthur, W., and Brugger, J. E. (1957). *Arch. Biochem. Biophys.* **71**, 326.

Kirk, J. T. O., and Tilney-Basset, R. A. E. (1967). "The Plastids." Freeman, San Francisco, California.

Knaff, D. B., and Arnon, D. I. (1969a). *Proc. Nat. Acad. Sci. U.S.* **63**, 956.

Knaff, D. B., and Arnon, D. I. (1969b). *Proc. Nat. Acad. Sci. U.S.* **63**, 963.

Knaff, D. B., and Arnon, D. I. (1969c). *Proc. Nat. Acad. Sci. U.S.* **64**, 715.

Knaff, D. B., and Arnon, D. I. (1971a). *Biochim. Biophys. Acta* **226**, 400.

Knaff, D. B., and Arnon, D. I. (1971b). *Biochim. Biophys. Acta* **245**, 105.

Kohl, D. H., Wright, J. R., and Weissman, M. (1969). *Biochim. Biophys. Acta* **180**, 536.

Kok, B., and Hoch, G. (1961). *In* "Light and Life" (W. D. McElroy, and B. Glass, eds.), pp 397–416. Johns Hopkins Press, Baltimore, Maryland.

Kok, B., Forbush, B., and McGloin, M. (1970). *Photochem. Photobiol.* **11**, 451.

Kouchkovsky, Y., de (1972). See Argyroudi-Akoyunoglou *et al.* (1972), Vol. I, pp. 233–246.

Kouchkovsky, Y., de, and Joliot, P. (1967). *Photochem. Photobiol.* **6**, 567.

Krause, G. H. (1972). See Elgersma (1972), p. 323.

Krey, A., and Govindjee (1964). *Proc. Nat. Acad. Sci. U.S.* **52**, 1568.

Krey, A., and Govindjee (1966). *Biochim. Biophys. Acta* **120**, 1.

Latimer, P., Bannister, T. T., and Rabinowitch, E. I. (1957). *In* "Research in Photosynthesis" (H. Gaffron, A. H. Brown, C. S. French, R. Livingston, E. I. Rabinowitch, B. L. Strehler, and N. E. Tolbert, eds.), pp. 107–112. Wiley (Interscience), New York.

Lavorel, J. (1959). *Plant Physiol.* **34**, 204.

Lavorel, J. (1962). *Biochim. Biophys. Acta* **60**, 510.

Lavorel, J. (1967). *J. Chem. Phys.* **47**, 2235.

Lavorel, J., and Joliot, P. (1972). *Biophys. J.* **12**, 815.

Lavorel, J., and Levine, R. P. (1969). *Plant Physiol.* **43**, 1049.

Levine, R. P. (1963). *Nat. Acad. Sci. Nat. Res. Council. Publ.* **1145**, 158.

Levine, R. P. (1968). *Science* **162**, 768.

Levine, R. P. (1969). *Ann. Rev. Plant Physiol.* **20**, 523.

Levine, R. P., and Gorman, D. S. (1966). *Plant Physiol.* **41**, 1293.

Levine, R. P., Gorman, D. S., Avron, M., and Butler, W. L. (1966). *Brookhaven Symp. Biol.* **19**, 143.

Litvin, F. F., (1972). See Elgersma (1972), Vol. I, p. 15,

Livingston, R., and Ke, C. L. (1950). *J. Amer. Chem. Soc.* **72**, 909.

Lozier, R., Baginsky, M., and Butler, W. L. (1971). *Photochem. Photobiol.* **14**, 323.

Malkin, S. (1968). *Biochim. Biophys. Acta* **153**, 188.

Malkin, S., and Kok, B. (1966). *Biochim. Biophys. Acta* **126**, 413.

Mar, T., Govindjee, Singhal, G. S., and Merkelo, H. (1972). *Biophys. J.* **12**, 797.

Mathis, P. (1969). *Photochem. Photobiol.* **9**, 55.

Mauzerall, D. (1972). *Proc. Nat. Acad. Sci. U.S.* **69**, 1358.

Mauzerall, D., and Malley, M. (1971). *Photochem. Photobiol.* **14**, 225.

McAlister, E. D., and Myers, J. (1940). *Smithsonian Misc. Collect.* **99**, 1.

Merkelo, H., Hartman, S. R., Mar, T., Singhal, G. S., and Govindjee (1969). *Science* **164**, 301.

Michel, J. M., and Michel-Wolwertz, M. R. (1969). *Progr. Photosynthesis Res.* **1**, 115.

Michel, J. M., and Michel-Wolwertz, M. R. (1970). *Photosynthetica* **4**, 146.

Mohanty, P. K. (1972). Ph.D. Thesis, Univ. of Illinois.

Mohanty, P. K., and Govindjee (1971). *Conf. Primary Photochem. Photosynthesis, Argonne, Illinois, Abstr.* p. 31.

Mohanty, P. K., and Govindjee (1973a). *Photosynthetica*, **7**, 146.

Mohanty, P. K., and Govindjee (1973b). *Plant Cell Physiol.* **14**, 611.

Mohanty, P. K., Munday, J. C., Jr., and Govindjee (1970). *Biochim. Biophys. Acta* **223**, 198.

Mohanty, P. K., Mar, T. and Govindjee (1971a). *Biochim. Biophys. Acta* **253**, 213.

Mohanty, P. K., Papageorgiou, G., and Govindjee (1971b). *Photochem. Photobiol.* **14**, 667.

Mohanty, P. K., Zilinskas-Braun, B., and Govindjee (1972). *FEBS Lett.* **20**, 273.

Mohanty, P. K., Zilinskas-Braun, B., and Govindjee (1973). *Biochim. Biophys. Acta* **292**, 459.

Morin, P. (1964). *J. Chim. Phys.* **61**, 674.

Munday, J. C., Jr. (1968). P.D. Thesis, Univ. of Illinois, Urbana.

Munday, J. C., Jr., and Govindjee (1969a). *Biophys. J.* **9**, 1.

Munday, J. C., Jr., and Govindjee (1969b). *Biophys. J.* **9**, 22.

Munday, J. C., Jr., and Govindjee (1969c). *Progr. Photosynthesis Res.* **2**, 913,

Murakami, S., and Packer, L. (1969). *Biochim. Biophys. Acta* **180**, 420.

Murakami, S., and Packer, L. (1970a). *Plant Physiol.* **45**, 289.

Murakami, S., and Packer, L. (1970b). *J. Cell. Biol.* **47**, 332.

Murakami, S., and Packer, L. (1971). *Arch. Biochem. Biophys.* **146**, 337.

Murata, N. (1969a). *Biochim. Biophys. Acta* **172**, 242.

Murata, N. (1969b). *Biochim. Biophys. Acta* **189**, 171.

Murata, N. (1970). *Biochim. Biophys. Acta* **205**, 379.

Murata, N. (1971a). *Biochim. Biophys. Acta* **226**, 422.

Murata, N. (1971b). *Biochim. Biophys. Acta* **245**, 365.

Murata, N., and Sugahara, K. (1969). *Biochim. Biophys. Acta* **189**, 182.

Murata, N., Nishimura, M., and Takamiya, A. (1966a). *Biochim. Biophys. Acta* **120**, 23.

Murata, N., Nishimura, M., and Takamiya, A. (1966b). *Biochim. Biophys. Acta* **112**, 213.

Murata, N., Tashiro, H., and Takamiya, A. (1970). *Biochim. Biophys. Acta* **197**, 250.

Myers, J. (1971). *Ann. Rev. Plant Physiol.* **22**, 289.

Nobel, P. S. (1967). *Biochim. Biophys. Acta* **131**, 1967.

Okayama, S., and Butler, W. L. (1972a). *Biochim. Biophys. Acta*, **267**, 523.

Okayama, S., and Butler, W. L. (1972b). *Plant Physiol.* **49**, 769.

Ornstein, L. S., Wassink, E. C., and Vermeulen, D. (1938). *Enzymologia* **5**, 110.

Packer, L. (1962). *Biochem. Biophys. Res. Commun.* **9**, 355.

Packer, L. (1972). See Duysens *et al.* (1972), p. 030.

Papageorgiou, G. (1968). PhD. Thesis, Univ. of Illinois, Urbana.

Papageorgiou, G., (1972). See Argyroudi-Akoyunoglou *et al.* (1972), Vol. II, pp. 1535–1544.

Papageorgiou, G., and Argoudelis, C. (1973). *Arch. Biochem. Biophys.* **156**, 134.

Papageorgiou, G., and Govindjee (1967). *Biophys. J.* **7**, 375.

Papageorgiou, G., and Govindjee (1968a). *Biophys. J.* **8**, 1299.

Papageorgiou, G., and Govindjee (1968b). *Biophys. J.* **8**, 1316.

Papageorgiou, G., and Govindjee (1969). *Progr. Photosynthesis Res.* **2**, 905.

Papageorgiou, G., and Govindjee (1971). *Biochim. Biophys. Acta* **234**, 428.

Papageorgiou, G., Isaakidou, J., and Argoudelis, C., (1972a). *FEBS Lett.* **25**, 139.

Papageorgiou, G., Argoudelis, C., and Isaakidou, J. (1972b). See Elgersma (1972), Vol. I, p. 385.

Park, R. B., and Sane, P. V. (1971). *Ann. Rev. Plant Physiol.* **22**, 395.

Park, R. B., Steinbach, K. E., and Sane, P. V. (1971). *Biochim. Biophys. Acta* **253**, 204.

Pirson, A. (1937). *Z. Bot.* **31**, 193.

Pirson, A., Tichy, C., and Wilhelmi, G. (1952). *Planta* **40**, 199.

Possingham, J. V., Vesk, M., and Mercer, F. V. (1964). *J. Ultrastruct. Res.* **11**, 68.

Pratt, L. H., and Bishop, N. I. (1968). *Biochim. Biophys. Acta* **153**, 664.

Rabinowitch, E. I. (1956) "Photosynthesis and Related Processes," Vol. 2, part 2, pp. 1375–1432. Wiley (Interscience), New York.

Richter, G. (1961). *Planta* **57**, 202.

Rurainski, H. J., and Hoch, G. E. (1972). See Argyroudi-Akoyunoglou *et al.* (1972), Vol. I, pp. 133–142.

Rurainski, H. J., Randles, J., and Hoch, G. E. (1971). *FEBS Lett.* **13**, 98.

Russell, G. K., Lyman, H., and Heath, R. L. (1969). *Plant Physiol.* **44**, 929.

Sato, V. L., Levine, R. P., and Neumann, J. (1971). *Biochim. Biophys. Acta* **253**, 437.

Schmid, G. H., and Gaffron, H. (1968). *J. Gen. Physiol.* **52**, 212.

Schliephake, W., Junge, W., and Witt, H. T. (1968). *Z. Naturforsch.* **24b**, 1038.

Schreiber, U., Bauer, R., and Franck, U. F. (1971). *Z. Naturforsch.* **26b**, 1195.

Schreiber, U., Bauer, R., and Franck, U. F. (1972). See Argyroudi-Akoyunoglou *et al.* (1972), Vol. I, pp. 169–179.

Senger, H., and Bishop, N. I. (1966). *Plant Cell Physiol.* **7**, 441.

Shavit, N., and Avron, M. (1967). *Biochim. Biophys. Acta,* **131**, 516.

Sineshchekov, V. A. (1972). See Elgersma (1972), Vol. I, p. 20.

Smillie, R. M., and Levine, R. P. (1963a). *J. Biol. Chem.* **238**, 4052.

Smillie, R. M., and Levine, R. P. (1963b). *J. Biol. Chem.* **238**, 4058.

Stacy, W. T., Mar, T., Swenberg, C. E., and Govindjee (1971). *Photochem. Photobiol.* **14**, 197.

Stern, O., and Volmer, M. (1919). *Phys. Z.* **20**, 183.

Stiehl, H. H., and Witt, H. T. (1968). *Z. Naturforsch.* **23b**, 220.

Stiehl, H. H., and Witt, H. T. (1969). *Z. Naturforsch.* **24b**, 1588.

Strehler, B. L. (1953). *Arch. Biochem. Biophys.* **43**, 67.

Sun, A. S. K., and Sauer, K. (1972). *Biochim. Biophys. Acta* **256**, 409.

Teichler-Zallen, D. (1969). *Plant Physiol.* **44**, 701.

Vater, J., Renger, G., Stiehl, H. H., and Witt, H. T. (1968). *Naturwissenschaften* **5**, 221.

Vaughan, W. M., and Weber, G. (1970). *Biochemistry* **9**, 464.

Volk, S. L., and Bishop, N. I. (1968). *Photochem. Photobiol.* **8**, 213.

Vredenberg, W. J. (1970). *Biochim. Biophys. Acta* **223**, 230.

Vredenberg, W. J., and Slooten, L. (1967). *Biochim. Biophys. Acta* **143**, 583.

Walz, D., Schuldiner, S., and Avron, M. (1971). *Eur. J. Biochem.* **22**, 439.

Wassink, E. C. (1951). *Advan. Enzymol.* **11**, 91.

Wassink, E. C., and Katz, E. (1939). *Enzymologia* **6**, 145.

Weber, G. (1972). *Ann. Rev. Biophys. Bioenerget.* **1**, 553.

Weller, A. (1961). *Progr. Reaction Kinet.* **1**, 187–214.

West, J., and Packer, L. (1970). *J. Bioenerget.* **1**, 405.

Whitten, D. G., Lopp, I. G., and Wildes, P. D. (1968). *J. Amer. Chem. Soc.* **90**, 7196.

Wilson, T., and Hastings, W. J. (1970). See Govindjee and Papageorgiou (1971), Vol. V, pp. 50–96.

Witt, H. T. (1971). *Quart. Rev. Biophys.* **4**, 365.

Wolff, C., and Witt, H. T. (1972). See Argyroudi-Akoyunoglou *et al.* (1972), Vol. II, pp. 931–936.

Wraight, C. A., and Crofts, A. R. (1970). *Eur. J. Biochem.* **17**, 319.

Yamamoto, H. Y., and Chichester, C. O. (1965). *Biochim. Biophys. Acta* **101**, 303.

Yamamoto, H. Y., and Takeguchi, C. A. (1972). See Argyroudi-Akoyunoglou *et al.* (1972), Vol. I, pp. 621–627.

Yamashita, T., and Butler, W. L. (1968a). *Plant Physiol.* **43**, 1968.

Yamashita, T., and Butler, W. L. (1968b). *Plant Physiol.* **43**, 2037.

Yamashita, T., and Butler, W. L. (1969). *Plant Physiol.* **44**, 435.

Zankel, K. L. (1971). *Biochim. Biophys. Acta* **245**, 323.

7

The Electron Transport Chain in Chloroplasts

M. Avron

ABBREVIATIONS

Chl	Chlorophyll
Cyt	Cytochrome
DAD	2,3,5,6-Tetramethyl-*p*-phenylenediamine

DBMIB	2,5-Dibromo-3-methyl-6-isopropyl-*p*-Benzoquinone
DCMU	3-(3,4-Dichlorophenyl)-1,1-dimethylurea
DCPIP	2,6-Dichlorophenol indophenol
E_0'	oxidation–reduction potential at pH 7
Fd	Ferredoxin
FRS	Ferredoxin reducing substance
HOQNO	2*n*-Heptyl-4-hydroxyquinoline *N*-oxide
NAD$^+$	Nicotinamide adenine dinucleotide
NADP$^+$	Nicotinamide adenine dinucleotide phosphate
P-ADPR	Phosphoadenosine diphosphate ribose
PC	Plastocyanin
Pi	Inorganic phosphate
PQ	Plastoquinone
PS I (II)	Photosystem I (II)

1. INTRODUCTION

Photosynthesis was recognized already by its very early investigators to be "a sensitized photochemical oxidation–reduction" (see Rabinowitch, 1945). However, the nature of the electron transport path was a subject of intense controversy and investigation throughout the modern period: from the early ideas of a light-mediated decomposition of CO_2 followed by interaction with water; to Van-Niel's unifying concept in which photoproduced reducing equivalents originating from water reduce CO_2 in green plant photosynthesis; to the very fundamental discovery by Robin Hill that isolated chloroplasts can perform the photochemical water-splitting and reducing-equivalent generation reaction independently of the carbon-fixation metabolism; to the relization that two photochemical reactions acting in series provide the best explanation of currently available data, to today's controversies regarding the number of photochemical acts involved and the order of the components linking them together. It is probably reasonable to assume that the general outline of the electron transport path which will be described in the following, has a lifetime which may be measured in units of decades. However, a detailed description of the electron transport path still requires a large amount of exploration, which we hope will be undertaken by the readers of this volume.

2. TWO PHOTOSYSTEMS

2.1. Energetic Considerations

Accepting for the moment the well-demonstrated fact that chloroplasts can easily reduce components with a potential as low as -0.55 V (see below), while producing at least one molecule of ATP, we need to supply photonic energy sufficient to drive the reaction:

$$H_2O + acceptor(-0.55 \text{ V}) + ADP + P_i \longrightarrow O_2 + reduced\ acceptor + ATP$$

$$(accepting\ 1/2H_2O \longrightarrow 1/4O_2 + H^+ + e^-, \qquad E_0, +0.8 \text{ V})$$

$$(ADP + P_i \longrightarrow ATP, \qquad E_0 \cong +0.3 \text{ V}/e^-)$$

(see Kraayenhof, 1969). Thus, 1.65 V/electron (0.8 + 0.55 + 0.3) are required.

A quantum of red light has sufficient energy to theoretically drive an electron across a gap of 1.82 V. However, taking into consideration some reasonable practical limitations on the utilization of this energy, it has been calculated that no more than 70% of this energy can be effectively used in photosynthesis (see Duysens, 1964) and therefore a maximum of 1.3 V are available to drive the process. Such simple calculations, rough as they are, immediately lead to the conclusion that the energy of at least 2 photons must be utilized in order to provide sufficient energy to drive an electron through the coupled photosynthetic electron transport system. The manner by which this is achieved was first hinted at in the observations of the red-drop phenomenon by Emerson and Lewis (1943).

2.2. The Red-Drop Phenomenon

Emerson and Lewis pointed out that photosynthesis became progressively less efficient in utilization of light energy when illumination of green plants was limited to wavelengths longer than 680 nm, despite the fact that Chl *in vivo* was still effective in absorbing this light. The concept of two photosystems acting in series (Fig. 1) predicts this result if one accepts the well-documented observation that most of the far-red absorbing Chl is effective only in sensitizing PS I-dependent reactions. Thus the drop of efficiency in the far-red region is due to an imbalance in the required equal distribution of quanta between the two photosystems in favor of PS I. This interpretation of the red-drop phenomenon was strongly supported by the further observation of the enhancement in photosynthetic systems (see Myers, 1971; Ben-Hayyim and Avron, 1972).

2.3. The Enhancement Phenomenon

In the enhancement phenomenon, also originally observed by Emerson and collaborators, the low quantum efficiency in the far-red region was shown to be markedly increased if simultaneously a second light beam (normally of red light around 650 nm) was provided. Thus, the rate of photosynthesis measured when the two beams were presented together was higher than the sum of the two beams presented separately. This, again is predicted by the scheme of Fig. 1 if we accept the suggestion that the red light excited mostly

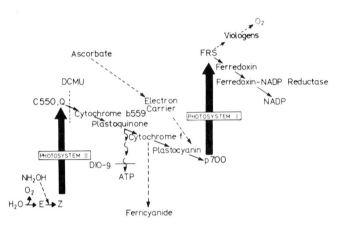

Fig. 1. Scheme summarizing current information on the electron transport path in photosynthesis. (From bottom left to the top right. the abbreviations in the diagram are: E, oxygen evolving complex; Z, primary electron donor of photosystem II; C550, a compound with a difference absorption maximum at 550 nm; Q, quencher of chlorophyll *a* fluorescence; DCMU, 3-(3,4 dichlorophenyl)-1, 1, -dimethyl urea, an inhibitor of electron flow; Dio-9, an antibiotic that inhibits energy transfer; P700, electron donor to photosystem I, a form of chlorophyll *a*; FRS, ferredoxin reducing substance; NADP, nicotinamide adenine dinucleotide phosphate. (The primary electron acceptor of photosystem I (X) may precede FRS.) The cyclic electron flow is not shown.

PS II. Thus, enhancement is the result of the balance in light distribution between the two photosystems obtained when a PS I favoring light is combined with a PS II favoring light.

2.4. Quantum Requirement of Partial Reactions

These interpretations were further reinforced by the elegant demonstration, originally by Hoch and Martin (1963), that chloroplast reactions sensitized exclusively by PS I proceed more efficiently in far-red light than in red light as expected from the above deductions. Thus, the transfer of electrons from ascorbate to $NADP^+$, or from ascorbate to diquat (Ben-Hayyim and Avron, 1971) proceeded with a quantum requirement of 1 quantum per electron transferred in the far-red region.

Recent data indicate that rather elaborate control mechanisms exist in isolated chloroplasts which control the relative distribution of absorbed light between the two photosystems and which respond to the content of the medium, the preillumination history, and the electron donor and acceptors present (see Myers, 1971; Ben-Hayyim and Avron, 1971; Duysens, 1972). Therefore, the wavelength dependence of the quantum yield of partial reactions cannot always be taken as an indication of the pigment content of the photosystem involved. Nevertheless, with reasonable assumptions the quantum yields of all partial reactions so far reported fit the general scheme presented in Fig. 1.

2.5. Oxidation–Reduction of Chain Components

Among the major experimental observations which led to the series-two-photosystem formulation, was the demonstration that under appropriate experimental conditions components of the electron transport system, such as Cyt f, are oxidized by light absorbed mainly by PS I and reduced by light absorbed mainly by PS II (Duysens and Amesz, 1962; Avron and Chance, 1966). This type of opposite response depending on the exciting wavelength has also been seen under appropriate conditions in P700, Cyt b, and Q (Fig. 1).

3. COMPONENTS OF THE ELECTRON TRANSPORT CHAIN

3.1. NADP$^+$

The final electron acceptor in the photosynthetic electron transport path, which provides the reducing equivalents necessary for the conversion of CO_2 to the level of carbohydrates, is generally considered to be NADP$^+$. The specificity for NADP$^+$, rather than NAD$^+$, is specified by the Fd-NADP$^+$ reductase which was demonstrated to be NADP$^+$-specific under most conditions (Shin, 1971; Forti, 1971). However, under some conditions NAD$^+$ can serve as an efficient final electron acceptor (Nelson and Neumann, 1969).

The photoreduction of NADP$^+$ from water has been observed to show both the red drop and the enhancement phenomena (see Avron, 1971), and for proceeding at maximal rates requires the addition of both Fd and Fd-NADP$^+$ reductase (see San Pietro and Black, 1965). It has therefore been concluded that both photosystems must be sensitized for this reaction to proceed.

3.2. Ferredoxin-NADP$^+$ Reductase

This enzyme is a flavoprotein containing FAD, and has been isolated in highly purified form from spinach leaves (see Bishop, 1971; Shin, 1971; Forti, 1971). The enzyme is rather specific for Fd or NADPH as electron donors, but rather non specific with regard to its electron acceptors, with dyes (ferricyanide, DCPIP), NAD$^+$ Cyt f, PC, NADP$^+$, and Fd all serving as reasonable electron acceptors. It has therefore been isolated under different names independently by several investigators.

Evidence for its functioning in the position suggested in Fig. 1 consists of (a) Its requirement for photoreduction of NADP$^+$ in chloroplasts depleted of its content (Davenport, 1963) (b) The rate of its photoreduction by isolated chloroplasts is linearly dependent upon the amount of Fd present in the chloroplasts (Avron and Chance, 1967), and (c) The oxidation of reduced Fd by NADP$^+$ in chloroplast preparations is fully inhibited by the addition of an antibody to the flavoprotein (see San-Pietro and Black, 1965; Avron and Chance, 1967).

The enzyme was recently shown to form a 1:1 complex with Fd, which completely and reversibly dissociates in high ionic strength media (see Shin, 1971; Boger, 1972). The fact that NADP$^+$ reduction in chloroplasts is inhibited by high-salt concentration and by pyrophosphate (Forti, 1970) may be related to this dissociation, but direct evidence on the obligate participation of the complex in the path of NADP$^+$ reduction is still lacking.

The best specific inhibitor of the enzyme so far described is phosphoadenosine diphosphate ribose (Ben-Hayyim *et al.*, 1967) with a K_i of about $1 \mu M$.

3.3. Ferredoxin

Originally identified under the misnomer photosynthetic pyridine nucleotide reductase (PPNR) by San Pietro (see San Pietro and Black, 1965), it was later isolated in crystalline form from a variety of sources (see Buchanan and Arnon, 1971). Ferredoxin is an iron-sulfur protein with a low redox potential ($E_0' = -0.42$ V). It is reduced by illuminated chloroplasts, or with H_2 + hydrogenase. Reduced Fd can serve as an electron donor to a variety of acceptors, including Cyt c, ferrihemoglobin, ferricyanide and its natural acceptor, the flavoprotein Fd-NADP$^+$ reductase.

Evidence for its functioning in the position indicated in Fig. 1 can be summarized as follows: (1) Its addition is required for NADP$^+$ reduction in depleted chloroplasts. (2) It is reduced by illuminated chloroplasts depleted of Fd-NADP$^+$ reductase under conditions which do not permit the reduction of NADP$^+$, Similarly, it is photoreduced even in the presence of an excess of the antibody to Fd-NADP$^+$ reductase which fully inhibits the reduction of NADP$^+$. (c) Its addition to depleted chloroplasts is sufficient for optimal rates of photoreduction of added Cyt c, while for NADP$^+$ reduction both Fd and the flavoprotein must be added. As mentioned above Fd forms a specific 1:1 complex with Fd-NADP$^+$ reductase, which may be functional in catalyzing electron flow.

3.4. Ferredoxin Reducing Substance (FRS)

Indirect evidence for the existence of a low-potential substance with a redox potential around -0.55 V between photosystem I and ferredoxin has been available for some time (see Avron, 1967, 1971). Recently Yocum and San-Pietro (1972) Trebst (1972) and Regitz and Oettmeier (1972) summarized the evidence for the presence of a factor which fulfills this position, and which San-Pietro termed ferredoxin reducing substance.

Chloroplasts depleted of FRS show an impaired ability to photoreduce $NADP^+$ in the presence of optimal amounts of Fd and the reductase. Addition of FRS restores much of the lost activity. Reduced FRS can reduce Fd in the dark. An antibody which reacts with the electron transport system between PS I and Fd has been described by Trebst and collaborators. The activity of this antibody can be preneutralized by reaction with isolated purified preparations of FRS.

Experimental difficulties prevented rapid progress in characterizing this important factor, but at present it is certainly the most well-characterized substance which precedes Fd in the electron transport path. Recently new factors have been described which were claimed to precede FRS, and so serve as the primary electron acceptor(s) of PS I (Hiyama and Ke, 1971; Malkin and Bearden, 1971).

3.5. P700

P700 is a Chl molecule(s) situated in a special environment which is responsible for the shift of its absorption peak to 700 nm. It accounts for about 0.25 % of the total Chl of higher plant leaves, and is detected by the bleaching of the absorption centered around 700 nm, due to its oxidation. It was originally described and thoroughly studied by Kok and collaborators (see Marsho and Kok, 1971). It has a redox potential of $E'_0 = +0.43$ V, is oxidized by light which sensitizes PS I and reduced by light which sensitizes PS II. The photooxidation proceeds down to liquid nitrogen temperature and when excited by a very intense flash of a very short duration was shown to be the fastest reaction observable in PS I. It is, therefore, generally assumed to occupy the position of the primary electron donor to PS I.

Recent experiments by Rurainski and Hoch (1972), in which the flux of electrons through P700 and into $NADP^+$ have been measured simultaneously cast doubt on the role of P700 as the exclusive electron donor to PS I. Under some conditions only a fraction of the electrons reaching $NADP^+$ seem to flow through P700.

3.6. Plastocyanin

Plastocyanin is a copper-protein originally isolated and thoroughly studied by Katoh and collaborators (see Katoh, 1971). It is present at a concentration equal to about 0.2 % of the total Chl, has a redox potential of $E'_0 = +0.37$ V and is easily photoreduced by isolated chloroplasts.

Its site of action is still a matter of some controversy. Most investigators place it in the position indicated in Fig. 1. This is based on evidence obtained from specific mutants (Gorman and Levine, 1966) in which it was shown that a PC-less mutant could still photoreduce but not photooxidize Cyt f, and that the latter activity could be restored in fragments by the exogenous addition of PC. Similar observations were also made in chloroplast preparations depleted of their PC (Hind, 1968; Avron and Shneyour, 1971). However, Knaff and Arnon (1969) claimed that in their hands a PC-depleted preparation could photooxidize Cyt f. Recently (Knaff, unpublished) these studies were extended, and it was shown that the photoreduction of Cyt f in these depleted preparations was accelerated by the addition of PC. Although these latter observations are difficult to explain in terms of the scheme presented in Fig. 1, it is nevertheless felt that the bulk of the evidence supports the position of PC indicated.

3.7. Cytochrome f

Cytochrome f was originally described, isolated and characterized by Hill and collaborators (see Bendall *et al.*, 1971). It has a redox potential of $E'_0 = +0.365$ V, and is again present at a concentration equal to about 0.25% of the total Chl.

It was the first component which was clearly shown to be located between the two systems (Duysens and Amesz, 1962). Its precise location is indicated by the data described above under PC which indicates that it precedes PC in the electron transport path, kinetic evidence indicating that a large pool of electron acceptors intervene between PS II and Cyt f (Avron and Chance, 1966), and inhibitor studies indicating that it is located beyond the site of action of DCMU and DBMIB (Böhme and Cramer, 1971; Gimmler and Avron, 1972; Bishop, 1972).

The site of ATP formation is also presently defined as preceding Cyt f and on less solid ground as following PQ (Avron and Chance, 1966; Ben-Hayyim and Avron, 1970a,b; Böhme and Cramer, 1972).

3.8. Plastoquinone

Plastoquinone was originally shown to occur in chloroplasts by Lester and Crane (1959). Its position between the two photosystems is indicated by its oxidation by light-sensitizing PS I and its reduction by light-sensitizing PS II. It is the only component present in an amount (5–10% of the total Chl) sufficient to account for the large pool of electron acceptors kinetically observed to occur between the photosystems and near PS II (Kok and Cheniae, 1966). Its exact location in Fig. 1 is based mostly on the data of Cramer *et al.* (1971) Cramer and Böhme (1972a), and Böhme and Cramer (1971), who showed that DBMIB, a PQ analog, inhibits the oxidation of Cyt b559 by PS I but not its reduction by PS II.

3.9. Other Cytochromes

Three other cytochromes have been described as playing a role in photo-synthetic electron transport, but their position has not been clearly defined as yet (see Bendall *et al.*, 1971; Wada and Arnon, 1971; Cramer *et al.*, 1971; Satoh and Katoh, 1972). These are the low and high potential Cyt b559 and Cyt b563. The former two seem to be interconvertible forms of the same cyto-chrome. Notably, the addition of FCCP converts the naturally most abundant high potential ($E_0' \cong +0.37$ V) Cyt b559 to the low potential ($E_0' \cong +0.06$ V) form. When so converted it behaves as if located between the two photo-systems in the position indicated in Fig. 1, but in its natural state its position is unclear.

Cytochrome b563 is the lowest potential cytochrome in photosynthetic material ($E_0' = -0.18$ V). It is easily observed in dithionite reduced minus ascorbate reduced difference spectra and equals in amount to the sum of the other cytochromes. Its function, however, is still very unclear. It has been suggested to be involved in cyclic electron flow around PS I (see Cramer and Böhme, 1972b; Hind and Olson, 1968), but the evidence is far from being complete.

3.10. Q,C550

Changes of Chl fluorescence yield in photosynthetic materials have been associated with the oxidation–reduction state of the primary electron accep-tor of PS II, termed Q with redox potential $E_0' = -0.035$ V (see Butler, 1966; Goedheer, 1972; Govindjee *et al.*, 1973). Assaying the redox state of Q by changes in fluorescence yield, it was shown to be reduced by excitation of PS II and oxidized by excitation of PS I. Its oxidation, but not its reduction is inhibited by DCMU.

Recent evidence indicates a possible identity between Q and a compound C550 showing an absorption change centered around 550 nm in chloroplasts (Knaff and Arnon, 1969; Erixon and Butler, 1971; Butler, 1972).

3.11. The Electron-Donating Side to Photosystem II

None of the components participating in the electron-donating side of PS II has been clearly defined, but several participants are strongly indicated.

Depletion of chloroplasts of chloride ions causes a marked reduction in their ability to catalyze photoinduced electron transport reactions with water as the electron donor, but not with a variety of other electron donors, which can be fully reversed by the addition of chloride ions (Bove *et al.*, 1963; Izawa *et al.*, 1969). It is, therefore, often assumed that chloride plays a role in the electron transport path donating electrons into PS II (see Cheniae, 1970).

Manganese ions have been recognized very early to be essential for photosynthetic activity (see Cheniae, 1970). However, successful removal and re-addition experiments have not been reported. The effect of manganese depletion has been clearly shown to be located on the electron donating side of PS II, since it did not affect photoreactions utilizing electron donors other than water.

3.12. Cyclic Electron Flow

In the presence of some electron carriers, such as phenazine methosulfate, pyocyanine, reduced DAD, high rate of phosphorylation are observed unaccompanied by any net electron flow (see Avron and Neumann, 1968). It is clear, therefore, that a cyclic electron flow is elicited by these electron carriers. The observations that this type of electron flow is insensitive to DCMU, and DBMIB Böhme and Cramer, 1971), shows a red-rise effect (Avron and Ben-Hayyim, 1969) and is inhibited in the absence of PC of P700, but not of Cyt f (see Levine, 1969), indicate that it is sensitized by only PS I and most probably involves an electron-carrying ability between a primary electron acceptor of PS I and PC. Some indication for the involvement of electron carriers not common to the noncyclic path, such as Cyt b_6, has already been mentioned, and is reinforced by data which indicates that the site of phosphorylation in the cyclic system may not be identical with that (one of those?) of the non-cyclic systems.

Evidence for the occurrence of cyclic electron flow *in vivo*, is abundant. However, it seems clear that it does not proceed via the same path as that occurring in the electron carrier mediated cyclic electron flow observed *in vitro*.

4. ELECTRON ACCEPTORS, DONORS, AND INHIBITORS

4.1. Electron Acceptors

In addition to the natural electron acceptor, $NADP^+$, discussed above, a variety of compounds can act as electron acceptors for the photosynthetic electron transport path.

As mentioned already, Cyt c, and ferrihemoglobin may serve as excellent electron acceptors receiving electrons from Fd. Depending on the preparative technique employed the electron accepting ability of these compounds can be shown to be completely or to a major extent dependent upon the addition of Fd.

Viologen dyes such as methyl viologen, benzyl-viologen, and diquat and other electron mediators such as anthraquinone-2-sulfonate (Trebst and Burba, 1967) and diaminodurene (Gromet-Elhanan and Redlich, 1970) act as electron acceptors at a position preceding Fd. When provided with an electron donor group such as ascorbate + DCPIP all these agents catalyze a DCMU-insensitive oxygen uptake in illuminated chloroplasts.

Ferricyanide and DCPIP seem to accept electrons mostly between the two photosystems, but under some conditions may also partially accept electrons from PS I. The former is indicated by the observation that with all electron donor systems tested to-date their reduction is DCMU-sensitive. However, under some conditions ferricyanide reduction has been claimed to show the enhancement effect (see Govindjee and Bazzaz, 1967, Myers, 1971) indicating the participation of both photosystems. Since DBMIB inhibits their reduction and their reduction is coupled to ATP formation, it is most likely that their normal site of electron acceptance is at the Cytochrome f–PC level (see Avron, 1967, and Gimmler and Avron, 1972).

4.2. Electron Donors

In addition to water which serves as the natural electron donor, a group of other electron donors may serve to compete or substitute for water as electron donors to PS II (see Cheniae, 1970).

Among them ascorbate (Ben-Hayyim and Avron, 1970b), hydroquinone, hydroxylamine, phenylenediamine, diphenylcarbazide (Vernon and Shaw, 1969), and manganous ions have been most commonly employed. The electron donation by these compounds is DCMU-sensitive, and they are most easily assayed in chloroplasts in which electron donation from water has been impaired (see below). Acceptors receiving electrons between the two photosystems or after PS I may be employed.

A variety of donors have also been employed which seem to donate electrons between the two photosystems. Electron donation by such donors is not impaired by DCMU, and they are mostly employed in the presence of fully inhibitory concentrations of DCMU. They can be assayed only in the presence of acceptors which receive electrons after PS I. Most commonly employed are ascorbate + DCPIP ascorbate + TMPD, ascorbate + DAD (see Gromet-Elhanan and Redlich, 1970).

At least one dismutation reaction has been described recently in which the same compound serves as both the electron donor and the electron acceptor (Shneyour and Avron, 1971; Vernon and Shaw, 1972). In a PS I-sensitized reaction 1,5-diphenylcarbazone is converted to a more oxidized and a more reduced product, simultaneously.

4.3. Electron Transport Inhibitors

Several inhibitors are available to more or less specifically block electron transport in chloroplasts (see Izawa and Good, 1972).

Phosphoadenosine diphosphate ribose, an analog of $NADP^+$, inhibits specifically the flavoprotein Fd-NADP reductase (Ben-Hayyim *et al.*, 1967). A specific antibody has also been prepared for this enzyme (see San Pietro and Black, 1965).

Disalicylidenepropanediamine(DSPD)and high salt concentrations inhibit at the Fd level (Trebst and Burba, 1967; Ben-Amotz and Avron, 1972).

DBMIB has recently been described as an inhibitor acting between Cyt b-559 and Cyt f, presumably as a competitor of PQ (Cramer *et al.*, 1971; Böhme and Cramer, 1971; Gimmler and Avron, 1972).

DCMU, Atrazine and HOQNO have been widely employed as specific inhibitors between Q and Cyt b559.

Few inhibitors and several treatments are available which inhibit the electron donating side to PS II. Hydroxylamine, and under some conditions other amines like ammonia or methylamine inhibit this part of the chain. However, since hydroxylamine serves also as a good electron donor beyond its site of inhibition it is not as useful as some of the other treatments.

Pretreatment of the chloroplasts with high concentrations of Tris is probably the most commonly employed procedure to inactivate the electron donating side of PS II, but removal of chloride ions, irradiation with ultraviolet light, and a mild heat treatment have also been employed (see Izawa and Good, 1972).

5. SUMMARY

Present information about the electron transport path in isolated chloroplasts fits into a linear sequence of steps which involve two photosensitized electron transport steps operating in series, and at least a dozen other electron carriers. These bridge across a potential gap of at least 1.3 V, and produce a reduced product with a redox potential of around -0.5 V. An impressive insight into the types and mode of action of the electron carriers involved has been gained during the last 15 years. Also, many experimental tools involving specific inhibitors, electron donors and acceptors, have been described. However, many unclear areas remain, awaiting further investigation.

REFERENCES

Avron, M. (1967). *Current Top. Bioenerg.* **2**, 1.
Avron, M. (1971). *In* "Structure and Function of Chloroplasts" (M. Gibbs, ed.), pp. 149–167. Springer-Verlag, Berlin and New York.
Avron, M., and Ben-Hayyim, G. (1969). *Progr. Photosynthesis Res.* **3**, 1185.
Avron, M., and Chance, B. (1966). *In* "Currents in Photosynthesis" (J. B. Thomas and J. C. Goedheer, eds.), pp. 455–463. Ad. Donker, Rotterdam.

Avron, M., and Chance, B. (1967). *Brookhaven Symp. Biol.* **19**, 149.
Avron, M., and Neumann, J. (1968). *Ann. Rev. Plant Physiol.* **19**, 137.
Avron, M., and Shneyour, A. (1971). *Biochim. Biophys. Acta* **226**, 498.
Ben-Amotz, A., and Avron, M. (1972). *Plant Physiol.* **49**, 244.
Bendall, D. S., Davenport, H. E., and Hill, R. (1971). *Methods Enzymol.* 327–344.
Ben-Hayyim, G., and Avron, M. (1970a). *Eur. J. Biochem.* **14**, 205.
Ben-Hayyim, G., and Avron, M. (1970b). *Europ. J. Biochem.* **15**, 155.
Ben-Hayyim, G., and Avron, M. (1971). *Photochem. Photobiol.* **14**, 389.
Ben-Hayyim, G., and Avron, M. (1972). *Methods Enzymol.* **24**, 293–297.
Ben-Hayyim, G., Hochman, A., and Avron, M. (1967). *J. Biol. Chem.* **242**, 2837.
Bishop, N. I. (1971). *Ann. Rev. Biochem.* **40**, 197.
Bishop, N. I. (1972). *Proc. 2nd. Int. Congr. Photosynthesis Res.*, (G. Forti *et al.*, ed.), pp. 459–468.
Böhme, H., and Cramer, W. A. (1971). *FEBS Lett.* **15**, 349.
Böhme, H., and Cramer, W. A. (1972). *Biochemistry* **11**, 1155.
Böhme, H., Reimer, S., and Trebst, A. (1971). *Z. Naturforsch.* **26b**, 341.
Bove, J. M., Bove, C., Whatley, F. R., and Arnon, D. I. (1963). *Z. Naturfosch* **18b**, 683.
Boger, P. (1972). *Proc. Int. Congr. Photosynthesis, 2nd.* (G. Forti *et al.*, ed.). pp. 449–458.
Buchanan, B. B., and Arnon, D. I. (1971). *Methods Enzymol.* 413–440.
Butler, W. L. (1966). *Current Top. Bioenerg.* **1**, 49.
Butler, W. L. (1972). *FEBS Lett.* **20**, 333.
Cheniae, G. M. (1970). *Ann. Rev. Plant Physiol.* **21**, 467.
Cramer, W. A., and Böhme, H. (1972a). *Biochim. Biophys. Acta* **256**, 358.
Cramer, W. A., and Böhme, H. (1972b). *Biochim. Biophys. Acta* **283**, 302.
Cramer, W. A., Fan, N. H., and Böhme, H. (1971). *Bioenergetics* **2**, 289.
Davenport, H. E. (1963). *Nature (London)* **199**, 151.
Duysens, L. N. M. (1964). *Progr. Biophys. Chem.* **14**, 1.
Duysens, L. N. M. (1972). *Biophys. J.* **12**, 858.
Duysens, L. N. M., and Amesz. J. (1962). *Biochim. Biophys. Acta* **64**, 243.
Emerson, R., and Lewis, C. M. (1943). *Amer. J. Bot.* **30**, 165.
Erixson, K., and Butler, W. L. (1971). *Biochim. Biophys. Acta* **234** 381.
Forti, G. (1970). *Biochem. Biophys. Res. Commun.* **140**, 107.
Forti, G. (1971). *Methods Enzymol.* **23**, 447–451.
Gimmler, H., and Avron, M. (1972). *Proc. Int. Congr. Photosynthesis, 2nd.* (G. Forti *et al.*, eds.), pp. 789–800.
Goedheer, J. C. (1972). *Ann. Rev. Plant Physiol.* **23**, 87.
Gorman, D. S., and Levine, R. P. (1966). *Plant Physiol.* **41**, 1648.
Govindjee, and Bazzaz, M. (1967). *Photochem. Photobiol.* **6**, 885.
Govindjee, Papageorgion, G., and Rabinowitch, E. (1973). *In* "Practical Fluorescence, Theory, Methods, and Techniques" (G. G. Guilbault, ed.), Dekker, New York. pp. 543–576.
Gromet-Elhanan, Z., and Redlich, N. (1970). *Europ. J. Biochem.* **17**, 523.
Hind, G. (1968). *Biochim. Biophys. Acta* **153**, 235.
Hind, G., and Olson, J. M. (1968). *Ann. Rev. Plant Physiol.* **19**, 249.
Hiyama, T., and Ke, B. (1971). *Proc. Nat. Acad. Sci. U.S.* **68**, 1010.
Hoch, G., and Martin, I. (1963). *Arch. Biochem.* **102**, 430.
Izawa, S. and Good, N. E. (1972). *Methods Enzymol.* **24**, 358–377.
Izawa, S., Heath, R. L., and Hind, G. (1969). *Biochim. Biophys. Acta* **180**, 388.
Katoh, S. (1971). *Methods Enzymol.* 408–413.
Knaff, D. B., and Arnon. D. I. (1969). *Proc. Nat. Acad. Sci. U.S.* **63**, 963.
Kok, B., and Cheniae, G. M. (1966). *Current Top. Bioenerg.* **1**, 1.
Kraayenhof, R. (1969). *Biochim. Biophys. Acta* **180**, 213.
Lester, R. L., and Crane, F. L. (1959). *J. Biol. Chem.* **234**, 2169.

Levine, R. P. (1969). *Ann. Rev. Plant Physiol.* **20**, 523.

Malkin, R., and Bearden, A. J. (1971). *Proc. Nat. Acad. Sci. U.S.* **68**, 16.

Marsho, T. V. and Kok, B. (1971). *Methods Enzymol.* 515–522.

Myers, J. (1971). *Ann. Rev. Plant Physiol.* **22**, 289.

Nelson, N., and Neumann, J. (1969). *J. Biol. Chem.* **244**, 1926.

Rabinowitch, E. (1945). "Photosyntheses and Related processes," Vol. I, pp. 51–56. Wiley (Interscience), New York.

Regitz, G., and Oettmeier, W. (1972). *Proc. Int. Congr. Photosynthesis Res., 2nd* (G. Forti *et al.*, ed.), pp. 499–506.

Rurainski, H. J., and Hoch, G. E. (1972). *Proc. Int. Congr. Photosynthesis 2nd* (G. Forti *et al.*, eds.), pp. 133–141. Junk, The Hague.

San Pietro, A., and Black, C. C. (1965). *Ann. Rev. Plant Physiol.* **16**, 155.

Satoh, K., and Katoh, S. (1972). *Plant Cell Physiol.* **13**, 807.

Shin, M. (1971). *Methods Enzymol.* **23**, 440–447.

Shneyour, A., and Avron, M. (1971). *Biochim. Biophys. Acta* **253**, 412.

Trebst, A. (1972). *Proc. Int. Congr. Photosynthesis, 2nd.* (G. Forti *et al.*, eds.), pp. 399–417. Junk, The Hague.

Trebst, A., and Burba, M. (1967). *Z. Pflanzenphysiol.* **57**, 419.

Vernon, L. P., and Shaw, E. R. (1969). *Plant Physiol.* **44**, 1645.

Vernon, L. P., and Shaw, E. R. (1972). *Plant Physiol.* **49**, 862.

Wada, K., and Arnon, D. I. (1971). *Proc. Nat. Acad. Sci. U.S.* **12**, 3064.

Yocum, C. F., and San Pietro, A. (1972). *Proc. Int. Congr. Photosynthesis, 2nd* (G. Forti *et al.*, eds.), pp. 477–489. Junk, The Hague.

8

Oxygen Evolution in Photosynthesis*

Pierre Joliot
and
Bessel Kok

ABBREVIATIONS

ANT 2p	2-(3-Chloro-4-trifluoromethyl)anilino-3,5-dinitrothiophene
CCCP	Carbonyl cyanide *m*-chlorophenylhydrazone
Chl	Chlorophyll
Cyt	Cytochrome
DCMU	3-(3,4-Dichlorophenyl)-1,1-dimethylurea
DCPIP	2,6-Dichlorophenol indophenol
P700	Chlorophyll a_1; reaction center of pigment system I
PS I (II)	Photosystem (pigment system) I (II)

*Supported in part by grants from the Atomic Energy Commission, Contract # AT(ll–1)–3326 and the National Science Foundation, Contract # NSF–C705.

1. INTRODUCTION

That in photosynthesis O_2 is evolved from water, as suggested by Wurmser (1930), Van Niel (1941), and others, became obvious when Hill and Scarisbrick (1940) showed that isolated chloroplasts can evolve oxygen while reducing ferrioxalate rather than CO_2. This hypothesis is supported by the experiments with H_2 ^{18}O (Ruben *et al.*, 1941).

As yet, little is known about the mechanism of the O_2 evolving system (cf. Cheniae, 1970). The oxidation of water, but not of artificial donors, requires the presence of chloride or other anions of strong acids (Izawa *et al.*, 1969), and bicarbonate ions (Stemler and Govindjee, 1973). Manganese appears to be a specific constituent of the O_2 evolving system; it occurs in an abundance of about six protein-bound atoms per trapping center. Mild heating or extraction procedures can remove four of these Mn atoms with a parallel loss of O_2 evolving capacity. However, the loss of this Mn fraction does not impair the photooxidation of electron donors other than water.

The formation of the O_2 evolving Mn catalyst requires light absorbed by PS II. Once photoactivated in a multiquantum process, it remains permanently active (Cheniae and Martin, 1967, 1970; Radmer and Cheniae, 1971).

The formation of an O_2 molecule from water requires the removal of 4 electrons:

$$2H_2O \longrightarrow 4e^- + 4H^+ + O_2 \qquad E_0' = 0.8 \text{ V} \qquad (1)$$

where E_0' is the oxidation reduction potential of the H_2O/O_2 couple at pH 7. In order to be able to oxidize water, the average potential of the four oxidizing equivalents must be higher than 0.8 V. The four individual steps, however, might well be at different potential levels. Since a single photoact transfers only one electron, this reaction requires the collaboration of four elementary photoacts, a most important problem to which much of this chapter will be devoted. Before this discussion of the kinetics of O_2 evolution, we will mention some aspects pertinent to the subject.

1.1. Photosystem II

Plant photosynthesis is driven by two series connected photoacts (see Chapter 7). One of these, PS II is directly connected with the O_2 generating system. In PS II, the energy of a red photon (~ 1.8 eV) is sufficient to carry out a charge separation which yields the strong oxidant Z^+ needed to oxidize water ($+0.8$ V). The primary electron acceptor of PS II, Q, presumably has a midpoint potential close to ~ 0 V, Cramer and Butler (1969), so that the energetic efficiency of the conversion would be $\sim 50\%$.

A small, flash-induced absorption decrease at 680 nm which rapidly returns in subsequent darkness ($t_{1/2} \sim 0.2$ msec) has been interpreted as the bleaching of Chl a_{II} which presumably sensitizes, but does not itself participate in, the primary charge transfer of PS II (Döring *et al.* (1967). (For further discussion, see Chapters 3, 5, 6, and 10 of this volume.)

An often used indicator of the early events in PS II is the yield of Chl a fluorescence. With all traps open, one finds a residual emission yield (F_0) which implies that not all singlet excitations reach the traps. With all traps closed (either Z or Q in the wrong redox state), the yield reaches a maximum (F_{max}) and under many conditions the yield of the "variable part" of the fluorescence (F-F_0) is inversely proportional to the degree of opening of the traps. Since, under most conditions strong oxidant Z occurs in the reduced state, the fluorescence yield (F-F_0) generally reflects the redox state of primary reductant Q (Duysens and Sweers, 1963). F becomes maximal (Q reduced) under reducing conditions and minimal (F_0) upon oxidation. (For other effects on Chl fluorescence, see Chapters 5 and 6 of this volume.)

On the reducing side of PS II, a pool (A) of about 10 equiv of plastoquinone per trapping center functions as the electron acceptor. This pool has been revealed by spectroscopic observation (Witt *et al.*, 1965; Amesz, 1965) as well as in other ways. One of these is responsible for a "gush" of O_2 which can be observed at the onset of illumination, under conditions of restricted PS I turnover (Blinks and Skow, 1938; P. Joliot, 1961).

Cytochrome b559 appears to be closely associated with PS II. Depending on conditions, a PS II-sensitized reduction *or* oxidation of this Cyt can be observed (see review by Hind and Olson, 1968). Its normally high midpoint potential of ~0.37 V (Bendall and Sofrova, 1971) is readily and irreversibly lowered to ~0.0 V by certain "uncouplers" (CCCP, antimycin A), NH_2OH, temperature shock, Tris extraction and detergents. No rapid turnover of this Cyt has as yet been observed and it may be located in a side path of the O_2 yielding reactions.

Photosystem II is intimately connected with the O_2 evolving system, as is suggested by the facts that both the fluorescence and chemiluminescence yields are affected by the state of the O_2 enzyme. Still, many agents have been reported to replace water as the electron donor—by reducing either Z^+ or one of the subsequent O_2 precursor states. The O_2-evolving mechanism is far more labile than the PS II trapping centers themselves (Döring *et al.*, 1969). Treatments such as aging, mild heating, high pH or poisons such as hydroxylamine destroy the capability to evolve O_2 while the photooxidation of artificial donors can still proceed with high efficiency (Cheniae and Martin, 1971b). Examples of such artificial electron donors are hydroxylamine, ascorbate, hydroquinone, phenylcarbazide, etc. (Trebst *et al.*; 1963, Yamashita and Butler, 1968; Habermann *et al.*, 1968; Vernon and Shaw, 1969). Other poisons, notably DCMU, specifically inhibit electron flow through PS II, presumably by blocking the oxidation of primary acceptor Q by its secondary acceptor.

1.2. Flash Yields of O_2; the Photosynthetic Unit

In their classical experiment, Emerson and Arnold (1932) measured the integrated rate of oxygen evolution induced by series of light flashes (~ 10 μsec). The amount of oxygen evolved per flash increased as a function of the energy of the flashes and of their spacing.

The maximum flash yield was about one O_2 molecule per 2000–2500 molecules of Chl. The authors concluded that a rate-limiting catalyst, present at low concentration, was involved. It later became clear that this rate-limiting enzyme was the photocenter itself (Gaffron and Wohl, 1936).

Emerson and Arnold's experiments thus led to the concept of the *photosynthetic unit*: a group of a few hundred associated Chl molecules which collect quanta into a trapping center where the photoconversion occurs. To compute the actual size of this unit, we must remember that four electrons are transferred per O_2 molecule produced and that two photoreactions are required for the transfer of one electron. Therefore the size of each PS I and each PS II unit is approximately $2000/8 \simeq 250$ Chl molecules.

1.3. Velocity Constants

In the Emerson–Arnold experiment the O_2 yields were half maximal at a dark spacing of ~ 40 msec between the flashes ($1°C$). This half-time is temperature-dependent and reflects the velocity of the reaction step which limits the overall rate of photosynthesis and which is located between the two photosystems ($P700^+ + Q^- \longrightarrow P700 + Q$). At room temperature, Kok *et al.* (1969), and Stiehl and Witt (1969) observed a half-time of ~ 10 msec for the photoreduction of the quinone pool which is responsible for the O_2 gush (~ 10 equiv). These results imply an overall reaction time of about a millisecond per equivalent moved from water to quinone at room temperature.

Vater *et al.* (1969) measured O_2 evolution and UV absorption changes at 335 and 265 nm in sequences of flash pairs. The spacing between pairs was long compared to the turnover time of the overall rate-limiting step and the dark time within each pair was varied. For both parameters a 0.6 msec half-time was observed for maximum effect of the second flash in each pair. Forbush and Kok (1968) reported a halftime of ~ 0.6 msec for the return of the fluorescence yield to F_0 after it had been raised to F_{max} by a flash. Presumably this time reflects the reoxidation of Q^- by the acceptor pool. P. Joliot *et al.* (1966), using modulated illumination, measured the phase delay of the evolved O_2. They computed a reaction time of 0.8 msec between the photoevent and the liberation of oxygen. Evidently the events in this system are rapid compared to the rate limiting steps between the two photosystems. Later in this chapter we will discuss some intermediate reaction steps in more detail.

1.4. Loss of O_2 Evolution Capacity in Darkness

The first measurements of the amounts of oxygen evolved by single flashes were performed by Allen and Franck (1955) who used a sensitive phosphorescence quenching method under strictly anaerobic conditions. No oxygen could be observed if a single (~ 1 msec) flash was given to algae which had been in dark for awhile. O_2 evolution did occur, however, after preillumination by a flash or by weak light. This phenomenon was confirmed by Wittingham and Bishop (1961, 1963) who interpreted their results as a cooperation between PS I and II.

P. Joliot (1960, 1961) studied the induction of oxygen evolution in dark-adapted material under more aerobic conditions. This author demonstrated a requirement of at least two photoacts for the evolution of oxygen after a dark period. This "priming" or "activation" was observed both in continuous light and in flashing light. The action spectrum of the activation process showed that only PS II was involved. For a more detailed discussion of these early experiments, see Kok and Cheniae (1966).

2. KINETICS OF O_2 EVOLUTION

2.1. Basic Observations and Hypothesis

The development of sensitive polarographic techniques (P. Joliot, 1960) and P. Joliot and A. Joliot, 1968) allowed precise measurement of the small amounts of O_2 evolved in very weak continuous illumination or by single, brief (10^{-5} sec) flashes. In principle a strong brief flash excites *all* trapping centers in the system once, so that all events occur stepwise and in synchrony, provided the system is uniform. Figure 1 shows an example of the yields of O_2 (Y_n: normalized to the yield after several flashes Y_{ss}) produced by a series of flashes given to a suspension of isolated chloroplasts which had been in darkness during 40 min. The basic features of the experiments are: the yields of the first two flashes (Y_1 and Y_2) are close to zero and the subsequent yields show a damped oscillation with a period of 4. The same basic features can be observed with all photosynthetic material capable of evolving oxygen. Figure 2 shows direct recordings of the rate of O_2 evolution induced by a modulated light beam of low intensity. Curve 1 (labeled O flash) is observed when the beam is given after a 10-min dark period. This time course reflects a lag which is due to the "activation" process discussed above.

Curves 2 and 3 were also observed after a 10 min dark period, but the modulated beam was preceded by one and two brief flashes, respectively. All three time courses reflect purely photochemical processes because the conversion is proportional to the number of quanta absorbed. In Expt. 1, the rate is initially zero, then increases and attains a steady state value after a small

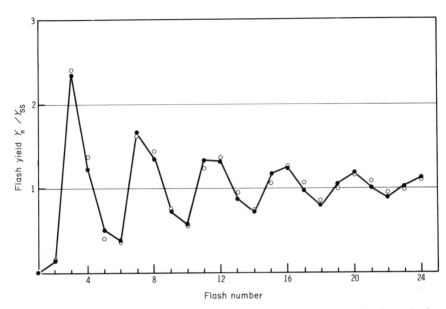

Fig. 1. (●) Flash yield sequence observed with isolated spinach chloroplasts after a 40-min dark period which followed a continuous illumination. (○) Prediction based on the following assumptions: $\alpha = 0.1$ for all steps, $\beta = 0.05$ for the conversions of S_0 and S_1, at time zero 25% of the centers are in the S_0 state and 75% in the S_1 state (see text; from Forbush *et al.*, 1971).

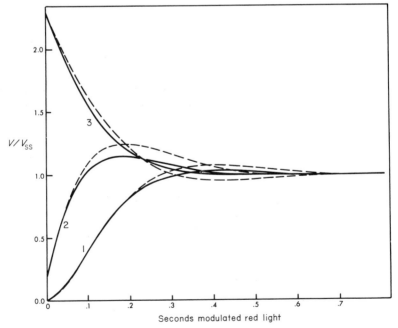

Fig. 2. (—) Time courses of the rate of O_2 evolution observed after 10 min dark following continuous light. The modulated rate-detecting beam was preceded by 0 (curve 1), 1 (curve 2), or 2 flashes (curve 3) (1 sec spacing). (––) Time courses predicted for $\alpha_{0.3} = 12$, $\beta_{0.1} = 0.05$. (From Forbush *et al.*, 1971.)

overshoot. Curve 2 shows a very low initial rate and a rapid rise to the steady state which includes an overshoot. Curve 3 shows that after 2 preilluminating flashes, the initial rate is ~ 2 times the steady state rate, which implies that initially the quantum yield of O_2 evolution is ~ 2 times the quantum yield during the steady state.

Note that in the experiment in Fig. 2 the initial rate after n preilluminating flashes is proportional to the amount of O_2 evolved in flash $(n + 1)$ in Fig. 1.

Several hypothesis have been proposed which are consistent with the observations of Figs. 1 and 2 (P. Joliot and A. Joliot, 1968; P. Joliot, 1968; P. Joliot et al., 1969; see Mar and Govindjee, 1972). All imply a mechanism to accumulate positive charges and four photoactive states of the reaction center. We will only discuss the proposal of Kok et al. (1970) simply because it appears to be the minimal hypothesis which can account for all presently available observations.

According to this hypothesis, the reaction chain which includes both the PS II (ZChlQ) photocenter and the O_2-evolving enzyme, can occur in four photoactive states (S) and at least 5 nonphotoactive states:

$$S_0 \xrightarrow[hv]{k_0 i} S_0' \longrightarrow S_1 \xrightarrow[hv]{k_1 i} S' \longrightarrow S_2 \xrightarrow[hv]{k_2 i} S_2' \longrightarrow S_3 \xrightarrow[hv]{k_3 i}$$

$$S_4 \longrightarrow S_0$$

$$2H_2O \qquad O_2 + 4H^+ \tag{2}$$

The states S_0, S_1, S_2, S_3, S_4 differ at least by the number of oxidizing equivalents in the system (as indicated by the respective indexes). The dark reactions $S_n' \longrightarrow S_{n+1}$ include reoxidation of the photochemical electron acceptor $(Q^- + A \longrightarrow Q + A^-)$ and the stabilization of the $+$ charge. The dark process $S_4 \longrightarrow S_0$ corresponds to the liberation of an O_2 molecule. To describe the observations in detail the following assumptions are made:

1. The reaction chains, each comprising a PS II trapping center and an associated O_2-evolving enzyme operate independently of each other, i.e., no exchange of positive charges or oxidants occurs between chains.

2. All four photosteps involve a 1-quantum process, have equal efficiency, and are sensitized by the same pigment system. The light-absorbing cross section and the photochemical rate constants are assumed to be equal for all centers.

3. The S_1 state is stable in the dark. States S_2 and S_3 are unstable and decay to state S_1. The lifetimes of S_2 and S_3 are long compared to the darksteps in Eq. (1), but less than ~ 15 min (the duration of the dark-adaptation period).

4. During each flash, a small fraction of the centers can undergo two transitions. $(S_n \longrightarrow S_{n+2})$. This implies that the duration of flashes as used in the experiment in Fig. 1 is not negligible compared to the duration of the intermediate dark steps in Eq. (1). The probability of the occurrence of these

"double hits" is defined as β. During each flash another fraction of the centers (α) does not undergo the transition $S_n \longrightarrow S_{n+1}$. Thus, starting from a state S_n, a flash induces a mixture of states S_n, S_{n+1}, and S_{n+2}. This disorder accounts for the damping of the oscillation of the flash yield.

In the above model the momentary distribution of the 4 S states can be computed from the observation of 4 subsequent flash yields. If coefficients α and β were 0, the first flash Y_1 of the series would measure the initial concentration of S_3; Y_2, Y_3, Y_4 would respectively measure the initial concentrations of S_2, S_1, and S_0. Introduction of the coefficient α leads to the following equations:

$$Y_1 = (1 - \alpha) S_3 \tag{3a}$$

$$Y_2 = (1 - \alpha)^2 S_2 + \alpha (1 - \alpha) S_3 \tag{3b}$$

$$Y_3 = (1 - \alpha)^3 S_1 + 2\alpha (1 - \alpha)^2 S_2 \tag{3c}$$

$$Y_4 = (1 - \alpha)^4 S_0 + 3\alpha (1 - \alpha)^3 S_1 \tag{3d}$$

Incorporation of the double hit coefficient β yields more complex equations which are not developed here.

With the above assumptions and using the values $\alpha = 0.1$ for all flashes and $\beta = 0.05$ for the first two flashes, satisfactory agreement could be obtained between the observations and predictions by computer modeling (Compare solid line with open circles in Fig. 1, and solid with dashed line in Fig. 2).

The distribution of the states at the beginning of the flash series was assumed to be: $S_0 = 0.25$, $S_1 = 0.75$, $S_2 = 0$, and $S_3 = 0$ (the sum of all states $= 1$).

The same initial distribution of the states was used to predict the activation curves in weak light without preilluminating flash (Fig. 2). The distributions of the S states after one and two flashes were computed from the data in Fig. 1.

Because the model of Kok *et al.* is a formal one, a great number of molecular representation can be, and have been proposed by Olson (1970), Renger (1971, 1972), Mar and Govindjee (1972). At this time, an insufficient number of specific chemical experiments has been performed to permit a choice among these models.

2.2 Discussion of the Basic Feature of Kok *et al.* Model

The agreement between the computed and experimental curves does not by itself prove that the various assumptions of the model are unique (cf. Mar and Govindjee, 1972). In the following sections, we will further scrutinize the assumptions and present quantitative information concerning the process described in Eq. (1).

2.2.1. *Independent Chains*

One of the main aspects of the model discussed above is that O_2 evolution occurs in independent centers. No diffusable intermediates are involved and the high energy states are stabilized in a quasi-solid state system. This assumption is supported by experiments in which the number of active reaction centers was decreased (Kok *et al.*, 1970). Partial inhibition with poisons (e.g., DCMU), which block a fraction of the reaction centers, decreases the flash yield of O_2 evolution. The same degree of inhibition was observed for all flash yields, regardless of flash number. In the experiment shown in Fig. 3, a relatively high concentration of DCMU was used. Although flash yields were decreased as much as 30-fold, the pattern of oscillation remained similar. The computed value for coefficient α and β are not significantly modified by addition of DCMU. Any type of cooperation between reaction centers would cause a severe distortion of the pattern due essentially to an increase of the "misses" coefficient.

Similar results were obtained when other methods were used to modify the number of reaction centers: UV irradiation, Mn deficiency (Kok *et al.*, 1970), and mutants with a low concentration of centers (Joliot *et al.*, 1973).

2.2.2. *Sensitization by System II*

P. Joliot (1965) observed that the wavelength dependence of the O_2 rate is constant during the activation process and corresponded to the sensitization

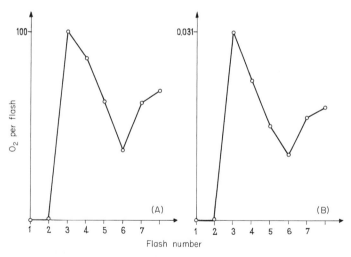

Fig. 3. Flash yield sequence observed with isolated chloroplasts in presence or absence of DCMU. Dark time between flashes 0.32 sec. (A) no DCMU; (B) DCMU 10^{-5} *M*. (P. Joliot, unpublished.)

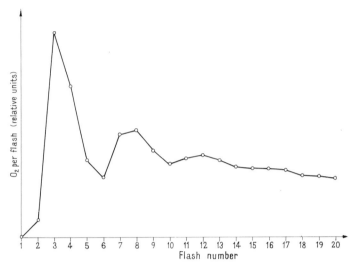

Fig. 4. Flash yield sequence obtained with *Chlamydomonas* mutant F_{14} which lacks PS I activity (no photoreduction of DCPIP, and no P700 oxidation). It does show a normal pool (A) of PS II acceptor. Dark time between flashes 0.32 sec. (Bennoun, unpublished.)

by PS II pigments. A similar conclusion was reached based on observations with mutants which lack PS I activity e.g., in Bishop's (1964) *Scenedesmus* mutant *8* (Kok, unpublished) or *Chlamydomonas* mutant F_{14} (Bennoun, unpublished). Figure 4 shows the oxygen yields of a series of flashes given to such a mutant. The first flashes of the sequence are normal. The decrease of the average O_2 yield and the abnormal damping are due to the depletion of the electron acceptor pool which in this mutant is not reoxidized by PS I.

2.2.3. Equal Quantum Yields of the Four Steps

The steady state oxygen production in weak light is attained when the rates of the four photochemical steps are equal. If k_n and α_n are the photochemical rate constant and the "misses" coefficient respectively, for the transition $S_n \longrightarrow S_{n+1}$ the steady state is defined by the equations:

$$k_0 i\,(1 - \alpha_0)\,S_0 = k_1 i\,(1 - \alpha_1)\,S_1 = k_2 i\,(1 - \alpha_2)\,S_2 = k_3 i\,(1 - \alpha_3)\,S_3 \quad (4)$$

Thus, in weak light, the steady-state distribution of states S depends upon the value of k_n which is proportional to the quantum yield of each photostep. In the case of series of short saturating flashes, the probability of the transition $S_n \longrightarrow S_{n+1}$ becomes independent of the quantum yield and dependent only upon the value of α_n.

The amount of O_2 evolved by a flash sequence after preillumination by a weak continuous light has been measured by P. Joliot *et al.* (1969), and more recently by Bouges-Bocquet *et al.* (1973). One observes only a small oscillation of the flash yield (less than $+5\%$) which proves that the steady state concentrations of S states attained in flashing light or continuous light are very similar. These results demonstrate that $k_0 \sim k_1 \sim k_2 \sim k_3$.

2.2.4. Double Hits and Misses

In order to explain the damping of the flash-yield oscillation, which suggests an increasing degree of disorder, Kok *et al.* (1970) introduced double hits (β) and misses (α). Double hits are predicted if the duration of the flash is not negligible compared to the dark relaxation time of the reaction $S'_n \longrightarrow S_{n+1}$. Experiments with saturating xenon flashes of different duration confirm this hypothesis (Fig. 5). If very brief flashes are used, no O_2 is evolved in the second flash (as is to be expected if all S_2 states disappear in the dark). In addition the amplitude of the oscillation is more pronounced and sustained longer with brief flashes than with longer flashes. Due to the "tail" following the main discharge, the duration of xenon flashes is difficult to evaluate precisely. Using very brief laser flashes, Weiss and Sauer (1970) (20–40 nsec) and Weiss *et al.* (1971) (2 nsec) found no O_2 evolution in the second flash, even with high flash energies.

Even when double hits are eliminated the flash yield oscillation is damped. The misses (α) cause the periodicity to be slightly higher than four, i.e., a delay in the oscillation. Such a phase retardation can be seen in the short flash experiment of Fig. 5. This point is discussed in detail by Forbush *et al.* (1971).

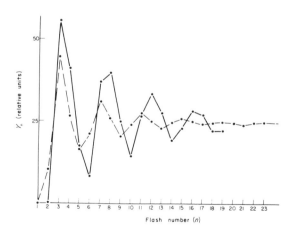

Fig. 5. Effect of the duration of the xenon flash on the flash yield sequence Y_n. Isolated spinach chloroplasts. (—) Short flash; (– –) long flash. Dark time between flashes 0.32 sec. Temperature: 20°C. Duration of the short flash: 2 μsec to one-third of the peak, 100 μsec including the tail of the flash. (From P. Joliot *et al.*, 1970.)

The number of misses remains constant even if the flash intensity is increased far above the saturation level. The misses can be explained by two possible mechanisms: (a) For a small fraction of the centers, the light transition

$$S_n \xrightarrow{\quad hv \quad} S'_n$$

does not occur, i.e., a fraction of S_n is not photosensitive. (b) A back-reaction annihilates the effect of the photoact.

It is interesting to note that in the other experiment in Fig. 5, using long flashes, the phase retardation due to misses is just about compensated by a phase advance caused by the double hits. Consequently, the periodicity remains very close to 4. We should stress that, in order to explain the damping of the oscillations, the double hits as well as the misses must occur randomly in all reaction centers. In the experiments shown in Fig. 5, the double hits were estimated to be $\sim 12\%$ for the long flash and $\leq 2\%$ for the short flash. The misses are generally of the order of 10–12% in isolated spinach chloroplasts, while *Chlorella* cells always show a higher percentage ($\sim 20\%$). At present, we do not know whether or not the probability of misses is different for the four steps.

2.2.5. Stability of the S_1 State

The hypothesis that the S_1 state is stable in the dark was introduced to explain why the yield of the third (and not the fourth) flash in a series is maximal. The simpler and therefore a priori more attractive hypothesis is that the centers deactivate completely to the ground state S_0. Models built on this assumption (P. Joliot, 1968a; Joliot *et al.*, 1969), however, were not supported experimentally. The Joliot's model is not able to interpret a variation of the ratio Y_4/Y_3 without significant change in value of Y_2. Also, this model cannot account for a value of Y_3 larger than 2 Y_{SS} which is also observed in Fig. 6.

Another possibility to explain the high value of Y_3, while assuming return to the S_0 state, is the assumption that the first flash after a long dark period induces two photoconversions in most of the PS II traps (Doschek and Kok, 1972). This assumption might be compatible with flashing light experiments as shown in Fig. 1. However, it predicts that the activation curve in weak light (Fig. 2) would essentially reflect a 4-quantum process—which it does not.

Probably the strongest argument for the stability of the S_1 state is the possibility to vary the ratio S_1/S_0 by appropriate preilluminations. Examples of this are given in Fig. 6 and are based on the following reasoning: During the steady state, the S states are distributed equally (25, 25, 25, 25, if the total number of states is 100). In subsequent darkness the S_2 and S_3 states decay to the S_1 state so that the distribution S_0, S_1, S_2, S_3 approaches 25, 75, 0, 0.

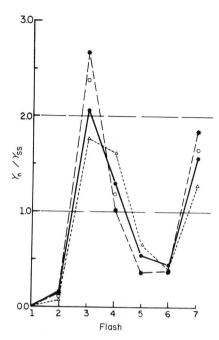

Fig. 6. Flash yield sequences observed with isolated chloroplasts after various pretreatments. ($\bullet$) Continuous light; ($*$) 1 flash; ($\bigcirc$) 2 flashes; ($\triangle$) 3 flashes. Observed after 30 min dark. Dark time between flashes in a sequence, 1 sec. In each experiment the procedure was: (1) 25 flashes to attain a steady state; (2) 5 min dark; (3) preillumination with (0, 1, 2, or 3) flashes; (4) 30 min dark; (5) 25 flashes; the yields of the first 7 of these are shown. (From Forbush *et al.*, 1971.)

When this distribution is reached, the chloroplasts are illuminated by 1, 2, or 3 flashes. During a new dark period of 10–20 min the system deactivates to different distributions of the states. (For instance, one would predict that one flash induces the distribution 0, 25, 75, 0 which in the dark will convert to 0, 100, 0, 0.) These new distributions can be determined by measuring a sequence of flash yields and then applying Eqs. (3a–3d). Since α and β are rather small, a variation of the ratio Y_3/Y_4 reflects a variation of the ratio S_1/S_0. Despite the uncertainties in α and β, it is evident that after one or two preilluminating flashes the ratio S_1/S_0 is higher than after deactivation from the steady state. On the other hand, preillumination with three flashes results in a ratio S_1/S_0 which is lower than the ratio attained after deactivation from the steady state.

As a first approximation, these results can be explained by assuming that (1) in darkness the S_2 and S_3 states deactivate to S_1, and (2) S_0 remains constant during the deactivation period. According to this hypothesis S_0 is made only in the light driven conversion

$$S_3 \xrightarrow{h\nu} S_4 \xrightarrow{H_2O} S_0 + O_2.$$

Hence, one or two preilluminating flashes on a deactivated system remove all S_0 states, so that in subsequent darkness S_2 and S_3 deactivate to S_1. Consequently the S_1/S_0 ratio will be very high. On the other hand, three preilluminating flashes produce a large number of S_0 states, so that after subsequent darkness the ratio S_1/S_0 will be low.

A careful scrutiny of the data in Fig. 6 reveals some second order but significant deviations from the model. For instance, one expects, but does not find, very similar distributions after one and two preilluminating flashes. This problem will be discussed later in more detail.

2.2.6. *Equilibration between* S_0 *and* S_1

Bouges (1971) studied the effects of dark times of longer duration than required for the complete loss of the S_2 and S_3 states. During such long dark periods, the ratio S_1/S_0 tends to approach a constant value rather independent of the preillumination. This result has been interpreted in terms of a dark equilibration between S_1 and S_0 which is slow compared to the disappearance of S_2 and S_3. Peculiarly, the equilibrium distribution (S_1/S_0) is close to the distribution (75/25) observed after deactivation from the steady state both in *Chlorella* and in isolated chloroplasts.

To check the interpretation that the two states equilibrate in the dark, Bouges-Bocquet (1973) has attempted to change the ratio S_1/S_0 by chemical oxidation and reduction. Figure 7 shows the effects of ferricyanide and of ascorbate + DCPIP on the oscillatory pattern in isolated chloroplasts. Computations based on these data suggest that the equilibrium distribution $S_1/S_0 = 80/20$ in this experiment is changed to 100/0 by ferricyanide and to 50/50 by reduced DCPIP. This allows us to estimate the midpoint potential of the reaction $S_1 + e^- \longrightarrow S_0$ to a value much lower than 0.8 V (potential of the O_2/H_2O couple). Thus, we conclude that a small amount of energy is fixed in the first oxidation step. This first oxidation step could be performed either slowly in the dark or by a photochemical transition. In contrast, the other three steps must conserve considerably more energy. Qualitative support for the view that much energy is stored in the S_2 and S_3 states, lies in the observation that the deactivation from both states is correlated with "luminescence" emission (see Chapter 5 of this volume).

2.2.7. *The Deactivation Process*

The disappearance of the S_2 and S_3 states in the dark can be monitored by measuring the yields of two flashes given after various dark intervals following a specific preillumination.

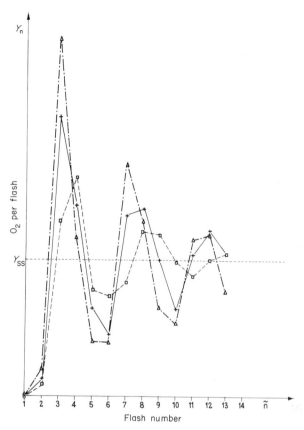

Fig. 7. Flash yield sequence observed with isolated spinach chloroplasts in the presence of oxidant or reductant. ($+$) Control; ($\square$) 0.1 mM DCPIP 1 mM ascorbate; ($\triangle$) 0.1 mM Ferricyanide (Bouges-Bocquet, 1973).

To best study the kinetics of the disappearance of S_2, a high concentration of this state can be generated by giving a single flash to dark adapted material. The two subsequent measuring flashes are given after various dark periods following this flash. Since the preilluminating flash does not generate S_3 states the concentration of S_2 can be computed using a simplified form of Eq. (2).

The deactivation of S_3 is best studied after a preillumination with two flashes, which produces a high concentration of this state. After various dark intervals a detecting flash is given which yields O_2 in proportion to the concentration of S_3. Figures 8 and 9 show the time courses of the deactivation of the states S_2 and S_3, as observed by P. Joliot *et al.* (1970) with *Chlorella* and chloroplasts, and by Forbush *et al.* (1971) with chloroplasts.

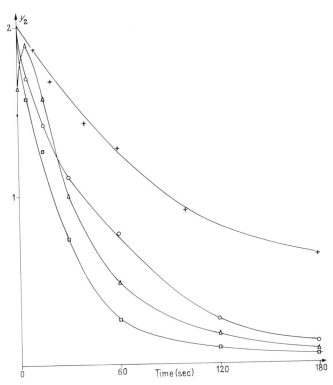

Fig. 8. Deactivation of state S_2 as a function of time. The concentration of S_2 is measured by the amount of O_2 (Y_2) evolved by the second detecting flash. (1) Preillumination by one flash (formation of state S_2 only): ($\square$) *Chlorella* cells (Joliot, *et al.*, 1970); ($\bigcirc$) spinach chloroplasts (Joliot *et al.*, 1970); ($+$) spinach chloroplasts (Forbush *et al.*, 1971). (2) Preillumination by two flashes (formation of states S_2 and S_3) ($\triangle$) *Chlorella* cells (Joliot *et al.*, 1970).

The data reveal the following points of interest: (1) Deactivation is faster in whole *Chlorella* cells than in isolated chloroplasts. In the latter material a considerable variation can be observed which is not readily correlated with the conditions of preparation. (2) In *Chlorella* cells, the half-life of S_3 is four times shorter than that of S_2. Such a large difference is not observed with isolated chloroplasts. (3) The decay of S_3 deviates considerably from first-order kinetics, especially in whole cells. In contrast the decay of S_2 is closer to first-order.

When deactivation of S_2 is studied after two preilluminating flashes which generate S_2 as well as S_3 (Fig. 8) a marked transient is observed which strongly suggest a one-step deactivation $S_3 \longrightarrow S_2$. Accordingly, Forbush *et al.* (1971) proposed a single-step deactivation mechanism $S_3 \longrightarrow S_2 \longrightarrow$ S_1. Again, the analysis encounters deviations between prediction and experi-

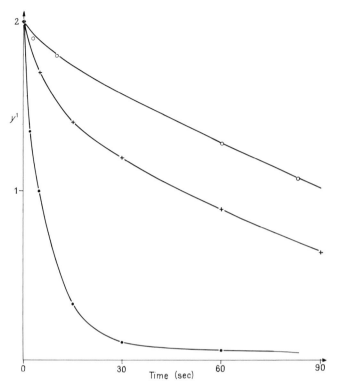

Fig. 9. Deactivation of state S_3 as a function of time. The concentration of S_3 is measured by the amount of O_2 evolved by the first detecting flash (Y_1). (●) *Chlorella* cells (Joliot *et al.*, 1970); (+) spinach chloroplasts (Joliot *et al.*, 1970); (O) spinach chloroplasts. (Forbush *et al.*, 1971.)

ment. These deviations have been interpreted by assuming that a small fraction of the S_3 or the S_2 states can yield S_0 states (P. Joliot *et al.*, 1970).

Nature of Reductant Involved in Deactivation. Deactivation involves the loss of oxidizing equivalents; thus the question arises which reductants acts as electron donors. One can a priori conceive two mechanisms:

1. The + charge is removed from S_2 or S_3 by a reductant which replaces water as the electron donor. This reductant can be an exogenous agent or a redox component of chloroplasts.

2. The + charge returns to the primary or secondary reductant of PS II, in a back-reaction of this photosystem.

Agents which operate according to the first mechanism are CCCP and derivatives (Renger, 1969, 1972; see Fig. 10). Some of these agents even at very low concentration (10^{-7} M), induce a rapid decrease of the positive charges from S_2 and S_3 so that deactivation can be accelerated drastically (up to 50-fold). Proof for this mode of action rests on the observation that in the

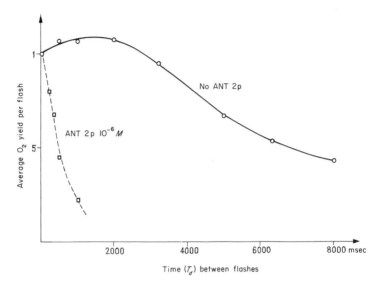

Fig. 10. Relative O_2 yield per flash as a function of dark time between flashes. Effect of 2-(3-chloro-4-trifluoromethyl)anilino-3,5 dinitrothiophene (Ant 2p) on the rate of deactivation (squares). ($\bigcirc$) Control experiment. (Renger *et al.*, unpublished.)

presence of these agents, the normal flash yield oscillation can be observed, provided the spacing between the flashes is very short. Recent observations (Renger *et al.*, 1973) suggest that these substances operate in a cycle between S_2 and S_3 and a chloroplast reductant which is not Q^-. It is not yet clear whether they act as catalysts undergoing a change of redox state, or modify some structural property of the chloroplast which favors a redox exchange between the electron donor of PS II and another component of the chain.

The second mechanism, a back reaction within PS II, is probably predominant in whole cells and in chloroplasts in the absence of exogenous electron donors. The rate of loss of the S_3 state depends upon the degree of reduction of the primary and secondary electron acceptors Q and A. Lemasson and Barbieri (1971) reported a faster decay of the S_3 state after preillumination with 650 nm light which reduces an appreciable fraction of the quinone pool, than after preillumination with 700 nm light which does not. An unexpected result of this experiment was that the loss of S_2 was not greatly affected by the redox state of the A pool. In chloroplasts without electron acceptor, most of the quinone pool becomes reduced upon illumination. Under these conditions the half life of the S_3 state is only 0.5 sec (i.e.. ~20 times shorter than under normal conditions; Kok, 1972).

2.2.8. Relaxation of the System after Flashes

The general principle used to analyze the relaxation time between each photostep is the use of a pair of flashes and varying the dark time between the two: If the dark time is short compared to the relaxation time of the system, the pair will act like a single flash. On the other hand, an increase of the spacing beyond a value which allows complete relaxation will have no further effect.

Two of the four relaxation times related to O_2 evolution are relatively easy to measure and compute: For $S'_1 \longrightarrow S_2$ the distance between the first two flashes is varied and the yield of the third flash (given a constant time after the pair) is measured. For the transitions $S'_2 \longrightarrow S_3$, the distance between the second and third flash is varied. The first measurements of Kok et al. (1970) showed clearly different relaxation times for these two steps (200 and 400 μsec, respectively, at room temperature). Zankel (see Kok, 1972) observed closely corresponding time courses for the return of the Chl fluorescence yield after the first and second flash. Bouges (1972) and Bouges-Bocquet

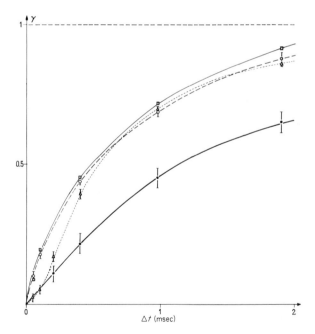

Fig. 11. Dark relaxation kinetics $S'_n \longrightarrow S_{n+1}$ observed with isolated spinach chloroplasts. $\gamma_n(\Delta t)$ represents the fraction of centers in the S'_n state which underwent the transition $S'_n \longrightarrow S_{n+1}$ during dark time Δt. ($\square$) $\gamma_0 (\Delta t)$; ($\bigcirc$) $\gamma_1 (\Delta t)$; ($\triangle$) γ_2 ($\bullet$) $\gamma_3 (\Delta t)$. (Bouges-Bocquet, 1973.)

(1973) performed similar experiments and also studied the relaxation times of the processes $S'_0 \longrightarrow S_1$ and $S'_3 \longrightarrow S_4$. Especially the latter study necessitated precise experiments and complex computations. The results, shown in Fig. 11, reveal about the same time constant for the transitions $S'_0 \longrightarrow S'_1$ and $S'_1 \longrightarrow S_2$ (half-time 400 μsec), neither reaction being first-order. Also, the energy of activation of this reaction proved to be very low in the initial phase and to increase in the final phase. The identical, rapid relaxations probably reflect the reoxidation of the acceptor Q^- of PS II. Bouges observed a somewhat longer time constant for the transition $S'_2 \longrightarrow S_3$, although shorter than the one reported by Kok *et al.* (1970). Typical for this transition is a lag of 100 μsec at the beginning. This lag indicates that at least two dark steps operate in series or in parallel.

The transition $S'_3 \longrightarrow S_4 \longrightarrow S_0$ is significantly slower than the others (half-time $= \sim 1$ msec) and proceeds close to first-order. This slow conversion reflects reaction with water molecule, the liberation of O_2 and 4 protons and the return of the system to the ground state. The computed half-time is very similar to the half-time of the limiting dark step in the formation of O_2 which was computed earlier by Joliot *et al.* (1966) from phase measurements in modulated light.

An important consequence of these measurements is that the double hit correction differs for the different transitions and is mainly important for states S_0 and S_1.

3. RELATED PROCESSES

3.1. Delayed Light Emission

The delayed light emission (Strehler and Arnold, 1951) is generally thought to be due to a back reaction of PS II (see Chapter 5 of this volume). Observations made in very different time ranges following illumination (10 μsec to 10 min) show clear correlations with the deactivation process. In the time interval 10 msec to 10 min, the luminescence intensity appears to be correlated with the *rate* of the conversions $S_3 \longrightarrow S_2$ and $S_2 \longrightarrow S_1$, the latter conversion being more important in the slower components of the emission (Barbieri *et al.*, 1970; P. Joliot *et al.*, 1970). In the shorter time range (< 1 msec), the predominant causes of luminescence emission are the back-reaction of S'_3 and S_4 (Zankel, 1971). Recent studies of Hardt and Malkin (1972) show that also the luminescence components which can be stimulated by salt, pH, or temperature jumps is correlated with the abundance of the more oxidized states. These phenomenon are discussed in more detail by Lavorel in Chapter 5 of this volume.

3.2. Fluorescence Emission

According to the hypothesis of Duysens and Sweers (1963), the Chlorophyle fluorescence yield is determined by the concentration of quencher Q, the electron acceptor of system II. However, this picture is now complicated by the fact that the yield is also a function of the distribution of the S states (Delosme, 1971; P. Joliot and A. Joliot, 1971).

Every photoreaction $S_n \longrightarrow S'_n$, $Q \longrightarrow Q^-$ is accompanied by an increase of the fluorescence yield. If this yield were determined solely by the state of Q, these increases should be independent of the flash number. This however, is not observed: the transitions $S_0 \longrightarrow S'_0$ and $S_1 \longrightarrow S'_1$ are accompanied by a large increase of the fluorescence yield, which approaches the increase which occurs in the presence of DCMU. On the other hand, the increase of the fluorescence yield in the transitions $S_2 \longrightarrow S'_2$ and $S_3 \longrightarrow S'_3$ is significantly smaller (Fig. 12). To interpret this, the authors have proposed that the photochemical center comprises two donor–acceptor couples with different quenching porperties, Q_1Z_1 and Q_2Z_2, each of which

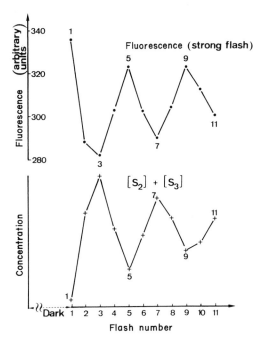

Fig. 12. Correlation between the flash-induced change of the fluorescence yield and the concentration of $S_2 + S_3$ at the onset of each flash in a sequence. The fluorescence yield is measured at the end of each flash in the sequence. $S_2 + S_3$ concentrations were determined in separate O_2 measurements during a series of flashes. (From Delosme, 1971.)

can store two positive charges. In the S_0 and S_1 states, the two photoacts involve Z_1Q_1, in the S_2 and S_3 states the photoacts involve Z_2Q_2.

This hypothesis is summarized by the following scheme in which the dark steps $(S'_n \longrightarrow S_{n+1})$ are not represented:

$$
\begin{array}{ccccccccc}
Z_1 \ \ Q_1 & & Z_1^+ \ \ Q_1 & & Z_1^{2+} \ \ Q_1 & & Z_1^{2+} \ \ Q_1 & & Z_1^{2+} \ \ Q_1 \\[2pt]
& \xrightarrow{h\nu} & & \xrightarrow{h\nu} & & \xrightarrow{h\nu} & & \xrightarrow{h\nu} & \\
\text{Chl} & & \text{Chl} & & \text{Chl} & & \text{Chl} & & \text{Chl} \\[6pt]
Z_2 \ \ Q_2 & & Z_2 \ \ Q_2 & & Z_2 \ \ Q_2 & & Z_2^+ \ \ Q_2 & & Z_2^{2+} \ \ Q_2 \\[4pt]
(S_0) & & (S_1) & & (S_2) & & (S_3) & & (S_3)
\end{array}
$$

$2\,H_2O \quad \nearrow$
$\quad \searrow O_2 + 4\,H^+$

(5)

At the basis of this model is the assumption that Z_1Q_1 reacts preferentially to Z_2Q_2, Q_1 being a more efficient quencher of fluorescence than Q_2. Only one of the two quenchers can be reduced during a short saturating flash.

3.3. Proton Liberation

In the reaction

$$2H_2O \longrightarrow O_2 + 4H^+$$

four protons are released per O_2 evolved. If we are to understand the mechanism of this process, it is important to know how these four protons are released: one in each S state transition, in pairs, or all four at the same time in a terminal "concerted mechanism."

Using a sensitive pH technique, Fowler (1973) monitored the acid produced by each flash in a sequence given after a dark period. He found the protons to be released in synchrony with oxygen (Fig. 13, full line). An interruption of the dark deactivation with 1, 2, or 3 flashes (as in experiment Fig. 6) changed the oscillation pattern in a way similar to the change of the O_2 oscillation pattern. The dashed line in Fig. 13 shows a three-flash preillumination experiment.

In experiment, Fig. 13, no electron acceptor was added; 10 mM methylamine was added to (1) annihilate the alkaline pH gradient due to the electron transport through plastoquinone, and (2) allow the protons, made in O_2 evolution, to escape from the thylakoid. This procedure demonstrated the oscillation of the proton yield most clearly. Essentially the same phenomena are observed using other uncouplers like gramicidin which make the membrane permeable for protons but do not necessarily suppress

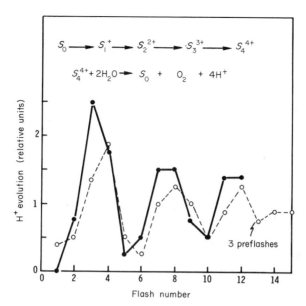

Fig. 13. Flash yield of proton liberation observed with isolated chloroplasts in the presence of methylamine. Flash spacing ∼1 sec. (—) Dark-adapted chloroplasts; (−−) after 5 min darkness following a preillumination with 3 flashes (Fowler, 1973).

the "classical" proton uptake. The oscillation then is superposed on this uptake.

It should be stated though, that, compared to the O_2 yield oscillation the first two flashes tend to behave anomalously, the first proton flash yield being somewhat erratic, the second consistently too high. The latter aspect could imply that one of the four protons is released in the step $S_2 \longrightarrow S_3$ but alternate, a priori more attractive, explanations for the high yields of the second (and other even numbered flashes) seem to be available. Thus for the moment we adhere to the concept of charge accumulation, followed by the decomposition of water in a final concerted event:

$$S_4 + 2H_2O \longrightarrow S_0 + O_2 + 4H^+ \qquad (6)$$

The fact that without uncouplers the oscillation of the proton yield is not observed implies that the proton liberation, and thus the O_2 evolution process, occur inside the thylakoid membrane.

The release of protons and O_2 inside the thylakoid membrane has been hypothesized earlier by Mitchell (1966) and other proponents of the chemiosmotic view of photophosphorylation (see Witt, 1971; also see Chapter 9, of this volume).

4. DONOR REACTIONS

Bennoun and A. Joliot (1969) were able to measure the rate of photo-oxidation of hydroxylamine with the modulated polarograph. This process is observed in the presence of high hydroxylamine concentrations (>0.4 mM) which irreversibly inhibit the O_2-evolving enzyme. Cheniae and Martin (1970) observed that such concentrations of hydroxylamine remove a fraction of the manganese associated with the PS II photocenters. In weak light, no activation lag is observed for the photooxidation of hydroxylamine. Also, when dark-adapted material is illuminated by a series of flashes, the amount of hydroxylamine oxidized in each flash does not vary with the number of the flash. These data therefore showed clearly that the 4-quantum mechanism is specific for the O_2 evolving mechanism. Bennoun (1972) and Bouges (1971) also studied the effect of lower concentrations of hydroxylamine (1 μM to 0.1 mM). At these concentrations the oxygen-evolving enzyme is not irreversibly inhibited, but at least one hydroxylamine molecule is bound to each reaction center. Studying the amounts of O_2 evolved during series of flashes B. Bouges observed that a low concentration of hydroxylamine induced a phase retardation of two flashes. These data can be interpreted in two ways:

1. Hydroxylamine reduces state S_1 to S_0 and one hydroxylamine molecule is bound to the center.

2. Hydroxylamine does not reduce S_1 to S_0 and two molecules of hydroxylamine are bound to the center. Measurements of the dark relaxation time between flashes performed recently by Bouges-Bocquet (1973) favors the second hypothesis. Bennoun (1972), observed that the binding of hydroxylamine at low concentration is highly specific. In contrast the inhibition observed at high concentration can be obtained with several substituted hydroxylamines (Cheniae and Martin, 1971a,b). A simple interpretation of the low concentration effect of hydroxylamine is to suppose that this molecule can replace water, being bound to a specific site of the PS II centers.

5. SUMMARY

Little is known about the biochemistry of the mechanism of O_2 evolution. For this reason, most of our present knowledge comes from kinetic information which lead to formal interpretations discussed here.

One of the functional problems is to understand how four photoreactions which correspond to the transfer of four electrons can cooperate in order to produce one O_2 molecule.

When dark-adapted photosynthetic material is submitted to a series of short (10^{-5} sec) saturating flashes, one observes that the amount of O_2 evolved per flash oscillates with a periodicity of four.

From these experiments, Kok *et al.* (1970) conclude that cooperation between the four photoreactions occurs at the level of the same photocenter, each one being independent of the others. One must assume that the electron donor of PS II centers can accumulate four positive charges or oxidizing equivalents before O_2 is evolved.

Different kinetics parameters have been precisely measured and discussed: e.g., lifetime of the oxidized form of the donor and dark time between photoacts.

Several other phenomena such as delayed light emission, fluorescence yield, and pH changes appear to be more or less directly related to the charge accumulation mechanism.

REFERENCES

Allen, F., and Franck, J. (1955). *Arch. Biochem. Biophys.* **58**, 510.

Amesz, J. (1965). *Biochim. Biophys. Acta* **79**, 257.

Barbieri, G., Delosme, R., and Joliot, P. (1970). *Photochem. Photobiol.* **12**, 187.

Bennoun, P. (1972). Thesis, Univ. of Paris, France.

Bennoun, P., and Joliot, A. (1969). *Biochim. Biophys. Acta* **189**, 85.

Bendall, D. S., and Sofrova, D. (1971). *Biochim. Biophys. Acta* **234**, 371.

Bishop, N. I. (1964). *Rec. Chem. Progr.* **25**, 181.

Blinks, L. R., and Skow, R. K. (1938). *Proc. Nat. Acad. Sci. U.S.* **24**, 413.

Bouges, B. (1971). These de Troisieme Cycle, Paris, France.

Bouges, B. (1972). *Biochim. Biophys. Acta* **256**, 381.

Bouges-Bocquet, B. (1973). *Biochim. Biophys. Acta* **314**, 250.

Bouges-Bocquet, B., Bennoun, P., and Taboury, J. (1973). *Biochim. Biophys. Acta.* **325**, 247.

Cheniae, G. M. (1970). *Ann. Rev. Plant. Physiol.* **21**, 467.

Cheniae, G. M., and Martin, I. F. (1967). *Biochem. Biophys. Res. Commun.* **28**, 89.

Cheniae, G. M., and Martin, I. F. (1970). *Biochim. Biophys. Acta* **197**, 219.

Cheniae, G. M., and Martin, I. F. (1971a). *Biochim. Biophys. Acta* **253**, 167.

Cheniae, G. M., and Martin, I. F. (1971b). *Plant Physiol.* **47**, 568.

Cramer, W. A., and Butler, W. L. (1969). *Biochim. Biophys. Acta* **172**, 503.

Delosme, R. (1971). *Proc. Int. Conf. Photosynthesis Res., 2nd, Stressa Italy* **1**, p. 189.

Döring, G., Stiehl, H., and Witt, H. (1967). *Z. Naturforsch* **22b**, 639.

Döring, G., Renger, G., Vater, J., and Witt, H. *Z. Naturforsch.* **24b** (1969).

Doschek, W. W., and Kok, B. (1972). *Biophys. J.* **12**, 832.

Duysens, L. N., and Sweers, H. E. (1963). *In* "Studies on Microalgae and Photosynthetic Bacteria," p. 353. Univ. Tokyo Press, Tokyo.

Emerson, R., and Arnold, W. (1932). *J. Gen. Physiol.* **15**, 391; **16**, 191.

Fowler, C. F. (1973). *Biophys. Soc. Abstr. 17th Annu. Meeting, Columbus, Ohio* p. 64a.

Forbush, B., and Kok, B. (1968). *Biochim. Biophys. Acta* **162**, 243.

Forbush, B., Kok, B., and McGloin, M. (1971). *Photochem. Photobiol.* **14**, 307.

Gaffron, H., and Wohl, K. (1936). *Naturwissenshaften* **24**, 81, 103.

Habermann, H. M. Handel, M. A., and McKellar, P. (1968). *Photochem. Photobiol.* **7**, 211.

Hardt, H., and Malkin, S. (1972). Paper presented at *Int. Congr. Photobiol., 6th, Bochum, Germany.*

Hill, R., and Scarisbrick, R. (1940). *Nature (London)* **146**, 61.

Hind, G., and Olson, J. M. (1968). *Ann. Rev. Plant Physiol.* **19**, 249.

Izawa, S., Heath, R., and Hind, G. (1969). *Biochim. Biophys. Acta* **180**, 338.

Joliot, P. (1960). Thèse de Doctorat es Sci. Phys., Paris, France.

Joliot, P. (1961). *J. Chim. Phys.* **58**, 584.

Joliot, P. (1965). *Biochim. Biophys. Acta* **102**, 116.

Joliot, P. (1968). *Photochem. Photobiol.* **8**, 451.

Joliot, P., and Joliot, A. (1968). *Biochim. Biophys. Acta* **153**, 625.

Joliot, P., and Joliot, A. (1971). *Proc. Int. Congr. Photosynthesis Res. 2nd, Stressa* **1**, 26.

Joliot, P., Hofnung, M., and Chabaud, R. (1966). *J. Chim. Phys.* **10**, 1423.

Joliot, P., Barbieri, G., and Chabaud, R. (1969). *Photochem. Photobiol.* **10**, 309.

Joliot, P., Joliot, A., Bouges, B., and Barbieri, G. (1970). *Photochem. Photobiol.* **12**, 287.

Joliot, P., Bennoun, P., and Joliot, A. (1973). *Biochim. Biophys. Acta* **305**, 317.

Kok, B. (1972). Paper presented at *Int. Congr. Photobiol., 6th, Bochum, Germany.*

Kok, B., and Cheniae, G. M. (1966). *Current Top. Bioenerg.* **1**, 1.

Kok, B., Joliot, P., and McGloin, M. (1969). *Progr. Photosynthesis Res.* **2**, 1042.

Kok, B., Forbush, B., and McGloin, M. (1970). *Photochem. Photobiol.* **11**, 457.

Lemasson, C., and Barbieri, G. (1971). *Biochim. Biophys. Acta* **245**, 386.

Mar, T., and Govindjee (1972). *J. Theor. Biol.* **36**, 427.

Mitchell, P. (1966). *Biol. Rev.* **41**, 445.

Olson, J. M. (1970). *Science* **168**, 438.

Radmer, R., and Cheniae, G. (1971). *Biochim. Biophys. Acta* **182**, 166.

Renger, G. (1969). *Naturwissenshaftern* **56**, 370.

Renger, G. (1971). *Z. Naturforsh.* **26b**, 149.

Renger, G. (1972a). *Eur. J. Biochem.* **27**, 259.

Renger, G. (1972b). *Biochim. Biophys. Acta* **256**, 428.

Renger, G., Bouges-Bocquet, B., and Delosme, R. (1973). *Biochim. Biophys. Acta* (in press).

Ruben, S., Randall, M. Kamen, M. D., and Hyde, J. L. (1941). *J. Amer. Chem. Soc.* **63**, 877.

Stemler, A., and Govindjee (1973). *Plant Physiol.* **52**, 119.

Stiehl, H. H., and Witt, H. T. (1969). *Z. Naturforsch.* **24b**, 1588.

Strehler, B. L., and Arnold, W. (1951). *J. Gen. Physiol.* **34**, 809.

Trebst, A., Eck, H., and Wagner, S. (1963). *In* "Photosynthetic Mechanisms of Green Plants" (B. Kok and A. J. Jagendorf, eds.), p. 174. National Academy of Science, National Research Council, Washington, D. C.

Van Niel, C. B. (1941). *Advan. Enzymol.* **1**, 263.

Vater, J., Renger, G., Stiehl, H. H., and Witt, H. T. (1969). *Progr. Photosynthesis Res.* **2**, 1006.

Vernon, L. P., and Shaw, E. R. (1969). *Biochem. Biophys. Res. Commun.* **36**, No. 6.

Weiss, C., and Sauer, K. (1970). *Photochem. Photobiol.* **11**, 495.

Weiss, C., Kenneth, J., Solnit, T., and Von Gutfeld, R. J. (1971). *Biochim. Biophys. Acta* **253**, 298.

Whittingham, C. P., and Bishop, P. M. (1961). *Nature (London)* **192**, 426.

Whittingham, C. P., and Bishop, P. M. (1963). Nat. Acad. Sci., Nat. Res. Council Publ. **1145**, 371.

Witt, H. T. (1971). *Quart. Rev. Biophys.* **4**, 365.

Witt, H. T., Döring, G., Rumberg, B., Schmidt-Mende, P., Siggel, U., and Stiehl, H. M. (1965). *Brookhaven Symp. Biol.* **19**, 161.

Wurmser, R. (1930). "Oxidations et Reductions," p. 30. Presses Univ. de France, Paris.

Yamashita, T., and Butler, W. L. (1968). *Plant Physiol.* **43**, 1978.

Zankel, K. L. (1971). *Biochim. Biophys. Acta* **245**, 373.

9

Mechanism of Photophosphorylation

André T. Jagendorf

ABBREVIATIONS

ADP	Adenosine diphosphate
ATP	Adenosine triphosphate
CCCP	Carbonyl cyanide *m*-Chlorophenylhydrazone
CF	"Coupling factor" protein of any organelle
CF_1	"Coupling factor" protein of higher plant chloroplasts
Chl	Chlorophyll
Cyt	Cytochrome
DCCD	Dicyclohexylcarbodiimide
DCMU	3-(3,4-Dichlorophenyl)-1,1-dimethylurea (the herbicide, Diuron)
DNP	2,4-Dinitrophenol
DTNB	Bisdithionitrobenzene sulfonic acid
DTT	Dithiothreitol
$\Delta\psi$	Membrane potential component of the protomotive force
EDTA	Ethylenediaminetetraacetic acid
FCCP	Fluorocarbonyl cyanide phenylhydrazone
NEM	*N*-Ethylmaleimide
P_i	Inorganic phosphate
pmf	Protonmotive force (electrochemical activity of hydrogen ions)
PP_i	Inorganic pyrophosphate
PS I (II)	Photosystem I (II)

1. COUPLING OF ELECTRON TRANSPORT AND PHOSPHORYLATION

1.1. Basic Observations

The transduction from the energy of electron transport in a membranous organelle to that of the dehydro bond of ATP was first discovered for mitochondria in 1937, but not until considerably later in spinach chloroplasts (Arnon *et al.*, 1954) or in chromatophores from photosynthetic bacteria (Frenkel, 1954; Geller, 1957). This chapter is concerned with this energy transformation, and also with the more recently discovered probable intermediate stage of proton and ion translocation.

The title of the chapter is premature in that the chemistry involved in ATP synthesis is not known, and the exact nature of the coupling between electron transport and ATP synthesis is not fully established. Approaches to these questions are well advanced, however. Any valid theory has to explain a number of familiar characteristics of ATP-synthesizing particles.

1. Only particles or vesicles are able to synthesize ATP from ADP and P_i, in the mode linked to electron flow through membrane-bound carriers. Over the years in the field of (mitochondrial) oxidative phosphorylation claims to have discovered active "soluble" systems have consistently come to grief. No such claims have yet been advanced for photosynthetic phosphorylation. Two kinds of vesicles will be under consideration here: those of bacterial chromatophores, and those derived one way or another from the

inner green membranes (thylakoids) of chloroplasts from algae or from higher plants. Unless otherwise specified work with higher plant chloroplasts will be under discussion. The term "chloroplasts" will be used, even if these have had their outer membranes and stroma material removed during preparation, or have been swollen and partially broken by suspension in hypotonic media. For a valuable discussion of the different degrees of preservation of isolated chloroplasts and appropriate terminology, see Hall (1972).

2. Electron transport is coupled to phosphorylation. No ATP is made unless electron flow occurs (a precept violated only by interposing an artificial ion gradient; see Section 3.4). *Vice versa*, there should be no electron transport unless ATP is synthesized at the same time, in the perfectly coupled situation. Isolated chloroplasts are generally found not to be perfectly coupled; a "basal" electron transport rate occurs at speeds between one-half and one-fourth those seen on adding ADP and P_i, with consequent control ratios* of 2 to 3.5 (Arnon *et al.*, 1958; West and Wiskich, 1968). With recent improvements in methods for isolating higher plant chloroplasts this control ratio is reported to reach values between 4 and 6 (Hall *et al.*, 1971).

3. A violation of paragraph 2 occurs in the presence of chemical "uncouplers." Electron transport speeds up to rates as fast or faster than those seen during phosphorylation, and ATP synthesis is prevented.

These and other observations have served as the basis for a variety of speculative proposals for the mechanism of phosphorylation. While most proposals were designed to explain oxidative phosphorylation, on the whole they are equally applicable to photosynthetic phosphorylation. All of them, of necessity, include some sort of common intermediate, linked in an obligate fashion to oxidation–reduction reactions on the one hand, and to the hydration–dehydration reactions resulting in ATP synthesis on the other. Basically there have been three kinds of intermediates proposed: (1) those involving a chemical covalent bond between one of the hydrogen or electron carriers and some adjoining compound; (2) a transmembrane gradient in electrochemical activity of hydrogen ions (pmf); and (3) varying combinations of charge redistributions, local proton concentrations, and protein conformational changes *within* the membrane structure rather than specifically from inside to outside a bounded vesicle.

1.2. Hypotheses for the Nature of Coupling

1.2.1. Chemical Coupling Hypotheses

The first mechanism of phosphorylation proposed involved a covalent bond between an electron carrier (cytochrome or others) and some other

---

* Control ratio is defined as the ratio of electron transport during active phosphosphorylation to the slower rate observed after ATP synthesis stops.

entity (see Chance and Williams, 1956, for instance). At times it has even been proposed that a complex is formed between one of the carriers and P_i (for a recent version see Tu and Wang, 1970 and Wang, 1972). However, it is generally observed that the exchange reactions associated with the phosphorylation mechanism are independent of the oxidation state of the redox chain, and this fact requires the postulation of at least one additional step prior to P_i participation. In addition, the pattern of oxygen exchange reactions (see Section 4.3.2) appears to be inconsistent with a phosphorylated electron carrier as a high-energy intermediate.

In these hypotheses a nearby entity (probably a CF but usually designated "X") forms a (presumably covalent) bond with an electron carrier at a low-energy level. This complex is the true electron donor to the next carrier in turn; and because of the oxidative step the original carrier complex bond is moved to a high energy level (i.e., from $-X$ to $\sim X$). (For a discussion of energy considerations, see Weber, 1972.) For coupling to occur, the original reduced carrier cannot be an electron donor *unless* it is present as the low energy complex with X. For electron transport to continue, therefore, the original electron carrier X complex must be broken down, as by adding first P_i then NADP across the bond to form ATP. These reactions are summarized:

$$AH_2 + X \rightleftharpoons AH_2 - X \tag{1}$$

$$AH_2 - X + B \rightleftharpoons BH_2 + A \sim X \tag{2}$$

$$A \sim X + P_i \rightleftharpoons X \sim P + A \tag{3}$$

$$X \sim P_i + ADP \rightleftharpoons ATP + X \tag{4}$$

In the complete set of reactions rapid electron transport is possible only when the bond between the electron carrier A and X is transformed into that of ATP. Similar release would occur, of course, by premature hydrolysis of either $A \sim X$ or of $X \sim P$, and the action of uncouplers has been postulated as inducing these hydrolyses.

In general the simpler forms of this kind of mechanism are no longer accepted as likely to be correct or sufficient. This is primarily because they neither require nor account for phenomena of membrane electric potentials (see Chapter 10 of this volume), of ion fluxes, or of membrane structural changes at both molecular and visible levels (see Chapters 2 and 11); all of which are now known to be directly associated with the coupling mechanism.

1.2.2. The Chemiosmotic Hypothesis

Based on earlier concepts to some extent, this hypothesis has been developed most extensively by Peter Mitchell (1961, 1966a, 1968, 1970, 1972; see also Skulachev, 1972). It derives from considerations of membrane physiology in addition to traditional biochemistry, and emphasizes the importance of vectorial reactions leading to group transfers (in this case, primarily of protons) across the membrane. This hypothesis can be broken down into five major postulates.

1. The relevant reactions occur on a membrane, poorly permeant to protons (or hydroxyl ions), which encloses an inner space.

2. The ordinary flow of reducing equivalents alternates from electron carriers (such as cytochromes) to hydrogen (atom) carriers (such as plastoquinone) and back again. Sequential reactions have to occur on opposite sides of the membrane, with a restricted geometry brought about by the specific transverse orientation of the electron and hydrogen carriers. When the reducing equivalents run from an electron to a hydrogen carrier the extra proton is picked up from the medium outside the chloroplasts; going from the hydrogen carrier back to an electron carrier the now unnecessary proton will be dumped inside the enclosed space. The net effect couples the flow of reducing equivalents down the chain to the translocation of hydrogen ions from outside to inside. (Note that the direction of proton flux is the reverse, in mitochondria.)

3. This net flow of hydrogen ions does two things: it makes the inside more acid than the exterior (i.e., creates a pH gradient) and makes the inside more electrically positive. To prevent excessive buildup of electric charge differences there must be some way—either by free diffusion of other ions, or by means of built-in carriers—for other cations to move out, or anions to move in, at the same time. The movement of other ions is electrophoretic in nature. To the extent that they fail to keep pace with the active movement of protons, some electric potential will build up.

4. The combination of the hydrogen ion concentration (pH) difference, and degree of positive charge on the inside, representing the electrochemical activity gradient for protons, is designated "protonmotive force" (pmf)

$$\text{pmf} = \Delta\psi + \frac{RT}{\mathscr{F}}(\Delta\text{pH}) \tag{5}$$

in which the pmf is expressed in millivolts, R and T are the gas constant and the absolute temperature, and $\mathscr{F}$ is the Faraday constant. This is the direct driving force for ATP synthesis, i.e., the coupling intermediate.

5. The pmf acts to form ATP via a membrane-bound enzyme (synthesizing ATP in the forward direction, an ATPase in the reverse direction) whose activities are strictly vectorial. Figure 1 is a diagram of the simplest form of Mitchell's proposal for the mode of action of this enzyme, working as an ATPase. The hydroxyl portion of the water needed to hydrolyze ATP (actually the crucial part of the reaction) in some unknown fashion must be obtained strictly from the inside of the vesicle, the proton from the outside. Thus the net effect is the apparent translocation of one hydrogen ion inward

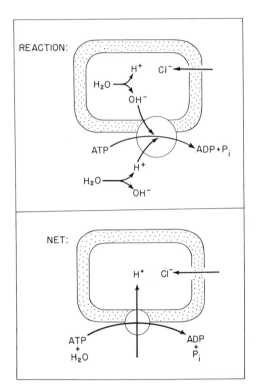

Fig. 1. Operation of a membrane-bound, vectorial ATPase according to the chemiosmotic hypothesis. The large rectangle represents the membrane vesicle, the small circle the ATPase. The top portion shows a presumed reaction in which the components of water are moved vectorially from opposite sides of the membrane to be used in ATP hydrolysis. The bottom portion indicates the net reaction of one proton moved inwards for each ATP hydrolyzed. Other modifications could be drawn to accomodate $2H^+$ moved per ATP hydrolyzed. ATP synthesis will consist of the reversal of all the arrows, including those of proton translocation hence a drainage of protons out of the interior.

as one ATP is split. With appropriate modifications a still fairly simple model can be diagrammed which will translocate two protons inward for every ATP split (Mitchell 1966a,b). Thus the consequence of ATP hydrolysis is an elevation of the internal pmf of the vesicle—raising the energetic status of the system, just as electron transport does.

ATP synthesis in this model will occur by a direct reversal of all the steps diagrammed in Fig. 1, including the reverse direction for proton and hydroxyl ion movement. Thus a preexisting condition of high internal acidity and positive electric charge, i.e., a high internal pmf, will pull ATP synthesis by reforming water as it combines with the hydroxyl ions removed from ADP plus P_i and transported inward.

The phenomena of coupling result from the fact that proton translocation is an obligate part of electron transport. Once the internal activity of protons has reached a high level, further reactions tending to drive protons inward (i.e., continued electron transport) will have to fight against the existing pmf. In that condition ATP synthesis, which translocates protons outward, lowers the pmf and so permits faster electron transport to resume. Uncouplers could act as before by catalyzing hydrolysis of intermediate high energy bonds (if they exist). They can also work, however, by making the membrane leaky to ions which would then discharge the pmf.

The chemiosmotic hypothesis requires a bounded and relatively impermeable vesicle, whereas the chemical theories ignore this requirement. On the other hand, the chemical theory postulates that the energy of the electron transfer reaction is first conserved in a chemical (covalent, most usually) bond, while the chemiosmotic theory puts an ion gradient stage as an obligate intermediate separating electron transfer from chemical bond steps.

1.2.3. Membrane Charge Distributions and Conformational Changes

A number of proposals have been made in recent years, tending to concentrate on events within the membrane. They differ in this respect from the earlier chemical hypotheses which in their original form essentially pay no attention to the existence of a membrane, and from the chemiosmotic hypothesis in paying less attention to the necessity of gradients across the membrane. One of the earliest and most comprehensive is the hypothesis advanced by R. J. P. Williams (1969). He proposed that electron transport can lead to the localization of protons *within* the hydrophobic membrane, where they will tend to combine with water removed from ADP plus P_i to form hydronium ions, which in turn are preferentially expelled. The net effect would be to facilitate dehydration reactions within the hydrophobic regions of the membrane. Both Dilley (1968a,b) and Lynn and Straub (1969a; 1969b) are also concerned with local proton concentrations, and the latter claim indirect evidence that the coupling factor itself (!) must be reduced and protonated, both, for ATP synthesis to occur.

Bennun (1971) speculates on charge redistribution resulting from electron transport leading to specific chemical group changes and eventual dehydration. The existence of charge redistributions within the membrane is also invoked by Chance (1972) to help explain the very interesting observations made in his and in Slater's laboratory on electron carriers (largely cytochromes) apparently showing two distinct midpoint redox potentials. Chance's comprehensive and impressive theory is concerned largely with the intimate nature of electron transfer through these components, which seem to connect clusters of carriers at discrete potential levels and may represent efficient energy-transducing regions. The heart of their activity is considered to depend on a rotational and translational mobility of the cytochromes within the membrane; hence conformational changes of the cytochromes become an important consideration. While the hypothesis is quite specific about the nature of the electron transfer reactions, very little is specified about the connection with ATP synthesis.

In chloroplasts the only variable potential electron carrier known so far is Cyt b559 (see Chapter 7 of this volume). Its low potential form appears only in damaged chloroplasts (Boardman *et al.*, 1971) or after adding electron transport inhibitors such as hydroxylamine or antimycin A, or the multi-functional uncoupler FCCP (Cramer and Böhme, 1972). The high potential form in undamaged chloroplasts does not appear to participate in normal electron flow between the two photosystems (Boardman *et al.*, 1971). On the whole it is difficult to say from present evidence whether even this Cyt fills the role postulated by Chance for energy-transducing regions of the electron transport chain.

Given charge redistributions within the membrane it is easy to imagine further effects on the conformational states of proteins embedded in the membrane. A specific suggestion for the role of conformational changes of proteins in the chemistry of ATP synthesis was made earlier by Boyer (1965). While many, if not most, of such changes would be difficult to detect, both with chloroplasts and mitochondria there seem to be further and major consequences which result in membrane alterations visible in the electron microscope, and alterations in light-scattering properties (see Chapters 2 and 11 of this volume). While one extreme proposal has placed these visible changes as the energetic intermediate between electron transport and ATP synthesis in mitochondria (Green *et al.*, 1968), this is probably not applicable to chloroplasts because the observable light-scattering and membrane conformational changes proceed too slowly (Hind and Jagendorf, 1965a and b; Chapter 11 of this volume). A refined version of the conformational hypothesis is presented by Green and Ji (1972) taking into account considerations of local proton activities and electric charge distributions within the membrane; and similar concepts are espoused by Young (1972). Current thinking based on electric and pH changes within the membrane also take into account the probably important role of conformational changes of membrane proteins (Williams, 1972; Azzone, 1972; Hatefi and Hanstein, 1972).

In general, these hypothesis place the energized "state" of the transducing membrane itself as the intermediate between electron transport and ATP synthesis. The energetic condition in most of these is a set of localized concentration differences (of protons or electrons) within the membrane rather than from one side to the other. Uncouplers in these hypotheses would be reagents capable of carrying protons to the lipophilic interior regions of the membranes, and the chemical activity of these introduced protons would be responsible for discharging the high-energy state.

The present chapter will tend to use the chemiosmotic hypothesis as a framework. This reflects the large amount of work inspired by it over the past several years, the surprisingly high proportion of its predictions that have been experimentally confirmed, and the consequent conviction that it has to be given serious consideration as providing a correct explanation for a number of the parts of the coupling mechanism, at a given level of analysis. In its original form it does not consider in any depth events at a deeper molecular level—neither the specific interactions between membrane macromolecules (including electron carriers) and the "energized state" of the membrane; nor the chemistry of ATP synthesis proper. As more extensive and sophisticated information becomes available about this level of events, the chemiosmotic hypothesis must of course be amplified and/or modified. To this extent it seems likely that many of the current proposals concerning membrane-located events are not contradictory, but in the long run will be amalgamated with some of the larger aspects of the chemiosmotic hypothesis.

The body of literature dealing with these interacting parameters—electron transport, ion fluxes and gradients, localized or transmembrane charge redistributions, subsequent readjustments between and within lipids and proteins, and ultimately the chemistry of ATP synthesis or hydrolysis—has become complex and far-reaching. This one chapter cannot attempt to outline all approaches or discuss all significant results. It is restricted primarily to evidence concerning proton movements and gradients (related to both electron transport and ATP synthesis) as measured with ordinary electrodes or some dyes, and then to the biochemistry of ATP synthesis proper. These topics should complement the chapter by Witt (Chapter 10), which deals with ion movements and electric fields in the membrane as studied by means of powerful biophysical techniques, especially rapid optical measurements; and those by Murakami et al. (Chapter 11), and Arntzen and Briantais (Chapter 2) which consider membrane alterations in more depth.

2. PROTON AND ION MOVEMENTS: RELATION TO ELECTRON TRANSPORT

2.1. Basic Observations

A major and reversible uptake of protons during light-driven electron transport, prevented or reversed by uncouplers, was first observed with swollen spinach chloroplast thylakoid membranes (Jagendorf and Hind, 1963; Neumann and Jagendorf, 1964). Similar observations were made with chromatophores from *Rhodospirillum rubrum* (Von Stedingk and Baltscheffsky, 1966), although with smaller pH changes, and subsequently were extended to almost every photosynthetic organelle examined *in vitro*. The number of protons disappearing from the medium in the case of chloroplasts can be on the order of 400 times the number of individual electron carriers such as Cyt f or plastocyanin.

In order to have a major amount of proton movement occur, it is clear that the internal pmf must not become excessive. Excessive rise of internal acidity (chemical concentration component) must be prevented by internal pH buffering groups; excessive rise of the membrane potential must be prevented by the appropriate flux of counterions. When these conditions are not met either the back-pressure from the internal pmf (according to chemiosmotic concepts) or abnormal conditions of internal acidity and membrane electric stress (for those who prefer alternative frameworks) will slow down electron transport and/or those processes moving protons inward.

2.1.1. Counterion Flux

The need to prevent excessive rise of the membrane potential is the more stringent of the two, since a very small number of entering protons can create unhealthy extremes of electric charge difference (depending on capacitance), whereas a much larger number is needed to change the pH (depending on internal volume and buffering capacity). Let us point out right away that discharge of the membrane potential can be accomplished, in the chemiosmotic framework, by proton movement caused by ATP synthesis, which is in the opposite direction to that caused by electron transport (see Fig. 1). However, in the absence of ATP synthesis, the membrane potential can be discharged by movement of cations in a direction opposite to that of the protons, or of anions in the same direction. These compensating ion fluxes destroy the membrane potential part of the pmf and represent work done in the form of movement of charge down an electrical potential gradient across a resistance; hence they are a way of dissipating energy. Section 2 deals for the most part with net proton fluxes measured in the absence of ATP synthesis.

Cation efflux in the light was described first by Dilley and Vernon (1965) who used ion-specific electrodes to measure K^+ and Mg^{2+} loss from chloroplasts. The sum of these apparently balanced the net amount of H^+ entry. The cation movements did not seem to be completely reversible in the dark, however, whereas the proton movements were. Alternatively, using labeled $^{36}CL^-$ and a centrifugation technique Deamer and Packer (1969) demonstrated chloride entry simultaneous with proton uptake, and Gaensslen and McCarty (1971) similarly measured chloride entry as amines moved into chloroplasts. In the latter case chloride entry matched the amount of amine entry at Cl^- concentrations above 20 mM only. Using the nonpermeating species polygalacturonate as the sole anion in the medium Cohen and Jagendorf (1972) obtained a proton uptake 40% as high as normal (as long as sucrose or mannitol was present to provide some osmotic support). We concluded that in the usual saline media about 40% of the proton uptake might be neutralized by cation extrusion, and the remainder by chloride entry. Using specific electrodes, Schröder et al. (1972) did indeed observe the simultaneous efflux of sodium ions and uptake of chloride ions in the light, with the sum of the two matching the amount of protons taken up.

Direct indications of the limitation on proton uptake by a rising $\Delta\psi$ are found in the stimulations of the amount of such uptake caused by providing a mobile counterion. For instance the combination of K^+ and valinomycin permits a passive, electrogenic flow of K^+; this addition stimulated net proton uptake by R. rubrum chromatophores (von Stedingk and Baltscheffsky, 1966) and by subchloroplast particles prepared with digitonin or by sonic oscillation (McCarty, 1970). These same R. rubrum chromatophores took up even larger amounts of H^+ when provided with any one of a series of highly artificial but permeant anions such as phenyl dicarbaundecane, or tetraphenylboron (Isaev et al., 1970). At the same time equivalent amounts of these anions entered the chromatophores. Uptake of SCN^- and of ClO_4^- by chromatophores was described by Gromet-Elhanan (1972), although in this case pH measurements do not seem to have been made. In at least some of these cases the anions were those of strong acids, so it could not have been the neutral protonated acid that was taken up.

2.1.2. Internal pH Buffers

A limit to the total amount of protons taken up based on excessive internal acidity is implied in the work of Rumberg et al. (1969) and of Avron and colleagues (see Avron, 1972). The rates of electron transport become progressively slower as (estimated) internal pH levels drop below 5. Slower electron transport results in slower inward pumping of H^+; given a steady rate of H^+ leakage in the light this becomes a self-limiting method of depressing the steady level of net H^+ uptake.

Striking experimental illustrations of the limitation by internal acidity are provided when the limitation is overcome by inserting an artificial buffer, able to diffuse into the thylakoid interiors. These have been organic bases with a low pK value so that their uncharged form predominates in the surrounding medium, and penetrates the thylakoids freely (Fig. 2). The sequence shown, with accumulation of the charged base-chloride and increases in net H^+ uptake to as much as 20 times higher than that normally seen, has been reported for tetramethylenediamine (Crofts, 1968), imidazole (Lynn, 1968), pyridine (Nelson et al., 1971), aniline, and p-phenylenediamine (Avron, 1972).

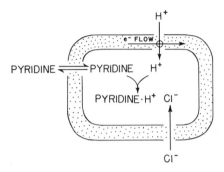

Fig. 2. Net uptake of pyridine by chloroplast vesicles in the light. The uncharged base enters freely and serves as buffer for continuing entry of many more protons than usual, providing there is simultaneous entry of the anion.

2.1.3. Conclusions on the Nature of the Proton Pump

Results described above are reasonably critical in deciding between some alternatives for the nature of the proton pump. It was often suggested, for mitochondria and inferentially for chloroplasts, that observed proton movements are largely secondary consequences of active and equally vectorial cation pumping (see Pressman, 1968, for instance). Now if this were the case in chloroplasts, there should be a cation pump extruding potassium or magnesium ions in the light. But this model cannot account for the near equivalence between uptake of protons and inward diffusion of artificial organic anions (Isaev *et al.*, 1970). For this to occur there must be either a proton pump, or one for artificial organic anions. The inward diffusion of organic bases such as pyridine could be due to active cation extrusion at first; but there would have to be a stoichiometry of charge movement, and as with protons the inward flow would cease when the supply of internal cations is exhausted. But we have seen that organic bases continue to accumulate to an enormous extent; to the point where very extensive osmotic swelling is induced, and this is obviously far beyond the original concentration of internal mobile cations.

Alternatively one might consider a possible inward-directed anion pump working in the light. But to account for the uptake of artificial anions as well as chloride it would have to possess an improbably broad specificity.

Furthermore it is possible at present to explain ATP synthesis under conditions of the acid-base experiment (see Section 3.4) in a consistent fashion on the basis of efflux of protons, left inside from the previous acid condition, to the now basic exterior. If there is no proton pump coupled to electron transport and energy conservation, but an anion pump instead, a more complex explanation must be sought for the energetic basis of acid–base ATP synthesis. Since an ion gradient of some sort is almost certainly responsible, at best it would have to be the outward diffusion of the particular buffer anion (succinate, phthalate, mesaconate, etc.) that might drive an anion pump backward to synthesize ATP. However, the experimental evidence shows a close correlation between the extent of ATP formation and the number of buffer anions trapped inside and failing to exit, i.e., the ones whose associated protons were able to leave by themselves.

On the whole, the combination of available evidence strongly favors the interpretation of an active inward movement of protons driven by light and electron transport, and a passive (electrophoretic) movement of any other ions to maintain electrical neutrality.

Liberman and Skulachev (1970) argue that the creation of a membrane potential in the light could be responsible for the observed proton fluxes. However, any charge redistribution would be sufficient to bring in those ions of opposite charge needed for neutralization, but not ions of the same charge at the same time. Further, the amount of proton entry needed to neutralize a membrane potential would be very small, probably almost nondetectable. Hence the observed massive entry of anions and protons together in the experiments of Isaev *et al.* (1970) or of anions and amines in those of Gaensslen and McCarty (1971) can only be due to the net active transport of protons to an interior region where they remain chemically active.

The existence of active proton translocation by thylakoids, virtually proven by these experiments, is a fact which is an integral part of the chemiosmotic hypothesis but a distracting irrelevancy for the others.

2.2. Geometry of Electron Carriers

A crucial part of the chemiosmotic hypothesis is the existence of proton-transferring "loops'. in the membrane, i.e., alternation of hydrogen atom and electron carriers, situated in the membrane to face specifically the interior or the exterior of the vesicle. This obligate transectional geometry for the carriers is not a necessary component of chemical hypotheses.

Evidence for specific transverse localization of carriers in the membrane is currently being obtained by means of both antibody studies, and the use of water-soluble chemical binding groups. The picture is most complete for mitochondria (Racker, 1970). However, even with chloroplasts it has been shown that CF_1, NADP reductase, ferredoxin, and unspecified components associated with PS I are exposed to the outside of thylakoid discs (Berzborn, 1968; Regitz *et al.*, 1970); while components associated with PS II are not exposed (see Chapter 2). The most interesting case is that of plastocyanin, which can participate in artificial electron flow through PS I when added to the outside, but must be placed on the inside of thylakoid discs (either endogenous, or by mild sonic oscillation) in order to support ATP synthesis (Hauska *et al.*, 1971).

The analysis of the localization of electron carriers with respect to membrane geometry is far from complete. Even at this point, however, the results show that the question is a meaningful one and when answered more fully may shed critical light on the chemiosmotic or membrane charge localization hypotheses.

2.3. Action of Ionophores and Uncouplers

The availability of specific ion-carrying chemicals and antibiotics with a mode of action demonstrated in both natural systems and model membranes (see Harold, 1970, for a valuable review) has been extremely useful in defining the physiological biochemistry of chloroplasts and chromatophores. Compounds are available to carry protons or metal cations, hydroxyl or Cl^- ions through membranes; and the net effect of the movement can either be neutral (no change in charge distribution); or electrogenic (net flow of charge as single ions diffuse).

2.3.1. Neutral Flux: Collapse of the pH Gradient

A neutral ion movement occurs when the carrier molecule (ionophore) such as nigericin is able to dissolve in the membrane only when complexing either with a cation or a proton; or only with an anion or a hydroxyl ion. In the case of nigericin both K^+ and H^+ are thus allowed to diffuse down their electrochemical activity gradients, in a neutral, one for one exchange (Fig. 3). An analogous exchange of Cl^- for OH^- ions is permitted by trialkyltin compounds (Selwyn *et al.*, 1970, Watling and Selwyn, 1970) in chloroplasts as well as other organelles and model membranes. The same neutral exchange was permitted by the lipid-soluble phenylmercuric acetate whose action would usually be thought of in terms of binding to $-SH$ groups; but not by its hydrophilic analog *p*-hydroxymercuriphenyl sulfonate. Note that these exchanges will tend to equalize the pH on the two sides of a membrane, and so collapse any pH gradient.

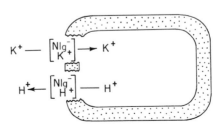

Fig. 3. Function of nigericin in permitting entry of K^+ for H^+ exchange across thylakoid vesicle membranes.

Probably related are the action of detergents such as Triton which uncouple at very low concentrations (Neumann and Jagendorf, 1965). These cause a rapid release of any stored protons, and undoubtedly change the membrane structure so that many ions permeate more easily. The effects are extremely dependent on minute variations in concentration and even on the method of addition of the detergent. A complete analysis as to whether neutral or electrogenic fluxes are engendered first, and which ions' movements are most affected, has not been reported.

A second sort of neutral flux is found in the passive entry of nonprotonated organic bases. It is both assumed and supported by indirect evidence (Hind and Whittingham, 1963) that the free base is the only permeating form. If the pH of the medium is alkaline relative to the pK of the base only the free base is present outside, and equilibration will occur without any change in energetic status. However, as in the case of pyridine with a pK of 5.2 (see Fig. 2) the now internal base molecules may serve as a buffer for entering protons after the internal pH in the light drops to the region of the pK of the base. This will permit more protons than usual to be taken up.

On the other hand, with a stronger base such as ammonia or many alkylamines, the usual pH of the medium is more acid than their pK and most of the base is present in the protonated form. In this case only the low concentration of free base actually present can equilibrate across the membrane in the dark. When the light goes on and protons start to move in, the internal pH will begin to drop below that of the outside. Accordingly a higher proportion of the internal free base will become protonated; the net concentration of free base inside will drop and so more can diffuse in from the outside. But as this in turn lowers the external concentration of free base, an equivalent amount from

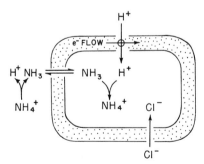

Fig. 4. Entry of ammonium chloride into thylakoid vesicles in the light. Mechanics of entry are the same as for pyridine (Fig. 2) except that NH_3 is a stronger base and exists to a greater extent as NH_4^+ at the pH used in chloroplast reactions. Consequently entry of NH_3 is accompanied by release of one H^+ on the outside, neutralizing the effect of each H^+ entering linked to electron transport. Internally, NH_3 as a relatively strong base binds all entering protons and maintains a high pH.

the reservoir of external protonated base will dissociate to maintain the usual equilibrium at the (constant or possibly rising) external pH. This dissociation liberates the equivalent number of protons on the outside (Fig. 4). Since the number of protons entering the thylakoids is matched by both the number of free base molecules entering and able to combine with them inside, and the number of protons liberated by the external dissociation of protonated base molecules, essentially no change will occur in either the internal or the external pH, and no pH gradient is created in the light.

These neutral fluxes, it is important to remember, do not affect any membrane potential brought about by a surplus of entering protons over entering Cl^- or leaving K^+ or Na^+. The extra positive charge is simply maintained by the K^+ or the NH_4^+ which has replaced the H^+.

2.3.2. Electrogenic Fluxes: Collapse of the Membrane Potential

Electrogenic movement of K^+ or NH_4^+ ions is engendered by the addition of antibiotics such as valinomycin (Fig. 5). This occurs because the antibiotic can exist in the membrane either without an ion, or complexed with K^+, etc. However it cannot bind protons. Therefore the K^+ or NH_4^+ ions carried across membranes by valinomycin cannot be exchanged for H^+ and their transport results in a change in the charge across the membrane.

A host of weak acids including the well-known uncouplers DNP and FCCP act as electrogenic proton carriers. This occurs because both the anion and the protonated, uncharged form dissolve in and can cross the membrane. Thus the protonated form can enter on one side, move across to the other and discharge its proton, then recross the membrane as an anion. The net unidirectional flow of protons (down their electrochemical activity gradient) is again able to cause a change in the membrane charge, i.e., this is an electrogenic flux. It should be noted that both the uncoupling weak acid and valinomycin are catalytic carriers, able to effect net transport of large numbers of ions without any net movement on their own part. [A caution: the specificity of any reagent should be checked carefully especially at higher

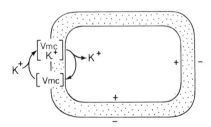

Fig. 5. Function of valinomycin in permitting electrogenic diffusion of K^+ ions across thylakoid vesicle membranes. Depending on the direction of the K^+ diffusion gradient this will either build up or dissipate an existing membrane potential.

concentrations. At 0.02 μM (Karlish et al., 1969) FCCP is probably a typical electrogenic proton carrier, but at 1.0 μM it can bind to sulfhydryl groups, inhibit electron transport, etc.]

Electrogenic ion flux also results from the diffusion of ions which are able to penetrate membranes without the addition of a special carrier. Quite a number of anions fall in this category, although the particular ones which are permeant vary with the particular organelle. For chromatophores of R. rubrum permeating anions include ClO_4^- and SCN^- (Gromet-Elhanan, 1972), SO_4^{2-} and picrate (Montal et al., 1970) as well as the artificial anions tetraphenylboron and phenyl dicarbaundecanoate (Isaev et al., 1970). With swollen chloroplasts in the light apparently Cl^- is a permeating anion, and studies of replacements for it have not been performed. Furthermore, it is possible that ammonium ions are permeant in whole chloroplast membranes, at least under some circumstances (N. Good, personal communication). Thus they could act in the same fashion that DNP does, as a catalytic shuttle for proton leakage, with the nonprotonated form entering and the protonated form leaving. This mechanism will collapse both a membrane potential and a pH gradient; whereas the one noted above (Section 2.3.1.) will collapse only the pH gradient.

2.3.3. Uncoupling Action in Thylakoid Systems

The role of ionophores and antibiotics predicted by the above considerations has largely been borne out in experiments with chloroplasts and chromatophores. Thus the net uptake of ammonia (Crofts, 1967; 1968) and of labeled amines (Gaennslen and McCarty, 1971) has been observed during electron transport in illuminated chloroplasts. The uncoupling effectiveness of amines is related to structural characteristics permitting solution in lipid membranes, and to the concentration of the nonprotonated form in solution (Hind and Whittingham, 1963; McCarty and Coleman, 1970). The action of nigericin, gramicidin, and other ionophores in inhibiting net H^+ uptake and substituting the uptake of K^+ ions, together with uncoupling, was shown in a number of studies (Shavit and San Pietro, 1967; Packer, 1967; Packer et al., 1968; Shavit et al., 1968a, 1970). Using nigericin, more K^+ could be accumulated, than H^+ in the absence of nigericin (Packer et al., 1968; Shavit et al., 1970). This should have been expected since the external concentration of K^+ used in these experiments, 1 mM, was 1000 times higher than that of external H^+ in the external solution at pH 6. Also accumulation of K^+ will not lead to acidification of the interior, hence the total amount taken up will not be limited by the availability of internal buffering groups as it is for H^+.

In general when using smaller particles, it is necessary to collapse both the ΔpH and $\Delta\psi$ components of the pmf for the most effective uncoupling. Thus for instance R. rubrum chromatophores require the simultaneous use of

(1) K^+ and nigericin to effect a collapse of the pH gradient, putting K^+ ions inside, and (2) valinomycin to permit electrogenic exit of the internal K^+ ions and so collapse the membrane potential. The membrane potential can be collapsed by the addition of a permeant anion, in the place of valinomycin. Similar results and conclusions were drawn by others with *R. rubrum* chromatophores (Jackson *et al.*, 1968; Thore *et al.*, 1968; Gromet-Elhanan, 1972) and for subchloroplast particles made by the use of digitonin or sonic oscillation (McCarty, 1969; Hauska *et al.*, 1970; Neumann *et al.*, 1970).

Ordinary ("whole") chloroplasts were thought to be different because they were uncoupled by ammonia, or by nigericin plus K^+. According to the first interpretation described above (Section 2.3.1) these reagents collapse only the pH gradient. This implied that little or no $\Delta\psi$ was present or needed during light-driven phosphorylation, at least in the presence of these reagents. Indeed, free entry of Cl^- anions, presumably compensating the electric charge of entering protons, was demonstrated during amine or ammonia uptake (Deamer and Packer, 1969; Gaennslen and McCarty, 1971). If this is the complete story then the difference between whole chloroplasts and the smaller particles would lie in a greater permeability of the former to Cl^- ions.

However, the second interpretation of ammonia as a proton shuttle (Section 2.3.2) is also quite possible; in which case ammonia would be collapsing both a pH gradient and a membrane potential (N. Good, personal communication). In this case the major difference between whole chloroplasts and the smaller particles would lie in a greater permeability of the former to NH_4^+ and possibly to K^+ cations.

Under other circumstances whole chloroplasts seem to demonstrate a clear requirement for maintenance of a membrane potential to synthesize ATP. At low light intensities both DNP and valinomycin alone appear to be effective uncouplers (Saha *et al.*, 1970) unlike their behavior at high light intensity. Also with a brief but repeated flashing light regime Junge *et al.* (1970) were able to drive ATP synthesis apparently on the basis of $\Delta\psi$ alone (no massive proton fluxes were observed); and in this ease valinomycin was a highly effective uncoupler. Again it had no effect under continuous high intensity light conditions. Thus at low light fluxes, it would appear that Cl^- is not a permeant anion, and in this case very small amounts of H^+ movement will be sufficient to create a meaningful $\Delta\psi$. It is also consistent that Cl^- is not a readily permeant anion in the dark. Among other things, it will not support the dark swelling of chloroplasts provided with a permeating cation (Karlish *et al.*, 1969).

The presumed increase in Cl^- permeability at high light intensities might be intrinsic, or it might be in large part due to the presence of ammonia, amines, or K^+ and nigericin. As indicated above, the total amount of cation (and therefore of anion) accumulated when using these reagents is not subject to the same limitations as for accumulation of protons. Water follows and osmotic swelling is ordinarily observed. This swollen condition could be partly responsible for the greater Cl^- permeation rates (see, for instance, the observation of increased K^+ and Na^+ permeability when red blood cells are in a swollen condition; (Parker and Hoffman, 1965; Poznansky and Solomon, 1972). Supporting this possibility is the fact that ammonia uncoupling is to a considerable extent counteracted by high osmotic concentrations of salts or mannitol in the reaction medium (Izawa, personal communication; Gaennslen and McCarty, 1971) which would tend to prevent the swelling. One might predict a synergistic interaction between NH_4^+ (or nigericin) and the electrogenic agent valinomycin to show up under these high osmotic strength conditions. Indeed this synergism was seen when chloroplasts were treated with limiting amounts of nigericin (Shavit et al., 1970), a result interpreted as showing the presence of at least some $\Delta\psi$ component for the pmf.

An anomalous observation is the failure of low concentrations of DNP or FCCP to act as uncouplers by themselves, but their ability to interact synergistically with valinomycin (Karlish et al., 1969). The rationale advanced was that DNP permits only electrogenic exit of H^+, and valinomycin at best would permit electrogenic exit of K^+. The two together, however, would reconstitute a nigericin-like exchange of potassium ions going in for protons moving out.

However, it would seem that low concentrations of DNP or FCCP by themselves really ought to be able to collapse a pH gradient as well as the membrane potential. Protons did enter in the first place, permitted to do so electrically by either the uptake of Cl^- or the exit of K^+ or Na^+ by passive diffusion. If these compensating-ions were permeant enough to permit H^+ entry in the first place, they should really be permeant enough to permit H^+ exit in the second place when DNP or FCCP are there. Either the compensating ions penetrate freely in one direction but not the other (an unlikely thought) or else a secondary phenomenon may be at work such as the presence of valinomycin plus K^+ facilitating entry of DNP or FCCP into the membrane. This possibility (previously suggested by P. Mitchell, 1970) has not been tested. It may be significant that more lipid-soluble substituted DNP's are uncouplers of chloroplast photophosphorylation at much lower concentrations than is DNP proper (Siow and Unrau, 1968).

[Inhibition by a few uncouplers has been studied as a function of light intensity (Avron and Shavit, 1965; Saha et al., 1970) or electron flow rate (Gross and San Pietro, 1969).]

2.4. Internal pH

With protons leaving the medium, a significant question is whether the internal pH actually falls, or whether the protons are only bound to internal buffering groups. Early attempts to measure the internal pH were by means of binding dyes such as bromthymol blue or methyl red (Chance *et al.*, 1966; Nishimura *et al.*, 1968; McEvoy and Lynn, 1970). However, it became apparent that the relatively slow bromthymol blue responses reflect some sort of binding-unbinding reponse, or other change in the nature of the membrane, rather than an internal pH change (Cost and Frenkel, 1967; Mitchell *et al.*, 1968; Nishimura and Pressman, 1969; Gromet-Elhanan and Briller, 1969). Since then two different methods have been used to show that the interior of thylakoid vesicles really does become acid during electron transport and H^+ translocation. The first, a quenching of Chl *a* fluorescence (Wraight and Crofts, 1970; see also Chapter 6 of this volume for a review) has been associated in a number of different kinetic situations with H^+ uptake (Wraight and Crofts, 1970; Cohen and Sherman, 1971; Cohen and Jagendorf, 1972). (However, see Mohanty *et al.*, 1972.) A quantitative relation between the extent of fluorescence quenching and actual internal pH has not been established yet.

The second way involves the classic approach of determining the partition of a weak base between inner and outer phases. The total concentration in each phase depends on the concentration of both the nonprotonated and the protonated form; the proportions of each of these depends on the relative pH in each phase. The method further depends on the fact that only the nonprotonated form can diffuse through the membrane in the time allowed. Using net uptake of ^{14}C-labeled amines at low concentrations, determined after centrifuging the chloroplasts, together with estimates of internal volume, Gaennslen and McCarty (1971) and Rottenberg *et al.* (1972) calculated the internal pH to be 2 units below the outside one. Somewhat surprisingly the internal pH reached its steady state level at lower light intensities and at somewhat more unfavorable external pH's then were needed to achieve the maximum amount of proton loss from the external medium (Rottenberg *et al.*, 1972).

The centrifugation technique is not ideal since chloroplasts shield each other from the light as they move into closer proximity and into the final pellet. Greater amounts of proton uptake are calculated in experiments where measurements can be made without removing chloroplasts from the medium. These include use of a cation-sensitive electrode to measure ammonium ions (Rottenberg and Grunwald, 1972) and of the fluorescent amine atebrin whose fluorescence is quenched when it enters the chloroplasts (Schuldiner and Avron, 1971). [Earlier work had shown that atebrin fluorescence is quenched in response to the "high energy state" of chloroplasts (Kraayenhof, 1970) or of *R. rubrum* chromatophores (Gromet-Elhanan, 1971) brought about by light or by ATP hydrolysis.] In applying these methods, Schuldiner *et al.* (1972) calculated a pH gradient of 3 to 3.5 units from outside to inside of illuminated chloroplasts. Uncouplers such as methylamine prevented the appearance of this pH difference in perfect parallel with their ability to inhibit phosphorylation.

Most significantly, these procedures resolved the long-standing superficial contradiction between a pH optimum of 6–6.5 for the total amount of H^+ uptake in the light, and one of 8–8.5 for photophosphorylation (Neumann and Jagendorf, 1964; Allen et al., 1958; Good et al., 1966). The pH gradient from outside to inside the thylakoids as measured by atebrin fluorescence quenching does indeed show an optimum between pH 8 and 9 (external), coincident with the pH optimum for photophosphorylation. This emphasizes again that the total extent of proton uptake measurable by a glass electrode reflects the internal buffering capacity of the chloroplasts, not the efficiency of the coupling mechanism (see Section 2.1.2). Internal buffering groups may have pK values in the region of 4–5, hence total proton uptake will be largest when a significant proportion are not protonated to begin with. In retrospect it is clearly necessary to distinguish between the amount of proton uptake which reflects only the buffer capacity under many circumstances, and the actual pH gradient which is the important factor for driving ATP synthesis. As yet no violation has been reported, using these procedures, in the correlation between a significant pH gradient and the ability to phosphorylate ADP.

However, it would be wise to retain some sense of caution about the conclusions derived from studies of the quenching of fluorescent amines. No particular rationale has been advanced for quenching in the thylakoid interior; the concentrations used are very small, and it may be that the physical basis for quenching lies in a particular sort of binding of the amines to the thylakoid membranes, on any face or in the middle. Independent lines of proof for their internal localization have not been presented. It is possible that the extent of such a binding would be quantitatively related to or caused by the internal protonation, and this could account for the good correlations (as far as they go) between net uptake of amines in quantities too large to represent binding, and the degree of fluorescence quenching of compounds such as atebrin. These uncertainties should be included in the evaluation of these measurements of internal pH.

2.5. Membrane Potentials

This topic is treated extensively in Chapters 5 and 10 of this volume, and only a few facets will be discussed here. Evidence has been obtained for the existence of $\Delta\psi$ by three means: band shifts of carotenoids (and other pigments to some extent); estimates of salt distributions across thylakoid membranes, and enhancement of Chl luminescence (delayed light emission).

The definitive association of carotenoid band shifts with membrane potentials in bacterial chromatophores came with the demonstration by Jackson and Crofts (1969) of these optical changes in the dark, identical to those seen in the light, by adding either K^+ or H^+ together with electrogenic carriers in such a way as to make the insides of the chromatophores more positive than the outside. Additions (such as valinomycin with no K^+ in the external medium) designed to make the inside negative reversed the direction of the band shift. Most importantly it became possible to quantitate the height of the optical change as a function of membrane potential created in the dark. The linear relation between height of the absorbance change and logarithm of the KCl concentration added meant that the response constituted a virtual membrane potential voltmeter. Application of these standards to shifts seen in the light indicated a spike of 420 mV when the light was first turned on, and a remarkable level of 249 mV for total steady state pmf (taking into account a net pH gradient estimated to be equal to 51 mV, under the same circumstances).

Analogous studies with chloroplasts have been more difficult due primarily to strong changes in light-scattering when salts are added. Recent work by Strichartz and Chance (1972) seems to have overcome these problems to some extent and a linear relation (with considerably more experimental scatter than for chromatophores) obtains again between the logarithm of the concentration of salt added to chloroplasts, and the change in absorbance at 518 nm. (This change, due to carotenoids and other pigments, has a broad band with a peak in the 515–530 nm range.) With the chloroplasts used in those studies the steady state membrane potential in the light was accordingly estimated at 30 mV positive on the inside.

Even lower estimates for the membrane potential come from studies of the distribution of radioactive chloride following centrifugation (Rottenberg *et al.*, 1972), using the Nernst equation and estimates of internal volume. The potential calculated in these studies was at most 9 mV more positive on the inside. Low values could be partly artifactual due to self-shielding, during centrifugation, with consequent lower light intensity and chloride loss (see earlier discussion). However, estimates of Na^+ and Cl^- partition between chloroplasts and the medium using electrode methods gave a similarly low value of 5 mV positive on the inside (Schröder *et al.*, 1972).

On the other hand, methods in which measurements are made rapidly after turning the light on seem to show much higher membrane potentials. Using techniques for rapid optical measurements Witt and colleagues (Schliephake *et al.*, 1968; Junge and Witt, 1968; Witt, 1971, 1972; and Chapter 10 this volume) have taken the 523 nm optical density change as a measure of the membrane potential (also see Larkum and Bonner, 1972), and provide a calibration based on the movement of only one electron and therefore only one charge unit in extremely rapid flashes of light, together with reasonable assumptions concerning the dielectric constant, thickness and area of the membranes involved. Their calculations then show membrane potentials of more than 100 mV in chloroplasts in the light. At least a part of the membrane potential seems to originate (indicating a charge separation) from each fundamental photoact (I and II).

It has not yet been shown that the quantitative relation between height of the 518 nm shift and the membrane potential is identical with chloroplasts in different physiological conditions. Some variation in this calibration might be suspected to account for the fact that "class I" chloroplasts retaining stroma show repeatable 518 nm shifts from 20 to 30 times as large as do "class II" (disrupted) chloroplasts from the same leaves (Larkum and Bonner, 1972). The basis of the shift lies in the solvent relations of the embedded carotenoids (Strichartz and Chance, 1972). Given a membrane with altered relations between the carotenoids and other lipids or proteins to begin with, an electric stress across it could easily produce a different degree of alteration in this parameter. The possibility of alterations in the quantitative relation between the 518 nm absorbance change and the membrane potential means that the conclusion that digitonin-prepared particles have a lower membrane potential than the original chloroplasts because their 518 nm shift is a bit less (Neumann et al., 1970) cannot be accepted as fully proven. This could be an important point for testing the chemiosmotic hypothesis, since these particles still phosphorylate but show much less net H^+ uptake than usual (although their internal pH does drop in the light, according to the criterion of atebrin fluorescence quenching).

The last method for estimating membrane potentials uses a changing external salt concentration for calibration, but is not affected by light-scattering changes. Illuminated chloroplasts reirradiate a little red light after the exciting light is turned off (see Chapter 5 of this volume). While this "luminescence" depends on the energy released by recombination of pools of oxidizing and reducing equivalents, its extent is affected by the existing electric potential across the chloroplast membranes. Thus sudden changes in salt concentration cause a spike of light emission (Miles and Jagendorf, 1969; Barber and Kraan, 1970), and the quantitative relations involved have been elucidated to provide a method for studying specific permeabilities of different ions (Barber and Varley, 1971, 1972). The effect of varying K^+ salt concentrations, and collapse of the luminescence by valinomycin, also permit establishing a calibration curve for the amount of luminescence vs membrane potential (Barber, 1972). Using flashing light, various chloroplast preparations showed membrane potentials between 75 and 105 mV with the inside positive, due to illumination. These results are thus in accord with those of Witt's group, using the 518 nm absorbance change.

2.6. Control of Electron Transport Rates

Considerations of the internal pH of chloroplast thylakoids have been used (Rottenberg *et al.*, 1972; Avron, 1972) to explain the shape of the pH curve for the rate of electron transport (i.e., the Hill reaction). These authors measured the rate of the Hill reaction and estimated the internal pH in the light, as a function of different external pH's. The highest electron transport rates were correlated with an inside pH between 5.0 and 5.5, and they postulate that this represents a true (internal) pH optimum. They point out that the external pH must be 8 or above, when using coupled chloroplasts, for the apparent internal pH to remain at 5 or above. If the external pH drops below 8 the internal pH will drop below 5, and therefore below the postulated optimum. On the other hand, added uncouplers diminish the extent of the internal acidification. In this case the internal pH might not get to 5 or below until the outside pH had dropped to 6 or 6.5, for instance. This hypothesis can thus account for a large part of the well-known broadening and shift in the (external) pH optimum toward the neutral or acid range every time uncouplers are added (see Neumann and Jagendorf, 1964, 1965, for early examples). Note that the concept of control of electron transport rates by internal pH alone is at variance with a chemiosmotic interpretation, which would depend on control by back pressure from an elevated internal pmf.

Control by the internal pH alone indeed seems inadequate to explain the 2- to 3-fold more rapid electron transport rates occurring when ADP and P_i are added to coupled chloroplasts, permitting phosphorylation to occur. These reagents change the estimated pH gradient to only a small extent (Avron, 1972). For this, the chemiosmotic explanation is more satisfactory: since ATP synthesis feeds on and causes a more rapid use of the internal pmf, it becomes possible for a more rapid electron transport rate to build it back up more rapidly. Essentially during phosphorylation one would expect a more rapid turnover of the protons causing the pmf, and this dynamic reason can explain the faster electron transport rates. Supporting this interpretation are the correlations observed by Rumberg *et al.* (1969), between the rates of electron transport as varied by a number of parameters, and the rates of proton turnover as deduced from proton efflux rates in the post-illumination darkness.

2.7. Alternative Roles for Proton Fluxes

The work discussed so far has gone a long way toward establishing the existance of measurable pH gradients and membrane potentials, arising from the more easily observed H^+ uptake in the light. While this is predicted by the chemiosmotic hypothesis, it is still somewhat difficult to critically rule out an alternative in which the pumping of protons, rather than being the direct coupling link between electron transport and ATP synthesis:

$$e^- \text{ flow} \xrightarrow[\text{H}^+ \text{ translocation}]{} \text{pmf} \underset{\text{ATPase}}{\overset{\text{reversible}}{\rightleftharpoons}} \text{ATP synthesis} \qquad (6)$$

is actually driven by a secondary "proton pump," taking energy from a true (chemical, configurational, etc.) high energy intermediate formed in some unkown fashion from the electron transport chain:

$$e^- \text{ flow} \xrightarrow[\text{mechanism}]{\text{coupling}} \sim X \underset{\text{ATPase}}{\overset{\text{reversible}}{\rightleftharpoons}} \text{ATP synthesis}$$

$$\begin{array}{c} \Updownarrow \text{ bound} \\ \text{proton} \\ \text{pump} \\ \text{pmf} \end{array} \qquad (7)$$

Adaptive significance would have to be found for the arrangement shown in Eq. (7), to explain its survival during evolutionary selection. This is not very hard to do, since proton and other ion fluxes can be shown to participate in a number of control functions.

A primary site could easily occur in the mechanism for "photosynthetic control," i.e., the inhibition of electron transport rates when ADP is not available and phosphorylation stops. Either the internal pH of the thylakoid discs or the internal pmf might often be limiting for electron transport (Section 2.6), and this could have some adaptive value.

A second role can easily be envisaged for indirect effects on photosynthetic carbon metabolism. The outside of the thylakoid discs is inside the stroma (mobile phase) where carbon dioxide fixation and other metabolic events occur. As proton pumping starts in the light, therefore, the stroma will become more alkaline and is likely to receive K^+ or Mg^{2+} formerly stored inside the thylakoid discs. Lin and Nobel (1971) provided direct measurements of Mg^{2+} ion levels in rapidly isolated pea chloroplasts, and found an apparent rise when the light went on due to loss of total water from the stroma. The consequences of this effect would be intensified by any Mg^{2+} ion movement from thylakoids to stroma, brought about by H^+ uptake into the thylakoids. The overriding importance of high Mg^{2+} levels in the dark reactions of CO_2 fixation were demonstrated in a dramatic way by recent findings of Jensen (1971).

Aside from adjustments of the internal ionic environment, light and the consequent proton fluxes lead to drastic shape changes of the whole chloroplast, and perceptible thinning of the thylakoid membranes (see Chapter 11). One functional effect of these changes has been demonstrated already in the quenching of chlorophyll fluorescence (Wraight and Crofts, 1970; Mohanty *et al.*, 1972; see also Chapter 6, this volume) and it would be surprising if there were no other consequences for control mechanisms.

However, if proton flux is entirely a secondary reaction involving a specific proton pump, this would mean that all of those uncouplers and ionophores whose action is now understood are acting at a secondary point of energy metabolism (Mitchell, 1972). Also, the need for a bounded vesicle with intact membranes would be imposed entirely because of the existence of this secondary pump, since its use of energy would be the point of damage caused by ion leakage. In that case any reagent that inhibited the proton pump, or any treatment that removed it, would (1) make phosphorylation insensitive to uncouplers, and (2) permit it to occur in a soluble, or at least in an open membrane system. There has been no sign over the years so far of any treatment or reagent which engenders these characteristics. Reversible inhibitors of proton translocation discovered so far have either been un-couplers, or substances like polygalacturonic acid at low osmotic strength (Cohen and Jagendorf, 1972) which inhibit both proton translocation and phosphorylation.

A most interesting role for a membrane potential in activating the CF_1 in some functional sense, prior to ATP synthesis, was suggested by Junge (1970). This would impose a requirement for ion fluxes in the mechanism of phosphorylation which might or might not involve stoichiometric turnover and continuing use of the pmf.

2.8. Proton/Electron Ratios

2.8.1. Number of Coupling (Proton-Translocating) Sites

The chemiosmotic hypothesis predicts that one proton will be translocated as one electron goes through a loop or coupling site of the electron transport chain. One way to test the hypothesis is to see if this prediction corresponds with the observed facts.

The first thing one really ought to know is where the coupling sites occur in the electron transport sequence. This has not been an easy job (see Chapter 7; Böhme and Cramer, 1972a) and will not be discussed here. Alternatively, it would do for the present purpose to know simply the number of coupling sites in a given overall electron transport sequence, compare the number of protons moved per electron transported through this sequence, and thereby see if Mitchell's prediction is verified.

Since these coupling sites are also the region of energy conversion supporting ATP synthesis, we can estimate their number from the ratio of ATP made to electrons moving through the chain. In view of the precedent with mitochondria it is generally assumed that it takes two electrons moving through one site to make one ATP; hence the $ATP/2e^-$ ratio is usually sought.

All early experiments showed one ATP formed per electron pair moving in the Hill reaction (Arnon *et al.*, 1958). Once Tris was no longer used as a buffer and the pH was raised to 8.5, repeatable ratios of 1.3 were found (Winget *et al.*, 1965). With recent improvements in methods for preparing and using chloroplasts, ATP/$2e^-$ ratios range from 1.5 to 2 (Hall *et al.*, 1971). Both of these results imply the existence of two distinct energy conversion sites. This number is reinforced by studies with inhibitors (Izawa and Good, 1968), in which the ratio of ATP synthesis inhibited/electron transport inhibited due to added increments of the inhibitor was constant at 2.0. The concept behind this calculation is that even in the presence of ADP, P_i, and Mg^{2+} a basal, nonphosphorylating electron transport rate continues and makes the gross ATP/$2e^-$ ratio artificially low by raising the denominator. The concept is not easy to test critically, but seems intuitively sound and is supported by finding the same answer with a wide variety of reagents that restrict phosphorylation. In any case it is clear that there is more than one phosphorylating site between water and ferricyanide in noncyclic electron flow; either 2, or some fractional number between 1 and 2.

The number of sites in cyclic electron flow with, for instance, pyocyanine added, is more difficult to measure because the number of electrons moving is not easily measurable. The number is at least one, and if the pathway includes both the site specific to the cycle and one on the chain between the photoacts, it could be two.

Since the chemiosmotic hypothesis shows one proton to be translocated as each electron goes through a coupling site, there should be between 1 and 2 protons translocated as each electron moves through the 1 to 2 coupling sites in the noncyclic pathway from water to ferricyanide (or $NADP^+$); and the same number as each electron moves around the cyclic pathway.

2.8.2. Problems of Measuring H^+/e^- Ratios

This has been more difficult than for mitochondria (Mitchell and Moyle, 1965, 1967a) primarily because the passive dark efflux of protons is more rapid in isolated chloroplasts. In mitochondria (which extrude protons during electron transport) it is possible to provide a pulse of oxygen and observe a very rapid lowering of the pH, then after the oxygen is used up the protons return slowly, with a half-time ($t_{1/2}$) of about 90 sec, back into the mitochondrion. The pH tracing is extrapolated back to the center of the period of the brief oxygen pulse, and this (when calibrated by titration) gives the total amount of protons released, to be compared to the known amount of oxygen added. The feature that makes this possible is the fact that the rate of proton ejection during active translocation is 10-fold or more faster than the rate of passive return afterward.

With chloroplasts, on the other hand, the passive diffusion of protons (in this case back out of the thylakoids) shows half-times between 1 and 10 sec, and active uptake is only 2 or 3 times as fast (see Table IV in Nishizaki and Jagendorf, 1971, for instance). With the back-reaction so close in rate to the forward one it is difficult to use the height of the pH tracing caused by a pulse of light as a valid estimate of the total amount of protons released, with or without extrapolation back from the decay curve. (From the chemiosmotic point of view this more rapid passive proton permeation is probably the cause of the lower control ratios for electron transport in chloroplasts than in mitochondria.) A return to the pulse methodology might be rewarding given chloroplasts with better control ratios (Hall et al., 1971) or using reagents that slow down passive efflux.

The available estimates of H^+/e^- ratios depend, instead, on measurements of the rates of electron transport and of proton translocation. One difficulty present from the start, but not recognized until much later, has to do with limitations of glass electrode methods for measuring rapid kinetics. All ordinary pH meters, for instance, have extremely slow response times; if these are used no really rapid reaction can be followed effectively. Glass electrodes themselves have a response time which is slowed down by adhering dirt such as a layer of the previous chloroplasts. Also, some buffer must be present or the electrode response is extremely slow (Mitchell and Moyle, 1967a; Nishizaki and Jagendorf, 1971). These factors undoubtedly limited the kinetic responses of earlier studies, which were performed without buffer in the medium. The use of fluorescent or absorbing pH-sensitive dyes with optical methods avoids these problems (and brings on a few others). Dyes provide reliable data as long as they do not bind to chloroplasts and can report the external pH (as in Grunhagen and Witt, 1970; Chance et al., 1970; Heath, 1972).

Even given an ideal method, it is not possible to measure proton turnover in the steady state, but only the initial rate when the light first goes on, or the efflux rate when the light is first turned off. A comparison has been made in a number of cases between this efflux rate, which is assumed to be the same as turnover rate in the steady state before the light was turned off, and the steady state rate of electron transport. However, the assumption is probably not correct. Rumberg and Schröder (1971) point out it is likely (if permeation of protons is at all faster intrinsically than that of the counterions) that a hydrogen ion diffusion potential will be created as protons begin to move out in the dark. This will reverse the polarity of the membrane, from positive inside in the light, to negative inside in the dark. These membrane potentials, which will change extremely rapidly since only a few extra ions need be involved, will tend to expel protons much more rapidly in the light than in the dark. An indication that this does occur comes from observations (Packer et al., 1968; Rumberg and Schröder, 1971) that the rate of efflux of protons in the dark is accelerated by the addition of valinomycin and K^+ to neutralize any membrane potentials.

Comparisons have also been made between the initial rate of proton uptake and the rate of electron transport. When the electron transport rate is measured at the steady state (Karlish and Avron, 1967; Lynn and Brown, 1967; Dilley, 1970) it leads to exaggerated H^+/e^- ratios, because the rate is not as fast during the steady state as that which occurs during filling up of the proton pool in the first 2 or 3 sec (Jagendorf and Hind, 1965; Crofts, 1968). An accurate comparison can only be made between the proton uptake and electron transport rates which occur at the same moment in time.

A final difficulty is a remarkable degree of variability in the observed kinetics of proton fluxes between successive runs, at least when using the glass electrode. Optimistic estimates of 25 % variation (Nishizaki and Jagendorf, 1971) and perhaps more realistic ones of 15–40 % (Telfer and Evans, 1972) can be found. The observed kinetics do follow apparent first-order rate laws (Karlish and Avron, 1968) so that rate constants can be derived; variation is apparent for these as well as the directly estimated initial rates.

2.8.3. Observed H^+/e^- and H^+/hv Ratios

For reasons such as the ones noted above, and no doubt others, the observed ratios have been highly variable. To begin with, ratios below 2.0 are not critical; they result from any degree of uncoupling of chloroplasts. They also occur when using the dark efflux rate of protons for the calculation, which are indeed lower than initial uptake rates (Dilley, 1970; Telfer and Evans, 1972, for instance); but as noted above these could be the result of inhibition by a proton diffusion potential positive on the outside.

Some studies with low-salt chloroplasts (Gross et al., 1969) showed especially low ratios, but these chloroplasts are in a highly abnormal state. Under low salt conditions the fine structure of chloroplasts may be drastically altered (Izawa and Good, 1966a,b), the chloroplasts are uniquely sensitive to electron transport inhibition by polycations (Brand et al., 1971), to proton pump inhibition by polyanions (Cohen and Jagendorf, 1972) and to uncoupling by ordinarily innocuous agents such as zwitterionic buffers (tricine, bicine) and quaternary ammonium compounds (choline) (Gross, 1971) which now tend to compete with and expel necessary divalent cations (Gross, 1972). The way in which detergents disrupt chloroplasts is different at high than at low ionic strength (Ohki et al., 1971); only at high ionic strength is a physical separation of the photosystems achieved. It might be well to remember that biological membranes are comprised of numerous molecules associated by noncovalent bonds, and these molecules are capable of rapid movements and reorientations within the membrane. Under drastically altered environmental conditions considerable reorientation of membrane components might lead to altered functional relations. At least one indication occurs in the fact that proton uptake by low-salt chloroplasts (Gross et al.,

1969), or ones in low-salt with polygalacturonate (Cohen and Jagendorf, 1972) is poorly reversible in the dark.

Others report H^+/e^- ratios close to the theoretical value of 2. These include Schwartz (1968) who used an especially rapidly responding glass electrode system, and Izawa and Hind (1967) who surmounted the slow response time of glass electrodes by using a flowing system. Telfer and Evans (1972) invented a clever and apparently effective way of correcting their observed pH tracings for the slow response of the glass electrode. With this procedure they obtained H^+/e^- ratios of close to 1.0; but this was for the dark proton efflux rate/steady-state electron transport rate. Similar measurements by Schröder *et al.* (1972), but using the pH-indicating dye phenol red rather than a glass electrode, gave H^+/e^- ratios of 1.0 also. These were increased by adding valinomycin up to a limiting value of 2.0.

Results presenting real discrepancies from the chemiosmotic concept are those which show H^+/e^- ratios much greater than 2. Some of these include reports of ratios as high as 4 to 6 (Karlish and Avron, 1967; Lynn and Brown, 1967; Dilley, 1970). In all three cases, however, the initial rate of proton influx was compared to the steady state rate rather than to the initial rate of electron transport, which, as noted above, is not a valid comparison. Thus it is not clear how serious these particular discrepancies are.

The most serious objection is found in two reports (Dilley and Vernon, 1967; Heath, 1972) of the quantum yield for H^+ uptake associated with cyclic electron flow. Dilley and Vernon used a glass electrode, and with 700 and 710 nm incident light calculated that up to 5 protons were transported per photon absorbed. The lower ratios at 650 and 685 nm they considered suspect since light absorbed by PS II may not have been functional in driving the cyclic electron transport permitted by the added dye (pyocyanine).

Heath (1972) used the much more rapidly responding optical method with bromcresol purple as external pH indicator, and took meticulous care in calibrations of the system. Again using pyocyanine the longer wavelengths were more effective than those between 640 and 690 nm. The H^+/hv ratio for initial uptake between 640 and 690 nm (with believable scatter between the experiments) was 3.5; above 700 nm it was 6.7. Both of these values, but especially the latter, are much too high to be consistent with the chemiosmotic proposal for the linkage of proton movements to electron transport, given the existence of no more than two coupling sites on the cyclic pathway.

2.8.4. Crude vs Simplified Systems

The simplest way to account for the results of Dilley and Vernon, and of Heath, might be to discard the concept of a direct connection between electron flow and proton translocation, and invoke a proton pump with potentially differing stiochiometry, running on hypothetical intermediates [as in Eq. (7)].

On the other hand, whole chloroplasts, or even subchloroplast particles now in use are still very complex, and really crude in any biochemical sense. With a complex system the simplest explanation does not always have to be the correct one. It is conceivable that chemiosmotic proton translocation does occur in the light, but that to it is added the proton-binding effect of some other reaction(s), perhaps, of an equilibrating sort. The reality of the extra H^+ translocated in the experiments of Dilley and Vernon and of Heath cannot be doubted. But since protonation and deprotonation are common to a vast proportion of biochemical reactions, their direct relation to electron transport may still be in question. Another significant question is whether these extra protons contribute in equal degree to the energetic status of thylakoid discs.

The complexity of proton movements in isolated chloroplasts is not a completely ad hoc assumption. The best published illustration is in the paper by Kahn (1971); using isolated *Euglena* chloroplasts, he documented carefully what appeared to be simultaneous proton uptake and extrusion in the light. Extrusion apparently was inhibited by DNP (in its presence the full extent of uptake appeared), and uptake was inhibited by ammonia or amines or Triton (in their presence the full extent of extrusion appeared). Light-induced proton extrusion by uncoupled chloroplasts has been reported occasionally by others (Izawa, 1970, low-salt chloroplasts; Uribe, 1972, EDTA-treated chloroplasts) and noted but not reported (R.E. McCarty, personal communication; our laboratory, some lots of spinach chloroplasts in the summer). These effects might be related to uptake of buffer cations, but have not been fully explained yet. They do lurk in the background as a quiet warning about coming to entirely firm conclusions based on the behavior of a nonspecific indicator (i.e., protons) in a crude system.

Given a complex system with the possibility of interfering reactions there are two possible routes to permit focusing on the particular reactions of interest. The first is the one used by Witt and colleagues (see Chapter 10 of this volume) the analysis of early events only using flashing light and powerful optical techniques, thereby avoiding distracting secondary events. These techniques have consistently yielded values of one proton per electron per site on the chain (for instance, see Grunhagen and Witt, 1970).

The second approach, necessary to supplement any spectroscopic analysis, is purification of the system to eliminate the distractions. This approach has hardly been started with chloroplasts. However, using components derived from mitochondria (Hinkle *et al.*, 1972) the basic validity of Mitchell's concept for the coupling of electron transport to H^+ translocation was demonstrated in a striking way. Hinkle *et al.* (1972) were able to recombine three known, purified components—Cyt c, Cyt oxidase, and phospholipids— in such a way as to form vesicles with the Cyt c on the outside. These tiny particles showed oxygen uptake when provided with an electron donor;

protons were translocated in the appropriate direction (outward), and the rate of electron transport was slow until accelerated by the addition of uncouplers. In other words they were able to reconstitute an electron-transporting, proton-translocating particle that showed respiratory control, and using known components with little room for inclusion of a mysterious "proton pump."

These experiments have yet to be duplicated with chloroplast components. Nevertheless, the direct demonstration of a working chemiosmotic model from purified components provides a strong foundation for continuing efforts to dissect fundamental processes of a similar sort (if they exist) out of the currently complex chloroplast system.

2.9. Subchloroplast Particles with "No" Proton Flux

In attempting to test the chemiosmotic hypothesis critically, several groups have developed systems which are able to phosphorylate ADP, but which have lost the ability to translocate detectable amounts of protons. These include phosphorylating particles (Wessels, 1963) made with the aid of digitonin (Nelson *et al.*, 1970) or the stroma lamellae vesicles formed after French Press treatment (Arntzen *et al.*, 1971). Also subchloroplast particles made by sonic oscillation (McCarty, 1968) showed a small amount of proton uptake; this was apparently abolished by ammonium chloride but phosphorylation was not uncoupled. And finally *R. rubrum* chromatophores (Shavit *et al.*, 1968b) had most or all of their apparent ability to catalyze a pH rise inhibited by nigericin, again with almost no effect on ATP synthesis or on the energy-linked transhydrogenase.

In each of these cases a premature conclusion was drawn that the proton pump is on a side path, irrelevant to the connection between electron transport and phosphorylation. Another explanation is that in each of these cases proton pumping occurs, but is limited to a very low (depending on the instrumentation, undetectable) level by either lack of internal buffers or lack of penetrating counterions to neutralize electric charge which builds up very quickly (see Sections 2.1.2, and 2.1.3). This explanation is indicated from the fact that further experiments in all of these cases showed that collapsing the membrane potential with electrogenic agents still caused uncoupling, even when no major amount of proton movement could be observed.

With *R. rubrum* chromatophores, observable H^+ uptake was prevented either by NH_4^+ salts, or by nigericin with K^+, and phosphorylation continued (Jackson *et al.*, 1968; Montal *et al.*, 1970; Briller and Gromet-Elhanan, 1970). But then a collapse of the membrane potential, by adding either permeating anions (Montal *et al.*, 1970) or valinomycin to carry out internal NH_4^+ (Briller and Gromet-Elhanan, 1970) or internal K^+ (Jackson *et al.*,

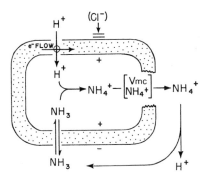

Fig. 6. Synergism between valinomycin and NH_4^+ in causing uncoupling in subchloroplast particles. NH_4^+ by itself does not uncouple, presumably due to a failure of Cl^- anions to penetrate or NH_4^+ to diffuse out, so only a few H^+ ions ever enter, ATP synthesis runs on a membrane potential, and the neutral entry of NH_3 does not dissipate the potential. Addition of valinomycin permits electrogenic exit of internal NH_4^+ which collapses the membrane potential; in the absence of a net pH gradient phosphorylation stops.

1968) caused uncoupling. Small particles made from lettuce chloroplasts with the aid of digitonin (Neumann *et al.*, 1970) or by sonicating spinach chloroplasts (McCarty, 1969) show the same behavior—a very small amount of H^+ uptake to begin with, completely abolished by NH_4Cl without destroying phosphorylation, but a synergistic interaction between ammonium salts and valinomycin in causing uncoupling (see Fig. 6).

As noted earlier (Section 2.3) the difference between chloroplasts and subchloroplast particles is likely to reside in a reduced permeability to either chloride or ammonium ions on the part of the latter. Thus a much smaller amount of H^+ uptake will lead rapidly to formation of a membrane potential, which has to be collapsed before uncoupling can occur. The smaller particles will also have less internal volume and probably a higher surface tension; accordingly they may be less able to accumulate ammonium chloride and swell the way whole chloroplasts do.

The proposed mechanisms seem forcibly confirmed by the otherwise inexplicable finding that amines, ordinarily uncouplers for whole chloroplasts, partially relieve the uncoupling caused by NH_4Cl plus valinomycin in subchloroplast particles (McCarty, 1969). The explanation lies in a competition between free base amines and ammonia for entry into chloroplasts (Crofts, 1968); the protonated amines replacing NH^{+4} internally cannot be carried out by valinomycin to collapse the membrane potential.

A further similarity between subchloroplast particles (McCarty, 1970) and *R. rubrum* chromatophores (Von Stedingk and Baltscheffsky, 1966) is that the amount of H^+ taken in is increased by valinomycin for both. This implies an inability of internal K^+ to move out in the absence of valinomycin, and again this will favor the rapid buildup of electric potential in the controls as the first amounts of H^+ move in.

In view of the above indications that H^+ uptake is limited due to rapid formation of a membrane potential, it was rather a surprise to see that estimates of internal pH for digitonin subchloroplast particles ran to 2.4 units below that of the medium (Rottenberg and Grunwald, 1972) or not so very much less than the 3.4 units found in whole chloroplasts. In all, these results suggest a combination of electrical and buffer capacity limitations on the uptake of protons by subchloroplast particles. It would be interesting to compare the stimulations of net H^+ uptake provided by valinomycin and K^+ (McCarty, 1970) with those which would probably be made possible by providing permeant anions, or pyridine, or other internal buffering bases (not yet tested).

3. PROTON GRADIENTS: RELATION TO ATP METABOLISM

3.1. Proton Uptake during ATP Hydrolysis in the Dark

The chloroplast latent ATPase, activated by a combination of light and — SH compounds (Petrack and Lipmann, 1961) continues active in the dark afterwards if supplied with ATP (Hoch and Martin, 1963; Marchant and Packer, 1963; Petrack *et al.*, 1965). If this hydrolysis truly represents a reversal of the enzyme ordinarily dehydrating ADP and P_i as would be postulated in the chemiosmotic hypothesis, it should move protons inward and build up the thylakoid pmf (see Fig. 1).

This proton translocation does occur. The first indications were the discoveries that ATP hydrolysis supports the uptake of ammonium salts (Crofts, 1966) and of amines (Gaennslen and McCarty, 1971). A direct demonstration was provided by Carmeli (1970) using phosphoenolpyruvate and pyruvate kinase to rephosphorylate the ADP formed, and thereby avoiding almost all of the complicating pH change associated with the differing pK values of P_i, ADP, and ATP.

A ratio of 1.7 amine equivalents were taken up per ATP split in the experiments of Gaennslen and McCarty (1971); and Carmeli's (1970) more direct measurements showed two protons taken up per ATP split. Thus chloroplasts appear to have an oriented proton-translocating reversible ATPase, showing a stoichiometry of $2H^+/ATP$ as predicted. It is true, as Carmeli (1970) points out, that the proton uptake could be due to operation of an unrelated proton pump, running on high energy intermediates created during ATP hydrolysis by the separate ATPase. A compelling rationale for the existence of this observed proton translocation driven by ATP hydrolysis exists within the framework of the chemiosmotic hypothesis; in alternative suggestions for the mechanism of ATP synthesis its place is largely that of a distracting side issue.

Chromatophores of *R. rubrum* in the absence of ADP form PP_i instead of ATP in the light, due to the presence of a membrane-bound, reversible, vectorial proton-translocating pyrophosphatase in addition to the usual ATPase (see Section 4.2.7.). As predicted from the chemiosmotic hypothesis these chromatophores do indeed show proton uptake during PP_i hydrolysis (Moyle *et al.*, 1972). The H^+/PP ratio was only 0.5, but this might have resulted from the presence of a second, noncoupled pyrophosphatase also present in the preparations.

3.2. Threshhold Effects in ATP Synthesis

The presence of intermediate steps between a driving reaction (electron transport in this case) and the driven reaction (ATP synthesis) should produce a finite lag in time between the onset of one and the other, depending on the pool size of the intermediates. The lag time might also be increased if there were any breakdown reactions of the intermediates competing with ATP synthesis (hydrolysis of $\sim X$; or nonuseful ion leaks).

The intermediate pool size should be very small according to chemical theories for photophosphorylation; it would be comprised of only one or two steps, stoichiometric with the passage of two electrons through a coupling site. In the chemiosmotic hypothesis the pool size would depend on whether the pmf were composed primarily of $\Delta \psi$ or of Δ pH components, quite small in the former case (since only a few charge equivalents can build a large membrane potential) and large only in the latter. In the case of Δ pH as the major driving force the extent of the lag should also depend on the concentration and pK values of internal buffering groups, and perhaps on the surface/volume ratio of the particles as well.

Lags have been observed both in the onset of phosphorylation with time (Kahn, 1962; Sakurai *et al.*, 1965) and in the steady-state rate of phosphorylation as a function of light intensity (Turner *et al.*, 1962; Schwartz, 1968; Dilley, 1968a; Saha *et al.*, 1970). On the basis of the timecourse lag Sakurai *et al.* (1965) calculated that a pool of 1 equiv for every 5 to 20 chlorophylls had to be filled before ATP synthesis started. The correlations between the onset of phosphorylation and a given steady-state level of proton uptake have been interpreted as showing a possible cause and effect relationship. Dilley (1968a) estimated that 1 H^+ has to be taken up per 5 chlorophylls for ATP synthesis to start, and Schwartz (1968) concluded that the ratio of H^+ flux to ATP synthesis had to reach a value of 2.0.

On the other hand, Saha *et al.* (1970) described almost negligible nonlinearities in the curves for phosphorylation rate vs light intensity even at the very lowest intensities, when using well-coupled chloroplasts. Only with poor chloroplasts, or in the presence of certain uncouplers, were nonlinearities detected. Now according to the chemiosmotic hypothesis the presence of a leak for H^+ as inducted by an uncoupler, should lead to a diminished pmf until faster inward H^+ pumping is accomplished by means of higher light intensities. The predicted nonlinearity in ATP synthesis rate vs light intensity was induced by some uncouplers (CCCP, DNP, gramicidin, valinomycin), but not by others (methylamine, atebrin, Triton). While it seems suggestive that the first group are all expected to collapse a membrane potential, and the second group are not, the basis for these discrepancies will have to be understood before the chemiosmotic hypothesis can be said to account for all experimental observations.

In elegant studies involving the use of groups of repeated light flashes, Junge *et al.* (1970) measured both the absorbance change at 523 nm as an indicator for the level of $\Delta\psi$ and its turnover rate, and the rate of ATP synthesis. The early indications were that a threshold level of $\Delta\psi$ is needed for phosphorylation but this interpretation seems to be superseded by later measurements (see Chapter 10 of this volume).

3.3. Diminution of the pH Gradient During ATP Synthesis

If the pmf drives ATP synthesis one might expect to see some decrease in the height of the pH gradient or membrane potential during phosphorylation. The decrease could not be a very large one, however, or the pmf would fall below the critical threshhold level needed as a driving force. On the other hand, if the pmf is a secondary parameter due to a proton pump on a side pathway [Eq. (7)] then phosphorylation would be expected to lower it to a drastic extent.

3.3.1. Amount of H^+ Uptake

The earliest indication of a lowered proton gradient due to phosphorylation was the observation (Crofts, 1966) of an inhibition of the uptake of NH_4^+ ions in the light due to the addition of ADP, P_i, and Mg. Direct demonstrations of H^+ uptake during phosphorylation are complicated by the buffering capacity of ADP and P_i (especially when used at nonlimiting concentrations) and the well-known rise in pH during ATP synthesis due to the disappearance of one of the ionized protons of P_i due to anhydride formation. The size of the pH rise can still be estimated from the decline in pH after turning of the light, however; and clear results can be obtained by using arsenate instead of P_i so that there is no net change in reactants.

One early report (Karlish and Avron, 1967) demonstrated an increase in the amount of proton uptake caused by adding ADP together with either phosphate or arsenate. The effect did not seem compatible with any previously postulated role for the pH gradient, and was rather a surprise. While a role for proton pumping in transporting ADP and phosphate across the chloroplast membranes was postulated, this concept has not been supported by later work. In retrospect it is significant that ADP and anion effects seen at that time were not affected by the energy transfer inhibitor, Dio-9. Similar insensitivity to Dio-9 of ADP plus phosphate effects on hydrogen exchange by the chloroplast coupling factor have led to postulating allosteric effector roles for these substrates (see Section 4.5.5.).

Some other groups have reported a decrease in the steady state level of H^+ uptake caused by adding ADP plus phosphate or arsenate (Schwartz, 1968; Dilley and Shavit, 1968; McCarty et al., 1971; Telfer and Evans, 1972; Schroder et al., 1972). The discrepancy with the results of Karlish and Avron (1967) seems to be a case of variability between chloroplast preparations whose origins are not yet known.

A complicating factor in these studies is the decreased rate of proton loss from membranes, and consequent higher steady state level in the light, when small amounts of ATP are present (McCarty et al., 1971; Telfer and Evens, 1972). Thus formation of $1-10$ μM ATP by the action of bound adenylic kinase on added ADP, or by photophosphorylation of ATP, can also increase the net amount of proton uptake. A decrease in net proton uptake due to simultaneous phosphorylation could be demonstrated by McCarty et al. (1972) only when strenuous efforts were made to remove ATP from chloroplasts: e.g., both arsenate to prevent new ATP from forming, and hexokinase and glucose to remove preformed ATP. On the other hand neither Schwartz (1968) nor Telfer and Evans (1972) had to take special precautions of that sort to observe some net decrease in proton uptake during photophosphorylation. The requirement for these precautions again may relate to a biologically varying precondition.

Interestingly, Schröder et al. (1972) detected less net proton uptake in chloroplasts treated only with valinomycin to permit ready proton efflux; removal of ATP as it formed was not needed.

With chromatophores of *Rhodopseudomonas capsulata* a clear increase in the extent of proton uptake occurs as phosphorylation is progressively inhibited by removal of the CF (Melandri et al., 1970). These chromatophores differ from chloroplasts in that removal of the CF does not make the membrane leaky to protons.

3.3.2. Rates of H^+ Efflux, and H^+/e^- Ratio

An acceleration of the rate of proton efflux due to simultaneous phosphorylation has been looked for in only a few cases. Dilley and Shavit (1968) found that Mg^{2+} ions by themselves will slow down the rate of proton efflux from chloroplasts, and no particular stimulation was obtained from ADP and P_i. The addition of arsenate did cause faster H^+ efflux for them. A complicating factor here is the decrease in H^+ efflux rates caused by traces of ATP, noted above.

According to the concept of Schröder *et al.* (1972) measurement of a dark proton efflux rate, which is close to that in the light, is only possible in the presence of valinomycin or some other reagent to relieve the outward proton diffusion potential. Using valinomycin they did find acceleration of proton efflux due to phosphorylation, and proportional to the rate of phosphorylation when the latter was varied by means of the CF_1 inhibitors Dio-9 or phlorizin. The linear relationship between extra rate of proton efflux and rate of ATP formation provided a ratio of $3H^+/ATP$. The same value had been derived earlier from the kinetics of decay of the 523 nm band shift in the presence or absence of ADP, P_i, and Mg^{2+} (Junge *et al.*, 1970).

Telfer and Evans (1972) point out that a difference between the formulations of Eq. (6) and (7) (pmf as an intermediate, or off on a side path) might be found in the variation in the H^+/e^- ratio depending on concurrent phosphorylation. If the pmf is on a side path, inhibiting ATP synthesis (by withholding substrates or by adding energy tranfer inhibitors) should permit more of the electron transport energy to be diverted to proton translocation, and accordingly cause a rise in the H^+/e^- ratio. If the pmf is on a direct line, nothing should change the H^+/e^- ratio.

In their own work Telfer and Evans (1972) observed a constant H^+/e^- ratio (within broad experimental limits) whether with or without ADP, ATP, Dio-9 or the complete phosphorylation mixture. This was true even when the total steady state uptake of protons was larger due to addition of one of these reagents. They concluded that the data supported the chemiosmotic hypothesis.

Dilley (1970) obtained a large number of variations in the apparent H^+/e^- ratios as a function of added ADP and P_i, ADP and AsO_4, polylysine or Dio-9. However the calculations were based on the initial rate of proton influx and the steady state rate of electron transport, which provides invalid ratios (see Section 2.8.1.). Telfer and Evans (1972) recalculated H^+/e^- ratios from Dilley's data using the maximal electron rates made possible when ADP and P_i were present in the appropriate mixture, and found relatively constant H^+/e^- ratios again.

3.4. ATP Synthesis Driven by pH Gradients

3.4.1. Postillumination ATP Synthesis

Interposition of pmf between electron transport and ATP synthesis [as in Eq. (6)] implies that the latter will continue after electron transport stops, to the extent that stored protons maintain an adequate membrane potential or proton concentration gradient. ATP synthesis after e^- transport stops is also expected if some high energy chemical intermediates are left over; however, the storage capacity could be much larger if it depends on internal protons. These predictions have been borne out in a series of experiments initiated independently by Shen and Shen (1962) and Hind and Jagendorf (1963a). In this work chloroplasts are preilluminated usually at pH 6 where the amount of proton uptake is highest, then synthesize ATP from substrates after the light is turned off. The stored material or condition responsible for ATP synthesis was initially called "X_E" but considerable subsequent work shows that it is really the store of internal protons. The ability to form ATP rises, in the light, with the same time course as that shown by proton uptake (Hind and Jagendorf, 1963b), dependent on the cofactor used. Correspondence of H^+ efflux and X_E decay time courses were also noted (Izawa, 1970).

The total amount of ATP formed even in the early experiments was greater than stoichiometric with electron carriers in the chain, amounting to as much as 40 or 50 moles of ATP per mole of Cyt f (Hind and Jagendorf, 1963a,b). Subsequently it was shown in a number of studies that the amount of ATP formed is entirely dependent on the amount of protons translocated in the previous light stage. These were varied by changing the pH of the light stage (Galmiche et al., 1967; Izawa, 1970) and by inhibition by polygalacturonate and its relief by NaCl (Cohen and Jagendorf, 1972). A significant net uptake of protons in subchloroplast particles occurred on adding valinomycin; at the same time postillumination ATP synthesis was stimulated 80% for sonicated particles and up to 7-fold for digitonin particles (McCarty, 1970). This procedure even permitted measuring X_E in R. rubrum chromatophores (McCarty, 1970).

The most dramatic instances of this correlation are found in experiments with pyridine (Nelson et al., 1971), aniline or p-phenylenediamine (Avron, 1972). Ten- to 20-fold stimulations of net proton uptake (see Section 2.1.3.) permitted 3- to 7-fold increases in the amount of postillumination ATP synthesis. Absolute values as high as 300–400 nmoles ATP/mg Chl were obtained, in place of the previous high values of 100.

An unexpected example of the same principle was illustrated in the discovery by Izawa (1970) that fairly high concentrations (0.3 mM) of phenazinium methosulfate (PMS) entered the chloroplasts at pH 8, were reduced there and contributed to internal acidification. This, as with pyridine, permitted high yields of postillumination ATP at pH 8. Ordinarily, pH 8 is not favorable due to both the smaller amount of total proton uptake, and a much more rapid decay of X_E at the higher pH.

Wang et al. (1971) present a few data showing decreased yields of postillumination ATP in the absence of added buffer. They assumed, for unknown reasons, that the amount of ATP formed was a measure of "efficiency of phosphorylation" rather than of storage capacity, and also decided to ignore the previously documented more rapid decay of X_E at higher pH's (these would occur due to the pH rise in the light in the absence of buffers). These faulty assumptions were included in an elaborate mathematical treatment. The consequent conclusion that X_E represents an accumulated chemical intermediate is highly questionable.

The method for measuring postillumination ATP formation allows the application of inhibitors separately to the dark (ATP-forming) and light stages (Hind and Jagendorf, 1965a,b) especially if ATP synthesis is accomplished in a highly diluted medium to remove the effectiveness of inhibitors applied to the light stage (Gromet-Elhanan and Avron, 1965). These studies showed that electron transport inhibitors (DCMU, 2n-nonyl-4-hydroxy-quinoline-N-oxide) acted on reactions of the light stage only; thus supporting the idea that ATP synthesis had truly been dissociated from electron transport.

Izawa (1970) presented some important measurements of the amount of ATP formed in relation to the number of protons stored. The molar ratio came out as 4 or 5 H^+/ATP at pH 6. However, some of these protons must have been lost by leakage; Izawa estimated 56% of the stored energy was captured as ATP. In that case it took 2.5 H^+ to synthesize one ATP, a number intermediate between those suggested by Schwartz (1968) and by Junge *et al.* (1970) or Schröder *et al.* (1972) on quite different grounds.

3.4.2. Acid to Base ATP Synthesis

A more definitive demonstration of ATP formation at the expense of a pH gradient came with the discovery of phosphorylation completely in the dark, when chloroplasts were placed first under acid (pH 4) then under basic (pH 8) conditions (Jagendorf and Uribe, 1966). Use of the same enzyme system as that functioning in the light was demonstrated by means of inhibition caused by the specific antibody to spinach chloroplast CF_1 (McCarty and Racker, 1966), which also inhibits photophosphorylation. Even the intimate chemistry of phosphate bond formation seemed identical, in that the same patterns of oxygen exchanges between ATP, P_i, and water were aroused by the acid–base transition as by light (Shavit and Boyer, 1966).

The acid–base transition substitutes for light in the —SH-dependent activation of both chloroplast ATPase (Kaplan *et al.*, 1967) and a weak P–ATP exchange reaction (Bachofen and Specht-Jurgensen, 1967; Kaplan and Jagendorf, 1968). These activations were thereby shown to require participation of the "high energy state" of chloroplasts rather than being directly linked to electron transport.

Evidence was obtained that an actual pH gradient is the driving force.

The pH curves for ATP synthesis, for instance, were shifted to a higher threshold pH and a higher optimum, if the acid stage was not sufficiently acid (Jagendorf and Uribe, 1966). A pH jump of 4 units was equally effective in activating the ATPase, whether from 4 to 8 or from 6 to 10 (although in each case expression of activity required pH 8) (Kaplan and Jagendorf, 1968).

Organic acids, required in the acid stage for the best yields of ATP subsequently, were proven to serve as buffers maintaining the internal pH of 4 during ATP synthesis. Effectiveness of the large number of dicarboxylic organic acids, including highly nonphysiological ones, coincided best with structures likely to penetrate at pH 4 and to be retained (as a doubly charged dianion) at pH 8 (Uribe and Jagendorf, 1967b). Using centrifugation techniques effective compounds were shown to penetrate at pH 4, and up to 20 % were retained after transition to pH 8 or 8.4 (Uribe and Jagendorf, 1967a). The kinetics of entry of succinate or other acids into this pool trapped at pH 8.4 corresponded very well with the rise time of the ability to phosphorylate ADP ("acid-poise") if moved to pH 8.4. Both the amount of acid entering and the ability to phosphorylate ADP were diminished by shrinking the thylakoids with osmotic concentrations of sucrose (Uribe and Jagendorf, 1968); apparently storing adequate amounts of organic acid buffers requires a large internal space provided in swollen thylakoid bubbles. It is not surprising, therefore, that subchloroplast particles with much diminished volume/surface area ratios are not able to make appreciable amounts of ATP this way (Nelson et al., 1970).

Freezing and thawing chloroplasts in an unfavorable medium results in a loss of the semipermeable character of thylakoid membranes (Heber, 1967; Uribe and Jagendorf, 1968). Although electron transport and fine structure is retained, these thylakoids do not repond osmotically to sucrose, fail to retain succinate after an acid–base transition, and are not able to make ATP either by acid–base or light-driven phosphorylation. This is strongly suggestive of the need for an intact, functioning osmotic membrane in these phosphorylating systems.

Two other hypotheses for the nature of the driving force after an acid–base transition were tested and largely ruled out. The energy might have come from returning cations in the base stage driving a specific, high-energy intermediate-using cation pump backward. However a variety of monovalent species (Na^+, K^+, Li^+, Tris, and choline) used as the sole cation in the activation period for the ATPase, all supported the activation process with almost equal effectiveness (Nishizaki and Jagendorf, 1969). This result is consistent with the lack of specificity for flux of counterions during proton translocation in the light.

Lynn (1968) raised the possibility that the acid–base transition changed redox potentials of internal carriers in such a way as to arouse continuing electron transport sufficient to drive some ATP formation. This was intrinsically unlikely in that (1) the amount of ATP made was highly dependent on the internal concentration of an artificial buffer, and (2) the amount made (up to 100 ATP/mole of Cyt f) was much larger than any known combination of electron-carrying and -accepting pools. Further proof (Miles and Jagendorf, 1970) came from the demonstration that a combination of three known inhibitors of electron transport (antimycin A, DCMU, and HOQNO) failed to inhibit acid–base ATP synthesis, and that efforts at both overloading internal e^- acceptors by anaerobic addition of reducing dyes, and draining internal e^- donors by oxidizing the chain with PS I light in the presence of O_2, methyl viologen, and DCMU, were similarly without effect.

A pH gradient is not an isolable intermediate. The sum of the above indirect experiments serves to prove its existence as the driving force for ATP synthesis in the acid–base experiment, and accordingly shows that chloroplasts contain the machinery for a transduction from ΔpH energy to that of the terminal bond of ATP, without returning through electron transport.

3.5. Considerations of Energetics

The central thermodynamic question for the chemiosmotic hypothesis is whether a proton electrochemical activity gradient has enough energy to drive the phosphorylation of ADP. The energy required for this act is measured by the free energy of hydrolysis of the terminal phosphate of ATP. Its actual value in any specific situation is a function of many parameters including pH, temperature, Mg^{2+} ion concentration, and ionic strength, as well as the better known facets of standard free energy and concentrations of all reactants and products:

$$\Delta G' = \Delta G'_0 - 1.36 \log \frac{(ATP)}{(ADP)(P_i)} \tag{8}$$

(note that in Eq. (8) the concentration of water is taken to be a constant, and so does not appear).

Estimates of actual $\Delta G'$ values for the last phosphate of ATP in reaction mixtures where chloroplasts were synthesizing ATP, were made by Kraayenhof (1969) following similar earlier work by Pressman and by Slater's group with mitochondria. In these studies efforts were made to find the point at which ATP synthesis stopped because the ratio of product to reactants in Eq. (8) reached too high a value, i.e., the "state 4–state 3" transition point. At this point the driving force available from electron transport, whether channeled through a chemiosmotic mechanism, or any other, would be stretched to its maximum ability. Under these specific conditions the (ATP)/

(ADP) ratio was about 10^4, and Kraayenhof calculated the $\Delta G'$ for the last ATP to be made as the extraordinarily high value of -15.5 kcal/mole. However, Kraayenhof's calculations necessarily included the published estimates for the $\Delta G'_0$ of ATP hydrolysis available at that time; and these have since been lowered some 2.33 kcal in more recent work by Rosing and Slater (1972) to about -7.5 kcal/mole. Hence Kraayenhof's work should be taken to indicate that chloroplasts can continue to make ATP until the $\Delta G'$ for ATP hydrolysis reaches -13 to -13.5 kcal/mole. This value, or something close to it, seems to be the challenge that has to be met by whatever driving force is invoked.

The energy available from a proton activity gradient can be represented (Mitchell, 1968) as:

$$\Delta G' = RT \ln \frac{\{H^+\}_{in}}{\{H^+\}_{out}} = 1.36 \log \frac{\{H^+\}_{in}}{\{H^+\}_{out}} \qquad (9)$$

where H^+ represents the electrochemical activity of hydrogen ions. Electrochemical activity represents the sum of chemical concentration (activity) and electrical forces acting on the ion, so that:

$$\log \{H^+\} = pH + \frac{\mathscr{F}}{RT}[\psi] = pH + \frac{\psi}{59} \qquad (10)$$

Where ψ represents the electric potential, $\mathscr{F}$ is the Faraday constant and R and T have the usual meanings. Substituting into Eq. (9):

$$\Delta G' = 1.36 \log \frac{[pH + \frac{\psi}{59}]_{in}}{[pH + \frac{\psi}{59}]_{out}} = 1.36[\Delta pH] + 1.36\left[\frac{\Delta \psi}{59}\right] \qquad (11)$$

The $\Delta G'$ term in Eq. (11) is the quantitative expression of the pmf term used all along. The energy driving ATP synthesis is that of the pmf times the number of protons consumed per ATP formed. Table I shows the value of the pmf required to synthesize ATP at three different possible values for its $\Delta G'$ of hydrolysis, depending on whether 2 or 3 protons are used per ATP. The pmf values are listed both in millivolts (directly applicable, if the pmf is entirely in the form of $\Delta \psi$), and in pH units. In any actual situation the total pmf would be made up partly one and partly the other.

Notice that the millivolt units for the pmf needed have the same values as the mV for the difference in redox potential between the electron donor and the electron acceptor at the phosphorylating site. If unreasonably high values of the $\Delta G'$ of ATP are calculated, they present difficulties for any proposed mechanism of phosphorylation using energy available in a given redox reaction. Thus Böhme and Cramer (1972b) presented evidence for a coupling

TABLE I
Protonmotive Force Requirements for ATP Synthesis

ATP $\Delta G'$	Required pmf in mV at given H^+/ATP		Required pmf in ΔpH units at given H^+/ATP	
(kcal)	2	3	2	3
-8.1	177	118	2.9	2.0
-13.0	283	188	4.8	3.1
-15.5	336	224	5.7	3.8

*Shown are required values for the protonmotive force, in millivolts and also in pH units, to synthesize ATP at the possible $\Delta G'$ values shown, depending on whether 2 or 3 protons are consumed per ATP synthesized.

site between plastoquinone and Cyt f; and the oxidation–reduction potential gap between these two was estimated at 250 mV. This is less than enough for the phosphorylation of ATP at -15.5 kcal/mole (Table I) no matter what the mechanism of conversion (assuming, as usual, 2 electrons per ATP). With appropriate assumptions about the steady state redox levels of the two components, it would be enough to drive ATP synthesis in the neighborhood of -13 kcal/mole, or below.

Since a pH gradient of 3.4 units has been estimated (Section 2.4) the data of Table I indicate that the stoichiometry (H^+/ATP ratio) could be quite important. If only two H^+ are consumed in making one ATP, the observed pH gradient will have to be supplemented with 83 mV of a membrane potential available at the same time in the steady state, to drive ATP synthesis at -13 kcal/mole (the value of -15.5 kcal/mole should be regarded as unrealistically high). However, if $3H^+$ are consumed in making ATP (a possibility still open since the mechanism is as yet unknown) then the observed pH gradients are sufficient at present levels, although with only a small margin for less than 100% efficiency. On the face of it these calculations do indicate a mechanism for ATP synthesis differing in detail than that originally proposed by Mitchell, in particular at the dehydration end of the sequence.

It seems too early to regard the thermodynamic arguments as being at all critical in evaluating the chemiosmotic or alternative hypotheses for ATP synthesis. For one thing any thermodynamic limitation is a constraint on all proposed hypotheses, and until alternatives are more clearly delineated it is not possible to know how serious the limitation is for them. And in view of the past history of fluctuations in the $\Delta G'_0$ for ATP, it is hard to be sure that even the present measurement will stand for all time.

These questions are intimately associated with the still contradictory data concerning the H^+/ATP ratio. Thus while Witt, Rumberg, and coworkers calculate a ratio of 3.0, Schwartz (1968) reported a value of 2.0 during ATP synthesis, and Carmeli (1970) demonstrated the flux of 2.0 H^+ for each ATP split. Most of these data might be resolved if the direct operation of the ATPase (backward or forward) translocates $2H^+$ (in alternative directions, respectively), but an extra proton derived from electron transport is needed during ATP synthesis to fulfill some accessory role in binding or effective concentrations of the substrates at the enzyme site.

The last statement might sound quite speculative, but there is an interesting analogy in the "energy-linked" transfer of reducing equivalents between NADH and $NADP^+$, occurring both in mitochondria and in *R. rubrum* chromatophores (Keister and Minton, 1969). The equivalent of one high energy bond is needed to drive electrons from NADH to $NADP^+$ in these systems, in spite of the fact that when measured in solution the two coenzymes have virtually identical oxidation–reduction potentials. Perhaps even more striking is the fact that electron transport from NADPH to NAD^+ can be shown to cause the "energization" of mitochondrial particles, including the net synthesis of ATP (Van de Stadt *et al.* 1971) and, as might be expected, the translocation of protons across the membrane (Moyle and Mitchell, 1973). While the synthesis of ATP is undoubtedly paid for in terms of concentrations of glucose-6-phosphate and acetaldehyde (the original electron donors and acceptor, respectively, in the system used), the phenomenon indicates that the effective potentials of the coenzymes are different from those found in solution, and that the machinery for proton translocation is involved in this altered energetic relationship. As with ATP synthesis a complete understanding of the mechanism is still to come.

3.6. Some General Interrelations

Before plunging into specific biochemical details it might be well to consider briefly the operation of chromatophores or chloroplasts in a larger, more integrated sense. The outstanding fact is that almost all the various aspects of energy metabolism have been shown to be in equilibrium with each other: the chromatophore or thylakoid vesicle is an integrated domain (Table II) showing almost complete reversibility.

3.6.1. Chromatophores

Energy input has been demonstrated not only from light, but also via the hydrolysis of either ATP or PP_i (Baltscheffsky, 1967; Keister and Yike, 1967; Baltscheffsky, 1969a,b; Jones and Vernon, 1969; Keister and Minton, 1969, 1971; Isaev *et al.*, 1970; Azzi *et al.*, 1971; Moyle *et al.*, 1972; Johansson *et al.*, 1972), from the energy-yielding transport of electrons from NADPH

TABLE II
Linked Reactions of Energy Metabolism in Chromatophores and in Chloroplasts

Organelle	Feasible energy inputs	Driven reactions or energy indicators
Chromatophores[b]	Light	ATP synthesis
	ATP hydrolysis	PP synthesis
	PP_i hydrolysis	Proton and anion uptake
	e^- transport from NADPH to NAD^+	Carotenoid bandshift
	(Proton gradient, using valinomycin)[a]	ANS fluorescence enhancement
		e^- transport from succinate to NAD^+
		e^- transport from NADH to $NADP^+$ (i.e., transhydrogenase)
		(Delayed light emission)[d]
Chloroplasts	Light	Proton uptake
	ATP hydrolysis	Activation of ATPase
	Proton gradient[e]	ATP synthesis
	Membrane potential[f]	Carotenoid band shift
		(Delayed light emission)[d]
		(CO_2 fixation)[g]

[a] See text for references.
[b] *Rhodospiriluum rubrum* primarily.
[c] Small extent so far.
[d] Facilitates light emission but is not the primary source of energy (see Chapter 5).
[e] Either postillumination, or an acid–base transition in the dark.
[f] With subchloroplasts particles; with whole chloroplasts in flashing light.
[g] Indirect evidence only.

to NAD^+ (Isaev *et al.*, 1970) and to a small extent from a proton gradient built up in the light with the help of valinomycin (McCarty, 1970). These have variously been shown to drive, aside from ordinary electron transport and phosphorylation in the light, the inward translocation of protons (Moyle *et al.*, 1972), reverse electron transport between bound cytochromes (Baltscheffsky, 1967, 1969b), ATP synthesis (Keister and Minton, 1971), the reduction of NAD^+ from donors such as succinate or ascorbate (Jackson and Crofts, 1968; Jones and Vernon, 1969; Isaev *et al.*, 1970; Govindjee and Sybesma, 1972), the energy-requiring reduction of $NADP^+$ by NADH (Keister and Minton, 1969), creation of a "high-energy state" or pmf as indicated by the carotenoid band shift (Baltscheffsky, 1969a) or by fluorescence of the added dye ANS (Azzi *et al.*, 1971; Johansson *et al.*, 1972), uptake of protons and added artificial anions (Isaev *et al.*, 1970), and facilitation of delayed light emission (Fleishman, 1970). Energy input from ATPase and

PPase has been shown for all of these reactions; the proton gradient has been used only for ATP synthesis so far, and energy from the transhydrogenase only for proton and permeant anion uptake. (Reversal of transhydrogenase leading to ATP synthesis has been demonstrated only in submitochondrial particles, so far.)

3.6.2. Chloroplasts

The range of biochemical activities is more limited here in that there is no membrane-bound hydrolase for PP_i and no energy-linked transhydrogenase. Energy inputs can come from light, from ATP hydrolysis, or from a proton gradient imposed in the dark. These inputs are used for proton (or amine) uptake, for ATP synthesis, and for activation of the ATPase. The high energy state facilitates delayed light emission; as with bacteria there is not sufficient energy in ATP or a proton gradient to drive emission of light at 40 kcal/einstein, but these inputs do help lower the activation energy for recombination of stored oxidants and reductants (see Chapter 5 of this volume).

A further role for the high-energy state might lie in facilitating CO_2 fixation by ribulose diphosphate carboxylase (Champigny and Migniac-Maslow, 1971; Migniac-Maslow and Champigny, 1971): inhibition of ^{32}P-fixing reactions by whole chloroplasts by antimycin A was shown to cause more rapid rates of CO_2 fixation into phosphoglyceric acid. The mechanism of this effect is not known.

An integration of different electron transport chains in the chloroplast was claimed by Siggel et al. (1972). From characteristics of DCMU inhibition at low light intensities or in flashing light, they estimated that up to 10 electron transport chains can trade electrons at the plastoquinone level.

The largest interacting unit is claimed to be the entire thylakoid comprising about 10^5 Chl molecules and correspondingly several hundred electron transport chains (Witt, 1971; and his chapter in this volume). The evidence is based on inhibitory effects by low concentrations of gramicidin D; essentially one molecule of the uncoupler per thylakoid is enough to cause a perceptible leakage of protons and loss of energy.

The opposite conclusion was reached by Kahn (1970) on the basis of experiments with tributyltin chloride and Euglena chloroplasts. Actually the experimental observation is essentially similar to those made by Witt with gramicidin D and spinach chloroplasts: chlorotributyltin continues to be an effective inhibitor of photophosphorylation at very low concentrations, and in this case even at sluggish electron flow rates. The difference lies in the interpretation of the action of the two compounds; Kahn (1968) had presented evidence that radioactive tributyltin is tightly bound to chloroplasts, and acts as an energy transfer inhibitor but not an uncoupler. Given this mode of action, then Kahn's results indicate that coupling sites are not shared in common over the whole-thylakoid discs. He estimated that each electron transport chain had the use of only 2 to 4 coupling sites. However the organotin compounds are the same ones that in other hands have also been shown to uncouple chloroplasts by facilitating free diffusion of anions (Watling and Selwyn, 1970). The potential for a residual uncoupling action of tributyltin over and above its binding reaction will have to be carefully evaluated, with respect to both *Euglena* and spinach chloroplasts.

4. BIOCHEMISTRY OF THE REVERSIBLE ATPASE

4.1. Coupling Factors—Removal and Reconstitution

4.1.1. Chloroplasts

Incubating chloroplasts in 1 mM EDTA at low ionic strength was shown by Avron (1963) to cause dissociation of a protein needed for phosphorylation, and thereby uncoupling. The protein could be rebound to chloroplasts and cause recoupling with the aid of Mg^{2+} ions. This protein was subsequently shown (McCarty and Racker, 1966) to be identical to a latent, trypsin-activated ATPase which was first extracted from chloroplasts by sonic oscillation in the presence of phospholipids (Vambutas and Racker, 1965). The protein before its activation to ATPase could reconstitute phosphorylation just as the EDTA-extracted protein did. Subsequently, further purification has been accomplished (Howell and Moudrianakis, 1967; Bennun and Racker, 1969; Farron, 1970; Lien and Racker, 1971a) and activity is shown to reside in a complex protein with molecular weight 325,000. The CF prepared by Lynn and Straub (1969a,b) was said to have a molecular weight of 250,000. The amino acid composition (Farron, 1970) is characterized by the absence of tryptophan, and the presence of only 12 —SH groups per mole. Dissociation in SDS followed by electrophoresis shows five different subunit peptide chains, three major and two minor ones (Racker *et al.*, 1972; Nelson, *et al.*, 1972a).

Electron microscopy of the isolated protein or of the one still on thylakoid membranes shows CF_1 to be a 90–100 Å particle, located on the outside of thylakoid vesicles (Vambutas and Racker, 1965; Howell and Moudrianakis, 1967). This conclusion is verified by the fact that CF_1 on the chloroplasts is accessible to antibody binding (Berzborn *et al.*, 1966; McCarty and Racker, 1966). The number of such 90 Å spheres has been estimated at 1 per 100 chlorophylls or about 3 to 4 per electron transport chain (Murakami, 1968). This estimate is close to that of Kahn (1968) for the number (1 per 60 chlorophylls in spinach, 1 per 120 chlorophylls in *Euglena*) of energy-conserving sites based on binding studies with chlorotributyltin.

The function of CF_1 is further corroborated by the *Chlamydomonas* mutant *F-54* (Sato, 1970) which is unable to phosphorylate, and lacks both the 90 Å particles on the surface of the thylakoid membranes, and the light-activated ATPase.

Although EDTA uncouples phosphorylation completely, this occurs with about half the CF_1 molecules remaining on the membranes (McCarty and Racker, 1967) as judged from residual trypsin-activatable Ca-dependent ATPase activity. Complete removal of CF_1 is possible either with the use of dilute (0.8%) silicotungstate solutions (Lien and Racker, 1971b) or by repeated extractions with EDTA. In both cases the reconstitution of phosphorylation by the residual membranes is difficult and incomplete; the more complete extraction seems to cause some so far irreversible damage to the system.

The nature of the function of those CF_1 molecules that rebind to depleted membranes is not entirely clear. The synthetic organic chemical DCCD can also restore phosphorylation as long as some endogenous CF_1 remains on the membranes (see Section 4.4.3(a)). DCCD is obviously not a catalyst of phosphorylation reactions, but does aid in restoring membrane impermeability to ions. The catalytic activity of these restored membranes must be satisfied by residual bound CF_1. In turn the critical part of the activity of CF_1 when reconstituting phosphorylation must lie in this structural role. But on the other hand the added CF_1 could be sharing the catalytic function with bound CF_1; since the catalytic function is not rate-limiting under these conditions additional evidence is required.

There is an intriguing possibility that the CF_1 of "Hatch-Slack" (or C_4) plant chloroplasts may have been modified during evolution. This is suggested by the report (Chen *et al.*, 1969) that phosphorylation by Bermuda grass chloroplasts has an apparent *Km* for P_i of 50 μM, whereas it is in the neighborhood of 600 μM for spinach chloroplasts (see for instance Ryrie and Jagendorf, 1971a). The Bermuda grass *Km* for ADP of 25 μM, however, is very close to that reported for spinach chloroplasts (Bennun and Avron, 1965; Ryrie and Jagendorf, 1971a).

4.1.2. Bacteria

Baccarini-Melandri *et al.* (1970) discovered that a CF could be removed from chromatophores of *Rhodopseudomonas capsulata* by a brief sonic oscillation in the presence of EDTA, even though EDTA washes alone were not effective. Phosphorylation was restored to the depleted particles by adding the (cold-liable, ATP-protected) supernatant protein back, together with Mg^{2+} ions. A similar procedure (Johansson, 1972) permitted the extraction of CF from *Rhodospirillum rubrum*. Simple extraction with tricine buffer at low ionic strength was sufficient to remove a CF from chromatophores of *Chromatium*, and rebinding with Mg^{2+} ions was demonstrated (Hochman and Carmeli, 1971).

Very little has been done yet with these isolated enzymes. However, the experimental advantages of working with bacteria may make these extremely useful systems in the long run. Zilinsky *et al.* (1971) selected arsenate-resistant mutants of *R. capsulata*, and among them found one strain with very much higher levels of the CF (as well as photoreaction centers and cytochromes) per unit membrane or per bacteriochlorophyll than in normal strains (Lien *et al.*, 1971).

An interesting aspect of photophosphorylation in many bacteria has always been that of its relation to oxidative phosphorylation in the dark. No effective separation seems possible of separate organelles for these two functions (Geller, 1963). *R. capsulata* grown heterotrophically in the dark yields particles capable of oxidative phosphorylation only; in the light the resulting chromatophores catalyze photophosphorylation. Melandri *et al.* (1971) showed that CF molecules derived from each of these particles could rebind and function in phosphorylation by the other. It is quite possible that the same protein functions in the two different systems, but this remains to be proven. These results contrast with the failure to trade coupling factors between spinach chloroplasts and rat heart mitochondria.

4.2. ATPase and PPase

4.2.1. CF₁ Bound to Chloroplasts

Activation of the chloroplast ATPase by cysteine and light was discovered by Petrack and Lipmann (1961). Other—SH compounds replace cysteine; DTT is especially effective. The chloroplast-bound activity requires Mg^{2+} ions. Rates of hydrolysis of ATP, GTP, and ITP are in the ratio of 3.0/1.0/0.65 (Bennun and Avron, 1965) which is consistent with the rates of phosphorylation of the respective nucleosides in the light. ADP inhibits ATPase, competitive with ATP, having a K_i of 40 μM (compared with *Km* values of 20–60 μM in phosphorylation). The activation of ATPase by light and DTT, and a number of other relationships, are diagrammed in Fig. 7.

Once activated in the light, ATPase continues to act in the dark afterward (Hoch and Martin, 1963; Marchant and Packer, 1963; Petrack *et al.*, 1965). Activation is also accomplished by acid–base transition in the dark in place of light (Kaplan *et al.*, 1967) or by prolonged dark incubation of chloroplasts with DTT in the absence of Tris (McCarty and Racker, 1968).

Other activations of the ATPase are accomplished by the use of trypsin. Ordinarily this leads to detaching CF₁ from the membranes and activating the Ca^{2+}-dependent function (McCarty and Racker, 1968). However, it appears that most commercial trypsin is contaminated with a small amount

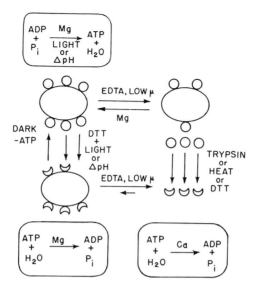

Fig. 7. The relations between phosphorylation activity of CF_1 (top left), Mg^{2+}-dependent ATPase of the chloroplasts (bottom left), detached CF_1 (top right) and Ca^{2+}-dependent ATPase of the soluble enzyme (bottom right) (see text; list of abbreviations).

of chymotrypsin. Lynn and Straub (1969a,b) first showed that very pure trypsin (commercially treated to inactivate chymotrypsin) activated the chloroplast ATPase without detaching it (see also Lien and Racker, 1971b). This bound ATPase again needs Mg^{2+} ions for activity.

The current concept (Nelson *et al.*, 1972b) for the nature of these various activations is that they represent different ways of removing—either by hydrolysis, or by some sort of conformational change—an inhibitory subunit of CF_1 which ordinarily binds to and obscures the active site of the ATPase. This concept is discussed further in Section 4.7.

It is intriguing that ATPase activity (and also a P–ATP exchange; see Section 4.3.1.) seems to be overt rather than latent in "class I" chloroplasts containing stroma and possibly some outer membrane material (Kraayenhof *et al.*, 1969; Frackowiak and Kaniuga, 1971). Some caution should be used in interpretation, since intact chloroplasts contain many more unrelated enzymes and are necessarily more crude than class II chloroplasts which are nothing but naked lamellar systems. In particular the presence of cytologically detectable ATPase localized on the chloroplast outer membrane *in vivo* (Sabnis *et al.*, 1970) is worrisome. However, the ATPase detected in class I chloroplasts resembles that of class II chloroplasts at least in sensitivity to the same inhibitors. If they are the same, it means that the inhibitory subunit is not bound to the active ATPase site of CF_1 when the enzyme is surrounded by its native stroma environment. In that case the latent state of ATPase in ordinary (class II) chloroplasts would be an artifact arising during chloroplast breakage and washing. Since most work has been done with these broken preparations, the activities of intact chloroplasts will not be discussed further.

The Mg^{2+}-dependent ATPase of chloroplasts shows rather complex behavior with respect to addition of ATP and of uncouplers. This is because of three salient facts: (1) activation, and continued activity of the ATPase requires the thylakoid membranes to be in the "high-energy state" (large internal pmf); (2) in the absence of the pmf, ATPase activity of the protein disappears with a half-time of 2–45 sec or more depending on the circumstances, and (3) the action of the enzyme results in the inward translocation of protons, thereby tending to maintain the pmf as long as ATP is present.

The activation process was noted above. The decay of ATPase is seen in experiments where ATP is withheld for varying periods after turning off the light, or after the acid–base transition. ADP causes a dramatic increase in the rate of disappearance of ATPase, acting at 1 μM concentrations (compared to 40 μM *Km* in phosphorylation). On the other hand P_i both prevents the ADP destabilization (Carmeli and Lifschitz, 1972) and stabilizes the ATPase by itself, also working at an extremely low concentration (20 μM, compared to 600 μM in phosphorylation). The ADP function here is more specific than in phosphorylation—GDP, UDP, and IDP have to be present at 100-fold greater concentrations to bring about 50% of the ADP effect. These effects of ADP and P_i probably occur as they bind to allosteric sites.

Uncouplers prevent the activation of the chloroplast ATPase, whether by light or by the acid–base transition. But addition of uncouplers afterward during ATP hydrolysis often stimulates the rate (Hoch and Martin, 1963; Petrack *et al.*, 1965; McCarty and Racker, 1968; Carmeli, 1969). This is probably an interesting analogy to the uncoupler stimulation of electron transport rates, since the ATPase is coupled to inward translocation of protons, when the internal pmf reaches too high a value the back pressure will slow down ATP hydrolysis. Since the ATPase needs a minimal level of the pmf to stay active, optimal activity requires a delicate balance of uncoupler concentration to avoid draining the pmf to too low a value. Indeed higher concentrations of uncouplers, or ordinary concentrations after prolonged exposure, do inhibit the ATPase (McCarty and Racker, 1968; Carmeli, 1969).

4.2.2. CF₁ Detached from Chloroplasts

On removal from chloroplasts by EDTA or by sonic oscillation CF_1 is still a latent ATPase. Its hydrolytic function can be aroused in three basic ways, with two variations (Table III): by trypsin treatment (Vambutas and

TABLE III
Properties of CF_1 Activated in Different Ways for Ca^{2+}-Dependent ATPase

Activation by	Ca^{2+}-ATPase	Mg^{2+}-ATPase[b]	Ability to bind to thylakoids	Loss of Ca^{2+}-ATPase on binding
Trypsin	+[c]	+	0	—
Treated typsin[d]	+	+	+	0
DTT	+	0	+	+
Heat	+	+	0	—
Heat and digitonin	+	+	+	?

[a]See text for references.

[b]With HCO_3^- at pH 8 or maleate at pH 6.

[c](+) indicates the property or activity is present, 0 means it is missing and (—) means it is not applicable.

[d]Chymotrypsin removed.

Racker, 1965), by incubation with DTT in the absence of Tris (McCarty and Racker, 1968) or by heat (Farron and Racker, 1970). In all cases the resulting ATPase, unlike the one on chloroplasts, is Ca^{2+}-dependent and inhibited by Mg^{2+} ions. These relations are included in Fig. 7.

Lynn and Straub (1969b) report a procedure for isolating CF_1 in the presence of Mg^{2+} ions which results in a protein whose ATPase activity is not expressed after treatment with trypsin. Further removal of the Mg^{2+}, however, does permit activity to show up.

Some efforts were made to elucidate changes in CF_1 when it was activated to the Ca^{2+}-ATPase state by heat (Farron and Racker, 1970). The only clear chemical change detected at that point was the appearance of 2 more $-SH$ groups per mole which could be titrated with DTNB, to a total of 4 (2 are exposed in the native protein and there are a total of 12 in the molecule after denaturation and reduction).

After activation with heat or by ordinary trypsin CF_1 can no longer rebind to EDTA-extracted chloroplast membranes and reconstitute photophosphorylation (Table III). However, if the heating is done in the presence of 0.5 % digitonin (Nelson *et al.*, 1972a,b) or if the trypsin is the special chymotrypsin-free variety, rebinding is still possible. The Ca^{2+}-ATPase activated by DTT can always rebind. These phenomena will have to be understood eventually in terms of a specific site on CF_1 which is needed for binding, being obscured or destroyed under some circumstances and being available under others.

4.2.3. Inhibitors

As with the bound enzyme, ADP inhibits ATP hydrolysis, but at the much higher levels of 300–700 μM (Nelson *et al.*, 1972a). With the soluble enzyme at pH 6 ADP is an allosteric effector; the curve for rate vs ATP concentration changes from hyperbolic to sigmoid when ADP is added, and the Hill plot (a mathematical treatment permitting estimates of the number of substrate binding sites of an enzyme) shows a cooperativity for ATP of $n = 2.3$.

The "energy transfer inhibitors" for chloroplasts are compounds which inhibit phosphorylation and the extra electron transport which accompanies it, but not basal electron transport. Their inhibitory effect is relieved by uncouplers. The classical inhibitor of this sort for mitochondria, oligomycin, has no effect on chloroplasts but acts in the usual way on *R. rubrum* chromatophores (Gromet-Elhanan, 1969). The following compounds are energy transfer inhibitors for chloroplasts; phlorizin (Izawa *et al.*, 1966; Winget *et al.*, 1969), Dio-9 (McCarty *et al.*, 1965), DCCD (McCarty and Racker, 1967), synthalin (Gross *et al.*, 1968), and chlorotri-*n*-butyltin (Kahn, 1968). DCCD shows a time-dependent inhibition implying a binding reaction, which goes faster with light on (Uribe, 1972). The extremely tight binding by chlorotributyltin (Kahn, 1968) is of special interest.

CF_1 is the logical target enzyme for a compound which is an energy tranfer inhibitor. A direct inhibition of the membrane-bound Mg^{2+}-ATPase or the soluble Ca^{2+}-ATPase or both, was shown for Dio-9 (McCarty *et al.*, 1965) DCCD (McCarty and Racker, 1967), phlorizin (Carmeli, 1969) and chlorotributyltin (Kahn, 1968). The data with the last compound are only preliminary, however. DCCD inhibits the bound, Mg^{2+}-ATPase at lower concentrations than it does the solubilized enzyme (McCarty and Racker, 1967). This behavior is close to that of oligomycin with mitochondria, where the ATPase is sensitive to the inhibitor only when it is attached to membranes.

4.2.4. Relation between ATPases and Phosphorylation

The concept that the same enzyme is concerned with ATP formation, the thylakoid-bound and the soluble ATPase activities (Fig. 7) is supported in several ways. The *Chlamydomonas* mutant (Sato, 1970) lacks both the coupling factor and the Mg^{2+}-ATPase activity. The DTT-activated Ca^{2+}-ATPase can rebind to the membranes and restore phosphorylation (McCarty and Racker, 1968). Compounds which act as energy transfer inhibitors for chloroplasts affect both phosphorylation and the ATPases. Perhaps, the most compelling reason is the inhibition of all three activities by the specific antibody to purified CF_1 (McCarty and Racker, 1966).

4.2.5. Enzyme Denaturation

CF_1 is a cold-labile enzyme, especially at high salt concentrations (0.3–0.8 M) and at pH 6 (McCarty and Racker, 1966; Lien et al., 1972). Inhibition of the enzyme when attached to thylakoid membranes by these conditions (Bennun and Racker, 1969) nicely explains an early observation of uncoupling by high salt (Krogmann et al., 1959). ATP protects the enzyme from cold denaturation at all stages of purification (it is more sensitive when pure).

Inactivation of CF_1 by heat was studied (Livne and Racker, 1969a); both RNA and lipids seemed to provide some protection. Acid inactivation was found to be very much accelerated in the presence of polyanions ranging from ferri- or ferrocyanide, to heparin and polygalacturonic acid (Polya and Jagendorf, 1970).

4.2.6. Ca^{2+} vs Mg^{2+} Dependency: The Allotopic Effect

The fact that the solubilized CF_1 becomes a Ca^{2+}-dependent ATPase, but the membrane-bound enzyme whether forming or hydrolyzing ATP requires Mg^{2+}, means that the soluble form is that much further away from serving as a good model of the reactions occurring during photophosphorylation. The modification of enzyme properties depending on binding to a membrane has been given the name "allotopic" (Racker, 1967). Modifying effects of the membrane are further demonstrated by the discoveries that (1) when EDTA is used to extract CF_1 from chloroplasts pretreated by light and DTT to activate the Mg^{2+}-ATPase, Ca^{2+}-ATPase already activated is found in the extract; and (2) the soluble CF_1 activated by DTT loses its Ca^{2+}-ATPase activity on rebinding to the membranes (Table III).

The property of imposing Mg^{2+} dependency for the ATPase activity is not unique to the thylakoid membranes. For one thing some aspect of the original enzyme structure is needed; when soluble CF_1 is activated by the chymotrypsin-free trypsin, it rebinds to the chloroplasts without losing its Ca^{2+}-ATPase activity (Lien and Racker, 1971b). On the other hand, there are occasional reports that ordinary solubilized CF_1 (Karu and Moudrianakis, 1969) especially if treated with DTT (Lien and Racker, 1971b) shows a perceptible level of Mg^{2+}-dependent activity. Surprisingly, phlorizin seems to activate Mg^{2+}-dependent ATPase of soluble CF_1 at the same time that it inhibits the Ca^{2+}-ATPase (Carmeli, 1969).

The most startling transformations of the soluble enzyme, however, are caused by dicarboxylic acids and by HCO_3^- ions (Nelson et al., 1972a). When activated by heat or by trypsin (not if by DTT) CF_1 will show rapid ATP hydrolysis with Mg^{2+} ions under two circumstances: (1) at pH 6, with 60 mM maleate or some other carboxylic acids, and 8 mM Mg^{2+};

(2) at pH 8, with 50 mM HCO_3^- and only 2 mM Mg^{2+} (excess inhibits). Itaconic, malonic, and phthalic acids are about half as effective as maleic.

The nature of these effects on the CF_1 ATPase is not known, but should bring one closer to the knowledge of its function in phosphorylation. They bring to mind associations with previous effects of similar compounds on photophosphorylation, the requirement for organic acids in the acid–base transition, and the stimulation of photophosphorylation in the spring season by HCO_3^- at a pH below the optimum (Punnett and Iyer, 1964; Batra and Jagendorf, 1965) which is duplicated weakly by organic acids (Punnett, 1965).

4.2.7. Bacterial ATPases

Bacterial chromatophores appear to demonstrate ATPase without the need for a special activation step. However, either a stimulation or an inhibition of ATPase by light can be noted, and Horiuti et al. (1968) report that rate of ATP hydrolysis shows a strong dependency on the redox potential of the environment (adjusted by adding varying concentrations of ascorbate, indophenol dyes, etc.). The ATPase was stimulated by DNP (Horio et al., 1965) which is probably an indication of its proton-translocating effect.

No overt ATPase activities were found in solubilized coupling factors from either *R. capsulata* (Baccarini-Melandri et al., 1970) or *Chromatium* D (Hochman and Carmeli, 1971) but latent ATPases may be present. Preliminary evidence was presented that the CF from *R. rubrum* may have Ca^{2+}-dependent ATPase when off the chromatophores, and Mg^{2+}-dependent ATPase when on (Johansson et al., 1972).

Surprisingly, the ATPase (and also PPase) activities of *R. rubrum* chromatophores were considerably inactivated by loss of phospholipids from the membranes. Klemme et al. (1971) digested chromatophores with phospholipase and showed about a 60% inhibition of ATPase (not as severe as the 95% inhibition of photophosphorylation) which was restored on adding back "soya" phospholipids. While it may be too early to say that ATPase activity itself is an allotopic property of this enzyme when bound to an intact membrane, the behavior is different than that seen with the equivalent chloroplast enzyme. The bacterial system certainly deserves more intensive investigation.

4.2.8. Bacterial PPase

A light-dependent formation of PP_i in place of ATP by *R. rubrum* chromatophores was discovered to occur when ADP was withheld (Baltscheffsky and von Stedingk, 1966). While at first it was thought this might represent a partial or interrupted reaction of the usual ATP-forming system, it has since become apparent that it is due to a separate, oligomycin-insensitive enzyme (Baltscheffsky and Von Stedingk, 1966). Although neither the PPase nor the ATPase have been isolated in pure form yet, two subsequent papers make it seem quite certain that different enzymes are responsible for PP_i and for ATP formation. Fisher and Guillory (1969) were able to remove ATPase and PPase activities separately and selectively, the first with 2 M LiCl and the second with 3.1 % butanol. The particles retained the alternate activity in each case.

The second, very interesting demonstration of separate pathways for ATP and PP_i formation was published by Keister and Minton (1971). They allowed $R.$ $rubrum$ chromatophores to hydrolyze ^{32}P-labeled PP_i and at the same time provided ADP and unlabeled P_i to permit ATP synthesis. These reactions did occur, but none of the labeled P from PP_i appeared in the ATP that was made. Thus PP_i hydrolysis did not lead to the production of a phosphorylated enzyme which could transfer the phosphate to ADP. This seems, rather, to be another case where hydrolysis of the energetic substrate leads to inward proton translocation (Moyle et $al.$, 1972; see Section 3.6.1.) and the resulting pmf drives the synthesis of ATP. The existence of two separate membrane-located, oriented proton-translocating reversible hydrolases on one electron-transporting organelle seems to emphasize the separate roles of electron transport and of the hydro–dehydro bond reactions (Mitchell, 1972).

4.3. Exchange Reactions

4.3.1. Chloroplasts—P_i and ADP Exchanges

The study of exchange reactions, often reporting parts of the overall reaction, has often been a valuable means of gaining insight into sequential steps of enzyme action. Originally chloroplasts appeared not to have any phosphorylation-related exchange reactions similar to those discovered earlier in mitochondria; but later careful use of more sensitive methods demonstrated that an exchange between the last phosphate of ATP and P_i occurs during net ATP synthesis in the light (Shavit et $al.$, 1967). Much faster rates of this exchange occur, however, under the conditions which arouse overt ATPase activity in the light (Carmeli and Avron, 1966, 1967; McCarty and Racker, 1967; Rienits, 1967). The function of CF_1 in this exchange reaction was clearly defined, and the same inhibitors were found to affect P-ATP exchange as inhibited either phosphorylation or ATP hydrolysis (McCarty and Racker, 1968). Chloroplasts illuminated with —SH compounds also demonstrate a perceptible rate of ADP–ATP exchange, although this is highly dependent on critical concentrations of the substrates (Stewart and Rienits, 1968). These exchanges can be activated by the acid–base transition in place of light (Bachofen and Specht-Jurgensen, 1967; Kaplan and Jagendorf, 1968).

The P-ATP exchange reaction requires the conservation of a high energy bond or the equivalent, as the inner ADP portion of ATP trades its un-labeled phosphate for a labeled one. One possible way to arrange this might be to form a high energy complex between the ADP portion of ATP and the enzyme, as an intermediate step. However, Carmeli and Lifschitz (1969) demonstrated a requirement for free ADP in the exchange reaction, as if it is actually a dynamic reversal of all the steps in ATP synthesis from ADP and P_i.

An important development for further analyses is the current reconstitution of Chl-deficient vesicles having P_i-ATP exchange activity (Carmeli and Racker, 1973). During dialysis of cholate extracts (i.e., the supernatant after sedimenting green particulate matter) vesicles were formed containing the 90 Å knobs on the outside. These had almost no Chl and only 15% of Cyt b_6 and f found in whole chloroplasts. However, they retained both ATPase (working with either Ca^{2+} or Mg^{2+} in this case) and an uncoupler-inhibited P-ATP exchange activity.

4.3.2. Oxygen Exchanges

Work with the heavy isotope of oxygen, ^{18}O, has provided definitive information about the pathway of oxygen in phosphorylation as well as data about exchange reactions which are helpful in delimiting various possibilities for the mechanism of phosphorylation.

ATP synthesis requires the formation of a new P—O bond, with consequent elimination of OH^- or HOH. However both P_i and ADP contain both P and O atoms, and the question is whether the electron-rich oxygen of P_i is bonded to the electrophilic P atom of ADP, or vice-versa. As in oxidative phosphorylation (Boyer, 1958) it was found that the "bridge" oxygen of ATP formed in photosynthetic phosphorylation is derived from ADP, not from P_i (Avron and Sharon, 1960; Schultz and Boyer, 1961; Avron *et al.*, 1965). That means that ADP is the last moiety to enter, and its oxygen atom is the one that bonds to the P atom of P_i in some activated form. The data rule out the formation of some X $\sim$ ADP covalent intermediate, and are consistent with but do not prove the sequence:

$$Y + P_i \rightleftharpoons Y-P \longrightarrow Y \sim P \qquad (12)$$

$$Y \sim P + ADP \rightleftharpoons Y + ATP \qquad (13)$$

where Y is a CF functional group at the active site.

In the light chloroplasts catalyze an exchange of oxygen atoms between those of water and those of ATP (Avron *et al.*, 1965; Shavit *et al.*, 1967) just as is found in oxidative phosphorylation by mitochondria (for a useful discussion, see Boyer, 1958). Smaller but quite perceptible rates of exchange occur between oxygen atoms of water and of P_i, and between P_i and ATP

(Shavit *et al.*, 1967). These results are considered to be most consistent with a complete, dynamic reversal of the steps in ATP synthesis, occurring at the same time as the light-driven phosphorylation.

The HOH $\rightleftharpoons$ P_i exchange of oxygen which occurs in mitochondria is completely dependent on added ADP (Jones and Boyer, 1969); that in chloroplasts after activation of the ATPase seems to require either ADP or ATP (Skye *et al.*, 1967). Taking this exchange to represent a dynamic reversal of the steps leading to at least a high energy phosphate compound of some sort, the results mean that P_i cannot form a high-energy intermediate in the absence of added ADP. As a matter of fact these results are highly consistent with those detailed in Section 4.5, involving studies both of phosphate analogs such as sulfate and arsenate, and of conformational changes in CF_1 as modified by phosphate and ADP. In all cases any significant activity of the anion requires the presence of ADP at the same time. These data could result most simply from an obligate order of addition of first ADP, then P_i, to substrate-binding sites of CF_1 prior to any reactions between them. However, this means that any (hypothetical) phosphorylated intermediate would only be formed after ADP was already in position, ready to convert it to ATP by nucleophilic attack of the future bridge oxygen. The chances of isolating the phosphorylated intermediate in this model seem remote indeed.

A most important point is that these exchange reactions occur in the dark and in the absence of electron transport, either following an acid–base transition (Shavit and Boyer, 1966; Shavit *et al.*, 1967) or in chloroplasts "modified" by illumination with — SH compounds to activate the ATPase (Skye *et al.*, 1967). The fact that these chemical reactions of ATP synthesis are quite separate from the oxidation–reduction reactions means that the otherwise intriguing "model reaction" studies of Tu and Wang (1970) are most probably quite irrelevant to photosynthetic phosphorylation.

Chaney and Boyer (1969) attempted to explore the nature of the chemical reactions of phosphorylation in more detail by measuring the relative rates of transfer of the two parts of the P_i ion, i.e., that of its oxygen atom into water and of its P atom into ATP. Any discrepancy in these rates could have been taken as a sign of the existence of a pool of some other intermediate in the process. No discrepancy was detected; either the pool does not exist, or its size is less than 3 per Cyt f (i.e., less than the number of CF_1 molecules present). While the results cannot be extended beyond the quantitative limits of the methods used, they are in harmony with a mechanism in which the oxygen from P_i goes directly to water without passing through any covalent compound.

Thus the present evidence from exchange reactions does not provide any clear evidence for the existence of a covalent $X \sim P$ intermediate of detectable stability. Indeed it is quite possible that no such intermediate exists, and enzyme-bound P_i and ADP react in a concerted mechanism during formation of the first chemical high energy bond (Boyer, 1958; 1965; or for a recent variant involving stereochemical considerations, see Korman and McLick, 1972). While there have been repeated claims of indirect evidence for the existence of a high-energy phosphorylated intermediate (recent examples include Wang *et al.*, 1971; Tyszkiewicz, 1972) their conclusions are far from compelling. Expecially in view of the negative indications from exchange studies the concept of a phosphorylated intermediate has to be considered as a speculation at the present time.

4.3.3. Bacteria

A P_i–ATP exchange reaction was reported as greatly enhanced (Horio *et al.*, 1965) or actually induced (Zaugg and Vernon, 1966) by light and electron transport in *R. rubrum* chromatophores. An indication of redox poise sensitivity was found for this reaction just as for ATPase and photophosphorylation (Horio *et al.*, 1965). Again as with ATPase and photophosphorylation, a requirement for phospholipid content of the vesicles is apparent (Klemme *et al.*, 1971). The P–ATP exchange and photophosphorylation functions were less easily reconstituted by adding exogenous phospholipids after lipase digestion, and more sensitive to removal of the lipids in the first place, than were the ATPase or PPase activities. It seems likely that the ATPase as such may require lipids in its structure in the same way that the isolated Na^+-K^+ ATPase of mammalian cells does (Hegyvary and Post, 1969;), but the reactions conserving a high energy bond have a further requirement for maintenance of an osmotically intact membrane of the chromatophores.

4.4. Coupling Factor—Membrane Interrelations

4.4.1. Nature of the Binding

Since coupling factors, both chloroplast and bacterial, come off the membrane with very mild treatments it is certain that no covalent bonds are involved. The need for EDTA in detaching CF_1 and the use of Mg^{2+} or other divalent cations in rebinding suggests (but does not prove) some sort of bridging function. The need for a low ionic strength (even 10 mM Na-EDTA itself is sufficient to prevent uncoupling; Jagendorf and Smith, 1962) is best interpreted in light of current understanding of the revolution in chloroplast membrane structure and function under very low-salt conditions (see Section 2.8.3). Either other aspects of the binding are loosened, or the Mg^{2+} bridging area must be exposed to the medium better, at low ionic strengths.

Some details of the binding process were studied by Livne and Racker

(1969b). The addition of CF_1 to membranes does not follow a sharp titration curve, and it takes two incubations with fresh depleted membranes to pick up all of the CF_1 from a given solution. It was inferred from these data that a reversible equilibrium exists between CF_1 on and that off the membranes.

The current working rationale is that there are specific binding sites on the membrane, and specific sites (areas or particular groups on particular subunits) on CF_1 involved in the binding. The goal will be to identify the nature of these binding sites. The chymotrypsin content of commercial trypsin, but not trypsin itself, must damage the CF_1 binding sites(s) (Lynn and Straub, 1969b; Bennun and Racker, 1969). Removal of the two minor, especially hydrophobic subunits (Section 4.7 below) does not prevent rebinding, so the sites may occur in the three major subunits.

4.4.2. Modification of CF_1 Function Due to Binding

The most striking effect is the change from Mg^{2+} to Ca^{2+} dependency of ATPas activity, discussed above in Section 4.2.6. Other effects include a greater stability toward cold inactivation (Bennun and Racker, 1969), toward heat denaturation (Livne and Racker, 1969a) and an increased sensitivity to low concentrations of DCCD (McCarty and Racker, 1967) and ADP when CF_1 is bound to chloroplast membranes.

4.4.3. CF_1 Function in Membrane Permeability

4.4.3.1. Effects of CF_1 Removal. Removal of CF_1 from the membranes of chloroplasts causes uncoupling—loss of ability to make ATP, and faster electron transport (Jagendorf and Smith, 1962; Avron, 1963). This is accompanied by and is almost certainly due to an increased permeability of the membrane to protons and to counterions, as seen in more rapid rates of proton efflux after illumination, and in lower net proton uptake (Jagendorf and Neumann, 1965) until at completely uncoupling levels none is perceptible. As noted in Section (4.1.1) this point occurs when only half the CF_1 molecules are removed.

On the other hand, the CF can be removed from chromatophores of *R. capsulata* to the point where phosphorylation is completely gone, and proton uptake is not only not inhibited, but is actually stimulated (Melandri *et al.*, 1970). The apparent nonessentiality of CF_1 per se for membrane permeability even for chloroplasts can be deduced from the fact that H^+ uptake is restored to EDTA-uncoupled chloroplasts by addition of some phosphorylation inhibitors. These include DCCD (McCarty and Racker, 1967; Uribe, 1972), chlorotributyltin (Kahn, 1968), and most surprisingly, silicotungstate anions (Girault and Galmiche, 1972). The last compound was used at concentrations that are said to cause complete removal of CF_1 from the membranes (Lien and Racker, 1971b), although the possibility of single subunits remaining attached in the presence of silicotungstate has not been evaluated.

The above experiments are analogous to results obtained in recoupling F_1-extracted mitochondria by oligomycin. The usual inference is that when CF_1 is removed from chloroplast membranes a functional "hole" is left behind through which protons and other ions can now move freely. (Note that these membranes still do not permit sucrose or the succinate dianion to penetrate; see Uribe and Jagendorf, 1968.) However, low ionic strength in itself leads to membranes which are extremely sensitive to uncoupling (Walz *et al.*, 1971) perhaps by otherwise innocuous buffers (Gross, 1971). It is possible that the uncoupling is an intrinsic property of thylakoid membranes suspended at unphysiologically low ionic strengths, and that restoration (by high-salt concentrations) of normal membrane structure and consequent proton impermeability is not possible unless sufficient CF_1 is present. In this model the proton leakage would not necessarily occur through unoccupied CF_1 binding sites.

The function of DCCD, silicotungstate, or chlorotributyltin in healing the gap left by CF_1 removal is not well understood. The fact that these are all reagents which interact with CF_1 in the first place suggests that they might affect the membrane by modifying residual CF_1 molecules (or its subunits for silicotungstate?) which have not been fully removed.

Restoration of the proton pump by DCCD is accompanied by restoration of some photophosphorylation, as long as the DCCD level is kept low (McCarty and Racker, 1967; Uribe, 1972) *and* the particles still contained some CF_1, or limiting amounts of CF_1 were added back. Further the antibody to CF_1 continued to inhibit that photophosphorylation made possible by DCCD (McCarty and Racker, 1967). Thus it is clear that DCCD restores ion impermeability, but does not really replace CF_1 in its catalytic function in phosphorylation.

4.4.3.2. CF_1 Modifiers and Proton Translocation. The question of CF_1 participation, directly or indirectly, in proton translocation driven by electron transport was attacked by applying antibodies to CF_1 obtained from mouse serum, sufficient to inhibit phosphorylation (McCarty and Racker, 1966). No inhibition of proton movements was observed. This seemed to rule out any direct participation of CF_1 in the light-driven proton pump [obviously the proton translocation caused by ATP hydrolysis (Carmeli, 1970) is another matter]. Similarly Dio-9, phlorizin, and tributyltin inhibit phosphorylation but not proton transport. However, more recent work with rabbit antisera to CF_1 has indeed demonstrated a partial inhibition, although the way in which this occurs is not understood.

A retardation of proton efflux caused by remarkably low concentrations (3 μM) of ATP and also by Dio-9, discovered independently by McCarty *et al.* (1971) and Telfer and Evans (1972) can be taken as a definite indication of membrane permeability modification due to altered functional states of the CF_1 bound to it. This conclusion is permitted because the effects of ATP were prevented by the use of a specific antibody against CF_1 (McCarty *et al.*, 1971), hence the particular protein must be involved. Telfer and Evans (1972) state that ADP is also effective; but McCarty *et al.* (1971) were able to suppress the ADP stimulation by a combination of arsenate plus glucose and hexokinase to remove contaminating ATP. As noted above (Section 3.3.1 and 2) the acceleration of dark proton efflux during phosphorylation which McCarty *et al.* (1971) brought about by scrupulously removing ATP, was probably induced by Schröder *et al.* (1972) by means of valinomycin instead; hence ATP could be decreasing permeability to K^+ or other compensating ions as its basic effect, and this would inhibit H^+ permeation secondarily.

The inhibition of proton turnover through the chloroplast membranes brought about by low concentrations of ATP can readily account for the partial inhibition of electron transport by ATP noted earlier (Avron *et al.*, 1958). The ATP effect is maximal at pH 8. The higher internal H^+ content caused by ATP is also seen to cause an increase in the extent of postillumination ATP synthesis from added ADP and P_i (McCarty *et al.*, 1971) which is otherwise very small at pH 8.

ATP itself (at 0.1 mM) inhibits phosphorylation by chloroplasts suspended in a low-salt medium (Walz *et al.*, 1971). However, these chloroplasts are very sensitive to loss of divalent cations, which can easily be displaced by otherwise innocuous buffer cations such as choline or triethylamine, for instance (Gross, 1972). Since ATP is a strong chelator of Mg^{2+} ions, this is most likely the basis for its inhibition of low-salt chloroplasts. The chelating action of ATP could also be involved in its apparent action as an energy transfer inhibitor when used with normal chloroplasts, in the 10–100 mM range (Shavit and Herscovici, 1970).

The decreased permeability to ions and inhibition of electron transport caused by ATP or by Dio-9 can be thought of as an enhanced degree of coupling of the thylakoid membranes. These effects were intensified when using chloroplasts suspended in a low-salt medium where they were somewhat uncoupled to begin with (Walz *et al.*, 1971).

4.5. ADP and Anion Effects on Coupling Factor Conformational States

There are a number of experimental observations which suggest that (1) the phosphate binding site of CF_1 may have a surprisingly broad specificity, and (2) that CF_1 can undergo one or more pronounced conformational changes, often reflected in some functional aspect, only when both ADP and P_i or an analog are present. These considerations are over and above the phenomena of ATPase activation, which were discussed above.

4.5.1. Arsenate Uncoupling

Arsenate is competitive with P_i in ATP synthesis (Avron and Jagendorf, 1959) and presumably binds at the same site as P_i does. In order for arsenate to cause faster electron transport, however, ADP must also be present. The half-maximal effect of ADP occurs at 20 μM, which is in the same range as for its function in ATP synthesis. This evidence is not consistent with the occurrence of an unstable high-energy arseno intermediate prior to ADP participation, nor by inference with a high energy phosphate intermediate. In view of other results (below) it is most easily explained by a concerted action of ADP and the anion (P_i or AsO_4) on CF_1 in which both substrates must be present simultaneously.

4.5.2. Inhibition by Sulfate

Sulfate is also competitive with phosphate in chloroplast phosphorylations (Asada *et al.*, 1968; Hall and Telfer, 1969). Over and above the direct competition with ATP synthesis, Ryrie and Jagendorf (1971a) noted that incubating chloroplasts with sulfate in the light led to a relatively irreversible inhibition, as measured during a second illumination in a fresh reaction mixture. The inhibition came to an endpoint at 50 %, both with increasing times in the light and with increasing sulfate concentrations. On the basis of a similar inhibition of both the trypsin- and DTT-activated Ca^{2+}-dependent ATPases, it was concluded the site of the inhibition was CF_1. All the conditions necessary for phosphorylation had to be met, for this inhibition to occur: both Mg^{2+} and ADP had to be present, and the onset of the inhibition was prevented by uncouplers or by the energy transfer inhibitor, phlorizin. It would seem that CF_1 in the high energy environment, further prepared by the presence of ADP, is strained into some semipermanent unnatural and inactive configuration by being forced to deal with sulfate in place of phosphate (see Fig. 9).

It was stated that lower concentrations of phosphate were effective in antagonizing this inhibition by sulfate than were needed to compete directly with sulfate in ATP synthesis, but unfortunately attempts were not made to estimate K_i values (e.g., for P_i as antagonist of the SO_4 effect). While K_m values for ADP were not measured, GDP and IDP failed to substitute for ADP in the SO_4 inhibition, which is unlike the specificity for phosphorylation. It is thus possible that an allosteric binding site rather than a substrate site for ADP is involved. From the point of the present thesis, the significant aspect is that the CF was not affected by (perhaps did not recognize) sulfate unless ADP was present.

4.5.3. Inhibition by N-Ethylmaleimide (NEM)

The light-dependent inhibition of photophosphorylation by NEM (McCarty et al., 1972) can best be explained by postulating a light-dependent conformational change in CF_1 to expose NEM-sensitive —SH groups. As with sulfate inhibition, that by NEM, demonstrated in a second light reaction after washing the chloroplasts, comes to an endpoint with both time and NEM concentration, at a point between 35 and 70%. Other examples of reagents inhibiting phosphorylation to an approximately 50% level include those by heptane (Laber and Black, 1969) and p-chloromer-curibenzoate (Izawa and Good, 1969). McCarty et al. (1972) suggest that these 50% effects might correlate with the fact that EDTA easily removes only 50% of the CF_1, and subsequent addition of DCCD restores phosphorylation. However, the specific participation of either the easily remove-able or the EDTA-resistant portion of the membranes' CF_1 population has not been tested yet.

Also as with the sulfate experiments, uncouplers prevented the NEM inhibition, i.e., the high energy state of the membranes had to occur. Inhibition was localized to CF_1 in that the trypsin- and DTT-activated AT-Pases were inhibited to the same degree as phosphorylation was. In later experiments (McCarty and Fagan, 1973) ^{3}H-labeled NEM was found to bind to CF_1 attached to chloroplasts to a greater extent in the light than in the dark. The conditions needed for the binding coincided with those needed to inhibit phosphorylation, and it seems assured that inhibition results from the NEM-binding. A number of NEM molecules (8 to 12) were attached to CF_1 in the dark (measured after extraction of CF_1 with EDTA). These did not interfere with phosphorylation. By using unlabeled NEM to saturate these sites in the dark first, the light-specific binding of labeled NEM (added in the light only) was measured as 0.6 moles per mole of CF_1. The label was further localized to the γ subunit of CF_1 after dissociation (see Section 4.7).

Onset of the NEM inhibition differed from that of sulfate in that it did not require, but rather was inhibited completely by ADP plus P_i. Although there was a synergistic effect of the two substrates (P_i was not effective by itself) either ADP or ATP alone were able to prevent about half the inhibition. ATP worked in this case at the lower concentration range with a half maximal effect at ca 5 μM. If increased sensitivity to NEM binding results from a conformational change exposing an otherwise internal —SH group (DTNB has similar effects, according to McCarty et al., 1972), then ADP plus P_i must prevent this change from occurring.

4.5.4. Inhibition by Permanganate

Incubation of chloroplasts with 50 μM permanganate for 10 sec has produced the same symptoms of inhibition as with sulfate (Datta and Jagendorf, in preparation). Since permanganate is a fierce oxidizing agent quite likely to damage other components than CF_1, the primary measurements have been of the trypsin-activated Ca^{2+}-dependent ATPase, which does not depend on maintenance of functional chloroplast membranes. Under the conditions chosen inhibition depends on light, on Mg^{2+} and ADP, and is prevented by P_i, by ASO_4, or by NH_4Cl. Again this seems to be a case where a sensitive group of CF_1 is not available to an anion, potentially a grim analog of P_i, unless ADP is present. Inhibition here comes to a plateau level which varies between experiments from 35 to 70%, with $KMnO_4$ concentrations between 50 and 300 μM. This probably corresponds to the more definite endpoint with sulfate or with NEM.

4.5.5. Conformational Changes: Evidence from Hydrogen Exchanges

Clear evidence for conformational changes of CF_1, attached to the membranes during illumination was obtained by Ryrie and Jagendorf (1971b) using hydrogen exchange techniques. The rationale for the experiment is diagrammed in Fig. 8. Chloroplasts are illuminated in a medium containing ^3H_2O in the hope that some hydrogen-exchanging groups of the protein, ordinarily hidden from the solvent, will now be exposed and incor-

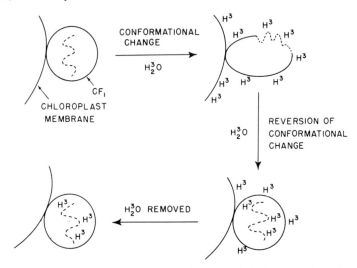

Fig. 8. Rationale for experiment incorporating ^3H atoms, by hydrogen exchange, into groups on the interior of CF_1 ordinarily hidden from the solvent but exposed during a light-dependent conformational change.

porate ^3H atoms. The experiment further depends on a return of these groups to their ordinarily hidden position at the inside of the refolded protein once the light is turned off. In that position they were able to survive steps including dilution and washing of the chloroplasts with unlabeled water, detaching the CF_1 by EDTA and purification through an ammonium sulfate and a column chromatography step. The final protein coming through the column had radioactive counts equivalent to 50 to 100 atoms per CF_1, and localization in the specific protein was shown by precipitation of the counts with a specific antibody to CF_1.

Kinetics of ^3H incorporation into these groups was expected to be complex, reflecting the rates of both the conformational change that makes it possible and the kinetics of a large series of hydrogen exchange reactions. Nevertheless, a large part of the exchange was complete in about 5 sec, and it reached an endpoint at about 90 sec in the light (Ryrie and Jagendorf, 1972).

The tritium incorporation was prevented by uncouplers, and the necessary high-energy states of the membranes could be supplied either as a proton gradient (chloroplasts in the light, or ATP hydrolysis in the dark, or an acid–base transition in the dark) or by a membrane potential (subchloroplast particles with ammonium chloride present to abolish the pH gradient). It was expected that ADP plus P_i might use up the energetic state, decrease the time the membranes stayed in the high-energy condition and so slow down ^3H incorporation. Added ADP plus P_i did slow down the rate, but unexpectedly also caused a 50% inhibition of the total amount of ^3H incorporated. A role for ADP and P_i other than that of substrates was most clearly shown by the fact that neither phlorizin nor Dio-9 prevented the partial inhibition of H-exchange caused by ADP plus P_i. Now phlorizin and Dio-9 certainly prevented ATP synthesis and therefore any drainage of the high-energy state that might have been caused by substrates. Thus the ADP plus P_i effect in this case has to be ascribed to preventing the conformational change simply because they bind, perhaps via some sort of allosteric effect. Again it is an instance where both substrates must be present to have a given effect on CF_1.

4.5.6. An Integrated Proposal

A speculative arrangement including a minimal number of conformational states of CF_1, attempting to integrate the results from P_i, arsenate, sulfate, permanganate, NEM, and hydrogen exchange experiments is shown in Fig. 9. In this formulation state S_0 (not to be confused with the "S" states involved in O_2 evolution, see Chapter 8 of this volume) is associated with the membranes' "ground" state in the dark. S_1 and S_2 are rearrangements occurring when the membranes are brought into a high internal pmf con-

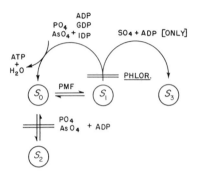

Fig. 9. Hypothetical conformational states of CF_1 (S_0 through S_3) bound to thylakoid membranes. S_1 and S_2 are achieved only via a net pmf (hydrogen ion electrochemical activity) across the vesicle membranes. For further explanation see text.

dition by light or by the acid–base transition; each one of them is responsible for 50% of the hydrogen exchange into otherwise cryptic groups. S_2 has one extra —SH group exposed to attack by NEM, but S_1 does not. It might be necessary to assume further that only about half the CF_1 molecules in a given chloroplast preparation are able to move into state S_2; this would account for the fact that NEM inhibits only 50%. Access to state S_2 would be forbidden by the presence of ADP plus P_i (or arsenate) by virtue of their binding only, not because they drain the high energy state (i.e., they work even in the presence of phlorizin).

State S_2 could not be crucial for phosphorylation, since it seems not to occur when ADP plus P_i are present. Thus state S_1 should be the form which reacts with adenylates plus anion substrate, to make ATP or other nucleotides. The evidence is clear (with arsenate, sulfate, and permanganate) that both the nucleoside and the anion must be present simultaneously. While this could be due to the formation of additional (anion-recognizing) configurational states following the binding of ADP or other nucleosides, further hypothetical conformers have been omitted in the interests of simplicity.

The reaction with the abnormal anion, sulfate, is possible only when ADP is present. The attempt to handle sulfate is indicated as putting a semipermanent crick in the structure leading to the inhibited state S_3. Reaction with permanganate will injure a group (probably —SH) which should be close to a binding site of phosphate, and would cause an artificially inhibited state S_4 (not shown).

Reactions involving phosphate (and presumably arsenate) are less demanding and occur when GDP or IDP are used instead of ADP. (Again, this could represent formation of slightly different conformer brought about by ADP vs GDP binding.) Reaction with phosphate yields ATP, with arsenate an unstable ADP-arsenate which decomposes. Only a complete return to state S_0 by one or another of these reactions will (in a manner not yet specified) allow full use of the 2 or 3 internal protons associated with ATP formation, and therefore more rapid electron transport.

It is unfortunate that at the present time data are lacking as to either the specificity of ADP (except in phosphorylation and in the sulfate inhibition) or the effective concentration range for ADP functions in most of these reactions. It is thus not yet possible to provide additional evidence for ADP or P_i actions at secondary, allosteric sites, even though these seem quite likely in most cases except for the S_1 back to S_0 transition.

The configurational state of CF_1 in which the inhibitory subunit is out of the way and the active site for Mg^{2+}-ATPase is exposed, is not shown. It ought to be derived from one of the activated states S_1 or S_2 (barring even further complications). The fact that S_2 has an extra $-SH$ group exposed to NEM attack suggests it as the point where DTT could act to bring on an active ATPase. However, the transition from S_1 to S_2 seems strictly forbidden by ADP plus P_i, whereas activation of the Mg^{2+}-ATPase is not (Carmeli and Lifshitz, 1972). Conversely the individual effects of ADP and P_i on deactivation of the ATPase have no counterparts in behavior of state S_2 as judged from hydrogen exchange data. Thus if ATPase is activated by an extension of the pathways shown, it should start out from state S_1.

Figure 9 must be considered a working hypothesis which will be replaced in time. Considerable data seem to be lacking from even the present experimental approaches, especially with respect to concentrations of nucleosides and anions which are effective. The data as they stand permit several alternative formulations; for instance, S_2 could as easily be derived from S_1 as from S_0. However, the diagram can help to bring together a number of the facts obtained so far, and may serve as a starting point for further speculation and experimentation.

4.6. Bound Adenylates

This topic represents the pioneering efforts of Roy and Moudrianakis (1971a,b) who described some interesting reactions between CF_1 and adenylates. Solubilized CF_1 was found to react quite slowly (1 to 2 hr time course) with added ^{14}C-labeled ADP to form an enzyme–ADP complex (Roy and Moudrianakis, 1971a). Two moles of ADP bind per mole of CF_1, probably at different concentration levels. These two undergo a transphosphorylation reaction forming AMP and ATP without ever leaving the enzyme surface.

Turning to chloroplasts: incubation in the light with labeled AMP (and ADP would not substitute) permitted the subsequent detection, after isolation of the enzyme, of a CF_1–ADP complex. These light reactions were faster than those seen with solubilized CF_1 and required light, P_i and AMP. Both sulfate and arsenate were inhibitory. Roy and Moudrianakis (1971b) propose a scheme in which AMP and P_i add on to the enzyme to form E-ADP which then transphosphorylates a neighboring ADP (i.e., the substrate brought in from the medium) to make the first ATP. The resulting AMP or at least ATP could dissociate, the AMP to begin the cycle all over again. More work is needed to decide if the reactions observed are on the main pathway of ATP formation, or are on a side pathway.

4.7. Subunit Structure of CF_1

With the participation of a protonmotive force in ATP synthesis largely assured (at the very least in model experiments; see Section 3.4), and the function of ATPase in causing proton translocation demonstrated directly (Section 3.1) the mechanism of action of the enzyme system able to effect the transduction between proton transport and gradients, and ATP synthesis or breakdown, becomes one of great interest. As with any other enzyme a firm knowledge of its structure at all levels and then of the nature of the interaction of this structure with substrates (in this case, including the thylakoid membranes and a transmembrane pH gradient) is a minimal prerequisite. And obviously if the speculations of Fig. 9 are ever to acquire concrete meaning the structure of the enzyme and its variations (probably via model reactions in solution) will have to be intensively explored.

The dissociation of pure CF_1 into five identifiable peptide chain subunits by 7 M urea has been accomplished recently (Racker *et al.*, 1972; Nelson *et al.*, 1973). There are three major bands (α, β, and γ) and two minor bands (δ and ε). The major subunits have molecular weights on the order of 50,000, and there are two each for a total of six for the complete molecule. McCarty *et al.* (1972) have identified the γ band as the site of attachment of NEM in the light, and Racker and colleagues have begun to identify the function of some of these subunits.

Nelson *et al.* (1973) were able to separate all the subunits and prepare rabbit antisera to each one, none of them cross-reacting. Inhibitory effects of these antisera discovered so far are summarized in Table IV. The antibodies to either α or γ subunits will inhibit phosphorylation; those to the β or δ subunits do not inhibit. No one antiserum will inhibit the Ca^{2+}-dependent ATPase, but a mixture of the anti-α and anti-γ is inhibitory. The ATP effect on membrane permeability, inhibiting proton efflux and therefore the rate of electron transport, is prevented by antibody to the α subunit.

Nelson *et al.* (1972a) identify one of the small (13,000), highly hydrophobic subunits as the inhibitor of ATPase. An assay for this function consisted

TABLE IV

Effects of CF$_1$ Subunit Antibodies on Specific Functions of the Native Enzyme

Function	Inhibited by antiserum to subunit(s)	
Photophosphorylation	α or γ	
Ca^{2+}–ATPase	(α and γ),	both required
ATP decrease of membrane permeability	α	
Inhibition of ATPase by ε subunit	γ	

[a]See text for references.

of dissociating one lot of CF$_1$ in 7 M urea, then applying a small aliquot to a second lot of CF$_1$ otherwise active as a Ca^{2+}-ATPase. Purification of the inhibitor started with dissolution of CF$_1$ in 50% pyridine then diluting the pyridine and use of ammonium chloride to precipitate the larger subunits, and column chromatography of the supernatant. The purified inhibitory subunit is not soluble in water, but is soluble in a Tris–urea mixture. It is appropriately highly sensitive to trypsin. Its hydrophobicity may account for heat-activated Ca^{2+}-ATPase failing to bind to membranes: the current feeling is that heating could remove the inhibitor from its normal site, permitting it to rebind to the (hydrophobic) portion of the remainder of CF$_1$ ordinarily concerned with binding to the hydrophobic thylakoid membrane. Heating CF$_1$ in digitonin apparently removes these small hydrophobic subunits so that they are discarded on subsequent column chromatography. The protein recovered from such a column consists of a slightly smaller CF$_1$ containing α, β, and γ subunits only. This preparation was shown to bind to chloroplast membranes, reconstitute phosphorylation, and maintain an active ATPase while bound. To the extent that this represents a partially simplified coupling factor, and provided that those CF$_1$ molecules which add back to depleted membranes perform a catalytic function, this should be the preferred material for further studies of the mechanism of photophosphorylation.

5. SUMMARY

The transformation of electron transport energy to that of the terminal phosphate (dehydro) bond of ATP is considered to go through one or more intermediate stages of still imperfectly understood nature. Hypotheses for the nature of the intermediate range from that of a distinct and therefore potentially isolable high energy chemical complex, to that of a transmembrane proton activity gradient according to the chemiosmotic hypothesis of Mitchell. A wealth of more recent suggestions tend to consider intramembrane electric potentials (charge redistributions) and conformational states of the entire membrane or its parts, as either additional or central components of the organelle high energy intermediate state.

Both chloroplasts and bacterial chromatophores, *in vitro*, demonstrate light-driven ion movements. The evidence indicates the primary event is the uptake of protons in the light, coupled to electron transport. Methods have been developed to estimate the height of both the pH gradient and the membrane potential created by proton translocation. The sum of these seems just adequate to drive ATP synthesis, but several assumptions are still involved. The action of many if not most uncouplers is best explained by their demonstrable ability to collapse either the proton gradient, the membrane potential, or both.

The formation of ATP causes some net decrease in the height of the light-driven gradients. Conversely ATP synthesis can also be driven by a stored proton gradient, present either by means of previous light and electron transport, or induced artificially in the dark by prior soaking of chloroplasts in acid. In these cases the chemistry of ATP synthesis is dissociated from reactions of electron transport.

After appropriate activation, chloroplasts hydrolyze ATP, and also catalyze exchange reactions not easily detectable before. The enzyme responsible for ATP hydrolysis has been extracted, purified, and identified by the use of inhibitors and antibodies as the same enzyme as that needed for ATP synthesis (i.e., the chloroplast CF). When ATP is hydrolyzed by the enzyme still bound to thylakoids, the reaction is coupled stoichiometrically to the inward translocation of protons across the membrane. The exchange reactions occur independent of electron flow. The pattern of oxygen exchanges and a dependence on added adenylates are not consistent with the occurrence of a high energy phosphorylated intermediate prior to ATP.

Some bacterial chromatophores can synthesize inorganic pyrophosphate in the light, in addition to ATP. This is due to the presence on their membranes of an additional enzyme capable of either forming or breaking down inorganic pyrophosphate, as well as the one metabolizing ATP. Hydrolysis of pyrophosphate in these chromatophores is similarly coupled to proton translocation.

All of the phenomena summarized above are consistent with and largely predictable from the chemiosmotic hypothesis. However, some other data, especially those concerning the stoichiometry between electron transport and proton flux, do not readily conform to its predictions.

The chemistry of the CF and the pattern of its reactivity while still on the membrane are under current study. After purification it is a protein weighing about 350,000 daltons, containing three major and two smaller kinds of subunit peptide chains. One of the smaller subunits is identifiable as the component ordinarily inhibiting ATPase activity either on or off the membrane. The protein seems able to bind 2 ADP molecules per mole at as yet unidentified sites.

While bound to thylakoid membranes, induction of the high-energy state causes the coupling factor to undergo a conformational change of some magnitude (as seen from H-exchange data), permitting one otherwise cryptic — SH group to react with NEM. The CF on the membranes appears to react with a variety of analogs of inorganic phosphate: with arsenate, as deduced from the increase in electron transport rates, and with sulfate and permanganate as deduced from the consequent induction of a partial inhibition of ATPase and of photophosphorylation. These interactions require both the high-energy state of the membranes, and the prior or simultaneous presence of ADP. At least in the case of sulfate and permanganate ADP may be filling an allosteric role. Other indications of conformational effects of the substrates are found in the prevention, by ADP plus phosphate, of both half the light-induced extra hydrogen exchange of the protein, and the exposure of the cryptic — SH group.

Intensive exploration of these and related aspects of the CF chemistry and reactivity should provide direct insight into the chemistry of ATP synthesis. The intriguing question for the future is how the action of this enzyme on its substrates is driven by the ion gradient-related electrochemical stresses on the membrane to which it is bound.

ACKNOWLEDGMENTS

I am very grateful to Drs. Racker, Nelson, McCarty, and Carmeli for permission to quote results from manuscripts in preparation or in press. Heartfelt thanks are due to Drs. Good, Racker, Nelson, Mitchell, Boyer, and Hill for critical reading in part or in whole, discussion and comments without which the number of errors would have been much larger. Support during the preparation of this manuscript came in part from grant GM 14479 from the National Institutes of Health.

REFERENCES

Allen, M. B., Whatley, F. R., and Arnon, D. I. (1958). *Biochim. Biophys. Acta* **27**, 16.
Arnon, D. I., Allen, M. B., and Whatley, F. R. (1954). *Nature (London)* **174**, 394.
Arnon, D. I., Allen, M. B., and Whatley, F. R. (1958). *Science* **127**, 1026.
Arntzen, C. J., Dilley, R. A., and Neumann, J. (1971). *Biochim. Biophys. Acta* **245**, 409.
Asada, K., Deura, R., and Kasai, Z. (1968). *Plant Cell Physiol.* **9**, 143.
Avron, M. (1963). *Biochim. Biophys. Acta* **77**, 699.
Avron, M. (1972). *Proc. Int. Photosynthesis Congr., 2nd, Stresa* (G. Forti, ed.), **2**, 861.
Avron, M., and Jagendorf, A. T. (1959). *J. Biol. Chem.* **234**, 967.

Avron, M., and Sharon, N. (1960). *Biochem. Biophys. Res. Commun.* **2**, 336.
Avron, M., and Shavit, N. (1965). *Biochim. Biophys. Acta* **109**, 317.
Avron, M., Krogmann, D. W., and Jagendorf, A. T. (1958). *Biochim. Biophys. Acta* **30**, 144.
Avron, M., Grisario, V., and Shavit, N. (1965). *J. Biol. Chem.* **240**, 1381.
Azzi, A., Baltscheffsky, M., Baltscheffsky, H., and Vainio, H. (1971). *FEBS Lett.* **17**, 49.
Azzone, G. F. (1972). *J. Bioenerg.* **3**, 95.
Baccarini-Melandri, A., Gest, H., and San Pietro, A. (1970). *J. Biol. Chem.* **245**, 1224.
Bachofen, R., and Specht-Jurgensen, I. (1967). *Z. Naturforsch.* **22b**, 1051.
Baltscheffsky, H., and von Stedingk, L.-V. (1966). *Biochem. Biophys. Res. Commun.* **22**, 722.
Baltscheffsky, M. (1967). *Biochem. Biophys. Res. Commun.* **28**, 270.
Baltscheffsky, M. (1969a). *Arch. Biochem. Biophys.* **130**, 646.
Baltscheffsky, M. (1969b). *Arch. Biochem. Biophys.* **133**, 46.
Barber, J. (1972). *FEBS Lett.* **20**, 251.
Barber, J., and Kraan, G. P. B. (1970). *Biochim. Biophys. Acta* **197**, 49.
Barber, J., and Varley, W. J. (1971). *Nature (London) New Biol.* **234**, 188.
Barber, J., and Varley, W. J. (1972). *J. Exp. Bot.* **23**, 216.
Batra, P. P., and Jagendorf, A. T. (1965). *Plant Physiol.* **40**, 1074.
Bennun, A. (1971). *Nature (London) New Biol.* **233**, 5.
Bennun, A., and Avron, M. (1965). *Biochim. Biophys. Acta* **109**, 117.
Bennun, A., and Racker, E. (1969). *J. Biol. Chem.* **244**, 1325.
Berzborn, R. J. (1968). *Z. Naturforsch.* **23b**, 1096.
Berzborn, R. J., Menke, W., Trebst, A., and Pistorius, E. (1966). *Z. Naturforsch.* **21b**, 1057.
Boardman, N. K., Anderson, J. M., and Hiller, R. G. (1971). *Biochim. Biophys. Acta* **234**, 126.
Böhme, H., and Cramer, W. A. (1972a). *Biochim. Biophys. Acta* **283**, 302.
Böhme, H., and Cramer, W. A. (1972b). *Biochemistry* **11**, 1155.
Boyer, P. D. (1958). *Proc. Int. Symp. Enzyme Chem., Tokyo and Kyoto, 1957* p. 301.
Boyer, P. D. (1965). *In* "Oxidases and Related Systems" (T. E. King, H. S. Mason, and M. Morrison, eds.), Vol. II p. 994. Wiley, New York.
Brand, J., Baszynski, T., Crane, F. L., and Krogmann, D. W. (1971). *Biochem. Biophys. Res. Commun.* **45**, 538.
Briller, S., and Gromet-Elhanan, Z. (1970). *Biochim. Biophys. Acta* **205**, 263.
Carmeli, C. (1969). *Biochim. Biophys. Acta* **189**, 256.
Carmeli, C. (1970). *FEBS Lett.* **7**, 297.
Carmeli, C., and Avron, M. (1966). *Biochem. Biophys. Res. Commun.* **24**, 923.
Carmeli, C., and Avron, M. (1967). *Eur. J. Biochem.* **2**, 318.
Carmeli, C., and Lifshitz, Y. (1969). *FEBS Lett.* **5**, 227.
Carmeli, C., and Lifshitz, Y. (1972). *Biochim. Biophys. Acta* **267**, 86.
Carmeli, C., and Racker, E. (1973). *J. Biol. Chem.* **248**, 8281.
Champigny, M. L., and Migniac-Maslow, M. (1971). *Biochim. Biophys. Acta* **234**, 335.
Chance, B. (1972). *FEBS Lett.* **23**, 3.
Chance, B., and Williams, G. R. (1956). *Advan. Enzymol.* **17**. 65.
Chance, B., Nishimura, M., Avron, and Baltscheffsky, M. (1966). *Arch. Biochem. Biophys.* **117**, 158.
Chance, B., Crofts, A. R., Nishimura, M., and Price, B. (1970). *Eur. J. Biochem.* **13**, 364.
Chaney, S. G., and Boyer, P. D. (1969). *J. Biol. Chem.* **244**, 5773.
Chen, T. M., Brown, R. H., and Black, C. C. (1969). *Plant Physiol.* **44**, 649.
Cohen, W. S., and Jagendorf, A. T. (1972). *Arch. Biochem. Biophys.* **150**, 235.
Cohen, W. S., and Sherman, L. A. (1971). *FEBS Lett.* **16**, 4.
Cost, K., and Frenkel, A. W. (1967). *Biochemistry* **6**, 663.
Cramer, W. A., and Böhme, H. (1972). *Biochim. Biophys. Acta* **256**, 358.

Crofts, A. R. (1966). *Biochem. Biophys. Res. Commun.* **24**, 725.

Crofts, A. R. (1967). *J. Biol. Chem.* **242**, 3352.

Crofts, A. R. (1968). *In* "Regulatory Functions of Biological Membranes" (J. Jarnefelt, ed.), Vol. 2, p. 247. Elsevier, Amsterdam.

Deamer, D. E., and Packer, L. (1969). *Biochim. Biophys. Acta* **172**, 539.

Dilley, R. A. (1968a). *Progr. Photosynthesis Res.* **3**, 1354.

Dilley, R. A. (1968b). *Biochemistry* **7**, 338.

Dilley, R. A. (1970). *Arch. Biochem. Biophys.* **137**, 270.

Dilley, R. A., and Shavit, N. (1968). *Biochim. Biophys. Acta* **162**, 86.

Dilley, R. A., and Vernon, L. P. (1965). *Arch. Biochem. Biophys.* **111**, 365.

Dilley, R. A., and Vernon, L. P. (1967). *Proc. Nat. Acad. Sci. U.S.* **57**, 395.

Farron, F. (1970). *Biochemistry* **9**, 3823.

Farron, F., and Racker, E. (1970). *Biochemistry* **9**, 3829.

Fisher, R. R., and Guillory, R. J. (1969). *FEBS Lett.* **3**, 27.

Fleischman, D. E. (1970). *Photochem. Photobiol.* **14**, 277.

Frackowiak, B., and Kaniuga, Z. (1971). *Biochim. Biophys. Acta* **226**, 360.

Frenkel, A. (1954). *J. Amer. Chem. Soc.* **76**, 5568.

Gaensslen, R. E., and McCarty, R. E. (1971). *Arch. Biochem. Biophys.* **147**, 55.

Galmiche, J. M., Girault, G., Tyszkiewicz, E., and Fiat, R. (1967). *C. R. Acad. Sci. Paris* **265**, 374.

Geller, D. M. (1963). *In* "Bacterial Photosynthesis" (H. Gest, A. San Pietro, and L. P. Vernon, eds.), p. 161. Antioch Press, Yellow Springs, Ohio.

Geller, D. (1957). Ph.D. Thesis, Harvard Univ., Boston, Massachusetts.

Girault, G., and Galmiche, J. M. (1972). *FEBS Lett.* **19**, 315.

Good, N., Izawa, S., and Hind, G. (1966). *Current Top. Bioenerg.* **1**, 75.

Govindjee, R., and Sybesma, C. (1972). *Biophys. J.* **12**, 897.

Green, D. E., Asai, J., Harris, R. A., and Penniston, J. T. (1968). *Arch. Biochem. Biophys.* **125**, 684.

Green, D. E. and Ji, S. (1972). *J. Bioenerg.* **3**, 159.

Gromet-Elhanan, Z. (1969). *Arch. Biochem. Biophys.* **131**, 299

Gromet-Elhanan, Z. (1971). *FEBS Lett.* **13**, 124.

Gromet-Elhanan, Z. (1972). *Biochim. Biophys. Acta* **275**, 125.

Gromet-Elhanan, Z., and Avron, M. (1965). *Plant Physiol.* **40**, 1053.

Gromet-Elhanan, Z., and Briller, S. (1969). *Biochem. Biophys. Res. Commun.* **37**, 261.

Gross, E. (1971). *Arch. Biochem. Biophys.* **147**, 77.

Gross, E. (1972). *Arch. Biochem. Biophys.* **150**, 324.

Gross, E., and San Pietro, A. (1969). *Arch. Biochem. Biophys.* **131**, 49.

Gross, E., Shavit, N., and San Pietro, A. (1968). *Arch. Biochem. Biophys.* **127**, 224.

Gross, E., Dilley, R. A., and San Pietro, A. (1969). *Arch. Biochem. Biophys.* **134**, 450.

Grünhagen, H. H., and Witt, H. T. (1970). *Z. Naturforsch* **25b**, 373.

Hall, D. O. (1972). *Nature (London) New Biol.* **235**, 125.

Hall, D. O., and Telfer, A. (1969). *Progr. Photosynthesis Res.* **3**, p. 1281.

Hall, D. O., Reeves, S. G., and Baltscheffsky, H. (1971). *Biochem. Biophys. Res. Commun.* **43**, 359.

Harold, F. M. (1970). *Advan. Microbiol. Physiol.* **4**, 45.

Hatefi, Y., and Hanstein, W. G. (1972). *J. Bioenerg.* **3**, 129.

Hauska, G. A., McCarty, R. E., and Olson, J. S. (1970). *FEBS Lett.* **7**, 151.

Hauska, G. A., McCarty, R. E., Berzborn, R. J., and Racker, E. (1971). *J. Biol. Chem.* **246**, 3524.

Heath, R. L. (1972). *Biochim. Biophys. Acta* **256**, 645.

Heber, U. (1967). *Plant Physiol.* **42**, 1343.

Hegyvary, C., and Post, R. L. (1969). *In* "The Molecular Basis of Membrane Function" (D. C. Tosteson, ed.), p. 519. Prentice-Hall, Englewood Cliffs, New Jersey.

Hind, G., and Jagendorf, A. T. (1963a). *Proc. Nat. Acad. Sci. U.S.* **49**, 715.

Hind, G., and Jagendorf, A. T. (1963b). *Z. Naturforsch.* **18b**, 689.

Hind, G., and Jagendorf, A. T. (1965a). *J. Biol. Chem.* **240**, 3195.

Hind, G., and Jagendorf, A. T. (1965b). *J. Biol. Chem.* **240**, 3202.

Hind, G., and Whittingham, C. P. (1963). *Biochim. Biophys. Acta* **75**, 194.

Hinkle, P. C., Kim, J. J., and Racker, E. (1972). *J. Biol. Chem.* **247**, 1338.

Hoch, G., and Martin, I. (1963). *Biochem. Biophys. Res. Commun.* **12**, 223.

Hochman, A., and Carmeli, C. (1971). *FEBS Lett.* **13**, 36.

Horio, T., Nishikawa, K., Katsumata, M., and Yamashita, J. (1965). *Biochim. Biophys. Acta* **94**, 371.

Horiuti, Y., Nishikawa, K., and Horio, T. (1968). *J. Biochem. (Japan)* **64**, 577.

Howell, S. H., and Moudrianakis, E. N. (1967). *Proc. Nat. Acad. Sci. U.S.* **58**, 1261.

Isaev, P. I., Liberman, E. A., Samuilov, V., Skulachev, V. P., and Tsofina, L. M. (1970). *Biochim. Biophys. Acta* **216**, 22.

Izawa, S. (1970). *Biochim. Biophys. Acta* **223**, 165.

Izawa, S., and Good, N. (1966a). *Plant Physiol.* **41**, 533.

Izawa, S., and Good, N. (1966b). *Plant Physiol.* **41**, 544.

Izawa, S., and Good, N. E. (1968). *Biochim. Biophys. Acta* **162**, 380.

Izawa, S., and Good, N. E. (1969). *Progr. Photosynthesis Res.* **3**, 1288.

Izawa, S., and Hind, G. (1967). *Biochim. Biophys. Acta* **143**, 377.

Izawa, S., Winget, G., and Good, N. E. (1966). *Biochem. Biophys. Res. Commun.* **22**, 223.

Jackson, J. B., and Crofts, A. R. (1968). *Biochem. Biophys. Res. Commun.* **32**, 908.

Jackson, J. B., and Crofts, A. R. (1969). *FEBS Lett.* **4**, 185.

Jackson, J. B., Crofts, A. R., and von Stedingk, L.-V. (1968). *Eur. J. Biochem.* **6**, 41.

Jagendorf, A. T., and Hind, G. (1963). *Nat. Acad. Sci. Nat. Res. Council Publ.* **1145**, 509.

Jagendorf, A. T., and Hind, G. (1965). *Biochim. Biophys. Res. Commun.* **18**, 702.

Jagendorf, A. T., and Neumann, J. (1965). *J. Biol. Chem.* **240**, 3210.

Jagendorf, A. T., and Smith, M. (1962). *Plant Physiol.* **37**, 135.

Jagendorf, A. T., and Uribe, E. (1966). *Proc. Nat. Acad. Sci. U.S.* **55**, 170.

Jensen, R. G. (1971). *Biochim. Biophys. Acta* **234**, 360.

Johannson, B. C. (1972). *FEBS Lett.* **20**, 339.

Johannson, B. C., Baltscheffsky, M., and Baltscheffsky, H. (1972). *Proc. Int. Photosynthesis Congr., 2nd* (G. Forti, ed.), **2**, 1203.

Jones, C. W., and Vernon, L. P. (1969). *Biochim. Biophys. Acta* **180**, 149.

Jones, D. H., and Boyer, P. D. (1969). *J. Biol. Chem.* **244**, 5767.

Junge, W. (1970). *Eur. J. Biochem.* **14**, 582.

Junge, W., and Witt, H. T. (1968). *Z. Naturforsch.* **23b**, 244.

Junge, W., Rumberg, B., and Schröder, H. (1970). *Eur. J. Biochem.* **14**, 575.

Kahn, J. S. (1962). *Arch. Biochem. Biophys.* **98**, 100.

Kahn, J. S. (1968). *Biochim. Biophys. Acta* **153**, 203.

Kahn, J. S. (1970). *Biochem. J.* **116**, 55.

Kahn, J. S. (1971). *Biochim. Bioyhys. Acta* **245**, 144.

Kaplan, J. H., and Jagendorf, A. T. (1968). *J. Biol. Chem.* **243**, 972.

Kaplan, J. H., Uribe, E., and Jagendorf, A. T. (1967). *Arch. Biochem. Biophys.* **120**, 365.

Karlish, S. J. D., and Avron, M. (1967). *Nature (London)* **216**, 1107.

Karlish, S. J. D., and Avron, M. (1968). *Biochim. Biophys. Acta* **153**, 878.

Karlish, S. J. D., Shavit, N., and Avron, M. (1969). *Eur. J. Biochem.* **9**, 291.

Karu, A. E., and Moudrianakis, E. N. (1969). *Arch. Biochem. Biophys.* **129**, 655.

Keister, D. L., and Minton, N. J. (1969). *Biochemistry* **8**, 167.

Keister, D. L., and Minton, N. J. (1971). *Arch. Biochem. Biophys.* **147**, 330.

Keister, D. L., and Yike, N. J. (1967). *Arch. Biochem. Biophys.* **121**, 415.

Klemme, B., Klemme, J.-H. and San Pietro, A. (1971). *Arch. Biochem. Biophys.* **144**, 339.

Korman, E. F., and McClick, J. (1972). *J. Bioenerg.* **3**, 147.

Kraayenhof, R. (1969). *Biochim. Biophys. Acta* **180**, 213.

Kraayenhof, R. (1970). *FEBS Lett.* **6**, 161.

Kraayenhof, R., Groot, G. S. P., and van Dam, K. (1969). *FEBS Lett.* **4**, 125.

Krogmann, D. W., Jagendorf, A. T., and Avron, M. (1959). *Plant Physiol.* **34**, 272.

Laber, L. J., and Black, C. C. (1969). *J. Biol. Chem.* **244**, 3463.

Larkum, A. W. D., and Bonner. W. D. (1972). *Biochim. Biophys. Acta* **256**, 396.

Liberman, E. A., and Skulachev, V. P. (1970). *Biochim. Biophys. Acta* **216**, 30.

Lien, S., and Racker, E. (1971a). *Methods Enzymol.* **23**, 547.

Lien, S., and Racker, E. (1971b). *J. Biol. Chem.* **246**, 4298.

Lien, S., San Pietro, A., and Gest, H. (1971). *Proc. Nat. Acad. Sci. U.S.* **68**, 1912.

Lien, S., Berzborn, R. J., and Racker, E. (1972). *J. Biol. Chem.* **247**, 3520.

Lin, D. C., and Nobel, P. S. (1971). *Arch. Biochem. Biophys.* **145**, 622.

Livne, A., and Racker, E. (1969a). *J. Biol. Chem.* **244**, 1332.

Livne, A., and Racker, E. (1969b). *J. Biol. Chem.* **244**, 1339.

Lynn, W. S. (1968). *Biochemistry* **7**, 3811.

Lynn, W. S., and Brown, R. (1967). *J. Biol. Chem.* **242**, 426.

Lynn, W. S., and Straub, K. D. (1969a). *Proc. Nat. Acad. Sci. U.S.* **63**, 540.

Lynn, W. S., and Straub, K. D. (1969b). *Biochemistry* **8**, 4789.

Marchant, R., and Packer, L. (1963). *Biochim. Biophys. Acta* **75**, 458.

McCarty, R. E. (1968). *Biochem. Biophys. Res. Commun.* **32**, 37.

McCarty, R. E. (1969). *J. Biol. Chem.* **244**, 4292.

McCarty, R. E. (1970). *FEBS Lett.* **9**, 313.

McCarty, R. E., and Coleman, C. H. (1970). *Arch. Biochem. Biophys.* **141**, 198.

McCarty, R. E., and Fagan, J. (1973). *Biochemistry* **12**, 1503.

McCarty, R. E., and Racker, E. (1966). *Brookhaven Symp. Biol.* **19**, 202.

McCarty, R. E., and Racker, E. (1967). *J. Biol. Chem.* **242**, 3435.

McCarty, R. E., and Racker, E. (1968). *J. Biol. Chem.* **243**, 129.

McCarty, R. E., Guillory, R. J., and Racker, E. (1965). *J. Biol. Chem.* **240**, PC 4822.

McCarty, R. E., Fuhrman, J. S., and Tsuchiya, Y. (1971). *Proc. Nat. Acad. Sci. U.S.* **68**, 2522.

McCarty, R. E., Pittman, P. R., and Tsuchiya, Y. (1972). *J. Biol. Chem.* **247**, 3048.

McEvoy, F. H., and Lynn, W. S. (1970). *FEBS Lett.* **10**, 299.

Melandri, B. A., Baccarini-Melandri, A., San Pietro, A.. and Gest, H. (1970). *Proc. Nat. Acad. Sci. U.S.* **67**, 477.

Melandri, B. A., Baccarini-Melandri, A., San Pietro, A., and Gest, H. (1971). *Science* **174**, 514.

Migniac-Maslow, M., and Champigny, M. L. (1971). *Biochim. Biophys. Acta* **234**, 344.

Miles, C. D., and Jagendorf, A. T. (1969). *Arch. Biochim. Biophys.* **129**, 711.

Miles, C. D., and Jagendorf, A. T. (1970). *Biochemistry* **9**, 429.

Mitchell, P. (1961). *Nature (London)* **191**, 144.

Mitchell, P. (1966a). Chemiosmotic Coupling in Oxidative and Photosynthetic Phosphorylation. Glynn Res., Bodmin.

Mitchell, P. (1966b). *Biol. Rev. Cambridge Phil. Soc.* **41**, 445.

Mitchell, P. (1968). Chemiosmotic Coupling and Energy Transduction. Glynn Res., Bodmin.

Mitchell, P. (1970). *In* "Membranes and Ion Transport" (E. E. Bittar, ed.), Vol. I, p. 192. Wiley, New York.

Mitchell, P. (1972). *J. Bioenerg.* **3**, 5.

Mitchell, P., and Moyle, J. (1965). *Nature (London)* **208**, 147.

Mitchell, P., and Moyle, J. (1967a). *Biochem. J.* **104**, 588.

Mitchell, P., and Moyle, J. (1967b). *Biochem. J.* **105**, 1147.

Mitchell, P., Moyle, J., and Smith, L. (1968). *Eur. J. Biochem.* **4**, 9.

Mohanty, P., Braun, B. Z., and Govindjee (1972). *Biochim. Biophys. Acta* **292**, 459.

Montal, M., Nishimura, M., and Chance, B. (1970). *Biochim. Biophys. Acta* **223**, 183.

Moyle, J., and Mitchell, P. (1973). *Biochem. J.* **132**, 571.

Moyle, J., Mitchell, R., and Mitchell, P. 1972. *FEBS Lett.* **23**, 233.

Murakami, S. (1968). *In* "Comparative Biochemistry and Biophysics of Photosynthesis" (K. Shibata, A. Takamiya, A. T. Jagendorf, and R. C. Fuller, eds.), p. 82. Univ Park Press, Baltimore, Maryland.

Nelson, N., Drechsler, Z., and Neumann, J. (1970). *J. Biol. Chem.* **245**, 143.

Nelson, N., Nelson, H., Naim, Y., and Neumann, J. (1971). *Arch. Biochem. Biophys.* **145**, 263.

Nelson, N., Nelson, H., and Racker, E. (1972a) *J. Biol. Chem.* **247**, 6506.

Nelson, N., Nelson, H., and Racker, E. (1972b). *J. Biol. Chem.* **247**, 7657.

Nelson, N., Deters, D., Nelson, H., and Racker, E. (1973). *J. Biol. Chem.* **248**, 2049.

Neumann, J., and Jagendorf, A. T. (1964). *Arch. Biochem. Biophys.* **107**, 109.

Neumann, J., and Jagendorf, A. T. (1965). *Biochim. Biophys. Acta* **109**, 382.

Neumann, J., Ke, B., and Dilley, R. A. (1970). *Plant Physiol.* **46**, 86.

Nishimura, M., and Pressman, B. C. (1969). *Biochemistry* **8**, 1360.

Nishimura, M., Kadota, K., and Chance, B. (1968). *Arch. Biochem. Biophys.* **125**, 308.

Nishizaki, Y., and Jagendorf, A. T. (1969). *Arch. Biochem. Biophys.* **133**, 255.

Nishizaki, Y., and Jagendorf, A. T. (1971). *Biochim. Biophys. Acta* **226**, 172.

Ohki, R., Kunieda, R., and Takamiya, A. (1971). *Biochim. Biophys. Acta* **226**, 144.

Packer, L. (1967). *Biochem. Biophys. Res. Commun.* **28**, 1022.

Packer, L., Allen, J. M., and Starks, M. (1968). *Arch. Biochem. Biophys.* **128**, 142.

Parker, J. C., and Hoffman, J. F. (1965). *Fed. Proc.* **24**, 589.

Petrack, B., and Lipmann, F. (1961). *In* "Light and Life" (W. D. McElroy and H. B. Glass, eds.), p. 621. Johns Hopkins Press, Baltimore, Maryland.

Petrack, B., Craston, A., Sheppy, F., and Farron, F. (1965). *J. Biol. Chem.* **240**, 906.

Polya, G., and Jagendorf, A. T. (1970). *Arch. Biochem. Biophys.* **138**, 540.

Poznansky, M., and Solomon, A. K. (1972). *Biochim. Biophys. Acta* **274**, 111.

Pressman, B. (1968). *Fed. Proc.* **27**, 1283.

Punnett, T. (1965). *Plant Physiol.* **40**, 1283.

Punnett, T., and Iyer, R. V. (1964). *J. Biol. Chem.* **239**, 2335.

Racker, E. (1967). *Fed. Proc.* **26**, 1335.

Racker, E. (1970). *In* "Essays in Biochemistry" (P. N. Campbell, and F. Dickens, eds.), vol 6, p. 1. Academic Press, New York.

Racker, E., Hauska, G. A., Lien, S., Berzborn, R. J., and Nelson, N. (1972). *Proc. Int. Photosynthesis Congr., 2nd Stresa* **2**, 1097.

Regitz, G., Berzborn, R. J., and Trebst, A. (1970). *Planta* **91**, 8.

Rienits, K. G. (1967). *Biochim. Biophys. Acta* **143**, 595.

Rosing, J. and Slater, E. C. (1972). *Biochim. Biophys. Acta* **267**, 275.

Rottenberg, H., and Grünwald, T. (1972). *Eur. J. Biochem.* **25**, 71.

Rottenberg, H., Grünwald, T., and Avron, M. (1972). *Eur. J. Biochem.* **25**, 54.

Roy, H., and Moudrianakis, E. N. (1971a). *Proc. Nat. Acad. Sci. U.S.* **68**, 464.

Roy, H., and Moudrianakis, E. N. (1971b). *Proc. Nat. Acad. Sci. U.S.* **68**, 2720.

Rumberg, B., and Schröder, H. (1971). *Eur. Biophys. Congr., 1st* (E. Broda, A. Locker, and H. Springer-Lederer, eds.), p. 57.

Rumberg, B., Reinwald, E., Schröder, H., and Siggel, U. (1969). *Progr. Photosynthesis Res.* **3**, 1374.

Ryrie, I. J., and Jagendorf, A. T. (1971a). *J. Biol. Chem.* **246**, 582.

Ryrie, I. J., and Jagendorf, A. T. (1971b). *J. Biol. Chem.* **246**, 3771.

Ryrie, I. J., and Jagendorf, A. T. (1972). *J. Biol. Chem.* **247**, 4453.

Sabnis, D. D., Gordon, M. and Galston, A. W. (1970). *Plant Physiol.* **45**, 25.

Saha, S., Izawa, S., and Good, N. E. (1970). *Biochim. Biophys. Acta* **223**, 158.

Sakurai, H., Nishimura, M., and Takamiya, A. (1965). *Plant Cell Physiol.* **6**, 309.

Sato, V. L. (1970). *J. Cell Biol.* **47**, 179a.

Schliephake, W., Junge, W., and Witt, H. T. (1968). *Z. Naturforsch.* **23b**, 1571.

Schröder, H., Muhle, H., and Rumberg, B. (1972). *Proc. Int. Photosynthesis Congr., 2nd Stresa* (G. Forti, ed.), **2**, 919.

Schuldiner, S., and Avron, M. (1971). *FEBS Lett.* **14**, 233.

Schuldiner, S., Rottenberg, H., and Avron, M. (1972). *Eur. J. Biochem.* **25**, 64.

Schultz, A. R., and Boyer, P. D. (1961). *Arch. Biochem. Biophys.* **93**, 335.

Schwartz, M. (1968). *Nature (London)* **219**, 915.

Selwyn, M. J., Dawson, A. P., Stockdale, M., and Gains, N. (1970). *Eur. J. Biochem.* **14**, 120.

Shavit, N., and Boyer, P. D. (1966). *J. Biol. Chem.* **241**, 5738.

Shavit, N., and Herscovici, A. (1970). *FEBS Lett.* **11**, 125.

Shavit, N., and San Pietro, A. (1967). *Biochem. Biophys. Res. Commun.* **28**, 277.

Shavit, N., Skye, G. E., and Boyer, P. D. (1967). *J. Biol. Chem.* **242**, 5125.

Shavit, N., Dilley, R. A., and San Pietro, A. (1968a). *Biochemistry* **7**, 2356.

Shavit, N., Thore, A., Keister, D., and San Pietro, A. (1968b). *Proc. Nat. Acad. Sci. U.S.* **59**, 917.

Shavit, N., Degani, H., and San Pietro, A. (1970). *Biochim. Biophys. Acta* **216**, 208.

Shen, Y. K., and Shen, G. M. (1962). *Sci. Sinica* **11**, 1097.

Siggel, U., Renger, G., Stiehl, H. H., and Rumberg, B. (1972). *Biochim. Biophys. Acta* **256**, 328.

Siow, K. S., and Unrau, A. M. (1968). *Biochemistry* **7**, 3507.

Skulachev, V. P. (1972). *J. Bioenerg.* **3**, 25.

Skye, G. E., Shavit, N., and Boyer, P. D. (1967). *Biochem. Biophys. Res. Commun.* **28**, 724.

Stewart, B. W., and Rienits, K. G. (1968). *Biochim. Biophys. Acta* **153**, 907.

Strichartz, G. R., and Chance, B. (1972). *Biochim. Biophys. Acta* **256**, 71.

Telfer, A., and Evans, M. C. W. (1972). *Biochim. Biophys. Acta* **256**, 625.

Thore, A., Keister, D. L., Shavit, N., and San Pietro, A. (1968). *Biochemistry* **7**, 3499.

Tu, S. -I., and Wang, J. H. (1970). *Biochemistry* **9**, 4505.

Turner, J. F., Black, C. C., and Gibbs, M. (1962). *J. Biol. Chem.* **237**, 577.

Tyszkiewicz, L. (1972). *Proc. Int. Photosynthesis Congr., 2nd, Stresa* **2**, 1303.

Uribe, E. (1972). *Biochemistry* **11**, 4228.

Uribe, E., and Jagendorf, A. T. (1967a). *Plant Physiol.* **42**, 697.

Uribe, E., and Jagendorf, A. T. (1967b). *Plant Physiol.* **42**, 706.

Uribe, E., and Jagendorf, A. T. (1968). *Arch. Biochem. Biophys.* **128**, 351.

Vambutas, V. K., and Racker, E. (1965). *J. Biol. Chem.* **240**, 2660.

Van de Stadt, R. J., Nieuwenhus, F. J. R. M., and van Dam, K. (1971). *Biochim. Biophys. Acta* **234**, 173.

Von Stedingk, L. -V., and Baltscheffsky, H. (1966). *Arch. Biochem. Biophys.* **117**, 400.

Walz, D., Schuldiner, S., and Avron, M. (1971). *Eur. J. Biochem.* **22**, 439.

Wang, J. H. (1972). *In* "Structural and Functional Aspects of Photochemistry," (V. C. Runeckles, and T. C. Tso, eds.), chapter 1. Academic Press, New York.

Wang, J. H., Yang, C. S., and Tu, S. -I. (1971). *Biochemistry* **10**, 4292.

Watling, A. S., and Selwyn, M. J. (1970). *FEBS Lett.* **10**, 139.

Weber, G. (1972). *Proc. Nat. Acad. Sci. U.S.* **69**, 3000.

Wessels, J. S. C. (1963). *Proc. Roy. Soc. London B* **157**, 345.

West, K. R., and Wiskich, J. T. (1968). *Biochem. J.* **109**, 527.

Williams, R. J. P. (1969). *Current Top. Bioenerg.* **3**, 79.

Williams, R. J. P. (1972). *J. Bioenerg.* **3**, 81.
Winget, G. D., Izawa, S., and Good, N. E. (1965). *Biochem. Biophys. Res. Commun.* **21**, 438.
Winget, G. D., Izawa, S., and Good, N. E. (1969). *Biochemistry* **8**, 2067.
Witt, H. T. (1971). *Quart. Rev. Biophys.* **4**, 365.
Witt, H. T. (1972). *J. Bioenerg.* **3**, 47.
Wraight, C. A., and Crofts, A. R. (1970). *Eur. J. Biochem.* **17**, 319.
Young, J. H. (1972). *J. Bioenerg.* **3**, 137.
Zaugg, W. S., and Vernon, L. P. (1966). *Biochemistry* **5**, 34.
Zilinsky, J. W., Sojka, G. A., and Gest, H. (1971). *Biochem. Biophys. Res. Commun.* **42**, 955.

Primary Acts of Energy Conservation in the Functional Membrane of Photosynthesis

H. T. Witt

ABBREVIATIONS

ADP	Adenosine diphosphate
ATP	Adenosine triphosphate
BChl	Bacteriochlorophyll
BTB	Bromthymol blue
BV	Benzyl viologen
Car	Carotenoid
(CH_2O)	Component of sugar
Chl	Chlorophyll
Chl a	Chlorophyll a
Chl b	Chlorophyll b
Chl a_I (P700)	Chlorophyll a_I (reaction center of pigment system I)
Chl a_{II} (P680)	Chlorophyll a_{II} (reaction center of pigment system II)
Cyt	Cytochrome
DCMU	3-(3,4-Dichlorophenyl)-1, 1-dimethylurea
DCPIP	2,6-Dichlorophenol indophenol
$\Delta\phi$	Electrical potential difference
F	Electrical field strength
Fecy	Potassium ferricyanide, $K_3Fe(CN)_6$
GmcD	Gramicidin D
i	Ion flux
$i_H{}^+$	H^+ flux
$i_K{}^+$	K^+ flux
Inc	Indigo carmine
$NADP^+$	Nicotinamide adenine dinucleotide phosphate
P	Inorganic phosphate*
PMS	N-Methylphenazonium sulfate
PQ	Plastoquinone
Saf	Safranine T
^1S	Symbol related to one turnover
S_{SS}	Symbol related to steady state conditions
τ	Half-life time and half-rise time, respectively
TIP	Thymol indophenol
UBF	Umbelliferone (7-hydroxycoumarin)
Vmc	Valinomycin
X-320	Probably plastosemiquinone

* *Note*: This symbol is used for inorganic phosphate in this chapter only.

1. INTRODUCTION

1.1. Some Basic Results

The basic events of photosynthesis take place in subcellular organelles, the chloroplasts. The inner system of the chloroplasts contains about 1000 small compartments, the so-called *thylakoids* (Menke, 1966) (see Fig. 15). Thylakoids are disk-shaped vesicles with a diameter of about 5000 Å. In the thylakoid membrane about 10^5 pigment molecules are embedded (for details see Chapter 2). In green plants the pigments are Chl *a* and *b* and Car. Light absorbed in the pigment system is channeled by excitation *energy migration* to photoactive centers (Emerson and Arnold, 1932; Gaffron and Wohl, 1936). The excited centers transfer electrons from H_2O via intermediates to $NADP^+$. $NADP^+$ can be replaced by artificial electron acceptors (Hill, 1939) (for details see Chapter 7). About 200 *electron transfer chains* are embedded in one thylakoid membrane. Each chain is surrounded by about 500 light energy-conducting pigment molecules. The electron transfer rate (O_2 evolution) as a function of the wavelength of light showed that far-red light (700–730 nm) produces a poor yield. The yield is, however, enhanced to normal efficiency when short-wavelength light (< 700 nm) is supplemented (Emerson, 1958). From this "two-color" effect Emerson and Rabinowitch (1960) proposed tentatively that *two pigment systems* drive the electrons: a long-wave Chl I and a short-wave Chl II (Govindjee and Rabinowitch 1960). A similar conclusion was reached by Blinks (1959). On the basis of Chl fluorescence measurements Kautsky *et al.* (1960) also proposed the operation of two light reactions (see Chapter 1). The described effects are, however, not a necessary consequence of two pigment systems (for details see review by Myers, 1971).

The electron transfer is accompanied by *phosphorylation*, i.e., by synthesis of ATP from ADP and P. This was discovered by Arnon *et al.* (1954) in green plants and by Frenkel (1954) in photosynthetic bacteria.

With the help of NADPH and ATP, absorbed CO_2 can be reduced into sugar (CH_2O) and "everything else" (see Fig. 1). The biochemical dark processes of the CO_2 cycle are well known and were discovered by Calvin and Benson (1948); also see Calvin (1962).

With respect to hypotheses efforts have been made to explain, first, how a cooperation of the two suggested pigment systems may occur within the electron transfer chain, and second, how the electron transfer may be coupled to phosphorylation: Hill and Bendall (1960) postulated that two light reactions may be arranged in series within the electron transfer chain and the coupling is effected by cytochromes. In an energy diagram

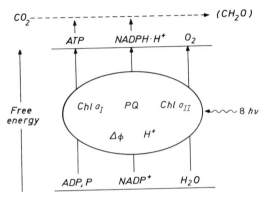

Fig. 1. Overall reaction of photosynthesis and four elements which are the basis for the bioenergetic concept developed in this chapter: Light reactions of chlorophylls (Chl a); charge translocation by plastoquinone (PQ); change of transmembrane potential ($\Delta\phi$); movement of protons (H^+). (For details see Section 1.1 and list of abbreviations.)

this concept results in the well-known Z-scheme. Coupling of two light reactions in series was also discussed by Kautsky *et al.* (1960).

The mechanism of phosphorylation is open to three hypotheses: (1) Mitchell (1961, 1966) postulated that electron transfer generates an electrochemical gradient of protons across a membrane. The energy of this gradient would drive the synthesis of ADP + P to ATP. (2) It has been suggested that the electron transfer generates an unidentified chemical intermediate (Lipmann, 1946; Slater, 1953, 1958). The energy of this intermediate should drive the synthesis of ATP. (3) It has been considered that electron transfer might induce conformational changes which may result in a mechanicochemical synthesis of ATP (Boyer, 1965; Green and Ji, 1972).

The problem of photosynthesis today is the question of the *molecular mechanism* of the transformation of light energy into the chemical energy of NADPH and ATP.

1.2. Molecular Events

Information on the mechanism of the molecular machinery can be obtained by direct measurement of the molecular events. These events are physical and chemical intermediate reactions which are in action between the beginning of light absorption and the end of the overall process terminated by the production of NADPH and ATP.

There exist several ways to obtain information on molecular events in photosynthesis. This chapter is devoted to results on primary acts obtained by pulse spectroscopic methods.

1.2.1. Characterization, Excitation, and Registration of Molecular Events

1. For the characterization of the molecular events which are reported in this chapter the change of *optical properties* has been used (see Section 1.2.2).

2. For excitation of the whole process light pulses much shorter than 0.6 msec, are necessary; this is the condition for single turnovers, i.e., for the transfer of only one electron from H_2O to $NADP^+$ without recycling (H. T. Witt *et al.*, 1966b; Vater *et al.*, 1968). Furthermore the pulses must be shorter than the reaction time of the considered event. This requirement has been realized in photosynthesis by using normal flashes (H. T. Witt, 1955a, b; H. T. Witt *et al.*, 1959), ultrashort flashes (Wolff *et al.*, 1967; Wolff and Witt, 1969) or laser giant pulses (Wolff *et al.*, 1967; Wolff and Witt, 1969).

3. For registration of the small signals within short times the *repetitive pulses spectroscopic method* (Rüppel *et al.*, 1962; Rüppel, 1962) is most suitable. The extremely high sensitivity of this technique enables the detection of very small signals. The high time resolution of this technique extended the analysis of biological events down to 20 nsec. Such techniques are described in detail elsewhere (H. T. Witt, 1967a; Rüppel and Witt, 1970).

The measurements have been made under *single turnover conditions* where insight into the relationship between the different events can more easily be obtained. In several cases two events have been measured with the same sample in double-beam instruments simultaneously to be comparable under exactly the same conditions.

1.2.2. Types of Molecular Events

Using the pulse spectroscopic methods the following events have been analyzed under single turnover conditions:
1. *Metastable states* of carotenoids
2. *Light reactions* of active chlorophylls
3. *Electron transfers* during the redox reactions of electron carriers
The events 1–3 have been measured by their characteristic intrinsic absorption changes.

4. *Electrical field* across the thylakoid membrane. This has been measured by absorption changes due to electrochromism (shift of absorption bands in an electric field).

5. *Ion fluxes* across the membrane. These have been measured by the derivative of the electrochromic absorption changes with regard to time.

6. *Proton transfer* across the membranes. This has been measured by artificial pH indicators, e.g., by UBF, which changes from a nonfluorescing into a fluorescing state during the H^+ transfer.

7. *Proton accumulation* in the inner space of the thylakoids. This has been measured by an intrinsic indicator for H_{in}^+ changes. The H_{in}^+ depending rate of Chl a_I^+ reduction has been used for this purpose.

8. *Phosphorylation.* At single turnovers ATP formation has been measured also by the pH indicator UBF. During the ATP generation, H^+ is consumed and can be measured by the corresponding irreversible fluorescence changes of UBF.

Properties of these eight events and their relationships are discussed in detail elsewhere (H. T. Witt, 1971).

1.2.3. Program and Restrictions

The following report is restricted to the "building blocks" which can bring out certain characteristic points of the bioenergetic concept of photosynthesis. As the necessary "essentials" for this concept four processes out of the different events cited in Section 1.2.2 have been recognized (see Fig. 1):

1. The light reactions of two active chlorophylls (Chl a_I, Chl a_{II})
2. The reaction of PQ
3. The change of transmembrane potential ($\Delta\phi$)
4. The movement of protons (H^+)

Results are discussed which give experimental evidence that the cooperation of the reaction of two excited chlorophylls and PQ leads to a vectorial electron transfer from H_2O to $NADP^+$ in such a way that the functional membrane is charged. It is shown that the discharging of the electrically energized membrane by H^+ is coupled with the generation of ATP.

2. LIGHT REACTIONS OF TWO CHLOROPHYLLS

2.1. Coupling of Two Pigment Systems

It has been assumed that two pigment systems may be engaged in photosynthesis and that these are coupled in series (see Section 1.1). One system should be excitable up to 700 nm, the other up to 730 nm. First experimental evidence for this coupling has been given independently by three spectroscopic phenomena observed by Kok, Duysens, and Witt and their coworkers. In all three experiments it was shown that an intermediate reaction of photosynthesis followed by absorption changes is "shuttled" from one direction to the contrary course of reaction when photosynthesis is triggered alternatively with < 700 and > 700 nm light. From these findings it was evident that two pigments systems are engaged and that these are coupled in series. (1) Kok and Hoch (1961) measured the intermediate reaction of a pigment P700. (2) Duysens *et al.* (1961) measured a reaction of a cytochrome. (3) Witt *et al.* (1961a) analyzed a cytochrome together with a reaction indicated at 515 nm. Kok and Duysens observed an oxidation of their compounds in far-red light and a reduction in red light. H. T. Witt *et al.* observed that in far-red light the extent of the 515 nm reaction was strongly decreased, the reduction of Cyt f$^+$, however, was not, but it was slow (1 sec). In red light the extent of the 515 nm reaction was increased, the extent of reduction of Cyt f$^+$ the same but fast (10^{-2}sec).

Further spectroscopic evidence which has been interpreted by a coupling of two pigment systems were given in the following years: (4) In far-red light PQ is oxidized. In red light it is reduced (Rumberg et al., 1963). (5) In far-red light the intensity of Chl fluorescence is decreased. In red light, however, it is increased (Duysens and Sweers, 1963).

The photoactive chlorophylls within the two coupled systems have been observed directly and are described in the following section.

2.2. Light Reaction of Chlorophyll a_I (P700)

A decrease of absorption at 700 nm during photosynthesis was first observed by Kok and attributed to a reaction of a pigment, P700 (Kok, 1957). P700 is oxidized in light (Kok, 1961). Its behavior as a radical follows from its ESR signal (Beinert and Kok, 1963). On the basis of a decrease at 430 nm in acetone-extracted chloroplasts it was assumed that P700 is a Chl a. The ratio of P700 to the bulk pigments is about 1 : 500 (Kok, 1961).

2.2.1. Kinetics

The timecourse of the changes has been recorded by the pulse techniques (Witt et al., 1963). In single turnover flashes the absorption decreases fast and returns to its original state in 10 μsec−20 msec (Fig. 2a). This time depends on the redox state of the electron donors (Haehnel et al., 1971). The rise time of the decrease has been measured by the repetitive laser technique (K. Witt and Wolff, 1970). It takes place in < 20 nsec (Fig. 2b).

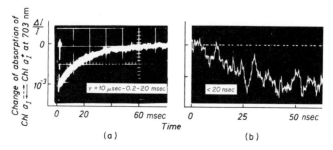

Fig. 2. Timecourse of the redox reaction of Chl a_I (P700) in chloroplasts of spinach. (a) Total trace (H. T. Witt et al., 1963). (b) Rise time (K. Witt and Wolff, 1970).

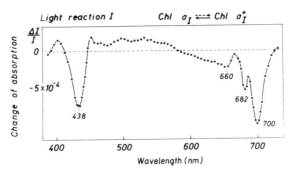

Fig. 3. Transient difference spectrum of the redox reaction of Chl a_I in chloroplasts of spinach Rise: $\leq$ 20 nsec; decay: 10µsec–20 msec. (Döring *et al.*, 1968a).

2.2.2. *Difference Absorption Spectrum*

The spectrum of the kinetics characterized in Section 2.2.1 is shown in Fig. 3. The bleaching at 700 and 438 nm agrees with the earlier results by Kok. Additionally a splitting in a double band with positions at 682 and 700 nm has been observed (Döring *et al.*, 1968a). This result was confirmed by Ke (1972). The magnitude and reaction half-life at 700 and 438 nm change in the same way when external conditions are varied (Rumberg and Witt, 1964). Since the Chl a absorption bands *in vivo* are located around 700 and 430 nm the spectrum must be caused by a photochemical reaction of Chl a. It is designated Chl a_I.*

The double band noted above may indicate that *two* or more Chl a_I molecules are combined in one reaction center with an oblique structure which would cause a symmetrical splitting of the original band (Döring *et al.*, 1968a). This result was supported recently by ESR measurements (Norris *et al.*, 1971) and dicroism analysis (Philipson *et al.*, 1972).

2.2.3. *Photooxidation of Chl a_I*

Three experiments give evidence that Chl a_I is oxidized in light.

(1) If Fecy is added to chloroplast suspensions the light-induced absorption change of Chl a_I is strongly decreased. Fecy obviously oxidizes Chl a_I in the dark. This means that Chl a_I is oxidized in light (Kok, 1961). The redox potential has been estimated as +0.45 V (Kok, 1961; Rumberg, 1964a).

(2) If reduced phenazonium methosulfate (PMS⁻) is added to aged chloroplast suspensions the recovery of Chl a_I absorption changes induced

---

*The use of Chl a_I and Chl a_{II} for the two photoactive chlorophylls should not be confused with the bulk chlorophylls of systems I and II, as used by other authors.

by flash light is strongly accelerated, i.e., Chl a_I is reduced in the dark by PMS^-, i.e., it is oxidized in the light (Rumberg and Witt, 1964).

(3) Photooxidation of pure Chl a in butanol with quinone as electron acceptor shows a difference absorption spectrum for Chl a which is, apart from a blue shift, comparable with that of chloroplasts in Fig. 3 (Seifert and Witt, 1968, 1969).

On the basis of the above we can interpret the fast rise of <20 nsec in Fig. 2b to represent the time of the oxidation of Chl a_I, and the slow return in the dark (10 μsec–20 msec) to the rereduction of Chl a_I^+ :

$$\text{Chl } a_I \; \underset{\text{10 }\mu\text{sec–20 msec}}{\overset{< 20 \text{ nsec}}{\rightleftharpoons}} \; \text{Chl } a_I^+$$

2.2.4. Chl a_I^+: A Primary Product

The extremely fast oxidation of Chl a_I within <20 nsec (see Fig. 2) suggests that Chl a_I^+ is probably a primary product. Evidence for this follows from trapping experiments at low temperature. Chl a_I can be oxidized in the light even at $-150°C$ (H. T. Witt et al., 1961b). The rereduction is, however, inhibited at this temperature. The spectrum of the irreversible change at $-150°C$ corresponds to the reversible spectrum of Chl a_I at 20°C in Fig. 3. Similar results have recently been obtained by Floyd et al. (1971) and K. Witt (1973a).

2.2.5. Electron Donor and Acceptor of Chl a_I

The electrons of Chl a_I are transferred ultimately to the electron acceptor $NADP^+$. A stoichiometry of 1 : 1 between Chl a_I–oxidation and $NADP^+$–reduction has been observed (Kok, 1963). An electron carrier (P430) between Chl a_I and $NADP^+$ and probably a primary electron acceptor of Chl a_I has been discussed by Hiyama and Ke (1971). The reduction of Chl a_I^+ is blocked by the addition of the poison DCMU (Rumberg, 1964a). Since DCMU blocks especially the O_2 evolution system, i.e., oxidation of H_2O (Bishop, 1958), H_2O must be the ultimate electron donor for Chl a_I^+ .

In the presence of an electron acceptor ($NADP^+$, BV, etc.) and addition of DCMU the anion PMS^- can act directly as an artificial electron donor because reduction of Chl a_I^+ blocked by DCMU is revived in full by PMS^- (see Fig. 5b) (Rumberg and Witt, 1964). In the absence of an electron acceptor Chl a_I^+ can be reduced by an intrinsic cyclic process (Rumberg, 1964a).

Conclusions that the linear electron transfer from H_2O to NADP[1] via PQ is not achieved by the action of Chl a_I and Chl a_{II} (Rurainski et al., 1971) have been rejected by results of Haehnel (1973) who showed that all electrons from PQ are transferred to oxidized Chl a_I^+. It is very likely that due to the slow response time of the apparatus used by Rurainski et al. (>1 msec) the faster electron transfer to Chl a_I^+ (see Section 2.2.2) escaped detection and therefore a lower rate of Chl a_I reaction was apparent as measured for the $NADP^+$ reduction.

2.2.6. *Coupling of Chl a_I with a Second Pigment System*

Actinic light with wavelength of <700 nm results in reversible absorption changes of Chl a_I (Fig. 2), i.e., oxidation is followed by rereduction. Far-red light between 700 and 730 nm, however, can only oxidize Chl a_I (Kok and Hoch, 1961). This is demonstrated by the irreversible decrease (I) of absorption (see dotted portion in Fig. 4a). Only a subsequent flash with <700 nm (638 nm) leads to the rereduction of Chl a_I^+, indicated by the increase (II) of absorption in Fig. 4a (Rumberg *et al.*, 1963, 1964). Since light of <700 nm causes an oxidation followed by a reduction, this quality of light might be channeled into two reaction centers: one part $h\nu_I$, is channeled to Chl a_I and causes its oxidation, and another part, $h\nu_{II}$ is channeled to a second pigment, where it provides electrons from H_2O for the reduction of Chl a_I^+. (This conclusion has been proved by a series of other experiments.) The increase (II) of absorption of Chl a_I indicates therefore qualities of the second pigment. The action spectra of both effects (I) and (II) in chloroplasts of spinach is depicted in Fig. 4b (Müller *et al.*, 1963; Junge and Witt, cited in H. T. Witt, 1967b). Measurements of this type were first carried out in blue-green algae by Kok and Gott (1960). The spectra show that Chl a_I can be excited with wavelengths <730 nm, but the second pigment only with wavelengths <700 nm. Chl a_I is therefore a long-wavelength Chl. The second pigment must correspond to a shorter wavelength Chl with an absorption band below 700 nm. Evidence for this active pigment is given in the next section.

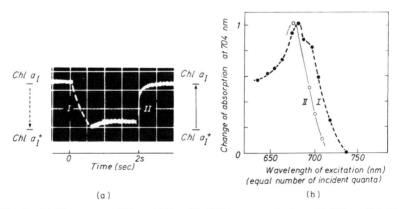

Fig. 4. (a) Timecourse of the oxidation (I) of Chl a_I on excitation with 720 nm light ($h\nu_I$) and rereduction (II) with 638 nm light ($h\nu_{II}$). (Rumberg *et al.*, 1963). (b) Action spectra of the oxidation (I) of Chl a_I and rereduction (II) of Chl a_I^+. (I) Subject: chloroplasts of spinach (Müller *et al.*, 1963; Junge and Witt, cited in H. T. Witt, 1967b).

2.3. Light Reaction of Chlorophyll a_{II} (P680)

Absorption changes with properties expected for the second active pigment in photosynthesis have a lifetime 100 times shorter than that of Chl a_I. Its signal is, therefore, masked by a 10 times greater noise. It is furthermore difficult to separate the fast signals in the red from the 10^6 times greater Chl fluorescence signal. These problems have, however, been overcome by the application of the repetitive technique (Döring *et al.*, 1967, 1968b, 1969; Govindjee *et al.*, 1970; Döring and Witt, 1972).

2.3.1. Kinetics

The timecourse of changes at 690 nm is depicted in Fig. 5a. The kinetics are biphasic. The slow phase of 20 msec indicates the reaction of Chl a_I (see Section 2.2), and the fast phase of 0.2 msec, the reaction of Chl a_{II}.* Because a product of the reaction of Chl a_{II}, the electrical field, is generated in < 20 nsec (see Section 4.1 and 4.6), the rise time of the Chl a_{II} reaction must be at least < 20 nsec.

2.3.2. Difference Spectrum

The spectrum of the fast phase is shown in Fig. 6. The same characteristic kinetics at all wavelengths and the same dependence on different parameters indicate that the spectrum is due to one substance (Döring *et al.*, 1968b). Since the maxima of the absorption change are at 682 and 435 nm, the position of the characteristic bands of Chl a, the difference spectrum must be caused by a photochemical reaction of this substance. It is designated

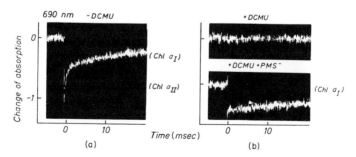

Fig. 5. (a) Timecourse of the Chl a_I and Chl a_{II} reaction at 690 nm. (b) top: in the presence of DCMU which blocks cleavage of H_2O; bottom: reactivation of the reaction of Chl a_I with the artificial electron donor PMS$^-$. Sample: chloroplasts of spinach (Döring *et al.*, 1967, 1969).

---

* An apparent absorption change caused by the change of the yield of the fluorescence (excited by the measuring beam) has been excluded (Döring and Witt, 1972).

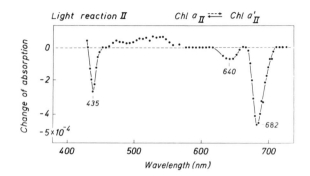

Fig. 6. Transient difference spectrum of the reaction of Chl a_{II}. Rise: $\leq$ 20 nsec; decay: 200 μsec. Sample: chloroplasts of spinach (Döring *et al.*, 1969).

Chl a_{II}. (Chl a_I and Chl a_{II} are probably not different in their chemical constitution but only in their special environment.)

The maximum of the absorption changes of Chl a_{II} on *whole* chloroplasts has recently been determined to be located at 687 nm (Döring, G., private communication) instead of 682 nm observed in system II enriched particles (Döring *et al.*, 1968b).

2.3.3 A New Component

Recently in whole chloroplasts in system II a new large absorption change at 690 nm with a decay time of 35 μsec was observed (Gläser *et al.*, 1974). Probably, this fast 35 μsec-component represents, together with the 200 μsec-component, the biphasic kinetics of one and the same Chl a_{II}. Assuming the same extinction coefficient for both Chl a_{II} components as for Chl a_I it can be estimated that about 1 molecule Chl a_{II} is active per electron chain, i.e. the ratio of Chl a_I to Chl a_{II} is about 1:1.

2.4. Coupling of Chlorophyll a_I with Chlorophyll a_{II}

The release of electrons from H_2O can be blocked by DCMU (Bishop, 1958.[†] Upon the addition of DCMU the reversible absorption changes of both Chl a_I and Chl a_{II} should disappear if both are coupled in series between H_2O and $NADP^+$ (Fig. 5b, top). In the presence of DCMU the artificial electron donor PMS^- can supply Chl a_I^+ with electrons (see Section 2.2.5).

[†] This is valid when the flash frequency is high enough (> 1 Hz) to prevent a removal of the block in a dark reaction ~ 10 sec (see Section 2.2.2).

If Chl a_I and Chl a_{II} are coupled, Chl a_I (slow phase) should be reactivated by PMS⁻ but not Chl a_{II} (fast phase) (Fig. 5b, bottom). Because both predictions are realized Chl a_I is very probably coupled with Chl a_{II} in series (see Fig. 7a) (Döring et al., 1967).

2.5. Properties of Chlorophyll a_I (P700) and Chlorophyll a_{II} (P680)

From the results in section 2.2.5 it follows that the reaction cycle of Chl a_I is separated from the overall electron-transfer chain by the addition of DCMU + PMS⁻. The separated cycle has been investigated as a function of different parameters. The characteristics are summarized in Table I.

The reaction of Chla_{II} can be separated from the overall electron-transfer chain simply by heating at 50°C (Döring et al., 1967, 1969). The separated Chl a_{II} cycle has been investigated as a function of different parameters. The characteristics are summarized Table I.

Precise measurements of the number of DCMU molecules which are necessary to block the reaction center II give the result that one DCMU molecule is sufficient to block two centers. This fact may be explained by the cooperation of two Chl a_{II} molecules (Siggel et al., 1972a,b). Trapping experiments at low temperature ($-200°C$) with respect to Chl a_{II} are ambiguous:

Floyd et al. (1971) observed negative changes (reversible?) at 680 nm. But it was not shown whether these changes belong to the main peak at 682 nm of P680 (see Fig. 6) or to minor peak at 682 nm of P700 (see Fig. 3). K. Witt (1973a) observed at $-160°C$ new positive absorption changes at 695 nm, which probably occur only at cryogenic temperature. In this case the negative

TABLE I

Properties of Separated Chlorophyll a_I and a_{II} Reaction Cycles [a]

Properties	Chl a_I (P700)	Chl a_{II} (P680)
Type of reaction	Electron donor	?
Characteristic	438, 660, 682–703 nm	435, 640, 687 nm
change of absorption	Band splitting	No band splitting
Rise time	≤ 20 nsec	≤ 20 nsec
Half-life time	20–0.2–0.01 msec [b]	200μsec $- 35\mu$sec
Redox potential	$+0.45$ V	?
Range of excitation	<730 nm	<700 nm
Range of pH	3–11	?
Temperature of inactivation	65°C	55°C
Sensitive to aging	No	$\tau \approx 95$ hr (0°C)
Sensitive to DCMU	No	$c \approx 10^{-7}$ M

[a] Data from H. T. Witt et al. (1963); Rumberg (1964a); and Döring et al. (1969).
[b] Depending on the redox state of the electron donors; see Section 2.2.1.

absorption changes of Chl a_{II} at 680 nm cannot be detected because they would be masked by the nonphysiological positive changes at 695 nm.

The molecular mechanism of the primary process of Chl a_{II} (photooxidation or photoreduction, etc.) is not yet known and must be the subject of future work.

In spite of the results in Sections 2 and 3 and Fig. 7, Arnon *et al.* (1970) proposed a scheme which includes only Chl a_I (P700) as light center and which performs only a cyclic electron flow. Two unidentified series light centers should drive in a separate linear system electrons from H_2O to $NADP^+$ via PQ. This means that photosynthesis is driven by three different light reactions. Such a system has been questioned recently by Knaff (1972). Results by Fork and Murata (1972) and Malkin and Bearden (1973) are also in contradiction to the result of Arnon and co-workers. Furthermore, a quantitative linear transfer of the electrons from PQ to Cyt f and Chl a_I has been demonstrated (Haehnel, 1973). This result excludes also the scheme of Arnon *et al.* (1970).

3. FUNCTION OF PLASTOQUINONE

Plastoquinone is discussed here because of its unique function as electron carrier, H^+ translocator, and mediator of a vectorial flow of hydrogen. The electron capacity of the PQ pool can furthermore be used as a valuable variable for evaluating quantitative relationships between different events.

3.1. Plastoquinone: Link between Chlorophyll a_I and Chlorophyll a_{II}

Bishop (1959) has shown that on extraction with petroleum ether chloroplasts lose the ability to oxidize water but that they regain their activity on subsequent recondensation of PQ. It was concluded, therefore, that PQ is involved in photosynthesis. With our finding of the transient ultraviolet difference spectrum of PQ during photosynthesis (Klingenberg *et al.*, 1962) the possibility was opened for a series of investigations on the function of PQ in more detail. Evidence was given that PQ is involved in a redox reaction. PQ is reduced to hydroquinone PQ^{2-} via excited Chl a_{II} and reoxidized by Chl a_I^+, i.e., PQ is located between Chl a_I and Chl a_{II}. Furthermore, it was shown that PQ exists as a pool of about five PQ molecules between the two light reactions (Rumberg *et al.*, 1963). Participation of PQ was also shown in other algae (Amesz, 1964).

3.2. Kinetics and Spectra of Plastoquinone and Its Precursor

In a single turnover flash one observes a timecourse of absorption change in the UV with a fast decrease at 265 nm (probably in 0.6 msec) followed by a return in 20 msec. The difference spectrum of these changes is depicted in Fig. 8b (Stiehl and Witt, 1968, 1969). The spectrum corresponds closely to the spectrum which is obtained when PQ is reduced chemically in ethanol to hydroquinone (Henninger and Crane, 1964).

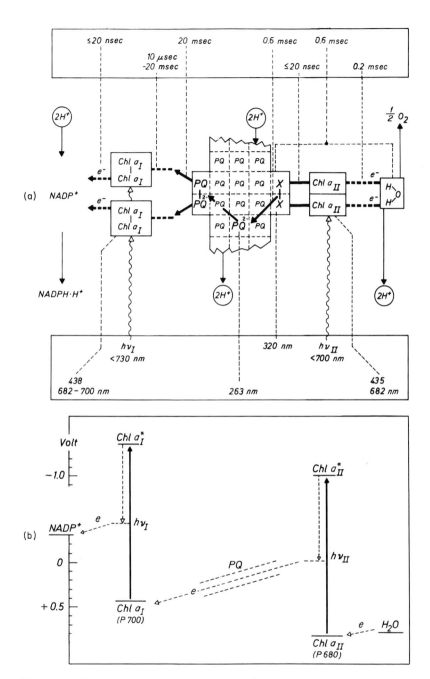

Fig. 7. (a) Electron transfer from H_2O to $NADP^+$ in chloroplasts of spinach. Results by pulse spectroscopic techniques. Electron carriers located on the dotted pathways as ferredoxin, plastocyanin, and cytochromes have been omitted. Kinetic data of the electron transfer times, and characteristic absorption changes of the intermediates are indicated (Stiehl and Witt, 1969). (b) Free energy diagram of the electron transfer from H_2O to $NADP^+$. The scale indicates the midpoint potentials. For details see Section 3.4.

The reduction of PQ to hydroquinone, even after a short flash in which only one electron is transferred through each chain (sec above), can be understood if two chains are assumed to be coupled parallel with each other. Each chain injects one electron to a PQ molecule. Plastosemiquinone (PQ^-) molecules, formed as a twin PQ^-–PQ^- by neighboring chains, can then dismutate in a first-order reaction into PQ^{2-} and PQ (see Fig. 8a). For a quantitative treatment see Stiehl and Witt (1969) and for further discussion see H. T. Witt (1971).

A substance which has the expected kinetics and spectral properties of PQ^-–PQ^- is described below as X-320.*

3.2.1. Rate Limiting Step.

PQ^{2-} is reoxidized via intermediates by Chl a_I^+ (for details see Stiehl and Witt, 1969). The oxidation time of PQ^{2-} (20 msec) represents the rate-limiting step in the whole electron-transfer chain (H. T. Witt, 1967b). This value agrees with the rate-limiting step in photosynthesis estimated by Emerson and Arnold (1932) from measurements of O_2 evolution in dependence on the flash frequency.

3.2.2. X-320. Semiquinone of PQ and Electron Acceptor of Chl a_{II}

The spectrum of X-320 is depicted in Fig. 8a. X-320 is formed rapidly ($<10^{-6}$ sec) and decays in a first-order reaction within 0.6 msec. Its characteristics are as follows.

1. The spectrum of X-320 (Stiehl and Witt, 1968) corresponds to that of PQ^- which Land *et al.* (1971) had produced artificially in butanol.

2. The ratio X-320 : Chl a_{II} is 1 (Stiehl and Witt, 1968).

3. X-320 is very probably a primary product because it can be trapped at $-160°C$ (K. Witt, 1973b).

4. The decay time of X-320 (0.6 msec) (Stiehl and Witt, 1968) has the same value as the overall transfer time of one electron from H_2O to the PQ pool (0.6 msec) (Vater *et al.*, 1968).

5. The formation of X-320 takes place in a single turnover flash also in the presence of DCMU. Its formation is, however, blocked when in the presence of DCMU the dark time of the following flash is shorter than 1 sec (K. Witt, 1973b). Such a time-dependent action of DCMU has been observed for the oxidation of H_2O (Bennoun, 1970; Mohanty *et al.*, 1971). Probably the reoxidation of X-320 (PQ^-) is blocked by DCMU, but can take place by a slow backflow of electrons from X-320 to the H_2O side in times >1 sec.

The characteristics 1–5 of X-320 are in accordance with the properties which are expected for the primary electron acceptor of Chl a_{II}.

---

*The use of X-320 for the primary electron acceptor of Chl a_{II} should not be confused with the use of X for the acceptor of Chl a_I by other authors.

Properties of an electron acceptor of Chl a_{II} derived from Chl fluorescence measurements (Duysen and Sweers, 1963) are discussed in Chapter 5. A substance C550 (absorption changes at 550 nm) considered by Knaff and Arnon (1969) to be the primary acceptor of Chl a_{II} shows no correlation with X-320 (also see Chapter 3). Also we were not able to observe in flashlight absorption changes at 550 nm with kinetics which are expected for a primary acceptor.

3.3. Pool and Strand of Plastoquinone

In a single turnover flash $\frac{1}{2}$ PQ per chain is reduced into hydroquinone (see Section 3.2). Exposed to long flashes the absorption changes of PQ can be increased up to tenfold. The absorption changes of all other electron carriers between H_2O and $NADP^+$ cannot be increased by a long flash. From this it has been concluded that a pool of about 5 PQ with an electron capacity of ~ 10 is located within each electron transport chain (Stiehl and Witt, 1968).

The pools of the different electron chains are combined with each other in the form of a strand. In this way at least 10 chains are electronically connected with each other in a parallel arrangement as indicated in Fig. 7. This has been concluded in principle for the following reasons. If most (90%) of the Chl a_{II} centers are blocked by addition of DCMU, all Chl a_I centers can

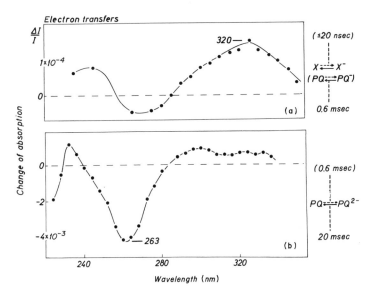

Fig. 8. (a) Transient difference spectrum of the redox reaction of X–320 (plastosemiquinone). Rise: $< 10^{-6}$ sec; decay: 0.6 msec. (b) Transient difference spectrum of the redox reaction of plastoquinone (Rise: 0.6 ms); decay: 20 msec. Sample: chloroplasts of spinach (Stiehl and Witt, 1968).

nevertheless carry out redox reactions because the remaining Chl a_{II} (10%) can supply after 10 turnovers all Chl a_I centers with electrons from H_2O via the assumed electron guiding PQ strands. For details see Siggel *et al.* (1972a,b).

3.4. Properties of the Charge Transporting Chain

3.4.1. Coupling in Parallel of Two-Electron Chains

Three findings gave evidence for a parallel cooperation of two-electron chains (see Fig. 7).

1. The formation of plastohydroquinone in a single turnover flash in which only one electron is transferred per electron chain can only be understood if two chains are joined parallel with each other (Section 3.2).

2. The observed first-order decay of X-320 can also only be explained by two parallel chains with two Chl a_{II} joined in one complex (Section 3.2.2).

3. The cooperation of two Chl a_{II} is finally supported by the observation that one DCMU molecule blocks the action of two Chl a_{II} centers (Section 2.5).

3.4.2. Proton Reactions

It can be expected that the electron transfer from H_2O to the ultimate electron acceptor $NADP^+$ is accompanied by proton uptake of 1 H^+ by 1/2 PQ^{2-} and 1 H^+ by 1/2 $NADP^+$ and a proton release of 1 H^+ by the cleavage of 1/2 H_2O and 1 H^+ by the reoxidation of 1/2 PQH_2 (see Fig. 7a). If the H^+ uptake occurs from the outside into the thylakoid membrane and the H^+ release from the membrane into the inside as proposed by Mitchell (1961) this would be equivalent to a *translocation of one proton at each light reaction* to the inside of the thylakoid. This has been measured as outlined in Section 4.7.

3.4.3. Energetic Considerations

On the basis of the light energy absorbed by Chl a_I and Chl a_{II} and the midpoint potentials of $NADP^+$, Chl a_I, PQ, and H_2O one can construct an energy diagram for the electron transfer as depicted in Fig. 7b. The energy of light absorbed by Chl a_I and Chl a_{II} is at the lower limit $hv_I = 1.80$ eV (700 nm) and $hv_{II} = 1.85$ eV (680 nm). According to thermodynamics, only about 70% is available as free energy, i.e., 1.19 eV and 1.23 eV (Duysens, 1959; Ross and Calvin, 1967; Knox, 1969). According to Fig. 7b in the redox span between Chl a_I (+0.45 eV) and $NADP^+$ or its precursor ferredoxin (−0.42 eV) ≥ 0.87 eV is used up. Therefore ≤ 0.32 eV remains available from hv_I for other work. In the redox span between H_2O (0.81 eV, pH_{out} 7) and PQ (∼ 0eV) or its precursor X-320 (− ?eV) ≥ 0.81 eV is used up. Therefore

≤ 0.42 eV remains available from hv_{II} light for other work.* The remaining energy could be stored if the reduction of $NADP^+$ occurs by vectorial electron transfer from H_2O to $NADP^+$ across a membrane from inside to outside. This would be equivalent to a transmembrane *electrical potenial difference*. This potential has been measured as outlined in Section 4. The electrical energy of this potential can be used for the formation of ATP. This is shown in Section 6.

4. ELECTRICAL FIELD GENERATION AND FIELD-DRIVEN ION FLUXES

Electrical potentials have been postulated by Mitchell (1966) in a remarkable hypothesis for the mechanism of phosphorylation (for details see Section 6).

Direct experimental evidence for electrical fields across membranes in bioenergetic systems came in 1967 from the observation of optical band shifts of membrane pigments in a field (electrochromism) (Junge and Witt, 1968; a first report of these results was published in H. T. Witt, 1967b).

Indication of electrical potential changes have later been derived from the enhancement of delayed light emission from Chl due to the recombination of charges (Barber and Kraan, 1970; Wraight and Crofts, 1971; Crofts, et al., 1971). (For details see Chapter 5.) In mitochondria analysis of salt distribution across membranes has been used for estimation of potential difference (Mitchell and Moyle, 1969). Recently potential changes in mitochondria and chromatophores of bacteria have been measured in an elegant way by observation of transmembrane fluxes of synthetic ions in the field (electrophoresis) (Bakeeva et al., 1970; Skulachev, 1972).

4.1. Kinetics of Absorption Changes Attributed to an Electrical Field

The timecourse of single turn over absorption changes $\Delta I/I = \Delta A$ which are interpreted to indicate an electrical field is shown in Fig. 9a. At 515 nm in the light an increase takes place faster than 10^{-5} sec (H. T. Witt et al., 1956) and in the dark a decay with a half-time of ~ 20 msec (H. T. Witt, 1955a). The decay time varies from 10 to 200 msec, probably depending on the proton concentration in the inner phase of the thylakoid membrane (see Section 4.8). The rise time is extremely fast and has been measured by the repetitive giant

---

*The calculation is based on infinitesimaly rates at thermodynamic equilibrium. For maximal *rates* part of the available energy is dissipated at the expense of a maximal yield.

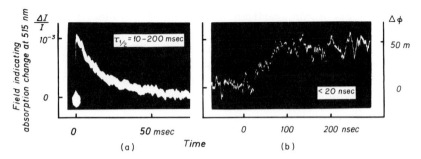

Fig. 9. Timecourse of the field indicating absorption change in chloroplasts. (a) Total trace (H. T. Witt, 1955a). (b) Rise time (Wolff *et al.*, 1967 cited in H. T. Witt, 1967b; Wolff *et al.*, 1969).

laser pulse technique (H. T. Witt, 1967b; Wolff *et al.*, 1969). Its value is of importance for the interpretation of the absorption change. A rise time of <20 nsec has been measured as documented in Fig. 9b.

It has been demonstrated: (1) The fast rise of the 515 nm change in the light indicates the formation of an electrical field and electrical potential difference $\Delta\phi$ across the thylakoid membrane (Junge and Witt, 1968). (2) The decay in the dark indicates the breakdown of the field by field-driven ion fluxes, i (Junge and Witt, 1968). (3) The indication is based on electrochromic changes of the bulk pigments in the thylakoid membrane (Emrich *et al.*, 1969; Schmidt *et al.*, 1971, 1972). These three statements are illustrated in Fig. 10, bottom. Support for the scheme is given in the following sections. An electrical analog is depicted in Fig. 10, top. The light reaction which shifts electrons across the membrane, is symbolized by a generator which charges the membrane capacitor. The electrochromic effect on the bulk pigments (Chl *a,b* and

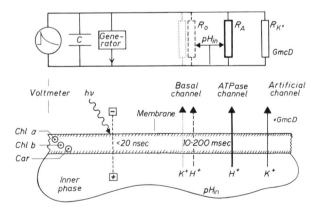

Fig. 10. Bottom: Scheme of the electrical field formation and ion translocation across the thylakoid membrane of chloroplasts. Top: Electrical analog. For details see Section 4.

Car) indicates the electrical potential changes, which are symbolized by a volt-meter. The discharge of the membrane capacitor occurs by ion fluxes through different channels. These are symbolized as resistors which are specified in the following sections (Junge and Witt, 1968).

The field-indicating absorption change $\Delta I/I = \Delta A$ is proportional to the potential change $\Delta \phi$ (see Section 4.9):

$$\Delta A(t) \sim \Delta \phi(t) \tag{1}$$

The derivative of the change with respect to time is proportional to the field-driven ion flux i (see Section 4.9):

$$\frac{d\Delta A(t)}{dt} \sim i(t) \tag{2}$$

In general the decay of ΔA with time t is biphasic and can be described ap - proximately by two first-order rate constants.

4.2. Kinetic Evidence for the Electrical Field

If the interpretation in Fig. 10 is correct it should be possible to change the field decay and the attributed absorption change into a faster one by arti-ficially increasing the membrane permeability for ions. This corresponds to a shunt parallel to the basal ion flux (see artificial channel in Fig. 10). This channel has been realized in three ways. (1) By treatment of thylakoids with osmotic shocks which produce holes for different types of ions (Cl^-, Mg^{2+}, etc.) (Junge and Witt, 1968). (2) By treatment with ionophores such as GmcD (Junge and Witt, 1968); GmcD increases the permeability specifically for alkali ions (Pressman, 1965; Chappell and Crofts, 1966). (3) By treatment with uncouplers as DCPIP (H. T. Witt and Müller, 1959); DCPIP increases the per-meability for H^+ (Karlish and Avron, 1967).

In Fig. 11a the treatment with DCPIP is shown (in 1959 this acceleration of τ (515) was assumed to be caused by a redox reaction). In Fig. 11b,c the speed of absorption decay at 515 nm is shown as a function of the concentra-tion (c) of the ionophore GmcD and of K^+. If the decay rate $(1/\tau)$ reflects the discharge of a loaded membrane by ions and if one type of ion is in excess, e.g., K^+, it is expected that:

$$1/\tau \, (+ \, \text{GmcD}) = \text{const} \, c_{K^+} \cdot c_{\text{GmcD}} + 1/\tau \, (- \, \text{GmcD}) \tag{3}$$

This relationship has been verified (Junge and Witt, 1968). The decay time $\tau(+\text{GmcD})$ decreases strongly with increasing concentration of GmcD (Fig. 11b) as well as with increasing concentration of K^+ (Fig. 11c). The type of ion which is engaged in the basal field driven ion flux and which cor-responds to $1/\tau(-\text{GmcD})$ in Eq. (3) is discussed in Section 4.8.

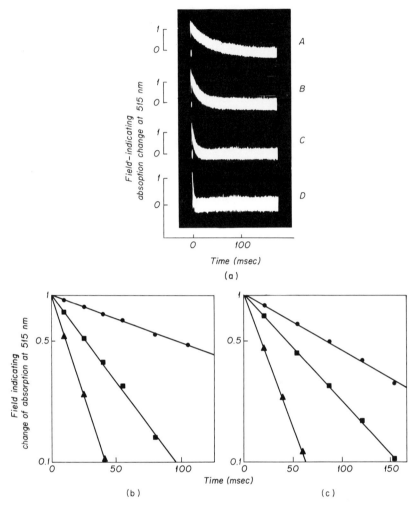

Fig. 11. Timecourse of the field indicating absorption change at 515 nm. (a) Independence on the concentration of the uncoupler DCPIP (H. T. Witt and Müller, 1959) (A = 8.5 × 10⁻⁵, B = 1.4 × 10⁻⁴, C = 203 × 10⁻⁴, D = 7.5 × 10⁻⁴ M. (b) Log plot of dependence on the concentration of the ionophore GmcD at constant K^+ concentration. (KCl = 4 × 10⁻² M). (▲) $c = 3 × 10^{-10}$; (■) $c = 10^{-10}$; (●) $c = 0$ M. (c) Log plot in dependence on the concentration of K^+ at constant GmcD concentration. (10⁻⁹ M). (▲) $c = 10^{-2}$; (■) $c = 10^{-3}$; (●) $c = 0$ M. Sample: chloroplasts of spinach (Junge and Witt, 1968).

The absorption change of the seven other events in photosynthesis which are listed in Section 1.2 and which are caused by formation of metastable states, light reactions, electron transfers, H^+ translocations, etc. are not accelerated by addition of ionophores. This is demonstrated in Table II, for instance for the rate of the overall electron transfer measured by O_2 evolution.

TABLE II
Rate of Electron Transfer and Decay Time, τ, of the Field-Indicating
Absorption Change as a Function of Gramicidin D[a]

GmcD ($\times 10^{-10}$ M)	0	5	10	50
Rel. electron transfer rate	100	100	100	100
$\tau(515)$ (msec)	51	12	5	1.3

[a]Data from Junge and Witt (1968).

4.3. Spectra of the Field-Indicating Absorption Changes

Because absorption changes attributed to other than electrical events are not accelerated by ionophores (see Section 4.2), one can discriminate between field-indicating absorption changes and the changes of all other events.

The spectrum of the absorption changes which can be accelerated by ionophores is presented in Fig. 12a (Emrich *et al.*, 1969). The dependence of the acceleration on the concentration of GmcD is in the limits of error the same at all wavelengths. This is indicated in Table III. The spectrum covers the whole spectral region where the bulk pigments—chlorophylls and carotenoids—absorb.

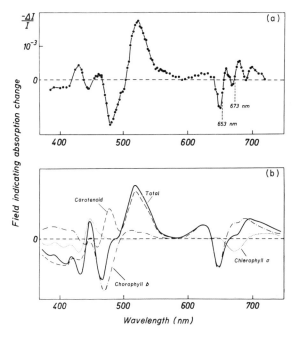

Fig. 12. (a) Transient spectrum of field-indicating absorption changes induced by light pulses in chlorplasts of spinach. Rise: $\leq$ 20 nsec; decay: 10–200 msec (Emrich *et al.*, 1969). (b) Transient spectrum of absorption changes of Chl *a*, Chl *b*, Car multilayers in a microcapacitor induced by an electrical field pulse of 10^6 V/cm (Schmidt *et al.*, 1971, 1972).

TABLE III

**Decay time, τ, of the Field-Indicating Absorption Change as a
Function of Gramicidin D at Different Wavelengths**[a]

GmcD ($\times 10^{-10}$ M)	1	2	5	10
$\lambda = 457$ nm	$\tau = 5.0$ msec	1.7 msec	0.7 msec	0.4 msec
478	5.1	1.5	0.8	0.3
515	5.1	1.5	0.8	0.3
648	5.1	1.5	0.8	0.3
685	3.3	—	0.6	0.3
700	3.2	1.7	0.6	0.2

[a]Data from Emrich *et al.* (1969).

If the spectrum in Fig. 12a is caused by an electrical field, it must be possible to explain the shape of the spectrum by interaction of the field with the bulk pigments. The interaction can be explained by different mechanisms.

In Section 4.9 it is shown that the electrical field set up across the membrane has a value of about 10^5 V/cm. Fields of this order can change: (1) the electronic absorption band of the bulk pigments incorporated in the membrane (electrochromism); (2) the chemical equilibrium between the bulk pigments in the membrane, e.g., monomer $\rightleftharpoons$ dimer; (3) the spatial orientation of the bulk pigments in the membrane, etc. All these changes are coupled with spectral changes. Very probably the spectrum in Fig. 12a is caused by the electrochromism (see below). In this case the theory predicts (see reviews by Labhart, 1967, and Liptay, 1969) that the frequency of the shift Δv depends on the field strength:

$$\Delta v = -\frac{1}{h}(\mu^* - \mu^0)F - \frac{1}{2h}(\alpha^* - \alpha^0)F^2 \qquad (4)$$

where μ^* and μ^0 are the vectors of the dipole moments in the excited and ground states, respectively, α^* and α^0 the polarizabilities, and F the vector of the field strength. At field changes of 10^5 V/cm the first term is predominant in dyes with permanent dipoles and causes a linear homogeneous shift of the absorption bands of the order of 0.1 Å. This would correspond to absorption changes of the order of 0.1 %. In the case of such small changes: (1) the spectrum of absorption changes of oriented molecules should follow the derivative of absorption bands; (2) the shape of the spectrum should not depend on the field strength; (3) the magnitude of the absorption changes should be proportional to the field strength.

The properties of (1)–(3) correspond closely to those of the spectrum in Fig. 12a (Emrich *et al.*, 1969): (1) In the red region the peaks at 648 and 660 nm are located antisymmetrically to a center at 653 nm. This shift is similar to the derivative of a Chl *b* band at 653 nm. The peaks at 668 and 680 nm are located antisymmetrically to a center at 673 nm. This shift is similar to the derivative of a Chl *a* at 673 nm. In the blue spectral range the interpretation is complicated due to the superposition of Chl *a*, Chl *b*, and Car shifts (carotenoids have three absorption bands in the blue). (2) The shape of the spectrum in Fig. 12a

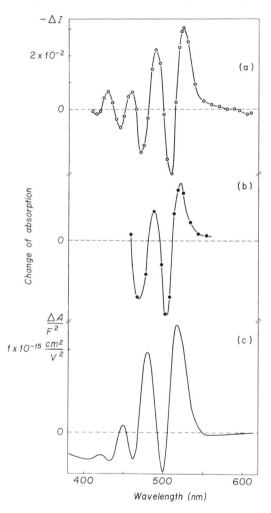

Fig. 13. (a) Transient spectrum of field-indicating absorption changes induced by light pulses in chromatophores of bacteria (Jackson and Crofts, 1969). (b) Transient spectrum of absorption changes induced with diffusion potentials set up with KCl gradients across the membrane of chromatophores of bacteria (Jackson and Crofts, 1969). (c) Transient spectrum of absorption changes of a carotenoid (lutein) in a microcapacitor induced by an electrical field pulse of 10^6 V/cm (Schmidt *et al.*, 1972).

is independent of the field strength, i.e., the location of the maxima does not shift with light intensity changes (unpublished results). (3) The amplitude of the spectrum is proportional to the field strength (see Section 4.9).

The findings of field-indicating absorption changes in green plants have been extended by Jackson and Crofts (1969) to corresponding changes in photosynthetic bacteria. The observed spectrum of these changes in bacteria are shown in Fig. 13a. The spectrum has been attributed to changes of carotenoids. The rise of these changes is extremely fast ($< 1 - 0.1 \,\mu$sec) (Baltscheffsky, 1969; Jackson and Crofts, 1969) as in the case of green plants (≤ 20 nsec; see Section 4.1). The rate of decay of these changes can be stimulated by ionophores such as Vmc (Jackson and Crofts, 1971) in the way shown for chloroplasts with GmcD in Fig. 11.

As in the case of chloroplasts the spectrum in bacteria is very probably caused by electrochromism. This argument is based again on the fact that the spectrum in Fig. 13a corresponds to a shift of the absorption band of the carotenoids, i.e., the spectrum is very similar to the derivative of the Car absorption bands *in vivo*. In bacteria the Car shift does not overlap with the field-induced shifts of the BChl bands because these are located in the far-blue and far-red spectral range. Therefore in this case the correspondence between absorption change and derivative of the absorption band has been demonstrated clearly (Schmidt *et al.*, 1972).

In this section some arguments have been discussed which make it likely that the shape of those light-induced spectra which are sensitive to ionophores, are caused by a field and are probably indicated through the mechanism of electrochromism. However, direct proofs for this assumption are discussed in the next section.

4.4. Spectroscopic Evidence for the Electrical Field

To substantiate the conclusion in green plants that (1) the spectral changes in Fig. 12a are field-indicating, and (2) the mechanism of the indication is due to electrochromism, multilayers of lipids with incooperated molecules of Chl a, Chl b, and Car were built up on glass slides (see Fig. 14). These layers were exposed in the dark between two electrodes to electrical fields of up to 10^6 V/cm and the resulting absorption changes caused by electrochromism were measured (Schmidt *et al.*, 1971, 1972). The superposition of these electrically induced changes is depicted in Fig. 12b. The agreement with the light-induced spectrum in chloroplasts in Fig. 12a is good. Only in the red spectral region is the spectrum in chloroplasts much more structured. This is due to the fact that physically different types of Chl a bulk pigments are present *in vivo*; on the multilayer, however, only one Chl a type is present. For an optimal fit one parameter was varied: the ratio of the pigment con-

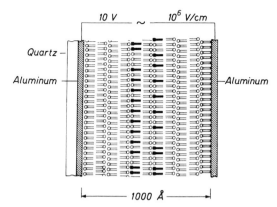

Fig. 14. Arrangement of monolayers of lipids with incorporated molecules of pigments in a microcapacitor which has been used for the experiment in Fig. 12b and Fig. 13c (Schmidt *et al.*, 1971, 1972). The monolayers have been built up according to the method of Blodgett and Langmuir (1937) and Kuhn *et al.* (1972). (○) Cd-arachidate; (■) Chl *a*, Chl *b*, and Car, respectively.

centrations. This was chosen as Chl a : Chl b : Car $= 1.7 : 5.2 : 1$. In chloroplasts the ratio of the pigments in the field-bearing lipids is unknown, only the ratio in toto is known; this value is $3.5 : 1.5 : 1$.

According to Fig. 12b the changes in Fig. 12a are a superposition of the band shift of all the three bulk pigments. However, the negative peak at 475 nm is due mainly to a Chl b shift, the positive peak at 515 nm to a Car, the negative peak at 648 nm to a Chl b shift.

The absorption changes observed in bacteria and shown in Fig. 13a have been analyzed by Jackson and Crofts (1969) in an elegant way by setting up an artificial diffusion potential across the inner membrane through gradient of ion concentrations. This was achieved by suspending potassium containing chromatophores of bacteria in potassium-free medium and then inducing a sudden potassium flux with Vmc. Spectral shifts of opposite polarity were then induced by the potassium pulses. The absorption change produced in the dark by such artificial potential changes corresponds very closely to the absorption changes which are produced when the bacteria are exposed to light (see Fig. 13b).

Corresponding salt jump tests with chloroplasts are complicated due to the large light-scattering effects on these particles. Changes at 520 nm induced in this way have been reported on recently (Strichartz and Chance, 1972).

To substantiate the conclusion in bacteria that (1) the spectral changes in Fig. 13a are band shifts of carotenoids in a field, and (2) the mechanism is due to electrochromism, the multilayer technique was also applied in this case. Layers of lipids incorporated with molecules of carotenoids (lutein) were built up on glass slides. In this case no concentration parameter had to be varied for an optimal fit (in chloroplasts the ratio of Chl *a*, Chl *b*, and Car was varied). When the lutein layers are exposed to electrical fields of up to 10^6 V/cm, absorption changes result as depicted in Fig. 13c (Schmidt *et al.*, 1972). The agreement with Fig. 13a,b is very good. Moreover, this agreement without any parameter variation justifies the application of the multilayer technique to the three-component system in chloroplasts as shown in Fig. 12b.

The field-indicating changes at 515 nm (rise $\leq$ 20 nsec, decay 200–10 msec) have been confused by some authors with absorption changes at 515 nm due to other events with different kinetics and different spectral changes in the blue and red:

1. Very slow optical changes at 515 nm (rise and decay in the range of seconds to minutes) are due to light scattering caused by osmotic-induced structural changes.

2. Slow changes at 515 nm (rise and decay in the range of seconds) may be due to deepoxidation of Car caused by pH changes in the inner space of the thylakoids (Hager, 1969).

3. Fast changes at 515 nm (rise < 20 nsec, decay 3 μsec) are due to the formation of metastable states of carotenoids (Wolff and Witt, 1969). Because the field-indicating changes can be seperated kinetically from others in flashlight and because they can furthermore be tested by the response to ionophores (see Section 4.2), it is always possible to distinguish clearly between field-indicating and the other three absorption changes at 515 nm.

The discussion on the pigment responsible for the absorption change at 515 nm arose after changes at 515 nm in the range of seconds were discovered by Duysens (1954): From the simultaneous changes at 648 nm it was concluded that Chl *b* causes the changes (Rumberg, 1964b). Govindjee and Govindjee (1965) speculated that the changes are due to redox reactions of Chl. From mutants with deficiency of special pigments Chance *et al.* (1967) concluded that Car contribute to the changes. Carotenoids have also been discussed by Fork and Amesz (1967) and Strichartz (1971). A decision was delayed because the question was focused mostly on the type of pigment responsible for the change at 515 nm and not to the event indicated. This was furthermore complicated because four events, not one, are signaled by the changes at 515 nm (see above).

4.4.1. Electrical Evidence for the Field-Indicating Absorption Changes

When in a chloroplast suspension two macroscopic electrodes are located at a distance of d ($\sim$ 1 cm) near the top and bottom of the cuvette and a nonsaturating flash is fired through one electrode (e.g., from the top), each thylakoid is slightly asymmetrically charged. This is because in the upper part of each thylakoid more light is absorbed than in the lower part. The charge difference induces a signal at the electrodes (Fowler and Kok, 1972, in Kok, 1974). The authors observed that the signal is diminished by one-half when one of the two light reactions is blocked. This confirms our result that each light reaction contributes one-half to the potential (see Section 4.6). The electrically measured signal decays in 10 μsec. The signal measured by the field-indicating absorption changes decays, however, in ~ 100 msec, i.e. 10^4 times slower. This discrepancy is explained as follows (Witt and Zickler, 1973). The electrically measured decay is caused by the equilibration of the asymmetrically charged thylakoid via ion fluxes in the water phase parallel to the plane of the membrane ($\tau_\parallel \approx 10$ μsec). However, the field-indicating changes decay by ion flux perpendicularly through the lipid phase of the membrane ($\tau_\perp \approx 100$ msec). Thus $\tau_\parallel \ll \tau_\perp$. To be able to measure $\tau_\perp$ by the electrodes it is necessary to prevent the equilibration via the membrane surface in 10 μsec. This holds if $\tau_\parallel \gg \tau_\perp$. A drastic increase of $\tau_\parallel$ has been realized by increasing the viscosity of H_2O from 1 to 10^3 cP through addition of $2M$ sucrose. $\tau_\perp$ has been decreased by addition of ionophores (Vmc). Under these conditions $\tau_\parallel \gg \tau_\perp$ and the electrically measured timecourse of $\Delta\phi$ is indeed in fair agreement with the timecourse of the field-indicating absorption changes (Witt and Zickler, 1973).

1. The agreement provides further support that the "field-indicating changes" indicate electrical potential differences.

2. The electrodes respond only to potential differences perpendicular to the membrane surface. Therefore, the agreement excludes the possibility that the "field-indicating changes" respond to field changes adjusted only in the plane of the membrane. This follows also from the response of $\Delta\phi$ to ionophores (see Section 4.2).

4.4.2. Field-Indicating Fluorescence Changes

The field indicating absorption changes are caused according to the results in Section 4.4 by a shift of the electronic states of the bulk pigments through the field. This shift should also change the prompt fluorescence of these pigments (Witt, 1971). Such fluorescence changes of chlorophyll have been observed. These changes also respond in the same way to ionophores as was shown for the absorption in Section 4.2 (Wolff, 1974).

4.5. Functional Membrane of the Electrical Field

The acceleration of the electrical field decay by GmcD can be achieved with extremely low concentrations (see Fig. 11). Data of Fig. 11 are listed in Table IV. Only one GmcD molecule per 10^5 Chl molecules causes more than 50% acceleration of the basal field decay. This means that about one GmcD molecule operates on a functional unit containing at least 10^5 Chl molecules. Such a number of Chl is spread over an area with a diameter of about 5000 Å

TABLE IV

Decay Time $\tau(515)$ of the Field-Indicating Absorption Change at
Different Ratios of Gramicidin D/Chlorophyll[a]

GmcD/Chl	0	$1:10^5$	$3:10^5$
$\tau(515)$ (msec)	100	30	10

[a]Data from Junge and Witt (1968).

(Kreutz, 1970b). A comparison with the different structural elements in chloroplasts shows that this area has the same order of size as that of so-called "thylakoids" (Fig. 15). From these results it has been concluded that the functional unit for electrical phenomena is the membrane of one thylakoid (Junge and Witt, 1968):

$$\Delta\phi - \text{unit} = 1 \text{ thylakoid} \tag{5}$$

An extreme treatment can be achieved when the concentration of GmcD is increased up to 10^{-6} M. At such concentrations membranes are perforated by holes and are permeable to different types of ions. Electron transfer and O_2 production should still operate even in such perforated membranes, because one electron transport chain covers an area of only about 10^5Å^2 including the energy conducting 500 bulk pigment molecules. On the other hand, the field-indicating absorption change should completely disappear because of the immediate breakdown of the field in such membranes. This is demonstrated in Table V. The result indicates that the field is bound to a whole thylakoid membrane but the electron transfers to a 100 times smaller area (H. T. Witt et al., 1968).

It would be of great interest to know whether the collapse of the field-indicating absorption changes by one ionophore per thylakoid is also valid

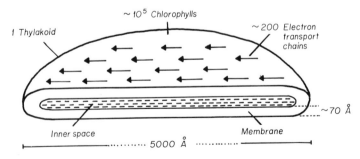

Fig. 15. Scheme of a thylakoid. The membrane is built up by lipids, proteins, and pigments (chlorophylls and carotenoids).

TABLE V

O_2 Evolution and Field-Indicating
Absorption Change ΔA (515) with and without
Gramicidin D[a]

GmcD ($\times 10^{-6}$ M)	0	10^{-6}
Rel. O_2 evolution	100	100
Rel. ΔA(515)	100	0

[a]Data from H. T. Witt *et al.* (1968).

when, instead of the thylakoids, vesicles with a size of quite different order are used, e.g., the chromatophores of photosynthetic bacteria. Indeed, the result of the extension of our titration experiment with chloroplasts to bacteria with the ionophore Vmc was that the collapse of the field-indicating absorption changes in the chromatophores takes place at a value of Vcm : BChl : 3×10^3 (Nishimura, 1970). 3×10^3 BChl molecules are spread over an area of a vesicle with a diameter of about 500 Å, which is the order of the size of one chromatophore. Nishimura has, however, not interpreted the changes at 515 nm as field-indicating.

4.6. Two Field Generators

When both light reactions are excited in one turnover flash the field-indicating changes have a magnitude as shown in Fig. 16a. When, however, the Chl a_I reaction is chemically eliminated with DCPIP + Fecy (Fig. 16c), the amplitude of the absorption changes diminishes by about one-half (Witt *et al.*, 1965; Schliephake *et al.*, 1968). This can also be demonstrated

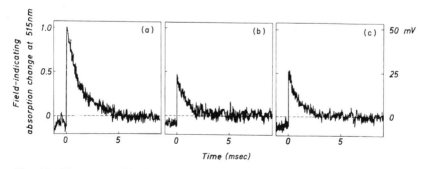

Fig. 16. Timecourse of field-indicating absorption changes by excitation of different light reactions. Addition of 10^{-9} M GmcD. (a) Light reactions I + II (+ TIP); (b) light reaction I (+ TIP + DCMU); (c) light reaction II (+ DCIP + Fecy). Sample: chloroplasts of spinach (Schliephake *et al.*, 1968).

TABLE VI

Decay Time $\tau(515)$ of the Field-Indicating Absorption Change as a Function of Gramicidin D
and Varying Light Reactions[a]

	Decay time (msec)			
GmcD ($\times 10^{-10}$ M)	0	1	3	10
Light reaction				
I + II	12	5	1.9	0.6
I	11.5	6.8	1.7	0.8
II	12.5	4.9	1.7	0.6

[a] Data from Schliephake et al. (1968).

when the Chl a_{I} reaction is saturated with 728 nm background light (which
excites only Chl a_{I}) (Rumberg, 1964b). When, on the other hand, the contribution of Chl a_{II} in light reaction II is prevented by deactivating Chl a_{II} with
DCMU, the amplitude of the absorption changes again diminishes by one-
half (Fig. 16b). This indicates that in a single turnover flash one-half of the
field or potential change is set up at each light reaction:

$$hv_{\mathrm{I}} \sim \tfrac{1}{2}{}^{1}\Delta\phi_0 \quad \text{and} \quad hv_{\mathrm{II}} \sim \tfrac{1}{2}{}^{1}\Delta\phi_0 \tag{6}$$

The decay time $\tau(515)$ in Fig. 16 is in all cases equal, no matter which of the
light reactions is active. This is the case even when $\tau(515)$ is shortened by different concentrations of GmcD (see Table VI). This substantiates the conclusion that at each light reaction the discussed absorption changes indicate
the same event, the electrical field (Schliephake et al., 1968). This conclusion
has been additionally supported by measuring the action spectra of the
absorption changes at 515 nm. Two spectra have been obtained which correspond to the action spectra in Fig. 4 (Schliephake et al., 1968).

4.7. Two Proton Pumps

pH phenomena in photosynthesis were discovered by Jagendorf and Hind
(1963). In continuous light large amounts of H^+ ions are translocated into
the inner phase of the thylakoids (Neumann and Jagendorf, 1964). The H^+
uptake would be stopped after a few H^+ transfer by electrostatic repulsion
forces if an equivalent number of other ions are not pushed across the
membrane in exchange for H^+ (e.g., by extrusion of K^+). Such a balancing ion
flux has been observed and is in its extent roughly equivalent to the uptake of
H^+ (Dilley and Vernon, 1965; Schröder et al., 1971). In Fig. 10 this flux has
been symbolized by a basal channel for K^+.

In saturating light, the pH value in the inner thylakoid space decreases
from pH_{in} 8–4.7. This corresponds to a maximal H^+ gradient across the
membrane of $\Delta pH_{ss} \approx 3.3$. This value has been measured first by an intrinsic

indicator for H_{in}^+. The H_{in}^+ depending rate of Chl a_I reduction has been used for this purpose (Rumberg and Siggel, 1969). Recently artificial indicators (amines) have been used for measuring pH_{in} (Rottenberg et al., 1971).

For the mechanism of the coupling between electron transfer, field generation and H^+ translocation and for the calibration of the field strength it is of interest to know how many H^+ ions are translocated per electron transfer. This ratio $\Delta H^+/\Delta e$ has been studied by several researchers in continuous light. An uptake of two protons per electron transfer was reported by Izawa and Hind (1967), Schwartz (1968a), and Rumberg et al. (1969), but also higher values between 2 and 6 protons have been found (Karlish and Avron, 1967; Dilley, 1970) (for details see Chapter 9). The values higher than 2 are, according to Rumberg et al. (1969), incorrect because in these cases the changes of the electron transfer rate at the beginning of illumination before the steady state has been reached have not been considered.

The complications of changes of electron transfer rates can, however, be completely avoided if the $\Delta H^+/\Delta e$ ratio is measured in a one-turnover flash in which only one electron is transported across the possible coupling site. By using pH indicators and by applying the repetitive pulse technique the corresponding changes can be analyzed with great accuracy. The values obtained with this independent and different method are listed in Table VII (Schliephake et al., 1968). For different electron acceptors it results a ratio of

$$^1\Delta H^+/1e \approx 2 \tag{7}$$

If the H^+ translocation is the result of proton reactions with the charge separated across the membrane (see Section 4.10) it can be concluded that the potential change is due to a separation of *two* elementary charges per electron chain. This value has been used for calibration of $^1\Delta\phi_0$ in Section 4.9.

The ratio $^1\Delta H^+/1e \approx 2$ may indicate the existence of one coupling site at each of the two light reactions or two at one light reaction. A discrimination can be achieved when Chl a_I in light reaction I has been separated by deactivating Chl a_{II} with DCMU (see Section 2.3.3). Figure 17 demonstrates that with DCMU the proton uptake is decreased by a factor of 2. This means

TABLE VII

Proton Uptake per Electron Transfer in the Presence of
Different Electron Acceptors[a]

Electron acceptor	BV		TIP		Saf		Inc	
$^1\Delta H^+/1e^-$	1.9	2.2	2.3	2.0	1.6	2.0	2.2	1.8

[a]Data from Schliephate et al. (1968).

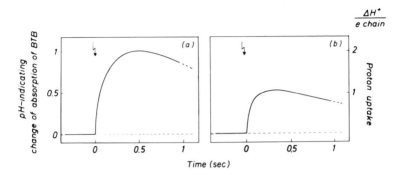

Fig. 17. pH-indicating absorption changes of bromthymol blue by excitation of different light reaction. (a) Light reactions I + II (+ TIP); (b) light reaction I (+ TIP + DCMU). Sample: chloroplasts of spinach (Schliephake *et al.*, 1968). The results have been confirmed also with other indicators.

that in a single turnover (^{1}) each of the two light reactions translocates one proton across the membrane (Schliephake *et al.*, 1968):

$$hv_I \sim 1H^+ \quad \text{and} \quad hv_{II} \sim 1H^+ \tag{8}$$

This result indicates that the potential change in a single turnover flash is due to a separation of one elementary charge at each light reaction. The field-indicating absorption change participates also in equal parts at each light reaction (see Section 4.6). This implies a *linear* relationship between field-indicating absorption changes ΔA and potential changes $\Delta \phi$. The linearity has, however, been demonstrated over a much wider range in Section 5. The linear response is the basis for the calibration of $\Delta \phi$ in Section 4.9.*

4.8. Field Decay and Proton Concentration

Results on chloroplasts of spinach in Fig. 18a indicate that the half-life time $\tau(515)$ of the potential decay decreases 20-fold with increasing flash frequency from 1 to 30 Hz. The reciprocal value $1/\tau(515)$ increases 20-fold in this range. The dependence on flash frequency (v) is linear (Boeck and Witt, 1972):

$$1/\tau(515) \sim v \tag{9}$$

---

* The pH increase in Fig. 17 is delayed due to the relative slow response time of the indicator. However, the pH *decrease* indicates the actual time of the H^+ efflux (≥ 1 sec). This H^+ efflux occurs in the case of Fig. 17 in exchange with the K^+ influx, after the field has already decayed to zero in ~ 100 msec predominantly by the efflux of K^+ (this has been checked by addition of V mc). This was due to a low H_{in}^+ concentration. If the field decays with the efflux of H^+, the pH decreases simultaneously with the decay time of the field (~ 100 msec).

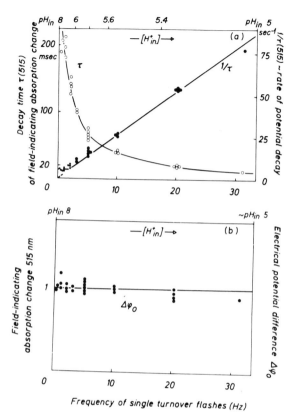

Fig. 18. (a) Decay time τ and decay rate $1/\tau$ of the field indicating absorption change at 515 nm as a function of the frequency of single turnover flashes and pH_{in}. (b) Extent of the initial field indicating absorption change at 515 nm and potential change $\Delta\phi_0$ as a function of the frequency of single turnover flashes and pH_{in}. Sample: chloroplasts of spinach, pH_{out} 8. 20°C (Boeck and Witt, 1972).

The measurements in Fig. 18 at different frequencies have been carried out when at a chosen frequency (v) stationary conditions have been reached. Similar results have also been observed with whole *Chlorella* cells (Gräber and Witt, 1974a).

With increasing frequency the concentration of H_{in}^+ is also increased because inward H^+ translocation caused by each flash starts before the H^+ efflux has finished. In exchange for H^+ an equivalent number of counterions, e.g., K^+, leave the inner phase of the thylakoid until steady state is reached. In the frequency range 0–30 Hz the pH_{in} values change from 8 to 5 (see Fig. 18a). pH_{out} is kept constant at pH_{out} 8.

The extent of the potential change $^1\Delta\phi_0$ caused by each flash is constant in the range between 1 and 30 Hz (see Fig. 18b):

$$^{1}\Delta\phi_0 = \text{constant} \tag{10}$$

If we multiply Eq. (10) by the capacity C of the thylakoid membrane, the product indicates a charge displacement. If we identify these charges as protons, the constancy indicates that at the indicated frequences (v) and pH_{in} values each single turnover flash translocates the same number of protons across the membrane:

$$^{1}\Delta\phi_0 C \sim {}^{1}\Delta H^+ = \text{constant} \tag{11}$$

The decrease of τ with increasing v can be explained by two different mechanisms (Gräber and Witt, 1974a).

1. The measured decay time τ is identical with the actual decay time τ_0 of $\Delta\phi_0$, i.e., $\tau = \tau_0$. It is assumed that τ_0 is a function of v due to the increase of H_{in}^+ with v. Provided that H_{in}^+ does not affect the permeability of the membrane for other ions, e.g., K^+, with increasing H_{in}^+ protons predominantly contribute to the discharge of the membrane. The averaged value of the corresponding H^+ efflux $\overline{\Delta\cdot\phi C} = \overline{i_{H^+}}$ is $\overline{i_{H^+}} \sim \Delta\phi_0 C/\tau_0$, or together with Eq. (9):

$$\overline{i_{H^+}} \sim v \tag{12}$$

For *Chlorella* cells this interpretation has been checked in an independent way. Different pH_{in} values have been established between pH_{in} 8 and 5 using continuous light with intensities between zero and saturating values. The decay time $\tau_0(515)$ of $\Delta\phi_0$ produced by one flash decreases at the different intensities from 1000 to 20 msec. When the constant light is switched off, the decay time τ_0 of $\Delta\phi_0$ induced by single test flashes increases again from 20 to 1000 msec probably due to the change from pH_{in} 5 to 8 (Gräber and Witt, 1974a).

2. The decrease of the measured decay time τ can also be explained, however, as follows. The actual decay time τ_0 is constant, i.e., independent of v. With increasing v thereby an underlying electrical potential is built up to $\Delta\psi$. The steady state value $\Delta\psi$ is approximately proportional to v, i.e., $\Delta\psi \sim v$. The measured lifetime τ of $\Delta\phi_0$ produced by each flash on the top of the underlying $\Delta\psi$ is $\tau \approx \tau_0\Delta\phi_0/\Delta\psi\ 2\ln 2$. Assuming again that the corresponding current is predominantly a H^+ efflux, in this case $\overline{i_{H^+}} = \Delta\phi_0 C/2\tau$. Because $\Delta\psi \sim v$ it follows of course also for this second mechanism.

$$\overline{i_{H^+}} \sim v \tag{13}$$

In isolated chloroplasts this second mechanism probably dominates.

The ability to change $\overline{i_{H^+}}$ with v is of value for the analysis of the mechanism of phosphorylation (see Section 6.2).

4.9. Calibration of the Electrical Potential Difference and Current Density

The field-indicating absorption changes have the specification of a molecular voltmeter which operates in spaces of 100 Å (thickness of the membrane of a thylakoid; see Section 4.5). In Sections 4.7 and 5 it is shown that in green plants these absorption changes $\Delta I/I = \Delta A$ depend linearly on the number of separated charges Q. In this case it follows:

$$\Delta\phi = Q/C = a\Delta A \tag{14}$$

where Q = charges translocated across the membrane during absorption changes (ΔA), C = capacity of the membrane, and a = proportionality factor. Because the derivative of the absorption changes with respect to time indicates the flux of charges, the former also have the specification of a molecular ammeter which operates in time ranges down to 10 nsec (see Fig. 9):

$$i = C\, d(\Delta\phi)/dt = Ca\, d(\Delta A)/dt \tag{15}$$

Absolute data for $\Delta\phi$ and i have been obtained in two different ways.

1. In a first approach the number of charges which are translocated across the membrane in one turnover were used for calibration. In Section 4.7 it was shown that *two* elementary charges per electron chain are translocated. One electron chain covers an area of $S \approx 10^5$ (Å)^2 (see Section 1.1). The thickness (l) of the isolating lipid layer of the membrane obtained by X-ray scattering is $l = 21$ Å (Kreutz, 1968). The result from electron microscopy is $l \approx 40$ Å (Mühletahler, 1969). Because of this uncertainty we use $l \approx 30$ Å. We assume that the effective dielectric constant of phospholipid membranes $(\epsilon \approx 2)$ is valid also for the thylakoid. Then it follows that, according to Eq. (14) in one turnover (^{1}) with the absorption change $^1\Delta A$ for the potential $^1\Delta\phi_0$ (Schliephake *et al.*, 1968):

$$^1\Delta\phi_0 = a\,^1\Delta A = \frac{Q}{C} = \frac{2e}{S\epsilon_0\epsilon/l} \approx \frac{2e}{1\mu F} \approx 50 \text{ mV} \tag{16a}$$

or

$$^1F \approx 2 \times 10^5 \text{ V/cm} \tag{16b}$$

Together with Eqs. (14) and (15) it follows for arbitrary values of ΔA:

$$\Delta\phi = \frac{50 \text{ mV} \cdot \Delta A}{^1\Delta A} \tag{17a}$$

and

$$i = 1\mu F \frac{50 \text{ mV}}{^1\Delta A}\, d(\Delta A)/dt. \tag{17b}$$

With Eq. (17b) and the data of Fig. 8 it follows that the onset of the field in $\leq$ 20 nsec corresponds with a current density in the field generator of $\geq$ 1 A/cm^2 and the field decay in 20 msec with a current density across the membrane of $\sim$1 μA/cm^2. For a current voltage analysis in thylakoids in the presence of ionophores see Junge and Schmid (1971).

The field-indicating absorption change increases as flash length is increased up to a factor 4: $\Delta A_{max}/^1\Delta A \approx 4$ (H. T. Witt, 1955b), i.e., the maximal voltage across the thylakoid membrane amounts to:

$$\Delta\phi_{max} \approx 200 \text{ mV} \tag{18a}$$

This increase occurs in a slow phase (1 msec) due to the filling up of the PQ pool with more electrons (H. T. Witt et al., 1965) (see Section 5). Larger transient changes of $\Delta\phi$ during the induction period (light on after long dark times) (H. T. Witt and Moraw, 1959b) may be due to ionic redistribution coupled with changes of the membrane capacity.

In saturating steady-state light (ss) in *Chlorella* cells $\Delta A_{ss}/^1\Delta A \approx 2$ (Witt and Moraw, 1959b), i.e.,

$$\Delta\phi_{ss} \approx 100 \text{ mV} \tag{18b}$$

In steady-state light this value may be subject to error because of slow changes associated with light scattering. However, this has been excluded in recent experiments with *Chlorella* cells (Gräber and Witt, 1974b).

2. In the second approach diffusion potentials have been set up across the membrane of chromatophores by salt jumps (see Section 4.4). This has been done in the presence of ion translocators which increase the membrane permeability specifically for cations or anions. The resulting absorption changes have been calibrated in volts by assuming that the Nernst equation is valid (Jackson and Crofts, 1969);

$$\Delta\phi = (RT/F)\ln K_{in}^+/K_{out}^+ \tag{19}$$

The results are shown in Fig. 19. The values are comparable with those of Eqs. (16)–(18). Figure 19 indicates that also in bacteria the field-indicating absorption changes depend *linearly* on the electrical potential change $\Delta\phi$.

Permanent Local Field

The electrochromic response of Car multilayers on slides shows a quadratic dependency on the field strength (Schmidt et al., 1971, 1972). In the case of oriented molecules this indicates the absence of permanent dipole moments. On the other hand, it has been shown that the field indication by the bulk pigments in the chloroplasts depends linearly on field changes F in the range of 2×10^5 V/cm (see above). This can be understood if the carotenoids are exposed in the membrane additionally to a high permanent local field F_p of

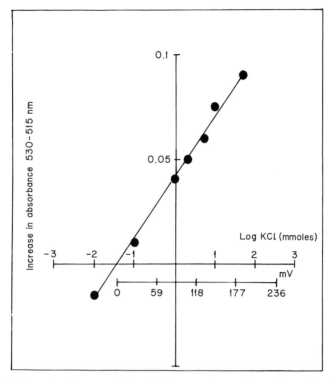

Fig. 19. Absorption changes at 515nm in chromatophores of bacteria as a function of the KCl gradient set up across the membrane of the chromatophores. The scale in millivolts is calculated from the Nernst equation (19) assuming that the intercept on the abscissa gives the internal K^+ concentration (Jackson and Crofts, 1969).

the order of 2×10^6 V/cm:

$$A = A_0 + b(F + F_p)^2$$

where A = absorption in the field, A_0 absorption without a field, and $b =$ constant factor. Relatively small changes of $F = 2.10^5$ V/cm would then cause a pseudolinear dependency:

$$\Delta A = b(F^2 + 2F_pF) \qquad \text{or} \qquad \Delta A \approx b'F$$

with $\Delta A = A - A(F = 0)$, $b' = b2F_p$, and $F^2 \ll 2F_pF$. Permanent local fields of 2.10^6 V/cm can be explained by ions such as phospholipids fixed in the thylakoid membrane (Schmidt et al., 1972; Schmidt and Reich, 1972).

5. VECTORIAL ELECTRON TRANSFER AND PROTON TRANSLOCATION

In this section the results reported so far are briefly summarized. The consequences of these results indicate a coupling between the different events

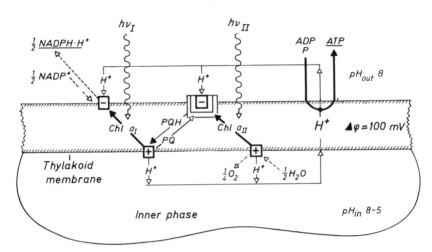

Fig. 20. Zigzag pathway of electrons and cycling of protons in the primary acts of photosynthesis derived from pulse spectroscopic results. The large square symbolizes the PQ pool. (For details see Section 5.)

in the form of a special zigzag scheme (see Fig. 20). This scheme has been supported by further results described at the end of this section.

1. It has been shown that the electrical events take place within the membrane of a thylakoid (see Section 4.5). The illustrated membrane element in Fig. 20 represents a segment of this thylakoid membrane in which one electron transport chain is embedded.

2. Two light reactions at Chl a_I 700 (see Section 2.2) and Chl a_{II} 680 (see Section 2.3) have been detected.

The energy of the excited states of Chl a_I and Chl a_{II} is used to transfer electrons from H_2O to $NADP^+$ in a series of two reactions (see Section 2.4).

3. A pool of PQ is the link between the two light centers (see Section 3).

4. Excitation of Chl a_I and Chl a_{II} leads to an electrical potential difference across the thylakoid membrane (see Sections 4.1–4.5). Potential differences occurring only in the plane of the membrane have been excluded (see Section 4.4.1). In a single turnover each of the two light reactions set up one half of the electrical potential change across the thylakoid membrane (see Section 4.6):

$$hv_I \sim {}^1\Delta\phi_0/2 \qquad \text{and} \qquad hv_{II} \sim {}^1\Delta\phi_0/2$$

5. The generation of ${}^1\Delta\phi_0$ takes place simultaneously with the photoact in <20 nsec (see Sections 2.2.1 and 4.1):

$$\tau(\text{Chl } a_I \rightarrow \text{Chl } a_I^+) \leq 20 \text{ nsec}$$
$$\tau(0 \longrightarrow {}^1\Delta\phi_0) \leq 20 \text{ nsec} \tag{20}$$

6. The results 1–5 have been explained by an electron shift at each light reaction with a component perpendicular to the membrane. The consequence is a vectorial zigzag flow of electrons from H_2O to $NADP^+$ via the pool of PQ (Junge and Witt, 1968). This is outlined in the zigzag scheme in Fig. 20. The scheme indicates an anisotropic organization of the thylakoid membrane as first proposed by Mitchell (1966).

7. In consecutive reaction steps at each of the two light reactions a proton translocation of $1H^+$ into the inner phase of the thylakoid takes place [see Section 4.7 and Eq. (8)]:

$$hv_I \sim 1H^+ \qquad \text{and} \qquad hv_{II} \sim 1H^+$$

In the presence of an artificial electron acceptor A this is explained if the protons are transferred as follows (see also Section 3.4). $1 H^+$ is taken up by $1/2 PQ^{2-}$ and $1 H^+$ by $1/2 A^{2-}$. $1 H^+$ is released by the cleavage of $1/2 H_2O$ and $1 H^+$ is released by the reoxidation of $1/2 PQH_2$ (see Fig. 20).

In toto this process is equivalent to a translocation of $2H^+$ to the inside of the thylakoid. Other mechanism of H^+ transfer may be discussed (Kreutz, 1970a). Whatever the mechanism is, the electronic charges of the field have been replaced by ions through chemical reactions. This charge conversion does not change the extent of the electrical potential difference across the membrane.

The reaction sequence as formulated by the zigzag scheme in Fig. 20 is supported by the following stoichiometries (8 and 9) between potential change, $\Delta\phi_0$, proton translocation, ΔH^+, and reduction of PQ:

8. In single turnover flash at light reaction II one $PQ^{2-}/2$ is produced and a potential change of $^1\Delta\phi_0/2$ is set up (see Sections 3.2 and 4.6). What happens if in a long flash with n turnovers $nPQ/2$ of the pool are reduced? This can be

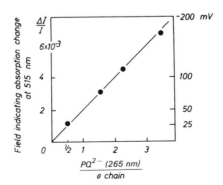

Fig. 21. Extent of the field-indicating absorption changes induced by light reaction II and its dependence on the amount of reduced PQ. Sample: chloroplasts of spinach (Witt et al., 1965, Reinwald et al., 1968).

answered by comparing the field-indicating absorption change at 515 nm accompanying light reaction II and that of PQ reduction at 265 nm while increasing the flash duration (Witt *et al.*, 1965; Reinwald *et al.*, 1968). In Fig. 21 the events are strictly proportional

$$n^1 \Delta\phi_0/2 \sim nPQ^{2-}/2 \qquad \text{(for } n = 1\text{–}7) \qquad (22)$$

The electrical potential change increases proportionally to the number of electrons injected into the pool of PQ. This substantiates the proposed transmembrane shift of electrons from H_2O (inside) to PQ (outside) as depicted in Fig. 20.

The relationship between electron transfer and field-indicating absorption change shows that over a much wider potential range than in Section 4.7 the relationship between membrane potential difference $\Delta\phi$ and its indication by the absorption changes ΔA at 515 nm is a *linear* one. This response was the basis for the voltage calibration in Section 4.9.

9. In a single turnover flash at light reaction II one $PQ^{2-}/2$ is produced (see Section 3.2), followed by the translocation of one H^+ (see Section 4.7). It is of interest to know whether this relationship holds if in a long flash with n turnovers $nPQ/2$ of the pool are reduced. Simultaneous measurements of both events between $n = 1$ and $n = 7$ are documented in Fig. 22 (Reinwald *et al.*, 1968). It was found that

$$nH^+ \sim nPQ^{2-}/2 \qquad \text{(for } n = 1\text{–}7) \qquad (23)$$

The proton uptake increases proportionally to the number of electrons which are injected into the pool of PQ. This substantiates the proposed H^+ translocation via H^+ uptake at PQ (outside) and H^+ release from H_2O (inside) as depicted in Fig. 20.

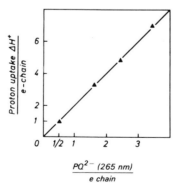

Fig. 22. Amount of proton uptake per electron chain induced by light reaction II and its dependence on the amount of reduced PQ. Sample: chloroplasts of spinach (Reinwald *et al.*, 1968).

10. The overall transfer time of one electron from H_2O to the PQ pool is 0.6 msec (Vater, *et al.* 1968) (see Section 3.2.2). When this electron transfer occurs vectorially across the membrane, i.e. generation of an electrical potential $^1\Delta\phi/2$ at light reaction II, the transfer time 0.6 msec should be identical with the passage time necessary before the potential can be generated a second time. The identity of the recovery time of the potential difference with the electron transfer time of 0.6 msec was observed (Witt and Zickler, 1974). This also supports the orientated electron shift from H_2O (inside) to PQ (outside) as depicted in Fig. 20.

We also note:

11. (a) The H^+ release from H_2O at light reaction II on the inner surface of the membrane is in accordance with pH measurements in single turn over flashes by Schwartz (1968b). He concluded from his results that cleavage of H_2O takes place on the inner surface of the membrane.

(b) The H^+ release at light reaction I on the inner surface of the membrane is in agreement with Hauska *et al.*, (1971) who located the oxidation of the electron donor of Chl a_I (plastocyanin) at that place.

(c) The polarity of the field with negative charges outside is in agreement with the observed passive efflux of cations (Dilley and Vernon, 1965) and passive influx of anions (Schröder *et al.*, 1971).

(d) The scheme has been confirmed and established by Junge and Ausländer (1974) through further analysis of the proton reaction on the inside of the thylakoids.

An alternative mechanism to $\Delta\phi$ generation by electron shifts as discussed above could be a primary H^+ shift at each light reaction in the opposite direction. This is, however, very improbable in view of the above-cited results and also from the fact that $\Delta\phi$ generation takes place simultaneously with the electron transfer at the active chlorophylls [see Eq. (20)].

12. The discharging of the loaded membrane is mainly caused by the efflux of protons (see Section 4.8):

$$\Delta\dot\phi = \Delta\dot\phi_{H^+} \quad \text{or} \quad i = i_{H^+}$$

With respect to phosphorylation the results of the following sections will demonstrate that the $\Delta\phi$ decay by the efflux of protons is coupled to the generation of ATP (see Fig. 20).

6. PHOSPHORYLATION IN THE ELECTRICAL FIELD

In 1966 an "intermediate" of photophosphorylation was detected which could be measured by light-induced absorption changes at 515 nm (Witt *et al.*, 1966a, Rumberg *et al.*, 1966). This was observed in the following way. With

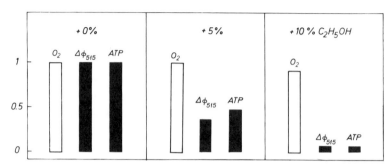

Fig. 23. Extent of the electrical potential change ($\Delta\phi_{515}$), phosphorylation (ATP) and electron transfer (O_2 evolution) as a function of alcohol concentration. Sample: chloroplasts of spinach (Rumberg *et al.*, 1966, Witt *et al.*, 1966a).

increasing concentration of added ethyl alcohol as well as with increasing concentration of desaspidin the extent of the absorption change at 515 nm was found to decrease parallel with the decrease of ATP production. The O_2 evolution, however, which indicates the oxidation of H_2O and the transfer of electrons, was practically unaffected (see Fig. 23). The changes at 515 nm were shown to indicate electrical potential changes across the thylakoid membrane (see Section 4). These results indicate therefore a relationship between an electrical potential change $\Delta\phi$ and phosphorylation.

Jagendorf and Uribe (1966) made the important discovery that if a pH gradient ΔpH 3 is artificially set up across the inner membrane of chloroplasts, ATP formation takes place in the dark. In the same way a light-induced pH gradient contributes to ATP generation as outlined in detail in Chapter 9. These results indicate a relationship between ΔpH and phosphorylation.

In the electrochemical hypothesis of Mitchell (1961, 1966) it is proposed that light-induced gradients of electrical potentials and pH drive protons through a membrane-bound ATPase enzyme. The energy released during the flux of n_i protons "down" the gradient is to be used for the synthesis of 1ATP. At $pH_{out} \approx 8$ it is:

$$1ADP + 1P + n_iH_{in}^+ + 1H^+ \rightleftharpoons 1ATP + n_iH_{out}^+ + 1H_2O \quad (24)$$

The ATPase enzyme was isolated by Racker (1970). Details of this enzyme are discussed in Chapter 9. In Fig. 10 the flux through the ATPase is symbolized by an ATPase channel. In the thermodynamic equilibrium, which requires a reversible ATPase, the electrochemical potential difference of H^+ equals the free energy of the ATP/ADP couple:

$$n_i(F \Delta\phi + 2.3RT \Delta pH) = \Delta G^{0\prime} + RT \ln (ATP/ADP \cdot P) \quad (25)$$

In the next sections the investigations are focused on the correlation between the electrical potential ($\Delta\phi$) and phosphorylation. We may formulate three interesting questions: What is the relationship (1) between the extent of the potential change $\Delta\phi_0$ and the yield of ATP, (2) between the rate of the potential decay $\Delta\cdot\phi$ and the rate of ATP formation, and (3) between the functional unit of the electrical events and the unit of phosphorylation?

At low frequencies the yield of the forward reaction in Eq. (24) (i.e., production of ATP) is very small and of the same order as the backward reaction (i.e., the hydrolysis of ATP). Therefore special precautions for the measurement of phosphorylation were observed.

The forward reaction has been measured by the chemically consumption of $1H^+$ during the esterification of 1P and 1ADP into 1ATP (Nishimura and Chance, 1962) [see Eq. (24)]. The ATP hydrolysis has been measured by the production of $1H^+$ per ADP. Consumption as well as production of H^+ have been measured by the sensitive pH-indicating fluorescence change of UBF*. (Günhagen and Witt, 1970). In comparison with the measurement of ATP by ^{32}P, the measurement by the UBF fluorescence, at low frequencies (≤ 1 Hz) where the yield of ATP-formation is comparable with ATP hydrolysis, has the advantage that the forward reaction in Eq. (24) in single turnover flashes can be estimated separately (Boeck and Witt, 1972).†

For this purpose the overall H^+ change (ATP-formation + ATP hydrolysis) *minus* H^+ production (ATP hydrolysis) was used for the calculation of ATP formation in Figs. 24–28.†

6.1. Amount of ATP Formation and Extent of Electrical Potential

The extent of the initial potential difference can be changed from $^1\Delta\phi_0 \approx$ 50 mV in a single turnover flash up to $\Delta\phi_{0\,max} \approx 200$ mV by using a longer flash. At the different $\Delta\phi_0$ values the flash frequencies have been adjusted so that ΔpH was always constant (ΔpH 2). Figure 24 shows that the yield of ATP formation per flash and per electron chain increases, within the margin of error, linearly with the initial $\Delta\phi_0$ value (Boeck and Witt, 1972);

$$\Delta ATP \sim \Delta\phi_0 \qquad \Delta pH = \text{constant} \qquad (26)$$

---

*The rate of the forward reaction is the total pH trace in the light minus the pH trace due to hydrolysis in the light. The trace of hydrolysis in the light has been estimated by addition of GmcD, which blocks the production of ATP (see Section 6.2). This technique has been controlled by subsequent addition of Dio-9, which poisons the ATPase. The separated trace attributed to hydrolysis in the light decreases as aspected to zero when Dio-9 is added.

†At higher frequencies (> 1 Hz) ATP hydrolysis is small in comparison with the forward reaction in Eq. (24). In this case the amount of ATP calculated on the basis of the change of H^+ concentration is identical with ATP measured by ^{32}P.

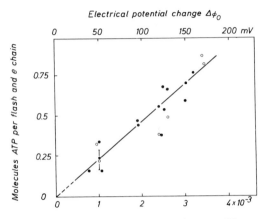

Fig. 24. Molecules of ATP per flash and electron chain in dependence on the extent of the field-indicating absorption changes and potential change $\Delta\phi_0$. The linearity is valid at a constant ΔpH value of 2. Simultaneous measurements of phosphorylation and $\Delta\phi_0$ by the repetitive pulse photometer with double beams. Sample: chloroplasts of spinach, pH_{out} 8, pH_{in} 6 (Boeck and Witt, 1972). The pH_{in} value was revised by Gräber and Witt (1973, unpublished).

The extrapolation to smaller $\Delta\phi_0$ values passes through the origin. Figure 24 shows that under these conditions any value of $\Delta\phi_0$ induced by the flash, no matter how small, shifts the ATP/ADP couple toward ATP.

The linear increase of ΔATP with $\Delta\phi_0$ allows a simple explanation for the causality between both events. At a constant electric capacity C of the thylakoid membrane, $\Delta\phi_0$ is proportional to the amount of charge translocated across the membrane. The above-discussed linearity then indicates that phosphorylation is stoichiometrically coupled to the amount of protons with which the membrane has been charged:

$$\Delta ATP \sim \Delta\phi_0 C \quad \text{or} \quad \Delta ATP \sim \Delta H^+ \qquad (27)$$

This result implies that $\Delta\phi$ is used for phosphorylation as has been proposed by Mitchell (1966).

The results reported here are different from earlier results from this laboratory in which a nonlinear dependence of $\Delta\phi$ and ΔATP characterized by a relatively large threshold of $\Delta\phi \approx 50$ mV had been observed (Junge et al., 1970). This had been interpreted by a triggering level which activates the phosphorylation center (Junge, 1970). In the case of an irreversible ATPase no equilibrium between ATP and ADP can be reached because the hydrolysis is hindered. In this case a certain value of $\Delta\phi$ (threshold effect) might overcome the block. However, this cannot be of fundamental importance because in the present results no irreversibility was ever observed.

During the preparation of this book we obtained the following results (Gräber and Witt, 1974a). The slope of the curve in Fig. 24 depends on the value of ΔpH. At ΔpH < 2 the curve is nonlinear and turns to a linear dependency at higher ΔpH values. At ΔpH 2.7 the slope is maximal (not shown in Fig. 24). At ΔpH 2.7 the slope indicates that 1 ATP per $4H^+$ is generated. This is in accordance with the earlier result of Rumberg (1974) measured in a different way (see Section 6.2).

A critical electrical potential has been excluded also in chromatophores of bacteria (Jackson, 1973). It was observed that the acceleration of the decay time of the $\Delta\phi$ which occurs under phosphorylating conditions (see Section 6.2) takes place the same as before when the extent of the flash intensity and thereby the potential change is reduced far below the single turnover value $^1\Delta\phi_0$.

When phosphorylation is measured in permanent light the existence of a critical potential of $\Delta\phi \approx 50$ mV should cause a relative large intensity range in which no phosphorylation takes place. This has, however, not been observed. Saha et al. (1970) observed a linear dependence of ATP production with light intensities even of values smaller than 100 erg/cm²-sec. This would be also in accordance with the absence of a critical potential.

6.2. Rate of ATP Formation and Rate of Electrical Potential Decay

Under conditions in which the potential difference decreases slowly in ≥ 200 msec due to a low basal permeability of the membrane, field-indicating absorption changes at 515 nm are observed as shown in Fig. 25. Under phosphorylating conditions the decay time $\tau(515)$ decreases 2- to 5-fold (Rumberg and Siggel, 1968). For the acceleration the presence of ADP, P, and Mg^{2+} are necessary. (The acceleration is reversed by addition of phloricin). These results have been confirmed and extended (Junge et al., 1970). An association of potential differences and phosphorylation has also been observed in chromatophores of bacteria but interpreted in a different way (Baltscheffsky, 1969). Recently, Jackson (1973) observed under phosphorylating conditions also in chromatophores an acceleration of the decay time of the field-indicating absorption changes.

The extra ion flux coupled to phosphorylation may be channeled through the oriented ATPase enzyme located in the membrane (Mitchell, 1966). This is outlined in Fig. 10. With this assumption the number of ions n_i which pass across the ATPase coupled pathway can be read out from the difference in the decay time $\tau(515)$ with and without phosphorylation. The conclusion is that the synthesis of one ATP molecule is coupled to the field-driven flux of about three ions (Junge et al., 1970). The same value has been obtained in continuous light, where predominantly ΔpH drives the protons (Schröder *et al.*, 1971). Recently this value has, however, been revised and specified to even 4 (Rumberg, 1974):

$$1\text{ATP} \sim 4\text{H}^+ \tag{28}$$

If it is accepted that $2e \sim 1\text{ATP}$ (see Chapter 9) and $2\text{H}^+ \sim 1e$ (see Section 4.7), one obtains $4\text{H}^+ \sim 1\text{ATP}$, which is in agreement with the direct measurement in Eq. (28).

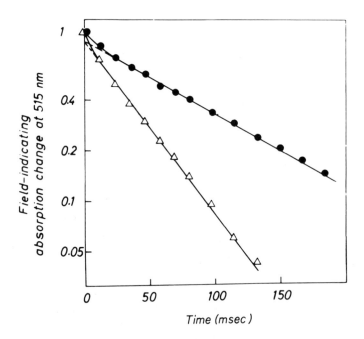

Fig. 25. Time course of the field-indicating absorption changes with ($\triangle$) and without ($\bullet$) phosphorylation in chloroplasts of spinach (plot in log scale) (Rumberg and Siggel, 1968).

Schwartz (1968a) reported values of $2\text{H}^+ \sim 1\text{ATP}$ under continuous light conditions. According to Schröder *et al.* (1971) this discrepancy can be quantitatively explained because, on the one hand, Schwartz did not discriminate between ATPase and dissipative pathways and, on the other hand, he did not correct for the retardation of H^+ flux caused by the low permeability of the coions.

Results imply that phosphorylation should be decreased when the ion flux is channeled through bypasses which can be set up by ionophores. This has been shown with Vmc (Junge et al., 1970) and also with GmcD (Boeck and Witt, 1972).

In Section 6.1 it was shown that the amount of synthesized ATP increases linearly with the extent of the initial electrical potential change $\Delta\phi_0$. The following experiments are focused on the question of whether there also exists a relationship between the rate of electrical potential decay $d\Delta\phi(dt = i/C$ and the rate of ATP formation.

In Section 4.8 it was shown that $\bar{i}$ increases linearly with v up to 30 Hz:

$$\overline{i_{H^+}} \sim v \tag{13}$$

The extent of the potential change $^1\Delta\phi_0$ is independent of v between 1 and 30 Hz:

$$^1\Delta\phi_0 = \text{constant} \tag{10}$$

Boeck and Witt (1972) have shown that the averaged rate of ATP formation increases linearly with v up to 30 Hz:

$$\overline{\text{ATP}} \sim v \tag{29}$$

The number of ATP molecules per single turnover flash and electron chain is constant between 1–3 and 30 Hz:

$$^1\Delta\,\text{ATP} \approx \text{constant} \tag{30}$$

Obviously in this range phosphorylation depends on v in the same way that the electrical values do [compare Eq. (13) and (10) with (29) and (30)]. Therefore, the relation $\overline{\text{ATP}} \sim \bar{i}_{H^+}$ might be valid. In Fig. 26 ATP was measured simultaneously with $1/\tau$ ($1/\tau$ was varied by changing v), ATP increases linearly with $1/\tau$ or $\bar{i}_{H^+}$:

$$\overline{\text{ATP}} \sim \bar{i}_{H^+} \tag{31}$$

For the linear increase of $\overline{A\dot{T}P}$ with $\overline{i}_{H^+}$ two mechanisms can be discussed:
1. The *actual* rate of ATP formation $A\dot{T}P(t)$ is controlled by the actual rate of field decay or $i_{H^+}(t)$, i.e.,

$$A\dot{T}P(t) \sim i_{H^+}(t) \tag{32}$$

2. The *actual* rate of ATP formation is constant $(A\dot{T}P_0)$, i.e., independent of i_{H^+} and v and its half-rise time shorter than 1/30 sec.

In both cases 1 and 2 it follows that the averaged rate $\overline{A\dot{T}P}$ within the time interval $t = 1/v$ is $\overline{A\dot{T}P} \sim v \sim \overline{i}_{H^+}$. Whether 1 or 2 is valid can be tested by the following experiments.

In the case of 1 and of Eq. (32) the slightest acceleration of the potential decay by additional ions such as K^+, i.e., the slightest artificial bypass i_{K^+} parallel to i_{H^+}, should deactivate phosphorylation. In the case of 2 a bypass should not deactivate phosphorylation. Because i_{H^+} can be changed 20-fold by varying H_{in}^+ and v, this criterion can be checked carefully over a wide range.

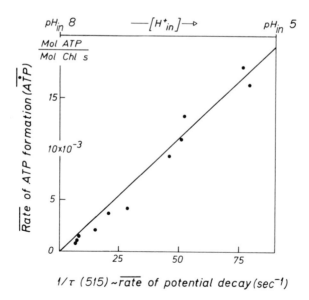

Fig. 26. Averaged rate of ATP formation $(\overline{A\dot{T}P})$ as a function of $1/\tau(515)$ and i_{H^+}. Simultaneous measurements of $1/\tau$ and $\overline{A\dot{T}P}$ by the repetitive pulse photometer with double beams. Sample: chloroplasts of spinach, $pH_{out}8, 20°C$ (Boeck and Witt, 1972).

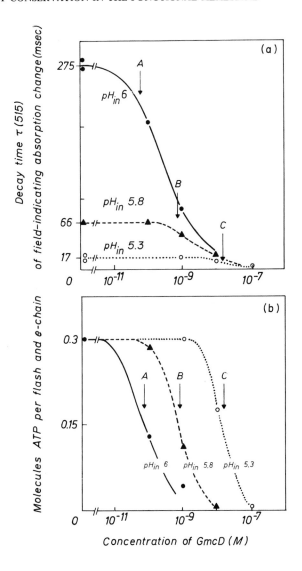

Fig. 27. (a) Decay time $\tau(515)$ of the field-indicating absorption changes as a function of the concentration of GmcD at different flash frequencies and pH_{in} values: (1 Hz, pH_{in} 6), (5 Hz, pH_{in} 5.8), (20 Hz, pH_{in} 5.3) pH_{out} 8. (b) ATP molecules per flash and electron chain as a function of the concentration of GmcD at different flash frequencies and pH_{in} values (see above). Sample: chloroplast of spinach. Simultaneous measurements of τ and ATP by the repetitive pulse photometer with double beams (Boeck and Witt, 1973). (See also Table VIII.)

A bypass i_{K^+} (shunt) has been induced by GmcD (Boeck and Witt, 1972). GmcD at $< 10^{-7} M$ (Chl $\approx 2.10^{-4} M$) is specific only for alkali ions, (Pressman, 1965). In Fig. 27a the effect of GmcD on the decay time $\tau(515) \sim 1/i$ is shown at different pH_{in} values. At the original large $\tau(515) = 275$ msec (1 Hz, pH_{in} 6 small i_{H^+}) a bypass induced with a GmcD concentration, as indicated by arrow A, decreases τ to about 25%. At $\tau(515) = 66$ msec (5 Hz, pH_{in} 5.8, medium i_{H^+}) a bypass induced with GmcD, as indicated by arrow B, is necessary to produce a corresponding decrease of τ. At $\tau(515) = 17$ msec (20 Hz, pH_{in} 5.3, large i_{H^+}) GmcD as indicated by arrow C is necessary to decrease τ to 25%. In Fig. 27b the effect of GmcD on the formation of ATP has been measured simultaneously. At different values of v and pH_{in} the yield of ATP per flash decreases beyond 50% at those GmcD concentrations (see arrows A, B, C) which are necessary in Fig. 27a to induce an effective bypass i_{K^+} parallel to i_{H^+}. The results are documentated in more detail in Table VIII. The results in Fig. 27 and Table VIII can also be discussed in the following way. At a certain bypass of i_{K^+} caused by a $10^{-9} M$ concentration of GmcD, phosphorylation is deactivated to 0.14 at 1 Hz and pH_{in} 6. At higher v and lower pH_{in} values i_{H^+} is markedly increased and competes with the artificial bypass i_{K^+}. Phosphorylation should increase again. Indeed at 5 Hz and pH_{in} 5.8 ATP formation is increased to 0.37. At 20 Hz and pH_{in} 5.3 i_{H^+} is so large that it outruns the artificial bypass i_{H^+} and phosphorylation reappears in full to 1 in the presence of 10^{-9} GmcD. The dependence of ATP from i_{H^+} and i_{K^+} is expressed in Fig. 10 (top) by two slide switches.

TABLE VIII

ATP Molecules (Relative) per Flash and Electron Chain and Decay Time τ of the Field-Indicating Absorption Change at Different Concentrations of Gramicidin D, flash Frequencies, and ΔpH values[a]

v	1 Hz		5 Hz		20 Hz	
GmcD (M)	ATP	$\tau(515)$ (msec)	ATP	$\tau(515)$ (msec)	ATP	$\tau(515)$ (msec)
0	1.0	275	1.0	66	1.0	18
10^{-10}	0.43	205	0.95	67	1.0	17
10^{-9}	0.14	85	0.37	48	1.0	16
10^{-8}			0.0	20	0.58	13
10^{-7}					0.0	2.4
pH_{out} 8	pH_{in} 6		pH_{in} 5.8		pH_{in} 5.3	
	ΔpH 2		ΔpH 2.2		ΔpH 2.7	

[a]Data from Boeck and Witt (1972, 1973). pH Values at 1 Hz were revised by Gräber and Witt, (1973, unpublished).

The above-discussed experiments support the interpretation formulated in Eq. (32): The actual rate of ATP formation is very probably controlled by the actual rate of the electrical potential decay ($\Delta \dot{\phi}_{H^+}$) and the efflux i_{H^+}. This relationship is expected if the hypothesis of Mitchell is valid.

Results by Neumann *et al.* (1970) are not in agreement with the results reported in this section. These authors compared in most of their experiments flash-induced absorption change at 515 nm with ATP yield in continuous light. However, under such conditions the above-discussed relationships would not be expected.

6.3 Absence of Phosphorylation in a pH Gradient Only

According to Fig. 27 and Table VIII at pH_{in} 6, pH_{in} 5.8 and pH_{in} 5.3 ATP formation is practically zero after addition of 10^{-9}, 10^{-8}, and 10^{-7} M GmcD, respectively. This has been explained by the breakdown of the field through the bypass i_{K^+}. However, the pH gradient across the thylakoid membrane is not quenched but ΔpH 2 (1 Hz), ΔpH 2.2 (5 Hz) and ΔpH 2.7 (20 Hz) (see Table VIII). (In the presence of GmcD, ΔpH is even higher than without GmcD; this is due to the release of K^+ instead of H^+). This means that ATP formation is zero even though gradients up to $\Delta pH \sim 2.5$ drive H^+ outward and this at rates which are on average as high as without addition of GmcD (Boeck and Witt, 1972). (The protons which have been translocated inward by each flash must have been released from the inner phase before the following flash: a necessary steady-state condition of periodical excitation). Since the ATPase is not deactivated in the presence of GmcD one has to assume that in isolated chloroplasts in the range of $\Delta pH \leq 2.5$ the H^+ efflux driven by a pH gradient alone is essentially not channeled through the ATPase but via basal pathways. Obviously H^+ efflux through the ATPase occurs under these conditions only when a field focuses H^+ to the ATPase (this is prevented in the presence of ionophores). The absence of phosphorylation at $\Delta pH \leq 2.5$ might be explained by results of Schröder and Rumberg (1971). They found in the absence of a field that H^+ efflux through the ATPase depends on the square of the proton concentration in the inner phase $(H_{in}^+)^2$ but basal H^+ efflux depends linearly on H_{in}^+. According to the value of the proportionality factors in the range up to $\Delta pH \leq 2.5$, the H^+ efflux driven by a pH gradient via the ATPase pathway is negligible as compared with the basal pathway. This also explains the finding of Jagendorf and Uribe (1966) that by artificially imposed pH gradients across the thylakoid membrane ATP formation does not take place when ΔpH is less than 3. The results probably also explain the observation that when pH gradients are artificially imposed across the thylakoid membrane, the yield of ATP formation can be increased considerably when a membrane potential (K^+ gradient) is additionally imposed. This has been observed by Uribe (1972) and Schuldiner *et al.* (1972).

6.4. Functional Unit of Phosphorylation

If according to the results discussed above the electrical events are inter-mediate steps in the synthesis of ATP, it is expected that the functional unit of phosphorylation is also the membrane of one thylakoid as it is for the electrical events. Therefore, it should be possible to titrate the functional unit of phosphorylation with GmcD, as has been done for the electric potential change (see Section 4.5). A definite result can only be expected at slow basal $\Delta\phi$ decay, i.e., at low frequencies of single turnover flashes ($v \leq 1H$). The reasons have been outlined in Section. 6.2.

At a ratio of GmcD: Chl of $1:10^5$ the decay time $\tau(515)$ of $\Delta\phi$ is decreased to 30 % through the shunt of GmcD (see Fig. 28a). This means that about one GmcD molecule operates on a functional unit containing 10^5 Chl molecules. Such a number of chlorophylls is spread over the membrane of a thylakoid. This confirms the results in Section 4.5. that a thylakoid is the unit of the electrical events. At a GmcD:Chl ratio of $1:10^5$, 30 % of the phosphory-lation is also deactivated (Fig. 28b). This clearly shows that one thyla-koid is also the functional unit of phosphorylation (Boeck and Witt, 1972).

$$\text{ATP unit} = 1 \text{ thylakoid} \tag{33}$$

In some chloroplast preparations GmcD:Chl values of $1:4 \times 10^5$ have been observed. This is probably due to varying sizes of the thylakoids or different

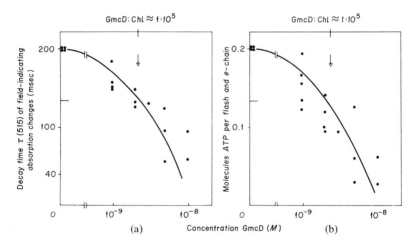

Fig. 28. (a) ATP molecules produced per single turnover flash and e-chain as a function of the concentration of GmcD (b) Decay time τ of the field-indicating absorption changes at 515 nm as a function of the concentration of GmcD. Simultaneous measurement of τ and ATP by the repetitive pulse photometer with double beams. Flash frequency 1 Hz. Chl $\approx 2 \times 10^{-4} M$. Sample: chloroplasts of spinach (Boeck and Witt, 1972).

interconnections between several thylakoids. In every case, however, the collapse of the electrical events and phosphorylation takes place at the same ratio of GmcD : Chl (Boeck and Witt, 1973).

The above results give evidence that phosphorylation cannot be understood as a phenomenon solely coupled to a chemical intermediate located within an electron chain. One primary component which initiates the act of ATP generation is bound to a whole thylakoid membrane: the transmembrane electrical potential difference $\Delta\phi$. (For more arguments see Section 7).

Kahn (1970) concluded from experiments with the inhibitor TBT (tri-n-butylin chloride) the absence of a common pool for phosphorylation. But due to a double function of TBT (increase of H^+ permeability and deactivation of the ATPase) Kahn's results can be interpreted also in a different way (Boeck and Witt, 1973).

6.5. Energetics of Phosphorylation

In respect to energetics we have to consider steady-state conditions in constant light because thermodynamic values are only available from this state. The actual free energy made available in photosynthesis for phosphorylation can be calculated when phosphorylation comes into thermodynamic equilibrium (reversibility of ATPase) in chloroplasts in constant light. From the values of ADP, P, and ATP in this state, in saturing light (ss) $\Delta G_{ss} \approx 13.5$ kcal/mole ATP (Kraayenhof, 1969) derived from $\Delta G_{ss} = \Delta G^{0\prime} + 1.36$ kcal log $(ATP/ADP.P)_{ss}$. For $\Delta G^{0\prime}$ the revised value 7.8 kcal/mole ATP (Rosing and Slater, 1972) has been used.

If one accepts the Mitchell mechanism, according to Eq. (25) the neccessary energy of 13.5 kcal must be obtained by a drop of n_i protons through an electrochemical potential span of about 580 mV/n_i. With $n_i = 4$ (or 2) (see Section 6.2) the transmembrane potential must be 145 mV (or 290 mV). The measured potential of ΔpH_{ss} and $\Delta\phi_{ss}$ is

$$\Delta pH_{ss} \; 3.3 \; (\approx 200 \text{ mV}) + \Delta\phi_{ss} \; (\approx 100 \text{ mV}) \approx 300 \text{ mV}$$

(see Sections 4.7 and 4.9) and would satisfy the energetic requirement, even with $n_i = 2$.*

Because both light reactions translocate protons, each of these centers must contribute to phosphorylation via the ATPase. This could explain the finding that at each light reaction system, one "ATP site" has been identified (Böhme and Trebst, 1969; West and Wiskich, 1973; Gould and Izawa, 1973).

---

*$\Delta\phi_{ss} \approx 100$ mV was measured in whole chlorella cells (see section 4.9).

7. SUMMARY AND CONCLUSIONS

The light-driven electron flow from H_2O to $NADP^+$ converts one part of the light energy into the form of the reducing power of NADPH. The electron transfer from H_2O to $NADP^+$ occurs in a series of two light reactions at Chl a_I 700 and Chl a_{II} 680. A pool of PQ is the link between the two light reaction centers. The transfer of the electrons is a vectorial transfer with a component perpendicular to the thylakoid membrane. In this way the membrane is charged and an electrical field across the membrane formed. The electrical energy of the charged membrane is an additional state into which light energy is converted. The electronic charges on the outside and inside of the thylakoid membrane are replaced by OH^- and H^+, respectively, through redox reactions. This corresponds to a proton translocation into the inner space of the thylakoid. Two field generators and two proton pumps have been recognized. This indicates in a different way the existence of the two active light reaction centers. These results are explained by the zigzag scheme of Fig. 20.

The discharging of the electrically energized membrane by H^+ efflux is coupled with the formation of ATP: The amount of ATP generated is proportional to $\Delta\phi_0$, i.e., to the number of charges Q (protons) by which the membrane has been loaded (see Section 6.1):

$$\Delta ATP \sim \Delta\phi C \quad or \quad \Delta ATP \sim \Delta H^+$$

For the stoichiometry $\Delta ATP:Q$ it has been found that the generation of 1ATP is probably coupled to four protons. The actual rate of ATP formation follows the actual rate of the electrical potential decay and H^+ efflux (see Section 6.2):

$$A\dot{T}P \sim \Delta\dot{\phi}_{H^+} \quad or \quad A\dot{T}P \sim i_{H^+}$$

The functional unit of phosphorylation as well as the unit of the electrical events is the membrane of one thylakoid (see Section 6.4):

$$ATP \text{ unit} = \Delta\phi \text{ unit} = 1 \text{ thylakoid}$$

Electrical Potential $\Delta\phi$ as the Primary Link between e-Transfer and Phosphorylation. The results indicate that the electron transfer e and phosphorylation are coupled by $\Delta\phi$ on the membrane of a thylakoid. This is in favor of the hypothesis of Mitchell. In a shorthand notation this can be expressed as written in Fig. 29a. The action of $\Delta\phi$ can subsequently induce special configurations between H^+, ADP, P, H_2O, and other components in the membrane, i.e., special intermediates and conformation are formed before ATP is completely synthesized. These unknown subsequent steps induced by $\Delta\phi$ are here symbolized by a double squiggle ($\approx$). Special proposals

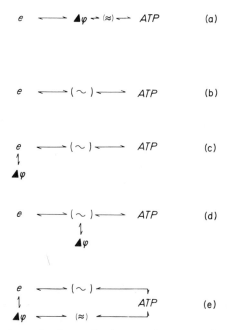

Fig. 29. Different schemes for the coupling between electron transfer e and ATP formation. For details see Section 7.

for such intermediates have been outlined by Green and Ji (1972), and Williams (1972).

An Intermediate (～) as Link between e-Transfer and Phosphorylation. Two major alternative hypotheses on the coupling between e-transfer and ATP generation are (1) the *chemical* and (2) the *conformational* hypothesis. Both can be written briefly as depicted in Fig. 29b. The squiggle (～) which is thought to be located at each electron chain represents either an unknown chemical intermediate directly produced by e-transfer and in which energy is conserved as covalent bond energy (Slater, 1953; 1958), or it represents conformational changes directly produced by e-transfer which conserve energy in the folding of numerous molecules (Boyer, 1965; Green and Ji, 1972).

The coupling by such a squiggle does not consider the existence of $\Delta\phi$ and is therefore in contradiction to our results. However, one can regard the existence of $\Delta\phi$ by assuming its action in a sidepath of the b-scheme (Wang, 1970; Bennun, 1971).

There are three possible sidepath coupling hypotheses as outlined in Fig. 29 c, d, and e. But all three alternatives can be excluded by the above-reported experimental results.

Δφ in a Sidepath at the level of e-Transfer.   The decay time of the electrical potential $\Delta\phi$ and the ion efflux can be adjusted to the same absolute values in the form of either a mainly H^+ efflux i_{H^+} or a mainly K^+ efflux i_{K^+} (see Table VIII).* In respect to scheme c the production of ATP should be independent of the type of ion fluxes. However, under conditions of a mainly K^+ efflux, i_{K^+}, the generation of ATP is exhibited (see Table VIII). Even the slightest change from i_{H^+} toward i_{K^+} (1 GmcD molecule per thylakoid) deactivates ATP generation (see Section 6.4). This excludes the c scheme.

Δφ in a Sidepath at the Level of the Squiggle.   In the d scheme (see Fig. 29d) the squiggle must be formed in $\leq$ 20 nsec because $\Delta\phi$ formation has been measured to take place in $\leq$ 20 nsec (see Section 4.1). The synthesis of the squiggle must therefore be faster than any e-transfer event apart from that coupled with the photoact. This is a very improbable assumption. Furthermore, $\Delta\phi$ generation is a direct result of the vectorial electron reaction at the chlorophylls (see Section 5), and therefore not a consequence of a squiggle. This makes the d scheme unlikely.

Δφ Parallel to the Squiggle Path.   In the case of two coexisting ATP generating pathways (see Fig. 29c), deactivation of one of them should still permit some ATP synthesis at a diminished rate via the other pathway. But 1 GmcD molecule per thylakoid deactivates ATP generation by a slight shift of i_{H^+} toward i_{K^+}. This means that the $\Delta\phi$ pathway "does all the work," i.e., a parallel squiggle path is very improbable.

In summary, it can be said that the above-discussed experimental results on phosphorylation are in favor of the hypothesis of Mitchell. This insight into the mechanism of phosphorylation is based on the zigzag scheme developed before and was supported by (1) restricting the experiments on the primary events; (2) focusing the experiments on the electrical potential; (3) running the investigation under single turnover conditions where insight into the relationship between the different events can more easily be obtained; and (4) arranging simultaneous measurements of two events. This method of investigation has required, on the other hand (1) special optical characterizations of the molecular events, and (2) the development of very sensitive and fast repetitive pulse spectroscopic methods working down to 20 nsec.

ACKNOWLEDGMENTS

I should like to express my gratitude to Dr. Govindjee, Dr. Fleischmann, Dr. P. Mitchell, Prof. W. Junge, and Prof. B. Rumberg for valuable criticism of the manuscript.

*For example, no GmcD. 20 Hz, τ = 18 msec and $+10^{-8} M$ GmcD, 5 Hz, τ = 20 msec.

I am most grateful to the members of the Max-Volmer-Institute for their collaboration. This work has been supported by grants of the Deutsche Forschungsgemeinschaft.

REFERENCES

Amesz, J. (1964). *Biochim. Biophys. Acta* **79**, 257.

Arnon, D. I., Allen, M. B., and Whatley, F. R. (1954). *Nature* **174**, 394.

Arnon, D. I., Chain, R. K., McSwain, B. D., Tsujimoto, H. Y., and Knaff, D. B. (1970). *Proc. Nat. Acad. Sci.* **67**, 1404.

Bakeeva, L. E., Grinius, L. L., Jasaitis, A. A., Kuliene, V. V., Levitsky, D. O., Liberman, E. A., Severina, L. I., and Skulachev, V. P. (1970). *Biochim. Biophys. Acta* **216**, 13–21.

Baltscheffsky, M. (1969). *Arch. Biochem. Biophys.* **130**, 646.

Barber, J., and Kraan, G. P. B. (1970). *Biochim. Biophys. Acta* **197**, 49.

Beinert, H., and Kok, B. (1963). *Natl. Acad. Sci-Natl. Res. Council Publ.* **1145**, 131.

Bennoun, P. (1970). *Biochim. Biophys. Acta* **216**, 357.

Bennun, A. (1971). *Nature (New Biol.)*, **233**, 5.

Bishop, N. I. (1958). *Biochim. Biophys. Acta* **27**, 205.

Bishop, N. I. (1959). *Biochem. N. Y.* **34**, 1696.

Blinks, L. R. (1959). *Plant Physiol.* **34**, 200.

Blodgett, K. B., and Langmuir, I. (1937). *Phys. Rev.* **51**, 964.

Boeck, M., and Witt, H. T. (1972). *Proc. Intern. Congr. Photosynthesis, 2nd, Stresa, 1971* **2**, 903.

Boeck, M., and Witt, H. T. (1973). Unpublished observations.

Böhme, H., and Trebst, A. (1969). *Biochim. Biophys. Acta* **180**, 137.

Boyer, P. D. (1965). *Oxidases Related Redox Syst.* **2**, 994.

Calvin, M. (1962). *Agew. Chem.* **74**, 165.

Calvin, M., and Benson, A. A. (1948). *Science*, **107**, 476.

Chance, B., Devault, D., Hildreth, W. W., Parson, W. W., and Nishimura, M. (1967). *Brookhaven Symp. Biol.* **19**, 115.

Chappell, J. B., and Crofts, A. R. (1966). *Biochim. Biophys. Acta* **7**, 293.

Crofts, A. R., Wraight, C. A., and Fleischmann, D. E. (1971). *FEBS Lett.* **15**, 89.

Dilley, R. A. (1970). *Arch. Biochem. Biophys.* **137**, 270.

Dilley, R. A., and Vernon, L. P. (1965). *Arch. Biochem. Biophys.* **111**, 365.

Döring, G. and Witt, H. T. (1972). *Proc. Intern. Congr. Photosynthesis, 2nd, Stresa, 1971* **1**, 39.

Döring, G., Stiehl, H., and Witt, H. T. (1967). *Z. Naturforsch.* **22b**, 639.

Döring, G., Bailey, J. L., Kreutz, W., Weikard, J., and Witt, H. T. (1968a). *Naturwissenschaften* **55**, 219.

Döring, G., Bailey, J. L., Kreutz, W., and Witt, H. T. (1968b). *Naturwissenschaften*, **55**, 220.

Döring, G., Renger, G., Vater, J., and Witt, H. T. (1969). *Natur forsch.* **24b**, 1139.

Duysens, L. N. M. (1954). *Nature* **173**, 642.

Duysens, L. N. M. (1959). *Brookhaven Symp. Biol.* **11**, 18.

Duysens, L. N. M., and Sweers, H. E. (1963). *In* "Studies of Microalgae and Photosynthetic Bacteria" p. 353. University of Tokyo Press, Tokyo.

Duysens, L. N. M., Amesz, J., and Kamp, B. M. (1961). *Nature* **190**, 510.

Emerson, R. (1958). *Ann. Rev. Plant Physiol.* **9**, 1.

Emerson, R., and Arnold, W. (1932). *J. Gen. Physiol.* **16**, 191.

Emerson, R., and Rabinowitch, E. (1960). *Plant Physiol.* **35**, 411.

Emrich, H. M., Junge, W., and Witt, H. T. (1969). *Z. Naturforsch.* **24b**, 1144.

Floyd, R. A., Chance, B., and Devault, D. (1971). *Biochem. Biphys. Acta* **226**, 103.

Fork, D. C., and Amesz, J. (1967). *Carnegie Inst. Yearbook* **66**, 160.

Fork, D. C. and Murata, N. (1972). *Proc. Intern. Congr. Photosynthesis, 2nd, Stresa, 1971* **1**, 39.

Frenkel, A. (1954). *J. Am. Chem. Soc.* **76**, 5568.

Gaffron, H., and Wohl, K. (1936). *Naturwissenschaften* **24**, 81.
Gläser, M., Wolff, C., Buchwald, H. -E., and Witt, H. T. (1974). *FEBS Lett.*, **42**, 81.
Gould, S. M., and Izawa, S. (1973). *Eur. J. Biochem.* **37**, 185.
Govindjee and Govindjee, R. (1965). *Photochem. Photobiol.* **4**, 675 R.
Govindjee and Rabinowitch, E. (1960). *Science* **132**, 355.
Govindjee, Döring, G., and Govindjee, R. (1970). *Biochim. Biophys. Acta* **205**, 303.
Gräber, P. and Witt, H. T. (1974a). In preparation.
Gräber, W., and Witt, H. T. (1974b). *Biochim. Biophys. Acta,* **333**, 389.
Green, D. E., and Ji, S. (1972). *Proc. Nat. Acad. Sci. U.S.* **69**, 726.
Grünhagen, H. H., and Witt, H. T. (1970). *Z. Naturforsch.* **25b**, 373.
Haehnel, W. (1973). *Biochim. Biophys. Acta* **305**, 618.
Haehnel, W., Döring, G., and Witt, H. T. *Z. Naturforsch.* **26b**. 1171.
Hager, A. (1969). *Planta* **89**, 224.
Hauska, G. A., McCarty, R. E., Berzborn, R. J., and Racker, E. (1971). *J. Biol. Chem.* **246**, 3524.
Henninger, M. D., and Crane, F. D. (1964). *Plant Physiol.* **39**, 598.
Hill, R. (1939). *Proc. R. Soc. B* **127**, 192.
Hill, R., and Bendall, D. S. (1960). *Nature* **186**, 136.
Hiyama, T., and Ke, B. (1971). *Arch. Biochem. Biophys.* **147**, 99.
Izawa, S., and Hind, G. (1967). *Biochim. Biophys. Acta* **143**, 377.
Jackson, J. B. (1973). Personal communication.
Jackson, J. B., and Crofts, A. R. (1969). *FEBS Lett.* **4**, 185.
Jackson, J. B., and Crofts, A. R. (1971). *Eur. J. Biochem.* **18**, 120.
Jagendorf, A. T., and Hind, G. (1963). *Natl. Acad. Sci-Natl. Res. Council Publ.* **1145**, 599.
Jagendorf, A. T., and Uribe, E. (1966). *Proc. Natl. Acad. Sci. U.S.* **55**, 170.
Junge, W. (1970). *Eur. J. Biochem.* **14**, 582.
Junge, W., and Ausländer, W. (1974). *Biochim. Biophys. Acta,* **333**, 59.
Junge, W., and Schmid, R. (1971). *J. Membr. Biol.* **4**, 179.
Junge, W., and Witt, H. T. (1968). *Z. Naturforsch.* **23b**, 244.
Junge, W., Rumberg, B., and Schröder, H. (1970). *Eur. J. Biochem.* **14**, 575.
Kahn, J. S. (1970). *Biochemistry* **3**, 116, 55.
Karlish, St. J. D., and Avron, M. (1967). *Nature* **216**, 1107.
Kautsky, H., Appel, W., and Amann, H. (1960). *Biochem. Z.* **332**, 277.
Ke, B. (1972). *Arch. Biochem. Biophys.* **152**, 70.
Klingenberg, M., Müller, A., Schmidt-Mende, P., and Witt, H. T. (1962). *Nature* **194**, 379.
Knaff, D. B. (1972). *FEBS Lett.* **23**, 142.
Knaff, D. B., and Arnon, D. I. (1969). *Proc. Natl. Acad. Sci. U.S.* **65**, 963.
Knox, R. S. (1969). *Biophys. J.* **9**, 1351.
Kok, B. (1957). *Acta Bot. Neerl.* **6**, 316.
Kok, B. (1961). *Biochim. Biophys. Acta* **48**, 527.
Kok, B. (1963). *Natl. Acad. Sci. Natl. Res. Council Publ.* **1145**, 35.
Kok, B. (1974)., *Intern. Congr. Photobio., 6th, Bochum, 1972,* in press.
Kok, B., and Gott, W. (1960). *Plant Physiol.* **35**, 802.
Kok, B., and Hoch, G. (1961). *In* "Light and Life" John Hopkins Press, p. 397.
Kraayenhof, R. (1969). *Biochim. Biophys. Acta* **180**, 213.
Kreutz, W. (1968). *Z. Naturforsch.* **23b**, 520.
Kreutz, W. (1970a). *Z. Naturforsch.* **25b**, 88.
Kreutz, W. (1970b). *Advan. Bot. Res.* **3**, 53.
Kuhn, H., Möbius, D., and Bücher, H. (1972). *Techniques of Chemistry* (A. Weissberger, ed.), Wiley-Interscience, p. 577.
Labhart, H. (1967). *Advan. Chem. Phys.* **13**, 179.
Land, E. J., Simic, M., and Swallow, A. J. (1971). *Biochim. Biophys. Acta* **226**, 239.
Lipmann, F. (1946). *Currents Biochem. Res.* **12**, 137.

Liptay, W. (1969). *Angew. Chem.* **81**, 195.

Malkin, R., and Bearden, A. J. (1973). *Biochim. Biophys. Acta* **292**, 169.

Menke, W. (1966). *Biochem. Chloroplasts* **1**, 1.

Mitchell, P. (1961). *Nature* **191**, 144.

Mitchell, P. (1966). *Biol. Rev.* **41**, 445.

Mitchell, P., and Moyle, J. (1969). *Eur. J. Biochem.* **7**, 471.

Mohanty, P., Mar. T., and Govindjee (1971). *Biochim. Biophys. Acta* **253**, 213.

Mühlethaler, K. (1966). *In* "Biochemistry of Chloroplasts" Vol. I, (T. W. Goodwin, ed.), p. 49, Academic Press, N.Y.

Müller, A., Rumberg, B., and Witt, H. T. (1963). *Proc. Roy. Soc. B* **157**, 313.

Myers, J. (1971). *Ann. Rev. Plant Physiol.* **22**, 289.

Neumann, J., and Jagendorf, A. T. (1964). *Arch. Biochem. Biophys.* **107**, 109.

Neumann, J., Ke, B., and Dilley, R. A. (1970). *Plant. Physiol.* **46**, 86.

Nishimura, M. (1970). *Biochem. Biophys. Acta* **197**, 69.

Nishimura, M., and Chance, B. (1962). *Biochim. Biophys. Acta* **59**, 177.

Norris, J. R., Uphaus, R. A., Crespi, H. L., and Katz, J. J. (1971). *Proc. Nat. Acad. Sci. U.S.* **68**, 625.

Philipson, K., Sato, V., and Sauer, K. (1972). *Biochemistry* **11**, 459.

Pressman, B. C. (1965). *Proc. Natl. Acad. Sci. U.S.* **53**, 1076.

Racker, E. (1970). In *Membranes of Mitochondria and Chloroplasts* (Racker, ed.) p. 127.

Reinwald, E., Stiehl, H. H., and Rumberg, B. (1968). *Z. Naturforsch.* **23b**, 1616.

Rosing, J., and Slater, E. (1972). *Biochim. Biophys. Acta,* **267**, 275.

Ross, R. T., and Calvin, M. (1967). *Biophys. J.* **7**, 595.

Rottenberg, H., Grünwald, T., and Avron, M. (1971). *FEBS Lett.* **13**, 71.

Rumberg, B. (1964a). *Z. Naturf.* **19b**, 707.

Rumberg, B. (1964b). *Nature* **204**, 860.

Rumberg, B. (1974). *Intern. Congr. Photobio. 6th.* Bochum, 1972, in press.

Rumberg, B. and Siggel, U. (1968). *Z. Naturforsch.* **23b**, 239.

Rumberg, B., and Siggel, U. (1969). *Naturwissenschaften* **56**, 130.

Rumberg, B., and Witt, H. T. (1964). *Z. Naturforsch.* **19b**, 693.

Rumberg, B., Schmidt-Mende, P., Weikard, J., and Witt, H. T. (1963). *Publ. Natl. Res. Council, Wash.* **1145**, 18.

Rumberg, B., Schmidt-Mende, P., and Witt, H. T. (1964). *Nature* **201**, 466.

Rumberg, B., Schmidt-Mende, P., Siggel, U., and Witt, H. T. (1966). *Angew. Chem.* **5**, 522.

Rumberg, B., Reinwald, E., Schröder, H., and Siggel, U. (1969). *Progr. Photosynthesis Res.* **3**, 1374.

Rüppel, H. (1962). Thesis, Universität Marburg.

Rüppel, H., and Witt, H. T. (1970). *Methods Enzymol.* **16**, 316.

Rüppel, H., Bültemann, V., and Witt, H. T. (1962). *Z. Elektrochem.* **66**, 760.

Rüppel, H., Bültemann, V., and Witt, H. T. (1964). *Z. Elektrochem.* **68**, 340.

Rurainski, H. I., Randles, B., and Hoch, G. E. (1971). *FEBS Lett.* **13**, 98.

Saha, S., Izawa, S., and Good, N. E. (1970). *Biochim. Biophys. Acta* **223**, 158.

Schliephake, W., Junge, W., and Witt, H. T. (1968). *Z. Naturforsch.* **23b**, 1571.

Schmidt, S., and Reich, R. (1972). *Ber. Bunsen Ges.* **76**, 589.

Schmidt, S., Reich, R., and Witt, H. T. (1971). *Naturwissenschaften* **8**, 414.

Schmidt, S., Reich, R., and Witt, H. T. (1972). *Proc. Intern. Congr. Photosynthesis 2nd*, Stresa, 1971, **2**, 1087.

Schröder, H., and Rumberg, B. (1971). *I. Eur. Biophy. Congr.* V, 57.

Schröder, H., Muhle, H., and Rumberg, B. (1971). *Proc. Intern. Congr. Photosynthesis 2nd, Stresa, 1971*, **2**, 919.

Schuldiner, S., Rottenberg, H., and Avron, M. (1972). *FEBS Lett.* **28**, 173.

Schwartz, M. (1968a). *Nature* **219**, 915.

Schwartz, M. (1968b). *RIAS Rept.* **49**.

Seifert, K. and Witt, H. T. (1968). *Naturwissenschaften* **55**, 222.

Seifert, K. and Witt, H. T. (1969). *Progr. Photosynthesis Res.* **2**, 750.

Siggel, U., Renger, G., Stiehl, H. H., and Rumberg, B. (1972a). *Biochim. Biophys. Acta* **256**, 328.

Siggel, U., Renger, G., and Rumberg, B. (1972b). *Proc. Intern. Congr. Photosynthesis, 2nd, Stresa 1971*, **1**, 753.

Skulachev, V. P. (1972). *J. Bioenergetics* **3**, 25.

Slater, E. C. (1953). Nature **172**, 975.

Slater, E. C. (1958). *Rev. Pure Appl. Chem.* **8**, 221.

Stiehl, H. H., and Witt, H. T. (1968). *Z. Naturforsch.* **23b**, 220.

Stiehl, H. H., and Witt, H. T. (1969). *Z. Naturforsch.* **24b**, 1588.

Strichartz, G. R. (1971). *Plant Physiol.* **48**, 553.

Strichartz, G. R., and Chance, B. (1972). *Biochim. Biophys. Acta* **256**, 71.

Uribe, E. G. (1972). *Plant Physiol.* **49**, 9.

Vater, J., Renger, G., Stiehl, H. H., and Witt, H. T. (1968). *Naturwissenschaften* **55**, 220.

Wang, J. H. (1970). *Science* **167**, 25.

West, K. R., and Wiskich J. T. (1973). *Biochim. Biophys. Acta* **292**, 197.

Williams, R. J. P. (1972). *J. Bioenergetics* **3**, 81.

Witt, H. T. (1955a). *Naturwissenschaften* **42**, 72.

Witt, H. T. (1955b). *Z. Elektrochem.* **59**, 981.

Witt, H. T. (1967a). *Nobel Symp.* **5**, 81.

Witt, H. T. (1967b). *Nobel Symp.* **5**, 261.

Witt, H. T. (1971). *Quart. Rev. Biophysics* **4**, 365.

Witt, H. T., and Moraw, R. (1959a). *Z. Phys. Chem.* **20**, 253.

Witt, H. T., and Moraw, R. (1959b). *Z. Phys. Chem.* **20**, 283.

Witt, H. T., and Müller, A. (1959). *Z. Phys. Chem.* **21**, 1.

Witt, H. T., and Zickler, A. (1973). *FEBS. Lett.* **37**, 307.

Witt, H. T., and Zickler, A. (1974). *FEBS Lett.* **39**, 205.

Witt, H. T., Moraw, R., and Müller, A. (1956). *Z. Elektrochem.* **60**, 1148.

Witt, H. T., Moraw, R., and Müller, A. (1959). *Z. Phys. Chem.* **20**, 193.

Witt, H. T., Müller, A., and Rumberg, B. (1961a). *Nature* **191**, 194.

Witt, H. T., Müller, A., and Rumberg, B. (1961b). *Nature* **192**, 967.

Witt, H. T., Müller, A., and Rumberg, B. (1963). *Colloq. Intern. Centre Natl. Rech. Sci.* **43**.

Witt, H. T., Rumberg, B., Schmidt-Mende, P., Siggel, U., and Skerra, B. (1965). *Angew. Chem. (Intern. Ed.)* **4**, 799.

Witt, H. T., Döring, G., Rumberg, B., Schmidt-Mende, P., Siggel, U., and Stiehl, H. H. (1966a). *Brookhaven Symp. Biol.* **19**, 161.

Witt, H. T.. Skerra, B., and Vater, J. (1966b). *Proc. Western Europe Conf. Photosynthesis, 2nd, 1965*, 273.

Witt, H. T., Rumberg, B., and Junge, W. (1968), *Colloq. Ges. Biol. Chem. 19th, Mosbach* 262.

Witt, K. (1973a). *FEBS Lett.* **38**, 112.

Witt, K. (1973b). *FEBS Lett.* **38**, 116.

Witt, K., and Wolff, C. (1970). *Z. Naturforsch.* **25b**, 387.

Wolff, C. (1974). *FEBS Lett.* in preparation.

Wolff, C., and Witt, H. T. (1969). *Z. Naturforsch.* **24b**, 1031.

Wolff, C., Buchwald, H. -E., Rüppel, H.. and Witt, H. T. (1967). *Nobel Symp.* **5**, 81 and 261.

Wolff, C., Buchwald, H. E., Rüppel, H., Witt, K., and Witt, H. T. (1969) *Z. Naturforsch.* **24b**, 1038.

Wraight, C. A., and Crofts, A. R. (1971). *Eur. J. Biochem.* **19**, 386.

11

Structure of the Chloroplast Membrane—Relation to Energy Coupling and Ion Transport

Satoru Murakami, J. Torres-Pereira, and Lester Packer

ABBREVIATIONS

ADP	Adenosine diphosphate
AMP	Adenosine monophosphate
ANS	8-Anilinonaphthalene-1-sulfonic acid
ATP	Adenosine triphosphate
CCCP	Carbonyl cyanide m-chlorophenylhydrazone
CF_1	Chloroplast coupling factor 1
Chl	Chlorophyll
Cyt	Cytochrome
DCMU	3-(3,4-Dichlorophenyl)-1,1-dimethylurea
DNP	2,4-Dinitrophenol
DCPIP	2,6-Dichlorophenol indophenol
DTE	Dithioerythritol
EDTA	Ethylenediaminetetraacetate
FCCP	Carbonyl cyanide p-trifluoromethoxyphenylhydrazone

NADP   Nicotinamide adenine dinucleotide phosphate
P_i Inorganic phosphate
PMA   Phenylmercuric acetate
PMS   Phenazine methosulfate
PS I (II)   Photosystem I (II) (or pigment system)

1. INTRODUCTION

Illuminated chloroplasts play a central role in the control of plant cell physiology. In this chapter we will examine the hierarchy of changes in structure and function following illumination of chloroplasts. In particular, the relations between the structural modifications that occur at the organelle, membrane, and molecular level will be emphasized. In recent years a large number of new methods have made it possible to scrutinize the structural modifications that arise following illumination from the molecular to the organelle level.

2. STRUCTURE AT THE ORGANELLE LEVEL

Changes in structure at the organelle level appear to arise mainly as a consequence of ion and osmotic flows driven initially by light energy and modified by the ionic environment of the plant cell. Thus ion movements that are chloroplast-dependent may affect ionic relations at the organismal level, e.g., control of leaf turgidity or the material flows that arise affect the translocation of various substances, such as products of photosynthetic metabolism. There is considerable evidence that chloroplasts (and also mitochondria) exhibit semipermeable properties and exchange ions at different rates, and thus behave as distinct compartments. The chloroplast is surrounded by an outer envelope composed of two membranes, and has three distinct compartments: (1) the space between the outer and inner membranes of its envelope; (2) the stroma, located between the inner membrane of the envelope and the thylakoids; and (3) the intraspace of the thylakoid (see Chapter 2 of this volume). The last two compartments play an important role in the regulation of metabolic and energy states of the chloroplasts by controlling the permeation of ions and other substances.

2.1. Ionic Composition of Chloroplasts

To obtain precise knowledge about the ion content of chloroplasts *in vitro*, the integrity of the chloroplasts' outer envelope is an essential, but sometimes difficult, condition to satisfy. Related studies with whole cells are an alternative approach, which has perhaps been most successfully employed in studies with giant algal cells. For example, Kishimoto and Tazawa (1965) studied the ion content of the cytoplasm in mature internodal cells of *Nitella flexilis*. They developed an elegant internal perfusion technique, to avoid contamination of the cell sap in the analysis of the cytoplasmic ionic concentration. When the vacuole of the internodal cell was rapidly perfused with 0.3 *M* mannitol or liquid paraffin, almost all flowing cytoplasm was lost with the perfusing solution, enabling the measurement of the ion concentrations in the chloroplast layer. Concentrations of K^+, Na^+, and Cl^- in the chloroplast layer were 110, 26, and 136 m*M*, and those of the flowing cytoplasm 125, 5, and 36 m*M*, respectively. It is interesting to note that the K^+ content in the flowing cytoplasm of the normal internodal cell is more or less of the same order as that in the cell sap, while the Na^+ and Cl^- contents of the internodal cell are only a fraction of those in the cell sap. K^+ is almost in electrical equilibrium in each phase, but there is a tendency of Na^+ movements from the flowing cytoplasm toward the vacuole and the external medium. Cl^- shows an overall tendency to move toward the external medium.

Using flame photometry and electrometric titration, Larkum (1968) studied the ion content of chloroplasts from mature internodal cells of *Tolypella intricata* var. *prolifera*. Whole chloroplasts were obtained using a non-aqueous isolation technique, involving density gradient separation. The concentrations of K^+, Na^+, Cl^-, Mg^{2+}, and Ca^{2+} were compared with those in chloroplasts of three angiosperms. Although it is questionable if the above-mentioned isolation technique was properly employed, because of the very high values reported, significant differences were found between the amount of ions and their ratio to one another in the chloroplasts of the four plants (Table I).

With the exception of *Limonium vulgare*, a salt-marsh plant, the magnesium and calcium contents are notably similar for all four plants. Using the Nernst equation*, it was possible to calculate the equilibrium potentials for Na^+, K^+, and Cl^- across the chloroplast envelope, the results indicating that at

---

*The Nernst equation expresses the relationship between the standard reduction potential of a given redox couple, its observed potential, and the concentration ratio of its electron donor and acceptor species. The Nernst equation is:

$$E_n = E_o' + \frac{2.303RT}{n\,\mathscr{F}} \log \frac{[\text{electron acceptor}]}{[\text{electron donor}]}$$

in which E_o' is the standard reduction potential (pH = 7.0, t = 25° C, all concentrations at 1.0*M*), E_n the observed electrode potential, R the gas constant, T the absolute temperature, n the number of electrons being transferred, and $\mathscr{F}$ the Faraday constant. The Nernst equation expresses mathematically the shape of a titration curve of a reductant by an oxidant. The Henderson–Hasselbalch equation expresses the same for titration of an acid with a base. For further details, see Lehninger (1970).

TABLE I
Ion Content of Chloroplasts Isolated by Nonaqueous Means from Four Different Plants[a]

	Ion content (nmoles/mg dry chloroplasts)				
Plants	K^+	Na^+	Cl^-	Mg^{2+}	Ca^{2+}
Tolypella intricata	620	65	620	103	39
Lagrosiphon major (*Elodea crispa*)	700	300	170	70	28
Limonium vulgare	540	960	1250	200	168
Beta vulgaris	500	390	130	120	26

[a]After Larkum (1968).

least Na^+ and K^+ are actively transported across such boundaries if the ions are in flux equilibrium.

Efflux and influx studies with intact chloroplasts from *Nitella opaca* have disclosed that the chloroplast envelope is a functional semipermeable membrane (Saltman *et al.*, 1963). There is a high concentration gradient between the external environment and the internodal cells for both K^+ and Na^+. However, ratios of internal K^+ to Na^+ as compared with those outside indicate a very strong preferential uptake of K^+. Direct measurements by flame spectrophotometry of the Na^+ and K^+ of the chloroplast, cytoplasmic, and vacuolar fractions of *N. opaca* showed that there is little, if any, apparent concentration gradient of the ions between cytoplasm and vacuole, but that the Na^+ concentration in the chloroplasts is lower than that found in the vacuole or cytoplasm. The internal ionic milieu of the chloroplasts is very low in Na^+ and high in K^+. A comparison of the ion content of chloroplast, cytoplasm, and vacuole in (A) *Nitella flexilis*, (B) *Tolypella intricata*, and (C) *Nitella opaca*, is given in Table II. Although intact *N. opaca* cells can accu-

TABLE II
Cellular Ionic Content of Three Algal Species

	Chloroplast (mM)[a]			Cytoplasm (mM)[a]			Vacuole (mM)[a]		
Ions	A	B	C	A	B	C	A	B	C
K^+	110	340	80.2	125	87–97	76	80	110–119	76.4
Na^+	26	36	3.6	2.8–9.1	4–22	13	28	8–39	12.4
Cl^-	136	340	—	2.0–28	23–31	—	136	116–136	—

[a]A = *Nitella flexilis* (after Kishimoto and Tazawa, 1965); B = *Tolypella intricata* (after Larkum, 1968); C = *Nitella opaca* (after Saltman *et al.* 1963).

mulate Na^+ against a concentration gradient, they preferentially accumulate K^+. An inward-directed chloride pump at the plasma membrane would establish the potential difference, which would suggest that the discrimination against Na^+ and a preferential uptake of K^+ must reside in an outward-directed Na^+ pump at the plasma membrane. K^+ would enter the cell either by passive diffusion as a counter ion or by an inward directed exchange mechanism coupled to Na^+ transport. A similar outward-directed Na^+ pump should also be operative at the chloroplast envelope.

On the other hand, passive permeability has been shown for Na^+, K^+, and Cl^- in isolated spinach (*Spinacia oleracea*) chloroplasts under conditions of minimal metabolic activity, in the dark at low temperature (Winocur *et al.*, 1968). The barrier to ion diffusion in chloroplasts appears to be quite loose as compared to that in cell membranes, as deduced from the low energy of activation and from apparent lack of specificity, since rapid efflux kinetics and pool sizes for Na^+, K^+, and Cl^- are similar. Hence the authors concluded that the chloroplast envelope has a relatively small selectivity for Na^+ as compared with K^+. However, it is not clear if the preparations used in this study were devoid of broken chloroplasts.

2.2. Permeability of Chloroplast Envelope to Ions and Other Photosynthetic Substances

It is very difficult to obtain completely homogeneous populations of class I chloroplasts or whole chloroplasts which retain outer envelopes and soluble stroma substances. In most cases, the described routine procedures of isolation yield preparations which contain chloroplasts without outer envelope and stroma, i.e., naked inner membrane (thylakoid) systems, the so-called class II chloroplasts. It is obvious that the use of heterogeneous populations of chloroplasts, in studies on the permeability of chloroplast outer membranes to ions and other substances, may lead to confusion, since nonhomogeneous chloroplast preparations exhibit a mixture of outer envelope and thylakoid membrane permeability properties. Therefore, it is necessary to perform studies on the translocation of ions and photosynthetic substances, as well as on photosynthetic reactions and metabolic control of energy levels, using well characterized populations of class I chloroplasts.

Chloroplast preparations containing a mixture of whole and envelope-free chloroplasts supported slower rates of cyclic photophosphorylation in comparison to those in which all chloroplasts had lost their outer membranes (Walker, 1965). The intact chloroplast envelope restricts or excludes the entry of exogenous reactants such as NADP, and other adenine nucleotides.

2.2.1. NADP

Heber and Santarius (1965) were able to demonstrate a compartmentation of pyridine nucleotides between chloroplast and cytoplasm of leaf cells of *Spinacia oleracea* and *Beta vulgaris*, which suggests that the chloroplast envelope is impermeable to pyridine nucleotides and that NADPH plays a main role in regulating photosynthesis. Chloroplastic NADPH is rapidly oxidized when the light is turned off. In the dark, NADPH continues to be reduced in the cytoplasm during respiration, and although the level of cytoplasmic NADPH is very high even in the dark, no, or little, cytoplasmic NADPH becomes available to the chloroplasts. Consequently, the reduction of 3-phosphoglycerate to triosephosphate is no longer possible in the dark and the photosynthetic carbon cycle stops. In contrast to nicotinamide cofactors, ATP, which is formed by oxidative phosphorylation in the dark, can penetrate the chloroplast envelope. Even in the dark, a moderately high level ATP is found in the chloroplasts, and would allow low rates of photosynthetic reactions in the chloroplasts if NADPH were available. The relatively high level of cytoplasmic NADPH does not cause 3-phosphoglycerate reduction in the cytoplasm, since NADPH-linked triosephosphate dehydrogenase is only present in the chloroplast.

On the other hand, there is some indication of the existence of either a specific mechanism of NADPH (internal) $NADP^+$ (external) transhydrogenation that resides in the limiting membrane of the chloroplast, or a specific mechanism for the entry of exogenous NADP (Harvey and Brown, 1969). With a preparation, obtained from a sucrose density gradient, in which 80 % of the chloroplasts were intact, these authors found that the nicotinamide cofactor content is very much greater than that in chloroplasts obtained by using aqueous techniques such as those used by Heber and Santarius (1965). Photosynthetic reduction of endogenous NADP occurs in the chloroplast stroma, but is closely associated with the inner membrane system. Exogenous NADP was reduced on illumination. Therefore, it is more likely that NADP is taken up by the chloroplast against a concentration gradient.

2.2.2. Other Adenine Nucleotides

As adenine nucleotides such as AMP, ADP, and ATP are also included in the metabolic and energy processes of the chloroplast, the translocation of these compounds across the chloroplast envelope also deserves consideration. However, as in the case of NADP and NADPH, this problem is very controversial. As reported by Santarius and Heber (1965), the ATP concentration in the chloroplasts increases upon illumination, being accompanied by a simultaneous increase in the ATP of the cytoplasmic fraction.

At the same time, ADP and AMP concentrations are reduced in both chloroplasts and cytoplasm. These changes are directly dependent upon the photosynthetic conditions. Since under light conditions ATP is formed in chloroplasts as a consequence of photophosphorylation, it is very likely that photosynthetically formed ATP must be translocated from the chloroplast into the cytoplasm across the chloroplast envelope. Addition of exogenous ADP, but not inorganic phosphate, strongly stimulates light-dependent oxygen evolution by intact chloroplasts which have an outer envelope (Robinson and Stocking, 1968). This also supports the view that the chloroplast envelope is relatively freely permeable to adenine nucleotides. However, Walker (1965) observed that intact pea chloroplasts exhibit a lower activity of photophosphorylation than broken chloroplasts. After exposing class I chloroplasts to an hypotonic medium, or with envelope-free chloroplasts (type II), a very marked stimulation of photophosphorylation can be attained. This seems to suggest that the permeability of the chloroplast envelope to exogenous ADP is rather limited.

More precise information of adenine nucleotide translocation has been obtained by assaying the exchange of endogenous and exogenous adenine nucleotides directly, using labeled compounds and intact *Acetabularia* chloroplasts (Strotmann and Berger, 1969). When the intact chloroplasts, separated from *Acetabularia* cytoplasm by a gentle centrifugation technique, are incubated in a medium containing [^{14}C]ATP or other adenine nucleotides, labeled nucleotides can be detected in the chloroplasts, though there is no change in the total concentration of adenine nucleotides in the chloroplast fraction. From these data, the chloroplast envelope again appears to be permeable to adenine nucleotides. Similar results were obtained by Heldt (1969) with spinach chloroplast preparations containing 70–80% intact chloroplasts, suggesting that the main role of adenine nucleotide translocation in chloroplasts is to deliver ATP, synthetized by glycolysis or respiration, to the metabolic sites inside the chloroplasts.

2.2.3. Photosynthetic Intermediates

Fixation of CO_2 can be stimulated by the addition of some intermediates of the carbon reduction cycle to broken chloroplasts or whole chloroplasts (Jensen and Bassham, 1966). Class I spinach chloroplasts can undergo CO_2 fixation for 6 min, following 3 min preillumination with a bicarbonate concentration of 3 mM, at a rate 60% of that shown by intact leaves without the addition of either cofactors (other than inorganic ions) or enzymes, as shown by the very high amounts of radiocarbon incorporated in intermediates of the carbon reduction cycle. Many of these compounds are transported, or diffuse, rapidly from the isolated chloroplasts to the suspending medium.

Walker *et al.* (1968), using chloroplasts with intact outer envelopes, also studied photosynthetic carbon assimilation by isolated chloroplasts and the oxygen evolution associated with it, and detected faster and higher rates of oxygen evolution with chloroplasts isolated in pyrophosphate-buffered medium than in those in which orthophosphate was used. In the latter case, the best rates are obtained only in the presence of added intermediates of the Benson–Calvin cycle. When ribulose 1,5-diphosphate was added, it was necessary to rupture the chloroplast envelope in order to release the ortho- phosphate inhibition. It seems probable that ribulose 1,5-diphosphate does not enter the intact chloroplast, or at least its ability to penetrate is drastically reduced.

2.2.4. Ions

Nobel (1967a) introduced an elegant, rapid (2 min) technique for the isolation of whole chloroplasts which are believed to closely reflect the situation occurring *in vivo*, thus providing a means of assessing the ongoing photosynthetic ability of the plant. Almost all (95%) of the chloroplasts obtained belong to class I. The number of osmotically active species in the chloroplasts can be calculated from the Boyle–Van't Hoff relation* which Nobel (1969a) reformulated to extend its applications to relatively simple measurements of osmotic responses of certain cells and organelles (Nobel, 1969a). There was a light-induced decrease *in vivo* of 32% in the content of osmotically active ions in the chloroplasts, when compared with that in chloroplasts from plants in the dark (Table III), which agreed with that predicted by the Boyle–Van't Hoff relation (Nobel, 1969b).

These results confirm that chloroplasts have a high concentration of K^+, which is the major cation, and Cl^-, which appears to be the principal anion. Na^+ is often present to about one-tenth the extent of K^+, and Mg^{2+} and Ca^{2+} are present in appreciable amounts, both perhaps near 15 mM when averaged over the osmotically active responding volume. Mg^{2+} and Ca^{2+} contents are approximately one-sixth that of K^+. Since K^+ is the most abundant cation in the chloroplasts *in vivo* and undergoes the greatest

---

*The membranes of cells and organelles exhibit a differential permeability, being generally much more permeable to water than to solutes and so behaving as osmometers when subjected to varied external osmotic pressures. This behavior is quantitatively expressed by the Boyle– Van't Hoff relation:

$$\pi(V - b) = nRT$$

Where π is the osmotic pressure of the external solution and by implication also of the internal solution, V is the volume of the cell or organelle; b is the nonosmotic volume or volume of a solid phase within volume V which is not penetrated by water; n is the apparent number of osmotically active moles in $V - b$. R is the gas constant and T the absolute temperature. For further details see Nobel (1969a).

TABLE III

Ion Content of *Pisum sativum*
Chloroplasts in Dark and Light Conditions[a]

Ions	Dark (μmoles/mg Chl)	Light (μmoles/mg Chl)
K^+	1.660 ± 0.018	1.214 ± 0.027
Na^+	0.155 ± 0.004	0.134 ± 0.003
Cl^-	1.704 ± 0.087	1.011 ± 0.071
Mg^{2+}	0.351 ± 0.013	0.450 ± 0.028
Ca^{2+}	0.242 ± 0.005	0.205 ± 0.010

[a] After Nobel, 1969b and Lin and Nobel, 1971.

variation in micromoles per milligram of Chl during the volume changes, it was interesting to find that the reversible light-induced K^+ efflux from chloroplasts had the same kinetic properties as the light-induced chloroplast shrinkage and its reversal in the dark (half-time near 4 min). Although it is not possible to consider a mechanism for the light-induced chloroplast shrinkage involving K^+ extrusion, K^+ and water loss may be correlated phenomena. Both the light-induced chloroplast shrinkage and the light-induced efflux of K^+ appear to depend on a high-energy intermediate, which is tentatively assumed to be a light-induced proton gradient.

2.2.5. Amino Acids

Using the above-mentioned rederivation of the Boyle–Van't Hoff relation, Nobel and Wang (1970) determined the reflection coefficients* of various amino acids, which is a useful parameter for estimating the penetrability of a particular solute crossing a specific biological membrane. It was found that the chloroplast envelope is highly permeable to amino acids with a very low reflection coefficient, such as glycine, DL-alanine, L-norvaline, DL-serine, DL-homoserine, DL-threonine, DL-phenylalanine, DL-methionine, and L-proline. Some other amino acids (DL-valine, L-leucine, and L-iso-leucine) and amides (acetamide and its methyl derivatives), which have intermediate reflection coefficient values, can penetrate the chloroplast envelope somewhat less easily. All these compounds contain methyl groups, which suggests the participation of these radicals in increasing reflection coefficients or decreasing compound penetrabilities. The results obtained

---

* Reflection coefficients are parameters which characterize a particular solute crossing a certain membrane. A species which cannot cross a specified membrane has a reflection coefficient of unity for that barrier. If the membrane does not distinguish or select between water and the solute, the reflection coefficient of the solute is zero. Reflection coefficients are useful for treating the many cases between the extremes of impermeability and nonselectivity. For further details, see Nobel (1969a) and Nobel and Wang (1970).

by Nobel and Wang (1970) are in accord with those previously presented by Ongun and Stocking (1965) and Aach and Heber (1967), who have shown that exogenous labeled amino acids are metabolically incorporated in chloroplast proteins, which suggests a relatively free permeability of the envelope to amino acids.

2.2.6. Sugars

The same approach was used for studies of the permeability of pea chloroplasts to alcohols and aldoses (Wang and Nobel, 1971). Methanol is freely permeable, ethylene glycol and glycerol are less permeable and *meso*-erythritol can hardly penetrate the chloroplast outer envelope. Alcohols become progressively less permeable as hydroxymethyl groups are added to their molecules. The chloroplast envelope is poorly permeable to various aldopentoses. D-Xylose, D-lyxose, and L-arabinose are more permeable than their optical isomers, i.e., L-xylose, L-lyxose, and D-arabinose. Aldohexoses such as D-glucose, D-galactose, D-mannose, and L-mannose have higher reflection coefficients and are therefore excluded from entering chloroplasts. Fairly good support for the contention that the outer envelope is impermeable to sugars has been recently presented by Heldt and Sauer (1971), whose experiments indicate that the outer membrane of the envelope is nonspecifically permeable to sugars (sucrose, sorbitol, glucose, and fructose) and sugar phosphates, but that these compounds do not enter the stroma space because they are unable to cross the inner membrane. The sucrose-permeable space, i.e., the space between the outer and inner membrane of the envelope, is also accessible for other low molecular weight compounds, including adenine nucleotides, inorganic phosphate, malate, glutamate, citrate, and acetate. The sucrose-impermeable space or stroma is accessible to certain metabolites such as 3-phosphoglycerate and malate.

2.2.7. Organic Acids

Nobel and Craig (1971) used electrical measurements and the Goldman equation* to obtain information about the permeability in palisade mesophyll

---

*Electrical potential differences across the membranes of cells and organelles generally result from the relative rates of passive diffusion of various ions. Such diffusion potentials can be quantitated using the Goldman equation:

$$E_m = (RT/\mathscr{F}) \ln (K° + \alpha/\beta j \, Aj° + \gamma)$$

where E_m is the electrical potential difference across the membrane, R is the gas constant, T is the absolute temperature, $\mathscr{F}$ is the Faraday constant, $K°$ is the external concentration of potassium, βj is the permeability of some monovalent anion Aj relative to that of potassium, $Aj°$ is the external concentration of Aj, while α and γ are parameters depending on the other ions that are present. For further details, see Nobel and Craig (1971).

cells of leaves of *Pisum sativum* to organic acids. Since organic acids are abundant products of photosynthesis, this approach should help to elucidate the problem of the permeabilities of organic acids *in vivo*. Assuming that the potentials across the cell membrane are diffusion potentials that can be analyzed using the Goldman equation which is based on the assumption of passive diffusion, it was determined that pyruvate, formate, butyrate, acetate, and bicarbonate can readily diffuse into the palisade cells, the permeability to organic acids being large compared with that to K^+ and even larger when compared with the permeability to Cl^-. These relative permeabilities might occur if the organic acid anions passively diffused across the membrane while bound to a carrier ("facilitated diffusion"). This would agree with previous studies by Nobel (1969c), in which a light-dependent uptake of the anions into pea leaf mesophyll cells was observed when the potassium salts of the above-mentioned organic acids were used. This uptake was sensitive to inhibitors, being an active process involving carriers. Therefore, the same carriers should be used passively in the dark to shuttle organic acids into the mesophyll cells, which would explain the high permeability to organic acid anions relative to K^+ and Cl^-.

2.2.8. Photosynthetic Control

West and Wiskich (1968), using preparations containing mostly class I chloroplasts obtained from *Pisum sativum* leaves, concluded that for the demonstration of "photosynthetic control"* morphological integrity was the critical factor. Their results seemed to correlate well with those presented by Walker (1965) and Jensen and Bassham (1966), suggesting that the rate of carbon dioxide fixation by isolated chloroplasts depends on the integrity of their outer envelopes. It was, however, found that chloroplasts broken by osmotic shock retain the ability to perform photophosphorylation with an ADP/O ratio of 1 and show photosynthetic control ratios† greater than 2

---

* In oxidative phosphorylation, the dependence of respiratory rates on ADP concentration is called *respiratory control* or *acceptor control*, because the rate of substrate oxidation is dependent on the concentration of phosphate acceptor present in the medium. In photosynthetic phosphorylation, where a light-dependent reduction of electron acceptors by chloroplasts is "coupled" to ATP production, it was also demonstrated that the rate of electron flow is not only stimulated by the addition of phosphate acceptor, but also that the rate is decreased on exhaustion of phosphate acceptor, and can be stimulated again by further addition. By analogy with *respiratory control* this phosphate acceptor dependence in photosynthetic electron flow is called *photosynthetic control*. For further details, see West and Wiskich (1968).

† In oxidative phosphorylation, the ADP-stimulated rate of oxygen consumption is called state III and the lower rate after all the ADP has been phosphorylated is referred to as state IV. The respiratory control ratios are derived from the state III/state IV ratio. By analogy, the photosynthetic control ratios are derived from the ADP-stimulated rate of oxygen evolution (state III) and the lower rate after all the phosphate acceptor has been phosphorylated (state IV). For further details, see West and Wiskich (1968).

(Telfer and Evans, 1971; Hall *et al.*, 1971). Integrity of the chloroplasts does not seem to be an essential requirement for photosynthetic control.

2.2.9. *Ion Translocation and Electrical Potential Difference*

Light leads to the building of electrical potential differences between the different compartments of plant cells (Nagai and Tazawa, 1962; Nishizaki, 1963; Barber, 1968). When the change in distribution of ions between two compartments separated by a membrane is a result of ion translocation, the electrical potential difference must be altered. Membrane resistance is closely associated with permeability properties of the membrane (Nishizaki, 1963; Hogg *et al.* 1968, 1969). Therefore, these two parameters were utilized to study the effect of light on ionic relations in plant cells.

In internodal cells of *Chara braunie* an increase in potential difference of about 50 mV between the vacuole and outer medium was observed upon illumination, which was abolished by 10 μM DCMU and enhanced by 0.2 M NaHCO$_3$ (Nishizaki, 1963). This suggests that the photoelectric response at the cell level is mediated by photosynthetic products depending on the electron transfer reactions associated with O$_2$ evolution.

Vredenberg (1969), using internodal cells of *Nitella translucens* and *Nitellopsis obtusa*, presented evidence showing that the ionic relation in a plant cell is established by ion transport in chloroplasts. The membrane potential increase upon illumination was inhibited completely by 1 μM of DCMU and 2 μM of CCCP, but not by Dio-9 (5 μg/ml), suggesting that membrane potential changes are directly dependent upon photosynthetic electron transport. By combining his results with previous data on H$^+$ uptake by chloroplasts *in vitro*, Vredenberg made the following interpretation: (1) H$^+$ uptake by chloroplasts in intact cells upon illumination results in a decrease in H$^+$ concentration in the cytoplasmic layer surrounding such organelles; (2) this would cause a flow of H$^+$ from the vacuole and the external medium toward the cytoplasm. A similar hypothesis was advanced by Lüttge and Pallaghy (1969) for photoresponses of membrane potential of anthophyta (*Atriplex spongiosa* and *Chenopodium album*) and bryophyta (*Mnium*). From experiments using DCMU (1–2 μM) and different wavelengths of actinic light to separate the two light reactions in photosynthesis, it was concluded that photosynthetic reaction II triggered the membrane potential changes, although the phytochrome system localized within the cell membrane could also have a possible participation.

It is clear that electrical potential studies emphasized the presence of correlated ionic relations between chloroplasts and other cell organelles, even suggesting a feedback mechanism which regulates cellular ionic states.

2.3. Light-Induced Ion and Proton Translocation

Chloroplasts and mitochondria undergo changes in structure that are closely associated with electron flow and energy-dependent reactions. The reactions leading to electron flow might also control the movement of both ions and water in these organelles (Packer, 1963; Jagendorf and Hind, 1963). Light-driven ion translocation was first documented for green plant cells. Hoagland and Davis (1923) observed light-induced increments of Cl^- and Br^- in *Nitella* cells. Jacques and Osterhout (1934) found that light increased the rate of K^+ entry into *Valonia macrophysa* and Arens (1936) demonstrated Ca^{2+} movement through *Elodea* leaves by a photo-dependent process. Cl^-, K^+, and PO_4^{3-} were found to be accumulated by *Elodea* leaves in the light. Subsequent investigations showed that light-dependent ion translocation is a general process in green plant cells (Kylin, 1960; Briggs *et al.*, 1963; Cummins *et al.*, 1966).

2.3.1. Intact Cells

In intact algal cells two independent mechanisms were found to be operative in light-dependent ion regulation: one for K^+, Cl^-, and H^+ influx, and the other for transporting Na^+ outward (Scott and Hayward, 1954; MacRobbie, 1965). Using a tracer analytical method, Cummins *et al.* (1966) were also able to demonstrate H^+ uptake by thalli of *Ulva*, paralleled by Na^+ extrusion. They concluded that several components of Na^+ transport existed within *Ulva*. The fairly slow rate of Na^+ transport in algae in the dark (0.15%/min) is attributable to the impenetrability of *Ulva* cell walls. The inhibitory effect exerted on this transport by ouabain, a specific inhibitor of the sodium pump according to Post *et al.* (1960), suggests that in the dark Na^+ transport is not entirely passive. Light causes a transient, rapid extrusion of Na^+ from the algae, equal to approximately twice the amount of Na^+ expected during a 2-min dark period. It is not possible to conclude how closely Na^+ extrusion is related to the primary reactions of photosynthesis, but it appears that with the onset of photosynthesis, chloroplasts, lying immediately at the interior cell wall, would take up H^+ in exchange for Na^+. This movement is not specific to Na^+, but also occurs with other cations, such as Rb^+ and Sr^{2+} (Cummins *et al.*, 1969). H^+ changes would be coupled to the movement of HCO_3^-, which supplies CO_2 for the dark reactions of photosynthesis (Atkins and Graham, 1971). Possibly H^+ would move with HCO_3^-, thus conserving charge and implying a function of the proton pump in HCO_3^- transport. As the overall result, a proton gradient would be established from the interior of the algae to the exterior medium.

Light-induced proton uptake in whole cells has been studied with uni-cellular algae. In *Dunaliella parva*, a unicellular halophilic alga, proton uptake is pH dependent with an optimum at pH 6.2 of 9–10 μEq H$^+$/mg Chl (Ben-Amotz and Ginzburg, 1969). This is in good correlation with the pH dependence of proton translocation activity in isolated spinach chloroplasts determined by Neuman and Jagendorf (1964) and Deamer *et al.* (1967). The proton transfer activity of *Chlamydomonas reinhardi Y-1* mutants was studied in relation to the light-induced formation of the chloroplast inner membrane system (Schuldiner and Ohad, 1969). The maximal rate of H$^+$ uptake of the light-grown normal cell in which chloroplasts have a normal inner membrane system is 1–2 μmoles H$^+$/mg Chl-min. The *Y-1* mutant lacks most of its proton uptake activity and Chl in dark-grown cells. Light-dependent H$^+$ activity is restored after 2 hr of light-induced formation of thylakoid membranes, which is about the same time it takes Cyt f to return. This study (Schuldiner and Ohad, 1969) was pertinent because it clarified the relationship between the development of H$^+$ activity and the formation of thylakoids. The proton uptake is not affected by the presence or removal of CO_2 in the reaction vessel, so that there is true H$^+$ uptake and no apparent side effects due to CO_2 fixation.

The difficulty of investigating ion transport in an intact plant cell arises from the existing organelle permeability barriers. A common approach has been the use of inhibitors in the study of ion distribution and transport. The action of phlorizin in *Hydrodictyon africanum* on light-stimulated ion transport, namely K$^+$ influx and Na$^+$ efflux, suggested that cyclic photophosphorylation is the most important mechanism for generating ATP in support of ion movements (Raven, 1968; Packer, *et al.*, 1970).

Penth and Weigl (1969) have also investigated the light-dependent anion uptake in detached leaves of *Limnophila gratioloides* as a function of external concentration of Cl$^-$ and SO_4^{2-}. As expected, nutritional and developmental states of the plant were critical with respect to the anion uptake. These authors also consider that the metabolic driving force was provided by ATP rather than by directed translocation powered by organelles. Another critical factor affecting anion transport is the composition of gas phases. The extent of Cl$^-$ uptake by *Elodea* leaves depends on light intensity levels (Jeschke and Simonis, 1967, 1969). Such activity saturates at low light intensities in pure N_2. In N_2 and 3% CO_2, on the other hand, there is much less Cl$^-$ uptake. CO_2 exerts a strong inhibitory effect upon Cl$^-$ uptake at low light intensities. This inhibition is almost undetectable at high intensities of white light. The CO_2 inhibition is wavelength-dependent and is greater at the optimal photosynthetic wavelengths. CO_2 also inhibits Cl$^-$ uptake at high levels of light intensity in the presence of trace amounts of DCMU. The action of CO_2 may arise from a competition for ATP between the CO_2 assimilation and Cl$^-$ uptake processes. At low light intensities, CO_2 appears to exert a regulation between noncyclic and cyclic electron flow and photophosphorylation. This would correlate with Heber's results (1969) showing an interaction of photoshrinkage with electron flow in the presence and absence of CO_2. Therefore, no direct mechanism for Cl^- uptake linked to photosynthetic electron transport can be anticipated in *Elodea densa*, and Cl^- uptake appears to be indirectly linked to electron transport activity. If this is so, the stimulation of PS II activity by Cl^- involves a different mechanism, according to Harvey and Brown (1969) and Hind *et al.* (1969). These studies need to be brought together with those of MacRobbie (1965), reporting that in *Nitella* Cl^- uptake is connected to PS II activity. Nevertheless, it is clear that a complex set of conditions, including the ionic environment, is critical to the control of electron flow through the photosystems in the green plant cell.

Nobel (1969c) extended his previous studies on light-dependent K^+ and Cl^- efflux in intact chloroplasts, using leaf fragments of *Pisum sativum*, and found that light-induced uptake of K^+ into the cells is partly balanced by H^+ release into the external solution. Moreover, HCO_3^- and Cl^- movements seem to balance the charge of the K^+ uptake which was not compensated for by H^+ release. His studies are relevant to the formulation of a general picture of ionic interactions within a cell, although knowledge of stoichiometries, specificity, and directionality of ion movements remains incomplete.

2.3.2. Intact (Class I) Chloroplasts

Light-induced H^+ uptake in intact pea chloroplasts was demonstrated by West and Wiskich (1968), to have the same direction of light-induced H^+ uptake as that in class II chloroplasts. This activity was inhibited by Triton X-100. West and Packer (1970) carried out careful experiments to investigate the magnitude of the light-induced proton movements in a population mainly constituted of class I chloroplasts, and obtained an average H^+ uptake of 0.53 μEq/mg Chl under cyclic electron flow, a much higher value than 0.02 μEq H^+/mg Chl obtained by West and Wiskich (1968). The low value obtained by the latter may well be due to the composition of their reaction medium which did not contain ions other than PMS. When the ionic balance in the compartment (chloroplast and thylakoid) is maintained by either influx of anions such as Cl^- or efflux of countercations (e.g., K^+) which are supplied from the medium, further H^+ uptake is to be expected. West and Packer (1970) also showed light-induced proton movements in glutaraldehyde-fixed class I chloroplasts, and Torres-Pereira and Packer (unpublished results) verified a proton uptake of 0.13 μEq H^+/mg Chl in glutaraldehyde-fixed class II chloroplasts, which did not show any light scattering changes, under conditions of light-induced basal electron transport. Mohanty *et al.* (1973) independently verified a proton uptake of 0.17

μEq. H$^+$/mg Chl in glutaraldehyde-fixed class II chloroplasts which did not show any PMS-induced Chl a fluorescence quenching related to scattering changes.

2.3.3. Energy-Dependent Processes

There is ample evidence for a very close correlation between the light-induced proton uptake or proton gradient established by illumination and the extent of ATP formation, supporting Mitchell's chemiosmotic hypothesis of phosphorylation (Mitchell, 1966; Mitchell and Moyle, 1965, 1967). Very pertinent considerations on this important problem are presented by Jagendorf in Chapter 9.

2.3.3.1. Divalent Cations Accumulation. Nobel and Packer (1964, 1965) demonstrated that Ca^{2+} and phosphate were taken up by isolated class II spinach chloroplasts in the presence of Mg^{2+} and ATP. This ion uptake is energized directly by cyclic electron flow (PMS) or light-triggered ATPase (Mg^{2+}-ATP, reduced thiol) (Petrack and Lipmann, 1961; Hoch and Martin, 1963; Marchant and Packer, 1963), and is sensitive to CCCP and NH$_4$Cl, being progressively reduced by increasing concentrations of the nonionic detergent Triton X-100. Ca^{2+} uptake is slow and not extensive, but the Ca^{2+} taken up is not removed by centrifugation. These results are similar to those obtained for light-induced phosphate uptake, and suggest the existence of insoluble deposits of these ions within the chloroplast. This was proved using electron microscopy by Nobel *et al.* (1966a, b) and Nobel and Murakami (1967). Upon illumination in the presence of P$_i$ and the chloride salts of strontium and calcium, ion accumulation is accompanied by the deposition of electron-dense particles associated with thylakoid membranes of chloroplasts. The lack of reversibility in the dark and the slow time course of accumulation in the light suggested that uptake of divalent cations, though energy-dependent, is not directly linked to the primary energy state resulting from electron flow in illuminated chloroplasts, and the slow light-induced swelling that is simultaneous with this accumulation suggests the possibility of membrane damage.

2.3.3.2. H$^+$ and Monovalent Cations Exchange. Light and electron transport-dependent proton uptake by class II chloroplasts was shown to be closely linked to K$^+$, Mg^{2+}, and water efflux (Dilley and Vernon, 1965). In the case of H$^+$-linked cation exchange, it was shown that a 1:1 stoichiometry is approached between total cations effluxed and H$^+$ taken up. Grana shrinkage or membrane conformation changes can occur as a result of this H$^+$-linked cation exchange. These results suggest that fixed-negative charge sites within the chloroplast thylakoid membrane are neutralized by K$^+$ and Mg^{2+} in the dark, and the H$^+$ taken up in the light replaces these cations and associates with the negative charges leading to structural modifications (Dilley and Rothstein, 1967). Since protons would be expected to bind groups such as $-COO^-$ covalently, as opposed to electrostatic binding by K^+ and Mg^{2+}, it appears that H^+ binding and K^+ loss would lead to dehydration, since the $-R^- \ldots K^+$ pair would be osmotically more active than the acid function ($-RH$) (Dilley, 1971). Deamer et al. (1967) and Deamer and Packer (1969) showed that, under conditions leading to chloroplast swelling, Cl^- is taken up, apparently to balance proton uptake.

The movement of ions and undissociated molecules across the thylakoid membranes and the consequences in terms of osmotic effect have been considered in close detail by Packer and Crofts (1967). From knowledge of the ready permeability of chloroplasts to organic anions in their undissociated form, they suggested that the concentration of these anions within chloroplasts might be affected by light-induced proton uptake. According to their hypothesis the fall in internal pH, due to H^+ uptake in chloroplasts, would be "buffered" by the presence of weak acid anions which on reaction with H^+ would give undissociated acid. The internal anion might then be displaced from illuminated chloroplasts in its uncharged form by virtue of the ability of the undissociated acid to equilibrate across the chloroplast membrane. The temporal sequence of events would be (1) in the dark the undissociated acid form of one weak acid anion would equilibrate across the chloroplast membrane; (2) upon illumination protons would be pumped into the chloroplasts, probably, at least in part, in exchange for other cations. This would shift the equilibrium of the dissociated acid anion within the chloroplast toward undissociated acid; (3) an efflux of undissociated acid would occur through a concentration gradient; (4) water loss would follow, because of the efflux of osmotically active material, resulting in shrinkage; (5) these events would be reversible in the dark. Therefore, weak acid anion movement in chloroplasts would lead to conformational changes and influence transport of other ions (Fig. 1,d).

Dilley and Vernon (1965) were the first to use ion-sensitive electrodes for detecting ion movements in chloroplasts and were able to show that about 0.22 nmoles K^+/mg Chl were released by chloroplasts upon illumination. They also investigated Mg^{2+} and Na^+ movements using atomic absorption spectrometry. Na^+ was found to move into the chloroplasts upon illumination, which is in agreement with the work of Nobel and Packer (1964). Dilley and Vernon (1965) concluded that ions are lost from chloroplasts to compensate for the inward movement of charge accompanying proton uptake, and suggested that the volume change observed on illumination results from the outward movement of Mg^{2+} and K^+. The movement of ions as measured by spectrometry was partially reversible. Crofts et al. (1967), using electrode techniques, partially confirmed these

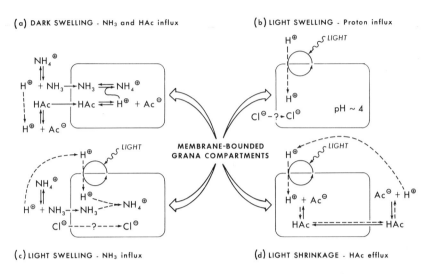

Fig. 1. Mechanisms of light-induced conformational changes in chloroplasts, mediated by ion transport. After Packer and Crofts (1967).

results, but failed to demonstrate reversibility of the ion movements. From the variability in content of K^+ in chloroplasts and the failure to demonstrate a significant amount and reversibility of the cation movements, it is apparent that K^+ movement cannot be commensurate with the magnitude of the water movements required by an osmotic mechanism.

Crofts (1966a,b, 1967) demonstrated that chloroplasts take up NH_4^+ on illumination in choline or Tris–chloride media. The observation of NH_4^+ uptake is of interest with regard to the powerful inhibition by amine salts of reactions associated with energy conservation in chloroplasts. Light-induced uptake of NH_4^+ increases as its concentration in the external medium increases and occurs at the expense of the proton uptake. Clearly, NH_4^+ uptake is closely coupled to electron flow. Proton uptake is not inhibited by increasing concentrations of KCl, but is markedly suppressed in the presence of increasing concentrations of NH_4Cl. The suppression of H^+ uptake is paralleled by an increased extent of NH_4^+ uptake. At 2 mM NH_4Cl, the rate of NH_4^+ uptake is some 20 times greater than that of H^+ uptake of the control. The pH dependence for uncoupling of the Hill reaction by amines shows that the uncharged amine is the primary active species (Good, 1960; Hind and Whittingham, 1963; Stiller, 1965). From kinetic data on NH_4^+ uptake at different pH values it was shown (Crofts, 1967) that, although the maximum velocity did not change markedly with pH, the K_m for NH_4^+ was much greater at lower pH. The calculated K_m for NH_3 varied less, suggesting that the active species in NH_4^+ uptake was $NH_3 \cdot NH_4^+$ uptake was found to be associated with a parallel inhibition of either H^+ uptake or photophosphorylation.

When chloroplasts are suspended in the presence of a weak-base cation such as NH_4^+ and other amines which can give rise to H^+ on dissociation, the light-induced H^+ uptake is inhibited, amine cation (at least in the case of ammonium and methylamine) is accumulated, and the chloroplasts swell. Izawa (1965) and Izawa and Good (1966a), as well as Deamer et al. (1967) and Crofts (1966a,b), demonstrated that uptake of amine cation and chloroplast swelling are closely associated. The temporal sequence of events can be summarized as follows: (1) NH_3 equilibrates across the chloroplast membrane as the uncharged species; (2) light-induced proton uptake displaces the equilibrium of NH_3 within the chloroplasts toward NH_4^+; (3) as the concentration of NH_3 within the chloroplasts decreases, more uncharged species come from the outside; (4) the overall result is an inward transport of NH_4^+, and consequently osmotic swelling since more water comes in. This mechanism (cf. Fig. 1c) explains why the addition of NH_4Cl rapidly dissipates light-induced proton gradients (Packer and Crofts, 1967).

It is clear from the above considerations that the nature of the ions present, in particular their state of association or dissociation, exerts control over the structural and energy states of isolated chloroplasts by interacting with light-dependent H^+ transport.

2.4. Gross Morphology Changes in Inner Membrane System

Packer (1962, 1963) first reported that a marked 90° light scattering change occurred in class II chloroplast suspensions upon illumination under conditions favorable for photophosphorylation or ATP hydrolysis. Subsequently, Itoh et al. (1963) reported that light caused a volume decrease in individual chloroplasts which could be detected both with Coulter counter and electron microscope techniques. Rapid progress in the field showed that photophosphorylation and ATP hydrolysis were not events primary to the light-induced structural modifications of chloroplasts, although they may contribute to and alter the process. The structural changes could be satisfactorily explained as an interaction of the light-dependent proton uptake in chloroplasts (Jagendorf and Hind, 1963) with the ionic environment (Deamer et al., 1967; Crofts et al., 1967). Ion transport mechanisms exert control on water movement in plant cells, and the major energy source for ion transport is the photosynthetic process in the chloroplast. Moreover, at the organismal level, structural changes in chloroplasts driven by ion movements may provide an important basis for understanding stomatal control and leaf movements. Furthermore, structural changes are obviously involved in regulatory mechanisms of photosynthesis. Structure and function are intimately related in chloroplasts *in vitro*, where the ionic environment and light are regulating factors in determining the size, configuration, and conformation of isolated chloroplasts.

The light scattering increments which occur upon illumination in sodium chloride and sodium acetate solutions are rapid, usually reaching a plateau within approximately 1 min. They are reversible in the dark and the light–dark cycle can be repeated several times. The ionic composition of the suspending medium determines the direction and extent of the scattering response. Spinach chloroplasts suspended in 0.1 M solutions of highly dissociated ions such as Na^+ and Cl^- show a scattering increase to 160% (dark level = 100%), accompanying swelling of the thylakoid system (Fig. 1b). This response has a pH optimum of about 6.0. However, in the presence of 0.1 M solutions of weak-acid anions such as acetate or phosphate, the response is enhanced by a factor greater than 2 (to 260%) and the pH optimum is spread between 6 and 8, falling off rapidly below 5.

The major effect of ammonium ions on chloroplasts is that the light-dependent high energy state is mainly used to drive their transport. If both weak-acid and weak-base ions are present, as in ammonium acetate, the original scattering level is greatly reduced and no further changes occur upon illumination (Fig. 1a). More information about the dissipation of light-induced proton gradients by amine cation accumulation was already presented in this chapter.

Photometric evidence for structural changes in chloroplasts *in vivo* during illumination was presented by Packer *et al.* (1967). Since air spaces within the leaf tissues of higher plants interfere with measurements of transmission and light-scattering changes, a vacuum infiltration technique is useful to remove air from the leaf and to place test substances at the cell surface. When the spinach leaf tissue is vacuum infiltrated in an acetate plus PMS solution at pH 6.5 and illuminated with red (600–700 nm) light, a large and reversible light-induced decrease of transmission at 546 nm which reaches a steady state in about 1 min is consistently observed. When the leaf is returned to darkness, the transmission decrease is reversed to its original dark level. The greatest response is observed between pH 5.5–6.5, corresponding to the same pH maxima seen for chloroplast suspensions *in vitro* (Deamer *et al.*, 1967). As with chloroplasts *in vitro*, the light-induced transmission decrease of the leaf tissue is dependent upon the presence of sodium acetate with a maximum light-induced response occurring in about 100 mM sodium acetate. Upon omission of PMS, a cyclic electron carrier, the magnitude of the response *in vivo* is much reduced. When PMS is replaced by ascorbate plus DCPIP and ferricyanide, a system which leads to noncyclic electron flow *in vitro*, a large light-induced transmission decrease also occurs.

Light-dependent changes in photometric parameters and structure of the thylakoid *in vivo* were further studied with thalli of *Ulva* and *Porphyra* by Murakami and Packer (1970a). The thalli of these algae are only two cell layers thick, which obviates the necessity of performing vacuum infiltration in order to bring nutrients and test solutions in contact with the cells. Moreover, the chloroplasts of *Ulva* and *Porphyra* are very large, so that the light-induced changes in the photometric signals reflect changes in the chloroplast structure *in vivo* which are proportionately very large with respect to the nonresponding background noise of unreacting cellular structures. Therefore, such marine algae offer certain advantages over leaf tissue. It was possible to verify that upon illumination of thalli in 0.1 M sodium acetate containing 2.5 mM potassium ferricyanide, a transmission decrease occurs (15 % decrease in 2 min). As the recording is continued, transmission slowly decreases somewhat further, which may reflect a secondary, perhaps osmotically induced process, involving either the chloroplast or another component of the cells. When the light is removed, a rapid increase of transmission occurs, but the response is largely though not fully reversible. The simultaneous recording of 90° light scattering showed an increase in the intensity of the light scattering upon illumination, which was reversed by extinguishing the red actinic light. DCMU, 10 μM, added to the incubation medium under conditions of continued illumination almost completely reversed both transmission and light scattering levels, and in the presence of DCMU no further responses of these two photometric parameters to illumination were observed. In blue-green algae, however, Mohanty and Govindjee (1973) showed a light-induced increase in transmission at 540 nm even in the presence of 15 μM DCMU; this increase was supressed by 5 μM salicylanilide, S-6. Under the photosynthetic conditions mentioned earlier and in the absence of DCMU, striking changes in the structure of chloroplast inner membrane system are observed.

Using high resolution electron microscopy and microdensitometry it was found that striking changes in the configuration of the thylakoids occur when the *Porphyra* thallus is illuminated by red light (Murakami and Packer, 1970a). Thylakoids assumed a more stretched and flattened form so that it became more difficult to resolve the two walls of a thylakoid. The distance between two thylakoid membranes of a single thylakoid of *Porphyra* (measured as center-to-center distance of the membrane) varies widely from 125 to 700 Å in the dark, whereas values for illuminated plastids range between 95 and 125 Å. Light-induced flattening of the thylakoid membrane system was demonstrated also for the chloroplasts of *Ulva*. In the dark, the thylakoids are in a swollen configuration and illumination induces drastic changes in chloroplast structure: the thylakoids are stretched, and stacks became very tightly packed. The stacking becomes so compact that intraspace of the thylakoids can hardly be resolved on the photographs even at higher magnifications. Spacing (center-to-center distance) of stacked thylakoid membranes decreased from 198 ± 10 Å in the dark to 144 ± 6 Å in the light. The thickness of the thylakoid membranes was also reduced. Photodensitometric curves revealed that in *Porphyra* plastids in the dark before illumination the thickness of the thylakoid membrane was 64 ± 8 Å, and this value was reduced to 49 ± 4 Å upon illumination, representing a 23% decrease. In *Ulva* chloroplasts, a reduction of the membrane thickness estimated at the partition (two opposed thylakoid membranes) was 20%. Return to darkness completely reverses the thylakoid flattening, resulting in a configuration of thylakoids similar to that seen in the dark before illumination. The membrane thickness measured with single thylakoid membranes was 70 ± 9 Å in the dark and 53 ± 7 Å for illuminated chloroplasts.

Miller and Nobel (1972) also observed flattening of pea chloroplast thylakoids *in vivo*: the average thickness of granal thylakoids decreased from 195 ± 4 Å in the dark to 152 ± 4 Å in the light. Miller and Nobel (1972) did not observe changes in the membrane thickness contrary to Murakami and Packer's (1970a) observations and interpreted this discrepancy as resulting from less physiological experimental conditions used by these authors.

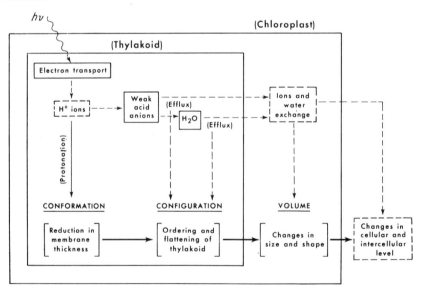

Fig. 2. Hypotheses for structural changes in thylakoid membrane *in vivo*. Dashed line boxes: proposed from *in vitro* and *in vivo* studies; closed boxes: observed in photometric studies; brackets: observed with electron microscope. After Murakami and Packer (1970a).

It is already known that in isolated chloroplasts transmission changes correlate more closely with the changes in spacing of the thylakoid membrane (Murakami and Packer, 1970b). A second type of configurational change is that in the shape and ordering of thylakoids as shown in algal chloroplasts *in vivo*. It now seems clear that gross morphology changes or configurational changes observed in chloroplasts *in vivo* and *in vitro* are mainly osmotic. Previously mentioned studies with chloroplasts *in vitro* and *in vivo* have demonstrated a light-dependent H^+ uptake and also movements of other ions. For example, illuminated chloroplasts show an efflux of weak acid anion-type substances, causing osmotic collapse of the thylakoid membranes. This type of mechanism, already described in this chapter, can be expected to occur both *in vitro* and *in vivo*, because weak acid anion substances are the common type of carbon compounds that would be expected to be present at appreciable concentrations within the chloroplasts. On this basis, Murakami and Packer (1970a) proposed that configurational changes have an osmotic basis, as illustrated in Fig. 2.

3. STRUCTURE AT THE MEMBRANE AND MOLECULAR LEVEL

3.1. Ionic Control of Structure and Energetic States

Cations affect structural and energetic states of the chloroplast environment at different levels. At the thylakoid level, interaction with membrane components, involving binding and/or releasing of cations, results in alteration of the various microenvironments located within the thylakoid membrane, thus leading to changes in its conformational and energetic states. At the organelle level, movement of cations (and anions) across the thylakoid membrane from inner to outer (stroma) space, or vice versa, results in osmotic shrinkage or swelling of the inner membrane system and an increase in concentration or dilution of cations responsible for cation-dependent biochemical reactions. Type and concentration of ions, and ionic strength are critical factors which significantly affect electron transport, photophosphorylation, ion movement, and structure of the photosynthetic apparatus. Energy states, as monitored by Chl fluorescence, delayed light emission and 515 mm absorption changes, are strictly influenced by various cations (see Section 3.3).

3.1.1. Energy States

Control of electron flow and phosphorylation by cations was noted by Jagendorf and Smith (1962), who observed that washing chloroplasts in low-salt medium or removing cations resulted in the loss of the chloroplasts' ability to phosphorylate. The inhibition of ATP formation was accompanied by an increase in the ability to reduce ferricyanide, thus showing an un-coupling effect which was prevented by a wide variety of 10^{-4} M divalent and 10^{-2} M monovalent cations. The authors suggested that the uncoupling effect could be due to a structural background of the electron transport mechanism but were unable to rule out the alternative of an internal break-down of a high-energy intermediate prior to ATP formation.

An extremely high rate of O_2 evolution in chloroplasts (10,000 μliters O_2/mg Chl-hr) was observed by Izawa and Good (1966b) when the mem-brane stacking had been disorganized in a low-salt medium. They proposed the view that membrane organization was not an important factor for Hill reaction viability. Shavit and Avron (1967) found that the Hill reaction and photophosphorylation rates, as well as the ratio of ATP produced to the electron flow, were strongly dependent on the solute concentration of the suspending medium. Addition of various cations stimulated the rate of ferricyanide reduction and efficiency of photophosphorylation. The presence of cations would induce a conformational change of the chloroplast struc-ture, which would favor electron transfer and coupled phosphorylation. Since a large part, but not all, of the requirement for $MgCl_2$ or phosphate in photophosphorylation could be replaced by $SrCl_2$ or other solutes, a dual role of Mg^{2+} in photophosphorylation is evident: a general one, replaceable by other cations, and a specific one, in phosphorylation. The induced conformational change would allow electron transport to proceed and the marked alteration in the $P/2e^-$ ratio would indicate a cationic function in producing a tighter coupling between ATP formation and electron transfer. The enhanced formation rate of a high-energy, nonphosphorylated inter-mediate could result from either one or both of these effects. The formation of such an intermediate was markedly increased when the cation was added during the light-activation stage.

Dilley and Shavit (1968) observed a Mg^{2+} stimulation of light-induced electron transport and proton uptake in chloroplasts. According to their interpretation, proton uptake would be a requirement for the ATP-forming mechanism and not an alternative use of energy. Hence, the esterification reaction which utilizes a proton, i.e.:

$$H^+ + ADP^{3-} + HPO_4^{2-} \longrightarrow ATP^{4-} + H_2O \tag{1}$$

would take place in a compartment in contact with that into which the light-driven H^+ transport mechanism deposits H^+. As the ATP was released into the suspension phase, H^+ would effectively be taken from the inner com-partment to the outer phase, thus lowering the internal concentration, the consequence of which was a decreased gradient, observable as a decrease in the extent and rate of the efflux of H^+ in the dark, relative to a non-phosphorylating control. Thus, one function of the H^+ internal pool would be to provide protons for ATP synthesis and would keep the pH of the internal space from being driven to high alkalinity, thus slowing down the esterification.

Another type of cationic effect on high-energy states of chloroplasts was discovered by Gross *et al.* (1969). It was found that very low concentrations (20 μM) of divalent cations, such as Mg^{2+}, Mn^{2+}, and Ca^{2+}, markedly reduced high rates of uncoupled electron flow in sucrose-washed chloroplasts. The same effect was observed with monovalent cations, but much higher concentrations were required to suppress the electron flow to the same extent. This inhibitory effect of divalent cations was reversed by uncouplers such as gramicidin. Higher rates of ATP formation were obtained under conditions in which electron flow was controlled by cations than in the absence of such control. Cations also increased the intensity of delayed light emission and the extent of the 515-nm absorption change, two parameters which in some way reflect the efficiency of energy conservation. In contrast, cations had little effect on the rate of uptake or decay of H^+ or on the total magnitude of the proton gradient, which suggests that they act on an earlier intermediate. As an interpretation of these results, Gross *et al.* (1969) proposed that divalent cations control electron flow by affecting the buildup (or decay) of high-energy states which would act as a driving force for proton translocation. Mg^{2+} could either promote the formation of ($\sim X$) by exerting an effect on k_1, or prevent its decay by a route leading to proton translocation, thus affecting k_2 [see Eq. (2)]. If this is so, the high-energy state is not identifiable with the proton gradient, probably being linked to a membrane-bound exchange carrier for protons and cations.

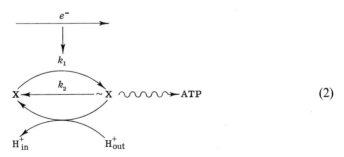

$$(2)$$

3.1.2. Structural States

Clearly, cations control the conformational and configurational states of the thylakoid membranes, through metabolism-independent binding and stabilization of the thylakoid association in the grana.

Gross and Packer (1967) studied the metabolism-independent binding of cations to thylakoid membranes, and verified that divalent species produce a greater decrease in volume on an equal osmolarity and ionic strength basis than did NaCl, for the concentrations tested. At 5 mM, Mn^{2+}, Mg^{2+}, Ca^{2+}, Co^{2+}, Cu^{2+}, and Ni^{2+} caused a marked decrease in the absorbance ratio (680/546 nm), whereas equal concentrations of the corresponding sodium salts had little effect. The divalent cations also caused an increase in the intensity of the scattered light, whereas NaCl did not. The authors were able to prove that the packed volume of grana suspended in 5 mM $MnCl_2$ was independent of the external NaCl concentration, indicating that the apparent inhibition of osmotic swelling was due to the nonosmotic action of $MnCl_2$. Thus, the volume changes observed in the presence of divalent cations should be of a nonosmotic nature. If the broken chloroplast preparations used (grana fraction) obeyed osmotic properties, the following criteria should be fulfilled: (1) their packed volume should obey the Boyle–Van't Hoff law as should (2) the volume changes; (3) volume changes should be independent of the nature of the external solute; (4) volume changes should be reversible by osmotic means alone. This was not the case because low concentrations of divalent cations were much more effective than equal concentrations of salts of monovalent cations, and the shrinkage was not reversible either by diluting the external concentrations of divalent cations or by the addition of equimolar concentrations of EDTA that is assumed to chelate the ions.

Binding of divalent cations to thylakoid membranes was suggested from the observation that lowering the pH of a chloroplast suspension to the isoelectric point (pH 4.7) results in the same degree of shrinkage as that obtained by extrapolation of Boyle–Van't Hoff plots (Dilley and Rothstein, 1967). This would suggest that chloroplast volume is in part determined by the degree of ionization of fixed dissociable charge groups of the membranes and possibly of nondiffusible polyelectrolytes within the chloroplast. Varying degrees of ionization of such groups would lead to changes in the amount of counterions within the structure, with concomitant changes in volume. The effect of many divalent cations (Mn^{2+}, Cu^{2+}, Co^{2+}, and Ni^{2+}) suggests that these cations may bind to fixed negative groups and induce contractions similar to those obtained by lowering the pH. The authors concluded that the fixed-charge groups which are present in chloroplasts can have a significant influence on the control of their volume and cation content apart from functional reactions.

Cations have another important role, namely, maintaining thylakoid association. The sensitivity of the stacking of grana thylakoid membranes to ionic conditions was studied by Izawa and Good (1966b). Chloroplasts isolated in low-salt media lose their grana, consisting almost entirely of swollen continuous sheets of thylakoids which are only loosely held together, primarily at the edges, by their material. This effect was reversed by adding salts to the grana-free chloroplasts in low-salt medium, thus providing strong interthylakoid attractions. This attraction apparently results in a stacking of the membranes which may be almost random but sometimes results in regular structures indistinguishable from the original grana. Therefore, the well organized stacking of grana thylakoid membranes is sensitive to ionic conditions.

Murakami and Packer (1971) systematically studied the structural changes caused by cations, using several parameters of membrane structure, such as light scattering and transmission, fluorescence of Chl and of the hydrophobic probe ANS (see Section 3.3 and Chapter 6 of this volume). The effect of cations in causing a reversible cementing of individual thylakoid membranes together into grana was relatively nonspecific, but divalent cations were much more effective than monovalent species on an ionic strength basis. Furthermore, the authors demonstrated that changes in Chl fluorescence were correlated with the structural states of the chloroplast membrane system which are controlled by the ionic composition of the medium. Under conditions in which membrane dissociation occurs, fluorescence yield at 734 nm markedly increased and that at 682 nm decreased. It is likely that cations might cause changes in the molecular organization or environment of components responsible for energy states within the thylakoid membranes, leading to changes in the fluorescence yield of two photosystems. For literature on the relation of Chl fluorescence to photosynthesis, see reviews by Govindjee *et al.* (1967), Govindjee and Papageorgiou (1971) and Goedheer (1972).

ANS fluorescence studies in glutaraldehyde-fixed chloroplasts suggested that since fixed chloroplasts do not manifest osmotic volume changes, the salt-induced changes in ANS fluorescence of fixed chloroplasts in distilled water indicated that changes in molecular structure and/or in hydrophobic environment occur in the membrane. Murakami and Packer (1971) also suggested that, since the electrostatic attraction between charged groups of membranes is weakened by the presence of electrolytes which stabilize the Debye–Hückel atmosphere* around these charged groups, this attraction force may not be dominant for cementing grana thylakoids. If it was, the association of membranes should be loosened in high ionic strength medium. As this was not found to be the case it is highly probable that hydrophobic interaction, rather than electrostatic interaction, is the dominant force in

---

* As a consequence of electrical attraction between positive and negative ions there are, on the average, more ions of unlike than of like signal in the neighborhood of any ion. Every ion may then be regarded as being surrounded by an ionic atmosphere of opposite charge (Debye–Hückel atmosphere) which reduces its mobility.

maintaining the association between grana thylakoid membranes. If so, the outer surface of the grana thylakoids, where the membranes are cemented tightly together, is of a more hydrophobic nature than the intergrana membrane surface.

3.1.3. Metabolic States

Cations exert a metabolic control at the organelle and cellular levels. It is now clear that light is not only necessary for photosynthesis, but also has side effects on the activity of chloroplasts, namely on their ionic content leading to a redistribution of cations and anions at the cellular level.

Nobel (1967a, 1968a,b,c, 1970) and Nobel et al. (1969) verified that chloroplasts isolated from previously illuminated Pisum sativum leaves were more efficient in photophosphorylation and CO_2 fixation than those isolated from pea leaves kept in total darkness. Dilley and Vernon (1965) had previously reported a light-induced Mg^{2+} efflux accompanying the light-induced shrinkage of spinach chloroplasts. Nobel (1967b) extended these studies and concluded that during light-induced chloroplast swelling there was a release of about 0.14 μmole Mg^{2+}/mg Chl in 30 min. Subsequently, Lin and Nobel (1971) determined that the endogenous photophosphorylation rate by class I chloroplasts isolated from Pisum sativum was doubled and the photosynthetic CO_2 fixation rate was increased 2- to 7-fold when plants were preilluminated. The addition of 5 mM of $MgCl_2$ to the incubation medium eliminated the photophosphorylation enhancement caused by illuminating the plants prior to chloroplast isolation, thus effectively suppressing the lag period. Moreover, Mg^{2+} considerably decreases the analogous enhancement in photosynthetic CO_2 fixation, while in the presence of ribose-5-phosphate and ATP the addition of 5mM of $MgCl_2$ causes the rate of CO_2 fixation to be the same for chloroplasts from plants in the light or the dark. Addition of other cations does not suppress the lag period, which suggests that the effect is due to Mg^{2+}. Chloroplasts from illuminated plants contain 0.45 μmole Mg^{2+}/mg Chl, while those from plants in the dark have only 0.35 (Cf. Table III). When concomitant volume changes are also taken into account, the light-induced rise in the amount of Mg^{2+} in the chloroplasts in vivo corresponds to a concentration increase of approximately 10 mM. Hence, it was found that the Mg^{2+} content in the chloroplasts actually increased, and there was a light-induced Mg^{2+} uptake by intact chloroplasts. Results showing light-induced release of Mg^{2+} may be attributed to contamination by broken isolated chloroplasts, since in such in vitro conditions thylakoid systems lose Mg^{2+}.

Lin and Nobel (1971) interpreted the reported enhancement in photophosphorylation and CO_2 fixation in the following manner: (1) Illumination would cause Mg^{2+} efflux from the internal space of thylakoids, release of bound Mg^{2+} from the thylakoid membranes and a decrease in the volumes of the aqueous stroma compartments in chloroplasts, since there is a 32% reduction in their water content, thus resulting in an increase in the concentration of Mg^{2+} in the chloroplasts of between 8 and 17 mM. (2) A movement of Mg^{2+} into the stroma compartment in the light would activate the carboxylation enzyme (ribulose-1,5-diphosphate carboxylase) and other Mg^{2+}-dependent enzymes such as phosphoribulokinase. Indeed Jensen (1971) has proved that increased Mg^{2+} levels not only affect the carboxylation enzyme (the light-induced increase in magnesium necessary to account for the observed changes in CO_2 fixation rate was about 6–10 mM), but also active photophosphorylation. (3) The overall result would be an enhancement of photophosphorylation and CO_2 fixation in the light.

At the cellular level, in daylight, chloroplasts are rich in Mg^{2+} and so photophosphorylation and CO_2 fixation are activated. In the dark, Mg^{2+} moves out to the cytoplasm and Mg^{2+}-dependent reactions in the cytoplasm are activated. It is clear that Mg^{2+} movements influence cytoplasmic enzymes in addition to playing a major role in regulating photosynthetic activity.

From the above considerations, it becomes evident that cations control structural and energy states of chloroplasts *in vivo* and *in vitro* at several different levels: the thylakoid membrane itself, membrane system, chloroplast, and cellular levels.

A possible sequence and mechanism for cation and/or light effects can be summarized in Scheme I.

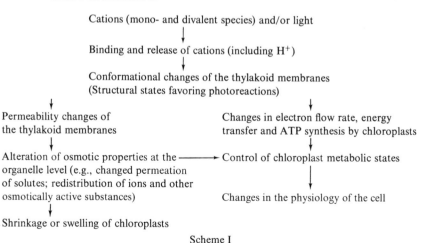

Scheme I

3.2. Protonation and Conformational and Energetic States in Thylakoid Membranes

The effect of protonation on the conformational state of the membrane has been investigated in several laboratories. Hind and Jagendorf (1963) were the first to report light-induced electron transport-dependent H^+ uptake. This proton uptake was shown to be closely linked to K^+, Mg^{2+} and water efflux (Dilley, 1964; Dilley and Vernon, 1965). These studies have already been described in the previous section. Dilley and Vernon (1965) and Dilley and Rothstein (1967) advanced the idea that cations bound to fixed-negative charges within the chloroplast thylakoid membrane are replaced by protons in the light, although not giving full details about such processes.

3.2.1. Protonation and Membrane Conformational State

Very pertinent findings about light-dependent chloroplast morphology and structural changes were revealed independently by Packer (1962, 1963) and Itoh *et al.* (1963). Deamer *et al.* (1967) studied ultrastructural changes mediated by proton transport in chloroplasts. They verified that chloroplasts suspended in NaCl solutions undergo structural modifications when illuminated by red actinic light in the presence of electron carriers. These changes were manifested by increments in light scattering and volume, and disruptive alterations of grana structure as demonstrated by electron microscopy. Simultaneously with these modifications, protons were transported into the chloroplasts in a light-dependent process. The authors postulated that the decrease in internal pH which would result from proton transport can cause many of the changes observed in chloroplasts during illumination. The kinetics of proton transport and light-scattering increments resemble each other both in onset and decay. The pH activation curves are similar for both processes, optimizing between 5.5 and 6.5. Lowering the external pH in the dark to levels which reproduce theoretical internal H^+ concentrations causes alterations in light scattering, volume, and ultrastructure similar to those induced by illumination. From these facts, it was concluded that light-induced modifications of chloroplast structure in NaCl solutions were satisfactorily explained by a decreased internal pH resulting from proton transport into the chloroplast. Shrinkage and swelling of chloroplasts has been extensively studied by Crofts *et al.* (1967) in relation to changes in ionic and protonic environment, using light scattering, Coulter counter, and electron microscopy. The proposed ionic mechanism for chloroplast shrinkage has been discussed at length elsewhere in this chapter, as well as ammonium uncoupling in swelling conditions.

The temporal sequence of events in the light-induced shrinkage of chloroplasts may well be: (1) binding of cation; (2) light-induced proton uptake;

(3) decrease in chloroplast internal pH; (4) decrease in chloroplast net negative charge; (5) loss of water; and (6) increase in the refractive index. Deamer *et al.* (1967) suggested a cationic dehydration of the thylakoid membrane. As light-induced proton uptake proceeds the pH is lowered and membranes lose their net charge, less water would be found and the refractive index of the membrane would increase, as can be monitored by light-scattering increments at low pH. The chloroplast membranes from a number of species do have an isoelectric point between 4.4 and 5.0. A similar effect should be induced by binding of polyvalent cations, since this would also reduce membrane charge. Mukohata *et al.* (1966) and Mukohata (1968) also studied the effect of acid and base conditions on the light scattering level of chloroplast suspensions and considered that changes in refractive index of the chloroplast thylakoid membrane were the main cause of light scattering responses, being closely related to the change in the internal pH of chloroplasts.

Murakami and Packer (1969, 1970a,b) thoroughly studied the microstructural state of thylakoid membranes during reversible contraction–relaxation processes, using such techniques as high resolution electron microscopy, microdensitometry, light scattering and transmission, and hydrophobic-seeking fluorescent (ANS) probe. Their results provided the experimental basis for the protonation concept of the membrane which was postulated earlier by Dilley and Rothstein (1967) and Deamer *et al.* (1967) as a cause of conformational changes of the thylakoid membrane; these changes are directly correlated with the energy state and appear to be dependent on the building of light-induced proton gradients both *in vivo* and *in vitro*, and are due to a primary event, protonation.

It was found that in class II chloroplasts in NaCl–PMA medium there is, upon illumination, not only a reversible decrease in thylakoid spacing of 32% (212 ±8 Å to 144 ±9 Å), but also a decrease in thylakoid membrane thickness of 21% (131 ± 10 Å to 104 ±6 Å), when the dimensions in the dark are taken as 100%. The same effects were observed in sodium acetate medium. In NaCl–PMA or sodium acetate medium, in the presence of either PMS or potassium ferricyanide, class II chloroplasts exhibit extremely amplified 90° light scattering responses to actinic light. Within the limits of error of the measurements, thickness and spacing changes closely followed the corresponding light scattering and transmission changes, respectively. Dark titration of the chloroplasts with H^+ also caused contraction of the thylakoid membranes, accompanying H^+ binding to the membranes in the pH range between 7 and 4.5. The amount of H^+ taken up was 4.5 μEq/mg Chl. The evidence in support of the idea that proonation is involved in the decrease in thickness (conformational change) of thylakoid membranes may be summarized as follows. (1) Several types of treatment (light, H^+, hypertonicity) decrease membrane spacing; all cause a decrease in membrane thickness, except osmotic force. These treatments, again except for osmotic changes, have in common a protonation resulting from illumination or acidification in the dark; (2) By using artificial means (titration with HCl, in the dark), the acidification of the membrane causes an increase in light scattering and a decrease in the thickness of the membrane in a reversible fashion if back-titration is done. Changes in internal structure of the membrane monitored by light scattering are closely correlated with the protonation of the membranes between pH 7.0 and 4.5. (3) Changes in ANS fluorescence by chloroplast membranes have been shown to occur at a faster rate than light-scattering changes and to be pH-dependent. In both these instances, the changes can be satisfactorily accounted for by a prior protonation. Moreover, changes in microstructure of the membrane can occur independently of changes in gross morphology and configuration, since light-induced ANS fluorescence intensity changes are retained in glutaraldehyde-fixed chloroplasts.

According to Murakami and Packer (1970a), protonation of negatively charged groups of molecules in the membrane (e.g., β- and γ-carboxyl, secondary phosphate and imidazole groups) would cause changes in conformation of the molecules and molecular interactions, as well as dehydration of the membrane, resulting in the decrease of membrane thickness. The authors also considered that protonation of negative charges within the thylakoid membrane would lead to an expulsion of attractive water dipoles localized in the vicinity of negatively charged groups, resulting in a decrease in membrane thickness as a consequence of the closer apposition of unit structures within the membrane and an increase of the membrane "hydrophobicity." Charge changes of the membrane as a result of protonation would alter the attractive and repulsive forces causing changes in quaternary protein structures and interaction between the molecules. Thus, it was concluded that the temporal sequence of events involved in the hierarchy of changes in structure following illumination probably is (1) protonation; (2) change in the environment within the membrane; (3) change in membrane thickness; (4) change in internal osmolarity accompanying ion movements with a consequent collapse and flattening of thylakoids; (5) change in the gross morphology of the inner chloroplast membrane system; and (6) change in the gross morphology of intact chloroplasts. A mechanism which relates membrane protonation to this hierarchy of changes in chloroplast membrane structure is presented in Fig. 3.

3.2.2. Protonation and Energy Coupling

Chance et al. (1970), using a fast and sensitive single-beam spectrophotometer and a variety of steady-state illumination sources, have been able to

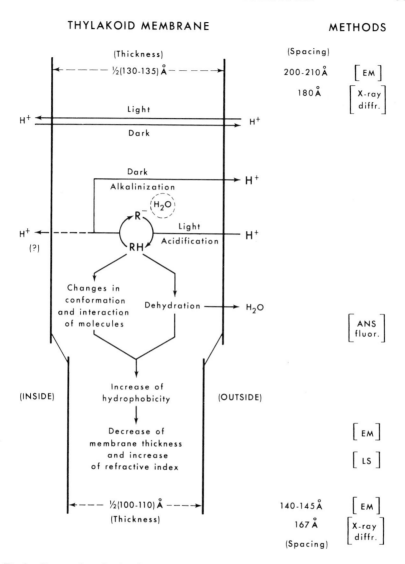

Fig. 3. Proposed mechanism for structural changes in chloroplast membrane by protonation. After Murakami and Packer (1970b).

distinguish bromocresol purple absorbance changes from background absorbance changes of *Chromatium* chromatophores, and identified these fast absorbance changes with rapid H^+ binding by chromatophores. A single turnover flash (10 μsec) by a xenon flash lamp of a total energy output of 8 J yields an H^+ : Chl ratio of 1 : 100, with a duration of the H^+ change of 800

μsec and a half-time of 400 μsec. This result suggested that the reduction of an electron carrier Y produces in 800 μsec an activated state of the membrane (a "membrane Bohr effect")* which is characterized by enhanced proton binding. This mechanism is believed to be analogous to the indirectly linked change of H^+ binding caused by oxygenation of hemoglobin (Bohr *et al.*, 1904). In chromatophores the component involved was not identifiable, but it could be a membrane-bound protein, phospholipid, carotenoid, or chlorophyll. The half-time for the bromocresol purple change would be characteristic of the transition between membrane states that were presumably elementary events in energy-coupling reactions. The insensitivity of the fast H^+ binding to uncouplers, electron transport inhibitors, and to the presence or absence of valinomycin suggests that the membrane transitions may represent a state that is necessary, but not sufficient, for energy-coupling reactions.

Jackson and Crofts (1969) had previously suggested that the rapid disappearance of H^+ from the medium indicated by the bromocresol purple change in the presence of chromatophores of *Rhodospirillum rubrum* would correspond to H^+ binding associated with the reduction of a H^+-carrying electron acceptor. Crofts *et al.* (1971) concluded that the distinction between the "membrane Bohr effect" mechanism and the "binding by H^+ carrier" mechanism will only be possible when and if the electron carrier is characterized. The rapid H^+ uptake observed by *R. spheroides* chromatophores has many features in common with the *Chromatium* system. Although the authors were not able to identify the process of H^+ binding, its close relation to electron flow was indicated by certain features: (1) the stoichiometry as indicated by the ratio of H^+/bacteriochlorophyll on excitation with a laser pulse was about doubled by the xenon flash, which allowed the reaction to turn over a second time; (2) the stoichiometry of H^+ uptake with electron flow on continuous illumination is close to $2H^+/e^-$. Valinomycin doubled the stoichiometric ratio following a flash, suggesting that the electrons activated in the flash passed through a second site for H^+ uptake when valinomycin was present; (3) antimycin inhibited part of the rapid phase of H^+ uptake and

---

*The oxygenation of hemoglobin results in a profound rearrangement of the electrons in the heme moiety. The strongly electrophilic O_2 extends this effect through to iron atom to the heme-linked imidazole group; the electrons binding the imidazolium-dissociable hydrogen are attracted and more closely held by the imidazole group, thereby loosening the dissociable hydrogen. Conversely, increased proton concentration, by repressing the dissociation of the imidazole, attracts electrons to the dissociable hydrogen and away from the O_2, thereby weakening the bond between O_2 and the iron atom. In a biological membrane, the reduction of an electron carrier Y by a hydrogen carrier X releases a proton, increasing the proton concentration in the membrane and leading to a more enhanced proton binding by an hypothetical membrane component. This would be the "membrane Bohr effect."

eliminated the slow phase. However, no inhibition of the initial rate of the phase was apparent.

These observations strongly suggest that H^+ uptake is coupled to electron flow at two sites, one before the antimycin block, and the other at or after the block. The first site is linked with the fast H^+ binding and an electron-flow event may be simultaneous and as rapid ($\tau_{1/2} = 250$–350 μsec) as the H^+ uptake. The second site could be associated with the slow phase of H^+ uptake, which suggests that electron flow at this site is much slower, with $\tau_{1/2}$ about 10 msec in the presence of valinomycin. Both the rapid and slow phase of the H^+ binding reflect the same process, the activity of an electrogenic H^+ pump and that any mechanism should account for a net electrogenic movement of H^+ across the chromatophore membrane, since the electrochemical nature of the gradient formed is well established (Crofts et al. 1971). In addition, the changes indicated by bromocresol purple would only reflect the chemical component of the H^+ movement. Both interpretations would agree with the concept advanced by Murakami and Packer (1970a,b) relating protonation with chloroplast membrane structure in vitro and in vivo. Moreover, Sundquist and Burris (1970) reported a light-dependent change in the distance between the lamellar membranes of isolated spinach chloroplasts, and verified that interlamellar distances in samples illuminated during 15–30 sec decreased by 18.6% (215 ± 19 to 175 ± 24 Å). The illumination time dependence of such structural changes, measured by electron microscopy, was comparable with the illumination time dependence of high-energy intermediate (Xe) formation in the two stage photophosphorylation, as reported by Hind and Jagendorf (1963). Hind and Jagendorf (1965) consider that Xe is a transmembrane pH gradient. Hence, the results presented by Sundquist and Burris (1970) also correlate well with the described investigations of Murakami and Packer (1970a,b) while showing a close correlation between structural states and the generation of high-energy intermediates in ATP formation.

3.2.3. Conformation of Membrane Components and Energy Coupling

Lynn (1968a,b) provided evidence that chloroplast thylakoid membranes have proton translocating oxidoreduction systems, which can create a pH gradient across them. Maintenance of both a high redox potential and a high H^+ concentration within the thylakoid compartments is required for phosphorylation. Proton gradients lead to ATP formation either in the light or by dark acid–base transition, as reported by Jagendorf (1967). Lynn's data strongly suggested that phosphorylation is the result of an energetic separation of charge, i.e., H^+ and OH^- in chloroplasts, and that the alteration of the pKa of fixed anionic sites (nature unknown) is an early and necessary event for conservation of energy in chloroplasts.

Lynn and Straub (1969) extended earlier concepts and verified that treatment of chloroplasts with trypsin activates a light-requiring ATPase whose properties are strikingly similar to those of the light-requiring ATP kinase of chloroplasts. Their observations suggest that there is in chloroplasts a single reducible enzyme which, in its reduced state, catalyzes the reversible reaction.

$$HPO_4^{2-} + ADP^{3-} + H^+ \rightleftharpoons ATP^{4-} + H_2O \tag{3}$$

By reduction and protonation of the catalytic site of this enzyme, light-induced electron transport would drive the reaction to the right. In the absence of light, in the presence of "uncouplers," or in the presence of certain electron acceptors which oxidize the internal electron acceptors, concentration of internal H^+ decreases, the catalytic site is neither reduced nor protonated and there is ATP hydrolysis. Activation of the enzyme requires light; it remains only active in the presence of ATP. Hydrolysis of all the ATP results in the inactivation of the ATPase. Lynn and Straub (1969) suggested that imidazole groups of the ATPase probably function in ATP synthesis and hydrolysis and may also undergo oxidoreduction during electron transport, being possibly involved in the protonation and dehydration of Pi. The possible participation of imidazole groups in membrane protonation had been previously suggested by Murakami and Packer (1970a).

According to Farron (1970), the terminal phosphorylation reaction in chloroplasts is catalyzed by CF_1, a latent, Ca^{2+}-dependent ATPase, which is a membrane-bound, high molecular weight protein having a multiple subunit structure (see Chapter 9 of this volume). Since previous work has well established the importance of either a proton gradient or membrane potential in the phosphorylating capacity of chloroplasts (cf. Walker and Crofts, 1970), what remains unclear is how energy conserved in this way is linked to anhydro bond formation. Boyer (1965) had suggested that the energy coupling in oxidative and photosynthetic phosphorylation may be mediated by conformational changes in proteins. Ryrie and Jagendorf (1971, 1972) tried to detect such conformational changes in CF_1 associated with phosphorylation, with the use of the method of hydrogen exchange with solvent water. A small amount of ^3H incorporated from a labeled water medium into CF_1 has been interpreted as indicating a conformational change in the protein. The incorporated ^3H represented at least 30 exchanged hydrogens per molecule of CF_1 and the "exchange-in" reactions of these hydrogens are presumed to occur only during the 40–60 sec of illumination, so that ^3H incorporation into this protein occurred in light but not in darkness. Moreover, this ^3H incorporation was markedly inhibited by NH_4Cl and CCCP. The authors considered that their results provided the first direct evidence suggesting the existence of conformational changes in a protein functioning in the phosphorylation reaction in chloroplasts. The ^3H labeling of CF_1 would reflect an energized state of the molecule itself, able to drive the phosphorylation of

ADP. Protonation would play a very important role in this process, since protein-bound hydrogens that could exchange in less than 1 min include the "instantaneously" exchanging hydrogens and those from peptide bonds in nonhelical regions of the chain.

Recently, Roy and Moudrianakis (1971a,b) studied the *in vitro* interaction of adenine nucleotides with CF_1 from spinach chloroplasts, and provided evidence that CF_1 forms a tight complex with ADP. Bound ADP undergoes a transphosphorylation reaction to give ATP and AMP. When spinach chloroplast thylakoids are illuminated in the presence of an electron acceptor and [^{3}H]AMP or [^{32}P]Pi, the enzyme can be recovered as a tight complex with [^{3}H]ADP or [^{32}P]ADP. The enzyme-bound ADP might serve as the phosphate donor to substrate ADP for phosphorylation. These findings indicate that the conformational state of the coupling factor protein (ATPase) which is affected by ADP binding and protonation is intimately correlated with the energy-transducing mechanism of photophosphorylation.

3.3. Probes of Conformational States of Thylakoid Membranes

As mentioned at the beginning of this chapter, a multitude of new methods has been used to derive information on structural modifications that occur in illuminated chloroplasts. Although in certain cases such evidence is still preliminary, it is now possible to build a comprehensive picture of the events related to conformational and configurational changes.

3.3.1. Light Scattering and Transmission

Light scattering at an angle of 90° and transmission (low-angle scattering) are very useful parameters to follow light-induced structural changes in chloroplasts, though of a complex nature as responses are included to: (1) structural changes of the membrane (conformational level); (2) volume changes; (3) shape changes; (4) heterogeneity; (5) changes in the refractive index of internal structure.

As previously described, increases in scattering occur both when the chloroplasts swell upon illumination or when they contract, as in the case when they are suspended in either NaCl or sodium acetate solutions. Thus, while transmission and light-scattering measurements are valuable as a means of assessing dynamic changes in chloroplast volume, only transmission changes are directly correlated with volume changes. Therefore, both parameters should be monitored simultaneously to provide more precise information about configurational and conformational changes of chloroplasts (Packer and Murakami, 1972).

Light scattering at 90° and transmission (low-angle scattering) are proportional to pigment content if chloroplasts contain less than 15 μg Chl/ml in-

cubation mixture. Above this concentration such proportionality disappears
and it is assumed that multiscattering of light occurs.

Simultaneous recording of light scattering and transmission changes of
chloroplasts incubated in sodium acetate solutions in the presence of PMS,
conditions under which shrinkage occurs upon illumination, leads to the
conclusion that the kinetics of light scattering and transmission changes are
closely correlated. Increases in light scattering and decreases in transmission
parallel one another in time and, moreover, when the light is extinguished,
the relaxation or reversal of the changes to initial conditions are likewise well
correlated, thus showing a general correspondence between changes in both
parameters when chloroplasts undergo extensive photoshrinkage *in vitro*.

However, when chloroplasts are suspended in a NaCl solution containing
small quantities of PMS and PMA, extensive contraction is observed upon
continuous illumination that can be reversed by adding a stoichiometric
quantity of a dithio reagent, such as DTE, although discrepancies in the
kinetics of light scattering and transmission changes are then found. While
there is a rapid decay in the light scattering, transmission changes show a biphasic response—there is an initial rapid increase of transmission which does
correlate with the light scattering change, followed by a slower and much

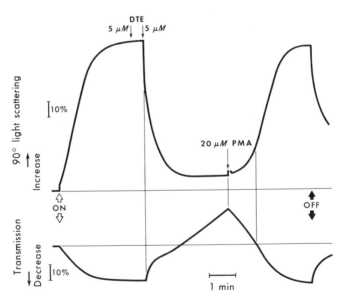

Fig. 4. Effect of DTE on light-induced 90° light scattering and transmission changes of
chloroplasts. Conditions: NaCl–PMA medium; at the points indicated, 10 μM DTE (dithio
erythritol) and 20 μM PMA (phenylmercuric acetate) were added to the incubation medium
during illumination. After Murakami and Packer (1970b).

larger increase in transmission. Upon adding a quantity of PMA larger than the amount of DTE present, the initial kinetics of light scattering and transmission changes show biphasic responses (Fig. 4). It has been concluded that the initial responses of light scattering and transmission changes under these conditions reflect the changes in the internal structure of the membranes themselves while the secondary larger, but slower, changes reflect changes in the spacing of thylakoid membranes and in the volume of the chloroplast as a whole (Murakami and Packer, 1970b). This provides further evidence that, in class II chloroplasts, transmission changes reflect changes in the chloroplast volume, whereas light scattering changes represent at first a change in inner membrane structure, which is subsequently masked by larger and slower volume changes reflecting the flattening of the thylakoid membrane. Therefore, light scattering changes seem to reflect more closely a structural, physical state within the membrane while changes in transmission reflects volume and/or shape changes. For theoretical predictions of the effects of particle volume changes in light scattering at various angles, see Latimer and Pyle (1972).

3.3.2. Electron Microscopy

It is reasonable to assume that many changes in cell structure which greatly affect the photometric properties of organelles might be induced upon illumination. Among these are changes in (1) location in the cell, (2) configuration (or shape), (3) volume (or size), and (4) internal membrane structures (conformation) of plastids or mitochondria, or both. The structural states of these organelles can be examined by electron microscopy and the results correlated with the corresponding metabolic states that are regulated by light conditions.

The membrane system of chloroplasts *in vitro* (Deamer *et al.*, 1967; Dilley *et al.*, 1967; Murakami and Packer, 1969, 1970b) and *in vivo* (Kushida *et al.*, 1964; Packer *et al.*, 1967; Nobel *et al.*, 1969; Murakami and Packer, 1970a; Miller and Nobel, 1972) becomes flattened (photoshrinkage) upon illumination, which is accompanied by changes in the internal structure of thylakoid membranes.

Murakami and Packer (1970b) utilized electron microscopy to detect reversible decrease in the thickness and spacing of thylakoid membranes occurring in class II chloroplasts upon illumination. Thickness and spacing of membranes can be estimated by two methods: (1) direct measurement on enlarged photographs at higher magnification, and (2) microdensitometry tracing of electron microscope negatives.

In order to obtain measurements that faithfully represent the real dimensions, certain precautions have to be taken. The grana portion of the membrane system is selected for measurement, since the grana membrane is extremely flat and the grana region is stacked with membranes at constant spacing. The grana region should be oriented perpendicularly to the direction of the electron beam. This orientation is crucial for examing the tightly packed grana structure, because if the orientation of the membranes within the section tilts by more than a few degrees in relation to the direction of the electron beam, it is difficult to clearly resolve individual membranes.

Densitometric curves can be obtained by tracings on electron microscope negatives at high magnifications by microbeam or a double-beam micro-densitometer. The size of the microbeam used has to be small enough to resolve the thickness and spacing of the membranes even when they are tightly packed. Analysis of samples by this method reveals a decreased thickness of the membranes upon illumination.

The presence of repetitive structures in grana affords a means of obtaining accurate information on dimensions if the profile of the membranes on an electron micrograph is in very sharp focus throughout the entire grana region. When, however, the dimensions of single intergranal membranes are considered, the exact orientation of membranes within the section are more difficult to ascertain. Therefore, numerous photodensitometric traces of single thylakoid membranes in the intergrana region have to be made to be able to conclude that all thylakoid membranes within the chloroplasts undergo the same changes upon illumination.

To investigate whether the kinetics of light scattering and transmission changes can be correlated with thylakoid thickness and spacing changes, Murakami and Packer (1970b) used glutaraldehyde fixation to rapidly stop structural modification during the actual course of the experiments. Their results show that the thickness changes correlate more closely with the kinetics of light-scattering responses and that the spacing changes are more closely monitored by the kinetics of transmission changes. As previously mentioned, chloroplasts may retain proton uptake activity upon illumination after being submitted to glutaraldehyde fixation under carefully controlled conditions (West and Packer, 1970; Mohanty *et al.*, 1973; Torres-Pereira and Packer, unpublished results), so that glutaraldehyde treatment is a very useful technique to stabilize structure, while some of the activity is retained.

Electron transport-dependent changes in the structure of thylakoid membranes occur in chloroplasts *in vivo* of *Ulva* and *Porphyra* thalli (Murakami and Packer, 1970a). These ultrastructural changes have been described earlier in this chapter and the experimental evidence thus obtained is in total agreement with previous results *in vitro*.

3.3.3. Extrinsic Fluorescence Probes

Extrinsic fluorescence probes have been used effectively to monitor structural states of membranes and molecules in relation to energy states. For literature on this subject, see reviews by Radda (1971), Brandt and Gohlke

(1972), and Weber (1972). Among the factors affecting fluorescence intensity are: (1) changes in the number of binding sites; (2) changes within the hydrophobic environment where the fluorochrome is embedded; and (3) energy transfer between extrinsic fluorescent probes and intrinsic fluorochromes such as tryptophan and chlorophyll. It is likely that both (1) and (2) are influential in altering fluorescence intensity in chloroplast thylakoid membranes. On the other hand, (3) has to be considered as an extremely critical factor for Chl-containing membranes. For example, there is some overlap between the emission band of ANS fluorescence and Chl absorption. It is clear that energy transfer between extrinsic and intrinsic fluorescence probes makes the situation somewhat complicated. However, it is possible to exclude this difficulty by using glutaraldehyde-fixed chloroplasts as a control, since these do not undergo structural changes (Murakami and Packer, 1970b). Alternatively, fluorochromes whose excitation and emission do not overlap with Chl absorption can be used. ANS is a very suitable probe to perform dark studies of pH and/or cation-induced conformational changes of the thylakoid membranes.

Ultraviolet-excited fluorescence of ANS increases when ANS molecules become associated with hydrophobic environments. This occurs when ANS binds to biological membranes or is placed in solutions of low dielectric constant (Stryer, 1965). When ANS is added to class II chloroplasts, the fluorescence increases by accompanying binding to the thylakoid membranes and is further enhanced upon illumination (Fig. 5). Class II chloroplasts treated with glutaraldehyde are no longer capable of undergoing changes in

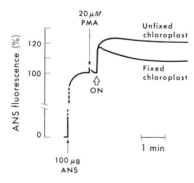

Fig. 5. Light-induced ANS (8-anilinonaphthalene-1-sulfonic acid) fluorescence changes of chloroplasts. 100 μg ANS was added to the chloroplasts equivalent to 14 μg Chl/ml in 50 mM Tris-HCL, pH 8.0, containing 175 mM NaCl and 15 μM PMS (phenazine methosulfate). When fluorescence yield in the dark reached the maximum level (100%) 20 μM PMA was added and then chloroplasts were illuminated by red light. For the experiment with fixed chloroplasts, chloroplasts were fixed with 0.1 M glutaraldehyde in the dark, thoroughly washed, and incubated. After Murakami and Packer (1970b).

volume, as measured by light scattering. Nevertheless, it was found that the light-dependent increases in the ANS fluorescence still occurs. This finding suggests that changes in microstructure of the membrane resist chemical fixation sufficient to prevent gross morphological changes and correlates well with the previously reported data on light-induced proton uptake in glutaraldehyde-fixed chloroplasts. The difference in ANS fluorescence observed between fixed and unfixed chloroplasts probably represents changes in binding sites or occupancy of ANS in the thylakoid membranes as a consequence of illumination. The timecourse of ANS fluorescence changes induced by illumination shows a rapid phase followed by a slow one. The rapid fluorescence change is not affected by glutaraldehyde fixation, but the slow phase is eliminated (Murakami and Packer, 1970b).

Light-induced ANS fluorescence increases in illuminated chloroplasts in the presence of weak acid anions (sodium acetate) or strongly dissociated ions (NaCl–PMA) show similar kinetics (Fig. 6). In both systems the speed of the ANS fluorescence changes upon illumination are faster than the corresponding light-scattering changes. Furthermore, the rate of the ANS fluorescence responses is similar for the two conditions, but the light-scattering kinetics

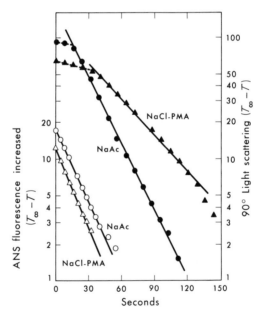

Fig. 6. Kinetics of light-induced 90° light scattering and ANS fluorescence changes in chloroplasts. Chloroplasts equivalent to 14 μg Chl/ml were suspended in either a NaCl-PMA or sodium acetate medium. Light scattering and ANS fluorescence increase upon illumination were expressed as the difference between the value at steady state and that at various times on a semi-log scale. After Murakami and Packer (1970b).

differ. The half-times for ANS fluorescence changes are 14 sec in the NaCl–PMA system and 15 sec in the sodium acetate system. These rates are considerably faster than light-scattering changes, where the half-times are 60 and 33 sec, respectively (Murakami and Packer, 1970b).

Faster and similar kinetics of ANS fluorescence in these two different incubation media are interpreted as indicating that structural changes at the conformational level occur more rapidly than can be monitored by light-scattering responses. It is likely that ANS fluorescence responses express more primary conformational changes in the thylakoid membranes. Hence, it is of considerable interest to understand how ANS fluorescence changes are related to structural changes in the thylakoid membrane. Since it was known that illumination of chloroplasts leads to an uptake of protons, Murakami and Packer (1970b) studied the effect of protonation of chloroplasts on ANS fluorescence (Fig. 7). Using pH titrations of the ANS fluorescence and light-scattering intensity changes these authors verified that the fluorescence change

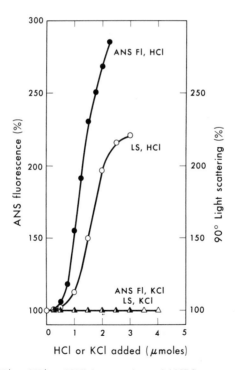

Fig. 7. Effect of H$^+$ and K$^+$ on 90° light scattering and ANS fluorescence intensity of chloroplasts. Chloroplasts were incubated in 100 mM sodium chloride, and 90° light scattering and ANS fluorescence were titrated with 0.05 N HCl or 0.05 N KCl. (After Murakami and Packer, 1970b.)

induced by protonation of the thylakoid membrane is more sensitive to pH change than light scattering. When a similar titration was done with K^+ (as KCl) no or a very weak increase in both ANS fluorescence or light scattering were detected. Similar negative results are obtained with other monovalent ions, but with divalent cations such as Ca^{2+} and Mg^{2+} rather marked increases can be observed.

It seems reasonable to advance the following explanations for ANS fluorescence enhancement: (1) Protonation of negatively charged groups on the thylakoid membranes results in an increase in the number of ANS binding sites. (2) Protonation would lead to an expulsion of attractive water dipoles adjacent to negatively charged sites, making the molecular environment more hydrophobic. (3) This would increase the attractive forces between the molecules, resulting in ANS fluorescence enhancement and decrease in membrane thickness as a consequence of the closer apposition of structural units within the membrane.

Auramine is a basic fluorochrome that binds strongly with bacterial chromatophore membranes and can be used as a fluorescence probe of photosynthetic membranes, since its fluorescence yield is very low in water, but markedly enhanced after binding to the membranes. Nishimura (1971) demonstrated a light-induced fluorescence intensity change of auramine O bound with chromatophores of *Rhodospirillum rubrum*. Kobayashi and Nishimura (1972) elegantly extended this work. Since auramine O fluorescence intensity decreases in the acid pH range and after addition of divalent cations, these authors concluded that the fluorochrome probably becomes attached to hydrophobic regions near the negatively charged groups on the membranes by electrostatic attraction. At low pH the negatively charged groups on the chromatophore membranes are protonated and hence binding of auramine O to the membranes is reduced, resulting in a decreased fluorescence intensity. Binding of cations has the same effect. These results and interpretations are in agreement with the work of Murakami and Packer (1970b) with ANS.

The auramine O fluorescence intensity change appears to be related with the energy state of the membrane which is controlled by ion translocation and phosphorylation. Auramine O in aqueous solution has a weak fluorescence and its emission maximum is about 510 nm. When this fluorochrome binds to *R. rubrum* chromatophores, the emission maximum shifts to a shorter wavelength (495 nm), with a much increased quantum yield of fluorescence. Fluorescence of auramine O bound to chromatophores changes reversibly by illumination of chromatophores with near-infrared light of wavelengths longer than 720 nm. The fluorescence intensity decreases rapidly at the start of illumination (rapid phase), and then slowly during illumination (slow phase). After the light is turned off, the fluorescence intensity increases gradually and reaches the preillumination level. The reversible fluorescence change can be repeated many times.

The rate and extent of light-induced proton transport and the extent of the light-induced decrease in auramine O fluorescence (rapid phase) is accelerated by addition of tetraphenyl boron or valinomycin plus KCl, suggesting that the rapid phase is associated intimately with light-induced proton uptake in the presence of K^+. However, the rapid phase is not a direct manifestation of the proton gradient formation itself, though it is a reflection of the physical state of the membranes caused by proton translocation, since NH_4Cl or valinomycin plus NH_4Cl, as compared with KCl or NaCl, moderately diminish the rapid phase of light-induced fluorescence change of bound auramine O. The redox level of the chromatophores is one of the factors which control the structural state monitored by light-induced fluorescence (Kobayashi and Nishimura, 1972). When the chromatophores are treated with ferricyanide, a mild oxidant which inhibits cyclic electron flow and other physiological activities of the chromatophores, the rapid phase is completely abolished. The inhibition is reversed by addition of a reductant such as ascorbate or dithiothreitol. The slow phase is resistant to ferricyanide treatment. The primary factor which determines the nature of the light-induced fluorescence change may be the change in the physical state of chromatophore membranes induced by protonation, which decreases the number of binding sites. However, the possibility of a change in hydrophobicity in the vicinity of the binding sites caused by protonation cannot be excluded. Abolishment of the rapid phase by ferricyanide treatment and restoration of the same phase by reductants indicate that the rapid phase is closely associated with the redox level of chromatophore membranes.

It would be pertinent at this stage to report some information on fluorescent and nonfluorescent indicators of pH changes in the outer and inner phases of biological membranes.

Chance and Mela (1966) considered that bromothymol blue, well known to bind tightly to a number of proteins and to indicate pH changes of opposite sign from those recorded by the glass electrode, would presumably indicate H^+ concentration changes on the "inside" of the vesicular membrane which separates the outside and the inside of mitochondrial membrane phases. The authors considered that bromothymol blue would bind to the hydrophobic portion of many types of membranes.

Jackson and Crofts (1969) found that *R. rubrum* chromatophores in suspension bind about 80 % of bromothymol blue and about 20 % of bromocresol purple in solution at low concentration. Binding of bromothymol blue is associated with a shift in pK_a of the indicator from 7.1 to 7.9, and with a decrease in extinction coefficient. No change of pK_a or extinction is found for bromocresol purple in the presence of chromatophores. On illumination of a suspension of chromatophores containing bromothymol blue, the indicator shows an absorbance change opposite in signal to that expected from the pH change in the medium. The absorbance change of bromocresol purple on illumination is indicative of the pH in the external medium. Increasing the buffering power of a medium containing chromatophores and bromothymol blue or bromocresol purple increases the response of bromothymol blue on illumination, but eliminates the response of bromocresol purple. The authors suggested that only a portion of the bromothymol blue present responds to external pH, while virtually all the bromocresol purple responds to changes in the external pH. It was concluded that, while portions of the bromothymol blue present in a suspension of chromatophores respond both to external and internal pH changes, the major proportion of the bromothymol blue change was a slow response to the energetic state reflecting neither internal nor external pH, nor kinetically, the onset of the energetic state. On the other hand, the absorbance change of bromocresol purple, at least during the first seconds of illumination, closely followed the pH change of the external medium. Therefore, it appears that bromocresol purple is not bound to the membrane, and can be used to measure H^+ changes in the medium. On the other hand, bromothymol blue does not seem to be in direct contact with the photosynthetic membrane because its indication is delayed (rise time, 70 msec). Therefore, it should not be used for kinetic studies, but is useful for measuring the relative pH (H^+) yield.

Using neutral red as a membrane-bound indicator of internal pH, Lynn (1968a) has demonstrated that the internal space of the thylakoids becomes more acidic upon illumination. Cations such as NH_4^+, PMS, and protonated neutral red, and also diffusible anions such as acetate and orthophosphate, which compete with H^+ ions for entry into thylakoid inner space, inhibit the buildup of pH gradients between chloroplasts and their external media.

Grünhagen and Witt (1970) used fluorescence changes of umbelliferone (7-hydroxycoumarin) as an indicator of pH changes in the outer phase of the thylakoid membrane. The excitation light has a λ_{max} around 366 nm, and the emission light has a λ_{max} close to 453 nm. Umbelliferone is a very sensitive indicator since the rise of the primary pH formation takes place in 20–30 msec, i.e., more rapidly than in the case of bromothymol blue. Moreover, this indicator is formed as a natural compound in many plants and concentrations of up to 5×10^{-5} M have no influence on the photosynthetic reactions, since the rates of electron transfer and ATP formation are not affected. Nor is any influence observed on the amplitude and on the time course of the absorption changes of Chl a_1 at 705 nm (see Chapter 10 of this volume) and on the amplitude (proportional to the field strength) and kinetics (proportional to ion fluxes) of the field indicating absorption changes (e.g., at 515 nm). The pK_a value of umbelliferone is ~ 8 and the indicator only indicates pH changes in the outer phase of functional membranes. The intensity of the fluorescence is pH-dependent, increasing between pH 5 and pH 10. These qualities make umbelliferone a suitable indicator for analytic measurements *in vitro*.

A possible interaction between the uncoupler atebrin and the "energized state" of chloroplasts was investigated by Kraayenhof (1970, 1971), who reported the existence of a stoichiometric relationship between the energy generation in spinach chloroplasts and the fluorescence quenching of atebrin. When atebrin is added to a chloroplast suspension, there is a small initial quenching of the fluorescence of atebrin (due to a little binding). Energization (either by light, ATP hydrolysis, or pH-jump) leads to a rapid quenching of the remaining fluorescence (and binding). It was assumed that the quenching reflects a binding of the uncoupler to certain structures related to the energy-conservation mechanism, or a change of the environment in which the uncoupler is situated. However, Schuldiner and Avron (1971) demonstrated that upon illumination atebrin was taken into the chloroplasts and that its distribution was determined by the H^+ gradient across the membrane. Thus, the fluorescent quenching of atebrin was useful to follow proton gradients generated across the chloroplasts, but would not have quantitative relation with the energized state. The quenching of the fluorescence of the atebrin molecules which have moved into the chloroplasts could be due to the low pH inside the chloroplast, where there would exist a highly quenching environment due to energy transfer to Chl or other components. It could also be attributed to the simple screening effect whereby the membrane-bound Chl would absorb a portion of the incoming exciting ultraviolet light before it reaches the atebrin molecules. Kraayenhoff *et al.* (1971) presented further evidence to corroborate that atebrin quenching is an energy-linked process, but admitted that more experiments were necessary to clarify this subject. Bazzaz and Govindjee (unpublished), using bundle sheath and mesophyll chloroplasts of maize, have shown that light-induced quenching of atebrin fluorescence cannot be correlated with the extent of proton uptake, but it can be with the ability to make ATP.

3.3.4. Chlorophyll Fluorescence

The fluorescence of Chl depends not only upon the redox state of the electron transport components but also upon the distance and orientation of the pigment molecules (see Chapter 6 of this volume). Hence, Chl fluorescence must be a good intrinsic marker of the structural state of the thylakoid membranes, since the properties of fluorescence emission directly reflect the state of the environment in which Chl molecules are embedded.

Papageorgiou and Govindjee (1968a,b) and Munday and Govindjee (1969a,b) studied the intensity dependence and the spectral changes during the rapid (sec) and the slow (min) transient of Chl *a* fluorescence yield, in

Anacystis, Chlorella, and *Porphyridium.* (The terms fluorescence induction, transient and change are used interchangeably by the Govindjee group.) Light-induced changes in the yield of Chl *a* fluorescence in intact green algae have charasteristic points in the fast induction: a low initial level *O*, hump *I*, followed by a dip *D*, a peak *P*, and a temporary steady-state low level *S* (Fig. 8). Characteristic points in the slow phase are: the low level *S*, followed by a rise to a maximum *M*, and then a final decline to a terminal steady state level *T* (see review by Govindjee and Papageorgiou, 1971). During the *SM* phase (slow induction) there is an increase in the rate of electron transport and O_2 evolution which reaches a constant level during the *MT* phase. In the blue green algae, there is a very slow decline from *M* to *T*. Poisons of electron transport like DCMU eliminate most of the slow changes in *Chlorella* and in *Porphyridium.* DCMU-poisoned cells retain only the fast *O–I* phase. However, blue green algae like *Anacystis* show prolonged induction of fluorescence even in the presence of DCMU and *o*-phenanthroline. This process may be related to the operation of cyclic electron transport of PS I *in vivo* and, perhaps, also the DCMU-resistant photodepression of O_2 uptake (Mohanty *et al.*, 1971b). It seems that the slow fluorescence changes in the case of *Porphyridium* and *Chlorella* need both cyclic and noncyclic electron flow. In the case of *Porphyridium*, unlike blue green algae, there is a limited extent of cyclic electron flow.

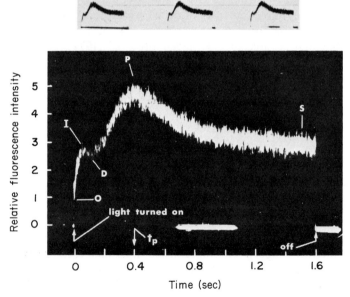

Fig. 8. The fluorescence transient in aerobic *Chlorella pyrenoidosa.* Wavelength of measurement: 685 nm; half-maximum bandwidth (BW) 6.6 nm. Excitation: 500 nm, BW 120 nm, incident intensity 1.5×10^4 ergs/sec-cm². The three small photographs were obtained in one experiment at 4 min dark intervals; the large photograph is an enlargement of the top right-hand photograph. The points *O*, *I*, *D*, *P*, and *S* are clear in each photograph. After Munday and Govindjee (1969a).

Uncouplers of photophosphorylation such as FCCP and atebrin abolish portions of the slow fluorescence changes in *Chlorella*, *Anacystis*, and *Porphyridium*. FCCP, even at low concentrations, slows down the *P* to *S* decline phase and supresses the slow *S* to *M* phase. The clear effect of uncouplers on the fluorescence yield, without significant change in the rate of oxygen evolution, suggests a relation between slow fluorescence change and phosphorylation. Higher uncoupler concentrations completely abolish the slow fluorescence transient. *Porphyridium* cells fixed with glutaraldehyde show no light scattering and no volume changes, nor ion transport, and do exhibit a response in the fast induction but not in the slow fluorescence change. These observations strongly suggest that electron transport per se contributes very little to the slow fluorescence changes in intact cells; the latter seem to be more intimately related to the energy coupling process than electron flow during photosynthesis (Mohanty *et al.*, 1971b; Mohanty and Govindjee, 1973). These authors believe that slow fluorescence induction is directly related to the energy-dependent structural changes of the thylakoid membranes. Other phenomena such as proton movement and phosphorylation would affect fluorescence via these structural changes. Although mentioning that West and Packer (1970) have shown that, upon illumination, glutaraldehyde-fixed chloroplasts can take up protons at a reduced rate, the authors consider that if fixed cells behave as fixed chloroplasts, it follows that proton uptake is not solely responsible for PSMT fluorescence yield transient in whole cells. This conclusion was recently confirmed for fluorescence changes induced by PMS in class II chloroplasts by Mohanty *et al.* (1973).

A light-induced reversible decrease in fluorescence yield of Chl *a*, designated as "fluorescence lowering," is observed upon illumination of isolated chloroplasts at high light intensities in the presence of PMS and DCMU, the latter being added to exclude the influence of photoreaction II on the yield of Chl *a* fluorescence (Govindjee and Yang cited in Govindjee *et al.*, 1967; Murata and Sugahara, 1969). The decrease in fluorescence yield amounts to 20–30%. There is involvement of the photophosphorylation system in the "fluorescence lowering" since this phenomenon is markedly suppressed by addition of uncouplers of photophosphorylation. Kinetic analyses showed that the "fluorescence lowering" commenced at about 0.2 sec and reached a half-level about 2 sec after the onset of illumination. The half-decay time of the recovery after cessation of illumination was approximately 5 sec (also see Mohanty *et al.*, 1973). A comparison of the timecourse of fluorescence and light-scattering changes disclosed that the light-scattering change is much slower than the change in fluorescence. The change in state of Chl *a* in the chloroplasts which may form the basis of "fluorescence lowering" may be different from the conformational changes responsible for the light-scattering change of chloroplasts, since the time responses of these two kinds of changes differed from each other

(Murata and Sugahara, 1969). However, the mechanisms of "fluorescence lowering" is still obscure, although these authors suggest that the high-energy state of the intermediate of photophosphorylation could play an important role.

The above-mentioned "fluorescence lowering" is also observed in intact cells in the presence of DCMU and PMS. Although a similar change in fluorescence yield may be operating in the absence of PMS in the intact cells of *Porphyra yezoensis*, its contribution seems small. The slow change in fluorescence yield which cannot be explained by the competition between fluorescence emission and photoreaction II is attributed to the change in distribution of excitation energy between the two pigment systems (Murata, 1970).

It has been established that cations affect the yield and properties of Chl *a* fluorescence in various photosynthetic organisms. This was previously observed with intact algal cells such as *Rhodosorus* (Giraud, 1963) and in *Euglena* chloroplast fragments (Brody, *et al.*, 1966). Subsequently, Murata (1969, 1971a,b) and Murata *et al.* (1970) extensively studied the behavior of Chl fluorescence in isolated spinach chloroplasts under different ionic conditions. Homann (1969) observed independently that a divalent cation (3.3 mM $MgCl_2$) enhanced the intensity of Chl fluorescence in the presence and absence of DCMU. All these authors assumed that cations affect Chl fluorescence by modifying the structural and energy states of photosynthetic membranes. Murata (1971b) further investigated the correlation of membrane structure with regulation of excitation transfer (spillover) between the two pigment systems in isolated spinach chloroplasts. Relatively low concentrations (3 mM) of divalent cations such as Mg^{2+}, Ca^{2+} and Mn^{2+} increased the intensities of both fluorescence yield at 684 nm and light scattering. The changes showed similar timecourses and concentrations dependence. Added Mg^{2+} also decreased the quantum yield of the reactions in PS I (DCPIPH$_2$ $\longrightarrow$ NADP), while increasing the quantum yield of PS I plus II (H$_2$O $\longrightarrow$ NADP). These facts suggest that there must be a close correlation between the regulation of excitation transfer between the two pigment systems and the membrane structure of chloroplasts. The excitation transfer between the two pigment systems depends upon the distance between them and upon their mutual orientation. Since the units of the two pigment systems lie on the thylakoid membrane, the supression of excitation transfer from PS II to PS I following the addition of cations suggests that there is a cation-induced conformational change of the thylakoid membranes, pulling apart the units of the two pigment systems. These concepts were discussed by several authors, namely Govindjee and Papageorgiou (1971), Myers (1971) and Duysens (1972).

An alternative interpretation was proposed by Sun and Sauer (1972).

Added divalent cations support the excitation transfer (spillover) from PS I to PS II, and might serve to increase significantly the quantum yield of the electron flow from H_2O to NADP in broken spinach chloroplasts. The optimum concentration of $MgCl_2$ or $MnCl_2$ was about 7.5 mM. Addition of salts of monovalent cations, such as NaCl, gave similar results, although at a much higher concentration. Assuming that the excitation transfer between the two pigment systems tends to equalize the rates of the two photoreactions, the increased quantum yield for the PS I plus PS II reaction can be explained if a spillover occurs more readily in the presence of divalent cations than in their absence. Spillover from PS I to PS II in the presence of divalent cations can account for (1) increased fluorescence yield; (2) increased quantum yield for the PS II reaction; and (3) decreased quantum yield for the PS I reaction. As was previously mentioned, divalent cations induce contraction of thylakoid membranes (Gross and Packer, 1967; Shavit and Avron, 1967; Murakami and Packer, 1971), which may result in a tighter packing of the active components of the two pigment systems. Thus, it is very likely that mono- and divalent cations may alter the microstructure at the inside of the thylakoid membranes, in a manner which would control the transfer of electronic excitation between the two pigment systems. Hydrophobic association is strengthened in the presence of electrolytes which decrease the water solubility of nonpolar regions of the membrane, resulting in a stabilization of the association between membranes. Thus cations appear to act by influencing hydrophobic interactions between thylakoid membranes. (Murakami and Packer, 1971). However, according to Mohanty and Govindjee (personal communication), there is the possibility that the observed tighter packing of membranes may still lead to the separation of PS I and II, leading to the decrease in spillover from PS II to PS I. This needs further experimentation.

3.3.5. Delayed Light Emission

Since the discovery of delayed light emission (delayed fluorescence, luminescence) in green plants by Strehler and Arnold (1951), the phenomenon has been detected in chloroplasts of various organisms, namely unicellular and thallophytic algae, higher green plants and in bacterial chromatophores (see Chapter 5).

Delayed light emission has been studied in relation to the primary events in photosynthesis and it is generally accepted that the delayed light emitted by chloroplasts originates mostly from the actively functioning PS II. The regeneration of excited Chl would be caused by a back reation of photoreaction II, namely, the oxidation of the reduced primary electron acceptor Q^- by the oxidized primary electron donor Y^+ that is produced by PS II during illumination (see Eq. 4).

$$\begin{array}{c} \xrightarrow{h\nu} \qquad\qquad e^- \qquad\qquad e^- \\ \boxed{\begin{array}{ccc} Q \cdot Chl \cdot Z & \xrightarrow{k_1} (Q^- \cdot Chl \cdot Z^+) \xrightarrow{k_2} (Q^- \cdot Chl \cdot Z) \\ & \text{Delayed} \\ & \text{light} \\ & \text{emission} \end{array}} \end{array} \qquad (4)$$

The intensity of delayed fluorescence emission depends on $(Q^- \cdot Chl \cdot Z^+)$, which is controlled by k_1 and k_2. Hence the amount of $(Q^- \cdot Chl \cdot Z^+)$ is regulated by redox state, O_2 evolving system and electron flow, which are related to phosphorylation coupling.

The spectrum of delayed light emission is identical to that of prompt fluorescence, and in algae the intensity of delayed fluorescence is proportional to the yield of prompt fluorescence. This indicates that the regenerated excited state of Chl is the same as the singlet state formed initially during light absorption (Clayton, 1969). Since prompt and delayed fluorescence both reflect the presence of singlet excitation quanta in the Chl associated with PS II, any factor (such as the state of the traps) that modifies the yield of the prompt fluorescence should have a corresponding effect on the intensity of delayed fluorescence (Lavorel, 1968). The delayed fluorescence of Chl in green plants is a sign that light energy has been stored, and then excitation quanta have been regenerated from this stored energy. The mechanism depends on the functioning of PS II. Therefore, it is likely that the regenerated quanta are formed in the component of Chl that communicates with the PS II traps.

The mechanism of delayed and stimulated delayed fluorescence of chloroplasts has been discussed by Kraan et al. (1970) in terms of a membrane potential and a proton gradient. Simultaneous measurements were made of prompt and delayed fluorescence of spinach chloroplasts upon acidification followed by a sudden jump of pH toward a higher value and upon a sudden increase in the ionic strength of the medium. It was found that the stimulation of delayed fluorescence caused by the above treatments is not paralleled by a significant increase of prompt fluorescence. This indicates that this stimulation is due to an increase in the rate of the back-reaction of primary photoproducts of photoreaction II which is not caused by an increase in concentration of reduced electron acceptor brought about by "reserved" electron flow. Experiments with mixtures of salts supported the hypothesis that the stimulation of delayed fluorescence upon addition of salt is due to the establishment of a diffusion potential which is positive on the inside with respect to the outside of the thylakoid membrane and which is caused by different permeabilities of cations and anions.

On the other hand, an increase in permeability to H^+ caused by damage to the thylakoid membrane or by addition of uncouplers of phosphorylation results in a stimulation of luminescence. Salt- and acid–base-induced stimulation of delayed light emission are tentatively explained by a model in which the reduced primary electron acceptor and oxidized donor of PS II are located on the outside and inside of the thylakoid membrane and react in a pH-dependent equilibrium:

$$Q^- + H^+ \rightleftharpoons QH, \text{ and } ZH^+ \rightleftharpoons Z + H^+ \tag{5}$$

Light emission occurs upon back-reaction of Q^- and ZH^+. This reaction is enhanced by a positive membrane potential and a higher proton concentration on the inside with respect to the outside of the thylakoid.

It was found that those treatments of chloroplasts and photosynthetic bacteria (uncouplers of phosphorylation, electron transport inhibitors, or the presence of a phosphate acceptor system), which would cause a decrease in the intermediates of photosynthetic phosphorylation, originate a decrease in the rapid components of delayed-light emission and that those treatments (electron acceptors or energy transfer inhibitors), which would yield an increase in the intermediates of photophosphorylation, cause an increase in delayed-light emission (Mayne, 1967a; Fleischman and Clayton, 1968). When the chloroplast phosphorylating system is intact, i.e., with no uncoupler, the amount of delayed-light emission is dependent on the electron flow associated with the Hill reaction. The addition of uncouplers of photosynthetic phosphorylation increases the rate of electron flow and inhibits delayed-light emission. Therefore, this phenomenon requires both electron flow and a competent coupling mechanism. On the other hand, inhibition of the phosphate acceptor (ATP-forming) system, either by adding inhibitors of this stage or by removing phosphate or ADP, stimulates delayed fluorescence. This suggests that the pool of metastable energy used for delayed-light emission is located in the coupling mechanism (Mayne, 1967a).

The timecourse of changes in delayed-light emission measured at 1.2 msec after each excitation flash during intermittent illumination of spinach chloroplasts shows at least two components (Itoh *et al.*, 1971). One is the initial rapid rise at the onset of illumination, which is suppressed by the inhibitors CMU, DCMU, *o*-phenanthroline, and high concentrations of CCCP. Assuming that the mechanism of delayed light emission is ultimately excited by a reverse reaction of the electron flow through PS II, the initial rapid-rise component should be a direct reflection of the rate of the reverse reaction of photoreaction II, i.e., the rate of recombination of the reduced primary electron acceptor, Q^-, and the oxidized primary electron donor, Z^+. (For detailed hypotheses, see Chapter 5 of this volume.) The other component developing slowly during illumination is selectively and thoroughly eliminated by addition of the uncouplers methylamine and CCCP, or by EDTA washing, sonication and addition of Triton X-100, being completely suppressed by DCMU and o-phenanthroline, which are electron transport inhibitors. This slowly developing component of delayed fluorescence should be related to an accumulation of the "high-energy state" of phosphorylation (Itoh et al., 1971). In view of this hypothesis, the initial rapid-rise component would be controlled by changing the oxidation–reduction levels of the primary electron donor and acceptor (Q and Z) of PS II. Itoh et al. (1971) made an extensive study of the initial rapid-rise component of delayed light emission and verified that electron donors of PS II (ascorbate, hydroquinone, p-phenylenediamine, and $MnCl_2$) decrease the magnitude of the initial rapid-rise component of delayed light without producing any significant change in the timecourse of fluorescence. Treatment of spinach chloroplasts with a high concentration of tris buffer, known to block the electron transport somewhere between photoreaction II and water, enhances the initial rapid-rise component but suppresses the second slow-increase component of delayed light. Addition of the electron donors to the Tris-washed chloroplasts suppresses the enhanced initial rapid-rise component of delayed light emission. The authors considered that these results indicate that the production of delayed light requires the oxidized form of the primary electron donor Z^+, in addition to the reduced form of the primary electron acceptor Q^- of photoreaction II. Similar conclusions were obtained independently by Bennoun (1970) and Mohanty et al. (1971a).

The relation between the "high-energy state" of phosphorylation and delayed fluorescence in spinach chloroplasts has also been studied by Wraight and Crofts (1971). The kinetics of the rise in intensity of light emitted 1 msec after illumination shows two distinct phases, a rapid one which is complete in less than 0.1 sec, and a sigmoidal slow phase with a half-rise time of about 0.1 sec. The maximum intensity and time course of emission vary with electron flow and the degree of coupling. With nigericin and with NH_4Cl, the slow phase of the intensity rise is abolished but the rapid phase is not affected. With valinomycin, the rapid phase is substantially inhibited, but the extent of the slow phase is increased to give the same maximum intensity as observed in the absence of ionophore. With FCCP, or a combination of valinomycin and nigericin, both phases are substantially inhibited. This spectrum of sensitivities has been interpreted as showing that the slow phase depends on the pH component and a part of the rapid phase on the electrical component of the electrochemical H^+ gradient. Addition of DCMU always inhibits emission. However, Mohanty et al. (1972) and Jursinic and Govindjee (1972) among others have observed the emission of DCMU–insensitive "slow" delayed light. The photochemical act of the system was considered by Wraight and Crofts (1971) to occur between acceptor and donor sites on opposite sides of the thylakoid membrane and in equilibrium with pools and midpoint potentials of which are dependent on the pH of the phase. In such a system the energy conserved in the oxidoreductive poise and in the electrochemical H^+ gradient would be additive and both would contribute to a decreased activation energy of delayed fluorescence. By dissipating the H^+ gradient, the activation energy requirement is increased so that the intensity of emission falls.

Murakami (1973) recently studied the effect on delayed-light emission of DNP and valinomycin, which affect H^+ transport and K^+ permeability across the thylakoid membranes, respectively (Karlish *et al.*, 1969), verifying that DNP and valinomycin affect the intensity and time course of the emission in a different manner. Valinomycin alone severely and selectively suppresses the initial rapid rise in the emission, while the slow one is relatively unaffected. Since valinomycin reduces electron transport to some extent and increases the permeability of photosynthetic membranes to K^+ ions, resulting in the K^+ gradient dissipation, the rapid component must be supported by a cation gradient, or a membrane potential which is established and controlled by the photosynthetic condition. On the other hand, the slow component of delayed fluorescence is only slightly reduced by DNP treatment alone, but is completely eliminated when DNP acts together with valinomycin. These results support Wraight and Crofts' (1971) views and more clearly demonstrate that the slow phase depends on the development of a pH gradient across the thylakoid membranes and the rapid phase relates to membrane potential. Moreover, contraction of the thylakoid membranes is also suppressed by the combination of valinomycin and DNP, as revealed by ANS fluorescence and electron microscopy. Thus, it is evident that the energized state and the contracted conformation of the thylakoid membranes are related to the light-induced proton gradient, which is accompanied by the light-dependent cation gradient. This implies that proton and cation gradients complementarily control the structural and energy states of the thylakoid membranes. Therefore, delayed light fluorescence is a good parameter to monitor the redox state of photochemical reactions and also coupling and high-energy states of the energy-transducing systems in thylakoid membranes.

An alternative approach to determine the relation of energy states, as monitored by delayed light emission, and structural states has been employed by Mayne (1967b, 1968). When preilluminated chloroplasts incubated in acid medium are transferred to base in the dark, a high-energy state is created in the dark (acid–base transition), which makes the chloroplasts able to form ATP, to activate ATP hydrolysis or to activate P_i–ATP exchange reaction (Jagendorf and Uribe, 1966; Jagendorf, 1967). Under similar conditions in the dark chloroplasts emit luminescence induced by the acid-base transition, which is inhibited by phosphorylating conditions and reactivated by phosphorylation inhibitors such as EDTA. DCMU and uncouplers also inhibit this luminescence. Luminescence is probably due to the formation of the same high-energy condition as that used in ATP formation. According to Mayne (1967b), this high-energy state would be a proton gradient across the grana membrane.

Miles and Jagendorf (1969) verified that by rapidly transferring chloroplasts from base to strong acid (below pH 3.2) luminescence is induced. Intensity was up to eight times that of acid–base induction. The delayed-light emission is also caused by a transition from a low to a high ionic strength environment. The salt induction of luminescence is in competition with the acid–base luminescence; whichever comes first prevents light emission upon going through the second process. This is probably due to competition in consuming energy stored at a common pool and agrees with the view that proton and cation gradients contribute complementarily to generate high energy states. On the other hand, the acid procedure does not seem to compete with acid–base transition or salt induction, since raising the pH or the salt concentration causes a further amount of luminescence even after the acid-induced burst (also see Malkin and Hardt, 1972). All three procedures share a sensitivity to phosphorylation inhibitors and to DCMU, and all require preillumination of the chloroplasts. Chloroplasts subjected to a low- to high-salt transition do not go into a high-energy state, as judged by the absence of either ATP formation or ATPase activation. Even in the case of acid–base transition the high-energy state appears not to be the "trigger" for light emission, though this treatment may act in some way as a "trigger" to produce recombination of stored electrons and holes. Barber and Kraan (1970) also verified that ionic treatments either with salts or by pH shifts can bring about luminescence from preilluminated chloroplasts. The magnitude and kinetics of the signals strongly suggest that they are controlled by a rate-determining membrane which shows a difference in cation and anion permeability. Whether the mechanism involves the induction of a proton efflux through a luminescence site is not certain. The key event may well be changes caused in membrane structural state by such transition, resulting in lowering the activation barrier for the emission.

The light-induced pH and electrical gradients have been considered to act in such a way as to decrease the activation energy necessary to lift electrons from the metastable state, created during the preillumination, to the first singlet of Chl (Wraight and Crofts, 1971). Based on this hypothesis, Barber and Varley (1971) assumed that the establishment of electrical gradients across thylakoid membranes by some means other than light-induced electron transport should change the intensity of millisecond delayed-light emission and used salt gradients to create membrane potentials. The magnitude of the potentials thus developed is a function of the concentration gradients and the relative rates of the penetration of the ions across the thylakoid membranes. Rapid addition of KCl to give a final concentration of 50 mM induces a transient increase in the intensity of 1.5 msec delayed light from a suspension of chloroplasts. Injection of other potassium salts also gives signals, but their sizes vary with the anion used. Sodium salts give transients comparable with those obtained with the corresponding potassium salts. The addition of $CaCl_2$ causes an inhibition rather than a stimulation of the intensity, possibly because Ca^{2+} is less permeable and probably binds to the membrane. The K^+-induced transients are enhanced after treating the chloroplasts with valinomycin, which has no effect on Na^+-induced signals. The K^+ signals are totally inhibited by DCMU even after valinomycin treatment. These results give strong support to the concept that the establishment of an electrical potential across the thylakoid membranes can increase the intensity of millisecond delayed light. It seems that the transients observed on the addition of potassium and sodium salts give a direct indication of the establishment and decay of a diffusion potential and that K^+ and Na^+ gradients stimulate delayed light by creating a positive potential on the inside of the thylakoids.

Barber and Varley (1972) extended their studies on the stimulation of delayed light emission by salt gradients and estimation of the relative ionic permeabilities of the thylakoid membranes, verifying that potassium salts of strong acids and zwitterions stimulate luminescence by creating a diffusion potential of the correct polarity across the thylakoid membranes. This stimulation of luminescence is, with some salts, due entirely to the electrical potential developed as the ions diffuse across the thylakoid membranes. By varying the magnitude of the concentration gradients and applying the Goldman voltage equation it has been possible to obtain quantitative estimates for the relative ionic permeabilities of the thylakoids. It seems that Cl^- permeates as much as 10 times more slowly than K^+. Salts of weak acids, such as benzoic and acetic acids, seem to have an additional capacity for stimulating luminescence, since it is difficult to explain their action totally by the diffusion potential hypothesis. The most likely explanation is that the undissociated acids in these salt solutions are able to penetrate the thylakoids and increase the internal acid reservoir (Crofts et al., 1967). Thus a change in the internal pH coupled to the establishment of a diffusion potential could be responsible for the particular effectiveness of the weak acid salts. The above reported evidence supports the model that delayed light emission may originate from a charge transfer complex specifically oriented in the thylakoid membrane system.

A quantitative analysis of the salt-induced millisecond delayed light emission in order to obtain a method of estimating the electrical potential created across the thylakoids of illuminated chloroplasts has recently been reported (Barber, 1972a). The establishment of an electrical gradient across the thylakoids of the correct polarity (inside positive) can decrease the activation energy for delayed light and can either result from light-induced charge transfer processes or from a diffusion potential created by a sudden salt addition to the chloroplast suspension. The size of the initial diffusion potential created after, for example, a rapid KCl addition can be estimated, assuming that K^+ and Cl^- are the main diffusion ions, from an equation derived from the Goldman theory. Barber (1972a) arrived at the conclusion that the maximum potential induced by a 50 mM KCl pulse is 64.5 mV and was able to draw up a membrane potential scale, thus proving that millisecond delayed-light emission can be used as a means of estimating the size of electrical potentials across the thylakoid membranes. His results give strong support to Witt's (1971) view that a membrane potential of about 100 mV (inside positive) can be developed across the thylakoids by light-induced charge transfer processes under steady-state conditions (see Chapter 10 of this volume).

The stimulation of millisecond delayed-light emission by KCl and NaCl gradients as a means of investigating the ionic permeability properties of the thylakoid membranes has been further pursued by Barber (1972b). The transient increases in millisecond delayed light brought about by the establishment of KCl or NaCl gradient across the thylakoid membranes of spinach chloroplasts have been interpreted in terms of a diffusion potential model and both the magnitude and kinetics of the salt-induced signals under various conditions can be explained by assuming the intensity of emission is an exponential function of the membrane potential developed. Treatment of the chloroplasts with valinomycin and gramicidin gives support to these concepts and further suggests that under adequate experimental conditions, membrane potentials of 50–110 mV (inside positive) can be created across the thylakoid membranes upon illumination.

3.3.6. 515-nm Absorption Change

The kinetics of the slow decay of the absorbance change at 515 nm of chloroplasts is one possible parameter to monitor the energetic state of thylakoid membranes. The 515 nm absorbance change has been considered to originate from Chl b (Witt, 1967; Junge and Witt, 1968; Rumberg and Siggel, 1968; Reinwald et al., 1968), but has also been regarded as a carotenoid absorption shift (Chance et al., 1967, 1969; Fork and Amesz, 1967). Finally, it was shown that all bulk pigments, i.e., Chl b, carotenoids, and Chl a, are responsible for the 515-nm absorbance change (Witt, 1971; see an earlier discussion by Govindjee and Govindjee, 1965).

The absorbance changes are generated as a result of charge separation at the thylakoid photocenters. The time course of 515 nm absorbance changes shows a rapid rise of 2×10^{-8} sec in the light, which indicates the formation of an electrical field across the membrane, and a slow decay in the dark, with a duration varying from 5 to 200 msec, dependent on pre-illumination, energetic state and ionic environment. This slow decay is due to a change of the proton concentration in the inner phase of the thylakoid membrane. The breakdown of the field (decay) is accelerated by osmotic (hyper- or hypotonic) shock which increases the permeability specifically for alkali ions, in the presence of K^+. Under phosphorylating conditions the decay time decreases 2- to 5-fold and it is assumed that phosphorylation is coupled to the discharging of the electrically energized membrane by a field-driven H^+ efflux, which might occur through an ATPase-coupled pathway as formulated by Mitchell (1966). Further details of these absorbance changes as a parameter for studying membrane energization are present in the relevant contribution by Witt (Chapter 10 of this volume).

4. SUMMARY

It becomes evident from this survey of the recent studies on structural and energy changes in chloroplast membranes and organelles by a variety of different approaches (ranging from the gross morphological and configurational to the molecular level), that a variety of changes in structure tightly coupled to its energy state occur at different levels. The initial conformational change of thylakoid membrane components induced by light excitation leads at first to a small structural change of the thylakoid, and eventually is amplified so that changes are observed at the organelle level.

It is clear that light causes molecular structural changes, changes in H^+ gradient and in protonation of the membrane and changes in the membrane potential. These events are supported by energy states of the membranes' components excited upon illumination. All of these effects have a cascade-like effect leading to alterations in the organization of the membrane, which in turn affect permeability processes, resulting in changes in the ionic distributions. Ionic distributions are influenced, of course, by the ionic milieu, within and without the chloroplasts. Many of the changes observed at the organelle level now seem to be reasonably predictable on the basis of knowledge that the effect of the ionic environment, the main area where further information is recorded, is on the structure of the membrane itself. Membrane lipids and protein, and lipoprotein interactions are involved in the events following illumination. The results summarized herein demonstrate that the use of a variety of different types of probes, both intrinsic and extrinsic, various photometric methods and electron microscopy, etc., when used in a coordinated fashion, lead to a better understanding of the organization of the membrane and, further, to a fuller comprehension of how light influences the structural organization and energy state of the membrane.

REFERENCES

Aach, H. G., and Heber, U. (1967). *Z. Pflanzenphysiol.* **57**, 317.

Arens, K. (1936). *Jahrb. Wiss. Bot.* **83**, 513.

Atkins, C. A., and Graham, D. (1971). *Biochim. Biophys. Acta* **226**, 481.

Barber, J. (1968). *Biochim. Biophys. Acta* **150**, 618.

Barber, J. (1972a). *FEBS Lett.* **20**, 251.

Barber, J. (1972b). *Biochim. Biophys. Acta* **275**, 105.

Barber, J., and Kraan, G. P. B. (1970). *Biochim. Biophys. Acta* **197**, 49.

Barber, J., and Varley, W. J. (1971). *Nature (London)* **234**, 188.

Barber, J., and Varley, W. J. (1972). *J. Exp. Bot.* **23**, 216.

Ben-Amotz, A., and Ginzburg, B. Z. (1969). *Biochim. Biophys. Acta* **183**, 144.

Bennoun, P. (1970). *Biochim. Biophys. Acta* **216**, 357.

Bohr, C., Hasselbalch, K. A., and Krogh, A. (1904). *Scand. Arch. Physiol.* **16**, 402.

Boyer, P. D. (1965). *In* "Oxidases and Related Redox Systems" (T. E. King, H. S. Mason, and M. Morrison, eds.), Vol. II, p. 994. Wiley, New York.

Brandt, L. and Gohlke, J. R. (1972). *Ann. Rev. Biochem.* **41**, 843.

Briggs, G. E., Hope, A. B., and Robertson, R. N. (1963). "Electrolytes and Plant Cells," p. 168. Davis, Philadelphia, Pennsylvania.

Brody, S. S., Ziegelmair, C. A., Samuels, A., and Brody, M. (1966). *Plant Physiol.* **41**, 1709.

Chance, B., and Mela, L. (1966). *J. Biol. Chem.* **241**, 4588.

Chance, B., Devault, D., Hildreth, W. W., Parson, W. W., and Nishimura, M. (1967). *Brookhaven Symp. Biol.* **19**, 115.

Chance, B., Kihara, T., Devault, D., Hildreth, W., Nishimura, M., and Hiyama, T. (1969). *Progr. Photosynthesis Res.* **3**, 1321.

Chance, B., Crofts, A. R., Nishimura, M., and Price, B. (1970). *Eur. J. Biochem.* **13**, 364.

Clayton, R. K. (1969). *Biophys. J.* **9**, 60.

Crofts, A. R. (1966a). *Biochem. Biophys. Res. Commun.* **24**, 127.

Crofts, A. R. (1966b). *Biochem. Biophys. Res. Commun.* **24**, 725.

Crofts, A. R. (1967). *J. Biol. Chem.* **242**, 3352.

Crofts, A. R., Deamer, D. W., and Packer, L. (1967). *Biochim. Biophys. Acta* **131**, 97.

Crofts, A. R., Cogdell, R. J., and Jackson, J. B. (1971). *In* "Energy Transduction in Respiration and Photosynthesis" (E. Quagliariello, S. Papa, and C. S. Rossi, eds.), pp. 883–901. Adriatica Editrice, Bari.

Cummins, J. T., Strand, J. A., and Vaughan, B. E. (1966). *Biochim. Biophys. Acta* **126**, 330.

Cummins, J. T., Strand, J. A., and Vaughan, B. E. (1969). *Biochim. Biophys. Acta* **173**, 198.

Deamer, D. W., and Packer, L. (1969). *Biochim. Biophys. Acta* **172**, 539.

Deamer, D. W., Crofts, A. R., and Packer, L. (1967). *Biochim. Biophys. Acta* **131**, 81

Dilley, R. A. (1964). *Biochem. Biophys. Res. Commun.* **17**, 716.

Dilley, R. A. (1971). *Current Top. Bioenerg.* **2**, 237.

Dilley, R. A., and Rothstein, A. (1967). *Biochim. Biophys. Acta* **135**, 427.

Dilley, R. A., and Shavit, N. (1968). *Biochim. Biophys. Acta* **162**, 86.

Dilley, R. A., and Vernon, L. P. (1965). *Arch. Biochem. Biophys.* **111**, 365.

Dilley, R. A., Park, R. B., and Branton, D. (1967). *Photochem. Photobiol.* **6**, 407.

Duysens, L. N. M. (1972). *Biophys. J.* **12**, 858.

Farron, F. (1970). *Biochemistry* **9**, 3823.

Fleischman, D. E., and Clayton, R. K. (1968). *Photochem. Photobiol.* **8**, 287.

Fork, D. C., and Amesz, J. (1967). *Photochem. Photobiol.* **6**, 913.

Giraud, G. (1963). *Physiol. Veg.* **1**, 203.

Goedheer, J. H. C. (1972). *Ann. Rev. Plant Physiol.* **23**, 87.

Good, N. E. (1960). *Biochim. Biophys. Acta* **40**, 502.

Govindjee, and Govindjee, R. (1965). *Photochem. Photobiol.* **4**, 675.

Govindjee, and Papageorgiou, G. (1971). *Photophysiology* **6**, 2.

Govindjee, Papageorgiou, G., and Rabinowitch, E. (1967). *In* "Fluorescence Theory, Instrumentation and Practice" (G. G. Guilbault, ed.), p. 511. Dekker, New York.

Gross, E. L., and Packer, L. (1967). *Arch. Biochem. Biophys.* **121**, 779.

Gross, E. L., Dilley, R. A., and San Pietro, A. (1969). *Arch. Biochem. Biophys.* **134**, 450.

Grünhagen, H. H., and Witt, H. T. (1970). *Z. Naturforsch.* **25b**, 373.

Hall, D. O., Reeves, S. G., and Baltscheffsky, H. (1971). *Biochem. Biophys. Res. Commun.* **43**, 359.

Harvey, M. J., and Brown, A. P. (1969). *Biochim. Biophys. Acta* **172**, 116.

Heber, U. W. (1969). *Biochim. Biophys. Acta* **180**, 302.

Heber, U. W., and Santarius, K. A. (1965). *Biochim. Biophys. Acta* **109**, 390.

Heldt, H. W. (1969). *FEBS Lett.* **5**, 11.

Heldt, H. W., and Sauer, F. (1971). *Biochim. Biophys. Acta* **234**, 83.

Hind, G., and Jagendorf, A. T. (1963). *Proc. Nat. Acad. Sci. U.S.* **49**, 715.

Hind, G., and Jagendorf, A. T. (1965). *J. Biol. Chem.* **240**, 3195.

Hind, G., and Whittingham, C. P. (1963). *Biochim. Biophys. Acta* **75**, 194.

Hind, G., Nakatani, H. Y., Izawa, S. (1969). *Biochim. Biophys. Acta* **172**, 277.

Hoagland, D. R., and Davis, A. R. (1932). *J. Gen. Physiol.* **6**, 47.

Hoch, G., and Martin, I. (1963). *Biochem. Biophys. Res. Commun.* **12**, 223.

Hogg, J., Williams, E. J., and Johnston, R. J. (1968). *Biochim. Biophys. Acta* **150**, 640.

Hogg, J., Williams, E. J., and Johnston, R. J. (1969). *Biochim. Biophys. Acta* **173**, 564.

Homann, P. H. (1969). *Plant Physiol.* **44**, 932.

Itoh, M., Izawa, S., and Shibata, K. (1963). *Biochim. Biophys. Acta* **66**, 319.

Itoh, S., Murata, N., and Takamiya, A. (1971). *Biochim. Biophys. Acta* **245**, 109.

Izawa, S. (1965). *Biochim. Biophys. Acta* **102**, 373.

Izawa, S., and Good, N. E. (1966a). *Plant Physiol.* **41**, 533.

Izawa, S., and Good, N. E. (1966b). *Plant Physiol.* **41**, 544.

Jackson, J. B., and Crofts, A. R. (1969). *Eur. J. Biochem.* **10**, 226.

Jacques, A. G., and Osterhout, W. J. V. (1934). *J. Gen. Physiol.* **17**, 727.

Jagendorf, A. T. (1967). *Fed. Proc.* **26**, 1361.

Jagendorf, A. T., and Hind, G. (1963). *Nat. Acad. Sci. -Nat. Res. Council Publ.* **1145**, 599.

Jagendorf, A. T., and Smith, M. (1962). *Plant Physiol.* **37**, 135.

Jagendorf, A. T., and Uribe, E. (1966). *Proc. Nat. Acad. Sci. U.S.* **55**, 170.

Jensen, R. G. (1971). *Biochim. Biophys. Acta* **234**, 360.

Jensen, R. G., and Bassham, J. A. (1966). *Proc. Nat. Acad. Sci. U.S.* **56**, 1095.

Jeschke, W. D., and Simonis, W. (1967). *Z. Naturforsch.* **22b**, 873.

Jeschke, W. D., and Simonis, W. (1969). *Planta* **88**, 157.

Junge, W., and Witt, H. T. (1968). *Z. Naturforsch.* **23b**, 244.

Jursinic, P., and Govindjee (1972). *Photochem. Photobiol.* **15**, 331.

Karlish, S. J. D., Shavit, N., and Avron, M. (1969). *Eur. J. Biochem.* **9**, 291.

Kishimoto, U., and Tazawa, M. (1965). *Plant Cell Physiol.* **6**, 507.

Kobayashi, Y., and Nishimura, M. (1972). *J. Biochem. (Tokyo)* **71**, 275.

Kraan, G. P. B., Amesz, J., Velthuys, B. R., and Steemers, R. G. (1970). *Biochim. Biophys. Acta*
    223, 129.
Kraayenhof, R. (1970). *FEBS Lett.* **6**, 161.
Kraayenhof, R. (1971). Uncoupling of Energy Conservation in Chloroplast and Mitochondrion,
    pp. 84–101. Thesis, Univ. Van Amsterdam. Mondeel-Offsetdrukkerij, Amsterdam.
Kraayenhof, R., Katan, M. B., and Grünwald, T. (1971). *FEBS Lett.* **19**, 5.
Kushida, H., Itoh, M., Izawa, S., and Shibata, K. (1964). *Biochim. Biophys. Acta* **79**, 203.
Kylin, A. (1960). *Bot. Notiser* **113**, 49.
Larkum, A. W. D. (1968). *Nature (London)* **218**, 447.
Latimer, P., and Pyle, B. E. (1972). *Biophys. J.* **12**, 764.
Lavorel, J. (1968). *Biochim. Biophys. Acta* **153**, 727.
Lehninger, A. (1970). "Biochemistry," pp. 365–368. Worth Publ. New York.
Lin, D. C., and Nobel, P. S. (1971). *Arch. Biochem. Biophys.* **145**, 622.
Lüttge, U., and Pallaghy, C. K. (1969). *Z. Pflanzenphysiol.* **61**, 58.
Lynn, W. S. (1968a). *J. Biol. Chem.* **243**, 1060.
Lynn, W. S. (1968b). *Biochemistry* **7**, 3811.
Lynn, W. S. and Straub, K. D. (1969). *Proc. Nat. Acad. Sci. U.S.* **63**, 540.
MacRobbie, E. A. C. (1965). *Biochim. Biophys. Acta* **94**, 64.
Malkin, S., and Hardt, H. (1972). *Proc. Int. Congr. Photosynthesis Res.* **1**, 253.
Marchant, R. H., and Packer, L. (1963). *Biochim. Biophys. Acta* **75**, 458.
Mayne, B. C. (1967a). *Photochem. Photobiol.* **6**, 189.
Mayne, B. C. (1967b). *Brookhaven Symp. Biol.* **19**, 460.
Mayne, B. C. (1968). *Photochem. Photobiol.* **8**, 107.
Miles, C. D., and Jagendorf, A. T. (1969). *Arch. Biochem. Biophys.* **129**, 711.
Miller, M. M., and Nobel, P. S. (1972). *Plant Physiol.* **49**, 535.
Mitchell, P. (1966). *Biol. Rev. Cambridge Phil. Soc.* **41**, 445.
Mitchell, P., and Moyle, J. (1965). *Nature (London)* **208**, 147.
Mitchell, P., and Moyle, J. (1967). *Nature (London)* **213**, 137.
Mohanty, P., and Govindjee (1973). *Biochim. Biophys. Acta* **305**, 95.
Mohanty, P., Mar, T., and Govindjee (1971a). *Biochim. Biophys. Acta* **253**, 213.
Mohanty, P., Papageorgiou, G., and Govindjee (1971b). *Photochem. Photobiol.* **14**, 667.
Mohanty, P., Braun, B. Z., and Govindjee (1972). *FEBS Lett.* **20**, 273.
Mohanty, P., Braun, B. Z., and Govindjee (1973). *Biochim. Biophys. Acta* **292**, 459.
Mukohata, Y. (1968). *In* "Comparative Biochemistry and Biophysics of Photosynthesis" (K.
    Shibata, A. Takamiya, A. T. Jagendorf, and R. C. Fuller, eds.), pp. 89–96. Univ. of Tokyo
    Press, Tokyo.
Mukohata, Y., Mitsudo, M., and Isemura, T. (1966). *Ann. Rep. Biol. Works, Fac. Sci., Osaka
    Univ.* **14**, 107.
Munday, J. C. Jr., and Govindjee (1969a). *Biophys. J.* **9**, 1.
Munday, J. C. Jr., and Govindjee (1969b). *Biophys. J.* **9**, 22.
Murakami, S. (1973). *In* "Organization of Energy-Transducing Membranes" (L. Packer and
    M. Nakao, eds.), pp. 291–313, Univ. of Tokyo Press, Tokyo.
Murakami, S., and Packer, L. (1969). *Biochim. Biophys. Acta* **180**, 420.
Murakami, S., and Packer, L. (1970a). *Plant Physiol.* **45**, 289.
Murakami, S., and Packer, L. (1970b). *J. Cell Biol.* **47**, 332.
Murakami, S. and Packer, L. (1971). *Arch. Biochem. Biophys.* **146**, 337.
Murata, N. (1969). *Biochim. Biophys. Acta* **189**, 171.
Murata, N. (1970). *Biochim. Biophys. Acta* **205**, 379.
Murata, N. (1971a). *Biochim. Biophys. Acta* **226**, 422.
Murata, N. (1971b). *Biochim. Biophys. Acta* **245**, 365.
Murata, N. and Sugahara, K. (1969). *Biochim. Biophys. Acta* **189**, 182.

Murata, N., Tashiro, H., and Takamiya, A. (1970). *Biochim. Biophys. Acta* **197**, 250.

Myers, J. (1971). *Ann. Rev. Plant Physiol.* **22**, 289.

Nagai, R., and Tazawa, M. (1962). *Plant Cell Physiol.* **3**, 323.

Neumann, J., and Jagendorf, A. T. (1964). *Arch. Biochem. Biophys.* **107**, 109.

Nishimura, M. (1971). *In* "Probes of Structure and Function of Macromolecules and Membranes" (B. Chance, C. P. Lee, and J. K. Blasie, eds.), pp. 227–233. Academic Press, New York.

Nishizaki, Y. (1963). *Plant Cell Physiol.* **4**, 353.

Nobel, P. S. (1967a). *Plant Physiol.* **42**, 1389.

Nobel, P. S. (1967b). *Biochim. Biophys. Acta* **131**, 127.

Nobel, P. S. (1968a). *Plant Cell Physiol.* **9**, 499.

Nobel, P. S. (1968b). *Biochim. Biophys. Acta* **153**, 170.

Nobel, P. S. (1968c). *Plant Cell Physiol.* **9**, 781.

Nobel, P. S. (1969a). *J. Theor. Biol.* **23**, 375.

Nobel, P. S. (1969b). *Biochim. Biophys. Acta* **172**, 134.

Nobel, P. S. (1969c). *Plant Cell Physiol.* **10**, 597.

Nobel, P. S. (1970). *Plant Cell Physiol.* **11**, 467.

Nobel, P. S., and Craig, R. L. (1971). *Plant Cell Physiol.* **12**, 653.

Nobel, P. S., and Murakami, S. (1967). *J. Cell Biol.* **32**, 209.

Nobel, P. S., and Packer, L. (1964). *Biochim. Biophys. Acta* **88**, 453.

Nobel, P. S., and Packer, L. (1965). *Plant Physiol.* **40**, 633.

Nobel, P. S., and Wang, C. (1970). *Biochim. Biophys. Acta* **221**, 79.

Nobel, P. S., Murakami, S., and Takamiya, A. (1966a). *Plant Cell Physiol.* **7**, 263.

Nobel, P. S., Murakami, S., and Takamiya, A. (1966b). *Int. Congr. Electron Microsc. 6th, Kyoto, Japan* pp. 373–374.

Nobel, P. S., Chang, D. T., Wang, C., Smith, S. S., and Barcus, D. E. (1969). *Plant Physiol.* **44** 655.

Ongun, A., and Stocking, C. R. (1965). *Plant Physiol.* **40**, 825.

Packer, L. (1962). *Biochem. Biophys. Res. Commun.* **9**, 12.

Packer, L. (1963). *Biochim. Biophys. Acta* **75**, 12.

Packer, L., and Crofts, A. R. (1967). *Current Top. Bioenerg.* **2**, 23.

Packer, L., and Murakami, S. (1972). *Methods Enzymol.* **24B**, 181.

Packer, L., Barnard, A., and Deamer, D. W. (1967). *Plant Physiol.* **42**, 283.

Packer, L., Murakami, S., and Mehard, C. W. (1970). *Ann. Rev. Plant Physiol.* **21**, 271.

Papageorgiou, G., and Govindjee (1968a). *Biophys. J.* **8**, 1299.

Papageorgiou, G., and Govindjee (1968b). *Biophys. J.* **8**, 1316.

Penth, B., and Weigl, J. (1969). *Z. Naturforsch.* **24b**, 342.

Petrack, B., and Lipmann, F. (1961). *In* "Light and Life" (W. D. McElroy and B. Glass, eds.), pp. 627–630. Johns Hopkins Press, Baltimore, Maryland.

Post, R. L., Merrit, C. R., Kinsolving, C. R., and Albright, C. D. (1960). *J. Biol. Chem.* **235**, 1796.

Radda, G. (1971). *Current Top. Bioenerg.* **4**, 81.

Raven, J. A. (1968). *J. Exp. Bot.* **19**, 712.

Reinwald, E., Stiehl, H. H., and Rumberg, B. (1968). *Z. Naturforsch.* **23b**, 1616.

Robinson, J. D., and Stocking, C. R. (1968). *Plant Physiol.* **43**, 1597.

Roy, H., and Moudrianakis, E. N. (1971a). *Proc. Nat. Acad. Sci. U.S.* **68**, 464.

Roy, H., and Moudrianakis, E. N. (1971b). *Proc. Nat. Acad. Sci. U.S.* **68**, 2720.

Rumberg, B., and Siggel, U. (1968). *Z. Naturforsch.* **23b**, 239.

Ryrie, I. J., and Jagendorf, A. T. (1971). *J. Biol. Chem.* **246**, 3771.

Ryrie, I. J., and Jagendorf, A. T. (1972). *J. Biol. Chem.* **247**, 4453.

Saltman, P., Forte, J. G., and Forte, G. M. (1963). *Exp. Cell Res.* **29**, 504.

Santarius, K. A., and Heber, U. (1965). *Biochim. Biophys. Acta* **102**, 39.

Schuldiner, S., and Avron, M. (1971). *FEBS Lett.* **14**, 233.
Schuldiner, S., and Ohad, I. (1969). *Biochim. Biophys. Acta* **180**, 165.
Scott, G. T., and Hayward, H. R. (1954). *J. Gen. Physiol.* **37**, 601.
Shavit, N., and Avron, M. (1967). *Plant Physiol.* **49**, 411.
Stiller, M. (1965). *Biochim. Biophys. Acta* **94**, 53.
Strehler, B. L., and Arnold, W. (1951). *J. Gen. Physiol.* **34**, 809.
Strotmann, H., and Berger, S. (1969). *Biochem. Biophys. Res. Commun.* **35**, 20.
Stryer, L. (1965). *J. Mol. Biol.* **13**, 482.
Sun, A. S. K., and Sauer, K. (1972). *Biochim. Biophys. Acta* **256**, 409.
Sundquist, J. E., and Burris, R. H. (1970). *Biochim. Biophys. Acta* **223**, 115.
Telfer, A., and Evans, M. C. W. (1971). *FEBS Lett.* **14**, 241.
Vredenberg, W. J. (1969). *Biochem. Biophys. Res. Commun.* **37**, 785.
Walker, D. A. (1965). *Plant Physiol.* **40**, 1157.
Walker, D. A., and Crofts, A. R. (1970). *Ann. Rev. Biochem.* **21**, 389.
Walker, D. A., Baldry, C. W., and Cockburn, W. (1968). *Plant Physiol.* **43**, 1419.
Wang, C., and Nobel, P. S. (1971). *Biochim. Biophys. Acta* **241**, 200.
Weber, G. (1972). *Ann. Rev. Biophys. Bioeng.* **1**, 553.
West, J., and Packer, L. (1970). *Bioenergetics* **1**, 405.
West, K. R., and Wiskich, J. T. (1968). *Biochem. J.* **109**, 137.
Winocur, B. A., Macey, R. I., and Tolberg, A. B. (1968). *Biochim. Biophys. Acta* **150**, 132.
Witt, H. T. (1967). *In* "Fast Reactions and Primary Processes in Chemical Kinetics" (S. Claesson, ed.), pp. 261–316. Wiley (Interscience), New York.
Witt, H. T. (1971). *Quart. Rev. Biophys.* **4**, 365.
Wraight, C. A., and Crofts, A. R. (1971). *Eur. J. Biochem.* **19**, 386.

12

Molecular Organization of Chlorophyll and Energetics of the Initial Stages in Photosynthesis

Felix F. Litvin and Vitaly A. Sineshchekov

1. INTRODUCTION

The aim of this article is to inform our readers of the research, published in the Russian language, on the mechanism of photosynthesis. Earlier, reviews of this kind have been published by Krasnovsky (1960, 1969, 1972). In accordance with the general subject of the book, the main attention is given to the energetics of the initial stages of photosynthesis and to the role of molecular structures in various energy-storage stages of the process (i.e., light absorption, energy migration and trapping, potential generation in membranes, net energy storage, etc.). Some related problems, such as electron transfer and oxygen evolution, are also discussed. Owing to the limitation of space, the authors could not, of course, present the whole picture and cover the work of all the authors. The traditional and fundamental investigations on pigment photochemistry are also omitted (see "Molecular Photonics" Nauka, Leningrad, 1970). Furthermore, the present review does not pretend to cover all the contributions of the authors cited here. The choice of the topics in this article is influenced by the scientific interests of the present authors and centers on the problems of the different native pigment forms. Investigations from countries other than the Soviet Union are cited in the corresponding chapters in this book.

2. NATIVE PIGMENT FORMS IN HIGHER PLANTS, ALGAE, AND PHOTOSYNTHETIC BACTERIA

2.1. Multiplicity of Pigment Forms *in Vivo*

It has been already established in earlier research that chlorophyll (Chl) incorporated into biological structures differs by its spectral and other properties from those of the solution. Krasnovsky and co-workers have pioneered a new approach in the investigation of native pigment forms. The main conclusion was that there exists two forms of Chl *a*; namely, a short wave, monomeric, photochemically labile form and a long wave, aggregated, less active form (see the above cited reviews by Krasnovsky).

In retrospect, it is of interest to compare this result with those obtained in the early works by Emerson and co-workers on the "red drop" of photosynthesis and the "enhancement" effect (see chapter by Govindjee and R. Govindjee, this volume). One can see that Krasnovsky's results would provide an explanation for the assumption of two photosystems in photosynthesis from the point of view of the pigment organization. The mutual connection of these two types of investigation became obvious later.

In the early works, the indications were already obtained that the number of the native chlorophyll forms is greater than two (Litvin and Krasnovsky, 1957, 1958; Litvin et al., 1959). Thus, at least three bands were found in addition to the 685 nm band in the low temperature (-150 to $-196°C$) fluorescence spectrum of leaves—the 730–735 nm maximum, sharply increasing upon cooling, a weak, complex band near 690–695 nm, and an infrared band at 812–815 nm. A large body of information concerning pigment forms was also obtained by C. S. French, J. Brown, E. Rabinowitch, Govindjee, S. S. Brody, J. B. Thomas, J. H. C. Goedheer and others (see chapter by Papageorgiou, this volume).

Complex fluorescence spectrum structure was found in greening and green higher plant leaves and algae at $-196°C$; the maxima and shoulders were located at 675, 682, 686, 696, 700–705, 710, 730, 740, and 760 nm. In a low-temperature absorption spectrum, other signs of complex structure, in addition to the maximum at 710 nm described by W. Butler, were also found (weak shoulders at 690–694, 700–705, 712–715, 730, 750–760, and 810 nm). The intensity of the long wave bands changed under the action of light, heating, or solvents. A similar complex structure was observed for concentrated Chl $(a + b)$ solutions in alcohol and petroleum ether (Litvin et al., 1962; Litvin and Gulyaev, 1966a,b; Gulyaev and Litvin, 1967). Considering the data obtained, the authors came to the conclusion that there were several native chlorophyll forms in a cell (aggregates with different degrees of packing).

From further investigations, the existence of a universal system of aggregated Chl forms was suggested in higher plants, algae, and photosynthetic bacteria (Litvin, 1965; Litvin and Gulyaev, 1966a; Gulyaev and Litvin, 1967). The experimental evidence and development of this thesis are presented below.

2.2. Absorption Spectra of the Native Forms of Chlorophyll and Other Pigments

The combination of the second derivative method with low-temperature spectroscopy and computer analysis of the curves thus obtained proved a useful approach (Litvin and Gulyaev, 1969a,b; Gulyaev et al., 1971).

The qualitative examination of the second derivative spectra confirmed that the broadening and red shift of the absorption band were only apparent effects (Litvin and Gulyaev, 1969c; Gulyaev and Litvin, 1970; Litvin *et al.*, 1970). In reality, they are caused by the appearance of many (not less than ten) narrow bands in the 640–740 nm region. The mathematical analyses of the spectrum (Fig. 1) allowed the authors to calculate the amplitude and half-widths of the bands (see Table I). The share of light absorbed by each form and its relative concentration were calculated from the area under the curves (Litvin and Gulyaev, 1969c; Litvin *et al.*, 1973a; Gulyaev *et al.*, 1973).

Chl *b* exists in the two forms with the maxima at 640–642 and 648–650 nm. The bulk of Chl *a* is found in the three short-wave forms. It is worthwhile to mention the first one, which is probably a monomeric pigment form absorbing at 661–662 nm. In spite of its relatively high concentration, it is not observed in normal absorption spectra. The second group, including three Chl *a* forms, absorbs in the region of the red (long-wavelength) slope of the

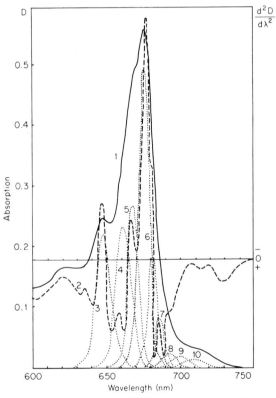

Fig. 1. Absorption bands of the native Chl forms in Phaseolus chloroplasts at $-196°C$: (1) absorption spectrum; (2) its second derivative; (3–10) Gaussian components (Gulyaev *et al.*, 1973).

TABLE 1

Spectral Characteristics of the Native Forms of Chlorophyll and Its Aggregates[a]

	In a cell			In a model system	
Chl form	Half band-width, (nm) (at −196°C)	Relative area of absorption bands (for Phaseolus)	Chl forms	Half band-width, (nm) (at −196°C)	Conditions (at −196°C)
		Chlorophyll b—narrow bands			
$Chl_{640}^{(645)^b}$	9	—	—	—	—
$Chl_{648}^{(653)^b}$	8	0.19	—	—	—
		Chlorophyll a—narrow bands			
Chl_{662}^{668}	10	0.18	Chl_{663}^{669}	10	Hexane 10^{-6-3} M
Chl_{668}^{674}	8	0.15	Chl_{670}^{676}	8–9	Hexane film
Chl_{676}^{680}	8	0.29	Chl_{677}^{681}	9	Hexane $10^{-5}M$
Chl_{682}^{686}	8	0.10	Chl_{682}^{687}	14	Hexane $10^{-5}M$ (20°C)
Chl_{687}^{692c}	7.5	0.04	$Chl_{(686)^b}^{691}$	9	Hexane 10^{-4} M film
Chl_{693}^{697}	7.5	0.02	$Chl_{690}^{694-696}$	9–11	Hexane cyclohexane film
Chl_{698}^{702}	7	—	Chl_{698}^{702}	7–8	Hexane $10^{-5}M$
$Chl_{703}^{706-707}$	8–10	0.02	Chl_{702}^{706}	Shoulder	Hexane $10^{-3}M$ $10^{-2}M$
Chl_{707}^{712}	10	<0.02	Chl^{712}	Shoulder	Hexane film
$Chl_{(712)^b}^{718}$	10–12	<0.01	$Chl_{715}^{(720)^b}$	16	Hexane $10^{-2}M$ (20°C)
$Chl_{(738)^b}^{743}$	9–11	<0.01	Chl^{740}	11–13	Hexane $10^{-5}M$
$Chl_{(755)^b}^{760}$	11–12	<0.01	$Chl_{(755)^b}^{760}$	10–12	Octane $10^{-2}M$
		Chlorophyll a—wide bands			
$Chl_{(702)}^{722}$	29	?	$Chl_{(700)}^{720}$	22–24	Hexane $10^{-5}M$
Chl_{704}^{726}	30	?	$Chl_{700-705}^{726}$	27–30	Octane $10^{-3}M$
$Chl_{(706)}^{732}$	32	?	Chl_{705}^{730}	30	Hexane $10^{-3}M$
Chl_{710}^{738}	36	?	Chl_{710}^{740}	38–35	Film on glass

[a] See Litvin et al. (1974a,b); Sineshchekov et al. (1973; 1974a,b).

[b] Found from the universal relationship between absorption and fluorescence spectra.

[c] Manifests well at 4°K.

band. The absorption by the long-wavelength forms is practically negligible. Nevertheless, migration of energy effects its transfer from the short-wavelength to the long-wavelength forms (see below). These data and also spectroscopic investigations of the Emerson enhancement effect prompt the authors to consider the weakly absorbing long-wavelength forms as the most important participants of the photosynthetic energy utilization process.

The half bandwidths of the forms are less even than those for the pigment in nonpolar solvents. It decreases by 1.2–2 times when temperature is lowered. The half bandwidth values and the effect of temperature on them are characteristic of different forms (see Table I).

The analysis of the derivative spectra for more than thirty species of higher plants and green, blue-green, brown, and red algae allowed the authors to confirm the universality of the Chl absorption spectrum *in vivo*. The position and half bandwidth are close to or practically coincide with one another (within an error of ±1–2 nm) for all species investigated. This thesis needs confirmation only for the longest wavelength bands because of less exact determinations of the parameters.

Such a confirmation is obtained from emission and exitation spectra of fluorescence measurements in algae (see Table I). In addition to this, wider fluorescence and absorption bands are also observed in these spectra (see below, p. 626).

The data obtained were used to calculate a reverse matrix, which allowed the estimation the optical density (or concentration) of each form directly from usual absorption spectra without computer calculations (Gulyaev *et al.*, 1973).

The analysis of low-temperature and derivative absorption spectra of a green bacterium, *Chlorobium ethylicum*, revealed a complex structure whose bands belonged to *Chlorobium* Chl (at 688, 705, 715, 741, 752, 762, 782, 790, 801, and 811 nm) and bacteriochlorophyll (BChl) (823 and 835 nm). In a purple bacterium (*Chromatium*), BChl bands are at 668, 675, 797, 808, 825, 840, 852, 865, 880, and 895–900 nm; additional bands are observed in some other species of purple bacteria. Upon lowering of the temperature from +20° to −196°C the half bandwidth of the narrow bands decreased from 12–16 nm to 8–10 nm (Litvin and Gulyaev, 1966b, 1969c). These bands are supposed to belong to the system of ten or more native bacterial pigment forms which are similar to that of the Chl *a* forms (see below) (Litvin and Gulyaev, 1966a).

Pigments of other chromophore structure can also exist as different native forms. Thus, in the intermediate spectral region, a number of bands are found that can be attributed to carotenoids (Litvin and Gulyaev, 1966b; Litvin *et al.*, 1965). Their comparison with the absorption spectra of films showed that the long wavelength maxima may belong to *cis*- and *trans*-β-carotene aggregates.

2.3. Fluorescence of the Native Forms and Their Role in the Initial Photophysical Processes of Photosynthesis

Second-derivative fluorescence spectra measurements of photosynthetic organisms at temperatures ranging from 4° to 297°K have confirmed the

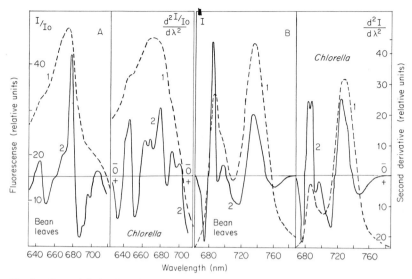

Fig. 2. Second derivative of excitation (A) and fluorescence (B) spectra: (1) usual spectra, (2) second derivative spectra (Litvin *et al.*, 1970; Sineshchekov *et al.*, 1974).

existence of the complex spectrum structure (Figs. 2B, 3B). It consists of not less than ten to twelve narrow bands. The number of the bands, their positions, and their half bandwidths practically coincided for the investigated species of higher plants and algae. The band positions did not depend on temperature. Their half bandwidths were different and decreased consider-

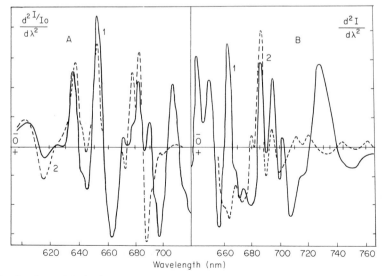

Fig. 3. Second derivative excitation (A) and fluorescence (B) spectra in algae at −196°C: (1) *Anacystis nidulans*; (2) *Hapolosiphon fontinalis* (Litvin *et al.*, 1974a).

ably upon cooling (for a narrow band, it decreased from 15 to 4 nm when temperature was lowered to 4°K) (Litvin *et al.*, 1971; Sineshchekov *et al.*, 1974).

In attempts to find out the relationship between the absorbing forms and their emission bands, the dependence of the fluorescence spectra on the wavelengths of monochromatic light and of the excitation spectra on the wavelengths of the registered monochromatic emission was investigated (Litvin *et al.*, 1964; Litvin and Sineshchekov, 1965, 1967a; Sineshchekov, 1972; Sineshchekov *et al.*, 1973). In the experiments on fluorescence excitation by very narrow monochromatic beams (1.5 nm) (650 nm $\leq \lambda_{ex} \leq$ 720 nm), it was found that the intensity of the short-wave fluorescence components decreased when the excitation wavelength was increased from 680 to 700 nm (Fig. 4A). At excitation wavelengths exceeding 700 nm, only one low-temperature fluorescence band is observed that differs from the others by its large half bandwidth. Its position is different in various species (722 nm, *Chlamydomonas*; 726 nm, *Chlorella Anabena*; 732 nm, *Nostoc*; 736–739 nm, *Phaseolus*). In *Euglena*, there are two wide bands at 720 and 730 nm.

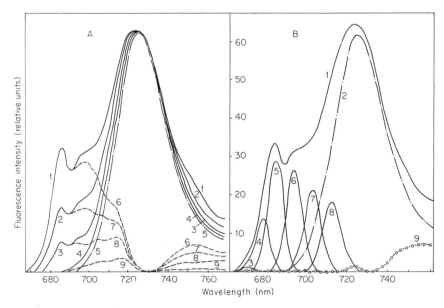

Fig. 4. Analysis of Chlorella fluorescence spectrum at −196°C. (A) Dependence of the spectrum on the excitation wavelength: (1) λ_{ex} = 676 nm, (2) 683 nm, (3) 686 nm, (4) 694 nm, (5) 705 nm; curves 6, 7, 8, and 9 are curves 1, 2, 3, and 4 minus curve 5, respectively.   (B) Analysis of the spectrum into a number of Gaussian curves: (1) fluorescence spectrum, λ_{ex} = 640 nm, (2) wide band, λ_{ex} = 705 nm, (3–8) narrow bands, (9) curve 1 minus curves 2–8. (Sineshchekov *et al.*, 1973.)

When temperature was raised to + 20°C, the intensity of the band decreased thirty-five to forty times and the halfwidth changed from 36 to 70 nm (for Phaseolus). Judging from the excitation spectra, the corresponding wide absorption band is at 710 nm. This group with wide bands belongs apparently to some other type of forms than those with narrow bands (Litvin et al., 1974a; Sineshchekov et al., 1973).

Fluorescence excitation spectra measurements pointed to the existence of a weak complex structure (the maxima and shoulders at 662–664, 668, 672, 680, 684, 693, 700, 710–715, and 722 nm) (Litvin and Sineshchekov, 1967a). Second-derivative measurements revealed that the positions of the bands were similar to those of the absorption bands (Litvin et al., 1971). The long wavelength forms were better manifested in the second-derivative excitation spectra, which were used for the determination of their parameters (Figs. 2A, 3A). The excitation spectra were investigated as a function of the registered monochromatic emission wavelength in the 670–740 nm fluorescence spectrum region (Fig. 5A) and of other factors (Sineshchekov, 1972;

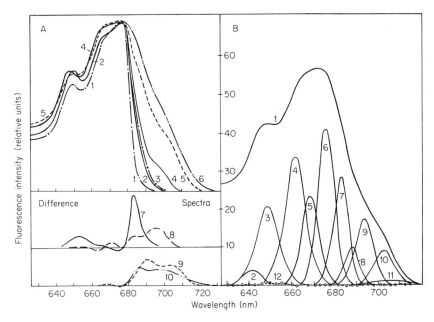

Fig. 5. Analysis of fluorescence excitation spectrum of *Chlorella* at −196°C. (A) Dependence of the spectrum on the wavelength of monochromatic emission: (1) $\lambda_m = 678$ nm, (2) 685 nm, (3) 695 nm, (4) 705 nm, (5) 713 nm, (6) 725 nm; curves 7, 8, 9, and 10 are difference spectra: (7) 2 minus 1, (8) 4 minus 3, (9) 5 minus 4, (10) 6 minus 5. (B) Analysis of the spectra into a number of Gaussian curves: (1) excitation spectrum, $\lambda_m = 713$ nm, (2–11) Gaussian curves, (12) curve 1 minus curves 2–11 and their vibrational satellites (the last ones are not shown in the figure). (Sineshchekov et al., 1973.)

Sineshchekov *et al.*, 1973). The system of the bands was the same for all investigated species within the error of the measurements (1.5–2 nm) (see Table I) (Sineshchekov *et al.*, 1974).

The above experiments allowed the analysis of the emission and excitation spectra of fluorescence into two sets of gaussian components (Figs. 4B, 5B). Along with the development of the methods, the positions of the maxima were found more precisely, and now the system of the native Chl forms can be presented (see Table I). Still, in some details, the system is not final (Litvin *et al.*, 1974a; Sineshchekov *et al.*, 1973).

Thus, the native forms can be considered as independent centers of electronic excitation. The energy migration between the forms does not eliminate their identity; they can play an independent role in the primary processes. This also suggests the relatively slow (resonance) energy transfer mechanism and the possibility of excitation energy stabilization within each form. (For a theoretical discussion of excitation energy migration, see Knox, this volume.)

2.4. The Role of Protein–Lipoid Matrix

The structure of the native pigment forms and the role of the pigment matrix in the organization of the pigment system was studied in the chromatophores of purple bacteria. Proteolytic enzymes (pronase, bacterial proteinase A) destroyed selectively the long wavelength forms $BChl_{890}$ and P_{890} (the energy trap). The lipolytic enzymes (phospholipase A) caused rearrangement of $BChl_{850}$ into a new form, $BChl_{830}$. The same change ($BChl_{850} \longrightarrow BChl_{830}$) was observed when chromatophores were treated by nonionic detergents and organic solvents and under the conditions of low pH or high ionic strength of a medium. After the elimination of these agents, the transformation $BChl_{850} \longrightarrow BChl_{830}$ was completely reversible. The reversibility was not observed in the presence of glutaraldehyde, which fixed the structure of the $BChl_{830}$ form (Erokhin, 1967; Erokhin and Sinegub, 1970a,b; Sinegub and Erokhin, 1971). The authors believe that the properties of the native BChl forms are due to pigment–pigment interaction. However, its character is determined by the structure and conformation of the protein bearers and membrane lipids.

One of the complexes obtained by the use of electrophoresis from *Chromatium* chromatophores contained $BChl_{890}$ and P_{890} (in a ratio of 50:1) and cyt c_{553} and c_{556} and was photochemically active, as suggested by the observation of P_{890} photo-bleaching and delayed light emission. The other one, photochemically inactive and lacking the electron-transport chain components, had only the light-harvesting BChl forms, $BChl_{800}$ and $BChl_{850}$. The molecular weights of the complexes were found to be 475,000 and 8000 for

BChl$_{890}$ and BChl$_{850}$, respectively; the numbers of protein molecules in the complexes were equal to 10–12 and 2 (Erokhin and Moskalenko, 1969, 1972).

Complex structure of the absorption band with maximum at about 680 nm has been revealed in the difference spectrum of oxidized minus reduced chloroplast fragments of photosystem I. The analysis of the experimental data permits Kochubei and Ostrovskaya (1972) to assume a relationship between observed spectrum shape alterations and changes in the pigment lipoproteid complex state.

2.5. Biosynthesis of the Native Chlorophyll Forms

Chl biosynthesis in etiolated and green plants was systematically investigated by Shlyk and coworkers. In particular, the Chl *b* synthesis from Chl *a*, the role of phytol-less pigments in the terminal stage of Chl biosynthesis, and other phenomena were elucidated (Shlyk, 1965, 1970). Recently Shlyk and co-workers have suggested the existence of the special sites of chlorophyll biosynthesis. By using a radioactive labeling method, preparative differential centrifugation, and polyacrylamide gel electrophoresis, it was found that "young," newly synthesized molecules (and the molecules of the precursor and of the Chl *a* into Chl *b* converting enzyme, as well) are concentrated in special fractions of subchloroplast particles (Shlyk *et al.*, 1966, 1969; Fradkin *et al.*, 1972a). The lability of a part of the biosynthesis centers is due to their links to young developing membrane sites (Fradkin *et al.*, 1972a).

Energy migration from carotenoid (lutein) to Chl was observed in model systems (Fradkin and Shlyk, 1967). However, in etiolated plants, such a transfer from carotene to protochlorophyll or to newly formed Chl is not observed (Fradkin *et al.*, 1969; Fradkin and Shlyk, 1967). There is, however, a low efficiency of energy transfer between protochlorophyllide and Chl. Analyzing these data, the authors came to the conclusion that protochlorophyllide was accumulated in specially isolated sites or biosynthesis centers (Fradkin *et al.*, 1968, 1969, 1972b; Fradkin and Shlyk, 1967). On the other hand, there are facts that indicate that synthesized Chl molecules are brought together and form groups. This is suggested by energy migration between Chl *a* and *b* formed from Chl *a* in dark in etiolated leaves after short illumination. (Fradkin *et al.*, 1966, 1969). The precursor molecules are also in juxtaposition.

The similarity between the fluorescence excitation spectrum of etiolated leaves and the absorption spectrum of protochlorophyllide aggregates in solution indicates that the long-wave protochlorophyllide form is an aggregated one (Krasnovsky and Bystrova, 1962; Fradkin *et al.*, 1969, 1972b).

The aggregated chlorophyll form with the fluorescence maximum at 720 nm appears practically at the beginning of illumination in etiolated leaves (Fradkin et al., 1970). The authors discuss the role of membrane structures in the formation of this aggregate by comparing the rates of the long wavelength fluorescence increase and Chl accumulation (Fradkin et al., 1970, 1972b). The correlation between the rise of the low temperature long wavelength fluorescence with that of afterglow (delayed light emission) in greening leaves points to the role of aggregated forms in energy storage (Litvin et al., 1966).

The sequence of the biosynthesis of the native chlorophyll forms was investigated by using low-temperature spectral measurments. Litvin and Krasnovsky (1957, 1959) have found the following train of the intermediate stages in the chlorophyll biosynthesis process (the figures are the low temperature flourescence maxima positions): P $Chl_{655} \longrightarrow Chl_{690-695} \longrightarrow Chl_{675-680} \longrightarrow Chl_{685-686}$.

The appearance and conversion of protochlorophyll (P) are also observed in green leaves; the Chl biosynthesis in green leaves also proceeds via the same chlorophyll forms (Litvin and Krasnovsky, 1959). Later, the terminal stages of the chlorophyll formation were shown to include not one but two consecutive photochemical reactions, in contrast to the existing data (Litvin, 1964, 1965). In the first reaction, the unstable product (Chl_{676}^{684}) is formed (step I in the scheme; the superscript refers to the emission wavelength, and the subscript to the absorption wavelength); in the second one (step II), it is converted in chlorophyllide Chl_{680}^{690}, considered earlier as a primary chlorophyll form. This form is converted into a longer wavelength one (step III) form (Chl_{685}^{695}). A parallel pathway of the intermediate transformation was found: the dark reaction (Ia), whose product did not undergo further transformations under action of light and was esterified by phytol, represented one of the resulting forms of chlorophyll. The photochemical stages of chlorophyll formation were found to correspond to two successive first-order reactions. At 20°C and light intensity of 2×10^5 erg cm^{-2} sec^{-1}, the rate constant for the first reaction (K_1) was 50 sec^{-1}, for the second (K_2), 200 sec^{-1}. Lowering the temperature to -40°C had little effect on K_1 but decreased K_2 to 15 sec^{-1} (Litvin and Belyaeva, 1968, 1971a, b; Litvin et al., 1974a).

The photochemical conversion of the protochlorophyllide into chlorophyll has been performed in a simple system (pigment solution) by Bystrova et al. (1966). The accumulation of the "adult" native chlorophyll forms was investigated by measuring derivative absorption and fluorescence spectra in greening leaves. The sequence of the accumulation of the chlorophyll forms corresponds, in general, to the sequence of the chlorophyll native forms put according to their absorption maxima positions. The comparison of the absorption, fluorescence, and excitation spectra shows the gradual develop-

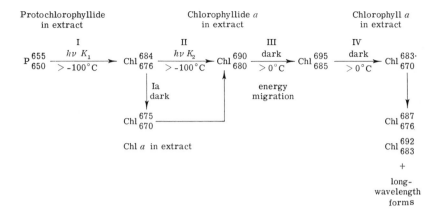

ment of the energy migration process among the Chl forms. This allows the conclusion that the pigment aggregates form even more complex molecular infrastructures, which are the pigment basis of a photosynthetic unit (Litvin et al., 1973a,b).

The biosynthesis of the native forms proceeds parallel to the development of the pigment function. In the first 2 hours of greening, maxima at 650 and 670 nm are observed in the action spectrum of oxygen uptake. Then, in the following period, maxima at 670, 690, and 715–720 nm appear in the action spectrum of oxygen evolution, maxima in the action spectra of the Emerson enhancement effect being at first at 700 nm, then at 715–720 nm (Litvin and He I-Tan, 1967; Gasanov et al., 1972).

The biosynthesis of tetrapyrrole compounds in Chlorella correlates with the activity of photosynthetic and oxidative phosphorylation processes (Gavrilenko et al., 1969; Gavrilenko and Budagovskaya, 1972; Chernavina et al., 1968). Under experimental conditions disturbing iron–proteid metabolism, cells retain their ability to accumulate energy by increasing the share of copper proteins in the electron transfer chain, in particular of plastocyanin connected with the cyclic system (Chernavina et al., 1968).

The problem of the regulation of chlorophyll biosynthesis in general, beginning from the very first stages, was investigated by Kaler (1972, 1973) and Kaler and Podchufarova (1968). The regulatory role of pigments and light in the early chemical steps of the formation of tetrapyrole compounds and of the PSU structures was established by examining the autoregulation schemes of the processes with the use of an analog computer.

2.6. Nature of the Native Chlorophyll Forms

The spectral similarity between the long wavelength forms of Chl and bacterial pigments and their aggregates in colloidal solutions and films has been already shown in the works of Krasnovsky and co-workers (see above).

Brody has concluded that the two maxima in the low-temperature fluorescence spectrum of *Chlorella* belong to the monomer and dimer of Chl *a* (see Brody, 1958, 1961). The complex structure of the low temperature fluorescence spectrum of a concentrated chlorophyll solution suggests, however, that there is a number of different aggregates that can be considered as a model of the native pigment forms.

More detailed investigations were made on solutions, monolayers, and films of the pigment. In contrast to preceding investigations it has been shown that there occurs not the broadening and shift of a single band, but the appearance of new bands when pigments are in a monolayer or in a film. As the pigment concentration increases in the above systems, maxima are observed at longer wavelength regions of absorption (Litvin, 1965; Litvin and Gulyaev, 1964, 1966b; Litvin *et al.*, 1965), excitation (Litvin *et al.*, 1965; Litvin and Sineshchekov, 1967a,b; Sineshchekov *et al.*, 1972) and fluorescence (Litvin *et al.*, 1964; Litvin and Gulyaev, 1964; Litvin and Sineshchekov, 1965; Sineshchekov *et al.*, 1972) spectra. The positions of the aggregate band correlate with the molecular packing densities of the layers (as evaluated from the average area occupied by a pigment molecule). The positions of the majority of the maxima and shoulders are close to or coincide with those observed *in vivo*.

In thick films, the predominant formation of one or another aggregate can be observed under different conditions of film preparation. The 678–680 nm maximum is a predominant one in films prepared from diethyl ether Their treatments by dioxane, ammonium, or methanol vapor are followed by the appearance of the maxima at 690, 720–730, or 730–740 nm respectively. Colloidal pigment solutions are characterized by maxima at 673 and 710–715 nm (water + 5% methanol) and 720 nm (water + 40% methanol) (Krasnovsky and Bystrova, 1967; Bystrova and Krasnovsky, 1967, 1968, 1971; Bystrova *et al.*, 1972). Some uncertainty in the position of the maxima is due to the fact that the main absorption bands include two or more subbands.

Synthetic pigment–protein–lipoid complexes have been investigated as a model of molecular organization. Under certain conditions, the strongly aggregated forms of the pigments are created in them (absorption maxima at 704 and 748–750 nm). Energy transfer between carotene and Chl was also observed in this model (Giller, 1972).

Basically, the spectral properties of films are not due to the long distance effects, but to the molecular structure of aggregates. As evidenced by spectroscopic measurements, there are the same aggregates in films and in concentrated solutions, and the effects of orientation or interaction of these aggregates, which could be attributed to the film as a whole, are not observed.

The appearance of Chl aggregates in concentrated nonpolar solutions and

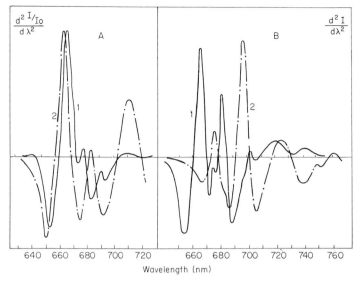

Fig. 6. Second derivative spectra of Chl aggregates in solution at $-196°C$. (A) Fluorescence excitation spectra in hexane: (1) $10^{-5} M$; (2) $10^{-3} M$. (B) Fluorescence spectra: (1) in hexane, $10^{-5} M$; (2) in octane, $10^{-5} M$. (Litvin *et al.*, 1974b.)

films analogous to the above Chl native forms can be followed by measuring second derivative spectra (Fig. 6) (Litvin *et al.*, 1974b; Sineshchekov *et al.*, 1974).

Apart from the similarity in the positions of maxima, a number of features common to the Chl aggregates and its native forms attract attention. The half bandwidth of the narrow aggregate bands is also small (7–10 nm at $-196°C$), but the bands do not turn into narrow exciton bands. The red shift can reach large values at high concentrations and the system of narrow bands can extend beyond 760 nm in the near infrared region. The temperature dependence of the aggregate half bandwidth and intensity is similar to that of the native forms.

A number of aggregates in films and solution have a different kind of spectrum, the half bandwidth of one of the fluorescence bands reaches up to 30–60 nm. The intensity of these bands in the region 720–760 nm sharply increases upon cooling (down to $-196°C$), the corresponding absorption maxima being at 700–715 nm. The aggregates of this type seem to be a model for the native forms with similar wide bands (Litvin *et al.*, 1974a,b; Sineshchekov *et al.*, 1973) (see above).

Proceeding from the McRae and Kasha chromophore interaction theory, an attempt has been made to estimate the number of molecules in a polymer (N), the distance between them (R) and the angle between an aggregate axis and a chromophore plane (α). Some estimations can be made in spite of the uncertainty due to the impossibility of determining independently one of the three parameters (Gulyaev and Litvin, 1967; Litvin and Gulyaev, 1969b). Thus, the shortest wavelength aggregate with the absorption maximum near 665 nm is most likely a dimer with a very weak interaction between chromophores ($N = 2, R = 15$ Å, $\alpha = 0°$). If the size of the aggregate is more ($N > 2$), then the aggregate should be more loose ($R \sim 20$ Å). The aggregate with the peak position near 670 nm is interpreted as a dimer in which chromophores are less separated ($R \sim 9.5$ Å, $\alpha = 50°$) or an oligomer with weak interaction. The 676 nm maximum is possibly associated with densely packed dimers ($R \sim 8.8$ Å, $\alpha = 55°$) or with pigment oligomers. The 684, 690, and 702 nm aggregates are not likely to be dimers, but are rather linear polymers ($N = 3-8$, $R \sim 7.8-6.5$ Å, $\alpha = 58-64°$). Perhaps, they are larger formations (two or three dimension crystal structures). Finally, the long wavelength aggregates absorbing beyond 700 nm are likely to be various pigment crystals.

The data obtained by means of infrared spectroscopy show that molecular chemical structure and the presence of addenda are important for aggregate formation (Krasnovsky *et al.*, 1969, 1972). The coordinate intermolecular bond is formed in the aggregation of the investigated Mg-complexes (Chl *b*, protochlorophyll, BChl, chlorobium Chl) as it is known for Chl *a*. The bond links the keto group at the ninth carbon atom of one molecule and the central magnesium atom of the other (see structure of Chl molecule in Sauer's chapter, this volume). In bacterial pigments aggregating more easily, all keto groups are bound, while in protochlorophyll, a considerable part of them remains free. In the last case, the bond can be formed between C-7 and C-10. The bond of this type is found also in bacterial pigment aggregates. The bond between a ketogroup and a carbon atom is formed via a low-molecular-weight addendum (methanol in particular). BChl aggregates are formed through the bonds acetyl $C = O \cdots Mg, \cdots = keto\, C = O \cdots HO \cdots Mg$. In chlorophyll, the bond formyl $C = O \cdots Mg$ plays an additional role.

In early works by Krasnovsky and co-workers, the stability against photobleaching was considered a test for the differentiation of the native pigment forms. It was shown later that in greening leaves, the long wavelength protochlorophyllide and chlorophyllide forms were the most photosensitive ones; they were oxidized in the presence of oxygen at room and liquid nitrogen temperatures. At the same time, the short wavelength forms of chlorophyll in chloroplasts are characterized by a higher degree of photosensitivity. At first stages of illumination, the 685 nm maximum in the low temperature fluorescence spectrum decreased, then the long wavelength form (the maximum is at 740 nm) became photobleached relatively faster. The illumination of a carotene maize mutant in the presence of O_2 caused the relative decrease of the long wavelength maximum intensity and the increase of that of the short wavelength one (Vorob'eva and Krasnovsky, 1968; Lang et al., 1968, 1969). The authors came to the conclusion, comparing the in vivo and in vitro data, that the in vivo photobleaching proceeds via the photodisaggregation of the pigment (Vorob'eva and Krasnovsky, 1967, 1969, 1970, 1972).

Krasnovsky and Brin (1968) and Brin et al. (1971) observed the short wavelength shift of the absorption spectrum and the simultaneous inhibition of the Hill reaction under the action of diethyl ether on chloroplast suspension. Both effects were reversible. These facts were interpreted to be due to the destruction and reversible "autoassembly" of the aggregated chlorophyll forms belonging to photosystem II and responsible for O_2 evolution.

Based on the above data, one can differentiate a number of common features determining the spectroscopic characteristics of photosynthetic pigments in the native and aggregated states, as compared with those of pigments in dilute solution. The main feature is the appearance of separate absorption and fluorescence bands (but not a gradual shift of a spectrum). The discrete character of the changes of the position of the electronic levels can be explained by the stability of only certain aggregate types. This, in turn, is the result of the formation of physicochemical bonds that allow only certain angles and distances between chromophores. On the other hand, this discrete character of the spectra is due to some selection rules of quantum mechanics that allow only discrete values of the interaction in an aggregate. In this connection, the similarity in the spectral band positions of the three chemically different pigments (Chl, BChl, chlorobium Chl) attracts attention.

The ratio of the red shift to the monomer half bandwidth changes significantly (from 0.5 to 10, and even more). This implies that chromophore–chromophore interaction varies from weak to strong in the series.

There exist two types of the forms with narrow and wide bands that differ in their temperature dependence. The narrow bands are found to be well described by the Stepanov relation characteristic for molecular spectra. It is possible that these two types of bands belong to two different aggregate types. The forms with the wide band in which the interaction between electronic and vibrational transitions is more pronounced are more likely to be "crystals" (Litvin et al., 1973b).

In search for the law determining the position of the spectra of the native forms, a simple empirical formula was found for the first six forms in the series (see Table 1) (Litvin et al., 1974a).

$$\nu_{cm}^{(N)} = \nu_0 - (N - 1)\,\Delta\nu$$

where N is the index number of the form in their sequence from the short wavelength to the long wavelength ones (e.g., for the "662" Chl form $N = 1$;

for the "668", $N = 2$; for the "676", $N = 3$, and so on), $\nu_0 = 15080\ cm^{-1}$ is close to the maximum position of the shortest wavelength form, and $\Delta\nu$ (a constant) is equal to $137\ cm^{-1}$.

A similar formula is also applicable to the bacteriochlorophyll forms, the interval constant $(\Delta\nu)$ being a multiple of or equal to $137\ cm^{-1}$. Thus, the electronic levels in the series of the forms are separated by $0.017\ eV$ interval or its multiple. One can assume that such a simple general law for different chromophores is due to the nature of monomer interactions in pigment aggregates or polymers.

3. ENERGY MIGRATION IN THE PHOTOSYNTHETIC UNIT (PSU) AND THE ROLE OF TRIPLET STATES IN THE PRIMARY STEPS OF PHOTOSYNTHESIS

3.1. Photophysical Properties of Pigments

Absorption and fluorescence investigations allow calculations of the probability of all main intermolecular transitions in the Chl a molecule in accordance with the Jablonsky scheme (Gurinovich *et al.*, 1968; Solov'ev *et al.*, 1972).

The measurements of triplet–triplet $(T-T^*)$ absorption dichroism for Chl a and some of its analogs indicate that the $T-T^*$ absorption oscillator is in the plane of the molecule. The porphyrin molecule oscillator remains rigidly fixed from 10^{-8} to 10^{-1} seconds (Semenova *et al.*, 1971). (This suggestion has been made from the comparison of the phosphorescence polarization, dichroism, and NMR spectra measurements of porphyrins and their films. The results pointed to the existence of NH tautomers with lifetimes of about 10^{-8}–10^{-1} seconds.) The results of systematic research on porphyrin spectroscopy are presented in a monograph by Gurinovich *et al.* (1968).

New opportunities in the spectroscopy of complex molecules are provided by the Shpolsky effect, i.e., the transformation of the usual wide spectrum into a quasilinear one with narrow (of several cm^{-1}) lines. Such spectra were obtained for porphyrins and phthalocyanins in n-octane solutions at $-196°C$ (Litvin and Personov, 1960). This method is useful for solving a number of problems in porphyrin spectroscopy. The Shpolsky effect could also be observed for Chl itself at $4°K$. The frequencies of the normal vibrations in the molecule, its exact electronic level, and the concentration dependence of the effect were determined (Litvin *et al.*, 1969).

The question of the participation of triplet and singlet molecules in the processes of energy migration and photosynthesis was the subject of a number of papers. The probability of Chl a and b triplet state formation in solution is very high in the absence of competing photochemical reactions. By the use of flash photolysis and fluorescence methods, the radiationless energy degradation of the excited singlet molecule is shown to proceed via triplet state formation in rigid solutions at room and low temperatures (Gurinovich *et al.*, 1968; Dzhagarov and Gurinovich, 1971; Solov'ev *et al.*, 1972).

Recently, the detailed spectroscopic investigations of pigment phosphorescence were made in solution. The position of Chl *a* phosphorescence in ethanol (980 nm for Chl *a* and 915 nm for Chl *b*) was found to differ from those reported earlier. The phosphorescence spectra were measured also for protochlorophyll (770 nm) and pheophytin *a* and *b* (940 nm and 910 nm, correspondingly). The delayed fluorescence of these pigments was found in solution (A. A. Krasnovsky, Jr. *et al.*, 1971, 1973).

3.2 Mechanism of Energy Migration and the Type of the PSU

In viscous solutions of Chl *a* and *b*, pheophytin *a*, and 4-vinyl-protochlorophyll, energy migration is shown to be due to inductive resonance interactions at concentrations up to 10^{-1} moles/liter (Losev and Zenkevich, 1968; Zenkevich *et al.*, 1972a). The energy migration characteristics are not affected by heterogeneity of pigment molecule distribution in micelles and oriented films. However, the effect of the molecular orientation on the concentration dependence of Chl *a* fluorescence polarization was found (Zenkevich *et al.*, 1972b).

It is shown that Chl fluorescence depolarization in greening etiolated leaves is due to energy migration between pigment molecules locally accumulated and incorporated into the cell structure (Losev and Gurinovich 1969). In system II, the local concentration of the fluorescence Chl *a* form is about 10^{-2} to 1×10^{-1} moles/liter. The conclusion is made that the energy migration in system II bulk Chl proceeds via singlet states and is due to inductive resonance interaction (with interaction energies of 1–6 cm^{-1}). The comparison of the polarization measurements *in vivo* and in stretched Chl films has shown that the specific orientation of antenna Chl in plants is low (less than 1–2%) (Zenkevich *et al.*, 1972b).

Tumerman and Sorokin (1967) and Sorokin and Tumerman (1971) established the proportionality between the lifetime and quantum yield of fluorescence during the induction period in intact *Chlorella* cells and chloroplast suspensions. (This was confirmed by Briantais, Govindjee, and Merkelo at the University of Illinois at Urbana.) It was suggested that PSU (also see chapters by Sauer, Knox, and Lavorel, this volume) was multicentral (statistical) by nature and that Chl fluorescence originated mainly in the molecules of the photosystem II chlorophyll matrix and was emitted during the time of the energy migration proceeding via the inductive resonance mechanism. However, in experiments on pea chloroplast (Sorokin, 1971b), the reverse transition of the multicentral PSU into the unicentral (separate units) was observed under the action of CMU. The data obtained (Sorokin, 1971a) allowed the calculation of the size of PSU (148 $\pm$ 4 Chl a and b molecules for a photosystem II reaction center) and the cross section of absorption at 436 nm, equal to 290 $\pm$ 90 Å^2.

Although the triplet molecules can potentially participate in the primary processes of photosynthesis, as suggested by the data on high yield of the generation of triplet states and their photochemical activity in solution (Chibisov, 1969, 1973; Dzhagarov and Gurinovich, 1971; Gurinovich *et al.*, 1968; Solov'ev *et al.*, 1972; Terenin, 1967), the energy migration in real biological structures occurs via singlet states. The above authors believe that energy is transferred in accordance with the "slow" inductive resonance mechanism (as it takes place in solutions).

However, there is a different approach to the problem. It is mainly based on the investigation of lifetimes and quantum yields of fluorescence as functions of the ratio of open to closed reaction centers (Borisov and co-workers). In purple bacteria, the experimentally found fluorescence intensity dependence on the share of the closed reaction centers corresponds neither to the unicentral nor multicentral model. But if constant background fluorescence intensity is subtracted, then the curve obtained is in rather good agreement with the multicentral model. The main argument, considered by the authors to be for the existence of the background fluorescence, is that the measured fluorescence lifetime does not increase but decreases, instead, upon the exciting light intensity increase. The level of the background fluorescence was so chosen that the remaining part (or photosynthetic fluorescence) fitted best in the theoretical Vredenberg–Duysens curve.

The increase of the photosynthetic fluorescence quantum yield is up to ten to thirteen times when reaction centers are photooxidized. This corresponds to the quantum yield of excitation energy trapping of 90–92%. The directly observed twofold fluorescence increase is explained by the authors as a result of the masking influence of the background fluorescence. The calculated relation of background and photosynthetic fluorescence is used by the authors for evaluations of their corresponding lifetimes from the data obtained by fluorimetric lifetime measurements. The calculated photosynthetic fluorescence lifetime is $1\text{--}5 \times 10^{-11}$ seconds in conditions of normal photosynthetic activity. The directly measured lifetime exceeds these values by approximately two orders of magnitude. This discrepancy, according to the authors' view, is due to the background fluorescence. Taking into account high values of calculated quantum yield of photosynthesis, the authors believe that the maximum quantum yield for triplet state formation does not exceed 10%. Thus, the participation of these states in energy transfer in purple bacteria can be neglected.

The above low lifetimes are not in conformity with the idea of the inductive resonance energy transfer mechanism via singlet states in purple bacteria. The energy transfer is suggested to follow the exciton mechanism with intermediate interaction energies ($\sim 10^{-2}$ eV) (Borisov and Godik, 1970, 1972a,b). (See Knox, this volume.)

The primary act of bacterial photosynthesis (P_{890} photooxidation) was shown to proceed within the times not exceeding 5×10^{-11} seconds after the absorption of a light quantum. This fact rules out the interaction of P_{890} with a diffusible primary electron acceptor. The primary charge separation should proceed in a rigid complex of molecules.

The exciton energy transfer mechanism takes place, according to the authors' view, in photosystem I of green plants. Here, the calculated lifetimes are less than 0.07 nsec and the interaction energies are more than 0.02 eV. (Borisov and Il'ina, 1971). The lighter subchloroplast fraction enriched in photosystem I is characterized by hyperbolic dependence of the rate of reaction center photooxidation on the portion of active P_{700}. The data obtained prove, according to the authors' view, the multicentral and connected models of PSU organization. The quantum yield of P_{700} photooxidation was estimated to be $\gg 0.75$ (Borisov and Godik, 1972b; Borisov and Il'ina, 1972).

Thus, according to Borisov and co-workers, the unusually high rates of singlet excitation energy transfer and the energy utilization in the primary act are characteristic for pigments *in vivo*. To confirm this point of view, which is based on a number of assumptions, it would be of interest to measure the decay times of the fast fluorescence components directly.

The problem of the triplet state formation *in vivo* also remains disputable. Borisov and co-workers tried to find triplets in purple bacteria by direct measurements. Photosynthesis was blocked by deep freezing or by the combined action of dithionite and saturating light. In experiments with flash photolysis, no absorption changes were found that can be attributed to the triplet–triplet transition; the calculations show that the quantum yield for triplet state formation is less than 1%. It might be higher, but in this case, the triplet lifetime could not exceed 10^{-5} seconds (Borisov et al., 1970a).

If the probability of a singlet–triplet interconversion *in vivo* is as high as in solution, then the observed absence of triplets *in vivo* points to the very high singlet state deactivation rates in the migration process or in the primary energy utilization act. Interconversion into triplet state, even with the rate constant of 10^9 seconds can not compete with such a fast process, and the triplets may be neglected. However the triplet state formation may not be observed because of short triplet lifetimes.

The direct measurements of *in vivo* pigment phosphorescence was performed only for porphyrins, which are not connected with the photosynthetic structure (Shuvalov and Krasnovsky, 1971). However, one of the components of the delayed light emission in green plants may be interpreted as delayed Chl fluorescence according to a number of its features, which may point to the existence of triplets (see below).

3.3. Energy Migration within Native Chlorophyll Forms

Energy migration is known to be one of the most important steps in the photophysical act of photosynthesis. The results below deal with energy transfer among the different spectroscopic Chl forms.

The red drop of the fluorescence quantum yield, its increase upon cooling of a sample, and dependence on background illumination were compared with the fluorescence spectra changes (Litvin and Sineshchekov, 1967a,b). This allowed us to distinguish three functional groups of the Chl forms differing by their ways of participation in the processes of absorption, emission, and transfer of quanta: the short-wave donor group (the first four forms in Table I, the intermediate (the following four forms) and the long-wave acceptor group (all the others in the series).

Based on the data on the position and shape of the spectra of discrete Chl forms spectra (see above), the theoretical calculations of the Förster overlapping integral were performed. It indicated that the downhill energy transfer (from the short-wave form to the long-wave ones) is possible within the whole system of the Chl forms, the uphill energy migration is effective between two spectroscopically adjacent forms.

Concentrations of the forms (evaluated from the analysis of the absorption spectra, see above) affect the energy distribution. This factor determines the mean distance between a donor and an acceptor and the ability of a form to compete for its share of excitation energy, if one assumes the statistical distribution of the centers. It increases the probability of the energy transfer among the main absorbing (short-wave) forms. At the same time, the energy flow to and among the long-wave forms is decreased because of their low concentration (Litvin *et al.*, 1971).

Another approach to the problem is based on the quantitative analysis of the experimental data. Analysis of the three types of spectra (absorption, excitation, and emission) allowed the derivation of the system of equations for balance and exchange of energy. Solving it, one can find the coefficients of energy transfer between the forms and the energy utilization coefficients for each of them. As a first step in this direction, mean values of these coefficients are obtained for the three groups of the Chl forms. These values confirm that the short-wave group is a donor group, the intermediate is a donor–acceptor group, and the long-wave is an acceptor group. It is likely that the forms with broad absorption and fluorescence bands (see Table I) are to play a substantial role as acceptors in the energy distribution process (Litvin et al., 1973b; Sineshchekov, 1972; Sineshchekov et al., 1973a).

The existence of several native forms may considerably speed the process of energy migration to the reaction centers. The computer calculations performed for ten different two-dimensional concentric models revealed that the number of excitation jumps required for efficient energy trapping diminished 2.6–10.4 times (to compare with that of the homogeneous model). The mean value, equal to 5, was used for calculating the in vivo jump times among BChl molecules (Borisov and Fetisova, 1971).

The main features of the energy transfer, which are characteristic for the chloroplast pigment layers, can also be obtained in the model systems. The energy migration from the short-wavelength forms to the long-wave ones appears upon the increase of the local chlorophyll concentration in a film or solution (Litvin et al., 1965; Litvin and Sineshchekov, 1967b; Sineshchekov et al., 1972).

When mixed monolayers (Chl a + carotenoid (i.e., cis- and trans- β-carotene), fucoxanthin, or zeaxanthin) are compressed so that the distance between the molecules is less than 20 Å, the energy transfer from carotenoid to Chl begins and proceeds possibly in a complex of the two pigments. The energy transfer efficiency increase correlates with the formation of the two-pigment complex and with the increase of the maxima of the long-wave Chl a aggregates. The observed picture is in general similar to the development of the energy transfer phenomena during chlorophyll accumulation in greening leaves and can possibly be considered as a model for the "autoassembly" of the pigment part of a PSU (Litvin et al., 1965; Sineshchekov et al., 1972).

Research on pigment photochemistry are of great importance for the understanding of the following and most important stage of photosynthesis, namely the transformation of quantum energy into chemical bond energy. To our regret, we are not able to discuss this special and very large field. Reference is made however, to a number of monographs and reviews (see Krasnovsky, 1960, 1969, 1972; Terenin, 1967; Evstigneev, 1963; "Molecular Photonics" Nauka, Leningrad, 1970).

New data on photoreduction and photooxidation are described and discussed by Evstigneev (1970), Evstigneev and Gavrilova (1972) Evstigneev and Shvedova (1971), Evstigneev et al. (1972), Chibisov (1969), Kiselev et al. (1970), Nazarova and Evstigneev (1971), Losev et al. (1972), Suboch et al. (1972) and Gurinovich et al. (1972).

4. DELAYED LIGHT EMISSION OF PHOTOSYNTHETIC ORGANISMS AND MODEL SYSTEMS—ENERGY TRAPS IN PHOTOSYNTHESIS

4.1. Afterglow (Delayed Light Emission) of Chloroplasts and Chromatophores

The energy migration in a PSU is followed by excitation energy trapping and its stabilization in the reaction centers. In the investigation of the unstable energy-rich products thus obtained, physical methods, afterglow (delayed light emission, DLE) measurements of cells and chloroplasts in particular, proved to be useful (see Lavorel this volume).

The DLE decay kinetics, light curves, temperature dependences of intensity and lifetimes, excitation and emission spectra, and action of photosynthetic inhibitors were systematically investigated (Litvin *et al.*, 1960b; Litvin and Shuvalov, 1968; Shuvalov and Litvin, 1969). As a result, five main components are found to constitute the whole process of the DLE. They differ by their spectroscopic and kinetic characteristics and by the mechanism of light quanta emission. Components I, III, IV, and V of the DLE are connected with light quanta absorption and emission by chlorophyll. (The less investigated component II is excited in the blue region of the spectrum and emits in the red.)

Component I is characterized by the short lifetime (1 msec). Its quantum yield and lifetime are constant in a temperature range from $-160°$ to $40°C$. The light curve of the emission is strictly linear at temperatures below $-20°C$, which points to the monomolecular character of the luminescence. But at $20°C$, the light dependence is quadratic. The luminescence is effectively inhibited by DCMU even at low temperatures.

It is found that DLE excited by $\lambda = 650$ nm at $20°C$ is inhibited by supplementary light ($\lambda = 700$ nm). This suggests that the emission is connected with photosystem II. In the lighter chloroplast particles, DLE is activated by ferricyanide, which oxidizes P_{700} completely in the dark.

Component I was supposed to be due to the reverse transition of an electron from the triplet level (similar to delayed fluorescence) (Litvin and Shuvalov, 1968). But this emission was found to weakly depend on O_2 concentration.

On the other hand, the afterglow in chloroplasts is shown to increase sharply at $-40°C$ in the presence of ferricyanide and other salts ($[Fe(CN)_6]_3$ or $Fe(NH_4)(SO_4)_2$, etc.). DCMU inhibits only DLE activated by these salts. The combined action of Mn^{2+} and Fe^{3+} is to inhibit DLE. Accordingly, DLE at $-40°C$ was interpreted as a result of the reversal of electron transfer between chlorophyll and Fe^{3+}. The authors believe that the above activation of DLE by the salts is due to the water molecule transfer to photoactive Chl via manganese or magnesium pores in the chloroplast membrane. The diameter of the pores are clearly determined from the maximum sum of anion and cation diameters (Shuvalov and Krasnovsky, 1972).

Thus, the emission (component I) is suggested to be due to the reversal of electron transfer and is a recombination by nature. It is possible that triplet Chl molecules participate in electron transfer.

Components III and IV (the latter is observed in the presence of DCMU) with $\tau_{1/2} = 1$ second are found only at temperatures above $-20°C$. The threshold of light saturation is close to that for photosynthesis, but photoproducts responsible for the emission can be formed even at $-196°C$. In this case the emission is thermoluminescence observed during heating. The maxima on the thermoluminescence curve are connected with the different DLE components. The emission is shown to be connected with energy traps in chloroplast (Litvin and Shuvalov, 1968; Shuvalov and Litvin, 1969). The depths of the traps for the singlet Chl levels are 0.35 and 0.9 eV (for components III and IV, respectively).

Energy barriers for the traps' depopulation in the direct photosynthesis process are about 0.27 and 0.42 eV for the two components, respectively, calculated from the lifetime activation energy. Spectral data show that component III belongs to photosystem II, and component IV to photosystem I. Component III is effectively excited by the short-wave Chl forms (Chl_{650}, Chl_{660}, and Chl_{675}). Component IV is excited at the maxima of 650 and 680 nm, and there is no quantum yield drop for it beyond 680 nm (Litvin et al., 1966; Shuvalov and Litvin, 1969). The effects of DLE enhancement and inhibition observed under the action of the two monochromatic light beams separated by a time interval of 20–40 seconds and the development of DLE in greening leaves were investigated by Litvin et al. (1966). These authors suggested a scheme for the participation of the chlorophyll forms in the afterglow and energy storage (Fig. 7B). The energy traps are in the main photosynthetic chain. They provide the efficient coupling of the primary physical stages with the following slower ones. High values of the trap population and depopulation constants and trapping cross-sections speak well for it. The trap with the depth of 0.35 eV has a crossection of 25 Å^2 and that with the depth of 0.9 eV, 100 Å^2. The high efficiency of energy trapping is suggested by the good correlation between the DLE (component IV) and the fluorescence quantum yield changes. The results allow the conclusion that the corresponding trap is the main energy collector from the excited singlet Chl molecules (Shuvalov and Litvin, 1973).

Rubin and Venediktov (1968, 1969) and Venediktov et al. 1969b, 1971) have established the relation of long-lived DLE components (with the lifetimes ranging from 0.5 seconds to several minutes and thermoluminescence to the photosystem I and II reactions. The component with the life-

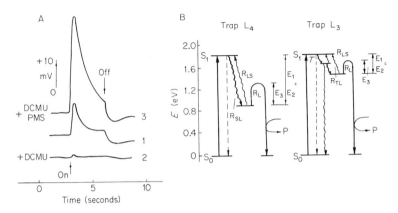

Fig. 7. (A) Photoinduced changes of the transmembrane potential: (1) potential measurements in a single illuminated chloroplast of *Piperomia metallica* by the use of a microelectrode; (2) the same, effect of DCMU, $10^{-6} M$; (3) the same as 1, effect of DCMU + PMS, $10^{-6} M$ (Bulychev *et al.*, 1971a,b). (B) Hypothetical scheme of the two types of the energy traps in a chloroplast manifesting in delayed light emission. L_4 is the trap of the long-wavelength system, $E_1 = E_2 = 0.9$ eV—activation energy of the reverse process obtained from the temperature dependence of afterglow (delayed light emission) and thermoluminescence; $E_3 = 0.42$ eV—activation energy of the direct process, as measured from the temperature dependence of the afterglow lifetime of delayed light emission; L_3 is the trap of the short-wavelength system; $E_1 = E_2 = 0.35$ eV, $E_3 = 0.27$ eV; P are the primary products of photosynthesis; $S_0 =$ singlet ground level, $S_1 =$ singlet excited level, $T =$ triplet level (Shuvalov and Litvin, 1969).

time of 0.5 seconds is generated in photosystem II; its intensity increases when the electron transfer between photosystems I and II is blocked. The component with the lifetime of several minutes is excited by near infrared light ($\lambda = 700$ nm), inhibited by DCMU and activated by pyocyanine. Proceeding from this, the authors explain the DLE component by the reverse electron flow from photosystem I to photosystem II through the chain of carriers. Its delay, due to the operation of the cyclic electron transport, causes the appearance of the maximum on the DLE kinetic curve. DCMU, as suggested, prevents not only the direct electron flow but also the reverse electron flow (from photosystem I to photosystem II).

It was found that the 10 msec delayed light emission in chloroplasts or *Chlorella* cells is inhibited by DCMU. It can be reactivated by the addition of electron donors for photosystem I. The lighter digitonin subchloroplast particles (photosystem I) do not emit this component. But in the presence of reduced dichlorophenol–indophenol or electron donors with redox potential less than 0.4 V, the particles are capable of emitting the 10 msec delayed light. These facts allow the conclusion that the millisecond afterglow is generated in photosystem I under conditions when pigment P_{700} can be reduced.

In the investigation of delayed light emission in photosynthesizing purple bacteria, it was found that for the millisecond-delayed light emission, the electron flow maintained by the photosynthetic reaction center activity is required. It is likely that the process is based on the reaction of the oxidized primary donor with the reduced acceptor. However, the delayed light can not be observed at 80°K. It is suggested that the electron flow creates an electrochemical proton gradient, which increases the probability of the electron recombination with oxidized BChl (Venediktov *et al.*, 1969a; Matorin *et al.*, 1971; Kononenko *et al.*, 1971b).

4.2. Afterglow of Pigments in Solution

Long-lived afterglow of the photosynthetic pigments in simple systems is a possible model simulating the photosynthetic DLE. The analysis of this phenomenon is one of the ways to elucidate the mechanisms of pigment photochemical reactions. Photochemiluminescence appearing in the reaction of O_2 with photoreduced chlorophyll was found (Litvin *et al.*, 1960b). Here, the emission connected with the oxidation of the pigment itself was also described.

Thermoluminescence of chloroplasts, extracts, and pigment solutions illuminated in the presence of oxygen was investigated by Litvin *et al.* (1960b) and by Tumerman and co-workers (1962). Later, the systematic investigation of the spectra, kinetics, and mechanism of pigment photochemiluminescence in solution under oxidation by oxygen were undertaken. The phenomenon is shown to be common for different types of molecules (Chl *a, b* protochlorophyll, *Chlorobium* chlorophyll, their pheophytins, porphyrins, synthetic dyes, and polycyclic hydrocarbons) (Litvin and Krasnovsky, 1967; Krasnovsky and Litvin, 1967, 1968, 1969, 1970). The experimental data allowed interpretations that lead to the mechanism of the stepwise process and a general scheme. According to this scheme, the emission follows the thermal destruction of labile pigment peroxides, which are formed in the reactions of the triplet pigment molecules with oxygen. The spectra show that terminal emitters are the singlet molecules of the pigments and products of their oxidation.

Belyakov and Evstigneev (1971) observed chemiluminescence in the course of the action of ascorbic acid on the products of reversible Chl *a* photooxidation by *p*-benzoquinone.

The relation of the photochemiluminescence in solution to the photosynthetic luminescence remains disputable. The combination oxygen–pigment–light, which is characteristic of photosynthetic DLE in green plants, is necessary also for the oxidative photochemiluminescence. Therefore, it is not unlikely that at least some components of the photosynthetic

DLE are due to a process similar to that in the *in vitro* system. These similarities are probably related to the long-lived (seconds) components, which are the oxygen-dependent components of the photosynthetic emission.

Low temperature thermochemiluminescence in chlorophyll solutions (Krasnovsky and Litvin, 1968, 1969, 1970) illuminated in the presence of oxygen can perhaps be considered as an analog to the low temperature thermoluminescence in leaves and cells. Its analysis allowed the calculation of the activation energy of the products responsible for the photochemiluminescence of pigments.

Electrochemiluminescence is an interesting model for the photosynthetic DLE appearing in the course of the electron transfer (Krasnovsky and Litvin, 1972). Chemiluminescence was found during the electrolysis of Chl in dimethylformamide. The character of the emission was shown to be dependent on the electrode potential and oxygen concentration. When the polarity of current was reversed the glow was observed; it was due to the recombination of the products of electroreduction and electrooxidation, probably of radicals or ion-radicals.

5. ELECTRICAL PHENOMENA IN MEMBRANE STRUCTURES CONNECTED WITH ENERGY STORAGE

5.1. Photoinduced Transmembrane Potential Changes in Chloroplasts

In recent years, the problem of energy relation between photosynthetic processes and ion transport in a cell and a chloroplast and the possible role of the electric potential in energy transformation have been receiving increasing attention by experimenters (see chapters by Lavorel, Witt, and Jagendorf, this volume). This problem was the subject of joint investigations held by Kurella and Litvin's research groups.

It has been established that the action of light, absorbed by the photosynthetic pigments, is followed by changes in the membrane potential in algae and plant cells. Action spectra measurements and experiments with inhibitors pointed to close links between the photoinduced changes and photosynthesis (Andrianov *et al.*, 1965, 1968a,b, 1969, 1970, 1971).

Recently, experiments on microelectrode insertion in a chloroplast and potential measurements were successfully performed both on a chloroplast inside an intact cell and on an isolated chloroplast of the higher plant *Peperomia metallika* (Bulychev *et al.*, 1971a,b, 1972a,b). Two types of photoinduced changes were found—slow and rapid. The slow photoinduced changes appear as the negative shifts of chloroplast potential (relative to cytoplasm). The fast photoinduced changes are observed as sharp (within 5–10 msec) bursts of the potential to the positive side.

The magnitude of the fast potential changes reached 75 mV. The action spectrum for the fast potential changes corresponds to the absorption spectrum of the photosynthetic pigments. Under the action of two monochromatic light beams (red and far red), the enhancement effect for the chloroplasts photoelectric response was observed. This is analogous to the Emerson enhancement effect for photosynthesis. It is interesting to mention that the effect was observed when there was a certain interval between the action of two beams, as was already shown for the DLE effect (see above). Temperature decrease of a medium down to 3°C has practically no effect on the fast photoinduced changes. In the presence of DCMU, light-induced changes are inhibited, but they can be restored by the addition of PMS (Fig. 7A) (see Bulychev et al., 1971a,b, 1972a,b). Thus, photosystem I under isolated excitation may be responsible for the potential generation. The experiments with inhibitors, electron transport cofactors, and uncouplers point to the direct coupling of the first steps of photosynthesis with the bioelectric potential generation and ion transport in a chloroplast. They can be explained by the Mitchell theory (see Witt and Jagendorf, this volume). The correlation between the photoinduced changes of pH and the potential indicates the important role of the proton transport.

The short time of the photoresponse and its weak dependence on temperature and other factors suggest a physical nature of the primary photoprocesses. It is assumed, that the charge separation may be carried out owing to the semiconductive properties of Chl incorporated in membrane structures of a chloroplast (see below).

The parallel action spectra measurements of photosynthesis and photoinduced changes of the resting and redox potentials in *Nitella* cells were made by the use of platinum (gold) quartz isolated microelectrodes (for results see Yefimtsev, 1972).

5.2. Photoinduced Transmembrane Potential Changes in Bacterial Chromatophores

Such measurements of the potential in bacterial cells or chromatophores are impossible at present, but the potential change can apparently be followed by using indirect methods.

Light energizing of chromatophore membranes in photosynthesizing purple bacteria results in a decrease in the rate of light-induced electron flow and the red shift of the major absorption band of bulk bacteriochlorophyll. These phenomena are inhibited by uncouplers, chloride or fluoride derivatives of phenylhydrazone, and atebrin and also at liquid nitrogen temperature (Kononenko et al., 1971a, 1972a,b; Kononenko, 1973).

It has been shown (Isaev et al., 1970) that the chromatophores of *Rhodo-*

spirillum rubrum accumulate the penetrating anions phenyl dicarbounde-carborane (PCB⁻) and tetraphenyl boron (TPB⁻), as well as iodide, if the electron carrier dipentafluorphenylmercury, is added. The anion accumulation can be supported by light-induced cyclic electron flow, by dark electron transfer (NADH ———→O$_2$, succinate ———→ferricyanide), by hydrolysis of ATP or PP$_i$, as well as by forward transhydrogenase reaction (NADPH ———→NAD⁺). The type of energy source influences only the extent of the anion accumulation process.

Switching on the light induces alkalinization of the medium containing chromatophores and penetrating anions PCB⁻ or TPB⁻ (Nazarenko *et al.*, 1971). Illumination results in acidification of intact bacterial cell suspensions in the presence of penetrating triphenylmethylphosphonium cations. Anions in experiments with bacteria and cations in experiments with chromatophores proved to be inefficient. Cessation of the energy supply brings about an efflux of the accumulated ions and reverses the pH changes of the medium. It is stated that the electric field is the driving force for ion transfer through the bacterial membranes against a concentration gradient.

Experimental evidence was provided for the membrane potential by the observations of energy-dependent carotenoid (Glinsky *et al.*, 1972) and bacteriochlorophyll (Barsky and Samuilov, 1972) absorption changes in *R. rubrum* chromatophores. Light-induced changes of bacteriochlorophyll absorption at 750–950 nm were found in *R. rubrum* chromatophores in the presence of oxidizable substrate. The observed absorption shifts were induced by the following treatments: illumination, ATP or PP$_i$ hydrolysis, NADH oxidation under aerobic conditions, exergonic transhydrogenase reaction of NADPH and NAD⁺, and generation of a K⁺ gradient. The energy-linked, uncoupler-sensitive changes of BChl absorption were found for the sulfur purple bacteria *Chromatium minutissimum* and *Ectothior-hodospira shaposhnikovii*. The authors consider the data obtained an indication that the shifts of bacteriochlorophyll absorption bands are due to the electrochemical potential generated in the chromatophore membranes.

5.3. Semiconductor Properties of Chloroplasts and Chromatophores

Hypotheses explaining the role of semiconductive processes in photosynthesis were proposed earlier by a number of authors. However, the experimental material on which they were based was rather limited. Owing to the improvement of the method, quantitative investigations of photoconductivity in chloroplasts and bacterial chromatophores were performed systematically with the use of a vacuum cell at room and low temperature (Litvin and Zvalinsky, 1967, 1968, 1971). From the measurements of action spectra, it was concluded that the photoconductivity was due to the light absorption by the photosynthetic pigments. However, the accessory pigments and the short wavelength Chl *a* forms participate, apparently, only in absorption and energy transfer. The charge separation process itself takes place in long-wavelength aggregated Chl *a* forms (the excitation spectrum maxima being 690, 700, 705, 712–715 nm). The width of the forbidden band calculated from the activation energies of dark conductivity (1.7 eV) coincides with the value obtained by measuring red drop of the photoconductivity. Thus, chloroplasts are semiconductors with intrinsic conductivity due to Chl. Aggregated Chl forms in films prepared from the pure pigment are found to possess similar features.

The analysis of an electric equivalent scheme with the use of a computor allowed the calculation of the quantitative characteristics of photoconductivity in chloroplasts (Litvin and Zvalinsky, 1971)—the local specific photoconductivity in films is 10^{-11} ohm^{-1} cm^{-1} (at light intensity of 10^{14} quantum cm^{-2} sec^{-1}), and the quantum yield of photocarriers is about 10^{-1} to 10^{-4}.

The characteristic of chloroplast films indicate that they are similar to molecular semiconductors such as phthalocyanines. The variable component of the photocurrent (0.05–0.5 seconds) was found to be connected with the long-wavelength photosystem, and the constant component suppressed by DCMU with the short-wavelength one. According to the authors' view, bulk polarization in chloroplast membrane structures plays a substantial role in these processes. It manifests, in particular, as photoconductivity appearing after switching off the applied voltage (the photocurrent due to the polarization electromotive force). By a number of features, the corresponding potential is analogous to the photoinduced potential measured in a whole chloroplast (see above). It is suggested that a part of the chloroplast membrane with the incorporated pigment is "a semiconductive electron pore" (Litvin *et al.*, 1973). The electron flow through it causes the appearance of the transmembrane potential, which is the driving force for ion flows. On the other hand, one cannot exclude the possibility that the electron transfer in aggregated Chl near the reaction center, where usual diffusion processes are slowed down, is one of the steps in the photosynthetic energy storage. An attempt to find free charges in intact chloroplasts with the use of photoconductivity measurements at high frequencies is undoubtedly of interest.

According to the authors' opinion photoconductivity measurements in photosynthetic samples at 10^{10} Hz point to the appearance of free or loosely bound charges in chloroplasts under illumination (Blumenfeld *et al.*, 1970).

The kinetic investigations of the narrow photoinduced singlet ESR signal in chloroplasts have shown the existence of not less than two types of different paramagnetic centers. The characteristics of photoinduced paramagnetic center formation are the following: the activation energies for the fast and slow stages of the process are 0.1 eV and 0.2 eV, respectively; the activation energy of the slow stage for the decay process is 0.18 eV.

The investigation of the narrow photoinduced ESR I signal dispersion allowed the separation of photoinduced paramagnetic centers according to their relaxation characteristics. The photoinduced ESR I signal was conditioned by five types of paramagnetic centers.

The average local concentration of the centers in primary photophysical stages of photosynthesis is about 10^{18}–10^{20} spins/cm^3, which is much higher than the mean spin concentrations in the sample (Blumenfeld *et al.*, 1968; Kafalieva *et al.*, 1969a,b).

The dependence of photoinduced ESR signals in plants and bacteria on the action of the inhibitors of photosynthesis, temperature and other factors was investigated by Timofeyev (1972) and Timofeyev and Rubin (1969); the connection between the ESR signals and the electron transport and P_{700} and $P_{870-890}$, conversion was confirmed. In purple bacteria, reversible ESR signals were found to appear at temperatures down to $-150°C$.

6. ENERGY STORAGE IN PHOTOPHOSPHORYLATION AND ITS COUPLING WITH ELECTRON TRANSFER

6.1. Photosynthetic Bacteria

By means of sensitive stationary and flash laser optical spectroscopy, it has been shown that light-induced redox reactions are observed in whole cells of purple sulfo-bacterium *Ectothiorhodospira shaposhnikovii* at 80°K. (Also see Sauer, this volume.) Under aerobic conditions, the reversible oxidation of the reaction center BChl P_{890} (Kononenko *et al.*, 1971a,b) and of a small part (less than 10%) of the high potential c_H cytochrome pool takes place, while under anaerobic conditions only the nonreversible oxidation of the low-potential c_L cytochromes is observed. At the same time, P_{890} reactions are greatly inhibited. The halftimes of both c_L and c_H cytochrome photooxidation, measured by means of pulsed laser spectroscopy, are approximately of I msec at room temperature.

According to the authors' view, the analysis of the results obtained supports the suggestion that there is only one photochemical system in *E. shaposhnikovii* (Kononenko *et al.*, 1971b; Grigorov *et al.*, 1970, 1971; Kononenko, 1973).

The rate of reduction of photooxidized cytochrome c_H was shown to depend on the concentration of endogenous photoreduced products. Effective cytochrome c_L oxidation by molecular oxygen is brought about by oxidase that is inhibited by potassium cyanide and salicylaldoxime. The halftime of

c_H cytochrome dark reduction varies greatly in contrast to that of P_{890} reduction (60 msec), which remains unaffected when light illumination period or exogenous electron donor concentration are changed. The fast reduction of photooxidized P_{890} is supposed to be carried out by photo-reduced electron acceptors, provided cytochrome c_H is oxidized or reduced slowly (Rubin *et al.*, 1968; Kononenko *et al.*, 1971b; Kononenko, 1973). Under these conditions, the process of P_{890} reduction is found to include both temperature-dependent and temperature-independent stages, the later with the characteristic halftime of 25–30 msec.

The data obtained allow the conclusion that in purple sulfo-bacteria the functional complex of photochemically active bacteriochlorophyll P_{890} and the primary electron acceptor is common for both the cyclic and noncyclic photosynthetic electron-transport systems. The high-potential c_H and the low-potential c_L cytochromes mediate the cyclic and noncyclic electron transfers, respectively (Rubin *et al.*, 1969; Kononenko *et al.*, 1971b; Kononenko, 1973).

The measurements of the difference spectra of purple bacteria and chromatophores revealed the increases of absorption at 865 and 905 nm in reducing conditions. Thus, Karapetyan and co-workers believe that these absorption changes are unlikely to be connected with conformation changes of a protein carrier.

The difference in the changes at 850 nm and 890 nm observed under the action of inactivating factors and redox agents suggests that the changes at 850 nm are due to photophosphorylation and can be considered an index of chromatophore activity (Karapetyan and Krasnovsky, 1968; Karapetyan, 1969; Karapetyan *et al.*, 1966, 1972).

To localize energy-coupling sites in chromatophores, the sensitivity of NAD^+ photoreduction to uncouplers was tested under conditions preventing the operation of alternative electron-transfer pathways. It is shown that photoreaction of NAD^+ in *R. rubrum* chromatophores occurs via an energy-dependent reversed electron transfer pathway. Hence, there is an energy-coupling site in the NADH dehydrogenase segment of the chromatophore redox chain (Borisov *et al.*, 1970b; Isaev *et al.*, 1970).

6.2. Green Plants

The comparison of the absolute efficiencies of photophosphorylation in isolated pea chloroplasts under the excitation of photosystem I or both the photosystems simultaneously leads to the conclusion that there are at least two sites of ATP formation in the electron transport chain between the two photosystems. These two coupling sites are located above and below the point of DCMU action. The study of photoinduced absorption changes at

520 nm shows the existence of a coupling site of the electron transfer and phosphorylation at the step between the artificial electron donor (reduced DCPIP) and NADP (Kononenko and Lukashev, 1972; A. Rubin and Krendeleva, 1971; Krendeleva *et al.*, 1971).

Experiments on destruction and photodestruction were made as an approach to the analysis of the pigment apparatus structure. The differences observed in chloroplast fluorescence-induction kinetics are interpreted to be associated with the loss of endogeneous reducing agents during the isolation of chloroplasts. These agents protect the electron transfer chain against the action of oxygen (Karapetyan *et al.*, 1971a,b). The photodestruction of the photosystem II centers in the presence of dithionite was found in chloroplasts illuminated by light absorbed in pigment system II. The action of light absorbed by pigment system I is reversible under these conditions. Irreversible loss of the fluorescence induction in chloroplasts under illumination by light absorbed by pigment system II is accompanied by the loss of their ability to perform the Hill reaction, and by the photobleaching of about 1% of the Chl. (However, the chloroplasts retain their ability of photoconversion of P_{700}).

In chloroplast fragments enriched in photosystem I, variable fluorescence was found; the fluorescence changes correlated with P_{700} photooxidation. The unicentral model of the photosynthetic unit is supposed to be applicable to these particles (Karapetyan *et al.*, 1973). The fluorescence induction is found in the lycopene and ζ-caroteneless maize mutants. The authors have concluded that the two photosystems exist in these mutants (Klimov *et al.*, 1972).

Preillumination of chloroplasts with pulses of a Q-switch ruby laser ($\lambda \geq 700$ nm) (Krendeleva *et al.*, 1972) in some conditions results in a considerable decrease of endogenous photophosphorylation as well as photophosphorylation mediated by pyocyanine, methylviologen, and reduced DCPIP. However, the rate of ATP formation induced by the two photosystems decreases by a smaller value. These facts imply the selective damage of photosystem I. Its efficiency in total ATP formation decreases by 70%, a 30% decrease is due to electron transport inhibition and a 40% decrease to uncoupling of ATP formation associated with photosystem I. The considerable differences of bioluminescence kinetic curves for the luciferin–luciferase system were observed when phosphorylation cofactors were added to illuminated chloroplasts. These differences are probably due to different availability of ATP formed at different sites of electron transport chain for the luciferin–luciferase system (Kukarskich *et al.*, 1972).

The photosynthetic electron transfer and possibly the formation of energy-rich chemical bond proceed via the tunnel effect (see Chernavsky *et al.*, 1972).

Ferredoxins from plants differing in their photosynthetic activity and physiological characteristics have different temperature dependencies of their various activities (Mukhin and Akulova, 1966, 1967; Mukhin *et al.*, 1970a; Mukhin and Gins, 1972). These features are attributed by the authors to differences in protein molecule conformations. The photochemical activity of $NADP^+$ reduction in chloroplasts is shown to be determined not only by the stage of water decomposition but also by a step near photosystem I and the intermediate electron transfer chain (Mukhin *et al.*, 1970b).

7. WATER DECOMPOSITION AND OXYGEN EVOLUTION IN PHOTOSYNTHESIS

The process of water decomposition and oxygen evolution is the unique process of biological oxidation and the most important stage of green plant photosynthesis (see Joliot and Kok, this volume). The oxidized form of chlorophyll is shown to participate in this process, according to investigations by a number of authors and by Kutyurin and his co-workers, in particular (see Kutyurin, 1970, 1972).

Precision mass-spectroscopy measurements have shown that the water molecule itself, not the hydroxyl ion, is oxidized during photosynthesis. Hence, the potential of an oxidation–reduction system oxidizing water molecules is not less than $+0.81$ V (Kutyurin, 1970).

Water is oxidized by Chl itself or by a system coupled to it. Oxidation–reduction properties of Chl and related compounds were studied using oscillopolarographic and other methods, and the consecutive character of Chl oxidation (Chl $\longrightarrow$ Chl$^+$ $\longrightarrow$ Chl^{2+}) has been shown. The stages of oxidation were reversible, and the potentials of corresponding systems were $E_0 = 0.77$ V and $E_0 = 0.90$ V for Chl a and 0.80 and 0.93 V for Chl b in acetone. These investigations and lifetime measurements have also shown that Chl$^+$ and Chl^{2+} are stable in the absence of electron donors and acceptors, their lifetimes being tens of seconds (Kutyurin, 1970; Korsun *et al.*, 1972a,b).

Detergents and high and low temperatures affect the rate of oxygen evolution much more than that of the electron transport from water. The activity of oxygen evolution depends on the content of structure-bonded manganese and is reversibly inhibited by CMU. The data obtained suggest that the two investigated processes, water oxidation and O_2 formation, are relatively independent (Kutyurin *et al.*, 1969; Grigorovich *et al.*, 1971; Zakrzhevsky *et al.*, 1972).

The long-wavelength Chl forms absorbing beyond 700 nm were also found to be effective in the process of oxygen evolution (Kutyurin *et al.*, 1968).

In the aspect of oxygen evolution modeling, the experiments on ultraviolet illumination of water suspensions (TiO_2, ZnO_2) are of interest. In the presence of electron acceptors (Fe^{3+} compounds) the authors observed oxygen evolution from water (A. A. Krasnovsky *et al.*, 1971).

8. NET ENERGY STORAGE AND GAS EXCHANGE IN PHOTOSYNTHESIZING PLANTS

Net storage of light energy by photosynthesizing plants, perhaps the most important aspect of photosynthesis, is usually assessed by indirect methods and primarily by measurements of O_2 evolution or CO_2 consumption. However, gas exchange measurements may yield erroneous results due to a number of causes. Hence, in studies of the energetics of photosynthesis direct energy measurements are not only desirable but in many cases obligatory (Bell, 1965). Bell and co-workers have proposed a photocalorimetric temperature curve method (Bell and Merinova, 1961, 1963); it is based on the assumption that at sufficiently high light intensities, corresponding to saturation of photosynthesis, energy storage is also saturated.

The values in the blue region of the spectrum of the energy yield of photosynthesis in *Chlorella* was found to exceed those calculated from quantum yields, i.e., from gas exchange measurements. It was found that the energy yield of weak red light, if measured on a background of considerably stronger blue light, could be enhanced on the average by 30–40% (Shuvalova and Bell, 1970). The high values of the energy yield cannot be attributed to dark respiratory anomalies produced by weak light, since the measurements were performed at light intensities exceeding 1500 erg/cm² second. Besides this, differential energy yields were measured. The authors believe that the primary processes of photosynthesis are independent of wavelength of light; the source of the discrepancy should probably be sought in secondary energy transformation processes. One possible explanation is photorespiration, but this remains to be proved. The enhanced energy storage phenomenon has been studied exclusively in *Chlorella*. If confirmed in other plants, it may signify, as the authors believe, that present-day concepts on the energy-storing processes *in vivo* may require a revision or extension (Bell, 1972).

9. SUMMARY

To summarize the above data, one can present the following picture (undoubtedly incomplete and hypothetical in several aspects). The energetics of the initial stages in photosynthesis is determined by the interaction of quanta with pigment molecules. However, these are not single isolated molecules in chloroplasts and chromatophores. With the increase in pigment content, the autoassembly of a number of the native forms of Chl and its analogs takes place. These forms are pigment aggregates or polymers by nature with special photophysical and, in particular, spectral properties. Spatial proximity of the aggregates causes the formation of even more complex structures. Such a system has new features, as compared with those of a simple set of molecules. The most important features of the system are strong overlapping of the spectra, special location of the electronic levels forming a stairway of levels, uneven pigment destribution between the Chl forms, and the existence of the conductivity zones. All this provides energy flow to several acceptor forms. The latter ones are probably in contact with reaction centers of different types. The collective properties of the Chl forms are supposed to be the basis of another function, namely, the generation of membrane potential (especially, if the latter proceeds via the semiconductive mechanism).

The energetics of a reaction center is of special interest. The fluorescence lifetime measurements and other data show that the primary process of photosynthesis takes place most likely in a complex of molecules. Such a complex is probably identical to the energy traps, investigated by measuring delayed light emission and thermoluminescence and plays a key role in photosynthesis.

The question of the participation of triplet and singlet states in the reaction center action still remains open. The fact that triplet states of chlorophyll are not observed in a living cell can be explained by the high values of the ratio of the rate constant of photochemical deactivation of excited singlet states to the interconversion rate constant *in vivo* (in comparison with the corresponding ratio for chlorophyll in solution). This, in turn, can be due to the increase of the photochemical deactivation rate constant (reaction in a complex of molecules?) or to the decrease of the interconversion rate constant.

The interpretation meets with difficulties because the interconversion constant *in vivo* is not known. Besides this, a constant for an aggregated native form may be different from that of molecule in solution. Here, the investigation of phosphorescence and photochemistry of aggregates and the use of laser techniques are suggested for future study of fast processes in reaction centers. They undoubtedly will enrich the store of information on the electron transfer mechanism and the coupling sites of the electron transfer with phosphorylation.

The problems of water decomposition and net energetics of photosynthesis, in spite of their long history, still remain unsolved and now seem to be the object of a new phase of intense research.

ACKNOWLEDGMENT

The authors express deep gratitude to all the investigators who contributed to the preparation of this chapter.

REFERENCES

Andrianov, V. K., Kurella, G. A., and Litvin, F. F. (1965). *Biofizika* **10**, 531.
Andrianov, V. K., Kurella, G. A., and Litvin, F. F. (196ºa). *Biofizika* **14**, 78.
Andrianov, V. K., Kurella, G. A., and Litvin, F. F. (1968b). *Abh. Deut. Akad. Wiss. Berlin, Kl. Med.* **4a**, 187.
Andrianov, V. K., Kurella, G. A., and Litvin, F. F. (1969). *Tsitologiya* **11**, 1014.
Andrianov, V. K., Bulychev, A. A., Kurella, G. A., and Litvin, F. F. (1970). *Biofizika* **15**, 190.
Andrianov, V. K., Bulychev, A. A., Kurella, G. A., and Litvin, F. F. (1971). *Biofizika* **16**, 1031.
Barsky, E. L., and Samuilov, V. D. (1972). *Biokhimiya* **37**, 1005.
Bell, L. N. (1965). *In* "Biochemistry and Biophysics of Photosynthesis (Biokhimiya i Biofizika Fotosinteza). (A. A. Krasnovsky, ed.), p. 252. Nauka, Moscow.
Bell, L. N. (1972). *In* "Theoretical Bases of Photosynthetic Productivity" (Teoreticheskie Osnovy Produktivnosti Fotosinteza) (A. A. Nichiporovich, ed.), p. 50. Nauka, Moscow.
Bell, L. N., and Merinova, G. L. (1961). *Fiziol. Rast.* **8**, 161.
Bell, L. N., and Merinova, G. L. (1963). *Fiziol. Rast.* **10**, 505.
Belyakov, V. A., and Evstigneev, V. B. (1971). *Biofizika* **16**, 544.
Blumenfeld, L. A., Chetverikov, A. G., Kafalieva, D. N., and Vanin, A. F. (1968). *Stud. Biophys.* **10**, 101.
Blumenfeld, L. A., Kafalieva, D. N., Lifshits, V. A., Solov'ev, I. S., and Chetverikov, A. G. (1970). *Dokl. Akad. Nauk SSSR* **193**, 700.
Borisov, A. Yu., and Fetisova, Z. G. (1971). *Mol. Biol.* **5**, 509.
Borisov, A. Yu., and Godik, V. I. (1970). *Biochim. Biophys. Acta* **223**, 441.
Borisov, A. Yu., and Godik, V. I. (1972a). *Bioenergetics* **3**, 211.
Borisov, A. Yu., and Godik, V. I. (1972b). *Bioenergetics* **3**, 515.
Borisov, A. Yu., and Il'ina, M. D. (1971). *Biokhimiya* **36**, 822.
Borisov, A. Yu., and Il'ina, M. D. (1972). *Progr. Photobiol., Proc. Int. Congr. 6th, 1972*, Book of Abstracts, p. 235.
Borisov, A. Yu., Godik, V. I., and Chibisov, A. K. (1970a). *Mol. Biol.* **4**, 500.
Borisov, A. Yu., Kondrat'eva, E. N., Samuilov, V. D., and Skulachev, V. P. (1970b). *Mol. Biol.* **4**, 795.
Brin, G. P., Krasnovsky, A. A., and Komarova, L. F. (1971). *Dokl. Akad. Nauk SSSR* **197**, 713.
Brody, S. S. (1958). *Science* **128**, 838.
Brody, S. S. (1961). *Nature* **189**, 547.
Bulychev, A. A., Andrianov, V. K., Kurella, G. A., and Litvin, F. F. (1971a). *Fiziol. Rast.* **18**, 248.
Bulychev, A. A., Andrianov, V. K., Kurella, G. A., and Litvin, F. F. (1971b). *Dokl. Akad. Nauk SSSR* **197**, 473.
Bulychev, A. A., Andrianov, V. K., Kurella, G. A., and Litvin F. F. (1972a). *Nature (London)* **236**, 175.
Bulychev, A. A., Andrianov, V. K., Kurella, G. A., and Litvin, F. F. (1972b). *Fiziol. Rast.* **19**, 443.
Bystrova, M. I., and Krasnovsky, A. A. (1967). *Mol. Biol.* **1**, 362.
Bystrova, M. I., and Krsnovsky, A. A. (1968). *Mol. Biol.* **2**, 847.
Bystrova, M. I., and Krasnovsky, A. A. (1971). *Mol. Biol.* **5**, 291.
Bystrova, M. I., Umrikhina, A., and Krasnovsky, A. A. (1966). *Biokhimiya* **31**, 83.
Bystrova, M. I., Lang, F., and Krasnovsky, A. A. (1972). *Mol. Biol.* **6**, 77.
Chernavina, I. A., Crendeleva, T. E., and Sverdlova, P. S. (1968). *Fiziol. Rast.* **15**, 1008.
Chernavsky, D. S., Chernavskaya, N. M., and Grigorov, L. N. (1972). *Proc. Int. Biophys. Congr., 4th. 1972*, Abstracts, Vol. 1, p. 317.
Chibisov, A. K. (1969). *Photochem. Photobiol.* **10**, 331.

Chibisov, A. K. (1973). *In* "Problems of Biophotochemistry" (A. B. Rubin and V. D. Samuilov, eds.), p. 12. Nauka, Moscow.

Dzhagarov, B. M., and Gurinovich, G. P. (1971). *Opt. Spektrosk.* **30**, 425.

Erokhin, Yu. E. (1967). *Stud. Biophys.* **5**, 171.

Erokhin, Yu. E., and Moskalenko, A. A. (1969). *All-Union Biochem. Congr., 2nd, 1969*, Abstracts, Sect. 19, p. 29.

Erokhin, Yu. E., and Moskalenko, A. A. (1972). *Proc. Int. Biophys. Congr., 4th, 1972.* Abstracts, Vol. I, p. 364.

Erokhin, Yu. E., and Sinegub, O. A. (1970a). *Mol. Biol.* **4**, 401.

Erokhin, Yu. E., and Sinegub, O. A. (1970b). *Mol. Biol.* **4**, 541.

Evstigneev, V. B. (1963). *Biofizika* **8**, 664.

Evstigneev, V. B. (1970). *Biofizika* **15**, 239.

Evstigneev, V. B., ānd Gavrilova, V. A. (1972). *Biokhimiya* **37**, 952.

Evstigneev, V. B., and Shvedova, T. A. (1971). *Biofizika* **16**, 25.

Evstigneev, V. B., Sadovnikova, N. A., Kostikov, A. P., and Kayushin, L. P. (1972). *Dokl. Akad. Nauk SSSR* **203**, 1343.

Fradkin, L. I., and Shlyk, A. A. (1967). *In* "Bioenergetics and Biological Spectrophotometry," p. 135. Nauka, Moskow.

Fradkin, L. I., Shlyk, A. A., and Koliago, V. M. (1966). *Dokl. Akad. Nauk SSSR* **171**, 222.

Fradkin, L. I., Faludi-Daniel, A., and Shlyk, A. A. (1968). *Dokl. Akad Nauk SSSR* **182**, 1420.

Fradkin, L. I., Shlyk, A. A., Kalinina, L. M., and Faludi-Daniel, A. (1969). *Photosynthetica* **3**, 326.

Fradkin, L. I., Kalinina, L. M., and Shlyk, A. A. (1970). *Dokl. Akad. Nauk SSSR* **194**, 201.

Fradkin, L. I., Koliago, V. M., and Shlyk, A. A. (1972a). *Dokl. Akad. Nauk SSSR* **207**, 453.

Fradkin, L. I., Kalinina, L. M., and Shlyk, A. A. (1972b). *Proc. Int. Congr. Photosyn. Res., 2nd, 1972*, Vol. III, p. 2298.

Gasanov, R. A., Abutalibov, M. G., Gasanchyan, R. M., and Ganieva, R. A. (1972). *Proc. Int. Biophys. Congr., 4th, 1972.* Abstracts, Vol. I, p. 366.

Gavrilenko, V. F., and Budagovskaya, N. V. (1972). *Dokl. Vysshei Shkoly, Biol. Nauki* **12**, 77.

Gavrilenko, V. F., Rubin, B. A., and Ermakova, L. P. (1969). *Dokl. Vysshei Shkoly, Biol. Nauki* **12**, 61.

Giller, Yu.E. (1972). *Proc. Int. Biophys. Congr., 4th, 1972.* Abstracts, Vol. I, p. 369.

Glinsky, V. P., Samuilov, V. D., and Skulachev, V. P. (1972). *Mol. Biol.* **6**, 664.

Grigorov, L. N., Kononenko, A. A., and Rubin, A. B. (1970). *Mol. Biol.* **4**, 483.

Grigorov, L. N., Zhivotchenko, V. D., Remennikov, S. M., Rubin, L. B., and Rubin, A. B. (1971). *Mol. Biol.* **5**, 744.

Grigorovich, V. J., Zakharova, N. J., Kutyurin, V. M., Rosonova, L. M., and Elpiner, J. E. (1971). *Biofizika* **16**, 260.

Gulyaev, B. A., and Litvin, F. F. (1967). *Biofizika* **12**, 845.

Gulyaev, B. A., and Litvin, F. F. (1970). *Biofizika* **15**, 670.

Gulyaev, B. A., Litvin, F. F., and Vedeneev, E. P. (1971). *Dokl. Vysshei Shkoly, Biol. Nauki* **4**, 49.

Gulyaev, B. A., Karneeva, N. V., Vedeneev, E. P., and Litvin, F. F. (1973). *Dokl. Vysshei Shkoly, Biol. Nauki.* **10**, 48.

Gurinovich, G. P., Sevchenko, A. N., and Solov'ev, K. N. (1968). "Spectroscopy of Chlorophyll and its Analoges" Akad. Nauk Byelorussian SSR, Minsk.

Gurinovich, I. F., Byteva, I. M., Chernikov, V. S., and Petsold, O. M. (1972). *Zh. Org. Khim.* **8**, 842.

Isaev, P. I., Liberman, E. A., Samuilov, V. D., Skulachev, V. P., and Tsofina, L. M. (1970). *Biochim. Biophys. Acta* **216**, 22.

Kafalieva, D. N., Blumenfeld, L. A., Solov'ev, I. S., Livshitc, S. A., and Darmanyan, A. P. (1969a). *Biofizika* **14**, 1117.

Kafalieva, D. N., Chetverikov, A. G., and Blumenfeld, L. A. (1969b). *Biofizika* **14**, 1119.

Kaler, V. L. (1972). *Proc. Int. Biophys. Congr., 4th, 1972*, Abstracts Vol. 1, p. 370.

Kaler, V. L. (1973). In "Problems of Biophotochemistry (A. B. Rubin and V. D. Samuilov, eds.), p. 243. Nauka, Moscow.

Kaler, V. L., and Podchufarova, G. M. (1968). *In* "Physico-chemical Basis of Autoregulation in Cells," p. 185. Nauka, Moscow.

Karapetyan, N. V. (1969). *Progr. Photosyn. Res., Proc. Int. Congr., 1968* Vol. 2, p. 778.

Karapetyan, N. V., and Krasnovsky, A. A. (1968). *Dokl. Akad. Nauk SSSR* **180**, 989.

Karapetyan, N. V., Krachmaleva, I. N., and Krasnovsky, A. A. (1966). *Dokl. Akad. Nauk SSSR* **171**, 1201.

Karapetyan, N. V., Klimov, V. V., Lang, F., and Krasnovsky, A. A. (1971a). *Fiziol. Rast.* **18**, 507.

Karapetyan, N. V., Klimov, V. V., Krachmaleva, I. N., and Krasnovsky, A. A. (1971b). *Dokl. Akad. Nauk SSSR* **201**, 1244.

Karapetyan, N. V., Krachmaleva, I. N., and Krasnovsky, A. A. (1972). *Mol. Biol.* **6**, 773.

Karapetyan, N. V., Klimov, V. V., and Krashovsky, A. A. (1973). *Photosynthetica* **7**, 330.

Kiselev, B. A., Kozlov, Ju. N., and Evstigneev, V. B. (1970). *Biofizika* **15**, 594.

Klimov, V. V., Lang, F., Karapetyan, N. V., and Krasnovsky, A. A. (1972). *Fiziol. Rast.* **19**, 151.

Kochubei, S. M., and Ostrovskaya, L. K. (1972). *Proc. Int. Biophys. Congr., 4th, 1972* Abstracts, Vol. 1, p. 327.

Kononenko, A. A. (1973). *In* "Problems of Biophotochemistry" (A. B. Rubin and V. D. Samuilov, eds.), p. 180. Nauka, Moscow.

Kononenko, A. A., and Lukashev, E. P. (1972). *Proc. Int. Biophys. Congr. 4th, 1972* Abstracts, Vol. 1, p. 375.

Kononenko, A. A., Venediktov, P. S., Lukashev, E. P., and Rubin, A. B. (1971a). *Stud. Biophys.* **28**, 9.

Kononenko, A. A., Venediktov, P. S., Lukashev, E. P., Matorin, D. N., and Rubin, A. B. (1971b). *Stud. Biophys.* **28**, 15.

Kononenko, A. A., Lukashev, E. P., Venediktov, P. S., and Rubin, A. B. (1972a). *Proc. Int. Congr. Photobiol., 6th 1972*, Abstracts, p. 253.

Kononenko, A. A., Lukashev, E. P., Rubin, A. B., and Venediktov, P. S. (1972b). *Biochim. Biophys. Acta* **275**, 130.

Korsun, A. D., Kutyurin, V. M., and Artamkina, I. Yu. (1972a). *Khim. Prir. Soedin.* **4**, 563.

Korsun, A. D., Kutyurin, V. M., and Artamkina, I. Yu. (1972b). *Khim. Prir. Soedin.* **4**, 567.

Krasnovsky, A. A., (1960). *Annu. Rev. Plant Physiol.* **11**, 363.

Krasnovsky, A. A. (1969). *Progr. Photosyn. Res. Proc. Int. Congr., 1968*, Vol. 2, p. 709.

Krasnovsky, A. A. (1972). *Biophys. J.* **72**, 749.

Krasnovsky, A. A., and Brin, G. P. (1968). *Dokl. Akad. Nauk SSSR* **179**, 726.

Krasnovsky, A. A., and Bystrova, M. I. (1962). *Biokhimiya* **27**, 958.

Krasnovsky, A. A., and Bystrova, M. I. (1967). *Dokl. Akad. Nauk SSSR* **174**, 480.

Krasnovsky, A. A., Bystrova, M. I., and Mal'gosheva, I. N. (1969). *Dokl. Akad. Nauk SSSR* **189**, 885.

Krasnovsky, A. A., Brin, G. P., and Aliev, Z.Sh. (1971). *Dokl. Akad. Nauk SSSR* **199**, 952.

Krasnovsky, A. A., Bystrova, M. I., and Mal'gosheva, I. N. (1972). *Dokl. Akad. Nauk SSSR* **204**, 1473.

Krasnovsky, A. A., Jr., and Litvin, F. F. (1967). *Mol. Biol.* **1**, 699.

Krasnovsky, A. A., Jr., and Litvin, F. F. (1968). *Biofizika* **13**, 146.

Krasnovsky, A. A., Jr., and Litvin, F. F. (1969). *Mol. Biol.* **3**, 282.

Krasnovsky, A. A., Jr., and Litvin, F. F. (1970). *Dokl. Akad. Nauk SSSR* **194**, 197.

Krasnovsky, A. A., Jr., and Litvin, F. F. (1972). *Biofizika* **17**, 764.

Krasnovsky, A. A., Jr., Shuvalov, V. A., Litvin, F. F., and Krasnovsky, A. A. (1971). *Dokl. Akad. Nauk SSSR* **199**, 1181.

Krasnovsky, A. A., Jr., Romanyuk, V. A., and Litvin, F. F. (1973). *Dokl. Akad. Nauk SSSR* **209**, 965.

Krendeleva, T. E., Kaurov, B. S., Sakharova, T. A., and Rubin, A. B. (1971). *Stud. Biophys.* **28**, 183.

Krendeleva, T. E., Nizovskaya, N. V., Ivanov, A. V., and Rubin, L. B. (1972). *Biokhimiya* **37**, 158.

Kukarskich, G. P., Krendeleva, T. E., and Rubin, A. B. (1972). *Biofizika* **17**, 85.

Kutyurin, V. M. (1970). *Biologiya* **4**, 569.

Kutyurin, V. M. (1972). *Proc. Int. Congr. Photosyn. Res., 2nd, 1971* Vol. 1, p. 569.

Kutyurin, V. M., Nazarov, N. M., and Anisinoma, I. N. (1968). *Dokl. Akad. Nauk SSSR* **181**, 1270

Kutyurin, V. M., Ulubekova, M. V., Matveeva, I. V., Shutilova, N. I., and Rosonova L. M. (1969). *Fiziol. Rast.* **16**, 181.

Lang, F., Vorob'eva, L. M., and Krasnovsky, A. A. (1968). *Dokl. Akad. Nauk SSSR* **183**, 711.

Lang, F., Vorob'eva, L. M., and Krasnovsky, A. A. (1969). *Progr. Photosyn. Res., Proc. Int. Congr., 1968* Vol. 7, p. 630.

Litvin, F. F. (1964). *Proc. Int. Photobiol. Congr., 4th, 1964* Abstracts, p. 126.

Litvin, F. F. (1965). *In* "Biochemistry and Biophysics of Photosynthesis" (A. A. Krasnovsky, ed.), p. 96. Nauka, Moscow.

Litvin, F. F., and Belyaeva, O. B. (1968). *Biokhimiya* **33**, 928.

Litvin, F. F., and Belyaeva, O. B. (1971a). *Biokhimiya* **36**, 615.

Litvin, F. F., and Belyaeva, O. B. (1971b). *Photosynthetica* **5**, 200.

Litvin, F. F., and Gulyaev, B. A. (1964). *Dokl. Akad. Nauk. SSSR* **158**, 460.

Litvin, F. F., and Gulyaev, B. A. (1966a). *Dokl. Akad. Nauk SSSR* **169**, 1187.

Litvin, F. F., and Gulyaev, B. A. (1966b). *Proc. Int. Microbiol. Congr., 9th*, Abstract, p. 28.

Litvin, F. F., and Gulyaev, B. A. (1969a). *Dokl. Vysshei Shkoly, Biol. Nauki* **2**, 118.

Litvin, F. F., and Gulyaev, B. A. (1969b). *Dokl. Vysshei Shkoly, Biol. Nauki* **5**, 130.

Litvin, F. F., and Gulyaev, B. A. (1969c). *Dokl. Akad. Nauk SSSR* **189**, 1385.

Litvin, F. F., and He I-Tan. (1967). *Dokl. Akad. Nauk SSSR* **167**, 1187.

Litvin, F. F., and Krasnovsky, A. A. (1957). *Dokl. Akad. Nauk SSSR* **117**, 106.

Litvin, F. F., and Krasnovsky, A. A. (1958). *Dokl. Akad. Nauk SSSR* **120**, 764.

Litvin, F. F., and Krasnovsky, A. A. (1959). *Izv. Akad. Nauk SSSR, Ser. Fiz.* **23**, 82.

Litvin, F. F., and Krasnovsky, A. A., Jr. (1967). *Dokl. Akad. Nauk SSSR* **173**, 451.

Litvin, F. F., and Personov, R. I. (1960). *Dokl. Akad. Nauk SSSR* **136**, 798.

Litvin, F. F., and Shuvalov, V. A. (1968). *Dokl. Akad. Nauk SSSR* **181**, 733.

Litvin, F. F., and Sineshchekov, V. A. (1965). *In* "Molecular Biophysics" (G. M. Frank, ed.), p. 191, Nauka, Moscow.

Litvin, F. F., and Sineshchekov, V. A. (1967a). *Biofizika* **12**, 433.

Litvin, F. F., and Sineshchekov, V. A. (1967b). *Dokl. Akad. Nauk SSSR* **175**, 1175.

Litvin, F. F., and Zvalinsky, V. I. (1967). *Dokl. Akad. Nauk SSSR* **173**, 703.

Litvin, F. F., and Zvalinsky, V. I. (1968). *Biofizika* **13**, 241.

Litvin, F. F., and Zvalinsky, V. I. (1971). *Biofizika* **16**, 420.

Litvin, F. F., Krasnovsky, A. A., and Rikhireva, G. I. (1959). *Dokl. Akad. Nauk SSSR* **127**, 699.

Litvin, F. F., Krasnovsky, A. A., and Rikhireva, G. I. (1960a). *Dokl. Akad. Nauk SSSR* **135**, 1528.

Litvin, F. F., Vladimirov, Ju. A., and Krasnovsky, A. A. (1960b). *Usp. Fiz. Nauk* **71**, 149.

Litvin, F. F., Rikhireva, G. I., and Krasnovsky, A. A. (1962). *Biofizika* **7**, 578.

Litvin, F. F., Sineshchekov, V. A., and Krasnovsky, A. A. (1964). *Dokl. Akad Nauk SSSR* **154**, 460.

Litvin, F. F., Gulyaev, B. A., and Sineshchekov, V. A. (1965). *Dokl. Akad. Nauk SSSR* **162**, 1184.

Litvin, F. F., Shuvalov, V. A., and Krasnovsky, A. A., Jr. (1966). *Dokl. Akad. Nauk SSSR* **168**, 1195.

Litvin, F. F., Personov, R. I. and Karataev, O. N. (1969). *Dokl. Akad. Nauk SSSR* **188**, 1169.

Litvin, F. F., Gulyaev, B. A., and Karneeva, N. V. (1970). *Dokl. Vysshei Shkoly, Biol. Nauki* **4**, 95.

Litvin, F. F., Gulyaev, B. A., and Sineshchekov, V. A. (1971). *Dokl. Akad. Nauk SSSR* **199**, 1428.

Litvin, F. F., Belyaeva, O. B., Gulyaev, B. A., and Sineshchekov, V. A. (1973). *In* "Problems of Biophotochemistry" (A. B. Rubin and V. D. Samuilov, eds.), p. 132. Nauka, Moscow.

Litvin, F. F., Belyaeva, O. B., Gulyaev, B. A., Karneeva, N. V. Sineshchekov, V. A., Stadnitchuk, I. N., and Shubin V. V. (1974a). *In* "The Chlorophyll" (A. A. Shlyk, ed.), p. 215. Nauka, Minsk.

Litvin, F. F., Shubin, V. V., and Sineshchekov, V. A., (1974b). *Biofizika.*

Losev, A. P., and Gurinovich, G. P.(1969). *Biofizika* **14**, 110.

Losev, A. P., and Zenkevich, E. I. (1968). *Zh. Prikl. Spektrosk.* **9**, 144.

Losev, A. P., Byteva, I. M., Dzhagarov, B. M., and Gurinovich, G. P. (1972). *Biofizika* **17**, 213.

Matorin, D. N., Venediktov, P. S., Kononenko, A. A., and Rubin, A. B. (1971). *Stud. Biophys.* **27**, 91.

Mukhin, E. N., and Akulova, E. A. (1966). *Dokl. Akad. Nauk SSSR* **169**, 699.

Mukhin, E. N., and Akulova, E. A. (1967). *In* "Bioenergetics and Biological Spectrophotometry," Nauka, Moscow.

Mukhin, E. N., and Gins, V. K. (1972). *Biokhimya* **37**, 1012.

Mukhin, E. N., Chruslova, S. G., and Gins, V. K. (1970a). *Fiziol. Rast.* **17**, 1193.

Mukhin, E. N., Chruslova, S. G., Egorova, E. E., and Shmeleva, V. L. (1970b). *Dokl. Akad. Nauk SSSR* **193**, 940.

Nazarenko, A. V., Samuilov, V. D., and Skulachev, V. P. (1971). *Biokhimiya* **36**, 780.

Nazarova, I. G., and Evstigneev, V. B. (1971). *Mol. Biol.* **5**, 826.

Rubin, A. B., and Krendeleva, T. E. (1971). *In* "Physiology and Biochemistry of Pathological and Normal Plants," p. 257. Mosk. Gos. Univ., Moscow.

Rubin, A. B., and Venediktov, P. S. (1968). *Fiziol. Rast.* **15**, 34.

Rubin, A. B., and Venediktov, P. S. (1969). *Biofizika* **14**, 106.

Rubin, A. B., Kononenko, A. A., Uspenskaya, N. Y., and Ivanov, I. D. (1968). *Izv. Akad. Nauk. SSSR, Biol.* **3**, 372.

Rubin, L. B., Rubin, A. B., Dubrovin V. N., and Shvinka, Y. E. (1969). *Mol. Biol.* **3**, 700.

Semenova, I. A., Sinyakov, G. N., and Gurinovich, G. P. (1971). *Opt. Spektrosk.* **31**, 368.

Shlyk, A. A. (1965). "Chlorophyll Metabolism in a Green Plant." Akad. Nauk Byelorussin SSR, Minsk.

Shlyk, A. A. (1970). *In* "Metabolism and Structure of the Photosynthetic Apparatus", p. 3. Nauka i Technika Publ., Minsk.

Shlyk, A. A., Fradkin, L. I., Chkanikova, R. A., and Sukhover, L. K. (1966). *Dokl. Akad. Nauk SSSR* **167**, 706.

Shlyk, A. A., Prudnikova, I. V., Fradkin, L. I., Nikolayeva, G. N., and Savchenko, G. E. (1969). *Progr. Photosyn. Res., Proc. Int. Congr., 1968.* Vol. 2, p. 572.

Shuvalov, V. A., and Krasnovsky, A. A. (1971). *Mol. Biol.* **5**, 698.

Shuvalov, V. A., and Krasnovsky, A. A. (1972). *Dokl. Akad. Nauk SSSR* **207**, 746.
Shuvalov, V. A., and Litvin, F. F. (1969). *Mol. Biol.* **3**, 59.
Shuvalov, V. A., and Litvin, F. F. (1973). *In* "Problems of Biophotochemistry" (A. B. Rubin and V. D. Samuilov, eds.), p. 148. Nauka, Moscow.
Shuvalova, N. P., and Bell, L. N. (1970). *Dokl. Akad. Nauk SSSR* **194**, 1223.
Sinegub, O. A., and Erokhin, Yu.E. (1971). *Mol. Biol.* **5**, 472.
Sineshchekov, V. A. (1972). *Proc. Int. Biophys. Congr., 4th, 1972* Abstracts, Vol. 1, p. 20.
Sineshchekov, V. A., Litvin, F. F., and Das, M. (1972). *Photochem. Photobiol.* **15**, 187.
Sineshchekov, V. A., Shybin, V. V., and Litvin, F. F. (1973). *Dokl. Akad. Nauk SSSR* **211**, 1226.
Sineshchekov, V. A., Stadnitchuk, I. N., and Litvin, F. F. (1974). Paper submitted to the Archives of the All-Union Institute of Scientific and Technical Information, August 1974.
Solov'ev, K. N., Gradyushko, A. T., Yegorova, G. D., and Tsvirko, M. P. (1972). *Proc. Int. Biophys. Congr., 4th, 1972* Abstracts, Vol. 1, p. 351.
Sorokin, E. M. (1971a). *Fiziol. Rast.* **18**, 473.
Sorokin E. M. (1971b). *Fiziol. Rast.* **18**, 1098.
Sorokin, E. M., and Tumerman, L. A. (1971). *Mol. Biol.* **5**, 753.
Suboch, V. P., Shulga, A. M., Gurinovich, G. P., Glazkov, Yu. V., Zhuravlev, Yu.V., and Sevchenko, A. N. (1972). *Dokl. Akad. Nauk SSSR* **204**, 404.
Terenin, A. N. (1967). "Photonics of Dye Molecules and Related Organic Compounds." Nauka, Leningrad.
Timofeyev, K. N. (1972). *Proc. Int. Biophys. Congr., 4th, 1972* Abstracts, Vol. 1, p. 339.
Timofeyev, K. N., and Rubin, A. B. (1969). *Biofizika* **14**, 995.
Tumerman, L. A., and Sorokin, E. M. (1967). *Mol. Biol.* **1**, 628.
Tumerman, L. A., Zavilgelsky, G. B., and Ivanov, V. I. (1962). *Biofizika* **7**, 21.
Venediktov, P. S., Kononenko, A. A., Matorin, D. N., and Rubin, A. B. (1969a). *Mol. Biol.* **3**, 592.
Venediktov, P. S., Matorin, D. N., and Rubin, A. B. (1969b). *Dokl. Vysshei Shkoly, Biol. Nauki* **2**, 46.
Venediktov, P. S., Matorin, D. N., Krendeleva, T. E., Nizovskaya, N. V., and Rubin, A. B. (1971). *Dokl. Vysshei Shkoly, Biol. Nauki* **11**, 56.
Vorob'eva, L. M., and Krasnovsky, A. A. (1967). *Biofizika* **12**, 240.
Vorob'eva, L. M., and Krasnovsky, A. A. (1968). *Biofizika* **13**, 456.
Vorob'eva, L. M., and Krasnovsky, A. A. (1969). *Dokl. Akad. Nauk SSSR* **189**, 420.
Vorob'eva, L. M., and Krasnovsky, A. A. (1970). *Dokl. Akad. Nauk SSSR* **195**, 731.
Vorob'eva, L. M., and Krasnovsky, A. A. (1972). *Dokl. Akad. Nauk SSSR* **205**, 233.
Yefimtsev, E. I. (1972). *Proc. Int. Biophys. Congr., 4th, 1972* Abstracts, Vol. 1, p. 383.
Zakrzhevsky, D. A., Rosonova, L. M., and Kutyurin, V. M. (1972). *Fiziol. Rast.* **19**, 1199.
Zenkevich, E. I., Losev, A. P., and Gurinovich, G. P. (1972a). *Proc. Int. Biophys. Congr., 4th, 1972* Abstracts, Vol. 1, p. 334.
Zenkevich, E. I., Losev, A. P., and Gurinovich, G. P. (1972b). *Mol. Biol.* **6**, 824.

Author Index

Numbers in italics refer to the pages on which the complete references are listed.

Subject Index

A

Absorbance change(s), 139
  in bromocresol purple, 587
  light-induced, 25, 26
  at 320 nm, 326, 329, 334, 508, 509
  at 515 (518) nm, 332, 520, 612, 613
  at 550 nm, 26, 27, 171, 172, 331-334, 381, *see also* C550
  at 680 nm, 325-328, 503-506, *see also* P680
  at 700 nm, 499-502, *see also* P700
Absorption, of light in photosynthesis, 15-17, 118-123
  band shift, 136
Absorption oscillators, molecular, 133
Absorption spectra, 120, 623
  broadening and red shift, 622
  of *Chlorobium*, 624
  of *Chromatium*, 624
  computer analysis of, 621, 622, 623
  at low temperature, 622
  of *Phaseolus*, 622, 623
  second derivative of, 621, 625
Acceptor (or respiratory) control, 565
Acceptor (electron), *see also* C550, Q and X
  primary, 147, 157, 164-172
  secondary, 157
Acid-base experiment for ATP synthesis, 424, 450-454
Acid-base transition
  activation of ATPase, 462, 464
  effect on delayed light, 254, 255, 610
  effect on exchange reaction, 471
  leading to ATP synthesis, 452, 468, 589, 609
Acidity, internal, 422

Action spectra
  of absorption change at 515 nm, 524
  of Emerson enchancement, 12
  of oxidation and reduction of P700, 502
Activation energy, 246, 247, 259, 293
"Activation phase," 337
Adenine nucleotides, 560
  concentration in chloroplasts upon illumination, 560
  translocation, 560, 561
Adenosine diphosphate (ADP), 449, 469, 470, 472
  in arsenate uncoupling, 476
  binding to CF_1, 481
  effect on proton uptake and efflux, 449, 450
  function in conformational change of CF_1, 478, 480, 481
  function in permanganate inhibition, 478
  function of, 464
  inhibition of CF_1 ATPase, 462, 464, 466
  in reversal of NEM (*N*-Ethylmaleimide) inhibition, 477
  in sulfate inhibition, 476
Adenosine monophosphate (AMP)
  binding to CF_1, 481
Adenosine triphosphate, 469, 476, *see also* ATP
  binding to CF_1, 481
  to $2e^-$ ratio, 438, 439
  effect on H^+ permeability, 449, 450
  to H^+ ratio, 539, 540
  inhibition to proton leakage, 475
  production of, forward reaction yield, 537
  regulatory effect of, 449, 450
  reversal of NEM (*N*-Ethylmaleimide) inhibition, 477

682

CELL BIOLOGY: A Series of Monographs

EDITORS

D. E. BUETOW

*Department of Physiology
and Biophysics
University of Illinois
Urbana, Illinois*

I. L. CAMERON

*Department of Anatomy
University of Texas
Medical School at San Antonio
San Antonio, Texas*

G. M. PADILLA

*Department of Physiology and Pharmacology
Duke University Medical Center
Durham, North Carolina*

Stuart Coward (editor). DEVELOPMENTAL REGULATION: Aspects of Cell Differentiation, 1973

I. L. Cameron and J. R. Jeter, Jr. (editors). ACIDIC PROTEINS OF THE NUCLEUS, 1974

Govindjee (editor). BIOENERGETICS OF PHOTOSYNTHESIS, 1975